CANCER GENE THERAPY

Past Achievements and Future Challenges

ADVANCES IN EXPERIMENTAL MEDICINE AND BIOLOGY

Recent Volumes in this Series

Volume 465
CANCER GENE THERAPY: Past Achievements and Future Challenges
Edited by Nagy A. Habib

Volume 466
CURRENT VIEWS OF FATTY ACID OXIDATION AND KETOGENESIS: From Organelles to Point Mutations
Edited by Patti A. Quant and Simon Eaton

Volume 467
TRYPTOPHAN, SEROTONIN, AND MELATONIN: Basic Aspects and Applications
Edited by Gerald Huether, Walter Kochen, Thomas J. Simat, and Hans Steinhart

Volume 468
THE FUNCTIONAL ROLES OF GLIAL CELLS IN HEALTH AND DISEASE: Dialogue between Glia and Neurons
Edited by Rebecca Matsas and Marco Tsacopoulos

Volume 469
EICOSANOIDS AND OTHER BIOACTIVE LIPIDS IN CANCER, INFLAMMATION, AND RADIATION INJURY, 4
Edited by Kenneth V. Honn, Lawrence J. Marnett, and Santosh Nigam

Volume 470
COLON CANCER PREVENTION: Dietary Modulation of Cellular and Molecular Mechanisms
Edited under the Auspices of the American Institute for Cancer Research

Volume 471
OXYGEN TRANSPORT TO TISSUE XXI
Edited by Andras Eke and David T. Delpy

Volume 472
ADVANCES IN NUTRITION AND CANCER 2
Edited by Vincenzo Zappia, Fulvio Della Ragione, Alfonso Barbarisi, Gian Luigi Russo, and Rossano Dello Iacovo

Volume 473
MECHANISMS IN THE PATHOGENESIS OF ENTERIC DISEASES 2
Edited by Prem S. Paul and David H. Francis

Volume 474
HYPOXIA: INTO THE NEXT MILLENNIUM
Edited by Robert C. Roach, Peter D. Wagner, and Peter H. Hackett

---

CANCER GENE THERAPY

Past Achievements and Future Challenges

Edited by

Nagy A. Habib

Imperial College School of Medicine
Hammersmith Hospital
London, United Kingdom

Kluwer Academic / Plenum Publishers
New York, Boston, Dordrecht, London, Moscow

Library of Congress Cataloging-in-Publication Data

Cancer gene therapy : past achievements and future challenges / edited by Nagy A. Habib.
p. ; cm. -- (Advances in experimental medicine and biology ; v. 465)
Includes bibliographical references and index.
ISBN 0-306-46191-9
1. Cancer--Gene therapy. I. Habib, Nagy A. II. Series.
[DNLM: 1. Neoplasms--therapy. 2. Gene Therapy--trends. 3. Neoplasms--genetics. QZ 266 C21434 1999]
RC271.G45 C36 1999
616.99'4042--dc21

99-044529

ISSN 0065-2598

ISBN 0-306-46191-9

233 Spring Street, New York, New York 10013

http://www.wkap.nl

10 9 8 7 6 5 4 3 2 1

A C.I.P. record for this book is available from the Library of Congress

Printed in the United States of America

To all the clinicians, scientists, and various other people who are passionate about cancer research and have dedicated their lives to the advancement of knowledge and the discovery of a cancer cure

PREFACE

With the coming of the new millennium we are witnessing a revolution in our understanding of cancer genetics. These are very exciting times. Today we have at our disposal the technology to diagnose abnormalities in our cancer genes and the means to correct the deficit and very soon we will have the complete sequence of the human genome.

With the use of gene chip technology the way doctors will be able to assess patients will change completely. Today we can diagnose abnormalities in ten thousand genes and within a short period of time we will be able to screen through our genome and discover potential abnormalities in our proto-oncogenes, tumour suppressor genes, differentiating genes, apoptotic genes and pro-inflammatory genes. In this book various authors have highlighted specific genes that could be expressed, overexpressed, neutralised or harnessed to achieve cancer control.

The problem of transferring the therapeutic gene into the cancer cell has been partly addressed with major developments in the field of naked plasmid DNA, adenovirus, retrovirus and adeno-associated viruses. However, further improvements are yet to be made to achieve significant gene transfer. Gene expression, in particular specificity of gene transfer, is obviously an important issue and one which is highlighted in this book by the use of specific promoter.

There are many potentially promising avenues to pursue to achieve cancer cell apoptosis or necrosis and some pre-clinical and clinical data are promising. Various clinical studies, including the use of Ad p53 and Ad E1B deleted, are progressing from phase II into phase III studies and are showing promise when their use is combined with chemotherapy.

Although progress in this field has been relatively slow, gene therapy has come of age. Already the first gene therapy drug has FDA approval for use in patients with CMV retinitis to save eyesight. Cancer is a more problematic disorder to solve as it involves many overexpressed oncogenes and several deleted or mutated tumour suppressor genes. Although cure is unlikely to happen tomorrow, cancer gene therapy is undoubtedly here to stay.

ACKNOWLEDGMENTS

I would like to acknowledge Sandra Doherty for putting this book together and for charming various authors for their manuscripts.

CONTENTS

III. Cell Cycle Control

IV. Apoptosis

V. Tumour Suppressor Genes

VI. Other Systems

VII. Antisense and Ribozymes

VIII. Immuno-Modulation

I

THE CLINICAL PROBLEM

1

MANAGEMENT PROBLEMS IN ONCOLOGY

Susan J. Cleator and Pat Price

Department of Clinical Oncology
Hammersmith Hospital
London

1. INTRODUCTION

Gene therapy is the introduction of nucleic acids into cells for the purpose of treating disease. Over the last decade the molecular basis of many sporadic and familial tumours has been elucidated. The development of the malignant phenotypes occurs as a result of loss of expression of recessive tumour suppressor genes and aberrant expression of dominant oncogenes. Somatic mutations tend to arise sequentially, with or without a predisposing germ line mutation. Research in this field has been parallelled by advances in the technology of mammalian gene transfer. This has enabled us to attempt to treat cancer at the molecular level.

1.1. Human Disease Models

The lead candidates for gene therapy are well-defined, monogenic diseases such as cystic fibrosis or adenosine deaminase deficiency which could be corrected, in principle, by replacing the "missing" gene. Much of the early gene therapy work focused on these diseases. However, most currently approved gene therapy protocols concern treatment of cancer. This approach can be readily tested in cancer patients and there is a recognised need for a novel treatment in this disease.

1.2. Gene Therapy Strategies

In principle, we could aim to develop strategies of either restoring tumour suppressor gene function or down-regulating oncogenic expression. However, carcinogenesis is generally a multistep process with multiple mutations. It would not be possible to identify and correct all the resultant abnormalities and gene function would need to be restored in all cells. These problems have led to the development of other strategies including the introduction of genes which (i) stimulate an immune response, (ii) which

Cancer Gene Therapy: Past Achievements and Future Challenges, edited by Habib
Kluwer Academic/Plenum Publishers, New York, 2000.

convert an inactive prodrug to an active toxin or (iii) which can render marrow stem cells resistant to chemotherapy drugs. Interest is growing in using gene therapy to interfere with tumour angiogenesis.

1.3. Gene Delivery

The success of gene therapy rests on our being able to insert genes into cells with sufficient efficiency and accuracy. Current approaches generally employ a vector. Broadly speaking there are two approaches to gene insertion. (i) *Ex vivo* techniques involve transducing target cells *in vitro* with a gene. These cells are then rendered replication incompetent and are then transplanted into the recipient. (ii) *In vivo* techniques introduce the vector directly into target tissue. Systemic injection of gene-bearing vector represents an attractive means of targeting disseminated disease, although it presents technical problems.

2. PROBLEMS WITH EXISTING ANTICANCER TREATMENT

Standard anticancer treatments include surgery, radiation and chemotherapy. These can be effective if disease is localised or if the disease is sensitive to treatment e.g. lymphomas and teratomas. However, 60% of patients presenting with their carcinoma are currently incurrable usually due to unresponsive metastatic disease. Current treatments are frequently associated with significant toxicity due to the non-selective nature of the approach. Strategies have been developed to "rescue" normal tissues such as peripheral stem cell transplantation or the addition of folinic acid. However, toxicity from intensive treatment is still significant.

Treatment can fail for a variety of reasons. Many tumours have a high metastatic potential and can escape local control. Cells within any one tumour are very heterogenous and therefore differ in sensitivity to radiation and cytotoxics. Tumours rapidly acquire drug resistance as a result of selection pressure on a genetically unstable population. Tumours usually harbor hypoxic cells within necrotic areas. Such cells are poorly accessed by blood-borne cytotoxics and are relatively radioresistant. The malignant phenotype often differs only subtly from the wild type resulting in a very narrow therapeutic window. Acute and late toxicity is particularly important in the setting of the adjuvant treatment of low to moderate risk patients and those patients with a heavy tumour load in whom treatment is often palliative.

Other therapeutic approaches include hormone therapy, cytokine therapy, antitumour antibodies and targeted chemotherapy. Hormone therapy, can be effective in breast and prostate cancer. The effect is often cytostatic rather than cytoxic. However, escape from hormonal control usually occurs after some time due to the growth of hormone irresponsive clones. Cytokine therapy has been disappointing as it is active only on a narrow spectrum of tumours. More targeted approaches to chemotherapy are aimed at increasing drug delivery to tumours and reducing exposure of normal tissue to the drug. Liposomal formulations of chemotherapy are an attempt to exploit the increased permeability of tumour vasculature resulting in preferential uptake within the tumour and relative sparing of normal tissue. So far however, this selectivity has not been sufficient to allow an increase in the dose of chemotherapy sufficient to significantly effect response rates. Anti-tumour antibodies continue to be developed. This strategy depends on tumour antigen targeting. In practice the approach has had limied success in bulk

Table 1. Summary of existing anticancer therapy

Treatment Modality	Advantage	Disadvantage
Radiotherapy	• Treatment completed in a few weeks. • Toxicity generally locoregional only. • Curative in some localised disease.	• Tumour cells vary in their • Treatment is usually locoregional. • Hypoxic cells are radioresistant. • Risk of second malignancy.
Chemotherapy	• Distant disease is targeted.	• Tumour cells vary in chemo-sensitivity. • Development of drug resistance. • Systemic toxicity. • Aberrant vasculature limits access.
Surgery	• Single treatment. • All malignant cells potentially removed immediately.	• Surrounding tissues and distant sites not treated. • May result in significant functional disruption.
Cytokines	• Systemic treatment.	• Flu-like symptoms common. • Narrow and often limited activity. • Neutralising antibodies may form.
Hormone manipulation	• Usually well tolerated. • Systemic treatment.	• Variable response. • Development of resistance. • Long-term treatment required.
Auto-antibodies	• Usually well tolerated. • Systemic treatment.	• Tumour antigenicity is heterogenous. • Normal cells may express same antigens. • Non-specific uptake in liver, spleen, bone marrow. • Poor uptake by tumour cells. • Development of humaral response to exogenous antibody.

disease probably due to physiological factors such as access and nonspecific binding. Currently there appears to be promise however in brast cancer and the adjuvant setting in colon cancer.

3. GENE THERAPY TREATMENT STRATEGIES

3.1. Replacing Defective Tumour Suppressor Genes

The tumour suppressor gene p53 is mutated in a wide variety of human tumours. (Hollstein *et al.*, 1991). Restoration of wild type p53 expression using a retroviral vector has been shown to suppress growth in human lung cancer cell lines (Cai *et al.*, 1993). Surprisingly, despite the multiplicity of genetic lesions within a cancer cell, studies suggest that correction of a single genetic lesion can be sufficient to suppress growth (Takahashi *et al.*, 1992). Furthermore, Cai's paper reports that in mixing experiments, retroviral wild-type p53 transduced cells could reduce the growth rate in nontransduced lung cancer cells, suggesting a bystander effect. However, to be clinically effective this approach demands a high transfection rate. Also unless stable expression is achieved, or unless transfection results in apoptosis, treatment would need to be given repeatedly. Specificity is not essential with this treatment as transfection of normal cells would not be expected to be harmful. P53 is probably important in the induction of apoptosis by some chemotherapeutic agents. One study has demonstrated that p53-deleted human lung tumours grown as subcutaneous xenographs in nude mice can be rendered more responsive to cisplatin by transduction with wild-type p53 (Fujiwara *et al.*, 1994).

3.2. Oncogene Inactivation

The ras oncogenes are among the most common activated oncogenes found in human cancers. They cause constitutive activation of signal transduction. Exogenous complimentary antisense oligonucleotides could be used to block transcription or translation of the oncogene. Systemic administration of oligonucleotides carries problems such as plasma degradation and selectivity of cellular uptake. To be effective all malignant cells would need to be targeted. There are also concerns of oligo homology to transcripts other than the target sequence. Antisense molecules would need to persist as down-regulation of an oncogene will not necessarily result in cell death.

3.3. Drug Sensitivity Genes

Selective transduction of tumour cells with a gene whose product can convert a prodrug to a cytotoxic drug within the cancer cell was one of the first gene therapy strategies to be pursued in oncology. The most commonly employed system is transduction with the herpes simplex virus thymidine kinase (HSV-TK) gene whose product phosphorylates gancyclovir to an active form. The advantage of this system is that expression need only be transient, and potant bystander effect will help to eliminate neighbouring untransfected cells (Freeman *et al.*, 1993). Normal cells will also be destroyed if they are successfully transduced so, in order to avoid toxicity, efforts must be made to target this treatment specifically to tumour cells.

3.4. Stimulating the Immune Response

Despite differing from their host, many tumour cells will escape immune control. For immunity to occur tumour antigens must be processed intracellularly and presented on the cell surface in the context of MHC class 1. Costimulatory signals are required, such as B7 on the surface of bone marrow-derived antigen presenting cells which interacts with CD28 on the cell surface. Cell kill occurs when the necessary cytokines activate cytotoxic lymphocytes. Immunity fails to occur for several reasons. Tumour cells may lack surface tumour specific antigens and are poor antigen presenters due to paucity of MHC class 1. Tumours are heterogeneous, and in particular, metastases may differ markedly from their primary. There is also some evidence to suggest that tumours exert an immune suppressive effect (Grimm *et al.*, 1988). Amongst the wealth of gene therapy protocols in this field are those involving immunisation of patients with irradiated autologous tumour cells which have been transduced *in vitro* with genes encoding cytokines (Abel-Wahab *et al.*, 1997). Patients have also been vaccinated with established human cell lines engineered to express cytokines or costimulatory molecules (Belli *et al.*, 1997). Tumour-infiltrating lymphocytes and fibroblasts have similarly been used to express cytokines. *Ex vivo* manipulation of cells has been avoided in recent protocols in which liposomes encapsulating plasmids encoding allogenic MHC moleculars are injected intratumorally (Hui *et al.*, 1997). Finally, dendritic cells, which represent potent, professional antigen presenters have been transduced with tumour-associated antigens (Wan *et al.*, 1997).

The immune approach is attractive because cytotoxic T cells can potentially target disease at many sites in the body. Expression need only be transient and there is an opportunity for response amplification. Of all gene therapy protocols, those involving the immune approach predominate. However, tumour antigenic diversity and tumour-induced anergy are causes of potential failure. Furthermore, many patients receiving gene therapy also receive standard cytotoxics which are immune suppressive.

3.5. Multidrug Resistant Genes

This approach involves transfer of genes conferring chemotherapy resistance into bone marrow or peripheral blood stem cells in order to allow higher doses of chemotherapy to be given. The best characterised gene of this type is the multidrug resistance gene which encodes P-glycoprotein. Many drugs however have additional dose limiting toxicity such as cardiac or neurological toxicity and so limit this approach. Also higher doses of chemotherapy may not translate to higher response rates. There is the additional risk of transducing cancer cells within the marrow.

3.6. Tumour Vasculature as a Target for Gene Therapy

Once tumours grow above a certain volume neovascularisation must occur. This process is referred to as angiogenesis and is controlled by the interplay of angiogenic and antiangiogenesis factors. The best characterised angiogenic mediator is vascular endothelial growth factor, (VEGF). Thrombospondin-1 is one of several inhibitors of the process. It is thought that metastasis from primary tumours is facilitated by angiogenesis (Fidler and Ellis, 1994). Attempts have been made to block the process of angiogenesis with gene therapy in the hope that tumour growth and metastasis will be inhibited. This has been approached by inserting genes for angiogenic or antiangiogeneic signals (Kong and Crystal, 1998).

4. CLINICAL PROBLEMS IN GENE THERAPY

4.1. Gene Delivery

Genes can either be inserted directly into tumour cells in situ or introduced into cells *ex vivo* before being reintroduced. Transfer of genes is generally facilitated by a vector which may be either viral or non-viral. The ideal vector for gene introduction would be safe and technically easy to produce in large quantities. It would be efficient at delivering genetic material, specific for its target and not recognised by the recipients' immune system. It would express the gene in a regulated fashion for as long as required. Viruses have evolved to infect cells and are obvious candidates.

Retroviruses are used widely in *ex vivo* techniques and have been employed in around 80% of gene therapy protocols. They become integrated into the genome and can give rise to stable expression, certainly *in vitro*. Unfortunately, the transduction rate is not high, partly because they tend to transfect dividing cells only. This, and the fact that the retroviruses lack the necessary receptors to enter some cells, whilst limiting efficiency of transduction, does offer some cell specificity. In fact attempts have been made to infuse live virus producer cells into the vicinity of the tumour allowing infection of cancer cells as they come into cycle (Culver *et al.*, 1992). If the surrounding normal cells are non-dividing they are left unaffected. Inevitably, integrated genes often fail to be expressed for various reasons, and there is a potential risk of insertion mutagenesis. The latter has never been demonstrated *in vivo* but it is a definite risk and targeted interaction is a focus of research.

The use of viruses carries the risk of recapture of deleted structural genes leading to development of replication-competent viruses. Many constructs undergo lysis in human serum as a consequence of complement activation. This limits expression of gene products but does provide protection against replication competent viruses. Adenoviruses, in contrast, have a broad host range, including non dividing cells. They have a high transduction rate and do not integrate but express their genes epichromosomally. This limits the duration of gene expression but means there is no risk of insertional mutegenesis. They are easier to grow in high titre and are often used for in situ work. Unfortunately, they tend to evoke a brisk immune response which further limits expression, and which could potentially trigger autoimmune disease. This is a particular concern when repeated treatments are planned. Ways to combat the immune system are therefore being investigated including the use of immunosuppressive drugs. Because of these limitations, new viral vectors are undergoing development.

Non viral methods of gene transfer avoid many of the safety issues and the technical complexities of viral transduction but all the expense of low efficiency. Most research is focusing on the use of plasmid DNA contained within cationic liposomes as they have low toxicity, are non immunogenic, simple to prepare in large quantity and are stable. There is also less restriction on the amount of DNA delivered. However transfection rates are lower and because expression is epichromosomal it tends to persist for a few weeks at the most. Dividing and non-dividing cells can be transfected, although the latter with less efficiency. Liposome-DNA complexes can be injected directly into a tumour site and there have been reports of intravenous injection of liposome-DNA complexes in mice leading to persistent reporter gene expression in widespread tissues (Zhu *et al.*, 1993). If we can achieve widespread gene delivery we have the possibility of treating multiple foci of metastatic disease.

4.2. Targeting

Targeting is important for the success of gene therapy in cancer. It may be achieved either at the level of gene transduction of gene expression. Injecting genetic material directly into tumours is the simplest approach. Viruses do possess some natural tissue tropism. In addition, retroviruses can be very specific for dividing cells which is very useful when the normal cells surround the cancer cells do not divide, as in the case of brain tumours. Viral envelopes and also liposomal surfaces can be modified to bear ligands which bind to tissue-specific, or ideally tumour-specific surface molecules. In terms of transcriptional targeting, tissue-specific promoters is one approach. In eukaryotic cells the regulation of gene expression is controlled by a promoter, enhancer, response elements, upstream activation elements and locus control elements. Several tissue specific genes have been isolated and used in the context of viral and lipsomal vectors as reviewed elsewhere (Roth and Cristiano, 1997). The ideal would be to employ tumour-specific promoters but in reality these are difficult to find. The timing and duration of gene expression can be controlled by inducible promoters. Radiation and heat inducible promoters have been studied but they would be suitable for treating disease in localised anatomical sites only.

Targeting treatment rests heavily on defining phenotypic differences between tumour and normal cells. Evidence is now growing that tumour vasculature differs from its normal counterpart in terms of surface molecules and these could potentially be used as a means of targeting gene therapy. Access to endothelial cells is simpler compared to tumour cells because of their contiguity with the circulatory system.

5. GENE THERAPY TO TREAT CANCER PATIENTS

There are currently over 100 approved clinical gene therapy trials world wide. One could argue that we are being premature in attempting to treat humans while we lack a more effective transfer vector. The drive to enrol patients in clinical trials has been fuelled by some very encouraging results in mouse tumours. However, it is clear that these tumours behave differently to their human counterparts making extrapolation to man difficult.

Gene therapy clinical trials require approval from ethics committees and regulatory bodies. It is clear that strict government controls are necessary. Protocols often require approval from several local and national bodies.

Design of trials and patient selection must be carefully considered. The standard approach of Phase I/II/III clinical trials may not be appropriate for the clinical development of gene therapy. Dosing to maximum tolerated dose may be inappropriate Proof of principle will probably need to be introduced at an early Phase I stage. Surrogate or complex endpoints will be needed. To refine techniques we will need means of assessing gene transduction and gene expression in target tissues. It is possible to do this by biopsy of accessible tissue but work is now in progress to develop a non-invasive technique using a marker substrate and marker gene. Herpes simplex virus type 1 thymidine kinase (HSV1-tk) expression can be imaged and monitored by single-photon emission tomography (SPECT), using a radiolabelled 2-fluoro-nucleoside (Tjurager *et al.*, 1996). If systemically administered gene therapy is to be developed, pharmacokinetic parameters will need to be defined. Chronic therapy trials may also be required.

6. THE FUTURE ROLE OF GENE THERAPY IN CANCER PATIENT MANAGEMENT

Clinically gene therapy is still in its infancy but its theory is based on extensive scientific knowledge and preclinical work is promising. At present its use in humans is limited to those with advanced malignancy but its future clinical application may lie in the adjuvant field. Its anticipated advantageous side effect profile may also provide a role in the palliative setting. Gene therapy should be well tolerated in combination with conventional treatments. In future it is conceivable that gene therapy will be applied to preinvasive disease, such as squamous metaplasia or in situ disease where reversal of adverse genetic changes will control disease. The genetic abnormalities in these conditions, compared to invasive disease, are fewer and therefore potentially easier to correct. Gene therapy may also have an important role in treatment of individuals who are discovered to carry a germ-line tumour suppressor gene mutation on genetic screening.

7. SUMMARY

Man has evolved sophisticated defence mechanisms over millions of years to combat insertion of foreign DNA into his cells. However, gene therapy carries huge potential for the treatment of cancer. The challenge is therefore to translate our scientific knowledge into a clinical reality.

REFERENCES

Abel-Wahab Z., Weltz C., Hester D., Pickett N., Vervaert C., Barber JH.R., Jolly D., and Seigler H.F., 1997. A Phase I clinical trial of immunotherapy with interferon-y gene-modified autologous melanoma cells. *Cancer* **80**:401–412.

Belli F., Arienti F., and Sule-Suso J., 1997. Active immunisation of metastatic melanoma patients with interleukin-2-transduced allogeneic melanoma cells: Evaluation of efficacy and tolerability. *Cancer Immunol. Immunother* **44**:197–203.

Cai D.W., Mukhopadhyay T., Lui Y., Fujiwara T., and Roth J.A., 1993. Stable expression of the wild-type p53 gene in human lung cancer cells after retrovirus-mediated gene transfer. *Hum Gene Ther.* **4**:617–624.

Culver K.W., Ram Z., Wallbridge S., Ishi H., Oldfield E.H., and Blaese R.M., 1992. *In vivo* gene transfer with retroviral vector-producer cells for treatment of experimental brain tumours. *Science* **256**:1550–1552.

Fidler I.J., and Ellis L.M., 1994. The implications of angiogenesis for the biology and therapy of cancer metastases. *Cell* **79**:185–188.

Freeman S.M., Abboud C.N., Whartenby K.A., Packman C.H., Koeplin D.S., Moolten F.L., and Abraham G.N., 1993. The "Bystander Effect": Tumour regression when a fraction of the tumour mass is genetically modified. *Cancer Res.* **53**:5274–5283.

Fujiwara T., Grimm E.A., Mukhopadhyay T., Zhang W.W., Owen-Schaub L.B., and Roth J.A., 1994. Induction chemosensitivity in human lung cancer cells *in vivo* by adenoviral-mediated transfer of the wild-type p53 gene. *Cancer Res.* **54**:2287–2291.

Grimm E.A., Crump W.L., Durett A., Hester J.P., Lagoo-Deenadalavan S., and Owen-Schaub L.B., 1998. TGF-beta inhibits the *in vitro* induction of lymphokine-activated killing activity. *Cancer Immumol. Immunother* **27**:53–58.

Hollestein M., Sidransky D., Vogelstein B., and Harris C.C., 1991, P35 mutations in human cancers *Science* **253**:49–53.

Hui K.M., Ang P.T., Huang L., and Tay S.K., 1997. Phase I study of immunotherapy of cutaneous metastases of human carcinoma using allogeneic and xenogeneic MHC DNA-liposome complexes. *Gene Ther.* **4**:783–790.

Kong H-L., and Crystal R.G., 1998. *J. Natl. Cancer Inst.* **90**:273–286.

Roth J.A., and Cristiano R.J., 1997. Gene therapy for cancer: What have we done and where are we going? *J. Natl. Cancer Inst.* **89**, No.1 21–39.

Takahashi T., Carbone D., Takahashi T., Nau M.N., Hida T., Linnoila I., Ueda R., and Minnay J.D., 1992. Wild type but not mutant p53 suppresses growth of human lung cancer cells bearing multiple genetic lesions. *Cancer Res.* **52**:2340–2343.

Tjuvajev J.G., Finn R., Watanabe K., Joshi R., Oku T., Kennedy J., Beattie B., Koutcher J., Larson S., and Blasberg R.G., 1996. Non invasive imaging of herpes virus thymidine kinase gene transer and expression: A potential method for monitoring clinical gene therapy. *Cancer Research* **56**:4087–4095.

Wan Y.H., Bramson J., Carter R., Grapham F., and Gauldie J., 1997. Dendritic cells transduced with an adenoviral vector encoding a model tumour-associated antigen for tumour vaccination. *Hum Gene Ther.* **8**:1355–1363.

Zhu N., Liggitt D., Liu Y., and Debs R., 1993. Systemic gene expression after intravenous DNA delivery into adult mice. *Science* **261**:209–211.

II

VECTORS

2

ADENOVIRAL VECTORS

Prem Seth*

Senior Scientist
Human Gene Therapy Research Institute
1415 Woodland Avenue, Suite 218
Des Moines IA 50309

In recent years there has been a tremendous interest in the use of adenoviral vectors for the gene therapy of cancer. The main reasons for this are the ease with which recombinant adenoviruses can be constructed in the laboratory; can be grown to high titers; and can infect a variety of target cells including both and non-dividing cells. Another advantage of using adenoviral vectors is the high level transient gene expression which appears to be sufficient for many cancer gene therapy purposes. The purpose of this chapter is to describe the biology of adenoviruses, the construction of the recombinant adenoviral vectors, and their important features. The problems associated with the use of adenoviral vectors and future considerations pertaining to vector development will be also be discussed.

1. ADENOVIRAL VECTORS-BIOLOGY

Adenoviruses are DNA-containing, non-enveloped viruses. There are 49 human serotypes which are classified into six sub groups (groups A to F) (Shenk, 1996). Adenoviruses have natural tropism for the epithelial cells of upper and lower respiratory tracts, the gastro-intestinal tract and ocular tissue. The two most commonly used adenoviruses for constructing recombinant vectors are Ad2 and Ad5, mainly because of our greater understanding of their genomes and because these viruses have never been shown to induce tumors even in animal models.

Adenoviral virions are icosahedral in shape, about 70 to 100 nm diameter. The double stranded DNA and core proteins are surrounded by a protein shell composed of capsid proteins. The capsid contains 252 surface projections called capsomers, with 12 vertices. There are eleven structural proteins numbered from protein II to XII (Fig. 1) (Shenk, 1996). The three capsid proteins which participate actively in the entry of virus

*Phone: 515-241-8727 (direct), 515-241-8787 (main office), Fax: 515-241-8788, e-mail: sethpk@ihs.org

Cancer Gene Therapy: Past Achievements and Future Challenges, edited by Habib
Kluwer Academic/Plenum Publishers, New York, 2000.

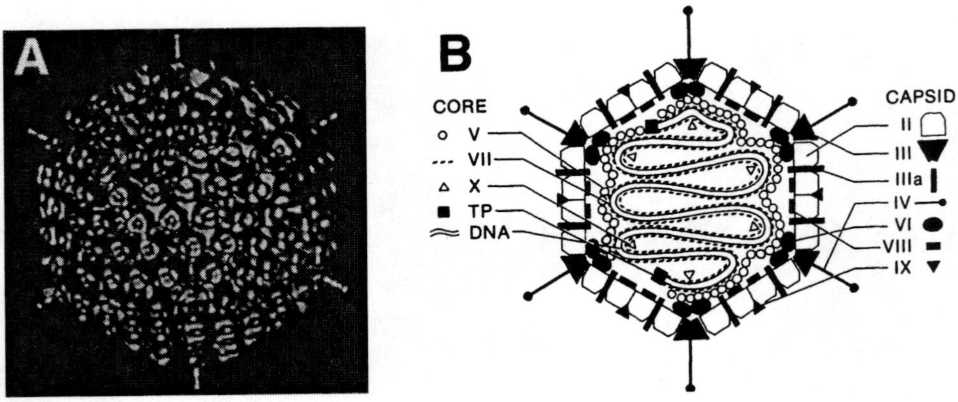

Figure 1. Adenovirus virion. **A.** The three-dimensional image reconstruction of the icosahedral capsid is viewed along the threefold axis. The major coat protein hexon has a triangular top with three towers. Fibers protrude from the penton bases at each fivefold vertex, but only the first third of the shaft is imaged as the remainder is washed out by the averaging method. Note that the capsid is rounded to provide a continuous protein shell. **B.** A stylized section summarizing current structural knowledge of the polypeptide components and the viral DNA. A real section through the icosahedral virion would not contain all these components (reprinted with permission from Field's Virology 3rd edition).

in the cells are fiber (protein III), penton base (protein IV), and hexon (protein II). There are at-least 4 inner core proteins which are known to participate mainly in the viral DNA replication.

The adenoviral genome consists of a linear double stranded DNA, approximately 36kb long. The genome is flanked by two short inverted terminal repeats, and is composed of various transcriptional units. These include an early region (E1 through E4), two delayed early units (IX and 1Va2), a late region (L1 through L5), and VA regions (Fig. 2). The E1 region encodes two proteins, E1a and E1b. E1a proteins are required for the regulation of other transcriptional units and also for initiation of viral DNA replication. E1b proteins induce the entry of the cells S-phase so that cellular machinery can be used for viral DNA replication. Another critical function of E1b proteins is to prevent the programmed cell death till the viral replication has been completed. E2 region encodes proteins required for viral DNA replication. E4 transcriptional region encodes for proteins which are involved in viral transcription, DNA replication, mRNA transport, and in turning off host protein synthesis. The region L1-L5 encode structural proteins including capsid proteins hexon, penton base, penton fiber, and protein IIIa (Shenk, 1996).

2. ADENOVIRUS REPLICATION

Electron microscopic examination of the adenovirally infected cells revealed that adenovirus enters the cells by receptor-mediated endocytosis (Seth *et al.*, 1986). The virus binding to the cellular receptor is mediated by its fiber protein. The main evidence for this is that free fiber competes for the binding of adenovirus to the cell surface. Scatchard analysis of the adenovirus binding to the cells has shown that there are about ten thousand adenoviral receptors per cell (Seth *et al.*, 1994). Competition experiments using various viruses had earlier suggested that perhaps adenovirus receptor falls in the same

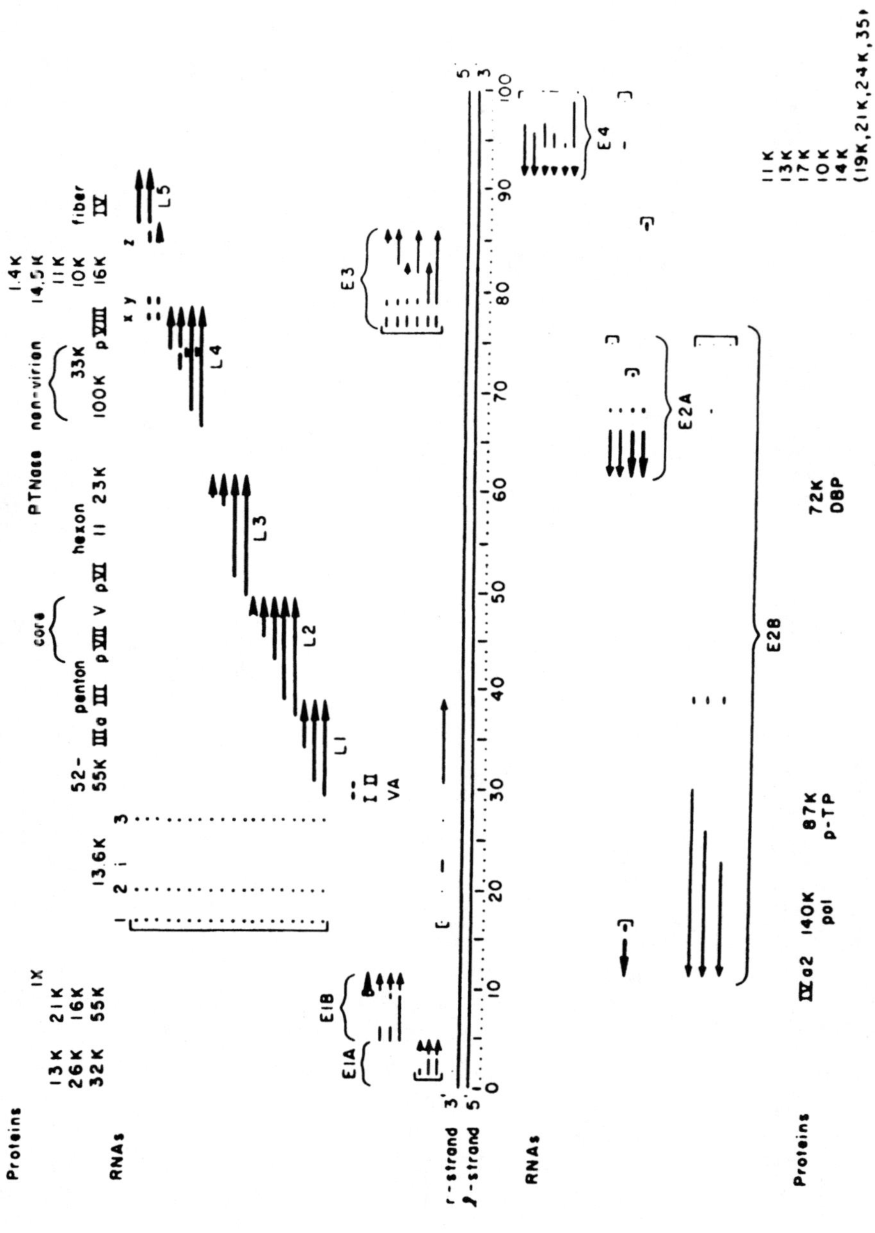

Figure 2. Ad5 transcription and gene products. The early and late mRNAs produced by Ad5 are shown along with identified and proposed protein products in relation to the genomic map units given in arbitrary units from 0 to 100. Transcription of E1, E3, VA RNA and late proteins is from the r strand and thus from left to right. Transcription of E2 and E4 is from the l strand and thus from right to left (reprinted with permission from Field's Virology 3rd edition).

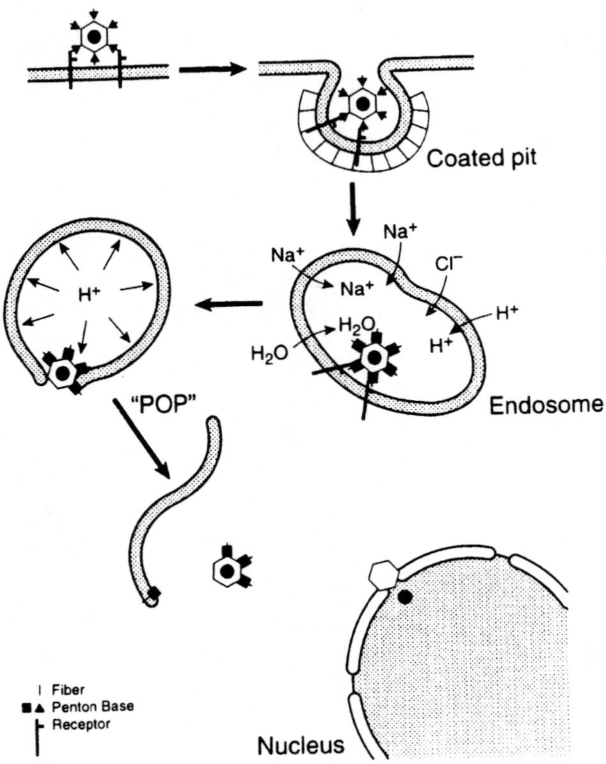

Figure 3. Model of adenovirus entry and vesicle disruption. Adenovirus binds to a cell surface receptor and moves into a coated pit, Soon thereafter, it appears in endosome. Low pH of the endosome causes the penton base protein to undergo a conformational change as a result of which penton base acquires amphiphilic characteristics which enables it to interact with the membrane of the endosome. Endosome, which is distended as a result of osmotic pressure, ruptures at the point where adenovirus penetrates the membrane (reprinted with permission from Cambridge University Press).

category as Coxsacki viruses. In fact, recent molecular cloning of the adenovirus receptor has suggested that it is a member of family of receptors used by Coxsackie B viruses, and therefore are appropriately called CAR (*C*oxsackie and *a*denovirus *r*eceptor) (Bergelson *et al.*, 1996). It has also been shown that an adenoviral capsid protein penton base possesses an amino acid sequences RGD. Such sequences are known to facilitate the binding of adenovirus to integrins alpha$_v\beta_3$ and alpha$_v\beta_5$. Thus it has been suggested that integrin molecules may function as the secondary receptors for adenovirus. After binding to its receptor, adenovirus-receptor complexes move to the clathrin coated pits, and then are subsequently internalized into the endocytic vesicles called endosomes. Adenovirus escape the endosomes by disrupting the membrane of the endosomes (Fig. 3). It has been shown that the adenoviral-mediated lysis of the endosome membrane requires the acidic pH of the endosome and the viral protein penton base plays a critical role in the lysis of the endosome membrane (Seth, 1994a, b). Adenovirus can also activate the membrane bound membrane bound ATPase, which causes ion fluxes across the endosome and cytoplasm, and the build-up of osmotic pressure inside the endosome which results in the rupture of the endosome membrane (Seth *et al.*, 1987) (Fig. 3). Adenoviruses then traverse toward the nucleus using the cytoskeletal microtubular network. Adenoviruses release their genome into the nucleus presumably through nuclear pores.

Adenoviral pre-terminal proteins possesses a nuclear localization signal, and therefore are responsible for the entry of adenovirus into the nucleus. During the process of adenovirus entry into the cells, the virus is uncoated by either shedding of the proteins from the nucleocapsid or by the degradation of the capsid proteins using an endogenous protease present in the adenovirus core (Shenk, 1996).

Immediately after entry of the adenoviral genome into the nucleus, viral early genes (E1 through E4) are transcribed. The protein products from these early regions help in the adenoviral DNA replication, and late transcription, and help the cells to evade the immune system. The E1a proteins, and E1b55 kd proteins can interact with cellular p53 protein thus forcing the cells to enter into S-phase (DNA synthesis phase). E1b19 kd protein by interacting with apoptotic machinery of the cells, prevents the host cells to undergo apoptosis. E2 region encodes proteins such as viral DNA polymerase, pre-terminal protein and DNA—binding proteins required for viral DNA replication. The initiation of DNA replication occurs at a specific DNA sequence at the origin of replication. Several cellular factors such as nuclear factor 1, and CAAT transcription factor participate in the viral DNA replication. Adenovirus DNA replication has been suggested to take place using a strand displacement model. Once viral DNA replication is underway, host protein mRNA transport and protein synthesis are shut off by the early gene proteins E1B-55 kd protein, E4orf6, VA RNA and tripartite leader. Once the early events of viral replication have been completed, the major late promoter located at 16 map units drives most of the subsequent viral transcription. During the late phase of viral replication large quantities of viral structural proteins (hexon, penton base, fiber, protein IIIa), and a few non-structural proteins required for virus assembly are produced. For viral assembly in the nucleus, it is believed that first the empty capsid structures are formed, followed by the entry of DNA into these capsids. Within 24 hr of viral entry into the cells several thousand viral particles can be produced per cell. Although E3 proteins are not directly required for viral DNA replication, they help the virus to escape the immune system. One of the E3 proteins (19 kd protein) helps the virus to subvert the host immune system by down regulating MHC class I expression. Once enough viral particles have been accumulated, another E3 proteins called death protein (a 14 kd protein) causes the lysis of the cells. Eventually the virion particles are released from the cell and begin a new cycle of viral replication in the neighboring cells of the infected organ (Shenk, 1996).

3. GENERATION OF THE RECOMBINANT ADENOVIRAL VECTORS

For gene therapy purposes it is generally desirable to use adenoviral vectors which are replication-deficient to minimize the toxic effects due to the viral vectors. Because E1 sequences are required for viral replication, replication-deficient adenoviral vectors can be easily generated by replacing E1 sequences by the foreign cDNA. Since adenovirus of upto 105% of the genomic size can be packaged, without removing any viral sequences, one can easily package sequences of upto 2 kb in adenoviral genome. To generate recombinant adenoviruses, a cDNA of choice coupled to an appropriate promoter, is cloned into a shuttle vector which provides the left adenovirus inverted terminal repeat (ITR), origin of replication, E1a enhancer, encapsidation signals and adenovirus homologous recombination sequences downstream of the E1 region. Adenovirus genome can be obtained either from restriction-digested adenovirus genome e.g. using Cla 1 and Xba1 sites located near the left end of the adenovirus genome dl309 (Strtford-Perricaudet *et*

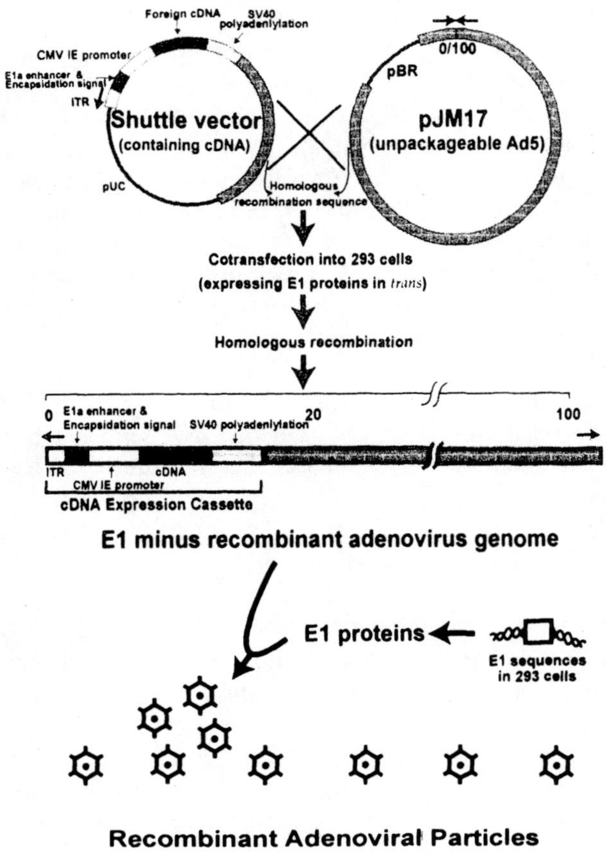

Figure 4. Construction of recombinant adenoviral vectors. cDNA of interest is cloned into a shuttle vector which provides cDNA expression cassette (adenovirus ITR, E1 enhancer, adenovirus encapsidation signal, CMV promoter, and SV40 a polyadenylation signal). Homologous recombination sequences are also cloned in this vector. Adenovirus genome (e.g. pJM17 shown in the figure) and the shuttle vector containing the cDNA are co-transfected in 293 cells. Intracellular homologous recombination between the two DNAs results in a $E1^-$ recombinant genome; the numbers 0, 20, 100 represent the approximate map units. This recombinant genome is replication defective. However, in the presence of E1 proteins (provided in *trans* by 293 cells), the recombinant genome will replicate and form adenoviral particles (reprinted with permission from Cambridge University Press).

al., 1992). Another approach utilizes circular plasmid DNA containing the adenovirus genome (e.g. pJM17) (Bett *et al.*, 1994). The large genomic adenoviral DNA and the shuttle vector are then co-transfected into a packaging cell line, 293, which is an adenovirus transformed cell line (using 10% of the left end of the adenoviral genome). This cell line provides the E1 proteins in *trans*. Homologous recombination between the two DNAs results in the generation of a recombinant adenovirus which is devoid of E1 sequences (Fig. 4). Individual viral plaques are then screened for E1 and the transgene sequences. The plaques which are devoid of E1 sequences and contain the transgene can be grown in 293 cells. Using these methods, in our laboratory we have constructed numerous recombinant adenoviral vectors containing foreign genome of up to 4.5 kb (Craig *et al.*, 1997, 1998; Katayose *et al.*, 1995a, b; Katayose *et al.*, 1997; Kim *et al.*, 1998; Li *et al.*, 1997; Rakkar *et al.*, 1999). To generate vectors with somewhat larger cDNA, one can use a viral backbone which is devoid of E3 region. Two types of such backbones are available. One is an adenovirus mutant dl327 which lacks E3 region. And the second is plasmid such as pBGH10 which is also devoid of E3 region. These viral genomes can be co-transfected with a shuttle vector containing the foreign cDNA as described above. Since E3 proteins are not required for viral replication, one can rescue such recombinant viruses also using 293 cells (Shenk 1996). Recombinant adenoviruses can be easily purified in a $CsCl_2$ gradient and stored, frozen, for an indefinite period of time.

A key problem associated with the first generation vectors described above is that they induce strong immune responses in the host. This problem is worsened by the fact that these viruses can undergo low level viral replication in the target cells (see below). To reduce this low level viral replication, several modification in the adenoviral vectors have been made. In one version adenoviral vectors expressing a mutant form of E2a protein (required for viral replication) was used. Recombinant adenoviruses totally devoid of E2 protein have also been constructed. The viruses have been further crippled by deleting in addition to E1 and E2 region, E4 (ORF4, PRF6). To rescue these viruses one needs packaging cells which also provide the missing proteins such as E2 and E4 proteins (reviewed in Seth, 1998).

Recently, recombinant adenoviruses in which all viral genes have been deleted such that they contain only viral ITRs and cis-packaging sequences, have been constructed. The construction of these viral vectors requires the helper adenovirus functions. One approach utilizes Cre-lox recombination system developed in several laboratories. In this method, first 293cre cells expressing Cre (a phage recombinase) enzyme are cloned. Simultaneously, *lox*P restriction sites are introduced into the helper virus located immediately after the left ITR. The infection of 293cre cells with the helper virus results in the cleavage of the helper virus, which therefore, can not package into the virion particles, but can still provide all the necessary helper functions needed for the replication of the recombinant adenoviral particles. Using these cloning methods, recombinant adenoviral particles containing upto 35 kb of foreign cDNA or small genomic fragments can be generated (Parks *et al.*, 1996; Fisher *et al.*, 1996; Hardy *et al.*, 1997).

4. ADENOVIRAL VECTORS: KEY FEATURES AND FUTURE DIRECTIONS

In recent years adenoviral vectors have become very popular vectors for cancer gene therapy. As discussed earlier, the key reasons for this are the ease with which recombinant viruses can be generated and grown to very high titers (up to 10^{12} pfu/ml) in the laboratory. Adenoviruses have other characteristics which favor their use as vectors which include the very efficient receptor-mediated endocytosis and upon infection of cancer cells exhibit high level of transgene expression. Because of their fairly large packaging capacity (upto 35 kb) adenoviruses can also be used to transfer larger genes than is possible with retroviruses or adeno-associated viruses. Another advantage of adenoviruses is that they can infect a variety of cell types; and produce the heterologous protein of interest in dividing as well as non-dividing cells.

Another feature of adenoviral vectors is that that they do not integrate into the host chromosomes or grow as episomal vectors, and therefore they are degraded by the host. This results in a short term expression of the transduced gene. However, as documented by numerous publications, short term gene expression could be sufficient to achieve the efficacy desired in the clinical setting of cancer gene therapy.

One problem associated with adenoviruses is that the host can induce a strong immune response against both the viruses as well as the transgene product. It has been suggested that these immunological responses to adenoviruses may also reduce the duration of expression of therapeutic gene (transgene) (Dai *et al.*, 1995). Immunological rejection is somewhat diminished by the use of anti-CTL4 IgGs, anti-CD4, or anti-CD8, to inhibit the T-helper cells; or cyclosporin to temporarily suppress the immune response (Kass-Eisler *et al.*, 1996) or cytokines such as IL-12 required for initiation of anti-viral antibody production. Some investigators have utilized different adenoviral serotypes for serial administration of adenoviral vectors in an attempt to circumvent this problem (Kass-Eisler *et al.*, 1996). Moreover, adenoviral vectors derived from E2a mutants, E4 deletion or adenoviral vectors in which all the essential adenoviral sequences have been deleted are expected to have diminished immunological responses.

Another problem associated with the first generation adenoviral vectors (E1-/E3-) is that very low level DNA replication can be observed when higher multiplicity of infections are used. Others have shown the emergence of replication-competent adenoviruses (RCA) in the viral stocks during the virus propagation in 293 cells (Seth, 1998). RCA appear to produce by homologous recombination between the E1 sequences present in 293 genome and the recombinant adenoviral constructs. However the use of cell lines which contain only the adenoviral sequences minus the viral ITRs can prevent the formation of RCA. The use of longer deleted vectors will also minimize the formation of RCAs.

Because adenoviruses like other viral vectors, lack cell and tissue specificity several approaches have been proposed to introduce cell and tissue specificity in adenoviral vectors. One approach to achieve targetable gene transfer is through receptor-mediated endocytosis. Work from many laboratories has now shown that if plasmid DNAs are co-internalized with adenovirus into the endosome, adenovirus can also increases the transfection efficiency of such DNAs into the cells by avoiding lysosomal degradation of the DNA. Monocationic and polycationic liposomes can further increase such viral-mediated transfection (Yoshimura *et al.*, 1993; Seth *et al.*, 1994; 1996). Moreover, by linking plasmid DNA and/or adenovirus with other ligands such as epidermal growth factor, transferrin. DNA can be delivered into cells through specific receptors (Seth, 1994c, De *et al.*, 1998). Attempts have been made to target adenoviruses to specific cell types by coupling ligands or antibodies to the adenovirus capsid proteins (Krasnykh *et al.*, 1996). Specificity of the transgene expression can be also introduced by using tissue specific such as carcinoembryonic antigen (CEA) for the treatment of pancreatic and colon cancers, mucin (MUC)-1 promoters for breast cancer cells, alpha-fetoprotein promoters for hepato cellular carcinoma, and tyrosinase promoter for melanoma, just to mention a few (Seth, 1998). Another approach to exploit the best properties of adenoviruses and other viral vectors is to use the hybrid vectors. Adenoviruses have been successfully linked with AAV, retroviruses, and herpes viruses. Another interesting approach to adenoviruses to selectively target certain cancer cells is the use of viral vectors which undergo conditional replication only in tumor cells but not in normal cells have recently been proposed (reviewed in Seth, 1998).

In spite of the very high level transgene expression in the adenovirally infected cells, direct administration of adenoviruses into the tumors can at best infect 5–10% of the cells in the tumor mass. Adenoviral vectors which induce a bystander effect, or replication-restricted vectors can at least in part overcome this problem, and hence research in this area is likely to be valuable. Methods will need to be developed whereby adenoviral vectors are modified in such a way that multiple administrations (if needed) will be feasible.

5. CONCLUSION

Since the original idea of vector-based gene therapy, a great deal of progress in adenoviral vector development has been made. Judging from the published work using the adenoviral vectors, it is apparent that the existing adenoviral vectors have provided excellent tools to learn the basic science of cancer gene therapy; and will most likely find some applications in the clinical setting. It has become clear that a high level transient gene expression through adenoviral vectors even in a limited number of the target cells could also be potentially beneficial. The most impressive progress in the vector development is in the area of generating the adenoviral vectors with complete deletion of viral sequences. These vectors if can be grown to high titers and have level transgene expression are likely to have minimum immunological responses the most common problem associated with adenoviruses. It is encouraging to note that in Phase I clinical trials, first generation adenoviral vectors have been found to be fairly safe.

REFERENCES

Bergelson, J.M., Cunningham, J.A., Droguett, G., Kurt-Jones, E.A., Krithivas, A., Hong, J.S., Horwitz, M.S., Crowell, R.L., and Finberg, R.W., 1997, Isolation of a common receptor for Coxsackie B viruses and adenoviruses 2 and 5, *Science (Washington DC)*, **275**:1320–1323.

Bett, A.J., Haddara, W., Prevec, L., and Graham, F.L., 1994, An efficient and flexible system for construction of adenovirus vectors with insertions or deletions in early regions 1 and 3, *Proc Natl Acad Sci (U S A)*, **91**:8802–8806.

Craig, C., Wersto, R., Kim, M., Ohri, E., Li, Z., Katayose, D., Lee, S.J., Trepel, J., Cowan, K., and Seth, P., 1997, A recombinant adenovirus expressing p27Kip1 induces cell cycle arrest and loss of cyclin-Cdk activity in human breast cancer cells, *Oncogene*, **14**:2283–2289.

Craig, C., Min, K., Ohri, E., Wersto, R., Li, Z., Katayose, D., and Seth, P., 1998, Consequences of adenoviral-mediated p16 protein expression on cell cycle arrest are determined by the endogenous Rb and p16 status in human breast cancer cells, *Oncogene*, **16**:265–272.

Dai, Y., Schwarz, E.M., Gu, D, Zhang, W.W., Sarvetnick, N., and Verma, I.M., 1995, Cellular and humoral immune responses to adenoviral vectors containing factor IX gene: tolerization of factor IX and vector antigens allows for long-term expression, *Proc Natl Acad Sci U S A*, **92**:1401–1405.

De, S.K., Venkateshan, C.N.S., Seth, P., Cajdusek, D.C., Marsh, J.W., and Gibbs, C.J., 1998, Adenovirus-mediated HIV-1 nef expression in human monocytes/macrophages and effect of Nef on down-modulation of Fcγ receptor, *Blood*, **91**:2108–2117.

Fisher, K.J., Choi, H., Burda, J., Chen, S.J., and Wilson, J.M. Recombinant adenovirus deleted of all viral genes for gene therapy of cystic fibrosis, *Virology*, **217**:11–22, 1996.

Hardy, S., Kitamura M, Harris-Stansil T, Dai, Y., and Phipps, M.L., 1997, Construction of adenovirus vectors through Cre-lox recombination, *J Virol.*, **71**:1842–1849.

Kass-Eisler, A., Leinwand, L., Gall, J., Bloom, B., and Falck-Pedersen, E., 1996, Circumventing the immune response to adenovirus-mediated gene therapy, *Gene Ther.*, **3**:154–162.

Katayose, D., Gudas, J., Nguyen, H., Srivastava, S., Cowan, K. and Seth, P.,1995a, Cytotoxic effects of a recombinant adenovirus expressing human wild type p53 on tumor and normal mammary epithelial cells, *Clinical Cancer Res.*, **8**:889–898.

Katayose, D., Wersto, R., Cowan, K.H., and Seth, P., 1995b, Effects of a recombinant adenovirus expressing WAF1/Cip1 on cell growth, cell cycle, and apoptosis, *Cell Growth Differ.*, **6**:1207–1212.

Katayose, Y., Kim, M., Rakkar, A., Li, Z., Cowan, K., and Seth, P., 1997, Promoting apoptosis: a novel activity associated with cyclin-dependent kinase inhibitor p27/Kip1, *Cancer Res.*, **57**:5441–5447.

Kim, M., Katayose, Y., Li, Q., Rakkar, A., Li, Z., Hwang, S., Katayose, D., Cowan, K., Trepel, J., and Seth, P., 1998, Recombinant adenovirus Von Hippel-Lindau-medaited cell cycle arrest is associated with the induction of cyclin-dependent kinase inhibitor p27 (kip1), *Biochem. Biophys. Res. Commm.*, **253**:672–677.

Krasnykh, V.N., Mikheeva, G.V., Douglas, J.T., and Curiel, D.T., 1996, Generation of recombinant adenovirus vectors with modified fibers for altering viral tropism, *J Virol.*, **70**:6839–6846.

Li, Z., Shanmugam, N., Katayose, D., Huber, B., Srivastava, S., Cowan, K., and Seth, P., 1997, Enzyme/prodrug gene therapy approach for breast cancer using a recombinant adenovirus expressing Escherichia coli cytosine deaminase, *Cancer Gene Ther.*, **4**:113–117.

Parks, R.J., Chen, L., Anton, M., Sankar, U., Rudnicki, M.A., and Graham, F.L., 1996, A helper-dependent adenovirus vector system: removal of helper virus by Cre-mediated excision of the viral packaging signal, *Proc Natl Acad Sci (U S A).*, **93**:13565–13570.

Rakkar, A., Katayose, Y., Kim, M, Craig, C., Choi, Y., Lee, S., Ohri, E., Li, Z., Cowan, K.H., and Seth, P., 1999, p53 mediated apoptosis is associated with up-regulation of Fas/CD95/Apo-1, *Cell Death and Differentiation* (in press).

Seth, P., Fitzgerald, D., Ginsberg, H., Willingham, M., and Pastan, I., 1996, Pathway of Adenovirus entry into cells, in : Virus Attachment and Entry into Cells (R. Crowell, R., and K., Lonberg-Holm eds), pp 191–195, American Society for Microbiology, Washington, D.C.

Seth, P.K., Pastan, I., and Willingham, M., 1987, Adenovirus dependent changes in cell membrane permeability: Role of Na^{+}, K^{+} ATPase, *J. Virol.*, **61**:883–888.

Seth, P., 1994a, Adenovirus-dependent release of choline from plasma membrane vesicles at an acidic pH is mediated by the penton base protein, *J Virol.*, **68**:1204–1206.

Seth, P., 1994b, Mechanism of adenovirus-mediated endosome lysis: role of the intact adenovirus capsid structure, *Biochem Biophys Res Commun.*, **205**:1318–1324.

Seth, P., 1994c, A simple and efficient method of protein delivery into cells using adenovirus, *Biochem Biophys Res Commun.*, **203**:582–587.

Seth, P., 1998, Vectors for cancer gene therapy, in Gene therapy for the treatment of cancer (B. Huber, and I. Magrath, eds) pp 41–77, Cambridge University Press, Cambridge, UK.

Seth, P., Brinkmann, U., Schwartz, G.N., Katayose, D., Gress, R., Pastan, I., and Cowan, K., 1996, Adenovirus-mediated gene transfer to human breast tumor cells: an approach for cancer gene therapy and bone marrow purging, *Cancer Res.*, **56**:1346–1351.

Seth, P., Rosenfeld, M., Higginbotham, J., and Crystal, R.G., 1994, Mechanism of enhancement of DNA expression consequent to cointernalization of a replication-deficient adenovirus and unmodified plasmid DNA, *J Virol.*, **68**:933–940.

Shenk, T., 1996, Adenovidae: The viruses and their replication in: Field Virology third edition, (B.N. Fields *et al.* eds), pp 2111–2148, Lippincot-Raven Publisher, Philadelhia.

Stratford-Perricaudet, L.D., Makeh, I., Perricaudet, M., and Briand, P., 1992, Widespread long-term gene transfer to mouse skeletal muscles and heart, *J Clin Invest*, **90**:626–630.

Yoshimura, K., Rosenfeld, M.A., Seth, P., and Crystal, R .G., 1993, Adenovirus-mediated augmentation of cell transfection with unmodified plasmid vectors, *J Biol Chem.*, **268**:2300–2303.

3

RETROVIRUS VECTORS

Yasuhiro Takeuchi and Massimo Pizzato

Chester Beatty Laboratories
The Institute of Cancer Research
237 Fulham Road, London SW3 6JB
England

1. INTRODUCTION

Retroviruses are enveloped viruses which contain 7–12 kb RNA genomes. After virus entry to cells via specific cell-surface receptors, their genomes are reverse transcribed into double-stranded DNA and subsequently integrated into the host chromosome in the form of provirus. The provirus replicates as the host chromosome replicates and is transmitted to all of the progeny of the host cells. This ability of retroviruses to stably transfer genetic information is an attraction to use retroviruses as gene transfer vectors. Because the retrovirus genome is relatively small and well-characterised, it is possible to engineer vector packaging systems which produce vectors encoding only transgenes and does neither produce replication competent viruses (RCV) nor transfer virus structural genes. Absence of RCV is required for safety, while the lack of expression of any viral protein in recipient cells could be advantageous in many preclinical and clinical settings as viral proteins may elicit undesirable immune responses. Most widely used retrovirus vectors to date are based on murine leukemia viruses (MLV). This chapter mainly summarises studies on MLV systems, but many principles can be applied for other retrovirus vector systems. Recent progress in the development of lentivirus vectors which, unlike MLV vectors, are able to infect non-dividing cells is also summarised. Detailed reviews on retrovirus vectors have been published elsewhere (Günzburg and Salmons, 1996; Miller, 1997; Collins and Porter, 1998).

2. DEVELOPMENT OF VECTOR SYSTEMS

2.1. Naturally Occuring Vectors

Acutely transforming retroviruses in birds and nonhuman mammals can be regarded as naturally occuring retrovirus vectors. These viruses have acquired cellular

Cancer Gene Therapy: Past Achievements and Future Challenges, edited by Habib
Kluwer Academic/Plenum Publishers, New York, 2000.

oncogenes which are transferred and spread by virus infection resulting in cell transformation and tumour formation. With the exception of some strains of Rous sarcoma virus which carry the *src* oncogene in addition to a full set of virus genes and therefore can replicate by themselves, most oncogenes are encoded by replication-defective virus genomes. Such oncogene-containing virus genomes lack all or part of virus structural genes but retain all necessary *cis*-acting sequences and can replicate as mixtures with non-transforming, nondefective "helper" viruses which provide virus proteins *in trans*. Thus, nature shows us a way to transfer non-viral genes to the cell using the mechanism of retrovirus replication. Development of packaging systems of retrovirus vectors started by mimicking and modifying this natural strategy. It may be interesting to note that the same strategy that has been used by "oncogenic" retroviruses is used in "cancer" gene therapy.

2.2. Design of Vector Genomes

2.2.1. cis-*Acting Retrovirus Elements.* Vector genomes include a number of *cis*-acting viral sequences which are critical for vector transduction (Fig. 1). Long terminal repeats (LTR) contain sequences to direct transcription initiation, polyadenylation at 3′ end of RNA genomes, template switch during reverse transcription and provirus integration. A packaging signal is essential for efficient encapsidation of the RNA genome and called Ψ in MLV and E for avian vector systems. A tRNA primer binding site (PBS) and a polypurine tract (PPT) are required for initiation of first and second strand DNA synthesis, respectively.

2.2.2. Insertion and Expression of Transgenes. Because the *cis*-acting sequences are localized in either 5′ or 3′ ends of virus genome, transgenes can be inserted in place of virus structural genes in the middle of virus genome. The most simple vectors have a single transgene whose expression is driven by the retroviral promoter in the LTR, whereas many recombinant retrovirus vectors are designed to express two genes one of which is often a selectable marker. While the first gene can be translated from genomic RNAs, the second gene requires an additional strategy for expression, such as splicing (Cepko *et al.*, 1984), transcription from heterologous promoters (Morgenstern and Land, 1990) and translation directed by an internal ribosome entry signal (IRES) (Adam *et al.*, 1991; Ghattas *et al.*, 1991). Although the usage of heterologous promoters may allow flexible control of transgene expression, such as tissue-specific expression, several studies have shown that the viral LTR promoter and heterologous promoters do not always work efficiently in combination due to promoter interference (Emerman and Temin, 1984; Emerman and Temin, 1986). To circumvent this, an entire expression unit of transgene with its own polyadenylation signal can be inserted in the opposite direction to virus genome transcription. An alternative strategy is to insert transgenes in U3 region of 3′-LTR. The heterologous promoter and transgene inserted in this region appear duplicated in both 5′ and 3′ LTR after transduction, hence such vectors are called "double copy vectors" (Hantzopoulos *et al.*, 1989).

2.2.3. Self Inactivating Vectors. Self inactivating vectors have been developed by modification of 3′-LTR U3 region. Deletion of the virus enhancer and/or promoter results in production of inactive proviruses, which lack transcription activity at both 5′- and 3′-LTR, in vector transduced cells whereas expression of transgenes from

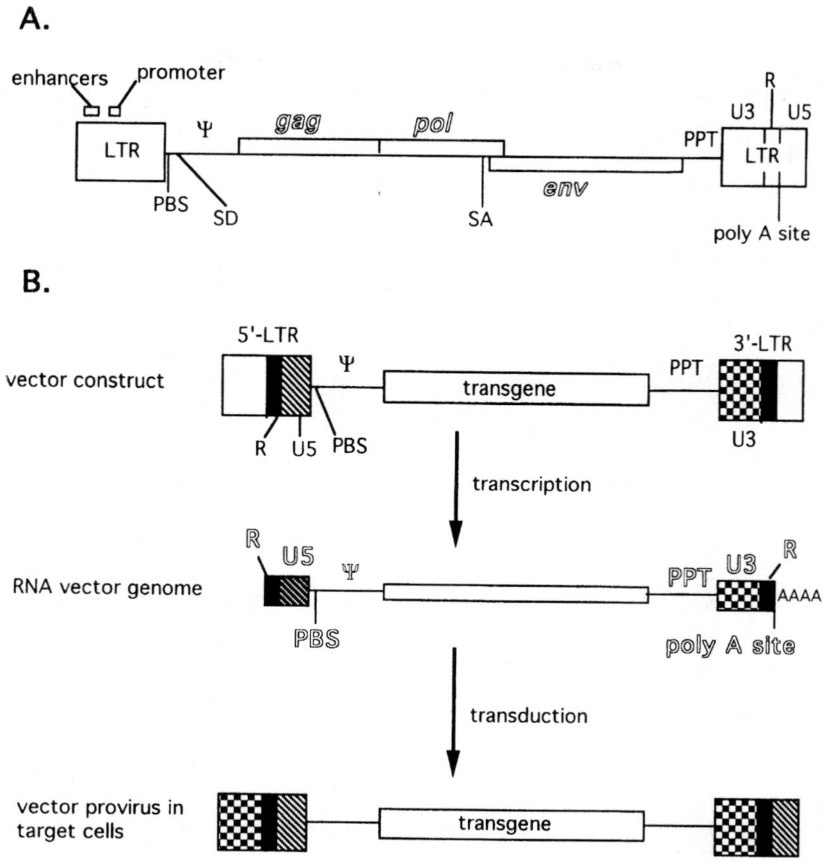

Figure 1. Genome structure of MLV and MLV based vector. A. MLV provirus. Structural genes, *gag*, *pol* and *env*, can be replaced by transgenes in vector genomes. SD and SA, splicing donor and acceptor to produce subgenomic mRNA coding for *env*. B. Vector construct, vector RNA genome and provirus in the target cell. Critical *cis* elements are shown on vector RNA genome. Transduction requires virus structural proteins, Gag for packaging, protease in *pol* for virus maturation, Env for virus entry to the cell and Pol for reverse transcription and provirus integration.

heterologous promoters remains active or may even be enhanced because of loss of promoter interference by viral promoters (Yu *et al.*, 1986). This inactivation should improve safety of the vector because lack of expression from 5′-LTR reduces the possibility of mobilisation of vector genome and that from 3′-LTR reduces the possibility of activation of cellular genes downstream of provirus integration sites. However, the titre of such self inactivating vectors tends to be low for unclear reasons. As an alternative strategy, self deleting vectors exploiting the Cre-loxP recombination system have been developed recently (Choulika *et al.*, 1996; Russ *et al.*, 1996b). In these systems, a single LTR containing a transgene expression unit, but not the other parts of vector genome, remain in the target chromosome after vector transduction and subsequent Cre-loxP excision.

2.2.4. Targeted Transgene Expression. Restricted delivery (see below, 2.3.4.) and expression of transgenes to specific types of target cell is critical for development of *in vivo* gene therapy protocols. Limited successes have been reported by incorporation of cell-type/tissue specific enhancers and promoters in U3 of 3′-LTR: Addition of the muscle creatine kinase enhancer between viral enhancer and promoter resulted in differentiation-specific expression in myogenic cells but was not entirely tissue-specific (Ferrari *et al.*, 1995); Replacement of the viral enhancer with the tyrosinase enhancer and promoter enabled specific expression in melanoma cells (Vile *et al.*, 1995; Diaz *et al.*, 1998). However, it is not clear whether it will ultimately be possible to incorporate all the control sequences to achieve faithful cell-type/tissue specificity. This area of research requires extensive studies on regulation of gene expression and largely empirical studies to incorporate cellular sequences into retrovirus vector genomes. It may be noteworthy that many cellular genes have sequences regulating their expression in their introns which are difficult to incorporate in retrovirus vectors and that many studies on gene regulation using transient transfection assays may be irrelevant for retrovirus vectors which stably integrate into host chromosomes.

2.2.5. Inducible Vectors. Timely or localised control of transgene expression may be useful in many clinical settings. Reversibly inducible vectors can be constructed using the tetracycline-controlled trans-activator-responsive promoter system (Gossen and Bujard, 1992; Paulus *et al.*, 1996). The Cre-loxP recombination system can be used to either activate (Russ *et al.*, 1996a) or delete (Westerman and Leboulch, 1996; Fernex *et al.*, 1997; Li *et al.*, 1997) transgenes irreversibly under controlled conditions. However, these strategies require much improvement before their clinical application.

2.3. Packaging Cells

2.3.1. Virus Proteins Essential for Packaging of Retrovirus Vector Genomes. Retroviral proteins encoded by *gag*, *pol* and *env* genes (Fig. 1) must be provided to package vector RNA genomes. *Gag* encodes structural proteins of the virus core, including matrix (MA), capsid (CA) and nucleocapsid (NC) proteins, whereas *pol* encodes enzymes necessary for virus replication, including protease (PRO), reverse transcriptase (RT) and integrase (IN) (Vogt, 1997). Envelope proteins which can be derived from a range of retroviruses or other viruses are necessary to direct virus entry into cells via their specific receptors.

2.3.2. Packaging Constructs. A classical way to produce helper-free retrovirus vectors is to introduce vector genome DNA into packaging cells which stably express viral proteins. The earliest examples of packaging cells for avian and murine systems contained proviruses as single packaging constructs which provided all the viral proteins, but lacked packaging signals (Linial *et al.*, 1978; Mann *et al.*, 1983; Watanabe and Temin, 1983). However, RCV can arise easily from such packaging constructs by inefficient processes of their co-packaging and recombination with vector genome to acquire a Ψ sequence. To reduce the occurrence of RCV, further deletion of the 3′-LTR and PPT (Fig. 1) which are necessary for initiation of the second DNA strand synthesis was introduced in the "second generation" packaging cell lines, such as the amphotropic PA317 (Miller and Buttimore, 1986). "Second generation" constructs occasionally give rise to RCV, presumably by two events of recombination with vector genomes to acquire Ψ, 3′-LTR and PPT. Packaging constructs were split to two plasmids, one for *gag-pol* and the other for *env* expression, in "third generation" packaging lines, such as ΨCRIP (Danos and Mulligan, 1988) and GP + EAM12 (Markowitz *et al.*, 1988). At least three recombination events are required to generate RCV from split packaging constructs. Thus, only one case of RCV occurrence from "third generation" packaging cells has been reported to date (Chong and Vile, 1996; Chong *et al.*, 1998) while many vectors produced by these cell lines have been rigorously tested for the presence of RCV. Improvement of viral protein expression has been achieved by further modifications of "third generation" split constructs: Ampli-GPE system uses episomal amplification of plasmids containing bovine papilloma virus sequence (Takahara *et al.*, 1992); FLY cells employ a strategy of translation re-initiation for selectable markers (Cosset *et al.*, 1995); ProPak-A uses a strong cytomegalovirus promoter to drive *gag-pol* and *env* genes (Rigg *et al.*, 1996).

2.3.3. Envelope Pseudotyping. A range of retroviral and non-retroviral envelope proteins can be incorporated into vector particles in order to direct entry of vectors into cells expressing corresponding receptors. Such complementation of virus cores with heterologous envelope proteins is called pseudotyping. The most widely used envelope protein for gene transfer to human cells derives from amphotropic MLV (MLV-A), typically pseudotyping vector cores derived from ecotropic Moloney MLV. Env proteins of gibbon ape leukemia virus (GALV) (Miller *et al.*, 1991), cat endogenous virus RD114 (Cosset *et al.*, 1995) and 10A1 MLV (Miller and Chen, 1996) allow vector entry to a wide range of human cell types at high titres and have been incorporated into packaging cell lines. Receptors for MLV-A, GALV and 10A1 have been identified as transmembranous proteins with function of phosphate transporter (O'Hara *et al.*, 1990; Kavanaugh *et al.*, 1994; Miller and Miller, 1994; Olah *et al.*, 1994; van Zeijl *et al.*, 1994), whereas RD114 receptor is yet to be identified. Among nonretroviral envelope proteins, vesicular stomatitis virus G protein (VSV-G) has been most extensively studied following the original observation of pseudotyping of a retrovirus with VSV-G by mixed infection (Weiss *et al.*, 1977). Since pseudotyping of MLV-based vectors using recombinant VSV-G was first demonstrated (Emi *et al.*, 1991), high-titre vectors with a wide host range including human cells have been prepared by concentration using ultracentrifugation and by improvement of VSV-G expression. Resistance of VSV-G pseudotyped vectors to ultracentrifugation is an advantage over retroviral Env pseudotypes which easily shed surface envelope subunits noncovalently associated to transmembrane subunits (Burns *et al.*, 1993). However because of cell toxicity of VSV-G, its expression systems to date are based on transient transfection (Yee *et al.*, 1994) or conditional induction by tetracyclin withdrawal (Yang *et al.*, 1995; Ory *et al.*, 1996), estrogen supplement (Chen *et al.*, 1996) or Cre-loxP recombination (Arai *et al.*, 1998). Quality control of vectors produced by these non-constitutive production systems is problematic. Another potential problem is that VSV-G pseudotypes can infect and replicate in the packaging cells during vector production because of lack of receptor interference (Arai *et al.*, 1998). Such vector amplification may cause genotoxicity and increases the risk of recombination of vector genomes with packaging constructs.

2.3.4. Envelope Targeting To achieve redirection or preferable infection of retrovirus vectors to specific tissues/cell types, several strategies have been explored to attach vector particles to cell surface molecules via their physiological ligands or antibodies against them. In early attempts, ecotropic MLV vectors could infect human cells, albeit at low titre, when they were linked to target cell surface molecules by bridge molecules which contained two antibody domains, one against the viral Env protein and the other against the targeted molecule (Roux *et al.*, 1989; Etienne-Julan *et al.*, 1992). A recent modification of this strategy has allowed efficient retargeting of avian leukosis virus (ALV) vectors. This method utilises a fusion protein of the soluble ALV receptor with EGF as a molecular bridge, allowing docking of the virus to the EGF receptor (Snitkovsky and Young,

1998). In this case, the soluble viral receptor may serve as not just a linker but also an activator of the fusogenicity of the viral glycoprotein. Env proteins have been genetically engineered to accommodate ligands to specific cell surface molecules. Insertion of single chain antibody fragments (scFv) to spleen necrosis virus (SNV) Env allowed SNV vectors to infect human cells bearing targeted molecules while the same cells were refractory to original SNV vectors without the scFv (Chu *et al.*, 1994; Chu and Dornburg, 1995; Chu and Dornburg, 1997; Jiang *et al.*, 1998). However, similar strategies gave only disappointing results using ecotropic MLV which may require the interaction with its own specific receptor to trigger efficient membrane fusion (Cosset and Russell, 1996) In contrast, the addition of a ligand at the N-terminus of the amphotropic MLV (MLV-A) Env protein, whose specific receptors are present on most human cells, led to preferential infection of cells expressing a melanoma associated antigen (Martin *et al.*, 1998). To further restrict MLV-A infection, two-step or multi-step targeting strategies which exploit protease-substrate interaction and conformational masking of Env fusion activity have been proposed (Cosset and Russell, 1996; Valsesia Wittmann *et al.*, 1996; Morling *et al.*, 1997; Valsesia Wittmann *et al.*, 1997). As an additional approach, chemical modifications of sugars made ecotropic MLV vectors infectious to human hepatocytes via asialoglycoprotein receptors (Neda *et al.*, 1991). Such an approach, however, may not be widely applicable as few tissue/cell type specific receptors for specific sugars are known. Overall, although the above studies have proved that targeted infection *in vitro* can be achieved, further improvement in specificity and transduction efficiency is required and *in vivo* efficacy has yet to be evaluated.

2.3.5. Inactivation of Vectors in Human Serum. Mammalian C type retroviruses including MLV were reported to be lysed and inactivated by human complement in late 1970's (Welsh *et al.*, 1975; Welsh *et al.*, 1976). It was proposed that a conserved region of transmembrane subunits of Env activates the classical pathway of complement cascade in an antibody independent manner (Cooper *et al.*, 1976; Bartholomew *et al.*, 1978), suggesting that a range of C type retrovirus strains produced by any cell line would be sensitive to human serum. MLV vectors produced by murine derived packaging cells were rapidly inactivated by fresh human serum, posing a problem for the usage of MLV vectors *in vivo*. However, certain cell lines including human cell lines were shown to produce viruses resistant to human serum (Takeuchi *et al.*, 1994). Further studies showed that the Gal(α1-3)Galactosyl (αGal) sugar structure on the retrovirus envelope is the major factor that causes virolysis by activation of complement via natural antibodies against this sugar structure (Rother *et al.*, 1995; Takeuchi *et al.*, 1996). Because humans, apes and old world monkeys lack a gene encoding αGal transferase, these species do not express αGal sugar, but instead recognise it as a xenoantigen and produce high amount of anti-αGal "natural" antibodies. Thus, packaging cell lines based on human, αGal-negative cells which produce serum-resistant MLV vectors have been developed (Cosset *et al.*, 1995). As lenti- and spumaviruses and VSV-G pseudotypes can also be sensitised to human serum by αGalactosylation, production of vectors based on these viruses for *in vivo* use also requires αGal negative packaging cells (Takeuchi *et al.*, 1997).

2.3.6. Cross-Packaging of Endogenous Retroviruses (ERV). The normal germ line DNA of most vertebrate species contains sequences related to infectious exogenous retroviruses (Boeke and Stoye, 1998; Herniou *et al.*, 1998). The presence of such ERV in retrovirus packaging cell lines poses a safety concern of unwanted transfer of ERV sequences to gene therapy targets. The murine endogenous VL30 sequences can be packaged in

MLV vector particles as efficiently as vector genomes (Scadden *et al.*, 1990; Patience *et al.*, 1998) and the transfer of VL30 and another ERV sequence, intracisternal A-type particle, has been reported for murine packaging cell lines (Hatzoglou *et al.*, 1990; Scadden *et al.*, 1990; Chakraborty *et al.*, 1994). Recombination between a vector genome, a packaging construct and murine ERV sequences gave rise to a replication competent virus in a third generation packaging cell line (Chong *et al.*, 1998), while the recombination of VL30 and MLV has been described (Itin and Keshet, 1983). In contrast several classes of human ERV transcripts, which are abundantly expressed in human packaging cell lines, were not packaged in MLV vector particles (Patience *et al.*, 1998). Moreover, random PCR amplification identified no ERV-like sequences in MLV vector particles produced by human packaging cell lines devoid of any vector genome (W Andrews, C Patience and Y Takeuchi, unpublished results). Therefore, human packaging cells appeared to produce MLV vectors far less contaminated with ERV RNA than murine packaging cells. Cross packaging of ERV into other retrovirus vector particles, such as spuma- and lentivirus vectors, has not been analysed extensively and is a subject of future research.

2.3.7. Transient Packaging Systems. Whilst stable packaging systems yield large batches of safety-controlled vector preparation, transient packaging systems can produce small batches of different vectors for comparison after transient DNA transfection in a time/labour-economical way. This approach is particularly useful to produce vectors encoding toxic genes or containing toxic virus structural proteins, such as VSV-G and some lentivirus accessory proteins. Human embryonal kidney 293 cells and their derivatives are currently used as producer cells in most studies, although COS-7 cells were also used in an early report (Landau and Littman, 1992). Some systems involve cotransfection of a vector genome and packaging constructs (Finer *et al.*, 1994; Soneoka *et al.*, 1995; Naldini *et al.*, 1996), whereas cells are available which harbour packaging constructs and can produce vectors upon transfection of a vector construct (Pear *et al.*, 1993). It should be emphasized that it is much more difficult to control quality of vectors produced by transient systems than those produced by stable packaging systems and further safety consideration is required before these systems are clinically used.

2.4. Lentivirus Vectors

One major drawback of MLV-based vectors for many gene therapy applications is that they are not able to infect non-dividing cells (Miller *et al.*, 1990). For some applications this may perhaps be used to advantage, for example to restrict transduction to proliferating tumour vascular endothelial cells as opposed to quiescent uninvolved cells. However, for gene delivery to tumour cells, only a small proportion of which will be dividing at the time of vector administration, transduction efficiency by MLV-based vectors will definitely be restricted and retroviral vectors without this limitation will be advantageous. Vectors based on lentiviruses including human immunodeficiency virus (HIV) are currently under development because lentiviruses are known to proliferate in non-dividing cells, such as macrophages.

A first generation vector based on human immunodeficiency virus type 1 (HIV-1) was described in 1996, which could transduce non-dividing cells *in vitro* and *in vivo* (Naldini *et al.*, 1996). Since then further modified HIV-1 vectors, lacking all but the Rev accessory protein, have been shown to be effective in many non-dividing cell types (Zufferey *et al.*, 1997; Kim *et al.*, 1998). Most HIV sequences apart from the *cis* elements required for nuclear export (RRE directing nuclear export of unspliced RNA vector genomes in concert with Rev), packaging, reverse transcription (2 modified LTR, tRNA primer binding site and purine rich sequence) and integration (LTR) were deleted from new generation vector constructs. In addition to such deletions of many parts of HIV-1 genome, the use of heterologous envelopes, such as VSV-G and MLV-A Env, and expression of Gag, Pol and Rev proteins by heterologous promoters reduces the risk of occurrence of unwanted, potentially harmful recombinant viruses. However, the process of vector production in most of the current HIV-1 vector systems includes transient transfection, which is difficult to standardize for yield and safety of the vector preparation and is therefore undesirable for the preparation of clinical reagents.

One potential further problem of HIV (HIV-1 and 2) vector systems is that HIV vector genomes can be packaged and mobilized by a replication competent (RC) HIV in patients already infected by HIV at the time of vector treatment as well as those who become HIV-infected after the vector treatment. Besides unwanted mobilization, recombination between vector and RC HIV genomes may in some circumstances give rise to a recombinant virus even more harmful than RC HIV. In addition, administration of large amounts of HIV antigens as parts of the vector particle may elicit anti-HIV antibodies in HIV negative patients. Such antibodies will produce false positive results in serological tests for HIV infection. To circumvent these problems, vectors based on non-primate lentiviruses including those of feline (Poeschla *et al.*, 1998) and equine (Olsen, 1998) origin are also under development.

3. APPLICATION OF RETROVIRUS VECTORS FOR CANCER GENE THERAPY

3.1. Applications to Date

Ex vivo treatments with retroviral vectors have been used to transduce immunomodulatory molecules (eg. IL-2 and TNF-α) into tumor cells (autologous tumor cell vaccines) and TIL (tumor infiltrating lymphocytes) or to transduce the multidrug resistance gene (MDR) into autologous bone marrow cells in order to induce higher tolerance to chemotherapy agents. Treatment of brain tumors represents the most frequent *in vivo* application of retroviral vectors. The neoplastic cells are the main population which actively replicates in the microenvironment and therefore can be selectively transduced by retroviruses. Either packaging cells or purified vectors can be stereotactically injected into the tumor mass to transduce a cytotoxic gene, such as HSV-Tk, alone or in combination with an immunomodulatory cytokine (eg. IL-2). In other cases, retroviral vectors have been used *in vivo* for treatment of melanoma, ovarian, breast, prostate and lung cancers (Roth and Cristiano, 1997).

3.2. Safety Control

Occurrence of replication competent recombinants (RCR) is the biggest safety issue. For MLV vectors produced by stable producer lines, standard procedures for detection of RCR have been established (Wilson *et al.*, 1997). In contrast, safety control procedures for vectors produced by transient DNA transfection or transplantation of packaging cells have yet to be established. Detection and characterisation of RCR which may arise from emerging vector systems including lentivirus vectors are still in early days.

Chimeric RCR which contain virus core and envelope proteins from different groups of viruses, e.g. HIV cores and MLV-A Env (Reiprich *et al.*, 1997), MLV cores and VSV-G, may be produced.

3.3. Towards in Vivo Application in Future

Further investigation is required to develop retrovirus vectors which work effectively by *in vivo* administration. Efficiency and specificity of delivery and expression of transgenes must be improved and there have not been sufficient pharmacokinetics studies following *in vivo* vector administration. In addition to studies discussed above, several strategies have been studied to enhance transduction efficiency by additives, including liposomes (Hodgson and Solaiman, 1996; Porter *et al.*, 1998) and calcium phosphate (Morling and Russell, 1995). Feasibility and efficacy of such approaches in different settings of *in vivo* treatment should be evaluated in preclinical trials. Producing retrovirus vectors *in vivo* instead of administration of ready-made vector particles has certain attraction to sustain vector input. Recent development of "hybrid" vector systems (Feng *et al.*, 1997; Savard *et al.*, 1997) which deliver retrovirus vectors and packaging functions transiently by other virus vectors may result in high efficiency of retrovirus transduction. However, such strategies to deliver retrovirus structural genes will pose challenging problems in safety.

REFERENCES

Adam, M.A., Ramesh, N., Miller, A.D., and Osborne, W.R., (1991), Internal initiation of translation in retroviral vectors carrying picornavirus 5′ nontranslated regions, *J Virol* **65**:4985–4990.

Arai, T., Matsumoto, K., Saitoh, K., Ui, M., Ito, T., Murakami, M., Kanegae, Y., Saito, I., Cosset, F.L., Takeuchi, Y., and Iba, H., (1998), A new system for stringent, high-titer vesicular stomatitis virus G protein-pseudotyped retrovirus vector induction by introduction of Cre recombinase into stable prepackaging cell lines, *J Virol* **72**:1115–1121.

Bartholomew, R.M., Esser, A.F., and Muller-Eberhard, H.J., (1978), Lysis of oncornaviruses by human serum. Isolation of the viral complement (C1) receptor and identification as p15E, *J Exp Med* **147**:844–853.

Boeke, J.D., and Stoye, J.P., (1997), Retrotransposons, endogenous retroviruses, and the evolution of retroelements, in *Retroviruses* (J.M. Coffin, S.H. Hughes, and H.E. Vermus eds.) pp. 343–436, Cold Spring Harbor Laboratory Press, New York.

Burns, J.C., Friedmann, T., Driever, W., Burrascano, M., and Yee, J.K., (1993), Vesicular stomatitis virus G glycoprotein pseudotyped retroviral vectors: concentration to very high titer and efficient gene transfer into mammalian and nonmammalian cells, *Proc Natl Acad Sci U S A* **90**:8033–8037.

Cepko, C.L., Roberts, B.E., and Mulligan, R.C., (1984), Construction and applications of a highly transmissible murine retrovirus shuttle vector, *Cell* **37**:1053–1062.

Chakraborty, A.K., Zink, M.A., and Hodgson, C.P., (1994), Transmission of endogenous VL30 retrotransposons by helper cells used in gene therapy, *Cancer Gene Ther* **1**:113–118.

Chen, S.T., Iida, A., Guo, L., Friedmann, T., and Yee, J.K., (1996), Generation of packaging cell lines for pseudotyped retroviral vectors of the G protein of vesicular stomatitis virus by using a modified tetracycline inducible system, *Proc Natl Acad Sci U S A* **93**:10057–10062.

Chong, H., Starkey, W., and Vile, R.G., (1998), A replication-competent retrovirus arising from a split-function packaging cell line was generated by recombination events between the vector, one of the packaging constructs, and endogenous retroviral sequences, *J Virol* **72**:2663–2670.

Chong, H., and Vile, R.G., (1996), Replication-competent retrovirus produced by a "split-function" third generation amphotropic packaging cell line, *Gene Ther* **3**:624–629.

Choulika, A., Guyot, V., and Nicolas, J.F., (1996), Transfer of single gene-containing long terminal repeats into the genome of mammalian cells by a retroviral vector carrying the cre gene and the loxP site, *J Virol* **70**:1792–1798.

Chu, T.H., and Dornburg, R., (1995), Retroviral vector particles displaying the antigen-binding site of an antibody enable cell-type-specific gene transfer, *J Virol* **69**:2659–2663.

Chu, T.H., and Dornburg, R., (1997), Toward highly efficient cell-type-specific gene transfer with retroviral vectors displaying single-chain antibodies, *J Virol* **71**:720–725.

Chu, T.H., Martinez, I., Sheay, W.C., and Dornburg, R., (1994), Cell targeting with retroviral vector particles containing antibody- envelope fusion proteins, *Gene Ther* **1**:292–299.

Collins, M., and Porter, C., (1998), Retroviral vectors. in *Blood Cell Biochemistry, Volume 8: Hematopoiesis and Gene Therapy*, (L.J. Fairbairn, and N.G. Testa, eds.) pp. 57–88, Plenum Press.

Cooper, N.R., Jensen, F.C., Welsh, R.M., Jr., and Oldstone, M.B., (1976), Lysis of RNA tumor viruses by human serum: direct antibody-independent triggering of the classical complement pathway, *J Exp Med* **144**:970–984.

Cosset, F.L., and Russell, S.J., (1996), Targeting retrovirus entry, *Gene Ther* **3**:946–956.

Cosset, F.L., Takeuchi, Y., Battini, J.L., Weiss, R.A., and Collins, M.K., (1995), High-titer packaging cells producing recombinant retroviruses resistant to human serum, *J Virol* **69**:7430–7436.

Danos, O., and Mulligan, R.C., (1988), Safe and efficient generation of recombinant retroviruses with amphotropic and ecotropic host ranges, *Proc Natl Acad Sci U S A* **85**:6460–6464.

Diaz, R.M., Eisen, T., Hart, I.R., and Vile, R.G., (1998), Exchange of viral promoter/enhancer elements with heterologous regulatory sequences generates targeted hybrid long terminal repeat vectors for gene therapy of melanoma, *J Virol* **72**:789–795.

Emerman, M., and Temin, H.M., (1984), Genes with promoters in retrovirus vectors can be independently suppressed by an epigenetic mechanism, *Cell* **39**:449–467.

Emerman, M., and Temin, H.M., (1986), Comparison of promoter suppression in avian and murine retrovirus vectors, *Nucleic Acids Res* **14**:9381–9396.

Emi, N., Friedmann, T., and Yee, J.K., (1991), Pseudotype formation of murine leukemia virus with the G protein of vesicular stomatitis virus, *J Virol* **65**:1202–1207.

Etienne-Julan, M., Roux, P., Carillo, S., Jeanteur, P., and Piechaczyk, M., (1992), The efficiency of cell targeting by recombinant retroviruses depends on the nature of the receptor and the composition of the artificial cell- virus linker, *J Gen Virol* **73**:3251–3255.

Feng, M., Jackson, W.H., Jr., Goldman, C.K., Rancourt, C., Wang, M., Dusing, S.K., Siegal, G., and Curiel, D.T., (1997), Stable in vivo gene transduction via a novel adenoviral/retroviral chimeric vector, *Nat Biotechnol* **15**:866–870.

Fernex, C., Dubreuil, P., Mannoni, P., and Bagnis, C., (1997), Cre/loxP-mediated excision of a neomycin resistance expression unit from an integrated retroviral vector increases long terminal repeat-driven transcription in human hematopoietic cells, *J Virol* **71**:7533–7540.

Ferrari, G., Salvatori, G., Rossi, C., Cossu, G., and Mavilio, F., (1995), A retroviral vector containing a muscle-specific enhancer drives gene expression only in differentiated muscle fibers, *Hum Gene Ther* **6**:733–742.

Finer, M.H., Dull, T.J., Qin, L., Farson, D., and Roberts, M.R., (1994), kat: a high-efficiency retroviral transduction system for primary human T lymphocytes, *Blood* **83**:43–50.

Ghattas, I.R., Sanes, J.R., and Majors, J.E., (1991), The encephalomyocarditis virus internal ribosome entry site allows efficient coexpression of two genes from a recombinant provirus in cultured cells and in embryos, *Mol Cell Biol* **11**:5848–5859.

Gossen, M., and Bujard, H., (1992), Tight control of gene expression in mammalian cells by tetracycline-responsive promoters, *Proc Natl Acad Sci U S A* **89**:5547–5551.

Günzburg, W.H., and Salmons, B., (1996), Retroviral Vectors. in *Gene Therapy*. (N.R. Limoine, and D.N. Cooper eds.) pp. 33–60, Oxford, BIOS Scientific Publishers Ltd.

Hantzopoulos, P.A., Sullenger, B.A., Ungers, G., and Gilboa, E., (1989), Improved gene expression upon transfer of the adenosine deaminase minigene outside the transcriptional unit of a retroviral vector, *Proc Natl Acad Sci U S A* **86**:3519–3523.

Hatzoglou, M., Hodgson, C.P., Mularo, F., and Hanson, R.W., (1990), Efficient packaging of a specific VL30 retroelement by psi 2 cells which produce MoMLV recombinant retroviruses, *Hum Gene Ther* **1**:385–397.

Herniou, E., Martin, J., Miller, K., Cook, J., Wilkinson, M., and Tristem, M., (1998), Retroviral diversity and distribution in vertebrates, *J Virol* **72**:5955–5966.

Hodgson, C.P., and Solaiman, F., (1996), Virosomes: cationic liposomes enhance retroviral transduction, *Nat Biotechnol* **14**:339–342.

Itin, A., and Keshet, E., (1983), Apparent recombinants between virus-like (VL30) and murine leukemia virus-related sequences in mouse DNA, *J Virol* **47**:178–184.

Jiang, A., Chu, T.H.T., Nocken, F., Cichutek, K., and Dornburg, R., (1998), Cell-type-specific gene transfer into human cells with retroviral vectors that display single-chain antibodies, *J Virol* **72**:10148–10156.

Kavanaugh, M.P., Miller, D.G., Zhang, W., Law, W., Kozak, S.L., Kabat, D., and Miller, A.D., (1994), Cell-surface receptors for gibbon ape leukemia virus and amphotropic murine retrovirus are inducible sodium-dependent phosphate symporters, *Proc Natl Acad Sci U S A* **91**:7071–7075.

Kim, V.N., Mitrophanous, K., Kingsman, S.M., and Kingsman, A.J., (1998), Minimal requirement for a lentivirus vector based on human immunodeficiency virus type 1, *J Virol* **72**:811–816.

Landau, N.R., and Littman, D.R., (1992), Packaging system for rapid production of murine leukemia virus vectors with variable tropism, *J Virol* **66**:5110–5113.

Li, L.P., Schlag, P.M., and Blankenstein, T., (1997), Transient expression of SV 40 large T antigen by Cre/LoxP-mediated site-specific deletion in primary human tumor cells, *Hum Gene Ther* **8**:1695–1700.

Linial, M., Medeiros, E., and Hayward, W.S., (1978), An avian oncovirus mutant (SE 21Q1b) deficient in genomic RNA: biological and biochemical characterization, *Cell* **15**:1371–1381.

Mann, R., Mulligan, R.C., and Baltimore, D., (1983), Construction of a retrovirus packaging mutant and its use to produce helper-free defective retrovirus, *Cell* **33**:153–159.

Markowitz, D., Goff, S., and Bank, A., (1988), Construction and use of a safe and efficient amphotropic packaging cell line, *Virology* **167**:400–406.

Martin, F., Kupsch, J., Takeuchi, Y., Russell, S., Cosset, F.L., and Collins, M., (1998), Retroviral vector targeting to melanoma cells by single-chain antibody incorporation in envelope, *Hum Gene Ther* **9**:737–746.

Miller, A.D., (1997), Development and application of retroviral vectors, in *Retroviruses* (J.M. Coffin, S.H. Hughes, and H.E. Vermus eds.) pp. 437–474, Cold Spring Harbor Laboratory Press, New York.

Miller, A.D., and Buttimore, C., (1986), Redesign of retrovirus packaging cell lines to avoid recombination leading to helper virus production, *Mol Cell Biol* **6**:2895–2902.

Miller, A.D., and Chen, F., (1996), Retrovirus packaging cells based on 10A1 murine leukemia virus for production of vectors that use multiple receptors for cell entry, *J Virol* **70**:5564–5571.

Miller, A.D., Garcia, J.V., von Suhr, N., Lynch, C.M., Wilson, C., and Eiden, M.V., (1991), Construction and properties of retrovirus packaging cells based on gibbon ape leukemia virus, *J Virol* **65**:2220–2224.

Miller, D.G., Adam, M.A., and Miller, A.D., (1990), Gene transfer by retrovirus vectors occurs only in cells that are actively replicating at the time of infection, *Mol Cell Biol* **10**:4239–4242.

Miller, D.G., and Miller, A.D., (1994), A family of retroviruses that utilize related phosphate transporters for cell entry, *J Virol* **68**:8270–8276.

Morgenstern, J.P., and Land, H., (1990), Advanced mammalian gene transfer: high titre retroviral vectors with multiple drug selection markers and a complementary helper-free packaging cell line, *Nucleic Acids Res* **18**:3587–3596.

Morling, F.J., Peng, K.W., Cosset, F.L., and Russell, S.J., (1997), Masking of retroviral envelope functions by oligomerizing polypeptide adaptors, *Virology* **234**:51–61.

Morling, F.J., and Russell, S.J., (1995), Enhanced transduction efficiency of retroviral vectors coprecipitated with calcium phosphate, *Gene Ther* **2**:504–508.

Naldini, L., Blomer, U., Gallay, P., Ory, D., Mulligan, R., Gage, F.H., Verma, I.M., and Trono, D., (1996), In vivo gene delivery and stable transduction of nondividing cells by a lentiviral vector, *Science* **272**:263–267.

Neda, H., Wu, C.H., and Wu, G.Y., (1991), Chemical modification of an ecotropic murine leukemia virus results in redirection of its target cell specificity, *J Biol Chem* **266**:14143–14146.

O'Hara, B., Johann, S.V., Klinger, H.P., Blair, D.G., Rubinson, H., Dunn, K.J., Sass, P., Vitek, S.M., and Robins, T., (1990), Characterization of a human gene conferring sensitivity to infection by gibbon ape leukemia virus, *Cell Growth Differ* **1**:119–127.

Olah, Z., Lehel, C., Anderson, W.B., Eiden, M.V., and Wilson, C.A., (1994), The cellular receptor for gibbon ape leukemia virus is a novel high affinity sodium-dependent phosphate transporter, *J Biol Chem* **269**:25426–25431.

Olsen, J.C., (1998), Gene transfer vectors derived from equine infectious anemia virus, *Gene Therapy* **5**:1481–1487.

Ory, D.S., Neugeboren, B.A., and Mulligan, R.C., (1996), A stable human-derived packaging cell line for production of high titer retrovirus/vesicular stomatitis virus G pseudotypes, *Proc Natl Acad Sci U S A* **93**:11400–11406.

Patience, C., Takeuchi, Y., Cosset, F.L., and Weiss, R.A., (1998), Packaging of endogenous retroviral sequences in retroviral vectors produced by murine and human packaging cells, *J Virol* **72**:2671–2676.

Paulus, W., Baur, I., Boyce, F.M., Breakefield, X.O., and Reeves, S.A., (1996), Self-contained, tetracycline-regulated retroviral vector system for gene delivery to mammalian cells, *J Virol* **70**:62–67.

Pear, W.S., Nolan, G.P., Scott, M.L., and Baltimore, D., (1993), Production of high-titer helper-free retroviruses by transient transfection, *Proc Natl Acad Sci U S A* **90**:8392–8396.

Poeschla, E.M., Wong-Staal, F., and Looney, D.J., (1998), Efficient transduction of nondividing human cells by feline immunodeficiency virus lentiviral vectors, *Nat Med* **4**:354–357.

Porter, C.D., Lukacs, K.V., Box, G., Takeuchi, Y., and Collins, M.K., (1998), Cationic liposomes enhance the rate of transduction by a recombinant retroviral vector in vitro and in vivo, *J Virol* **72**:4832–4840.

Reiprich, S., Gundlach, B.R., Fleckenstein, B., and Uberla, K., (1997), Replication-competent chimeric lenti-oncovirus with expanded host cell tropism, *J Virol* **71**:3328–3331.

Rigg, R.J., Chen, J., Dando, J.S., Forestell, S.P., Plavec, I., and Bohnlein, E., (1996), A novel human amphotropic packaging cell line: high titer, complement resistance, and improved safety, *Virology* **218**:290–295.

Roth, J.A., and Cristiano, R.J., (1997), Gene therapy for cancer: what have we done and where are we going?, *J Natl Cancer Inst* **89**:21–39.

Rother, R.P., Fodor, W.L., Springhorn, J.P., Birks, C.W., Setter, E., Sandrin, M.S., Squinto, S.P., and Rollins, S.A., (1995), A novel mechanism of retrovirus inactivation in human serum mediated by anti-alpha-galactosyl natural antibody, *J Exp Med* **182**:1345–1355.

Roux, P., Jeanteur, P., and Piechaczyk, M., (1989), A versatile and potentially general approach to the targeting of specific cell types by retroviruses: application to the infection of human cells by means of major histocompatibility complex class I and class II antigens by mouse ecotropic murine leukemia virus-derived viruses, *Proc Natl Acad Sci U S A* **86**:9079–9083.

Russ, A.P., Friedel, C., Ballas, K., Kalina, U., Zahn, D., Strebhardt, K., and von Melchner, H., (1996a), Identification of genes induced by factor deprivation in hematopoietic cells undergoing apoptosis using gene-trap mutagenesis and site-specific recombination, *Proc Natl Acad Sci U S A* **93**:15279–15284.

Russ, A.P., Friedel, C., Grez, M., and von Melchner, H., (1996b), Self-deleting retrovirus vectors for gene therapy, *J Virol* **70**:4927–4932.

Savard, N., Cosset, F.L., and Epstein, A.L., (1997), Defective herpes simplex virus type 1 vectors harboring gag, pol, and env genes can be used to rescue defective retrovirus vectors, *J Virol* **71**:4111–4117.

Scadden, D.T., Fuller, B., and Cunningham, J.M., (1990), Human cells infected with retrovirus vectors acquire an endogenous murine provirus, *J Virol* **64**:424–427.

Snitkovsky, S., and Young, J.A., (1998), Cell-specific viral targeting mediated by a soluble retroviral receptor-ligand fusion protein, *Proc Natl Acad Sci U S A* **95**:7063–7068.

Soneoka, Y., Cannon, P.M., Ramsdale, E.E., Griffiths, J.C., Romano, G., Kingsman, S.M., and Kingsman, A.J., (1995), A transient three-plasmid expression system for the production of high titer retroviral vectors, *Nucleic Acids Res* **23**:628–633.

Takahara, Y., Hamada, K., and Housman, D.E., (1992), A new retrovirus packaging cell for gene transfer constructed from amplified long terminal repeat-free chimeric proviral genes, *J Virol* **66**:3725–3732.

Takeuchi, Y., Cosset, F.L., Lachmann, P.J., Okada, H., Weiss, R.A., and Collins, M.K., (1994), Type C retrovirus inactivation by human complement is determined by both the viral genome and the producer cell, *J Virol* **68**:8001–8007.

Takeuchi, Y., Liong, S.H., Bieniasz, P.D., Jager, U., Porter, C.D., Friedman, T., McClure, M.O., and Weiss, R.A., (1997), Sensitization of rhabdo-, lenti-, and spumaviruses to human serum by galactosyl(alpha1-3)galactosylation, *J Virol* **71**:6174–6178.

Takeuchi, Y., Porter, C.D., Strahan, K.M., Preece, A.F., Gustafsson, K., Cosset, F.L., Weiss, R.A., and Collins, M.K., (1996), Sensitization of cells and retroviruses to human serum by (alpha 1–3) galactosyltransferase, *Nature* **379**:85–88.

Valsesia Wittmann, S., Morling, F.J., Hatziioannou, T., Russell, S.J., and Cosset, F.L., (1997), Receptor co-operation in retrovirus entry: recruitment of an auxiliary entry mechanism after retargeted binding, *Embo J* **16**:1214–1223.

Valsesia Wittmann, S., Morling, F.J., Nilson, B.H., Takeuchi, Y., Russell, S.J., and Cosset, F.L., (1996), Improvement of retroviral retargeting by using amino acid spacers between an additional binding domain and the N terminus of Moloney murine leukemia virus SU, *J Virol* **70**:2059–2064.

van Zeijl, M., Johann, S.V., Closs, E., Cunningham, J., Eddy, R., Shows, T.B., and O'Hara, B., (1994), A human amphotropic retrovirus receptor is a second member of the gibbon ape leukemia virus receptor family, *Proc Natl Acad Sci U S A* **91**:1168–1172.

Vile, R.G., Diaz, R.M., Miller, N., Mitchell, S., Tuszyanski, A., and Russell, S.J., (1995), Tissue-specific gene expression from Mo-MLV retroviral vectors with hybrid LTRs containing the murine tyrosinase enhancer/promoter, *Virology* **214**:307–313.

Vogt, V.M., (1997), Retroviral virions and genomes, in *Retroviruses* (J.M. Coffin, S.H. Hughes, and H.E. Vermus eds.) pp. 27–70, Cold Spring Harbor Laboratory Press, New York.

Watanabe, S., and Temin, H.M., (1983), Construction of a helper cell line for avian reticuloendotheliosis virus cloning vectors, *Mol Cell Biol* **3**:2241–2249.

Weiss, R.A., Boettiger, D., and Murphy, H.M., (1977), Pseudotypes of avian sarcoma viruses with the envelope properties of vesicular stomatitis virus, *Virology* **76**:808–825.

Welsh, R.M., Jr., Cooper, N.R., Jensen, F.C., and Oldstone, M.B., (1975), Human serum lyses RNA tumour viruses, *Nature* **257**:612–614.

Welsh, R.M., Jr., Jensen, F.C., Cooper, N.R., and Oldstone, M.B., (1976), Inactivation of lysis of oncornaviruses by human serum, *Virology* **74**:432–440.

Westerman, K.A., and Leboulch, P., (1996), Reversible immortalization of mammalian cells mediated by retroviral transfer and site-specific recombination, *Proc Natl Acad Sci U S A* **93**:8971–8976.

Wilson, C.A., Ng, T.H., and Miller, A.E., (1997), Evaluation of recommendations for replication-competent retrovirus testing associated with use of retroviral vectors, *Hum Gene Ther* **8**:869–874.

Yang, Y., Vanin, E.F., Whitt, M.A., Fornerod, M., Zwart, R., Schneiderman, R.D., Grosveld, G., and Nienhuis, A.W., (1995), Inducible, high-level production of infectious murine leukemia retroviral vector particles pseudotyped with vesicular stomatitis virus G envelope protein, *Hum Gene Ther* **6**:1203–1213.

Yee, J.K., Miyanohara, A., LaPorte, P., Bouic, K., Burns, J.C., and Friedmann, T., (1994), A general method for the generation of high-titer, pantropic retroviral vectors: highly efficient infection of primary hepatocytes, *Proc Natl Acad Sci U S A* **91**:9564–9568.

Yu, S.F., von Ruden, T., Kantoff, P.W., Garber, C., Seiberg, M., Ruther, U., Anderson, W.F., Wagner, E.F., and Gilboa, E., (1986), Self-inactivating retroviral vectors designed for transfer of whole genes into mammalian cells, *Proc Natl Acad Sci U S A* **83**:3194–3198.

Zufferey, R., Nagy, D., Mandel, R.J., Naldini, L., and Trono, D., (1997), Multiply attenuated lentiviral vector achieves efficient gene delivery in vivo, *Nat Biotechnol* **15**:871–875.

4

TARGETABLE GENE DELIVERY VECTORS

Paul L. Hallenbeck and Susan C. Stevenson

Genetic Therapy Inc., A Novartis Company

1. CANCER THERAPY: TARGETING

The Holy Grail of cancer therapy, and a host of many therapies for major diseases, is to elicit a desired and selective biological effect on a specific cell type(s). In essence, cancer therapy is targeting. The key aspect of this therapy is in achieving a high rate of killing of cancer cells vs. normal cells. Accomplishing this has been extremely difficult for many reasons including the wide array of cell types involved, the systemic dissemination of cancer cells due to metastases, and the narrow biological differences between normal and cancer cells. While tremendous progress has been made in targeting, much still needs to be done, as current cancer therapies are clearly inadequate.

1.1. Conventional Therapy

In the past surgeons have tried to physically remove the tumor surgically without harming normal tissue. Even complete removal of a primary tumor does not ensure survival since earlier metastases to unknown sites in the body are left undetected. There is also some research, which even suggests that surgical intervention may enhance the growth of distant metastases due to removal of tumor cells producing angiogenesis inhibitors (O'Reilly *et al.*, 1994). Finally, in many cases the tumor grows back at the original site after surgical removal. Radiation aims to selectively destroy the most rapidly proliferating cells at the expense of the others. However, tumor cells can escape radiation therapy either by becoming resistant or by being in a non-dividing state during treatment. In addition, radiation is not always selective in that many normal cells are actively dividing and killed by the treatment. (Gastrointestinal cells, hair follicles, etc). Like radiation, chemotherapy is not completely selective and thus destroys many normal cells, and does not kill all tumor cells due to drug resistance and/or division state of the cell.

1.2. Immunotherapy

Immunotherapy is essentially the induction of the immune system to selectively expand a subset of cells and/or antibodies, which can target tumor cells. A variety of

Cancer Gene Therapy: Past Achievements and Future Challenges, edited by Habib
Kluwer Academic/Plenum Publishers, New York, 2000.

approaches are currently in preclinical and clinical studies to attempt to boost the number and specificity of T cells, which will track and destroy tumor cells (For review see Rosenberg, 1998).

One approach, the use of monoclonal antibodies specific for tumor cells, was thought to be the great promise for cancer therapy approximately 20 years ago. However, most antibodies were generated in mice and elicited immune responses in humans, which effectively eliminated the therapeutic antibodies. However, they are now again showing great promise, as methods have been developed to humanize these antibodies. Of significant note are the clinical studies utilizing a humanized form of a monoclonal antibody (Herceptin) developed at Genetech that targets the HER2/neu receptors which are over-expressed on approximately 30% of breast cancer patients. Phase II trials have demonstrated a 65% increase in delay of disease progression in those patients receiving Herpcetin and standard therapy compared with standard therapy alone (Robertson, 1998). This therapy again relies on the fact that systemically injected antibody binds selectively to tumor cells, and kills tumor cells via either inclusion of a toxin moiety fused to the antibody or in some cases the antibody may elicit a toxic effect through binding alone. Newer and more potent antibody-toxin fusions are becoming available, some of which can even extend the killing effect to surrounding tumor cells (Niculescu-Duvas, 1996 and references therein).

1.3. Barriers to Targeting

Tumors have an intricate and unusual blood supply network which renders binding of a molecule to the underlying tumor cell and support cells difficult. Tumors have high interstitial pressure, uneven and changing blood flow throughout the tumor, and an endothelial basement membrane barrier, all of which make the binding of any molecule to a significant fraction of tumor cells difficult (for review see Jain, 1997). In addition, tumors are composed of a multitude of cell types, which contribute to the overall ability of a tumor to thrive. Even the tumor cells within one patient can vary in genetic makeup and biological characteristics. Finally, any therapy must have the appropriate pharmacological profile *in vivo*, taking into account many parameters such as distribution and half-life.

2. CANCER GENE THERAPY: VECTORS

Gene therapy is essentially any method, which can deliver a desired gene(s) to a desired site(s), i.e. tumors, most often delivered to the site by gene therapy vectors. A vector is the substance, which carries the DNA to the desired state (nucleus of target cells) and can be synthetic, and/or viral based.

In vivo gene therapy for cancer for the most part has not involved inclusion of cell targeting but instead has focused on intra-tumoral or *in situ* injection of the vector. Some methods have been able to partially target the gene therapy vector by injecting into partially confined sites harboring metastases. These include the peritoneum (Deshane *et al.*, 1997), pleural space (Sterman, 1998), and metastases in the liver (Anderson *et al.*, 1998). However, even treatment for partially localized tumor may not be efficient since metastases will spread and localize to many various organs throughout the body. Thus vector development must focus on systemically deliverable vectors which can achieve selective and reasonable transduction of tumor and/or tumor blood vessels.

Of course many of the barriers to chemotherapy and monoclonal antibody therapy will also hold true for viral or synthetic vectors. The vector must not be inactivated by the immune system or any other facet of the therapy and should be inert with the exception of binding to the desired cell type(s) within the tumor. Ideally the vector should remain in circulation only long enough to selectively transduce cells within any tumor mass. Once selective transduction is achieved then a multitude of cytotoxics and expandable therapies exist which should be able to be incorporated into vectors for effective therapy. These include HSV-Tk, cytosine deaminase, carboxypeptidase, p53 or other complementing genes, E1a, restriction of vector replication approaches, and the induction of immune responses (for review see Cusack *et al.*, 1998).

3. GENE DELIVERY VECTORS-PAST AND PRESENT ACHIEVEMENTS

Targeted gene delivery has been the emphasis in many gene delivery systems including retroviral, adenoviral, and synthetic vectors. Specific gene delivery to desired tissues or cell types can be "targeted" on multiple levels including transcriptional control through regulatable promoters and manipulation of the receptor specificity of the viral or non viral system. Recently, interest has focused on evaluating approaches for viral vector targeting by modifying the viral attachment protein in order to redirect the receptor specificity of the vector particles.

3.1. Adenoviral Vectors

Recently, work has focused on the development of targeted adenoviral vectors. Adenoviral vectors are useful for both *in vivo* and *in situ* gene delivery testing in animal tumor models and appear to be promising for potential therapeutic applications in humans including cancer and cardiovascular gene therapy. However, caution must be utilized in comparing efficiencies of vector targeting and efficacy in mice since mice are not the natural hosts of human adenoviruses. Adenoviral vectors when delivered systemically by intravenous administration in mice show a marked preferential transduction of liver cells although detectable levels of vector DNA can be found in most other tissues examined (Smith *et al.*, 1993). For reasons of safety and efficacy it will be important to control the cellular specificity of gene delivery by adenoviral vectors. One way to direct a vector to a specific target tissue or cell type is by selection of the route of administration (e.g. intratumoral, intranasal, or intramuscular delivery). In addition to the physical localization of gene delivery vectors, the transduction efficiency of these localized target cells may be increased by the addition of specific ligands and this may increase vector potency. Several approaches have been described to target adenoviral vectors and gene expression and are described below.

3.1.1. Transcriptional Regulation for Targeted Gene Expression. The use of regulatable or tissue specific promoters represents an additional strategy which will be of value in restricting the expression of the foreign gene to a desired cell type of tissue, particularly for cancer (for review see Clary *et al.*, 1998). A multitude of vectors have now been described utilizing tumor specific promoters to express cytotoxic genes for cancer therapy, including those from our own laboratory (Kaneko *et al.*, 1995). This general approach is widely applicable as specific promoters could be incorporated into all currently used gene delivery systems and new promoters are rapidly being identified.

3.1.2. Targeting Adenoviral Vectors Via Fiber Modifications. While both the delivery route and tissue specific promoters will contribute to the cellular specificity which will be required for safe and effective therapies, a key factor in determining susceptibility of a particular cell to viral infection is the expression of receptors which permit viral attachment and entry. The capacity of human adenoviruses to bind to and infect a broad range of cultured cell lines and primary tissues from different species indicates that the adenovirus type 5, group C receptor is widely distributed and evolutionarily conserved. The identity of the group B coxsackievirus and group C adenovirus receptor (CAR) in human and mouse provides evidence for this observation (Bergelson *et al.*, 1997; Tomko *et al.*, 1997). One approach to selective cell transduction is to manipulate the adenovirus capsid in such a way as to redirect or change the receptor specificity and target the vector to a specific cell type or tissue. Targeting can be achieved by constructing an altered viral capsid which is no longer capable of binding to the normal cellular receptor, CAR but has acquired the ability to bind a specific cell type via a new receptor which has a defined and restricted pattern of expression. Selective transduction of tumor cells for example would be achieved by constructing a vector which selectively binds to a receptor which is overexpressed on the surface of tumor cells.

Attachment of the adenovirus particle to the cell is mediated by a high affinity interaction between the fiber protein and CAR (Wickham *et al.*, 1993). Following binding, virion cellular entry is mediated by an interaction between RGD peptide sequences in the penton base and cell surface integrins, which act as coreceptors (Wickham *et al.*, 1993). As a first step in the cellular transduction process, the interaction between the fiber protein and the cell is a logical target for controlling the cell specificity for transduction by adenoviral vectors. The other major adenoviral capsid proteins including hexon (Roy *et al.*, 1997) and penton (Wickham *et al.*, 1995), in addition to fiber (Karnasykh *et al.*, 1996; Stevenson *et al.*, 1997; Krasnykh *et al.*, 1997), have been modified to incorporate specific antigenic epitopes or receptor ligands.

The fiber protein is responsible for attachment of the virion to CAR cell surface receptors and various strategies have been attempted to incorporate novel receptor ligands into the fiber protein structure while still allowing for trimerization of the protein and subsequent capsid association. It was initially shown that the receptor specificity of an adenoviral vector could be changed from the Ad5 CAR receptor to the Ad3 serotype receptor by switching of the fiber head domains (Stevenson *et al.*, 1995; 1997). Also, adding a series of positively charged, lysine amino acid residues onto the C-terminus of the fiber protein allowed for the specific interaction with cell surface heparin sulfated proteoglycans and infection of cells which did not express CAR on the cell surface (Wickham *et al.*, 1996). Modified fiber proteins with novel receptor specificities can also be constructed by replacement of the fiber head domain with other trimeric proteins including head replacements from other naturally occurring adenovirus serotypes that utilize different cellular receptors (Stevenson *et al.*, 1995; 1997), fusion of peptide sequences to the fiber protein C-terminus (Michael *et al.*, 1995, Wickham *et al.*, 1996), or additions of peptide ligands within exposed loop regions of the fiber head domain (Xia *et al.*, 1994, Krasnykh *et al.*, 1998).

Successful targeting will depend on the identification and use of ligands with sufficient binding characteristics. Depending on the target cell type and the ligand selected, it may be necessary to ablate the native receptor tropism of the adenovirus fiber protein.

The crystal structure of the fiber knob domain has been reported (Xia *et al.*, 1994). Subsequent structure:function studies of the adenovirus fiber protein have demonstrated that many mutations will destroy the homotrimeric structure of the functional fiber which is required for assembly onto the viral capsid (Hong and Engler, 1996; Novelli and Boulanger, 1991). Recent studies have demonstrated that mutations of the fiber knob domain can be created that reduce fiber binding to CAR while still maintaining the ability to trimerize. A series of 6 individual amino acid substitutions in the fiber knob, DE loop between residues 460 to 471 reduced receptor binding by 100 fold (Robert Gerard, personal communication). Another approach to ablate the receptor-binding domain is to replace the fiber knob domain with a smaller peptide sequence that will drive trimerization of the fiber protein (Krasnykh *et al.*, 1998). Subsequent incorporation of these receptor binding negative fiber mutants into adenoviral capsids is still relatively new and untested but is a promising concept.

Multi-component systems have also been used to redirect the receptor tropism of adenoviral vectors. Bifunctional conjugates were constructed which consisted of a blocking anti-adenoviral fiber knob Fab linked to basic fibroblast growth factor 2 (Goldman *et al.*, 1997) or to folate (Douglas *et al.*, 1996). These adenoviral complexes successfully transduced cells that overexpressed either the FGF or folate receptor and which were previously refractory to adenovirus transduction. These systems while demonstrating effective retargeting and proof of concept have the disadvantage of not being amenable to large-scale production in preparation for human clinical trials.

Trans-complementation of a fiber-deleted adenoviral vector is an alternative approach to incorporate fiber proteins with novel receptor specificities into adenoviral vectors (Von Seggern *et al.*, 1998). This approach has several potential advantages compared to genetic modification of the adenovirus genome, including ease of switching fibers by preparing multiple cell lines that express novel fiber proteins and the ability to rapidly construct and incorporate multiple targeting ligands into adenoviral vectors. Preliminary studies utilizing this Ad5 fiber expressing cell line, 211B demonstrate successful incorporation of modified fibers into virion capsids. Both fiber genetic modification and trans-complementation may lead to the development of customized adenovirus vectors that selectively target specific cell types.

3.2. Targeted Retroviral Vectors

Modifications of the retroviral receptor ligand have been carried out to incorporate novel receptor binding domains to target specific cell types. The envelope protein of retroviral vectors has been modified to include various targeting ligands including ones for the human epidermal growth factor receptor (Han *et al.*, 1995), erythropoietin receptor (Kasahara *et al.*, 1994) and single chain antibody fragments against the LDL receptor (Somia *et al.*, 1995). Hall *et al.* (1997) have developed retroviral vectors which target and increase the transduction efficiency of vascular lesions through the incorporation of a high-affinity collagen-binding domain derived from von Willebrand clotting factor. There are many examples of incorporation of specific targeting ligands into retroviral vectors, however, due to the complex biology of the retroviral envelope protein, each modification of the retroviral envelope must preserve the additional functions associated with the envelope protein which produces infectious retroviral particles. Unlike adenovirus, where the fiber protein functions in only the initial attachment of the virion to cell surface receptors, the retroviral envelope gp70 protein is involved in the virion cell fusion process that must be maintained for successful targeted gene transfer.

3.3. Non-Viral Vector Gene Delivery

When administered systemically, non-viral vector complexes naturally target certain tissues including lung and tumor beds. Local administration of non-viral complexes has yielded the greatest success so far and many studies have been performed in tumor bearing animals (Yang *et al.*, 1998). Taking the field of drug delivery into account, many approaches for the use of targeting ligands for non-viral systems have been utilized and from these studies there are two major issues to be addressed for targeting (Woodle and Storm, 1998). First, there is a need to reduce "non-specific" interactions of the vector with unwanted biological cells and tissues. This then provides a foundation for addition of ligands for the desired specificity. Interestingly, just establishing vectors with reduced "non-specific" interactions can yield vectors that can accumulate in tumors, sites of infection, or sites of inflammation. The addition of ligands to such systems still has value in some cases but in and of itself can have sufficient added value to improved therapeutic responses. Sosnowski *et al.* (1996) for example, using FGF has demonstrated efficient ligand-mediated targeting of DNA to proliferating cells which overexpress the FGF receptor.

4. GENE DELIVERY VECTORS: FUTURE PROMISES AND CHALLENGES

4.1. Tumor Endothelial Cells as A Target

As has been previously discussed, there are several barriers to entry into tumors even when using high affinity ligands. For a variety of reasons, angiogenic endothelial cells may be the best choice for targeting and subsequent killing. Endothelial cells within tumors differ from normal quiescent vascular endothelial cells in many aspects including a dramatically increased proliferation rate, overexpression of a variety of proteins such as $\alpha_v\beta_3$ integrins and theses cells are directly assessable to systemically delivered "targetable" vectors. In addition, the elimination of tumor endothelial cells is likely to amplify anti-tumor therapy since tumors cannot grow beyond 1–2 mm^3 without a blood supply (for review see Folkman, 1996). In addition to targeting the vector to the angiogenic endothelial cells, several groups have reported promising therapy by expressing anti-angiogenic proteins systemically (for review see Folkman, 1998). Adenoviral vectors have been shown to systemically express therapeutic proteins into the circulation for at least one year in mice (Connelly *et al.*, 1998 and references therein). The adenoviral-mediated gene expression of anti-angiogenic proteins may result in sustained levels of therapeutic anti-angiogenic proteins that may selectively kill proliferating tumors cells systemically.

4.2. Expandable Vector Systems to Increase Therapeutic Efficacy

There are now a host of vector systems, in which the payload can be effectively amplified following the initial transduction of the tumor cell and/or endothelial cell. Chimeric vectors have recently been developed which can effectively increase the number of cells transduced with a vector via the utilization of an adenoviral vector which is capable of expressing retroviral genomes (Feng *et al.*, 1997). Tumor specific replication restricted vectors (TSRRV) are rapidly increasing in utility and use since they amplify both the number of cells transduced and the amount of therapeutic gene product at the desired site. Replication restriction adenoviral vectors have been developed by our laboratory (Hallenbeck *et al.*, 1997) and others (Rodriguez *et al.*, 1997) by expressing a gene necessaary for viral replication under the control of a tumor specific promoter. Another type of TSRRV involves the ability of particular cancer cells to complement specific gene defects in adenovirus (Bischoff *et al.*, 1996) or Herpes (Yazaki *et al.*, 1995). Clinical trials are currently ongoing which utilize the E1B deleted adenoviral vector (ONYX015) which has been shown to replicate in and selectively kill tumor cells which lack a functional copy of p53 *in vitro* and *in vivo* (Bischoff *et al.*, 1996). Another strategy involves utilizing an adenoviral vector, which expresses high affinity receptors at the tumor site. Following expression of the receptor, tumor cells are selectively eliminated utilizing radioactvely labeled antibodies specific for the receptor (Raben *et al.*, 1996). There could be combinations of enzyme/prodrugs which again expand the therapeutic index of one vector. Finally, the use of tissue specific promoters and gene regulatory systems will permit high level but controlled gene expression. All approaches for the payload will be combined in various ways and are likely to be very effective once systemic and selective delivery of the vector to the tumor site is achieved.

5. IMMUNE EVASION

One of the most daunting challenges for the use of systemically delivered vectors is the avoidance of immune-mediated neutralization of the vector. Delivery of Ad5 based adenoviral vectors may especially be problematic since at least 50% of normal humans have neutralizing antibody. Nevertheless, there have been reports that suggest that these problems may be overcome by changing capsid components through altering immunogenic regions of capsid proteins (Roy *et al.*, 1998), by immunosuppression (Smith *et al.*, 1996), and/or by partially localized delivery. In addition, a recent report (Molnar-Kimber *et al.*, 1998) indicates that despite the existence of significant and preexisting humoral and cellular immunity to Ad5 in patients with mesothelioma, gene transfer of HSV-tk was not prevented when an adenoviral vector expressing HSV-tk was delivered intrapleurally.

6. SUMMARY

Adenoviral vectors, which have targeting ligands for tumor cells on the capsid, no natural tropism, and carry a therapeutic payload should be constructed soon and tested in pre-clinical models. Nevertheless, there are still important considerations for the design and therapeutic use of targetable vectors. Perhaps the single greatest challenge in the future, as it was in the past, will be finding ligands that have a higher apparent affinity for tumor and/or tumor endothelial cells then normal cells. However, the advent of many rapidly advancing technologies and information including the sequencing of the human genome, *in vivo* and *in vitro* phage display, rapid analysis of gene and protein expression in any context, and new cellular targets such as angiogenic endothelial cells, may provide many opportunities for the discovery of novel and useful ligands. In addition, the interests in targeting vectors are rapidly growing with new journals and meetings solely devoted to this subject increasing annually. Within the next 5 years, we should have meaningful clinical data on targetable vectors to reassess our progress.

ACKNOWLEDGMENTS

The authors would like to thank Drs. Suzanne Forry-Schaudies and Sheila Connelly for critical review of the manuscript and Dr. Martin Woodle for advice and imput.

REFERENCES

Anderson, S.C., Johnson, D.E., Harris, M.P., Engler, H., Hancock, W., Huang, W.M., Wills, K.N., Gregory, R.J., Sutjipto, S., Wen, S.F., Lofgren, S., Shepard, H.M., and maneval, D.C., 1998. P53 gene therapy in a rat model of hepatocellular carcinoma:intra-arterial delivery of a recombinant adenovirus. *Clin. Cancer. Res.* **7**:1649–1659.

Bergelson, J.M., Cunningham, J.A., Droguett, G., Kurt-Jones, E.A., Krithivas, A., Hong, J.S., Horwitz, M.S., Crowell, R.L., and Finberg, R.W. 1997. Isolation of a common receptor for coxsackie B viruses and adenoviruses 2 and 5. *Science* **275**:1320–1323.

Bischoff J.R., Kirn D.H., Williams A., Heise C., Horn S., Muna M., Ng L., Nye J.A., Sampson-Johannes A., Fattaey A., and McCormick F., 1996. An adenovirus mutant that replicates selectively in p53-deficient human tumor cells. *Science*. **274**:373–376.

Clary, B.M., and Lyerly, H.K., 1998. Transcriptional targeting for cancer gene therapy. *Surg. Oncol. Clin. N. Am.* **7**:565–574.

Connelly, S., Andrews, J.L., Gallo, A.M., Kayda, D.B., Qian, J., Hoyer, L., Kadan, M.J., Gorziglia, M.I., Trapnell, B.C., McClelland, A., and Kaleko, M., 1998. Sustained phenotypic correction of murine hemophilia A by in vivo gene therapy. *Blood*. **91**:3273–3281.

Cusack, Jr., J.C., and Tanabe, K.K., 1998. Cancer gene therapy. *Surg. Oncol. Clin. N. Am.* **7**:421–469.

Deshane, J., Siegal, G.P., Wang, M., Wright, M., Bucy, R.P., Alvarez, R.D., and Curiel, D.T., 1997. Transductional efficiency and safety of an intraperitoneally delivered adenovirus encoding an anti-erbB-2 intracellular single-chain antibody for ovarian cancer gene therapy. *Gynecol. Oncol.* **64**:378–385.

Douglas J.T., Rogers B.E., Rosenfeld M.E., Michael S.I., Feng M., and Curiel D.T. 1996. Targeted gene delivery by tropism-modified adenoviral vectors. *Nat. Biotechnol.* **14**:1574–1578.

Feng, M., Jackson, W.H., Goldman, C.K., Rancourt, C., Wang, M., Dusing, S.K., Siegal, G., and Curiel, D.T., 1997. Stable in vivo gene transduction via a novel adenoviral/retroviral chimeric vector. *Nat. Biotechnol.* **15**:866–870.

Folkman, J., 1996. Fighting cancer by attacking its blood supply. *Sci. Am.* **275**:150–154.

Folkman, J., 1998. Antiangiogenic gene therapy. *Proc. Natl. Acad. Sci. U.S.A.*, **95**:9064–9066.

Goldman C.K., Rogers B.E., Douglas J.T., Sosnowski B.A., Werbin Y., Siegal G.P., Baird A., Campain J.A., and Curiel D., 1997. Targeted gene delivery to Kaposi's sarcoma cells via the fibroblast growth receptor. *Cancer Research* **57**:1447–1451.

Hall F.L., Gordon E.M., Wu L., Zhu N.L., Skotzko M.J., Starnes V.A., and Anderson W.F., 1997. Targeting retroviral vectors to vascular lesions by genetic engineering of the MoMLV gp70 envelope protein. *Hum. Gene. Ther.* **8**:2183–2192.

Hallenbeck, P.L., Chang, Y-N., Hay, C., Golightly, D., Stewart, D., McGarrity, G., and Chiang, Y., 1996. Novel tumor specific replication competent adenoviral vector for gene therapy of cancer. *Abstracts of the fifth International Conference on gene Therapy of Cancer., November, 1998.*

Han, X., Kasahara N., and Kan, Y.W., 1995. Ligand-directed retroviral targeting of human breast cancer cells. *Proc. Natl. Acac. Sci. USA*. **92**:9747–9751.

Hong J.S., and Engler J.A., 1996. Domains required for assembly of adenovirus type 2 fiber trimers. *J. Virology* **70**:7071–7078.

Kaneko S., Hallenbeck P., Kotani T., Nakabayashi H., McGarrity G., Tamaoki T., Anderson W.F., and Chiang Y.L., 1995. Adenovirus-mediated gene therapy of heptacellular carcinoma using cancer-specific gene expression. *Cancer Res.* **55**:5283–5287.

Kasahara, N., Dozy, A.M., and Kan, Y.W., 1994. Tissue specific targeting of retroviral vectors through ligand-receptor interactions. *Science* **266**:1373–1376.

Krasnykh, V.N., Mikheeva, G.V., Douglas, J.T., and Curiel, D.T., 1996. Generation of recombinant adenovirus vectors with modified fibers for altering viral tropism. *J. Virology* **70**:6839–6846.

Krasnykh, V.N., Dmitriev, I., Mikheeva, G., Miller, C.R., Belousova,N., and Curiel, D.T., 1998. Characterization of an adenovirus vector containing a heterologous peptide epitope in the HI loop of the fiber knob. *J. Virology* **72**:1844–1852.

Krasnykh, V.N., Douglas, J.T., and Curiel, D.T., 1998. Tropism-modified adenoviral vectors for cell-specific gene delivery. *Abstracts of Cold Spring harbor Symposium "Gene Therapy", September, 1998.*

Jain, R.K., 1997. The Eugene M. Landis award lecture 1996. Delivery of molecular and cellular medicine to solid tumors. *Microcirculation* **4**:1–23.

Michael, S.I., Hong, J.S., Curiel, D.T., and Engler, J.A., 1995. Addition of a short peptide ligand to the adenovirus fiber protein. *Gene Therapy* **2**:660–668.

Molnar-Kimber K.L., Sterman D.H., Chang M., Kang E.H., Elbash M., Lanuti M., Elshami A., Gelfand K., Wilson J.M., Kaiser L.R., and Albelda S.M., 1998. Impact of preexisting and induced humoral and cellular immune responses in an adenovirus-based gene therapy phase I clinical trial for localized mesothelioma. *Human Gene Therapy* **9**:2121–2133.

Niculescu-Duvaz, I., and Springer, C.J., 1996. Development of prodrugs for antibody-directed enzyme prodrug therapy (ADEPT). *Exp. Opin. Invest. Drugs.* **5**:289–308.

Novelli A., and Boulanger P.A., 1991. Deletion analysis of functional domains in baculovirus-expressed adenovirus type 2 fiber. *Virology* **185**:365–376.

O'Reilly, M.S., Holmgren, L., Shing, Y., Chen, C., Rosenthal, R.A., Moses, M., Lane, W.S., Cao, Y., Sage, E.H., and Folkman, J., 1994. Angiostatin: a novel angiogenesis inhibitor that mediates the supression of metastases by a Lewis lung carcinoma. *Cell.* **79**:315–328.

Rabin, D., Buchsbaum, D.J., Khazaeli, M.B., Rosenfeld, M.E., Gillespie, G.Y., Grizzle, W.E., Liu, T., and Curiel, D.T., 1996. Enhancement of radiolabeled antibody binding and tumor localization through adenoviral transduction of the human carcinoembryonic antigen gene. *Gene Ther.* **3**:567–580.

Robertson, D., 1998. Genetech's anticancer Mab expected by November. *Nature Biotechnology* **16**:615.

Rodriguez, R., Schuur, E.R., Lim, H.Y., Henderson, G.A., Simons, J.W., and Henderson, D.R., 1997. Prostate attentuated replication competent adenovirus (ARCA) CN706: a selective cytotoxic for prostate-specific antigen-positive prostate cancer cells. *Cancer Research* **57**:2559–2563.

Rosenberg, S.A., 1998. New opportunities for the development of cancer immunotherapies. *Cancer J. Sci. Am. 4 Suppl.* **1**:S1–4.

Roy, S., Shirley, P.S., McClelland, A., and Kaleko, M., 1998. Circumvention of immunity to the adenovirus major coat protein hexon. *J. Virol.* **72**:6875–6879.

Smith T.A., White, B.D., Gardner, J.M., Kaleko, M., and McClelland, A., 1996. Transient immunosuppression permits successful repetitive intravenous administration of an adenvirus vector. *Gene Ther.* **3**:496–502.

Somia, N.V., Zoppe, M., and Verma, I.M., 1995. Generation of targeted retroviral vectors by using single-chain variable fragment: An approach to in vivo gene delivery. *Proc. Natl. Acad. Sci. USA.* **92**:7570–7574.

Smith, T.A.G., Mehaffey, M.G., Kayda, D.B., Saunders, J.M., Yei, S., Trapnell, B.C., McClelland, A., and Kaleko, M., 1993. Adenovirus mediated expression of therapeutic plasma levels of human factor IX in mice. *Nat. Genet.* **5**:397–402.

Sosnowski B.A., Gonzalez A.M., Chandler L.A., Buechler Y.J., Pierce G.F., and Baird A., 1996. Targeting DNA to cells with basic fibroblast growth factor (FGF2). *J. Biol. Chem.* **271**:33647–33653.

Sterman D.H., Kaiser, L.R., and Albelda, S.M., 1998. Gene therapy for malignant pleural mesothelioma. *Hematol. Oncol. Clin. North. Am.* **12**:553–568.

Stevenson, S.C., Rollence, M.L., White, B., Weaver, L., and McClelland, A., 1995. Human adenovirus serotypes 3 and 5 bind to two different cellular receptor via the fiber head domain., *J. Virology* **69**:2850–2857.

Stevenson, S.C., Rollence, M.L., Marshall-Neff, J., and McClelland, A., 1997. Selective targeting of human cells by a chimeric adenovirus vector containing a modified fiber protein, *J. Virology* **71**:4782–4790.

Tomko R.P., Xu R., and Philipson L., 1997. HCAR and MCAR: The human and mouse cellular receptors for subgroup C adenoviruses and group B coxsackieviruses. *Proc. Natl. Acad. Sci. U.S.A.*, **94**:3352–3356.

Von Seggern, D., Kehler J., Endo R.I., and Nemerow G.R., 1998. Complementation of a fibre mutant adenovirus by packaging cell lines stably expressing the adenovirus type 5 fibre protein, *J. General Virology.* **79**:1461–1468.

Wickham, T.J., Mathias, P., Cheresh, D.A., and Nemerow, G.R., 1993. Integrins avb3 and avb5 promote adenovirus internalization but not virus attachment. *Cell.* **73**:309–319.

Wickham, T.J., Carrion, M.E., and Kovesdi, I., 1995. Targeting of adenovirus penton base to new receptors though replacement of its RGD motif with other receptor-specific motifs. *Gene Therapy* **2**:750–756.

Wickham, T.J., Roelvink, P.W., Brough, D.E., and Kovesdi, I., 1996. Adenovirus targeted to heparin-containing receptors increase its gene delivery efficiency to multiple cell types. *Nature Biotechnol.* **14**:1570–1573.

Wickham, T.J., Segal, D.M., Roelvink, P.W., Carrion, M.E., Lizonova, A., Lee, G.M., and Kovesdi, I., 1996. Targeted adenovirus gene transfer to endothelial and smooth muscle cells by bispecific antibodies. *J. Virology.* **70**:6831–6838.

Woodle, M.C., and Storm, G. eds., 1998. *Long circulating liposomes, old drugs, new therapeutics*. Spinger-Verlag Berlin-Heidelberg and Landes Bioscience, Georgetown, Texas.

Yang, J., 1999. Direct intratumoral injection of naked DNA, in *Gene Therapy of cancer: methods and protocols*, (W. Walther and U. Stein, eds.), in press.

Yazaki, T., Manz, H.J., Rabkin, S.D., and Martuza, R.L., Treatment of human malignant meningiomas by G207, a replication-competent mulitmutated herpes simplex virus 1. *Cancer Res.* **55**:4752–4756.

Xia D., Henry L.J., Gerard R.D., and Deisenhofer J., 1994. Crystal structure of the receptor binding domain of adenovirus type 5 fiber protein at 1.7 A resolution. *Structure*. **2**:1259–1270.

5

HUMAN α-FETOPROTEIN TRANSCRIPTIONAL REGULATORY SEQUENCES

Application to Gene Therapy

Taiki Tamaoki

Department of Biochemistry and Molecular Biology
University of Calgary
Calgary, Alberta T2N 4N1
Canada

INTRODUCTION

One of the main characteristics of hepatocellular carcinoma (HCC) is the production of α-fetoprotein (AFP). This protein is produced abundantly in fetal liver but its synthesis decreases rapidly after birth and is hardly detectable in adult life. However, the AFP gene is frequently reexpressed in HCC (Abelev, 1971). This suggests that the AFP regulatory sequences may be used to express a cytostatic gene in an HCC-specific manner to eliminate tumor cells without harming normal cells. Over 70% of HCC are reported to be AFP-positive and therefore this therapeutic approach should be applicable to the majority of HCC patients. This article will review the main features of human AFP regulatory sequences and their applications to HCC gene therapy. Discussions below are largely confined to the human AFP gene.

1. HUMAN AFP TRANSCRIPTIONAL REGULATORY REGIONS

Transcription of the human AFP gene is controlled by three regulatory regions, promoter, enhancer and silencer, in the 5′-flanking sequence (Sakai *et al.*, 1985; Watanabe *et al.*, 1987; Sawadaishi *et al.*, 1988; Nakabayashi *et al.*, 1989, 1991) (Fig. 1).

1.1. Promoter

The AFP promoter is a 200-bp region immediately upstream of the AFP gene (Feuerman *et al.*, 1989; Papaconstantinou *et al.*, 1990; Bois-Joyneux and Danan, 1994;

Cancer Gene Therapy: Past Achievements and Future Challenges, edited by Habib
Kluwer Academic/Plenum Publishers, New York, 2000.

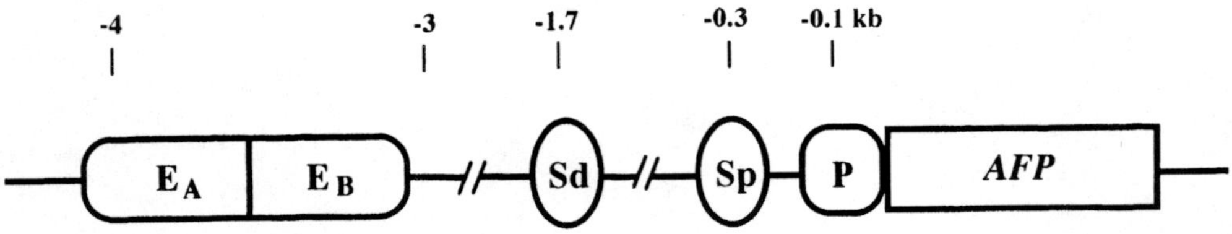

Figure 1. Transcriptional regulatory regions of the human α-fetoprotein gene. E_A, enhancer domain A; E_B, enhancer domain B; Sd, distal silencer; Sp, proximal silencer; P, promoter

Nakabayashi *et al.*, 1989). There are two binding sites of hepatocyte nuclear factor 1 (HNF1) in the promoter, both of which are indispensable and play a key role in tissue-specific transcription (Fig. 2). In addition, nuclear factor 1 (NF1) and CCAAT/enhancer binding protein (C/EBP) have been shown to bind to sites overlapping with the distal HNF1 site in the promoter. The human AFP promoter is up-regulated by dexamethasone through a glucocorticoid responsive element (GRE) at –0.17 kb (Nakabayashi *et al.*, 1989). Overall, AFP promoter sequences are similar among humans, mice and rats.

The AFP promoter is a typical tumor-specific promoter highly suitable for HCC gene therapy. Ido *et al.* (1995) have shown specific growth inhibition of AFP-producing HCC cells *in vitro* by expressing the herpes simplex virus thymidine kinase (HSV-tk) gene under the control of the AFP promoter (–230 to +29) followed by treatment with ganciclovir (GCV) (see below). However, transcriptional activity of the AFP promoter is relatively weak and, for the purpose of gene therapy, augmentation of its activity by a suitable enhancer is likely required.

1.2. Enhancer

The human AFP enhancer is present between 3 and 4.9 kb upstream of the transcription initiation site of the AFP gene (Watanabe *et al.*, 1987). Six transcription factor-binding sites have so far been identified in the human AFP enhancer region, one each for HNF1 and HNF4 and two for C/EBP in domain A, and one each for HNF1 and HNF3 in domain B (Fig. 3). The human AFP enhancer shows liver-specificity which may be due largely to the action of HNF1 (Watanabe *et al.*, 1987). However, the affinity of the HNF1 protein to the HNF1 site in domain B has been reported to be one-tenth of that to the *bona fide* HNF1 site (Feuerman *et al.*, 1989), suggesting that the domain B HNF1 site may not be exclusive for HNF1. In fact, a transcription factor, termed ATBF1, has been shown to bind to this site in competition with HNF1 to down-regulate AFP enhancer activity (Morinaga *et al.*, 1991; Yasuda *et al.*, 1994). No HNF1 site has been

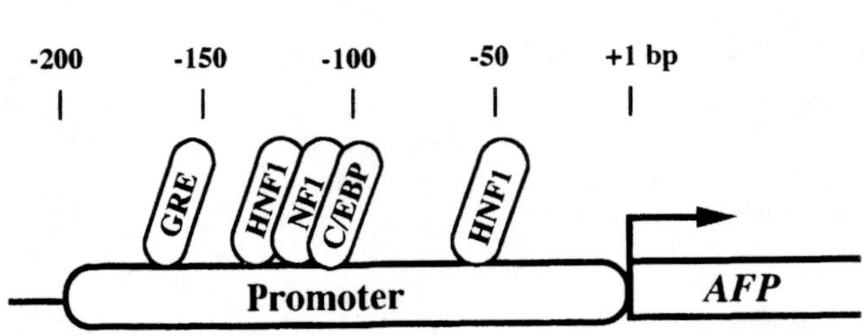

Figure 2. Transcription factors regulating the human α-fetoprotein promoter.

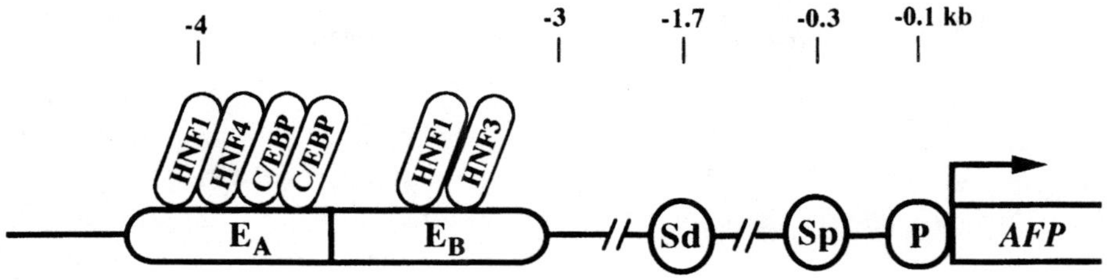

Figure 3. Transcription factors regulating the human α-fetoprotein enhancer. E_A, enhancer domain A; E_B, enhancer domain B; Sd, distal silencer; Sp, proximal silencer; P, promoter.

found in mouse and rat AFP enhancers, although all other transcription factor binding sites are shared by the human, mouse and rat AFP enhancers. Thus, the presence of the HNF1 site is a characteristic feature of the human AFP enhancer. The rodent AFP enhancers also differ from the human AFP enhancer in that they are located in three widely separated regions which exhibit different tissue specificities (Godbout *et al.*, 1988; Wen *et al.*, 1991).

In fetal liver and HCC, the enhancer is believed to play a major role in determining the level of AFP expression as the AFP promoter is a relatively weak promoter. Most past studies of gene therapy of HCC have used the human AFP enhancer to stimulate the AFP promoter. Bui *et al.* (1997) have used the MER-I enhancer, one of the three mouse AFP enhancer regions having liver specificity, with the mouse AFP promoter to target IL-2 gene expression in AFP-producing human HCC cells (HepG2, Hep3B) *in vitro* and *in vivo*. No studies have so far been reported in which nonspecific enhancers, such as SV40 and cytomegalovirus, are used in HCC gene therapy.

1.3. Silencer

The AFP silencer is a negative *cis*-acting element with a consensus sequence, 5′-CTTCATAACTAATACTT-3′ (Nakabayashi *et al.*, 1991). Two silencer regions have been identified upstream of the human AFP gene (Fig. 1). The proximal silencer is located at −0.31 kb and contains one copy of the consensus sequence. The distal silencer is located at −1.75 kb and carries four copies of consensus sequence. The latter exhibits much stronger suppressive activity than the former. The function of the AFP silencer is to prevent the AFP enhancer from stimulating the AFP promoter. This inhibitory effect is manifested only when the silencer is situated downstream of the enhancer and upstream of the promoter. The AFP silencer functions in either orientation and in an enhancer-nonspecific manner, i.e. it blocks the activity of the SV40 enhancer as well as AFP enhancer, as long as it is situated between the enhancer and the promoter.

In transcription assays using the chloramphenicol actetyltranserase (CAT) gene under the control of the AFP 5′-flanking sequence, the deletion of the silencer regions from the AFP 5′-flanking sequence resulted in increased expression of the linked CAT reporter gene in low AFP-producing huH-1 cells but little changes in high AFP-producing HuH-7 cells (Nakabayashi *et al.*, 1991). This indicates that the silencer activity is inversely correlated with the level of AFP expression, suggesting that the silencer is involved in determining the level of AFP expression in HCC.

In mice and rats, AFP silencer activities have been located between −838 and −250 (Vacher and Tilghman, 1990) and between −742 and −87 (Wen *et al.*, 1991), respectively. In either case, the precise sequence of the silencer element has not been identified.

The rodent silencers, like the human counterpart, function in a position-dependent manner. The mouse silencer activity is low in the fetal liver and high in the adult liver (Camper and Tilghman, 1989; Vacher and Tilghman, 1990), indicating that the silencer plays a major role in suppressing AFP expression during liver ontogeny, although this is not the only mechanism involved in developmental regulation of the AFP gene (Emerson *et al.*, 1992). The possibility exists that the silencer may serve as a molecular switch to re-direct the action of the AFP enhancer to stimulate the albumin promoter as the liver matures (Nakata *et al.*, 1992). Little information is available at present regarding protein factors that bind and regulate the silencer element. In gene therapy of HCC, the use of the silencer is desirable as it will provide additional safeguard against damage of normal cells.

2. MODULATION OF TRANSCRIPTIONAL ACTIVITY OF THE AFP REGULATORY SEQUENCES

2.1. Hormones, Growth Factors and Oncogene Products

Activities of the human AFP promoter and enhancer are modulated by a variety of agents, including hormones (Nakabayashi *et al.*, 1991) and growth factors (Nakata *et al.*, 1992; *Nakao et al.*, 1991) and oncogene products (Nakao *et al.*, 1990; Saegusa *et al.*, 1994). Dexamethasone stimulates the human AFP promoter. Epidermal growth factor (EGF) and 12-*0*-tetradecanoylphorbol-13-acetate (TPA) synergistically suppress AFP expression primarily through the HNF1 element in the enhancer (Nakata *et al.*, 1992). The HNF1 element in the promoter is also affected by EGF and TPA, although to a lesser degree. Conversely, v-Ras suppresses the activity of the AFP promoter HNF1 element more than that of the enhancer HNF1 element (Nakao *et al.*, 1990). AFP promoter activity is also down-regulated by transforming growth factor β1 (Nakao *et al.*, 1991), hepatocyte growth factor (Hatano *et al.*, 1992) and prostaglandin Δ^{12}-PGJ2 (Mitsuoka *et al.*, 1992) although the exact elements involved have not been identified. It is possible that the silencer activity is subject to similar modulation but little information is available at present.

As discussed below, dexamethasome, EGF and TPA regulate AFP regulatory sequences introduced into cells by viral vectors in a manner similar to the endogenous regulatory sequences (Ido *et al.*, 1995; Arbuthnot *el al.*, 1995).

2.2. Mutation of the AFP Promoter

There is a naturally occurring mutation of the human AFP promoter that increases transcriptional activity. Members of a family showing persistent elevation (average 23-fold) of AFP production in adult life contain a G to A point mutation at –109 (McVey *et al.*, 1993). This is within the overlapping binding site for HNF1, NF1 and C/EBP in the promoter. The G to A substitution results in an increase in the sequence similarity to the consensus HNF1 recognition site and in the affinity to HNF1. This may lead to reduced binding of NF1 and C/EBP which appear to function negatively for the AFP promoter (Papaconstantinou *et al.*, 1990; Thomassin *et al.*, 1992; Bois-Joyeux and Danan, 1994).

In theory enhancer activity can also be modulated by changing specific nucleotides. No naturally occurring mutations of the AFP enhancer leading to increased enhancer activity have been reported.

2.3. Use of Cre-*loxP* System

The Cre-*loxP* system has been used to express a gene in a tissue-specific manner (Sauer and Henderson, 1989; O'Gorman *et al.*, 1991; Gu *et al.*, 1994) or to enhance the level of gene expression without losing the tissue specificity (Kanegae *et al.*, 1995; Sato *et al.*, 1998). Sato *et al.* (1998) have applied this system to gene therapy of HCC. In this work, two recombinant adenoviral vectors were constructed. One contained the Cre recombinase gene linked to AFP regulatory sequences to express Cre in HCC cells producing AFP and the other carried the lacZ gene linked to a powerful hybrid promoter termed CAG (Niwa *et al.*, 1991). A stuffer sequence sandwiched by the *loxP* element was inserted between the CAG promoter and the lacZ gene to prevent the CAG promoter from initiating transcription of the lacZ gene. When AFP-producing HCC cells, e.g. HuH-7, were infected with these two types of recombinant viruses Cre recombinase was expressed under the control of the AFP regulator. The resultant Cre recognized the *loxP* sequences to delete the stuffer sequence to allow the CAG promoter to initiate transcription of the lacZ gene. The level of lacZ expression in HuH7 cells doubly infected with these viruses was 50-fold higher than that in cells infected with recombinant viruses carrying the lacZ gene under the direct control of the AFP regulatory sequences. In cells not producing AFP, Cre was not expressed and consequently no appreciable lacZ expression was observed. Thus, this system enables enhanced expression of lacZ in AFP-positive HCC without losing the tissue-specificity.

3. APPLICATION OF AFP REGULATORY SEQUENCES TO GENE THERAPY OF HCC

3.1. Construction of Gene Therapy Vectors

The human AFP regulatory sequences are present in the 4.9-kb region upstream of the AFP gene. Most viral and nonviral vectors currently available can accommodate DNA of this size. If shorter sequences are required, the enhancer and silencer regions are isolated individually and combined with the promoter in single or multiple copies. For example, an AFP regulatory construct containing two copies of the AFP enhancer domain A and B (–4.1 to –3.3 kb), six copies of the 31-bp AFP silencer core sequence and the 0.2-kb AFP promoter in a total size of less than 2 kb has been constructed and successfully used to drive the lacZ gene in an HCC-specific manner (Sato *et al.*, 1998).

Delivery vectors so far used include retrovirus (Ido *et al.*, 1995; Arbuthnot *et al.*, 1995; Xu *et al.*, 1996), adenovirus (Kaneko *et al.*, 1995; Kanai *et al.*, 1996, 1997; Bui *et al.*, 1997), adeno-associated virus (Su *et al.*, 1996, 1997) and HVJ-liposome (Hirano *et al.*, 1998a,b). Genes introduced under the control of human AFP regulatory sequences include HSV-tk (Macri and Gordon, 1994; Ido *et al.*, 1995; Kaneko *et al.*, 1995; Kanai *et al.*, 1996; Su *et al.*, 1996, 1997), varicella-zoster virus thymidine kinase (VZV-tk) (Huber *et al.*, 1991), cytosine deaminase (Kanai *et al.*, 1997), p53 (Xu *et al.*, 1996), and an ecotropic retrovirus receptor (Igarashi *et al.*, 1998). The interleukin-2 gene has been introduced under the control of mouse AFP promoter and enhancer (Bui *et al.*, 1997). Table 1 shows a list of various vectors carrying the human AFP regulatory sequences so far used in gene therapy of HCC.

Table 1. Vectors Carrying Human AFP Regulatory Sequences Used for Gene therapy of HCC

AFP Regulatory Sequences	Genes Expressed	Delivery Vehicles	Assay Systems	References
5.1 kb	VZV-tk	Retrovirus	*in vitro*	Huber *et al.* (1991)
0.3 kb	HSV-tk	Retrovirus	*in vitro*	Ido *et al.* (1995)
4.9 kb	HSV-tk	Adenovirus	*in vitro*, *in vivo*	Kaneko *et al.* (1995)
$E_AE_B \cdot 0.9$ kb	p53	Retrovirus	*in vitro*	Xu *et al.* (1996)
$E_AE_B \cdot 0.2$ kb	HSV-tk, lacZ	Adenovirus	*in vitro*	Kanai *et al.* (1996)
$E_AE_B \cdot 0.2$ kb	CD	Adenovirus	*in vivo*	Kanai *et al.* (1997)
$E_B \cdot 0.3$ kb*	HSV-tk	Adeno-associated virus	*in vitro*	Su *et al.* (1996)
$E_B \cdot 0.3$ kb*	HSV-tk	Adeno-associated virus	*in vitro*	Su *et al.* (1997)
$[E_AE_B]_2 \cdot [S]_6 \cdot 0.2$ kb	Cre	Adenovirus	*in vitro*	Sato *et al.* (1998)
4.9 kb	HSV-tk	HVJ-liposome	*in vitro*, *in vivo*	Hirano *et al.* (1998a,b)
$E_AE_B \cdot 0.3$ kb	Retrovirus receptor	Adenovirus	*in vitro*	Igarashi *et al.* (1998)

Figures under AFP Regulatory Sequences indicate the lengths of the human AFP 5′-flanking sequence. E_AE_B, human AFP enhancer domain A and B (–4.0 to –3.3); E_B, human AFP enhancer domain B (–3.7 to –3.3); S, 31-bp human AFP silencer core sequence; *, human albumin promoter; VZV, varicella zoster virus; tk, thymidine kinase; HSV, herpes simplex virus; CD, cytosine deaminase; Cre, Cre recombinase; HVJ, hemagglutinating virus of Japan.

3.2. Examples of HCC Gene Therapy Using AFP Regulatory Sequences

3.2.1. Retroviral Vectors. Ido *et al.* (1995) infected several human HCC cell lines with varying levels of AFP production with amphotrophic recombinant retroviruses carrying the HSV-tk gene under the control of the human AFP promoter (–230 to +29 bp). Two highly AFP-producing cell lines, HuH-7 and HepG2, were most sensitive to subsequent treatment with GCV, moderately AFP-producing PLC/PRF/5 cells less sensitive, and low AFP-producing huH-1 cells least sensitive. Growth of HeLa cells was hardly affected by this treatment. These results show that the cell survival is inversely proportional to the level of AFP production. However, this relationship is not strict since the infectivity of retrovirus and the toxicity of GCV are influenced by cellular mitotic activity.

Huber *et al.* (1991) infected various cell lines with amphotrophic retroviruses containing a chimeric VZV-tk gene under the control of the 5.1-kb 5′-flanking sequence of the human AFP gene. AFP-positive cells (HepG2, Hep3B, HuH-7) selectively expressed the VZV-tk gene and became sensitive to 6-methoxypurine arabinonucleoside.

Igarashi *et al.* (1998) have reported a two-step procedure for highly efficient, HCC-specific retroviral gene transfer. In this procedure, AFP-positive (HepG2) and -negative (HeLa, HLE) cells were transduced with an adenoviral vector containing the ecotropic retrovirus receptor gene under the control of the AFP promoter and enhancer. This resulted in expression of ecotropic retrovirus receptors in HepG2 cells but not in HeLa and HLE cells. Subsequent transduction with an ecotropic retroviral vector containing the lacZ gene under the control of the SV40 early promoter resulted in stable transfer and expression of the lacZ gene in HepG2 cells but not in HeLa and HLE cells. The efficiency of targeted gene transfer by the two-step procedure is comparable to that of tissue non-specific gene transfer with an amphotropic retroviral vector.

As discussed above, the transcriptional activity of the AFP regulatory sequences can be modulated by hormones and growth factors. When HepG2 cells were infected with recombinant retroviruses carrying the HSV-tk under the control of the AFP promoter and then treated with aciclovir (ACV) in the presence or absence of 3×10^{-3}M dexamethasone, growth inhibition was about 2-fold greater in the presence of dexamethasone than in its absence (Ido *et al.*, 1995). Arbuthnot *et al.* (1995) have shown that, in the context of an integrated provirus, the transcriptional activity of the 3-kb rat AFP regulatory sequence was down-regulated by EGF and TPA. These results show that the AFP promoter and enhancer introduced retrovirally are subject to regulatory control by hormones and growth factors as those of the endogenous gene.

3.2.2. Adenoviral Vectors. Several groups have used adenoviral vectors to introduce cytotoxic genes under the control of various AFP promoter constructs. Kaneko *et al.* (1995) constructed recombinant adenovirus carrying the HSV-tk gene under the control of the 4.9-kb AFP 5′-flanking sequence. Transduction of HuH-7 and SK-Hep-1 cells with this recombinant virus followed by GCV treatment resulted in killing of HuH7 but not SK-Hep-1 cells. The effect of adenovirus-mediated transduction *in vivo* was tested using tumors arising from HuH-7 and SK-Hep-1 cells transplanted in nude mice. When the recombinant virus was directly injected into the tumors and then GCV was injected into mice intraperitoneally, HuH7-derived tumors regressed completely but SK-Hep-1 tumors did not. Kanai *et al.* (1996) reported specific killing of AFP-positive HCC cells *in vitro* using a recombinant adenovirus vector harboring the HSV-tk gene under the control of the 0.2-kb AFP promoter and one copy of AFP enhancer domain A and B (–4.0 to –3.3 kb). They also reported regression of PLC/PRF/5-derived tumors *in vivo* by direct injection of recombinant adenoviruses carrying the cytosine deaminase gene under the control of the above regulatory sequences into the tumors followed by treatment of mice with 5-fluorocytosine (Kanai *et al.*, 1997).

3.2.3. Adeno-associated Viral Vectors. The adeno-associated virus (AAV) has been used as vector for the delivery of the HSV-tk gene under the control of the human albumin promoter and the human AFP enhancer domain B (3.7 to –3.3 kb) *in vitro* and *in vivo*. Infection of human HCC cell lines with the recombinant AAV followed by GCV treatment *in vitro* caused death of AFP-positive HCC cells but not AFP-negative cells (Su *et al.*, 1996). The dose required to kill the AFP-positive cells was inversely proportional to the level of AFP expression. Subcutaneous tumors generated in nude mice by implanting PLC/PRF/5 cells previously transduced with the recombinant AAV shrank dramatically after treatment with GCV (Su *et al.*, 1997). Growth of tumors derived from untransduced PLC/PRF/5 cells in nude mice was also retarded by direct injection of the recombinant AAV into tumors followed by intraperitoneal injection of GCV.

3.2.4. Liposomes. Hirano *et al.* (1998a) have embodied the lacZ gene under the control the 4.9-kb AFP 5′-flanking sequence in liposome containing hemagglutinating virus of Japan (HVJ-liposome) and injected into tumors derived from HuH-7 and a colon cancer cell line, LS180, growing subcutaneously in mice. LacZ expression was observed in HuH-7 tumors but not in LS180 tumors. To test the effect of the HVJ-liposome-mediated transduction on tumor growth *in vivo*, HVJ-liposomes carrying the HSV-tk gene under the control of the 4.9-kb AFP 5′-flanking sequence were injected into HuH-7 tumors in nude mice. The mice were then injected with GCV intraperitoneally. After this procedure of gene transfer and GCV treatment was repeated ten times complete regression of HuH-7 tumors resulted. HVJ-liposome was also used to examine whether HCC-specific gene expression is enhanced by using the Cre/*lox*P system (Hirano *et al.*, 1998b). Two plasmid vectors, one carrying the AFP-Cre gene and the other the CAG-stuffer-lacZ gene (see 2.3), were embodied in HVJ-liposome and injected into HuH-7 and LS180 tumors growing in nude mice. LacZ expression was observed in HuH-7 tumors at a level 20 times higher than that in tumors received the lacZ gene directly controlled by the AFP regulatory sequences. No lacZ expression was observed in LS180 cells.

SUMMARY/CONCLUSION

The AFP regulatory sequences are among the best known tumor-specific transcriptional regulators. A number of groups have demonstrated that a variety of genes can be expressed in an HCC-specific manner under the control of the AFP regulatory sequences *in vitro* and *in vivo*. It would appear that, with the development of a suitable delivery system, HCC-directed gene therapy using the AFP regulatory sequences holds a promising future.

REFERENCES

Abelev, G.I., 1971, Alpha-fetoprotein in ontogenesis and its association with maligant tumors. *Adv. Caner Res.* **14**:295–358.

Arbuthnot, P., Bralet, M.-P., Thomassin, H., Danan, J.-L., Brechot, C., and Ferry, N., 1995, Hepatoma cell-specific expression of a retrovirally transferred gene is achieved by α-fetoprotein but not insulin like growth factor II regulatory sequences, *Hepatology* **22**:1788–1796.

Bois-Joyeux, B., and Danan, J.-L., 1994, Members of the CAAT/enhancer-binding protein, hepatocyte nuclear factor-1 and nuclear factor-1 families can differentially modulate the activities of the rat α-fetoprotein promoter and enhancer, *Biochem. J.* **301**:49–55.

Bui, L.A., Butterfield, L.H., Kim, J.Y., Ribas, A., Seu, P., Lau, R., Glaspy, J.A., McBride, W.H., and Economou, J.S., 1997, *In vivo* therapy of hepatocellular carcinoma with a tumor specific adenoviral vector expressing interleukin-2, *Hum. Gene Ther.* **8**:2173–2181.

Camper, S.A., and Tilghman, S.M., 1989, Postnatal repression of the α-fetoprotein gene is enhancer independent, *Genes Devel.* **3**:537–546.

Emerson, J.A., Vacher, J., Cirillo, L.A., Tilghman, S.M., and Tyner, A.L., 1992, The zonal expression of α-fetoprotein transgenes in the livers of adult mice, *Devel. Dynamics* **195**:55–66.

Feuerman, M.H., Godbout, R., Ingram, R.S., and Tilghman, S.M., 1989, Tissue-specific transcription of the mouse α-fetoprotein gene promoter is dependent on HNF1. *Mol. Cell. Biol.* **9**:4204–4212.

Godbout, R., Ingram, R.S., and Tilghman, S.M., 1988, Fine-structure mapping of the three mouse α-fetoprotein gene enhancers, *Mol. Cell. Biol.* **8**:1169–1178.

Gu, H., Marth, J.D., Orban, P.C., Mossmann, H., and Rajewsky, K., 1994, Deletion of a DNA polymerase beta gene segment in T cells using cell type-specific gene targeting, *Science* **265**:103 106.

Hatano, M., Nakata, K., Nakao, K., Tsutsumi, T., Ohtsuru, A., Nakamura, T., Tamaoki, T., and Nagataki, S., 1992, Hepatocyte growth factor down-regulates α-fetoprotein gene expression in PLC/PRF/5 human hepatoma cells, *Biochem. Biophys. Res. Comm.* **189**:385–391.

Hirano, T., Ueki, T., Okamoto, E., Kaneda, Y., Saito, I., Kaneko, S., Tamaoki, T., and Fujimoto, J., 1998a, Gene therapy for hepatoma by repetitive transgene of HSV-tk gene under the control of α-fetoprotein gene promoter, Abst. 441, American Society of Gene Therapy meeting, Seattle, 1998.

Hirano, T., Ueki, T., Okamoto, E., Kaneda, Y., Saito, I., Tamaoki, T., and Fujimoto, J., 1998b, Cre/*lox*P switching system using HVJ-liposome enhanced specific gene expression under the control of AFP promoter in hepatocellular carcinoma *in vivo*, Abst. 594, American Society of Gene Therapy meeting, Seattle, 1998.

Huber, B.E., Richards, C.A., and Krenitsky, T.A., 1991, Retroviral-mediated gene therapy for the treatment of hepatocellular carcinoma: An innovative approach for cancer therapy, *Proc. Natl. Acad. Sci. USA* **88**:8039–8043.

Ido, A., Nakata, K., Kato, Y., Nakao, K., Murata, K., Fujita, M., Ishii, N., Tamaoki, T., Shiku, H., and Nagataki, S., 1995, Gene therapy for hepatoma cells using a retrovirus vector carrying herpes simplex virus thymidine kinase gene under the control of α-fetoprotein gene promoter, *Cancer Res.* **55**:3105–3109.

Igarashi, T., Suzuki, S., Takahashi, M., Tamaoki, T., and Shimada, T.A., 1998, A novel strategy of cell targeting based on tissue specific expression of the ecotropic retrovirus receptor gene, *Hum. Gene Ther.* in press.

Kanai, F., Shiratori, Y., Yoshida, Y., Wakimoto, H., Hamada, H., Kanegae, Y., Saito, I., Nakabayashi, H., Tamaoki, T., Tanaka, T., Lan, H.-H., Shiina, S., and Omata, M., 1996, Gene therapy for α-fetoprotein-producing human hepatoma cells by adenovirus-mediated transfer of the herpes simplex virus thymidine kinase gene, *Hepatology* **23**:1359–1368.

Kanai, F., Lan, K.-H., Shiratori, Y., Tanaka, T., Ohashi, M., Okudaira, T., Yoshida, Y., Wakimoto, T., Hamada, H., Nakabayashi, H., Tamaoki, T., and Omata, M., 1997, *In vivo* gene therapy for α-fetoprotein-producing hepatocellular carcinoma by adenovirus-mediated gene transfer of cytosine deaminase gene, *Cancer Res.* **57**:461–465.

Kanegae, Y., Lee, G., Sato, Y., Tanaka, M., Nakai, M., Sakaki, T., Sugano, S., and Saito, I., 1995, Efficient gene activation in mammalian cells by using recombinant adenovirus expressing site-specific Cre recombinase, *Nucl. Acids Res.* **23**:3816–3821.

Kaneko, S., Hallenbeck, P., Kotani, T., Nakabayashi, H., McGarrity, G., Tamaoki, T., Anderson, W.F., and Chiang, Y.L., 1995, Adenovirus-mediated gene therapy of hepatocellular carcinoma using cancer-specific gene expression, *Cancer Res.* **55**:5283–5287.

Macri, P., and Gordon, J.W., 1994, Delayed morbidity and mortality of albumin/SV40 T-antigen transgenic mice after insertion of an α-fetoprotein/herpes virus thymidine kinase transgene and treatment with ganciclovir, *Hum. Gene Ther.* **5**:175–182.

McVey, J.H., Michaelides, K., Hansen, L.P., Ferguson-Smith, M., Tilghman, S., Krumlauf, R., and Tuddenham, G.D., 1993, A G to A substitution in an HNF1 binding site in the human α-fetoprotein gene is associated with hereditary persistence of α-fetoprotein, *Hum. Mol. Genet.* **2**:379–394.

Mitsuoka, S., Ohtsuru, A., Nakao, K., Tsutsumi, T., Hamasaki, K., Hatano, M., Sato, Y., Nakayama, T., Shima, M., Nakata, T., Ishii, N., Koji, T., Tamaoki, T., and Nagataki, S., 1992, Inhibitory effect of prostaglandin Δ^{12}-PGJ2 on cell proliferation and α-fetoprotein expression in HuH-7 human hepatoma cells, *Prostaglandins* **43**:189–197.

Morinaga, T., Yasuda, H., Hashimoto, T., Higashio, K., and Tamaoki, T., 1991, A human α-fetoprotein enhancer-binding protein, ATBF1, contains four homeodomains and seventeen zinc fingers. *Mol. Cell. Biol.* **11**:6041–6049.

Nakabayashi, H., Watanabe, K., Saito, A., Otsuru, A., Sawadaishi, K., and Tamaoki, T., 1989, Transcriptional regulation of α-fetoprotein expression by dexamethasone in human hepatoma cells, *J. Biol. Chem.* **264**:266–271.

Nakabayashi, H., Hashimoto, T., Miyao, Y., Tjong, K.-K., Chan, J., and Tamaoki, T., 1991, A position-dependent silencer plays a major role in repressing α-fetoprotein expression in human hepatoma, *Mol. Cell. Biol.* **11**:5885–5893.

Nakao, K., Lawless, D., Ohe, Y., Miyao, Y., and Tamaoki, T., 1990, c-Ha-ras down-regulates the α-fetoprotein gene but not the albumin gene in human hepatoma cells, *Mol. Cell. Biol.* **10**:1461–1469.

Nakao, K., Nakata, K., Mitsuoka, S., Ohtsuru, A., Ido, A., Hatano, M., Sato, Y., Nakayama, T., Shima, M., Kusumoto, Y., Koji, T., Tamaoki, T., and Nagataki, S., 1991, Transforming growth factor β1 differentially regulates α-fetoprotein and albumin in HuH-7 human hepatoma cells, *Biochem. Biophys. Res. Comm.* **174**:1294–1299.

Nakata, K., Motomura, M., Nakabayashi, H., Ido, A., and Tamaoki, T., 1992, A possible mechanism of inverse developmental regulation of α-fetoprotein and albumin genes. Studies with epidermal growth factor and phorbol ester, *J. Biol. Chem.* **267**:1331–1334.

Niwa, H., Yamamura, K., and Miyazaki, J., 1991, Efficient selection for high-expression transfectants with a novel eukaryotic vector, *Gene* **108**:193–200.

O'Gorman, S., Fox, D.T., and Whal, G.M., 1991, Recombinase-mediated gene activation and site-specific integration in mammalian cells, *Science* **251**:1351–1355.

Papaconstantinou, J., Rabek, J.P., and Zhang, D.-E., 1990, Molecular mechanisms of liver specific albumin and α-fetoprotein gene regulation: A review, *Devel. Growth Diff.* **32**:205–216.

Saegusa, M., Ido, A., Nakabayashi, H., Sakai, M., and Tamaoki, T., 1994, Down-regulation of human α-fetoprotein promoter by c-jun and c-fos, *J. Tumor Marker Oncol.* **9**:29–33.

Sakai, M., Morinaga, T., Urano, Y., Watanabe, K., Wegmann, T.G., and Tamaoki, T., 1985, The human α-fetoprotein gene: Sequence organization and the 5′ flanking region, *J. Biol. Chem.* **260**:5055–5060.

Sato, Y., Tanaka, K., Lee, G., Kanegae, Y., Sakai, Y., Kaneko, S., Nakabayashi, H., Tamaoki, T., and Saito, I., 1998, Enhanced and specific gene expression via tissue-specific production of Cre recombinase using adenovirus vector, *Biochem. Biophys. Res. Comm.* **244**:455–462.

Sauer, B., and Henderson, M., 1989, Cre-stimulated recombination at *lox*P-containing DNA sequences placed into the mammalian genome, *Nucl. Acids Res.* **17**:147–161.

Sawadaishi, K., Morinaga, T., and Tamaoki, T., 1988, Interaction of a hepatoma-specific nuclear factor with transcription regulatory sequences of the human α-fetoprotein and albumin genes, *Mol. Cell. Biol.* **8**:5179–5187.

Su, H., Chang, J.C., Xu, S.M., and Kan, Y.W., 1996, Selective killing of AFP-positive hepatocellular carcinoma cells by adeno-associated virus transfer of the herpes simplex virus thymidine kinase gene, *Hum. Gene Ther.* **7**:463–470.

Su, H., Lu, R., Chang, J.C., and Kan, Y.W., 1997, Tissue-specific expression of herpes simplex virus thymidine kinase gene delivered by adeno-associated virus inhibits the growth of human hepatocellular carcinoma in athymic mice, *Proc. Natl. Acad. Sci. USA* **94**:13891–13896.

Thomassin, H., Hamel, D., Bernier, D., Guertin, M., and Belanger, L., 1992, Molecular cloning of two C/EBP-related proteins that bind to the promoter and the enhancer of the α_1-fetoprotein gene. Further analysis of C/EBP$_\beta$ and C/EBP$_\gamma$, *Nucl. Acids Res.* **20**:3091–3098.

Vacher, J., and Tilghman, S.M., 1990, Dominant negative regulation of the mouse α-fetoprotein gene in adult liver, *Science* **250**:1732–1735.

Watanabe, K., Saito, A., and Tamaoki, T., 1987, Cell-specific enhancer activity in a far upstream region of the human α-fetoprotein gene, *J. Biol. Chem.* **262**:4812–4818.

Wen, P., Groupp, E.R., Buzard, G., Crawford, N., and Locker, J., 1991, Enhancer, repressor, and promoter specificities combine to regulate the rat α-fetoprotein gene, *DNA Cell Biol.* **10**:525–536.

Yasuda, H., Mizuno, A., Tamaoki, T., and Morinaga, T., 1994, ATBF1, a multiple homeodomain zinc finger protein, selectively down-regulates AT-rich elements of the human α-fetoprotein gene. *Mol. Cell. Biol.* **14**:1395–1401.

Xu, G.W., Sun, Z.T., Forrester, K., Wang, X.W., Coursen, J., and Harris, C.C. 1996, Tissue specific growth suppression and chemosensitivity promotion in human hepatocellular carcinoma cells by retroviral-mediated transfer of the wild-type p53 gene, *Hepatology* **24**:1264–1268.

6

TUMOR-TARGETED *SALMONELLA*

Highly Selective Delivery Vectors

David Bermudes[1]*, Brooks Low[2], and John Pawelek[3]

[1]Vion Pharmaceuticals, Inc.
4 Science Park, New Haven
Connecticut 06511 USA and
Yale University
School of Medicine
Departments of Therapeutic Radiology[2] and Dermatology[3]
333 Cedar Street, New Haven
Connecticut 06520 USA

1. BACKGROUND

The development of novel cancer therapies presents a formidable challenge to find means to selectively target and eradicate neoplastic cells while sparing healthy tissue. Cancerous cells arising from normal cells in the body have few and often subtle differences from their normal counterparts. Vasculature supplying the tumor is generally irregular and positive pressure within the tumor combines to create a physical barrier to penetration, often reducing the effectiveness of therapeutic agents (Boucher *et al.*, 1996; Jain, 1994). Thus the mandate for new therapeutic agents is not only to exploit differences that tumors may exhibit whereby the agent used is more toxic to cancerous cells than to normal cells, but to overcome the barriers that limit their delivery to the tumor.

One approach has been to develop methods to selectively bind to tumor cells using antibodies. Highly specific monoclonal antibodies directed to tumor-associated antigens such as Lewis y or HER2/neu, have the ability to be administered sysemically and to preferentially accumulate at the site of the tumor cells to which they bind. The anti-cancer activity of these antibodies has been enhanced in several ways. Such antibodies have been used to target cells overexpressing certain antigens and carry enzymes that activate non-toxic prodrug forms of chemotherapeutic agents, inducing selective toxicity at the site of

* To whom correspondence should be sent: DBermude@Vionpharm.com, Tel: (203) 498-4210 ×304, Fax: (203) 498-4211

Cancer Gene Therapy: Past Achievements and Future Challenges, edited by Habib
Kluwer Academic / Plenum Publishers, New York, 2000.

accumulation. These antibodies have also been directly coupled to toxic agents, or used in combination with them, including small molecule chemotherapeutics such as doxirubicin (Trail *et al.*, 1993; Baselga *et al.*, 1998) or protein toxins such as pseudomonas endotoxin (Friedman *et al.*, 1993; Siegall, 1995).

Yet while these or similar agents hold promise for providing selective delivery to tumors, they are subject to the same limitations posed by the fundamental barriers encountered by other agents requiring both the vasculature and diffusion to accomplish delivery into poorly vascularized tumors having positive pressure. Thus, antibody agents would be expected to be limited to the periphery of tumors and the selective accumulation in tumors predicted to be relatively low (Jain, 1994).

Gene therapy agents directed towards cancer are based on the principle of altering the cancer cells by introduction of genetic material in the form of modified viruses, plasmids, ribozymes, and other nucleic acid-based vectors. Specificity for cancerous cells is achieved by several means, including dependency on p53 mutations (Bischoff *et al.*, 1996), tumor specific promoters (e.g., Gotoh *et al.*, 1998), or direct intratumoral injection. By and large, these agents lack ability as systemically delivery vehicles, limiting their usage to local-regional disease.

Liposomes have also been developed for use as gene therapy delivery vehicles. Modification of the spheroids with antibodies provides some selectivity towards the target tissue (Allen *et al.*, 1995). These agents can be administered systemically but encounter difficulty in achieving the level of delivery that enables them to be effective.

Overall, current approaches to tumor-targeted therapy remains promising, yet most of the approaches require further development.

2. TUMOR-TARGETED BACTERIA

Systemically administered bacteria are a little-studied approach that offers surprising advantages over other systems. Clostridia were long ago tested as anti-tumor agents (Parker *et al.*, 1947; Möse and Möse, 1964; Carey *et al.*, 1967) and continue to be investigated (Fox *et al.*, 1996; Lemmon *et al.*, 1997). These bacteria, it was recognized, have a built-in specificity for tumors in that they are strict anaerobes, and can flourish within the anaerobic regions of large tumors. Smaller tumors, however, are not targeted. These bacteria multiply within the necrotic area, and achieve levels of 10^5–10^6/gm tumor tissue (Lemmon *et al.*, 1997).

We have investigated *Salmonella* as tumor-targeting agents (Pawelek *et al.* 1997). These bacteria are facultative anaerobes, growing well in oxygenated as well as hypoxic regions of the tumor. *Salmonella* are motile, and thereby overcome diffusion and pressure gradients that exist in tumors. They are also sensitive to a wide range of antibiotics, and can be engineered to exhibit many of the properties desired in an antitumor delivery vehicle.

Salmonella typhimurium is a common cause of food poisoning frequently contracted by eating undercooked chicken, eggs, and other foods. In humans, disease is usually self-limiting, confined to the lower intestines. Occasionally, the bacteria escape the confines of the gut into circulation, resulting in bacteremia and in some cases septic shock and death if left untreated. Mice are highly sensitive to *S. typhimurium*, and relocation of the bacteria from the lower intestines into circulation occurs at a high frequency similar to the etiology of thyphoid fever in humans caused by *Salmonella typhi*. In mice, after orally ingested bacteria make their way into circulation, these bacteria are found disseminated throughout the body, but are particularly prevalent in the liver, spleen and bone marrow (Carter and Collins, 1974). The bacteria traverse both the vasculature and lymphatic system (Carter and Collins, 1974), and are resistant to complement (Joiner, 1988). Wild type *Salmonella* are also resistant to macrophages (Groisman and Saier, 1990) and tend to persist in tissues in which macrophages reside. While these attributes make these bacteria effective as pathogens, they also predispose them for use as delivery vectors with access to a large portion of the human body.

The initial experiments on the biodistribution of *Salmonella* in mice bearing tumors were performed by Pawelek *et al.* (1997). Using the wild type *S. typhimurium* and a purine auxotrophic clone selected to be hyperinvasive *in vitro* (clone YS72), the phenomenon of differential replication within tumors was observed (Table 1). Tumor types that were shown to be targeted by *Salmonella* included melanoma, lung, colon, breast, kidney and liver. Further study by Kops *et al.* (1997) confirm the ability of the bacteria to spread throughout tumors, apparently overcoming the physical barriers of tumors that limit other agents.

Studies of attenuation of virulent *Salmonella* by mutations in metabolic pathways including purines and amino acids suggested that the major difference between the wild type and the YS72 was due to the purine deficiency. Early studies by Bacon *et al.*, (1950; 1951) on purine auxotrophic *Salmonella* indicated that such mutations limit the pathogenesis of these bacteria. Amino acid auxotrophy has also been shown to strongly attenuate *Salmonella* pathogenic capabilities (Hoiseth and Stocker, 1981). In light of this, a reasonable interpretation of the data in Table 1 is that the restricted growth within the liver by the stain YS72 indeed indicates some form of attenuation, and therefore probably due to the purine auxotropy. However, although this strain was attenuated compared to the wild type strain, it was nonetheless too virulent for long term antitumor studies in sensitive mice such as C57B6.

This finding on attenuation was further exploited by making strains of bacteria that were polyauxotrophic (Fig. 1). Selection for polyauxotrophy reduced the toxicity of the strains, increasing the time to death of tumor-bearing mice in a single dose challenge of bacteria (Table 2). Polyauxotrophic strains of *Salmonella* retain high-level replication within tumors and some exhibit tumor to normal tissue rations of greater than 2,000:1 with a total accumulation of greater than 10^9 c.f.u./g (Table 3). This high-level replication within tumors in highly attenuated strains suggests that the acquisition of nutrients abundant in tumors forms a basis for maintaining tumor specificity.

Table 1. Tumor and liver distribution of *Salmonella* at 5 h, and 2 d post i.p. inoculation of 10^6 c.f.u. Points are for C57B6 mice-bearing B16F10 melanoma tumors, representing avg ± S.D. for n = 3 (5 h) and n = 5 (2 d) animals. Modified from Pawelek *et al.* (1997)

	Salmonella/g tissue		
Time Post-Inoc.	Tumor	Liver	Tumor : Liver
5 h			
Wild type	$7.1 \pm 2.2 \times 10^4$	$1.3 \pm 7.6 \times 10^5$	1:2
YS72	$1.2 \pm 1.3 \times 10^5$	$5.5 \pm 4.8 \times 10^4$	2:1
2 d			
Wild type	$6.5 \pm 6.8 \times 10^9$	$2.4 \pm 2.8 \times 10^7$	270:1
YS72	$1.7 \pm 1.2 \times 10^9$	$1.9 \pm 2.3 \times 10^5$	9,000:1

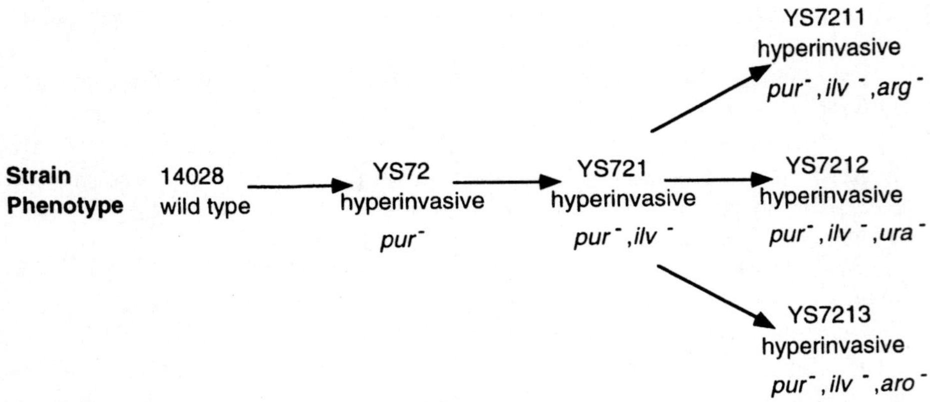

Figure 1. Sequential mutation of the wild type (ATCC 14028) bacteria inducing auxotrophic and polyauxotrophic mutations.

Table 2. Survival of C57B6 mice after introduction of wild type, auxotrophic and polyauxotrophic *Salmonella*. Modified from Pawelek *et al.* (1997)

Strain	Phenotype	Mean Survival (days)
14028	Wild type	3
YS72	*pur*⁻	6
YS721	*pur*⁻, *ilv*⁻	30
YS7211	*pur*⁻, *ilv*⁻, *arg*⁻	>30
YS7212	*pur*⁻, *ilv*⁻, *ura*⁻	>30
YS7213	*pur*⁻, *ilv*⁻, *aro*⁻	>30

Table 3. Tumor and liver distribution of *Salmonella* at 2 d and 4 d post i.p. inoculation of 10^6 c.f.u. Points are for C57B6 mice-bearing B16F10 melanoma tumors, representing avg ± S.D. and n = 5 (2 d, 4 d) animals. Modified from Pawelek *et al.* (1997)

	Salmonella/g tissue		
Time Post-Inoc.	Tumor	Liver	Tumor : Liver
2 d			
YS72	$1.7 \pm 1.2 \times 10^9$	$1.9 \pm 2.3 \times 10^5$	9,000 : 1
YS721	$8.7 \pm 3.1 \times 10^8$	$4.2 \pm 3.6 \times 10^6$	210 : 1
YS7211	$3.3 \pm 3.0 \times 10^7$	$8.1 \pm 8.4 \times 10^5$	41 : 1
YS7212	$3.9 \pm 7.3 \times 10^7$	$1.1 \pm 0.8 \times 10^6$	35 : 1
YS7213	$1.5 \pm 2.8 \times 10^8$	$4.0 \pm 3.1 \times 10^5$	375 : 1
4 d			
YS72	moribund/dead		
YS721	$3.2 \pm 1.5 \times 10^9$	$4.7 \pm 6.9 \times 10^6$	680 : 1
YS7211	$1.6 \pm 2.2 \times 10^9$	$6.3 \pm 9.9 \times 10^6$	253 : 1
YS7212	$1.1 \pm 7.4 \times 10^9$	$5.1 \pm 8.6 \times 10^5$	2,200 : 1
YS7213	$1.3 \pm 2.5 \times 10^9$	$2.2 \pm 6.9 \times 10^5$	5,900 : 1

Lipid A modification has now been shown to offer an additional means to attenuate *Salmonella* (Low *et al.*, 1998; Kahn *et al.*, 1998). *Salmonella* deficient in terminal myristolization of lipid A are greatly reduced in their ability to elicit TNFα, the cytokine that initiates septic shock. Yet strains of *Salmonella* bearing this mutation retain both tumor-targeting and antitumor activity. This mutation seems particularly promising in furthering the potential use of systemically administered gram-negative bacterial antitumor agents.

2.1. *Salmonella* Antitumor Activity

These highly attenuated strains of *Salmonella* possess innate antitumor activity (Pawelek *et al.*, 1997). When subcutaneously implanted B16F10 melanoma is staged to day 8 and the bacteria administered systemically by i.p. injection, there is a dramatic slowing of tumor progression (Fig. 2). With one strain YS7212, the treated group survived more than twice as long as the untreated group.

In addition, it has been shown that this same strain inhibits tumor metastasis *in vivo* (Zheng *et al.*, 1997). Since metastatic disease is one of the major causes of death in patients with solid tumors, this aspect of antitumor *Salmonella* is also particularly encouraging.

2.2. *Salmonella* Expression of Prodrug-Converting Enzymes

Bacteria are well developed as protein expression systems. Tumor-targeted *Salmonella* extend the utility of bacteria to include both the expression and the delivery of anticancer therapeutic proteins directly within cancerous tissue. While bacteria do not perform mammalian glycosylation and other protein modifications, there are many effector proteins in which such modifications are unnecessary. The herpes simplex thymidine kinase (HSV TK) is an example of a prodrug converting enzyme that is functionally expressed in bacteria (Garapin *et al.*, 1981). This enzyme activates nucleoside analogues

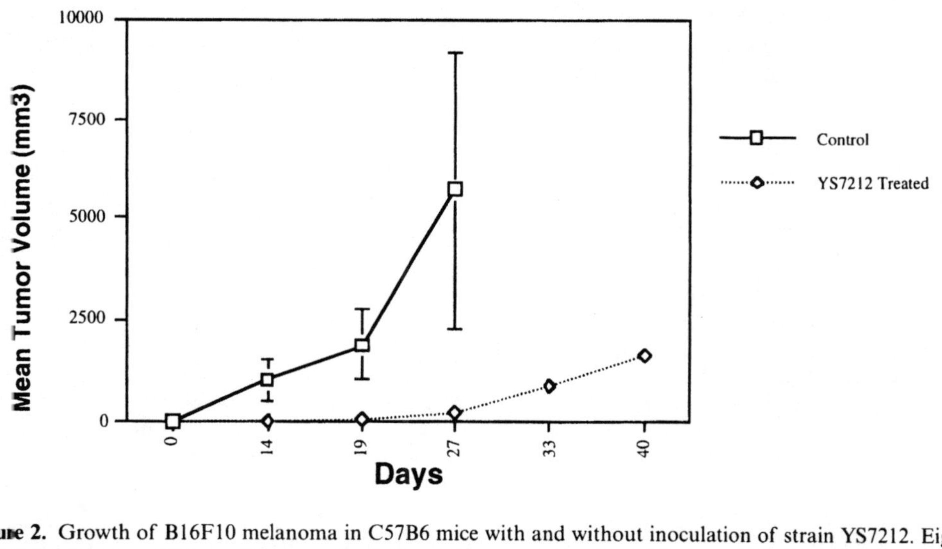

Figure 2. Growth of B16F10 melanoma in C57B6 mice with and without inoculation of strain YS7212. Eight days following introduction of tumor cells, the treated group received 4 × 10^6 c.f.u. injected i.p. Each point represents the mean ± S.E. for n = 5 animals.

such as acyclovir (ACV) and ganciclovir (GCV). When Pawelek *et al.* (1997) used a strain of *Salmonella* expressing a secreted form of HSV TK in a B16F10 subcutaneous melanoma model, they found 1) that the presence of the plasmid lessened the innate antitumor activity of the bacteria, and 2) that when these bacteria were coadministered with GCV the tumors were 2.5 times smaller than without added GCV. They concluded that these bacteria were capable in delivering a prodrug converting enzyme effective in activating a compound into its chemotherapeutic form.

Another example of prodrug conversion by tumor-targeted *Salmonella* has been studied by King *et al.* (1998). This study demonstrated that *Salmonella* expressing cytosine deaminase effectively convert 5-flurocytosine (5-FC) to 5-fluorouricil (5FU) within tumors and resulted in antitumor activity against subcutaneously implanted B16F10 melanoma.

Taken together, these studies indicate that *Salmonella* may be useful in expressing a variety of prodrug converting enzymes directly within tumors resulting in antitumor effects when given together with their appropriate prodrug, and therefore applicable in treatment of a wide range of tumor types.

3. SUMMARY

Genetically engineered *Salmonella* offer an intriguing new approach to selectively target solid tumors, including melanoma, lung, colon, breast, kidney and liver. These bacteria target tumors after systemic administration and selectively replicate within them. Specificity for tumors is often more than 1,000 times greater than for any other tissue. Auxotrophic mutations make these bacteria highly safe and form the basis for maintaining tumor specificity. An altered lipid greatly reduces the potential for septic shock yet also retains the antitumor properties of these bacteria. These bacteria have innate antitumor activity towards both primary and metastatic tumors and the ability to deliver proteins capable of activating chemotherapeutic agents directly within tumors. The delay in tumor growth results in mice that survive up to twice as long. These bacteria are susceptible to a wide range of antibiotics, allowing external control of the vector after administration. The combination of these features within a single vector seems especially surprising considering their unlikely source.

REFERENCES

Allen, T.M., Hansen, C.B., and Zalipsky, S. 1995. Antibody-targeted stealth liposomes. In: *Stealth Liposomes*, D.D. Lasic (ed), CRC Press, Boca Raton, pp. 233–244.

Bacon, G.A., Burrows, T.W., and Yates, M. 1950. The effects of biochemical mutation on the virulence of *bacterium typhosum*: The virulence of mutants. *Br. J. Exp. Path.* **31**:714–724.

Bacon G.A., Burrows T.W., and Yates M. 1951. The effects of biochemical mutation on the virulence of *bacterium typhosum*: The loss of virulence of certain mutants. *Br. J. Exp. Path.* **32**:85–96.

Baselga, J., Norton, L., Albanell, J., Kim, Y.-M., and Mendelsohn, J. 1998. Recombinant humanized anti-HER2 antibody (Herceptin™) enhances the antitumor activity of paclitaxol and doxorubicin against HER2/neu overexpressing human breast cancer xenografts. *Cancer Res.* **58**:2825–2831.

Bischoff, J.R., Kirn, D.H., Williams, A., Heise, C., Horn, S., Muna, M., Ng, L., Nye, J.A., Sampson-Johannes, A., Fattaey, A., and McCormick, F. 1996. An adenovirus mutant that replicates selectively in p53-deficient human tumor cells. *Science* **274**:373–376.

Boucher, Y., Leunig, M., and Jain, R.K. 1996. Tumor angiogenesis and interstitial hypertension. *Cancer Res.* **56**:4264–4266.

Carey, R.W., Holland, J.F., Whang, H.Y., Neter, E., and Bryant, B. 1967. Clostridial oncolysis in man. *Eur. J. Cancer* **3**:37–46.

Carter, P.B., and Collins, F.M. 1974. The route of enteric infection in normal mice. *J. Exp. Med.* **139**:1189–1203.

Fox, M.E., Lemmon, J.J., Mauchline, M.L., Davis, T.O., Giaccia, A.J., Minton, N.P., and Brown, J.M. 1996. Anaerobic bacteria as a delivery system for cancer gene therapy: in vitro activation of 5-fluorocytosine by genetically engineered clostridia. *Gene Therapy* **3**:173–178.

Friedman, P.N., McAndrew, S.J., Gawlak, S.L., Trail, P.A., Brown, J.P., and Siegall, C.B. 1993. BR96 sFv-PE40, a potent single-chain immunotoxin that selectively kills carcinoma cells. *Cancer Res.* **53**:334–339.

Garapin A.C., Colbère-Garapin, F., Cohen-Solal, Horondniceanu, F., and Kourilsky, P. 1981. Expression of herpes simples virus type I thymidine kinase gene in *Escherichia coli. Proc. Natl. Acad. Sci. USA* **78**:815–819.

Gotoh, A., Ko, S.C., Shirakawa, T., Cheon, J., Kao, C., Miyamoto, T., Gardner, T.A., Ho, L.J., Cleutjens, C.B., Trapman, J., Graham, F.L., and Chung, L.W. 1998. Development of prostate-specific antigen promoter-based gene therapy for androgen-independent human prostate cancer. *J. Urol.* **160**:220–229.

Groisman, E.A., and Saier, M.H. Jr. 1990. *Salmonella* virulence: new clues to intramacrophage survival. *Trends Biochem. Sci.* **15**:30–33.

Hoiseth, S.K.J., and Stocker, B.A.D. 1981. Aromatic-dependent *Salmonella typhimurium* are non-virulent and effective as live vaccines. *Nature* **291**:238–239.

Jain R.K. 1994. Barriers to drug delivery in solid tumors. *Scientific American* **271(July)**:58–65.

Joiner, K.A. 1988. Complement evasion by bacteria and parasites. *Ann. Rev. Microbiol.* **42**:201–230.

Khan, S.A., Everest, P., Servos, S., Foxwell, N., Zähringer, U., Brade, H., Rietschel, E. Th., Dougan, G., Charles, I.G., and Maskell, D.J. 1998. A lethal role for lipid A in *Salmonella* infections. *Mol. Mircobiol.* **29**:571–579.

King, I., Feng, M, Luo, X., Lin, S., Bermudes, D., and Zheng, L.-M. 1998. Tumor-targeted *Salmonella* expressing cytosine deaminase converted 5-fluorocytosine to 5-fluorouracil and inhibited tumor growth in vivo. *Proc. Amer. Asoc. Can. Res.* **39**, p. 512, Abstract 3484.

Kops, S.K., Luo, X., Fischer, J., Le, T., Bermudes, D., Bolognia, J.L., Carmichael, E., Key-Yen, A., King, I., Low, K.B., Pawelek, J.M., Sodi. S.S., and Zheng, L.-M. *Salmonella typhimurium* as an anti-cancer vector: Localization within solid tumors. *Proc. Amer. Asoc. Can. Res.* **38**, p. 7, Abstract 46.

Lemmon, M.J., van Zijl, P., Fox, M.E., Mauchline, M.L., Giaccia, A.J., Minton, N.P., and Brown, J.M. 1997. Anaerobic bacteria as a gene delivery system that is controlled by the tumor microenvironment. *Gene Therapy* **4**:791–796.

Low, K.B., Ittensohn, M., Le, T., Platt, J., Sodi, S., Amoss, M., Ash, O., Carmichael, E., Chakraborty, A., Fischer, J., Lin, S.L., Luo, X., Miller, S.I., Zheng, L.-M., King, I., Pawelek, J.M., and Bermudes, D. 1998. Disruption of the *Salmonella msbB* gene suppresses virulence and TNFα induction yet retain tumor-targeting *in vivo*. *Proc. Amer. Asoc. Can. Res.* **39**, p. 60, Abstract 409.

Möse, J.R., and Möse, G. 1964. Oncolysis by Clostridia. I. Activity of Clostridium butyricum (M-55) and other nonpathogenic clostridia against the Ehrlich carcinoma. *Cancer Res.* **24**:212–216.

Parker, R.C., Plummer, H.C., Siebenmann, C.O., and Chapman, M.G. 1967. Effect of histollyticus infection and toxin on transplantable mouse tumors. *Proc. Soc. Exp. Biol. Med.* **66**:461–467.

Pawelek, J., Low, K.B., and Bermudes, D. 1997. Tumor-targeted *Salmonella* as a novel anti-cancer vector. *Cancer Research* **57**:4537–4544.

Siegall, C.B. 1995. Targeted therapy of carcinomas using BR96 sFv-PE40, a single-chain immunotoxin that binds to the Le(y) antigen. *Semin Cancer Biol.* **6**:289–295.

Trail, P.A., Willner, D., Lasch, S.J., Henderson, A.J., Hostead, S., Casazaza, A.M., Firestone, R.A., Hellström, I., and Hellström, K.E. 1993. Cure of xenografted human carcinomas by BR96-doxorubicin immunoconjugates. *Science* **261**:212–215.

Zheng, L.-M., Luo, X., Fischer, J., Le, T., Bermudes, D., Low, B., Pawelek, J.M., and King, I. 1997. Attenuated *Salmonella typhimurium* inhibited tumor metastasis *in vivo*. *Proc. Amer. Asoc. Can. Res.* **38**, p. 9, Abstract 60.

7

MUTANT ADENOVIRUSES SELECTIVELY REPLICATION-COMPETENT IN TUMOR CELLS

Makoto Sunamura*

First Department of Surgery
Tohoku University School of Medicine Sendai
Japan

INTRODUCTION

The development of a malignant tumor is a multi-step process resulting from the mutation of several specific genes involved in the control of cell growth and programmed cell death.

Most retroviral and adenoviral vectors containing a deletion in an essential gene required for viral propagation are unable to replicate in cells. Rather than destroying tumor cells through viral replication, these vectors rely on the delivery of therapeutic genes to destroy tumor cells. One of the major difficulties with this approach is the daunting goal of delivering genes to every cancer cell of a tumor.

The idea of using viruses as antitumor agents has been emerged from different groups.

The adenovirus relies on cellular replication factors for efficient viral DNA replication. Thus, the virus must induce cell cycle progression for productive infection in normal growth-arrested cells. Adenovirus E1 gene products, in addition to transactivating the other early gene promoters, prepare the cellular environment for optimal viral replication by associating with a number of key cell cycle proteins. Selectively-replicating viruses may overcome the limitation of gene transfer of adenoviral vectors. Virus replication in a small fraction of the tumor cells leads to amplification and extension of the antitumor effect. Cell killing is due to viral replication and cell lysis exclusively.

Phone: +81-22-717-7205, Fax: +81-22-717-7209, E-mail: msun-thk#umin.u-tokyo.ac.jp

Cancer Gene Therapy: Past Achievements and Future Challenges, edited by Habib
Kluwer Academic/Plenum Publishers, New York, 2000.

E1B MUTATED ADENOVIRUS

The *p53* gene is mutated in more than half of human tumors (Greenblatt *et al.*, 1994) indicating that it plays a key role as tumor suppressor. In support of this notion, is the evidence that members of the cancer-prone Li-Fraumeni families carry germline *p53* mutations (Malkinet *et al.*, 1990) and *p53* null mice frequently develop tumors (Donehower *et al.*, 1992). On the other hand, reconstitution of the wild-type *p53* gene expression suppresses the growth of different human and mouse tumor cells in culture through the induction of cell-cycle arrest and/or apoptosis (Baker*et al.*, 1990; Yonish-Rouach *et al.*, 1991).

p53, tumor suppressor protein, mediates cell cycle arrest and/or apoptosis in response to DNA damage caused by chemotherapy, radiation or foreign DNA synthesis during viral replication. The p53 protein is normally present in minute levels and is probably inactive. However, when cells are exposed to DNA damage or start to divide aberrantly, p53 levels rise and the protein production is switched on. Activated *p53* gene acts to block cell division and induce programmed cell death of the damaged cell.

As *p53* is functionally inactivated in many human tumors, one strategy to impede their growth is through transduction of wild-type *p53* gene to cancer cells using a virus vector. But one of the limitations of this strategy and other cancer gene therapies is the low efficiency at which therapeutic genes can be delivered in vivo, especially into solid tumors.

In order to replicate themselves, the small DNA tumor viruses encode proteins, which shut down *p53* gene. In adenovirus, this is the task of the E1B encoded proteins, E1B 19kDa and E1B 55kDa (Fig. 1). The former acts downstream of *p53* to prevent apoptosis, whereas the latter physically associates with *p53* to prevent p53-mediated transactivation (Chinnadurai, 1998). One example is that, DNA tumor viruses such as adenovirus, SV40 and human papilloma virus encode for proteins that inactivate p53, resulting

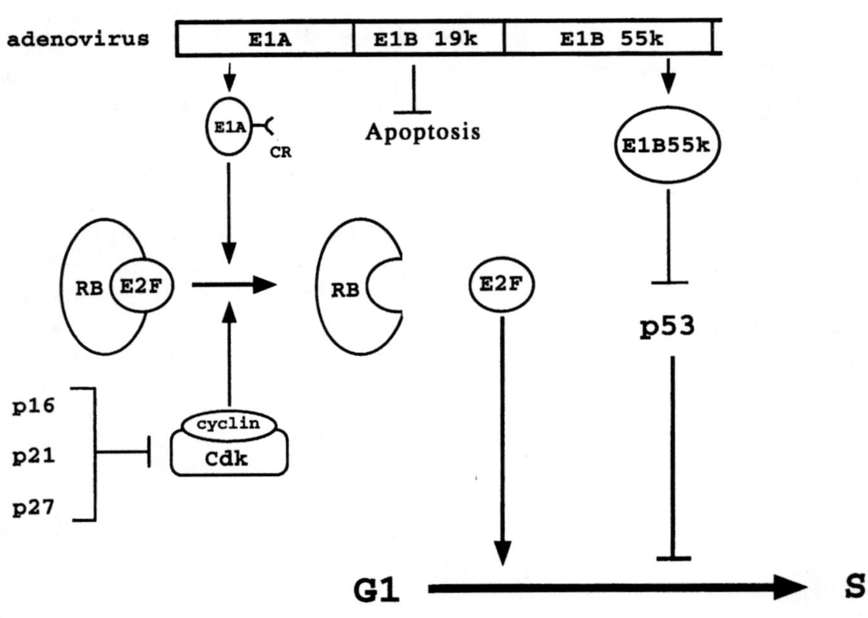

Figure 1. The role of E1A, E1B 19K and E1B 55K proteins in controlling cell cycle.

in efficient viral replication. E1B-region 55 kDa protein is reported to bind p53 followed by the inactivation of p53 function (Barker and Berk, 1987). E1B co-operates with E1A to produce a completely transformed cell. Essentially, E1A pushes on the accelerator and E1B cuts the brake cables. The E1B 55 K and 19 K proteins act independently to inhibit growth arrest and apoptosis, respectively. The E1B 55 KDa protein binds to p53 and prevents it from stimulating the promoters of growth arrest genes, such as p21 and GADD45 (Ricciardi, 1995; Beck *et al.*, 1998). Mechanistically, E1B 55 KDa protein complexes with the amino-terminal activation domain of p53 while allowing p53 to remain bound to its DNA recognition site on the promoter. Interestingly, E1B 55 K may function as a dimer that can actually increase the DNA binding affinity of p53 to the promoter. Even in the absence of p53, E1B 55 K protein exhibits an inherent transcriptional repression activity, which may be regulated by phosphorylation, when it is tethered to a promoter as a GAL4-fusion protein (Teodoro and Branton *et al.*, 1997). Thus, E1B 55 K may not only block p53 mediated transcription by masking its activation domain but also may use the surface of DNA-bound p53 as a strategic site to exert active repression.

Recent studies demonstrated the usefulness of restricted replication-competent adenovirus, ONYX-015 (ONYX Pharmaceuticals, Richmond, CA) (Bischoff *et al.*, 1996). A mutant adenovirus can replicate in and lyse p53-deficient human tumor cells but not cells with functional p53. ONYX-015 is an attenuated adenovirus type 2/5 chimera (dl 1520) with two mutations in the early region E1B, 55 kD gene. Since p53 function must be blocked in order to allow efficient virus replication, an adenovirus lacking E1B 55 kD gene expression can not replicate in normal cells holding normal p53 function. Tumor cells, which lack p53 function, allow the replication of mutant virus lacking E1B, 55 kD gene and this is followed by destruction of host cells.

Moreover, the released virus infect neighbouring cells and their subsequent proliferation results in a destructive cycle (Fig. 2). This character of the mutant adenovirus distinguishes the strategy from other gene therapy approaches, which are limited due to the non-replicating character of conventional virus vectors.

The hypothesis is attractive, but what is really interesting is that the strategy actually works out well on several tumor models. ONYX-015 induces cytopathic effects identical to that of wild-type adenovirus, in tumors lacking functional p53, whereas these cytopathic effects are remarkably reduced in the tumor line harboring normal p53.

How p53 prevents viral replication in normal cells remains mysterious. Although p53 promotes apoptosis in cells expressing adenovirus E1A (Lowe and Ruley *et al.*, 1993), apoptosis presumably is not the mechanism that prevents ONYX-015 replication within these cells because the virus is not toxic in normal cells. ONYX-015 efficiently destroys certain tumor cells that acquire resistance to chemotherapeutic drugs (Heise *et al.*, 1997). These results bring about the exciting possibility that mutant adenoviruses may be able to shrink tumors in cases where standard therapies have failed.

Phase I clinical testing of ONYX-015 begun in April 1996 in patients with head and neck cancer and continued on those with pancreatic cancer, liver metastasis of colorectal cancer and ovarian cancer. More than 150 cancer patients have been treated to date with ONIX-015. Toxicities of the treatment combining ONYX-015 with two chemotherapeutic drugs, Cisplatin™ and 5-FU, observed in these studies are no more than those associated with the administration of Cisplatin™ and 5-FU alone. In studies of ONYX-015 alone in head and neck cancer treatment has been well tolerated with no remarkable dose-limiting toxicities. Among the 26 patients with head and neck cancer treated with intratumoral injection of ONYX-015 and chemotherapy employing Cis-

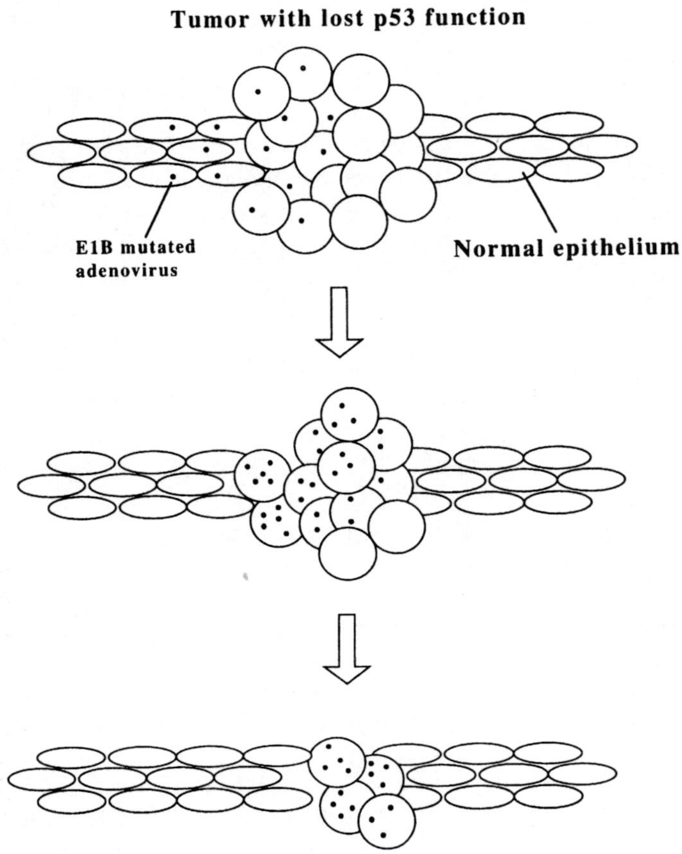

Figure 2. Propagation of the mutant adenovirus after transfection within tumors lacking p53 function.

platin™ and 5-FU, 16 patients have experienced tumor regressions of more than 50%, this equals 61% of the total number of patients compared to 35% historically obtained with chemotherapy alone. Furthermore, a complete response was recorded in six patients (23%) compared to less than 10% with chemotherapy alone. The median time to tumor progression for all patients in the study was five months and was expected to increase further. This is in contrast to the median time to tumor progression of 3 months historically for chemotherapy alone. Two of the 15 patients who entered the study with a single tumor mass and developed a second tumor and were then treated at the site of the new tumor. Both of the treated patients experienced another complete response in their second tumor sites. This result is exciting though remains to be clarified and indicates that the immunity to the virus is not preventing efficacy in patients who are retreated after achieving a response at the initial tumor site. Encouraging results of the efficacy of ONYX-015 (Kirn *et al.*, 1998) are expected to emerge from the Phase II combination studies in pancreatic cancer and Phase I/II study in liver metastasis of colorectal cancer.

In order to develop further strategies for gene therapy of cancer using adenoviruses, we constructed E1B 55K deficient adenovirus (AxE1AdB) and examined the combination of this mutant adenovirus with other adenovirus vectors on pancreatic cancer line (Motoi *et al.*, 1999). Coinfection of E1-deficient adenovirus expressing reporter lacZ gene together with AxE1AdB resulted in the replication of both viruses and marked increase in reporter gene expression in pancreatic cancer cells without p53 function. Pancreatic cancer cells coinfected with E1B 55K deficient adenovirus and adenovirus vector for human interleukin-2 (AxCAhIL-2) produced 110 times more IL-2 than those infected with AxCAhIL-2 alone. Moreover, injection of AxE1AdB with AxCAhIL-2 resulted in a complete regression of the established tumors. These data suggest that this E1B 55K deficient adenovirus is not only the strong weapon itself but also a carrier of genes for anti-tumor factors as virus vector specific to tumors without normal p53 function.

E1A MUTATED ADENOVIRUS

A family of transcription factors, altogether called E2F, is required for transcription of several genes involved in DNA and deoxynucleotide synthesis, inducing G1-S transition of cell cycle (Adams and Kaelin *et al.*, 1995). Binding of two proteins, one called pRB and another related protein called p107, to E2F inhibits its ability to activate transcription (Ricciardi, 1995; Beck *et al.*, 1998). *RB* was initially identified as the product of the prototype tumor suppressor gene *Rb* (Weinberg, 1995). The underphosphorylated form of pRB may act as a growth suppressor by blocking exit from the G0 or G1 phase. Phosphorylation of pRB inhibits the growth suppression function of pRB, allowing the cell to enter S phase.

The mechanisms of action of the Rb gene have been clarified through studies of the early region 1A (E1A) oncogene of human adenovirus type 5 (Fig. 3). The binding of proteins, such as adenovirus E1A or SV40 large-T-antigen gene products, to pRB negates the requirement for pRB phosphorylation and allows quiescent cells to enter the cell cycle.

Transformation seems to require specific interactions between E1A polypeptides and key cellular proteins. Several cellular proteins where E1A proteins bind have been noted (Yee and Branton, 1985; Harlow *et al.*, 1986). Two of these have molecular masses of 105 kDa and 300 kDa (p105 and p300, respectively). P105 has been identified as the product of the Rb gene (Whyte *et al.*, 1988). Interestingly, E1A mutants, whose proteins fail to bind to p105-RB also fail to transform cells. Furthermore, mutant E1A products that bind p105-RB normally but fail to bind p300 are also defective for transformation (Egan *et al.*, 1988; Jelsma *et al.*, 1989; Whyte *et al.*, 1989).

P300 is closely related to CBP (CREB binding protein), a nuclear factor that binds selectively to the protein-kinase-A-phosphorylated form of CREB (cAMP-regulated enhancer protein) (Chrivia *et al.*, 1993; Kwok *et al.*, 1994). p300/CBP can arrest the cell cycle by functioning as an co-activator of p-53 mediated transcription of p21, the cyclin-dependent kinase inhibitor. Thus, by associating with E1A, the p300/CBP both induction of tissue specific gene expression and growth arrest are abolished.

The E1A coding region comprises multiple discrete domains that physically target different cellular factors. The CR1, CR2 and amino-terminal domains of E1A are essential for transformation and regulate transcription of genes involved in cellular growth by combining to critical factors such as pRB and p300, resulting in the loss of their abilities to govern the cell cycle (Ricciardi, 1995; Mymryk, 1996).

Recently another mutant adenovirus with mutation in E1A region was reported to be the alternative strategy for cancer therapy. There is hypothesis that an adenovirus with a deletion in the pRB-binding (CR2) region of E1A or with mutations in the p300-

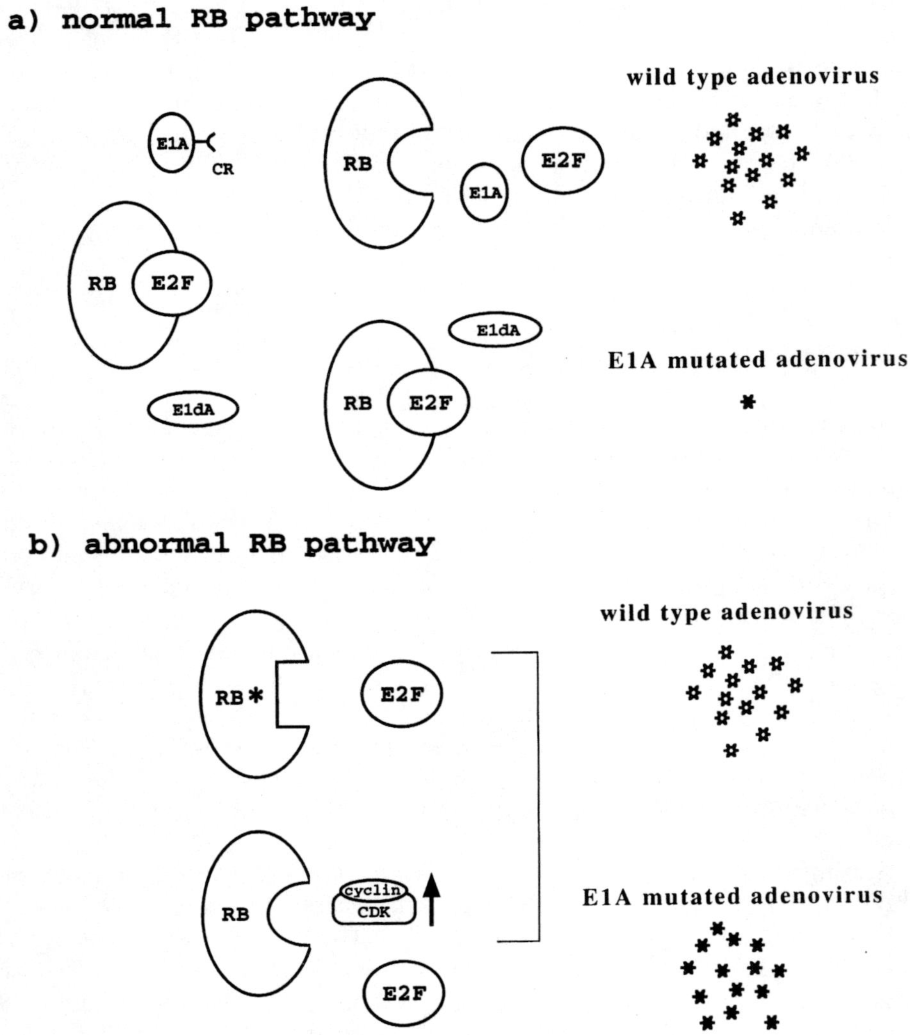

Figure 3. The fate of mutated and wild type adenovirus in cells with normal, a) and abnormal, b) Rb pathways.

binding region (CR1) of E1A might selectively replicate in cancer cells with defects in the pRB pathway (e.g., RB mutation, p16 loss, cyclin D amplification). Researchers at Onyx Pharmaceuticals, Inc. have developed a second viral therapeutic, ONYX-838 that uses its deficiency in RB (retinoblastoma) gene pathway to replicate and destroy cancer cells while leaving resting cells unharmed. The newly developed virus is meant to target a mutation present virtually in all human cancers. ONYX-838 is currently tested in preclinical studies and it is expected to enter soon clinical trials. Adenoviruses with deletion in the pRB-binding (CR2) region of E1A or with mutations in the p300-binding region of E1A (CR1) selectively replicate in and lyse a wide variety of tumor cells. This may be another promising anti-cancer strategy for many tumors that have aberrant pRB pathway.

REFERENCES

Adams P.D., and Kaelin W.G. Transcriptional control by E2F. Semin Cancer Biol 6, 99–108, 1995.

Baker S.J., Markowitz E.R., Fearon J.K., Willson K.V., and Vogelstein B. Suppression of human colorectal carcinoma cell growth by wild-type p53. Science 249, 912–915, 1990.

Barker D.D., and Berk A.J. Adenovirus proteins from both E1B reading frames are required for transformation of rodent cells by viral infection and DNA transfection. Virology 156, 107–121, 1987.

Beck G.R., Zerler B.R., and Moran E. Introduction to DNA tumor viruses: adenovirus, simian virus 40, and polyomavirus. In: McCance D.J. ed. Human Tumor Viruses. Washington DC: Am Soc Microbiol, 1998.

Bischoff J.R., *et al.* An adenovirus mutant that replicates selectively in p53-deficient human tumor cells. Science 274, 373–376, 1996.

Chinnadurai G. Control of apoptosis by human adenoviruses. Semin in Virol 8, 399–408, 1998.

Chrivia J.C., Kwok R.P., Lamb N., Hagiwara M., Montminy M.R., and Goodman R.H. Phosphorylated CREB binds specifically to the nuclear protein CBP. Nature 365:855–859, 1993.

Donehower L.A., *et al.* Mice deficient for p53 are developmentally normal but susceptible to spontaneous tumors. Nature 353, 215–221, 1992.

Egan C., Jelsma T.N., Howe J.A., Bayley S.T., Ferguson B., and Branton P.E. Mapping of cellular protein-binding sites on the products of early-region 1A of human adenovirus type 5. Mol Cell Biol 8, 3955–3959, 1988.

Greenblatt M.S., Bennet W.P., Hollstein M., and Harris C.C. Mutations in the p53 tumor suppressor gene: Clues to cancer etiology and molecular pathogenesis. Cancer Res 54, 4855–4878, 1994.

Harlow E., Whyte P., Franza B.R. Jr, and Schley C. Association of adenovirus early-region 1A proteins with cellular polypeptides. Mol Cell Biol 6, 1579–1589, 1986.

Heise C., *et al.* ONYX-015, an E1B gene-attenuated adenovirus, causes tumor-specific cytolysis and antitumoral efficacy that can be augmented by standard chemotherapeutic agents. Nat Med 3, 639–645, 1997.

Jelsma T.N., Howe J.A., Mymryk J.S., Evelegh C.M., Cunniff N.F., and Bayley S.T. Sequences in E1A proteins of human adenovirus 5 required for cell transformation, repression of a transcriptional enhancer, and induction of proliferating cell nuclear antigen. Virology 171, 120–130, 1989.

Kirn D., Hermiston T., and McCormick F. ONYX-105: Clinical data are encouraging. Nat Med 4, 1341–1342, 1998.

Kwok R.P., Lundblad J.R., Chrivia J.C., Richards J.P., Bachinger H.P., Brennan R.G., Roberts S.G., Green M.R., and Goodman R.H. Nuclear protein CBP is a coactivator for the transcription factor CREB. Nature 370(6486):223–226, 1994.

Lowe S.W., and Ruley H.E. Stabilization of the p53 tumor suppressor is induced by adenovirus E1A and accompanied apoptosis. Genes Dev 7, 535–545, 1993.

Malkin D., *et al.* Germ line p53 mutations in a familial syndrome of breast cancer, sarcomas, and other neoplasias. Science 250, 1233–1238, 1990.

Motoi F., Sunamura M., Sadata A., Yoshida Y., Matsuno S., and Hamada H. Effective gene therapy for pancreatic cancer by cytokines mediated by restricted replication-competent adenovirus. (submitted for publication).

Mymryk J.S. Tumor suppressive properties of the adenovirus 5 E1A oncogene. Oncogene 13, 1581–1589, 1996.

Ricciardi R.P. Transformation and tumorigenesis mediated by the adenovirus E1A and E1B oncogenes. In: Barbanti-Brodano G., Bendienelli M., Friedman H., eds. DNA Tumor Viruses: Oncogenic Mechanisms. New York: Plenum Press, 195–225, 1995.

Teodoro J.G., and Branton P.E. Regulation of p53-dependent apoptosis, transcriptional repression, and cell transformation by phosphorylation of the 55-kilodalton E1B protein of human adenovirus type 5. Virol 71, 3620–3627, 1997.

Weinberg R.A. The retinoblastoma protein and cell cycle control. Cell 81, 323–330, 1995.

Whyte P., *et al.* Association between an oncogene and an anti-oncogene: the adenovirus E1A proteins bind to the retinoblastoma gene product. Nature 334, 124–129, 1988.

Whyte P., Williamson N.M., and Harlow E. Cellular targets for transformation by the adenovirus E1A proteins. Cell 56, 67–75, 1989.

Yee S.P., and Branton. Detection of cellular proteins associated with human adenovirus type 5 early region 1A polypeptides. Virology 147, 142–153, 1985.

Yonish-Rouach E., *et al.* Wild type p53 induces apoptosis of myeloid leulemic cells that is inhibited by interleukin-6. Nature 353, 345–347, 1991.

8

POLYOMA AND PAPILLOMA VIRUS VECTORS FOR CANCER GENE THERAPY

Nina Krauzewicz and Beverly E. Griffin

Department of Infectious Diseases (7th Floor)
Imperial College School of Medicine
Du Cane Road, London W12 ONN
UK

1. INTRODUCTION

Gene therapy is becoming increasingly accepted as a strategem for treating human disease. Over 60% of all therapies approved for clinical trials are directed towards cancer targets, the majority of which have used retroviral or adenoviral vectors to deliver genes to the target tissues (Verma and Somia, 1997; Anderson, 1998). Viral vectors are engineered to carry a gene expression cassette and only be capable of replication in a helper system. Results with these vectors have been encouraging, however, concerns for safety and adverse immunological responses *in vivo* may ultimately limit their use in the clinic. Lipid and colloid based non-viral vectors are chemically better defined than viral vectors, but lack the specificity required to target cancer cells (Cooper, 1996). The limitations of the currently available "vectorology" has led several groups to focus on developing alternative systems, which retain the targetting and cell uptake properties of viruses, whilst having the potential to be totally synthesised *in vitro*. One such approach uses viral like particles, or pseudocapsids.

2. PSEUDOVIRIONS AND PSEUDOCAPSIDS

2.1. Early Studies

In the early 1970s, Aposhian and co-workers proposed that "empty" particles from the non-enveloped mouse virus, polyoma, could be co-assembled with heterologous DNA in a cell free system and used as carriers of genes into mammalian cells (Osterman *et al.*, 1971; Qasba and Aposhian, 1971). In their experiments, capsids devoid of DNA and composed only of protein were isolated by density gradient centrifugation from infected

Cancer Gene Therapy: Past Achievements and Future Challenges, edited by Habib
Kluwer Academic/Plenum Publishers, New York, 2000.

cells. These particles could be encouraged to create polyoma-like particles (PLPs), by incubating the mixture at low ionic strength with supercoiled polyoma viral DNA, and these were able to enter rat cells, resulting in viral protein expression (Slilaty and Aposhian, 1983). Subsequently, it was shown that, using the closely related simian polyomavirus, SV40, empty capsids could be dis-assembled under mild reducing conditions and reformed around foreign DNA (Colomar *et al.*, 1993). These experiments demonstrated that: first, it is possible stably to interact DNA with viral coat protein assemblies; secondly, polyoma-like particles could be assembled from the basic subunits of the virus and DNA and, thirdly, these complexes could transfer DNA to cells in a biologically active form.

2.2. *In Vitro* Expression Systems

A second line of research, developing in parallel with the reconstitution studies, centred around generating viral coat protein genes in *in vitro* expression systems to analyse coat protein functions and assembly pathways. Such studies included protein production in bacteria and in eukaryotic systems and have been applied, in particular, to the analysis of mouse polyoma virus structure and assembly. Particles of this virus are composed of one major coat protein (VP1), which forms the bulk of the coat of the virus, and two minor coat proteins (VP2 and VP3), surrounding a double-stranded circular DNA/histone core (Garcea and Liddington, 1997). This structure represents one of the simpler solutions to forming an icosahedral protein coat, the main framework comprising, predominantly, a single protein species. Expression of VP1 in bacteria yielded high levels of protein in pentameric complexes (Leavitt *et al.*, 1985), which under appropriate salt and calcium ion conditions could be made to assemble into higher order structures that resembled viral capsids (Salunke *et al.*, 1986). When VP1 was expressed from a recombinant baculovirus in insect cells, however, it was found efficiently to self assemble in the nuclei of the infected cells, and could easily be purified by density gradient centrifugation (Montross *et al.*, 1991; Forstová *et al.*, 1993). The structure of these higher order assemblies, called pseudocapsids or viral-like particles (VLPs), is shown in Fig. 1A. They can be purified to near homogeneity, using a series of gradient centrifugation steps, as illustrated by the purity of VP1 in the preparations when analysed by denaturing gel electrophoresis (Fig. 1B). VLPs have also now been produced from SV40 VP1 (Sandalon and Oppenheim, 1997), and other members of the polyomavirus family (Rodgers *et al.*, 1994; Pawlita *et al.*, 1996), including the human polyomavirus JC, which has been expressed both in insect cells (Chang *et al.*, 1997) and in monkey Cos 7 cells (Shishido *et al.*, 1997).

2.3. Papillomavirus Like Particles

Papillomaviruses, being part of the papovavirus family, are related to polyomaviruses and have a similar structure that includes one major coat protein (L1), one minor coat protein (L2) and a slightly larger DNA/histone core (reviewed in Howley, 1990). This virus, unlike the polyoma viruses, has many human subtypes, a number of which have been implicated in contributing to the development of human malignancies, such as cervical cancer (reviewed in Shah and Howley, 1990). Papillomavirus VLPs can also be synthesised *in vitro*, from, for instance, recombinant vaccinia virus (Hagensee *et al.*, 1993; Unckell *et al.*, 1997), or a recombinant baculovirus in insect cells (Volpers *et al.*, 1994; Touze *et al.*, 1998).

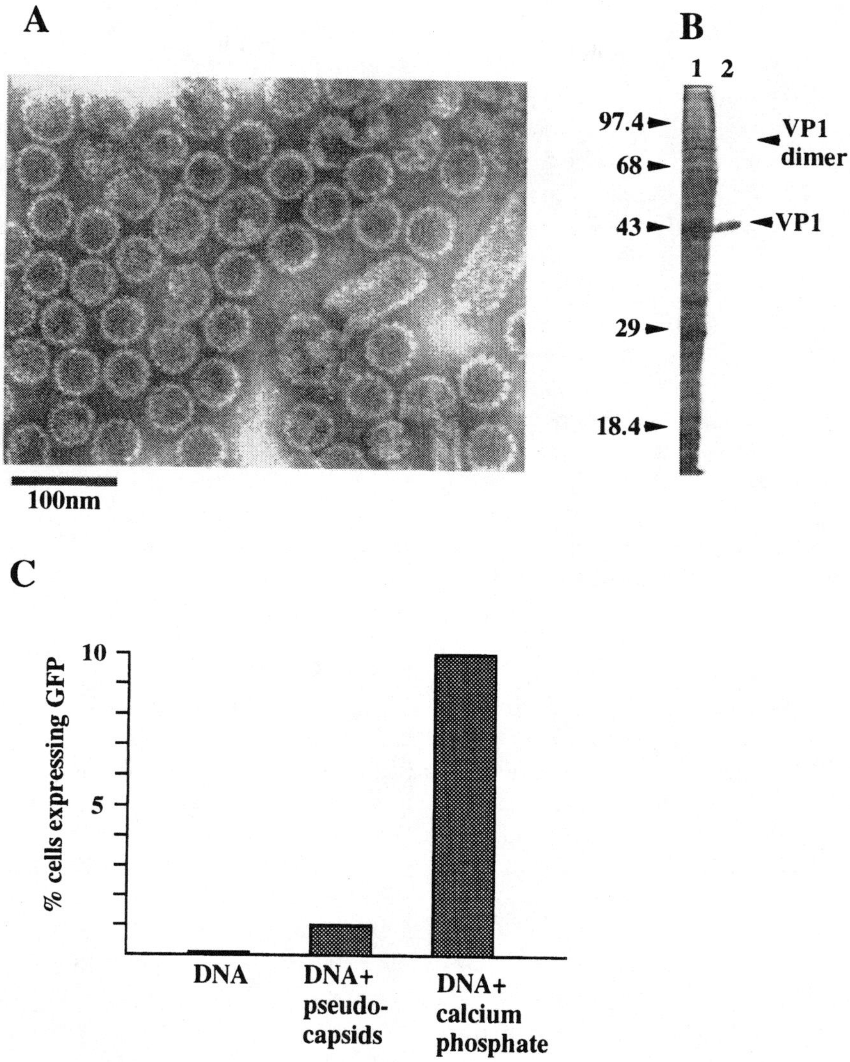

Figure 1. Structure and DNA transfer properties of mouse polyoma virus VP1 pseudocapsids. A. Electronmicrograph of polyoma virus pseudocapsids negatively stained with phosphotungstic acid. B. Coomassie blue stained SDS-polyacrylamide gel of protein from cells infected with a recombinant virus expressing polyoma VP1 (track 1) and density gradient purified pseudocapsids (track 2). Positions of VP1, VP1 dimer and molecular weight markers (in kD) are indicated. C. Cos 7 monkey cells incubated with DNA alone or with mouse polyoma VP1pseudocapsids mixed with plasmid DNA carrying the gene for green fluorescent protein (GFP), or DNA-calcium phosphate precipitates (1 µg DNA per 5×10^4 cells). 24 hours later, cells expressing the GFP gene were scored by UV light microscopy and are presented as a percentage of the total cell population.

2.4. Pseudocapsids as Gene Carriers

2.4.1. In Vitro Studies. Following the isolation and characterisation of VLPs/pseudocapsids, an obvious question arises, that is, can they be used as reagents for gene transfer? Interaction of murine polyoma pseudocapsids with exogenous DNA resulted in complexes in which the DNA was protected from DNaseI activity. These complexes, when incubated with cells in tissue culture, mediated DNA transfer of a polyoma viral oncogene to produce transformed foci in rodent (rat 2) cells. It was also possible to transfer DNA to cells of human origin (Forstová *et al.*, 1995). This work demonstrated that pseudocapsids could act as a DNA carrier, and that transfer could be achieved using a number of human cell lines. In monkey Cos 7 cells, approximately 1% cells expressed detectable levels of a marker gene (green fluorescent protein) in a transient assay, compared to 10% for calcium phosphate transfected DNA (Fig. 1C). As noted earlier for the former (Forstová *et al.*, 1995), although only low levels of DNA appeared to be transferred, stable expression was achieved, indicating that pseudocapsids may retain some of the properties of the natural virus. Recently, a modified packaging method has been described (Touze and Coursaget, 1998), in which papilloma VLPs assembled from the major capsid protein, L1, were disassembled under reducing conditions and then reassembled in the presence of marker plasmid. In these experiments capsids were used to transfer a marker gene to a range of cell lines with relatively high efficiency. Preliminary experiments to disassemble polyoma VP1 pseudocapsids and re-form them in the presence of plasmid DNA have not proved successful (J. Forstová, Charles University, Prague, personal communication). It is unclear whether this represents an inherent difference between papilloma and polyoma pseudocapsids, or a failure to identify the required conditions for the process in the case of the mouse virus protein. Further studies on the nature of pseudocapsids and their interaction with DNA should answer this question.

2.4.2. In Vivo Studies. Encouraged by the *in vitro* data, two pilot *in vivo* experiments have been carried out to determine whether pseudocapsids could be used in gene therapy. The first experiment was performed in athymic nude mice (nu/nu), to obviate possible complications from the host immune system. In this study, mice were treated with pseudocapsids complexed to plasmid DNA (pCMVβ, encoding the gene for β-galactosidase under the control of the cytomegalovirus immediate early promoter), administered either subcutaneously, intraperitoneally, intravenously or, to capitalise on the natural route of infection of the virus, intranasally. Mice were also treated with DNA and buffer alone, as controls. Single animals were harvested at one and seven weeks and organs analysed for expression of β-galactosidase by histochemical staining, or use of the polymerase chain reaction to measure the presence of DNA or RNA, using primers selected to amplify only sequences derived from the plasmid, in tissues of treated animals.

The results, summarised in Table 1, demonstrate that pseudocapsids are capable of enhancing DNA uptake into cells in tissues, and also targetting non-dividing cells. At one week, expression was detected at varying levels in most tissues tested, both in animals treated with DNA/pseudocapsids complexes, or with DNA alone. At seven weeks, however, no evidence of the gene could be identified in animals treated with DNA alone, whereas with pseudocapsid/DNA complexes both the presence and expression of the gene could be found. Of particular note was the fact that, although at one week no expression was detected in the brain of pseudocapsid treated animals, at seven weeks a significant level of the gene was found, particularly in the animal where complexes had been administered intranasally (Fig. 2). On the basis of these promising results, a second study was undertaken in immunocompetent animals. Six week old female balb c mice were treated intranasally with DNA, or pseudocapsid/DNA complexes, and analysed at one, seven and fourteen weeks post inoculation, as described above. Results, similar to those from the nude mouse experiments, were obtained with regard to gene expression and presence of the DNA in tissues, despite the fact that a considerable immune response was

Table 1. Summary of in vivo β-galactosidase expression in nude mice

	DNA[a] (1 week)	DNA + P[b] (1 week)	DNA[a] (7 weeks)	DNA + P[b] (7 weeks)
High level expression[c]		spleen		heart
Intermediate expression	spleen	heart kidney		brain kidney spleen
Low to no expression	heart kidney lung parotid liver skin brain	lung liver parotid skin brain	parotid heart kidney lung liver spleen skin brain	lung liver parotid skin

[a]mice treated with DNA (5 μg) alone.
[b]mice treated with pseudocapsid/DNA complexes (150 μg/5 μg [w/w]).
[c]high, intermediate and low to no expression refer to detection of >20, 20 – 2, or <2 positively stained areas per tissue.

raised in the treated animals. Further, it is noteworthy, that no ill effects were observed in the treated animals (our unpublished observations).

These preliminary experiments demonstrate that it should be possible to adapt pseudocapsids, or VLPs, for *in vivo* gene therapy applications. Although, in the case of mouse polyoma VP1 pseudocapsids, they are capable of mediating gene delivery in the presence of a competent immune system, the fact that they generate a significant antibody response (C. Cox and B. Clark, unpublished work in our laboratory), at

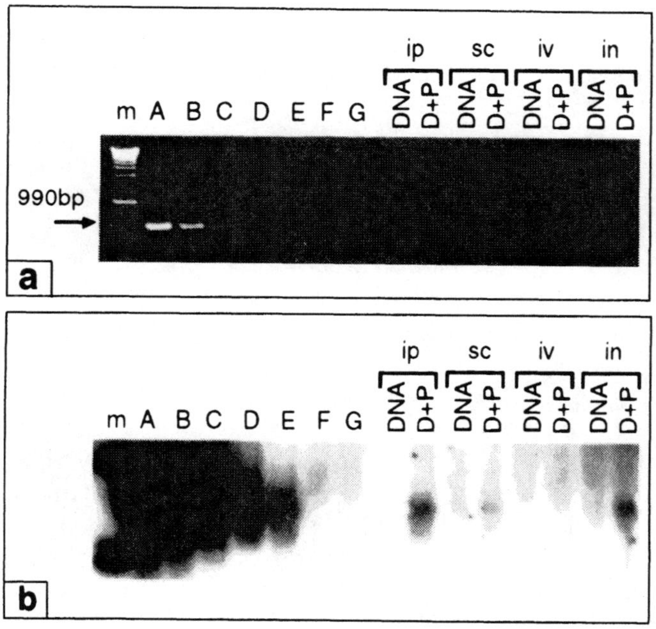

Figure 2. PCR of DNA samples prepared from brain tissue of mice, assayed at seven weeks. Panel (a), PCR products separated by gel electrophoresis and visualised with ethidium bromide; Panel (b), autoradiograph of Southern blotted products in Panel (a) hybridised with a radiolabelled DNA probe. Tracks A–F contain PCR products from a mixture of 500 ng DNA from untreated nude mouse brain with 10^6, 10^5, 10^4, 10^3, 10^2 and 0 added copies pCMVβ, respectively, and track G, contains no template. PCR on products from 500 ng brain DNA from mice with either pseudocapsid-DNA complexes (D + P) or DNA alone (DNA), introduced by different routes (ip, intraperitoneally; sc, subcutaneously; iv, intravenously; in, intranasally), loaded as indicated. After hybridisation and autoradiography, specific products were observed in the D + P ip, in and sc PCRs only. Molecular weight markers are shown in tracks labelled m. The position of the expected 990 bp product is indicated by an arrow.

least in the mouse, may be problematic with regard to certain applications. However, given that in cancer therapy simulation of the immune system may play an important role in killing tumour cells (Wels *et al.*, 1995), this apparent drawback could prove advantageous.

2.4.3. Future Development. Sudies described above indicate that pseudocapsids/VLPs can act as gene transfer vectors. In general, however, transfer is less efficient than generally observed with the parental viruses. Several obvious differences may account for this discrepancy. Firstly, heterologous DNA is not packaged in the same way as in the virion and, secondly, the minor capsid proteins are either not present in VLP/pseudocapsids, or, where incorporated, may not adopt the same configuration as in the natural capsid. Whereas it is possible to achieve virus-like packaging by generating pseudovirions (where heterologous DNA is introduced into the cells making the capsid; Strayer, 1996; Strayer and Milano, 1996; Sandalon *et al.*, 1997; Strayer *et al.*, 1997; Rund *et al.*, 1998) this depends on carrying out the whole process in the cell in the presence of viral sequences, and thus presents no advantage over the other currently available viral systems.

A major challenge is to provide a completely defined system for producing VLP/pseudocapsid-DNA complexes. DNA binding studies with mouse polyoma VP1 pseudocapsids have shown that there is a preference for interaction with DNA structural motifs, such as origins of replication, or single-stranded DNA, rather than double-stranded nucleotide sequences (our unpublished data, also J. Forstová, personal communication). In the case of SV40 it has been suggested that transcription factors may play a role, since AP1 and Sp1 binding sites in plasmid DNA appear to promote packaging into pseudovirions (viruses containing heterologous DNA synthesised in a helper packaging cell; Dalyot *et al.*, 1996). For papilloma VLPs, both the major and minor capsid proteins, L1 and L2, were found to be necessary for DNA packaging, as were unspecified papilloma DNA sequences (Zhao *et al.*, 1998). Given, however, that non-viral sequences have been reported to be packaged into both natural virions (Michel *et al.*, 1967; Winocour, 1967; Winocour, 1969) and pseudocapsids (Pawlita *et al.*, 1996; Gillock *et al.*, 1997), efficient packaging of completely heterologous sequences remains a real possibility. In order to assist this process, and possibly the entry of complexes into cells to provide maximum efficiency, it may be advantageous to make particles that also contain minor coat proteins.

The three-dimensional structures of several non-enveloped icosahedral virus particles, including mouse polyoma virus (Stehle *et al.*, 1994), simian virus 40 (Stehle *et al.*, 1996) and human papillomavirus (Hagensee *et al.*, 1994) have been solved to high resolution by X-ray crystallography. Potential receptors have been identified for a number of these viruses (Griffiths and Consigli, 1986; Breau *et al.*, 1992; Liu *et al.*, 1998). Pseudocapsid/VLPs presently display a broad tropism with regard to the types of cells with which they can interact (Volpers *et al.*, 1995; Soeda *et al.*, 1998; our unpublished data), therefore, information regarding receptors and capsid surfaces provides the basis for making informed alterations to their structures, either to increase uptake efficiency, or to alter cell/tissue tropisms. These changes are feasible given that the structure of the capsid subunits consists of a β sheet/α helical core, which present a series of loops on the outer surface of the particle. These loops are involved in a number of key interactions with cell surface receptors, including, in the case of polyoma virus, sialic acid residues on glycosylated proteins (Stehle *et al.*, 1994; Stehle and Harrison, 1996).

2.5. Alternative Approaches

Besides using various components of papovavirus genomes in therapeutic vectors, such as employing SV40 gene regulation sequences (polyadenylation, splicing and promoter motifs), or using the SV40 large T antigen nuclear localisation signal to target DNA through the nuclear pore (Morris *et al.*, 1997), a number of alternative methods for utilising papovaviruses in gene delivery or tumour targetting are being developed.

2.5.1. Pseudovirions. Several groups have investigated the use of SV40 pseudovirions for gene delivery. In one series of experiments, the multidrug resistance gene (MDR1) was cloned between the SV40 origin and polyadenylation sequences that had been engineered into a bacterial plasmid. After excision of the bacterial sequences, the DNA was recircularised and cotransfected into Cos cells with wild type SV40 DNA, to provide viral functions *in trans*. The resulting viral progeny were used to infect mouse haemopoietic cells and human bone marrow cells. Between 73 to 96% of cells were found to be expressing the gene as determined by a fluorescence dye exclusion assay, indicating high efficiency gene transfer by this route (Rund *et al.*, 1998). This system has a major drawback, however, in that large quantities of wild type virus are also produced during the process.

Using a similar rationale, in another series of experiments, sequences encoding the regulatory, large T antigen of SV40 (LT; approximately 2 kbp in size) were excised from the genome and replaced with a heterologous gene (luciferase). The engineered genome was then recircularised and transfected into cells that provided the deleted LT functions, *in trans*. In this study, little or no wild type viral progeny were produced (Strayer *et al.*, 1997), and the resulting pseudovirions were used both in *in vitro* and *in vivo* (in mice) with encouraging results (Strayer, 1996; Strayer and Milano, 1996; Strayer *et al.*, 1997). Expression of the luciferase gene was detected in a range of tissues, particularly in peripheral blood nucleated cells following intravenous injection, where significantly greater than control levels were observed for at least six months. Interestingly, in this experiment, expression was not detected immediately following injection, but rather emerged after two to three weeks, in a manner analogous to that observed for delivery of genes with mouse polyoma VP1 pseudocapsids (see above). In both cases, this has been attributed to the integration of the gene into the host chromosome. In the SV40 system, no significant immune response was detected, as judged by the failure of the of immune system to eliminate "infected" cells and the absence of inflammatory responses (Strayer and Milano, 1996). Both SV40 pseudovirion approaches generated vectors that provided significant levels of expression of the input gene, but, in each case viral sequences (including LT antigen, a potentially powerful oncogene), or the possibility of reversion, reduce the value and final applicability of the approach. On the positive side, they demonstrate, however, that, given appropriate development, pseudocapsids or VLPs have the potential to be very efficient DNA transfer vehicles.

2.5.2. Immunisation Therapy. A quite different approach has been taken in the papilloma virus field, based on the fact that human papilloma viruses, far from being "harmless" to their host *in vivo* (as is the case with most polyoma viruses) are strongly implicated in the development of malignancies, particularly cervical cancer (WHO, 1995). This field has capitalised on the strong immune response directed towards papilloma VLPs to effect immunisation against papilloma virus infection (Schiller and Roden, 1996). To increase the viral antigen targets seen by the host, chimeric VLPs were created that present the

E7 and E2 non-structural gene products of the virus on a background of the L1 major capsid protein. These constructions are still able to form VLPs and elicidate a strong antibody response (Muller *et al.*, 1997; Greenstone *et al.*, 1998). Further, they have been shown to be able to protect mice from tumour challenge even in the absence of adjuvant, presumably via a class 1-restricted cytotoxic lymphocyte route (Muller *et al.*, 1997; Greenstone *et al.*, 1998). The same potential may exist for the polyoma Vp1 pseudocapsids.

3. CONCLUSIONS

The studies carried out to date using polyoma and papillomavirus based carriers demonstrate their considerable promise, although numerous aspects of the systems require further attention. Present vectors vary widely in efficacy of DNA transfer, and parameters important in regulating this process need to be more clearly defined. Following establishment of a generalised prototype, modifications could then be made to target specifically to tumours, for instance by encoding tumour specific epitopes within the capsids, or crosslinking them to their surfaces. Further, delivery of prodrug activating enzyme genes to kill tumour cells, once targetted, could be explored. If it proves possible reliably to generate efficient and targetted pseudocapsid/VLP carriers, further approaches to anticancer therapy could be considered. For instance, delivery of wild type anti-oncogenes, such as p53, could be attempted, or where viruses (for example, papillomaviruses, Epstein Barr virus, or herpesvirus-8) have been implicated in tumour formation, an approach aimed at inhibiting expression of identified oncogenic functions could be explored. These methods, coupled with stimulation of T cell killing by eliciting an immune response against tumour antigens expressed on the VLP/pseudocapsid surface, could produce a very efficient therapy vector.

REFERENCES

Anderson, W.F., 1998, Human gene therapy, *Nature* **392**:25–30.

Breau, W.C., Atwood, W.J., and Norkin, L.C., Class I major histocompatibility proteins are an essential component of the simian virus 40 receptor, *J Virol* **66**:2037–2045.

Chang, D., Fung, C.Y., Ou, W.C., Chao, P.C., Li, S.Y., Wang, M., Huang, Y.L., Tzeng, T.Y., and Tsai, R.T., 1997, Self-assembly of the JC virus major capsid protein, VP1, expressed in insect cells, *J Gen Virol* **78**:1435–1439.

Colomar, M.C., Degoumois, S.C., and Beard, P., 1993, Opening and refolding of simian virus 40 and *in vitro* packaging of foreign DNA, *J Virol* **67**:2779–2786.

Cooper, M.J., 1996, Noninfectious gene transfer and expression systems for cancer gene therapy, *Semin Oncol* **23**:172–187.

Dalyot, H.N., Ben-nun-Shaul, O., Gordon, S.A., and Oppenheim, A., 1996, The simian virus 40 packaging signal ses is composed of redundant DNA elements which are partly interchangeable, *J Mol Biol* **259**:69–80.

Forstová, J., Krauzewicz, N., Sandig, V., Elliott, J., Palková, Z., Strauss, M., and Griffin, B.E., 1995, Polyoma virus pseudocapsids as efficient carriers of heterologous DNA into mammalian cells, *Hum Gene Ther* **6**:297–306.

Forstová, J., Krauzewicz, N., Wallace, S., Street, A.J., Dilworth, S.M., Beard, S., and Griffin, B.E., 1993, Cooperation of structural proteins during late events in the life cycle of polyomavirus, *J Virol* **67**:1405–1413.

Garcea, R.L., and Liddington, R.C., 1997, Structural Biology of polyomaviruses, in: *Structural Biology of Viruses* (W. Chiu, R.M. Burnett, and R.L. Garcea, eds), pp 187–208, Oxford University Press, Oxford.

Greenstone, H.L., Nieland, J.D., de, V.K., De, B.M., Kirnbauer, R., Roden, R.B., Lowy, D.R., Kast, W.M., and Schiller, J.T., 1998, Chimeric papillomavirus virus-like particles elicit antitumor immunity against the E7 oncoprotein in an HPV16 tumor model, *Proc Natl Acad Sci USA* **95**:1800–1805.

Griffiths, G.R., and Consigli, R.A., 1986, Cross-linking of a polyomavirus attachment protein to its mouse kidney cell receptor, *J Virol* **58**:733–781.

Hagensee, M.E., Yaegashi, N., and Galloway, D.A., 1993, Self-assembly of human papillomavirus type 1 capsids by expression of the L1 protein alone or by coexpression of the L1 and L2 capsid proteins, *J Virol* **67**:315–322.

Hagensee, M.E., Olsen, N.H., Baker T.S., and Galloway, D.A., 1994, Three dimensional structure of vaccinia produced human papilloma type 1 capsids, *J Virol* **68**:4503–4505.

Howley, P.M., 1990, Papillomavirinae and their replication, in: *Virology*, Volume 2, 2nd ed., (B.N. Fields, ed), pp 1625–1650, Raven Press, Ltd., New York.

Leavitt, A.D., Roberts, T.M., and Garcea, R.L., 1985, Polyoma virus major capsid protein, VP1. Purification after high level expression in Escherichia coli, *J Biol Chem* **260**:12803–12809.

Liu, C.K., Wei, G., and Atwood, W.J., 1992, Infection of glial cells by the human polyomavirus JC is mediated by an N-linked glycoprotein containing terminal alpha(2–6)-linked sialic acids, *J Virol* **72**:4643–4649.

Michel, M.R., Hirt, B., and Weil, R., 1967, Mouse cellular DNA enclosed in polyoma viral capsids (pseudovirions), *Proc Natl Acad Sci USA* **58**:1381–1388.

Montross, L., Watkins, S., Moreland, R.B., Mamon, H., Caspar, D.L., and Garcea, R.L., 1991, Nuclear assembly of polyomavirus capsids in insect cells expressing the major capsid protein VP1, *J Virol* **65**:4991–4998.

Morris, M.C., Vidal, P., Chaloin, L., Heitz, F., and Divita, G., 1997, A new peptide vector for efficient delivery of oligonucleotides into mammalian cells, *Nucleic Acids Res* **25**:2730–2736.

Muller, M., Zhou, J., Reed, T.D., Rittmuller, C., Burger, A., Gabelsberger, J., Braspenning, J., and Gissmann, L., 1997, Chimeric papillomavirus-like particles, *Virology* **234**:93–111.

Osterman, J.V., Waddell, A., and Aposhian, H.V., 1971, Gene therapy systems: the need, experimental approach, and implications, *Ann NY Acad Sci* **179**:514–519.

Pawlita, M., Muller, M., Oppenlander, M., Zentgraf, H., and Herrmann, M., 1996, DNA encapsidation by viruslike particles assembled in insect cells from the major capsid protein VP1 of B-lymphotropic papovavirus, *J Virol* **70**:7517–7526.

Qasba, P.K., and Aposhian, H.V., 1971, DNA and gene therapy: transfer of mouse DNA to human and mouse embryonic cells by polyoma pseudovirions, *Proc Natl Acad Sci USA* **68**:2345–2349.

Rodgers, R.E., Chang, D., Cai, X., and Consigli, R.A., 1994, Purification of recombinant budgerigar fledgling disease virus VP1 capsid protein and its ability for *in vitro* capsid assembly, *J Virol* **68**:3386–3390.

Rund, D., Dagan, M., Dalyot, H.N., Kimchi, S.C., Schoenlein, P.V., Gottesman, M.M., and Oppenheim, A., 1998, Efficient transduction of human hematopoietic cells with the human multidrug resistance gene 1 via SV40 pseudovirions, *Hum Gene Ther* **9**:649–657.

Salunke, D.M., Caspar, D.L., and Garcea, R.L., 1986, Self-assembly of purified polyomavirus capsid protein VP1, *Cell* **46**:895–904.

Sandalon, Z., Dalyot, H.N., Oppenheim, A.B., and Oppenheim, A., 1997, *In vitro* assembly of SV40 virions and pseudovirions: vector development for gene therapy, *Hum Gene Ther* **8**:843–849.

Sandalon, Z., and Oppenheim, A., 1997, Self-assembly and protein-protein interactions between the SV40 capsid proteins produced in insect cells, *Virology* **237**:414–421.

Schiller, J.T., and Roden, J.B.S., 1996, *Papillomavirus-like particles: basic and applied studies*. Leeds, UK, Leeds Medical Information. 101–112.

Shah, K.V., and Howley, P.M., 1990, Papillomaviruses, in: *Virology*, Volume 2, 2nd ed., (B.N. Fields, ed), pp 1651–1676, Raven Press, Ltd., New York.

Shishido, Y., Nukuzuma, S., Mukaigawa, J., Morikawa, S., Yasui, K., and Nagashima, K., 1997, Assembly of JC virus-like particles in COS7 cells, *J Med Virol* **51**:265–272.

Slilaty, S.N., and Aposhian, H.V., 1983, Gene transfer by polyoma-like particles assembled in a cell-free system, *Science* **220**:725–727.

Soeda, E.S., Krauzewicz, N., Cox, C., Stokrová, J., Forstová, J., and Griffin, B.E., 1998, Enhancement by polylysine of transient, but not stable, expression of genes carried into cells by polyoma VP1 pseudocapsids, *Gene Therapy* **5**:1410–1419.

Stehle, T., Gamblin, S.J., Yan, Y., and Harrison, S.C., 1996, The structure of simian virus 40 refined at 3.1 A resolution, *Structure* **4**:165–182.

Stehle, T., and Harrison, S.C., 1996, Crystal structures of murine polyomavirus in complex with straight-chain and branched-chain sialyloligosaccharide receptor fragments, *Structure* **4**:183–194.

Stehle, T., Yan, Y., Benjamin, T.L., and Harrison, S.C., 1994, Structure of murine polyomavirus complexed with an oligosaccharide receptor fragment, *Nature* **369**:160–163.

Strayer, D.S., 1996, SV40 as an effective gene transfer vector in vivo, *J Biol Chem* **271**:24741–24746.

Strayer, D.S., Duan, L.X., Ozaki, I., Milano, J., Bobraski, L.E., and Bagasra, O., 1997, Titering replication-defective virus for use in gene transfer, *Biotechniques* **22**:447–450.

Strayer, D.S., Kondo, R., Milano, J., and Duan, L.X., 1997, Use of SV40-based vectors to transduce foreign genes to normal human peripheral blood mononuclear cells, *Gene Ther* **4**:219–225.

Strayer, D.S., and Milano, J., 1996, SV40 mediates stable gene transfer in vivo, *Gene Ther* **3**:581–587.

Touze, A., and Coursaget, P., 1998, *In vitro* gene transfer using human papillomavirus-like particles, *Nucleic Acids Res* **26**:1317–1323.

Touze, A., Dupuy, C., Mahe, D., Sizaret, P.Y., and Coursaget, P., 1998, Production of recombinant virus-like particles from human papillomavirus types 6 and 11, and study of serological reactivities between HPV 6, 11, 16 and 45 by ELISA: implications for papillomavirus prevention and detection, *Fems Microbiol Lett* **160**:111–118.

Unckell, F., Streek, R.E., and Sapp, M., 1997, Generation and neutralisation of pseudovirions of human papillomavirus type 33, *J Virol* **71**:2934–2939.

Verma, I.M., and Somia, N., 1997, Gene therapy—promises, problems and prospects, *Nature* **389**:239–242.

Volpers, C., Schirmacher, P., Streeck, R.E., and Sapp, M., 1994, Assembly of the major and the minor capsid protein of human papillomavirus type 33 into virus-like particles and tubular structures in insect cells, *Virology* **200**:504–512.

Volpers, C., Unckell, F., Schirmacher, P., Streeck, R.E., and Sapp, M., 1995, Binding and internalization of human papillomavirus type 33 virus-like particles by eukaryotic cells, *J Virol* **69**:3258–3264.

Wels, W., Moritz, D., Schmidt, M., Jeschke, M., Hynes, N.E., and Groner, B., 1995, Biotechnological and gene therapeutic strategies in cancer treatment, *Gene* **159**:73–80.

WHO, 1995, Human Papillomaviruses, *IARC monographs on the evaluation of carcinogenic risks to humans*. Lyons, France, WHO.

Winocour, E., 1967, On the apparent homology between DNA from polyoma virus and normal mouse synthetic RNA, *Virology* **31**:15–28.

Winocour, E., 1969, Some aspects of the interaction between polyoma virus and cell DNA, *Adv Virus Res* **14**:153–200.

Zhao, K.N., Sun, X.Y., Frazer, I.H., and Zhou, J., 1998, DNA packaging by L1 and L2 capsid proteins of bovine papillomavirus type 1, *Virology* **243**:482–491.

9

COCHLEATES

Lipid-Based Vehicles for Gene Delivery-Concept, Achievements and Future Development

Leila Zarif[1] and Raphael J. Mannino[2]

[1]BioDelivery Sciences Inc
c/o UMDNJ, MSB A691
185 South Orange Ave, Newark
New Jersey 07103, USA
[2]UMDNJ MSB A691
185 South Orange Ave, Newark
New Jersey 07103, USA

1. INTRODUCTION

Gene therapy represents one contemporary approach for the treatment of congenital and acquired diseases. In the last decade extensive research has been conducted on the potential for gene therapy in correcting genetic diseases, including, cystic fibrosis, diabetes, hematologic diseases, immune system disorders, arthritis, asthma, hemophilia, autoimmunity, neurodegenerative disorders and others. Gene therapy is also being examined as a therapeutic for acquired diseases, including, cancer, AIDS, infectious diseases, and cardiac and vascular disease.

In order to achieve a successful gene therapy in mammalian systems, a safe, non-toxic, non-inflammatory, and efficacious delivery system is required. This system should mediate intracellular delivery and expression of exogenous DNA with high efficiency. Various approaches are currently under investigation by many laboratories developing gene therapy. These include the use of facilitating agents, such as calcium phosphate (Wigler *et al.*, 1977), DEAE dextran (Ishikawa and Homcy, 1992); microinjection of DNA (Capecchi, 1980), electroporation (Paquereau and Cam, 1992), viral vectors (Ferry *et al.*, 1991; Culver *et al.*, 1992), ballistic guns, and receptor mediated gene transfer. A wide variety of non-viral gene delivery systems has been developed for gene therapy (Felgner, 1990; Wagner *et al.*, 1990; Behr, 1993). Liposomes have been used for this purpose (Felgner *et al.*, 1987; Mannino *et al.*, 1988; Legendre *et al.*, 1992, Behr *et al.*, 1989; Smyth-Templeton *et al.*, 1997). All these approaches have advantages but also limitations

Cancer Gene Therapy: Past Achievements and Future Challenges, edited by Habib
Kluwer Academic / Plenum Publishers, New York, 2000.

such as toxicity and/or inefficient delivery *in vivo*. Cochleates, a safe and effective lipid-based delivery system, has the potential to overcome many of these limitations.

2. COCHLEATE DELIVERY VEHICLES

2.1. Background

Cochleates were first described by Papahadjoupoulos and coworkers in 1975 (Papahadjoupoulos, 1975). They are highly stable lipid bilayer structures that can be stored in buffer at 4 °C or lyophilized to a powder and stored at room temperature. In studies in which cochleates have been used to formulate ("encochleate") vaccine antigens, the cochleates themselves have been shown to be nontoxic and non-inflammatory. Multiple administrations of cochleate formulations to the same animal do not result in either the development of an immune response to the cochleate material, or to any side effects relating to the cochleate vehicle.

Cochleate formulations of DNA plasmids, oligonucleotides, peptides and proteins have been made and tested in animal models. These studies have demonstrated that cochleates consistently deliver biologically active materials whether they are administered by oral, intranasal or parenteral routes (Mannino *et al.*, 1997). Cochleate formulations of small molecule anti-infective drugs have also been prepared and are being tested directly against microorganisms in culture and in infected animals (Zarif *et al.*, 1998).

2.2. Composition and Structure

Cochleates are stable phospholipid-cation precipitates composed of naturally occurring materials, for example phosphatidylserine and calcium. They have a multilayered structure consisting of a continuous, solid, lipid bilayer sheet oftentimes rolled up in a spiral, with little, or no, internal aqueous space. The tightly packed unique structure of cochleates provides protection from degradation for molecules encompassed within the lipid bilayers. Since the cochleate crystalline structure is composed of a series of solid layers, biologically active molecules within the interior of the structure remain intact, even when the structure is exposed to acid pH or enzymes. Encochleated DNA plasmids have been stored in calcium-containing buffers for up to two years (Fig. 1).

Biological activity and morphology of DNA cochleate preparations are maintained after long term storage. In addition, cochleate formulations can be lyophilized, without altering the cochleate structure and the intactness of encochleated DNA molecules, and stored as a dry powder at room temperature (Fig. 2).

Divalent cation concentration *in vivo* in serum and mucosal secretions are such that the cochleate structure should be maintained. Hence, the majority of cochleate associated molecules are present in the inner layers of a solid, stable, impermeable structure. Therefore, it is anticipated that there will be little or no free DNA *in vivo* in the circulation accompanying cochleate delivery.

2.3. Methods of Preparation of DNA Cochleates, Characterization

We have developed procedures for the preparation of cochleate formulations containing a wide variety of biologically important molecules, including proteins, peptides, DNA plasmids, antisense oligonucleotides, drugs, and lipid-soluble compounds. These

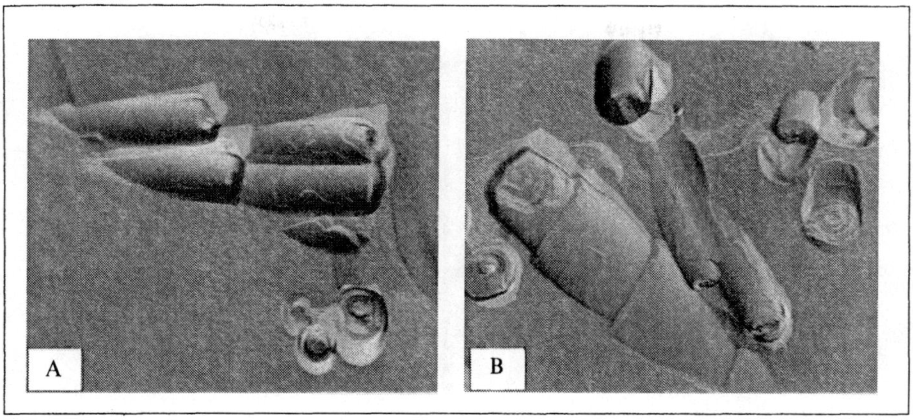

Figure 1. Freeze-Fracture Electron Micrographs of DNA-cochleate formulations before (A) and after (B) long term storage.

procedures employ mild conditions in order to maintain the physical integrity and biological activities of the formulated molecules.

A typical process to encochleate DNA is shown in Fig. 3. In this figure, the DC method (I) consists of adding a detergent solution of lipids in buffer to the DNA, dialyzing the detergent out, then dialyzing against Ca-buffer to obtain the DNA-cochleate; the trapping method (II) begins directly with sonicated or microfluidized liposomes in buffer, DNA suspension in buffer is added and followed by a slow addition of calcium. In process I, upon removal of detergent, liposomes form spontaneously. In both methods I and II, the addition of divalent metal ions to liposomes composed predominantly of negatively charged lipids induces the collapse and fusion of the liposomes into large sheets composed of lipid bilayers, which spontaneously roll up into cochleate cylinders. The cochleates are purified to remove unencochleated material, resuspended in a buffer containing divalent metal ions, and then either stored as a suspension or lyophilized.

We investigated various aspects of the formulation process, including pH, type of metal ion, ratio of lipid to material, rate of addition of cation, and other parameters, in order to optimize the formulation process for each specific material.

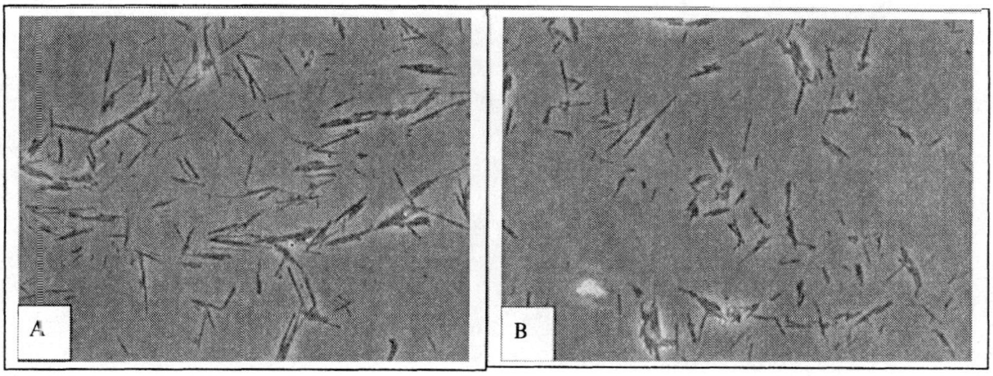

Figure 2. Stability of DNA-cochleate to lyophilization: before (A) and after lyophilization (B).

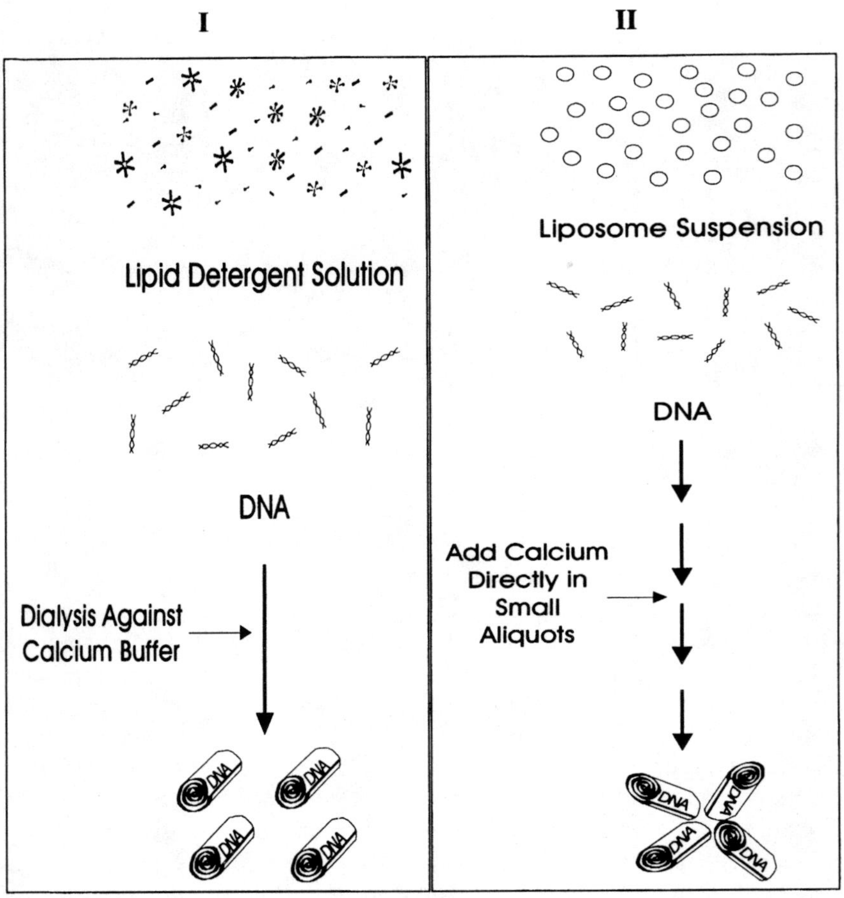

Figure 3. Processes developed to encochleate DNA plasmids.

These procedures have been applied successfully to encapsulate several DNA's such as DNA plasmid encoding for the env, tat and rev genes of HIV-1 IIIB, and a DNA plasmid encoding the glycoprotein B envelop gene of herpes simplex virus type I, (HSV-1 gB), with high entrapment efficiency, (40 to 80 %). The formulated DNA-cochleates are characterized by Freeze Fracture scanning electron microscopy. Cochleate particle size can be evaluated routinely either by Laser Light Scattering, (LLS, not shown), or using optical microscopy to capture digitalized images and, using a Scion computer program, determined the areas and lengthes. An example is illustrated in Fig. 4.

2.4. Biocompatibility of the Cochleate Lipid Vehicle

The fundamental components of the cochleate delivery vehicle are phosphatidylserine (PS) and calcium. Phosphatidylserine is a natural component of all biological membranes, and is most concentrated in the brain. Clinical studies conducted to evaluate the potential of phosphatidylserine as a nutrient supplement indicate that PS is very safe and may play a role in the support of mental functions in the aging brain (Crook *et al.*, 1991). We evaluated the *in vivo* tolerance of mice to multiple intraperitoneal administrations of cochleates (Fig. 5).

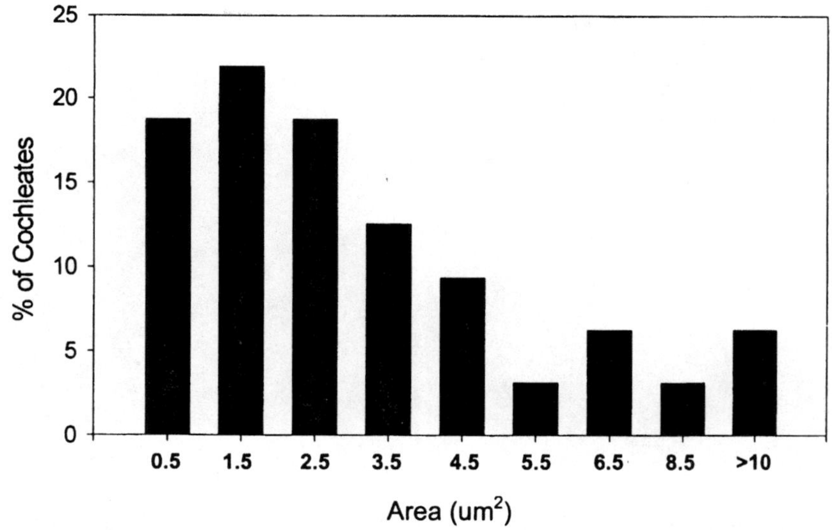

Figure 4. Particle size distribution of cochleates cylinders measured using Scion computer program.

Multiple administrations of cochleate formulations to the same animal do not result in either the development of an immune response to the cochleate matrix, or to any side effects relating to the cochleate vehicle. This is a major advantage over viral vector-based delivery systems where prior or induced immune responses to the carrier reduce the effectiveness of subsequent administrations. Cochleate formulations are also superior to some lipid-based systems (particularly cationic) where inflammatory activities of the lipid and lack of biodegradability are known to be serious limitations.

2.5. Biodistribution of Cochleates

The biodistribution of ^{3}H-cholesterol labeled cochleates has been examined following intraperitoneal (i.p.), intramuscular (i.m.), oral, and intranasal administration. Their widespread biodistribution varies with the route of administration. Only the intramuscular and the oral routes will be discussed briefly below.

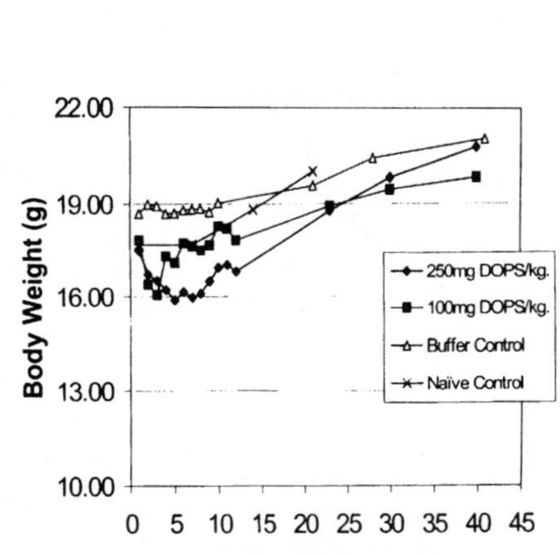

Figure 5. Safety of cochleate vehicle assessed by the change in body weight of mice after multiple injections of empty cochleate formulation prepared with dioleoylphosphatidylserine (DOPS).

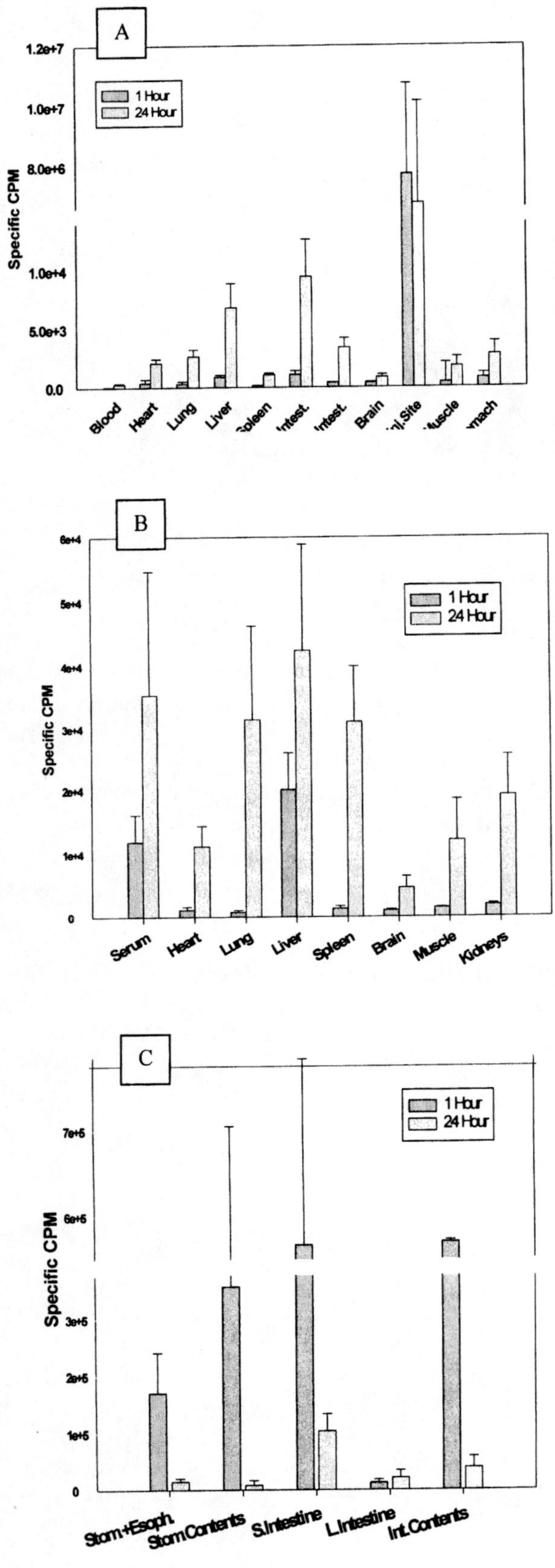

Figure 6. Biodistribution of ^{3}H-chol labeled DC cochleates following intramuscular (A) and oral (B,C) administration.

DC type ^{3}H-cholesterol labeled plain cochleates were administered to Balb/c mice (10 million CPM/dose, 3 mice per group) intramuscularly (50 ul). Mice were sacrificed at 1 and 24 hours and blood and organs were removed, solubilized in soluene, and assayed for radioactivity.

Following intramuscular administration, it was shown that the majority of radiolabeled cochleate material was localized at the injection site forming a depot. Twenty four hours after administration, the amount of radioactive cochleate material decreased at the site of injection while radioactivity increased in the organs examined. In particular, the liver, small intestine and lungs showed relatively high levels of radiolabeled cochleate accumulation.

The same dose of DC type ^{3}H-cholesterol labeled plain cochleates were administered to Balb/c mice orally (50 ul). Twenty four hours after oral administration, it was shown that there was a general increase in the level of radiolabeled cochleate material in all organs except the esophagus, stomach and small intestine. A relatively large accumulation of radiolabeled cochleate material was detected in the serum, lung, liver, spleen and kidneys after 24 hours. It was interesting to observe the presence of radiolabeled material in the brain and that the level of this material increased over time until the 24 hour time point.

2.6. Mechanism of Delivery of Encochleated DNA

Cochleates deliver material into cells. This was illustrated by the high immunological response generated after i.m. and oral administration of DNA- or antigen cochleates. The proposed mechanism of this delivery, illustrated for a DNA cochleate in Fig. 7, is through fusion of the outer layer of the cochleate with the cell membrane.

The interaction of calcium with negatively charged lipids has been extensively studied by many investigators. Several naturally occurring membrane fusion events involve the interaction of calcium with negatively charged phospholipids (generally phosphatidylserine and phosphatidylglycerol). Calcium induced perturbations of membranes containing negatively charged lipids, and the subsequent membrane fusion events, are an important mechanism in many natural membrane fusion processes. Hence, cochleates can be envisioned as membrane fusion intermediates.

This postulated fusion would result in the delivery of a small amount of the encochleated material into the cytoplasm of the cell. The cochleate could then break free

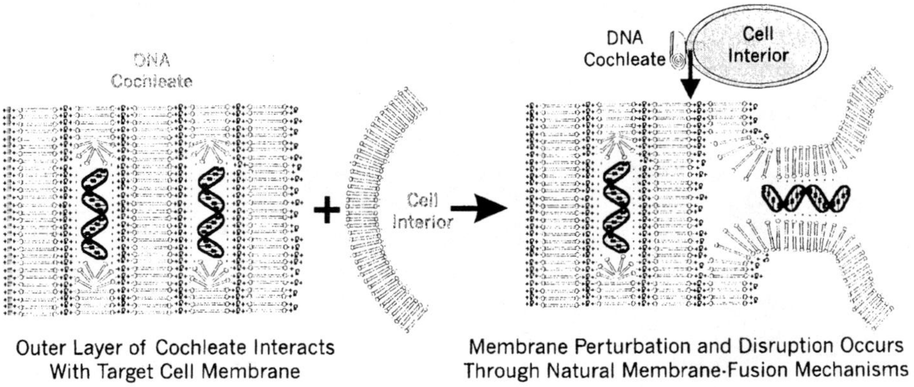

Figure 7. Proposed Mechanism of Delivery of Encochleated DNA.

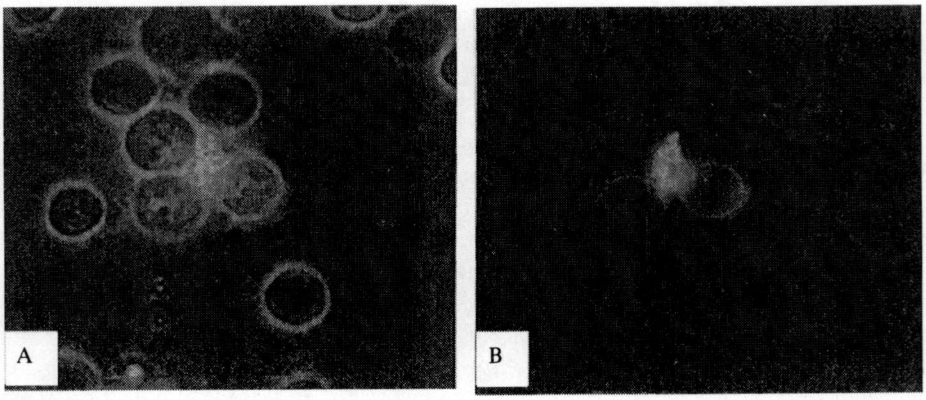

Figure 8. Cochleate Delivery to Mouse Splenocytes as seen by phase contrast (A) and fluorescence (B) optical microscopy.

of the cell and be available for another fusion event. Alternatively, cochleates could be taken up by endocytosis and deliver their contents to the cytoplasm by fusion with endocytic vesicle membranes.

Cochleates prepared with 0.1%w/w fluorescent rhodamine labeled phosphatidylethanolamine were added to mouse splenocytes. The splenocytes can be seen using phase contrast microscopy. When observed with fluorescence microscopy, cochleates which are in contact with splenocytes, can be seen transferring their orange-red fluorescent lipid to the surface membrane of the target cells (Fig. 8).

2.7. Cochleates in Gene Transfer

2.7.1. Immune Response. The ability of cochleates to deliver DNA vaccines using two systems: (1) an 11 kb plasmid coding for the env, tat and rev genes of HIV-1 IIIB driven by the cytomegalovirus promoter and (2) a DNA plasmid encoding the glycoprotein B envelope gene of herpes simplex virus type 1. The HIV studies showed that strong cellular responses to proteins encoded by the DNA were induced by a single dose of 3 μg or 15 μg (im) or 25 μg (po) of encochleated DNA per animal. These responses were significantly greater in animals administered encochleated DNA compared to animals receiving either naked DNA (50 μg or 100 μg) or cochleates that did not contain DNA. Cellular responses were measured by assessing lymphocyte cytotoxicity to target cells using chromium release assays, as well as by antigen-specific splenocyte proliferation assays. Antigen-specific cellular immunity was shown to persist for at least four months in animals that had received a single intramuscular injection.

In the HSV-1 studies, DNA plasmids encoding the gB protein were formulated into cochleates and administered to mice on days 0 and 10 by various routes: intranasal, intragastric, intramuscular, and intraocular. Control groups received naked DNA by intragastric or intramuscular injection. Mucosal and humoral immunity were assessed by determining levels of IgG or IgA antibody in serum, and fecal and vaginal wash samples taken 15 days after the second immunization. Encochleated HSV-1 DNA stimulates HSV-1 specific IgG and IgA antibodies following intranasal, intragastric, intramuscular or intraocular administration. Encochleated DNA induces substantially more efficient delivery than naked DNA (given at 200 μg), as smaller doses of encochleated DNA, ranging from 5–25 μg, were needed to generate equal or superior responses. In comparison, previous experiments demonstrated that naked DNA at 5–25 μg induced no measurable antibody. Lethal dose live virus challenge assays were also performed to assess whether protective immunity had been conferred to the animals given HSV-1 cochleates. The immunity induced by DNA cochleates was shown to protect mice against a lethal intravaginal challenge with HSV.

2.7.2. In Vitro *Transfection.* NIH3T3 fibroblast cells were used to assay the transfection efficiency of DNA-cochleate formulations. A serial dilution of DNA-cochleates was made with 10 ul and added to the wells, and the cells were incubated for 16 hours at 37C. After transfection, cells were washed with fresh medium and allowed to grow for 48–72 hours. Cells were fixed and the x-gal development solution was added to the cells. Calcium phosphate precipitated DNA was used as a positive control and cochleates with no DNA as negative control. These assays indicated that DNA-cochleates (lipid : DNA ratio of 10 : 1) at a concentration 100 : 1, containing only 32 ng DNA, resulted in obvious gene transfection (Fig. 9).

Our working hypothesis for the mechanism by which DNA-cochleates achieve cell transfection, is that the DNA macromolecule is localized between the cochleate bilayers with calcium bridging between the DNA and the phospholipid phosphate groups. We assumed that in this fibroblast transfection assay, the transfection induced by DNA cochleates is distinguished from any effect of the calcium phosphate-DNA precipitate that might form during the preparation of DNA-cochleate. In our hands, 32 ng of DNA per well in a calcium phosphate precipitate form do not induce any gene transfection in the cells while the same amount in a DNA-cochleate form results in obvious transfection.

2.7.3. In Vivo *Transfection.* The potential of cochleate technology to perform as gene therapy vectors can be inferred by the successful delivery of DNA cochleates as vaccines. Biologically active DNA capable of expressing antigenic gene products is effectively delivered when DNA cochleates are given by oral, intranasal and intramuscular routes.

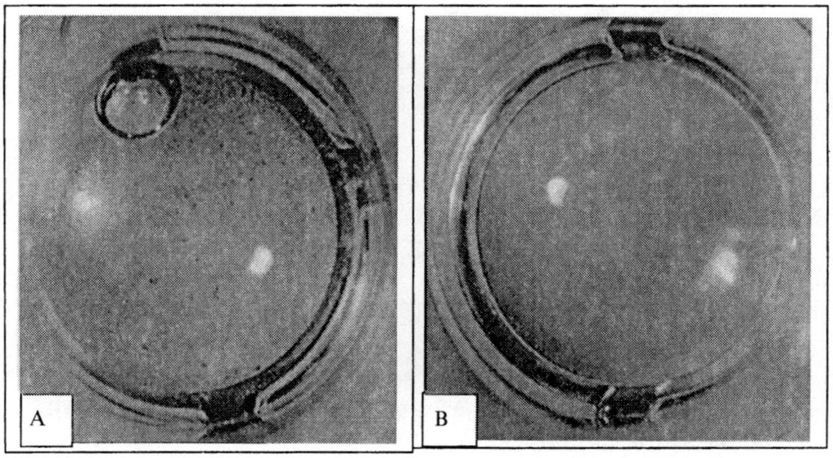

Figure 9. Cells transfected with DC type cochleates at 1 : 100 dilution (A) and untransfected cells as negative control (B).

In studies in mice, we have observed the ability of cochleates to deliver a plasmid encoding the enzyme β-galactosidase. In these studies, a 7.4 kb plasmid containing the gene for β-galactosidase, driven by the promoter RSV-LTR, was encochleated and delivered by intramuscular and intranasal routes. Strong expression of active enzyme was demonstrated in liver and spleen tissues in mice dosed by the intranasal route; mice administered the DNA cochleate IM also showed gene expression at the site of injection. This series of experiments showed that effective *in vivo* delivery of gene therapy vectors can be achieved through the use of plasmid cochleates.

We are conducting additional studies relating to gene therapy to assess the potential of the cochleate technology in both the *in vivo* and *ex vivo* delivery of genes.

2.8. Cochleates in Drug Delivery

The cochleate delivery system may represent a new technology platform for the formulation and administration of efficacious but problematic therapeutic molecules.

Proof-of-principle studies of the cochleate delivery vehicle for systemic and oral delivery of small therapeutic molecule are being carried out with amphotericin B (AmB), a potent antifungal agent. AmB is the treatment of choice for systemic candidiasis but suffers from extreme toxicity resulting in severe side effects, including permanent renal insufficiency. Three different lipid-based AmB formulations were developed in the last decade in other laboratories. These formulations displayed two major limitations. They do not deliver AmB orally and the efficient doses are approximately 5 times higher than conventional AmB preparations. Therefore, a need for new AmB formulations that combine low toxicity, high efficacy and economical price exists. AmB-cochleates were developed to address these limitations.

AmB-cochleates appear to be promising as a new, broad spectrum, antifungal formulation. Cochleates based on dioleoylphosphatidylserine and calcium were shown to associate a high amount (more than 99%) of AmB. These formulations are stable for several months at 4 °C. AmB cochleates were found to have low toxicity both *in vitro* and *in vivo* compared to Fungizone, the most clinically used AmB antifungal preparation. The AmB-cochleates were efficient in inhibiting the fungus growth *in vitro*. The Minimal Inhibitory Concentration (MIC) of AmB-cochleates *in vitro* against *Candida albicans*, *Aspergillus fumigatus*, and *Cryptococcus neoformans* were found to be equivalent to that of Fungizone. Low doses of AmB-cochleates (0.5 mg/kg) were sufficient to achieve 100% of survival in immunocompetent ICR mice infected with *Candida albicans* versus 60% survival when mice received Fungizone at the same dose.

3. CONCLUSION & PERSPECTIVES

Cochleates appear to have potential in delivering genes as well as small therapeutic molecules. Toxicity is an important issue in the development of a delivery vehicle. All studies to date indicate that the negatively charged lipid-based cochleate structures safely mediate the delivery of macro- and small therapeutic molecules. Future efforts in cochleate development will focus on enhancing the delivery of therapeutic molecules by engineering the cochleate formulations: Monitoring the particle size, enhancing the stability by modification of the nature of the lipid, (the major constituent of the cochleate), controlling the release rate of therapeutic molecules from cochleates by modifying the cation involved in the membrane fusion process, and modifying the surface of cochleates to eliminate aggregation and achieve targeting that will address specific clinical requirements.

ACKNOWLEDGMENTS

The authors aknowledge their collaborators whose names are cited in the references. Special aknowledges to Dr Fan Zhang, Dr Dan Brois, Dr. Zei Wei Chen, Mr. Anthony Scolpino, Ms. Diane Hyra, Dr. Barry Rouse, Dr Massoud Daheshia and Dr. Susan Gould-Fogerite.

REFERENCES

Behr, J.P., 1993, Synthetic gene-transfer vectors, *Acc. Chem. Res.* **26**:274–278.

Behr, J.P., Demeinex, B., Loeffler, J.P., and Perez-Mutul, J., 1989. Efficient Gene Transfer into Mammalian Primary Endocrine Cells with Lipopolyamine-coated DNA, *Proc Natl Acad Sci USA*. **86**:6982–6986.

Crook, T.H., Tinkenberg, J., Yesavage, J., Petrie, W., Nunzi, M.G., and Massari, D.C. 1991. Effects of Phosphatidylserine in age-associated memory impairment, *Neurology*. **41**:644–649.

Culver, K.W., Ram, Z., Wallbridge, S., Ishii, H., Oldfield, E.H., and Blaese, R.M., 1992. In vivo gene transfer with retroviral vector-producer cells for treatment of experimental brain tumors. *Science*, **256**:1550–1552.

Felgner, P.L., Gadek, T., Holm, M., Roman, R., Chan, H.W., Wentz, M., Northrop, J.P., Ringold, G.M., and Danielsen, M., 1987. Lipofection: a Highly efficient, lipid-mediated DNA Transfection Procedure. *Proc Natl Acad Sci USA* **84**:7413–7417.

Felgner, P.L., 1990. Particulate systems and polymers for in vitro and in vivo delivery of polynucleotides. *Adv. Drug Delivery Rev* **5**:163–187.

Ferry, N., Duplessis, O., Hopussin, D., Danos, O., and Heard, J.-M., 1991: Retroviral-mediated gene transfer into hepatocytes in vivo. *Proc Natl Acad Sci USA* **88**:8377–8381.

Ishikawa, Y., and Homcy, C.J., 1992, High efficiency gene transfer into mammalian cells by a double transfection protocol. *Nuc Acids Res*. **20**:4367.

Legendre, J.Y., and Szoka, F.C., 1992, Delivery of Plasmid DNA into mammalian cell lines using pH-sensitive liposomes: Comparison with Cationic Liposomes, *Pharm Res* **9**:1235–1242.

Smyth-Templeton, N., Lasic, D.D., Frederik, P.M., Strey, H.H., Roberts, D.D., and Pavlakis, G.N., 1997. Improved DNA: liposome complexes for increased systemic delivery and gene expression. *Nature Biotech* **15**:647–652.

Mannino, R.J., and Gould-Fogerite, S., 1988. Liposome mediated gene transfer, *BioTechniques*, **6**(7):682–690.

Mannino, R.J., and Gould-Fogerite S., 1997, Antigen cochleate preparations for oral and systemic vaccination, in: New Generation Vaccines (M.M. Levine, G.C. Woodrow, J.B. Kaper, and G.S. Cobon, eds). pp 229–37. Marcel Dekker, Inc. New York.

Papahadjopoulos, D.,1975. "Cochleate lipid cylinders: formation by fusion of unilamellar lipid vesicles," *Biochim Biophys Acta* 394:483.

Paquereau L., and Cam, A.L., 1992, Electroporation-mediated gene transfer into hepatocytes: preservation of a growth hormone response. *Anal Biochem*, **204**:147–151.

Wagner, E., Zenk, M., Cotton, M., Beug, H., and Birnsteil, M.L., 1990, Transferrin-polycation conjugates as carriers for DNA uptake into the cells. Proc Natl Acad Sci USA **87**:3410–3414.

Wigler, M., Silverstein, S., Lee, L.-S., Pellicer, A., Cheng, Y.-C., and Axel, R., 1977. Transfer of purified herpes virus thymidine kinase gene to cultured mouse cells. *Cell* **11**:223–232.

Zarif L., Scolpino, A., Hyra, D., Jin, T., Mannino, R.J., Perlin, D., Lambros, C., and Graybill, J.R., 1998. Amphotericin B cochleates as a novel delivery system for the treatment of fungal infections. *Abstract Drug Delivery Technologies, Preclinical and Clinical Study Results and Novel Technologies*, IBC December 3–4, 1998, Coronado, CA.

10

THE USE OF SKELETAL MUSCLE TO EXPRESS GENES FOR THE TREATMENT OF CANCER

Stephen Coe, Michael Harron, Marc Winslet, and Geoffrey Goldspink

Department of Anatomy and Developmental Biology and
Department of Surgery The Royal Free and
University College Medical School
University of London, UK., Royal Free Campus
Rowland Hill St. London NW3 2PF

1. INTRODUCTION

The available evidence shows that skeletal muscle is an appropriate target tissue for the transfer of DNA for the treatment of a number of diseases. These include those in which a systemic protein is absent or defective as well as muscle diseases per se such as Duchenne muscular dystrophy. It has also been shown that it is an appropriate target tissue for the introduction of vaccine DNA although the requirements for effective gene therapy procedures are somewhat opposite to those in which the main purpose is to elicit an immune response.

This review deals with the design of gene constructs for the treatment of cancer including the vectors and the regulatory elements required for optimisation of expression following introduction of the relevant cDNA by intramuscular injection. The relative merits and problems associated with each type of vector including the immunogenic responses they elicit are discussed. The latter is relevant even when introducing vaccine genes as the immune response is required from the anti cancer cDNA not from other parts of the gene construct.

Direct intramuscular injection of plasmid DNA was shown by Jon Wolff's group (Wolff *et al.*, 1990) to result in a small percentage of the fibres taking up and expressing the cDNA. This expression has been shown to persist for a considerable time (Wolff *et al.*, 1992). Levels of expression were improved by using different regulatory sequences to drive the expression of the introduced cDNA (Hansen *et al.*, 1991; Novo *et al.*, 1995; Skarli *et al.*, 1998) and by introducing the plasmid constructs into young muscle (Wells and Goldspink, 1992). Plasmids are very useful for transfecting muscle by intramuscular injection. However, the possibilities of using other vectors are discussed. Hence this

Cancer Gene Therapy: Past Achievements and Future Challenges, edited by Habib
Kluwer Academic / Plenum Publishers, New York, 2000.

review also deals with the immune response to the vectors as well as the gene product of the introduced cDNA as this is one of the crucial issues in gene therapy.

2. DESIGN OF VECTORS AND GENE CONSTRUCTS FOR APPROPRIATE EXPRESSION IN MUSCLE

In any gene therapy protocol one needs to consider the construction of the gene cDNA or genomic DNA to be transferred, the vector backbone, as well as host factors in determining immune responses. Consideration needs to be given to individual components as any one of them can influence the therapeutic efficacy of a gene therapy approach. In this review these factors will be discussed. However, it is clear that additional factors such as the host immune status (Michou *et al.*, 1997), dose of vector (Svenson *et al.*, 1997), and mode of delivery all affect the effector immune response but are beyond the scope of this review.

2.1. Gene Constructs for Intramuscular Injection

For intramuscular injection, consideration needs to be given to whether the desired product is secretable or non-secretable. For exocytosis from the muscle fibres it may be necessary to include an appropriate signal sequence. Also the antigen presentation to the immune system of these responses will differ depending on which MHC pathway is involved. It is known that MHC class I will present intracellularly derived peptides whereas secreted proteins may be processed via MHC II pathways. Most of the early published studies were performed with non-secretable reporter genes such as lacZ whose gene product (β galactosidase) is non-secretable. Another problem that hampered early gene therapy studies was the availability of species-specific transgenes; most of the available cloned transgenes for early studies were human and these were tested in non-human animal hosts. Recently, many more animal cDNA's have been cloned which will allow appropriate testing of gene therapy strategies in the homologous animal.

2.2. Regulatory Elements

For effective gene transfer into muscle, tissue-specific expression of the introduced gene is desirable. This can be achieved by the use of promoters and enhancers specific for the tissue of interest. In the case of muscle, creatine kinase (Wang *et al.*, 1997), a-actin (Bergsma *et al.*, 1986; Draghia-Akli *et al.*, 1997), and myosin heavy chain (Skarli *et al.*, 1998) have been used. The necessity for the development of vector bearing muscle-specific regulatory elements is further emphasised by recent studies which have shown that intramuscular injection of genes results in the uptake of a proportion of the vector by the motorneurones (Keir *et al.*, 1995; Ribota *et al.*, 1997). Retrograde transport of the vector is a possible way of treating neuroblastomas, particularly as it is necessary to transfect only some of the neurones by injection of the the muscle they innervate. There is a "bystander effect", in other words the anticancer factor produced by some of the neurone cell bodies have a paracrine.

Engineered muscle type genes that are under the control of myosin DNA regulatory elements have several potential advantages. Myosin isoforms are differentially regulated by physical activity and by hormones and this, therefore, offers the possibility for developing inducible vectors. Also muscle-specific elements used to drive DNA are more likely perceived as safer than viral promoters. It is worth noting that viral and ubiquitous promoters present another safety problem as they may result in expression of the introduced gene in the germ cells and early embryo. Thus in the next generation the gene product will not be regarded as foreign and this may increase susceptibility rather than immunise against cancer.

3. VECTORS

The choice of vector is crucial for the outcome of the immune response; construction of vectors that addressed these problems in early studies involved non-viral as well as viral delivery methods.

3.1. Non-Viral Delivery: A Plasmid Approach

For non-viral vectors, such as injection of naked plasmid DNA, there is now evidence to suggest that the actual plasmid backbone acts as an adjuvant to the stimulation of a host immune response. It is clear that this immune response is both cell mediated and humoral. The immune response is associated with the presence of non-coding immunostimulatory sequences (ISS) within the plasmid backbone which are centered around hypomethylated CpG base pairs (Sato *et al.*, 1996). These motifs are encountered frequently in eukaryotic DNA. Less than 5% of the cytosine's in the CpG base pairs in prokaryotic DNA are methylated, but the frequency of methylated cytosine's in CpG base pairs in eukaryotic DNA is 70–90% (Tighe *et al.*, 1997). These hypomethylated motifs rapidly stimulate the innate immune response with production of IFN-g by NK cells and IFN-a and b, Il-12 and Il-18 by macrophages. This response is important phylogenetically as it is the host's first line of defence against bacterial infection. In the adaptive immune response bacterial DNA favours the development of a TH1-driven response and secretion of IFN-g favours immunoglobulin class switching to the IgG2a subtype. Because of the strong and persistant cell mediated and humoral immune response to the plasmid backbone and encoded transgene, the use of DNA vaccines has an enormous potential against infectious diseases, allergy, and cancer. However, for use in gene transfer to generate a secretable protein this strong immune response is counter productive (see below). Therefore, plasmid vectors for gene therapy strategies should be designed to lack these ISS sequences. Nevertheless, for vaccine-based strategies the incorporation of these ISS endowed with adjuvant properties would appear beneficial.

3.1.1. Plasmid Construct; Factor VII. Some of the early in vivo attempts using muscle as a target cell in our laboratory, consisted of direct injection of a plasmid construct encoding the factor VII gene (this was chosen at the time because of availability and ease of manipulation in cloning and expression). The engineered construct consisted of the FVII cDNA under the control of a myosin heavy chain promoter from which negative regulatory elements were removed. A myosin enhancer was also included and this has later been shown to increase the level of expression several fold. The product of the internal gene was detected systemically and shown to have biological activity (at day 4) although this was short lived and disappeared by day 7 (Miller *et al.*, 1995). Subsequent follow up experiments revealed that although factor VII expression was detectable at a tissue level, the muscle sections revealed dense inflammatory infiltrates (consisting of CD4 cells, CD8 cells and macrophages) around the areas of tissue expression (Fields *et al.*, 1998a). An antibody was also detected in the serum of the injected animals to the human factor VII antigen. Similar experiments were performed following injection of a plasmid containing the human Factor IX cDNA into mouse muscle and again this demonstrated little plasma elevation of systemic factor antigen but the presence of an antibody isotype IgG2a to the transgene. This would be in keeping with TH1 driven cell response occurring in the setting of plasmid mediated gene transfer for a secretable protein (Fields *et al.*, 1998b). These experiments illustrated the importance of selecting a species-specific transgene in a gene therapy setting. They also demonstrated that although the concept of in vivo gene therapy for haemophilia using muscle is a viable approach, the efficiency of gene transfer is low. Although expression was demonstrable, plasmid-mediated gene transfer was short lived due to the induction of the host immune response. New evidence suggested that although plasmid mediated-gene transfer is very inefficient for expression of a secretable protein, its immunostimulatory adjuvant effects are highly desirable in various vaccine strategies against infectious and malignant diseases.

4. VIRAL DELIVERY

4.1.1. Adenoviral Vectors. The first vectors to be tested in gene therapy approaches were based on the adenovirus; adenoviral vectors still have advantages and are currently being used particularly in the field of haemophilia.

Advantages include the ease with which they can be prepared in high titre, their wide host range and their ability to transduce non-dividing target cell such as liver and muscle. However, one of the early problems encountered with the use of these vectors was a rapid decline in gene expression although the early expression levels attained were high; re-administration of the adenoviral vector was precluded because of immunogenicity; expression of the transgene was finally lost. Subsequent experiments revealed that the decline in expression appears to arise from the elimination of the therapeutically-transduced cells (both muscle and liver) by the host immune response (Dai *et al.*, 1995, Yang *et al.*, 1994a, Yang *et al.*, 1994b). This resulted from a cytotoxic T lymphocyte (CTL) response induced by the adenoviral proteins. The immune response also appears to mediate the failure of expression following repeated administration of the vector. Because of these problems a number of immunomodulatory strategies are being developed. Engineering of less immunogenic adenoviral vectors (so called gutted adenoviral vectors) is being developed so that the vectors contain only the minimal cis elements required for replication and packaging but are devoid of any regions encoding viral proteins (Fisher *et al.*, 1996, Hardy *et al.*, 1997). Immunomodulatory strategies have involved the use of immunosupressive drugs such as cylophosphamide and cyclosporin, and the blocking of costimulatory pathways involved in T cell activation (Kay *et al.*, 1995). These strategies have met with varying degrees of success but once optimisation can be reached it would be a major advance in adenovirus-mediated gene transfer. Because of the problems encountered with these two approaches, investigators have sought alternative vectors for use in gene transfer strategies in muscle. One such candidate vector for gene transfer is the adeno-associated virus (AAV). Significantly, AAV-mediated gene transfer in muscle is not highly immunogenic as is the case for adenovirus.

Two types of viral vectors have been mainly used in gene transfer studies for muscular dystrophies. Retroviral vectors have been initially used but their efficiency in vivo is extremely limited because their uptake by cells requires mitotic division. Therefore their use in muscular dystrophy gene therapy has been limited to myoblast transfer studies (Fassati *et al.*, 1997). More recently research has focused on the use of adenoviral vectors which are taken up very efficiently by muscle fibres (Lochmuller *et al.*, 1996). However, first-generation adenoviral vectors induced a cellular immune and inflammatory response that precluded long term expression of the gene (Stratford-Perricaudet *et al.*, 1990; Li *et al.*, 1993). This immune response was thought to arise from the presence of viral sequences in the vector. Recently a new adenoviral vector has been developed with a capacity for 28 kb of foreign DNA. This vector is devoid of virtually all viral coding sequences and has been successfully used to transduce fibres in vivo (Clemens *et al.*, 1996; Kochanek *et al.*, 1996; Chen *et al.*, 1996, 1997; Floyd *et al.*, 1998). A number of other viral vectors have been used in gene transfer studies in skeletal muscle. Of these adeno-associated viral constructs have been shown to be effective. However, their limited capacity for foreign DNA (about 4.5 kb) makes them unsuitable for use with dystrophin. They may, however, be suitable for other forms of gene transfer, for example, for haemophilia B. Herpes simplex virus has also been used in muscle but uptake by muscle fibres is very limited (for review see Huard *et al.*, 1997). One way to increase uptake by muscle fibres is to damage the muscle by inducing necrosis and regeneration by chemical means. Various agents have been used such as bupivacaine (Davis *et al.*, 1993), notexin, and barium chloride. It has been shown that regenerating muscle exhibits increased uptake of DNA, which is taken up mainly by the newly formed fibres. The use of viral vectors is also limited by safety factors. Since helper virus is used to produce the virus in the encapsulated form that can be taken up by cells, the construct to be transferred has to be highly purified. This is difficult to achieve especially on a large scale and there is always the possibility of contamination by helper virus. In addition recombinant events may occur which could lead to the activation of oncogenes; this is especially important when regenerating muscle models are used. Another approach for the treatment of muscular dystrophy has been the use of isolated satellite cells stably transformed with dystrophin genes that are then reintroduced into the host organism. This approach is hindered by the fact that there is limited integration between the introduced cells and the existing muscle fibres, by the development of immune responses and the limited penetration of satellite cells though muscle. There have been a number of studies on animal models (Rando and Blau, 1994; Rando *et al.*, 1995) but clinical trials have shown that there is a negligible therapeutic value in this approach (Miller *et al.*, 1997).

4.1.2. Adeno-Associated Virus (AAV) Vectors. Recently, AAV vectors have shown great promise in a gene therapy setting of haemophilia using muscle as a target tissue, both in small and large animal models of haemophilia B. Several studies have documented that AAV vectors can direct persistent expression of reporter genes in muscle fibres of immune competent animals (Xiao *et al.*, 1996; Kessler *et al.*, 1996; Fisher *et al.*, 1997). One early report documented expression of therapeutic levels of erythropoietin following intramuscular injection of an AAV vector expressing erythropoietin (Xiao *et al.*, 1996). AAV vectors have certain advantages that make them particularly attractive for muscle-directed gene therapy. These include their relative non-pathogenicity in normal individuals (up to 80% of humans are infected with parvovirus), their ability to infect non-dividing cells and a broad range of host recipients (primates, canine and murine models). Recombinant AAV vectors contain the engineered expression cassette for the transgene flanked only by the inverted terminal repeats and is therefore devoid of any viral coding sequences. Transduction of muscle with recombinant AAV is very efficient, and expression is stable and long-lived without eliciting the cellular immune responses characteristically seen in muscle after adenovirus-mediated gene transduction (Xiao *et al.*, 1996; Snyder *et al.*, 1997). Various ex vivo approaches have targeted muscle in the past for factor IX gene expression; it is known that factor IX produced in myotubes in vitro is biologically active (Dhawan *et al.*, 1991). With the use of AAV vectors it is now possible in an in vivo gene therapy setting to transduce muscle in a highly efficient way. Stable expression of therapeutic plasma levels of human factor IX (1 year) has been demonstrated after intramuscular injection of AAV-factor IX to Rag-1 immmunodeficient mice; levels of 4–7% (200–350 ng/ml FIX Ag) of normal levels of FIX in a human being were attained (Herzog *et al.*, 1997a). When this experiment was repeated in an immunocompetent mouse, no factor IX could be detected in mouse plasma due to the presence of an antibody to the human transgene. When this experiment was scaled up to a canine haemophilia B model a similar result was obtained (Monahan *et al.*, 1998); furthermore data from the canine model indicated that antibody formation against the transgene could be avoided if species specific transgene boundaries are not transgressed (Herzog *et al.*, 1997b). The results achieved so far with these vectors are very encouraging. Optimisation of these gene vectors has been achieved in rodent models resulting in persistent expression of therapeutic plasma levels of clotting factor IX with reduced or absent cellular immune responses against the transduced muscle cells. A scale up of these approaches to larger animal models of haemophilia will form the basis for future human clinical trials.

4.2. Retroviral

Many of the initial viral vectors tested for a gene therapy approach in haemophilia B were retroviral. This involved an ex vivo approach in transfecting cell lines and subsequently transplanting them back into rodent animal models. Some early success was achieved transfecting murine myoblasts with a retroviral construct expressing factor IX, with factor IX being expressed for up to six months, although at low levels (Dai *et al.*, 1992; Yao *et al.*, 1992). This was presumed to arise from inefficient expression cassettes (Dai *et al.*, 1992) and a cross species specific transgene evoking a host immune response. Studies were then performed in large animal models but the results were less successful with only short-term expression at levels that were clinically insignificant (Verma *et al.*, 1994). Although some limited success was derived from the use of retroviral vectors there were still many problems to be resolved linked to these vectors. These related to low viral copy numbers, requirement for target cell division, size limitation of the transferred gene, and the concern about potential long term safety of these vectors with respect to their tropism, random integration and oncogenic potential. Because of these early problems with retroviruses investigators began working with adenoviruses; some of the problems associated with adenoviruses have been discussed above.

5. POSSIBLE TARGETS FOR CANCER GENE THERAPY

As our understanding of the molecular genetics underlying the causes of cancer improves, the scope for molecular medicine and gene therapy as a novel therapeutic approach is increasing. The direct transfer into skeletal muscle of engineered genes using gene promoters and enhancers, which are specific to the muscle, as far as expression is concerned (Fig. 1). However, the gene product can be delivered systemically. The object

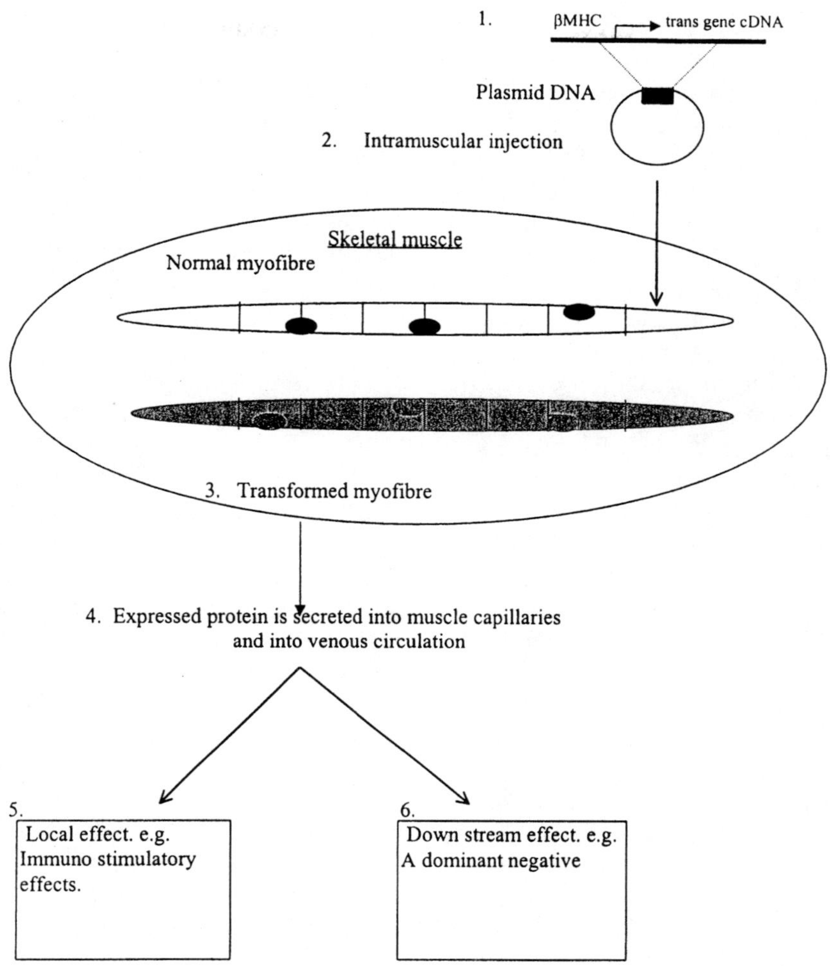

Figure 1. A schematic representation of muscle based cancer gene therapy. 1. Trans gene is under the control of a muscle specific promoter i.e. β myosin heavy chain (βMHC) and downstream enhancer (not shown) 2. Plasmid is injected directly into skeletal muscle 3. Muscle fibres take up the plasmid containing the trans gene 4. Gene product is secreted into the muscle capillaries and passes into the circulation leading to 5. A local immuno stimulatory response or 6. A downstream dominant negative response.

here is not to review all the possible antisense, antibody or vaccine cDNAs that could be included in a muscle type gene cassette for the control of tumorigenesis and malignancy but to describe some of the targets that we are currently working on in conjunction with the intramuscular route of gene delivery.

5.1. Insulin like Growth Factor-1 (IGF-1) and the IGF-1 Receptor

The rate of division and growth of cells during normal post-natal growth of tissues is to a large extent regulated by the growth hormone/insulin-like growth factor axis. Also insulin-like growth factor (IGF-1) is known to trigger cell division and growth of transformed cells. Indeed it is the commonly used growth factor to test if cells have mutated and become tumorigenic. However, IGF-1 exists as different isoforms and these have different binding proteins, which to a large extent determine their actions. Therefore this seems to be a good starting point to look for anticancer therapies.

IGF-1 is a protein consisting of 70 amino acids having approximately 70% homology with proinsulin (Daughady and Rotwein, 1989). It mediates the effects of growth hormone and has metabolic effects as well as mitogenic. The major source of IGF-1 is the liver, however, it is also produced in other tissues (D'Ercole *et al.*, 1984; Lund *et al.*, 1986; Schlechter *et al.*, 1986; Murphy *et al.*, 1987) and has paracrine/autocrine modes of action (Lund, 1994).

IGF-1 exerts its effect by binding to a heterotetrameric transmembrane receptor (IGF-1R) which has intrinsic tyrosine kinase activity (Ullrich *et al.*, 1985 and 1986). Binding leads to the inhibition of apoptosis by activating multiple signal transduction pathways which include the Phosphadityl Inositol 3′ Kinase and the MAP kinase pathways (Resincoff *et al.*, 1995; Parrizas *et al.*, 1997). The type 1 IGF-1R has also been implicated in tumorigenesis with over expression of the receptor protecting the cells from apoptosis. Therefore binding of IGF-1 to elevated levels of receptors may prevent a cell from undergoing programmed cell death, which it might otherwise have done, in the absence of IGF-1.

The tumor suppressor gene p53 is mutated in many human cancers (Hollstein *et al.*, 1991) one of the functions of the p53 protein is as a transcription factor binding specifically to DNA. Wild type p53 has been shown to suppress promoter activity of the IGF-1R gene, whilst tumor derived mutant forms of p53 stimulate the promoter activity (Werner *et al.*, 1996). Therefore, mutations of p53 lead to the over expression of the IGF-1R by its failure to repress transcription. Saos-2-DH4 is an osteosarcoma cell line, which expresses a temperature sensitive p53. Incubating the cells at the permissive temperature leads to wild type p53 expression which reduces IGF1-R expression by 69% when compared with controls (Ohlsson *et al.*, 1998). When not at the permissive temperature a mutant p53 is expressed which leads to increased expression of IGF1-R by 85% compared with the wild type. The IGF-1R is not the only component of the IGF1 system that is regulated by p53. The activity of IGF-1 is modulated by seven IGF binding proteins (IGFBP's) (Rechler *et al.*, 1994). Of these binding proteins, IGFBP-3 has shown to be induced by p53 (Buckbinder *et al.*, 1995). This binding protein is the most common form of IGFBP found in serum. The reduced transcription of this gene decreases IGFBP-3 levels and increases the total free pool of IGF-1 able to bind the receptor. Clearly the IGF-1 system in tumorigenesis is an excellent target for muscle based gene therapy.

5.2. Strategies Targeting the Type 1 IGF1-R

A dominant negative of the type 1 IGF-1 receptor 486/stop, has a frame shift mutation in the RNA leading to a stop codon at residue 486. The RNA is of normal length but it codes for the first 481 amino acids of the α sub unit plus four additional amino acids which are not found in the original sequence (D'Ambrosio *et al.*, 1996). This mutant receptor lacks the transmembrane domain and it is therefore secreted. Reiss *et al.*, (1998) reported that when 486/stop is stably expressed it inhibits growth in five tumor cell lines (two prostate, colon, ovarian and breast). The injection of stably transfected cancer cell lines into nude mice resulted in a marked inhibition of tumor growth. In 3 out of the 4 cell lines used the majority of mice did not develop any tumors. In the fourth cell line the tumors were much smaller than controls. Another important observation was that expression of 486/stop inhibits colony formation in soft agar. A bystander effect was illustrated when cells expressing 486/stop were coinjected with tumor forming cell lines. This led to an inhibition in tumor formation and also showed inhibition of established tumors in nude mice. In some experiments carried out by Reiss *et al.*, there was a delayed appearance of tumors in cells which had been expressing 486/stop when compared with the parental cell line. This was explained by the fact that the cells had eliminated the plasmid and therefore the inhibition afforded by the protein was no longer apparent. As dividing cells will eliminate a plasmid, tumor cells may not be the most suitable target for a therapy based upon this strategy.

As muscle cells are non-dividing cells they are therefore a suitable target for a construct containing the cDNA for 486/stop under the control of a muscle specific promoter/enhancer which could be administered intramuscularly. The muscle could then secrete the protein to produce a therapeutic effect down stream. The use of muscle would negate the need to transform the tumor cells directly and the systemic production of 486/stop could not only lead to regression of the tumor but could also prevent metastases. The advantage of this method over that described by Reiss *et al.*, (1998) is that high levels of expression can be obtained by the muscle fibres. Since the muscle of an adult are post-mitotic, expression of the introduced gene will persist for long periods. This should mean that patients do not require repeat injections at frequent intervals. The findings of Reiss also show that the effects are dose dependent, the greater effects are seen as the number of cells that have been transformed increases. This leads from inhibition to abrogation of tumor growth and would indicate that there is a critical concentration over which leads to tumor cell death. By far the most important observation for this therapy is that it appears to affect tumor cells more than normal cells (Baserga *et al.*, 1997), indicating the suitability for a systemic application of this therapy whilst minimising toxicity

6. CARCINOEMBRYONIC ANTIGEN, IMMUNO THERAPY AND INTRAMUSCULAR IMMUNISATION IN MICE

Immunotherapy is one of the most active areas of cancer gene therapy with many efforts being concentrated at enhancing the immunological response to tumors. The carcinoembryonic antigen (CEA) is a 180 KDa glycoprotein the properties of which has been extensively studied (Shively and Beatty, 1985; Thompson *et al.*, 1991). It is commonly expressed in human colon, breast and non-small cell lung carcinoma (Shively and Beatty, 1985; Thompson *et al.*, 1991) however, it is also found less extensively in normal colonic mucosa. Conry *et al.*, (1995) constructed a plasmid containing the full-length cDNA for CEA under the control of a cytomegalovirus promoter/enhancer. When this plasmid was injected intramuscularly into mice there was an immune response leading to "CEA specific lymphoblastic transformation, lymphokine release and an antibody response". The data indicated that this response was dose dependent in which 50 μg of the plasmid led to a maximum effect with no significant increase in effects above this dosage. The results also indicated that the schedule in which the plasmid is injected will also determine the response. Not suprisingly a thrice-weekly 50 μg plasmid injection elicits a stronger CEA immune response within 43 days than does a weekly injection. The injection of the plasmid intramuscularly led to subsequent immunoprotection. Immunised mice were challenged with MC-38-CEA-2 colon carcinoma cells 3 weeks after immunisation. Of the 20 mice inoculated none had developed tumors after 42 days, compared with 7 out of 10 naïve mice inoculated with MC-38-CEA-2, which had developed measurable tumors over the same period. The immunised mice were monitored for a further 10 weeks and evidence of tumor growth was found in only one mouse.

6.1. Immunisation of Non Human Primates Against Carcinoembryonic Antigen

A continuation of the work above is reported by Conry *et al.*, (1998) in preparatory work for a clinical trial using a vaccine developed against CEA. Non human primates (pig-tailed macaques) were inoculated by intramuscular injection or by particle bombardment (plasmid coated gold beads which are delivered by gene gun (Eisenbraum *et al.*, 1993)) into the inguinal skin. A plasmid containing the cDNA for CEA and a hepatitis B surface antigen under the control of two separate cytomegalovirus promoters/enhancers was constructed. The hepatitis B antigen was included to act as an internal positive control in which an immune response to this antigen would give an indication of the efficacy of these methods of gene transfer. The dual transfer with positive control was required, as human CEA is highly homologous with that of the monkey.

Both methods resulted in similar CEA specific lymphoproliferation and anti-CEA antibody responses. Animals that had been immunised by i.m injection exhibited anti-CEA antibodies both prior to and after a intradermal boost injection of CEA protein, only one animal immunised by particle bombardment showed anti-CEA antibodies prior to the boost injection. Moderate levels of antibodies could be detected in both groups 74 weeks after a single protein boost.

Only monkeys immunised by intramuscular injection were protected when subsequently challenged by syngeneic CEA-expressing carcinoma cells. The immune responses elicited by particle bombardment and intramuscular injection were different. Particle bombardment was exclusively IgG1, whereas intramuscular administration showed a substantial IgG2a response. The presence of the IgG2a antibodies is compatible with a T helper1 response. This is the principal effector of delayed type hypersensitivity, which was observed in 2 out of the 3 monkeys treated by i.m injection. Delayed type hypersensitivity was totally absent from the animals immunised by particle bombardment.

The anti-CEA antibody response required approximately 30 weeks to reach what is described as a moderate titre after primary immunisation, with maximum levels being reached at between 34 to 40 weeks. There was little difference between the two methods in the levels of antibody achieved. The time scale involved in establishing an adequate antibody response would suggest that this may be an important application in the immunisation against distant micrometastases, however, for patients with established metastases its uses would be extremely limited due to the limited life expectancy of many of these patients.

Experimental evidence favours the use of i.m injection over particle bombardment as an efficient means of gene delivery and subsequent immunisation in the non-human primate and has been chosen by Conry *et al.*, in initial clinical trials for patients with colon carcinoma.

The technique employed above was free of toxicity. Non essential viral DNA sequences were removed to reduce the immunogenecity of the plasmid, with none of the animals showing measurable anti-DNA antibodies. No inflammation or tenderness at the sites of application was reported.

The observations of Conry *et al.*, have shown that the muscle is both a suitable and efficient site for eliciting humoral and cytolytic T cell immune responses. The use of the cytomegalovirus promoter/enhancer however may raise some safety concerns (Gorecki and Simons, 1999). These issues must be addressed particularly in light of recent evidence suggesting that injecting genes into muscle results in the uptake by motorneurones (Keir *et al.*, 1995; Ribota *et al.*, 1997). Possible solutions to these safety concerns would be the use of muscle specific promoters and enhancers, ensuring muscle selective expression.

7. ANTI-ANGIOGENESIS GENE THERAPY

Angiogenesis, the formation of new blood vessels, is a complex physiological process essential for embryonic development, wound healing, and tissue or organ regeneration. Angiogenesis is essential for tumors to continue growing above a volume of 1–2 mm^3 (Folkman, 1995), and without adequate angiogenesis primary tumors lie dormant with a balance between proliferation and apoptosis. Extensive data exists that tumors express genes coding angiogenic mediators and it is believed that these mediators act in a paracrine fashion. This has lead to anti-angiogenic strategies as a method of suppressing tumor growth. Various non-gene therapy techniques including antiangiogenic drugs, polypeptides, and monoclonal antibodies are under evaluation in an attempt to starve the tumor of blood supply (Kong and Crystal, 1998). Targeting angiogenesis is a relatively new concept in anti-cancer gene therapy. Virtually all cancer gene therapy trials target the cancer cell including: preventing or correcting genetic defects, immunogene therapy and increasing tumor cell sensitivity to chemotherapeutic agents. Conventional antiangiogenic treatment has several disadvantages including repeated dosing, indefinate treatment period, it is cumbersome and expensive. Angiogenic gene therapy may overcome these pharmaceutical problems. Furthermore gene transfer may allow tumor-based angiogenesis to be targeted without systemic effects on physiological angiogenesis.

Using gene transfer to disrupt the angiogenic system *in vitro* has been shown to decrease the tumor forming abilities of cancer cell lines subsequently inoculated into nude mice. Using retroviral and adenoviral vectors to express an antiangiogenic protein, platelet factor 4(sPF4), Tanaka *et al.* (1997) where able to produce marked slowing of RT2 rat glioma growth *in vivo* associated with hypovascular tumors. Bouvet *et al.* (1998) transfected human colorectal cancer cell lines KM12L4 and SW620 with wild type p53 tumor suppressor gene using an adenoviral vector. This decreased Vascular Endothelial Growth Factor (VEGF the angiogenic factor which seems to be associated with neovascularisation in colon cancer) production. This decreased neovascularisation *in vivo* of treated cell lines tumors inoculated into nude mice (Bouvet *et al.*, 1998).

Local deliver of gene therapy by intratumoral injection has been shown to slow tumor growth. Using liposome-mediated gene transfer, Wang and Becker (1997) injected malignant human melanomas grown subcutanaseously in nude mice with episomal vectors containing antisense to basic fibroblast growth factor (bFGF) and fibroblast growth factor receptor-1 (FGFR-1). The expression of these genes is essential in sustaining the proliferation of malignant melanomas *in vitro*. The growth of tumors injected with these constructs was completely arrested or the tumors regressed due to blocked angiogenesis. Tanaka *et al.* (1997) prolonged animal survival by injecting established intracerebral RT2 rat gliomas with sPF4 carried in an adenoviral vector. Griselli *et al.* (1998) showed a significant arrest of growth in established rat C6 glioma or human MDA-MB-231 pre-established in nude mice with an intratumoral injection of an adenovirus vector expressing angiostatinK3 (an angiogenesis inhibitor). Furthermore systemic injection delayed C6 glioma establishment and growth.

Kong *et al.* (1998) used an adenoviral vector to transfer cDNA encoding a secreted form of the extracellular domain of the *flt-1* VEGF receptor to established CT26.CL25 splenic tumors with liver and lung metastases. This was administered on a regional basis (intrajugular for splenic and liver tumors, intratracheal for lung deposits) and significantly reduced growth in the liver, spleen and lung deposits.

Tie2 is an endothelium-specific receptor tyrosine kinase known to play a role in tumor angiogenesis. Systemic (intravenous) administration of an adenoviral vector with recombinant, soluble Tie2 receptor (Lin *et al.*, 1998), which is capable of blocking Tie2 activation, significantly inhibited growth of established murine mammary carcinoma and murine melanoma. Furthermore the animals appeared well, without histological abnormalities in the normal vasculature suggesting inhibition of physiological angiogenesis was minimal.

7.1. Future Prospects for Antiangiogenic Therapy

Several authors have suggested that antiangiogenic therapy should be localised (Kong and Crystal, 1998) or transient (Paillard, 1998), due to fears of damage to physiological angiogenesis. Evidence of systemic side-effects in uterine tissues and pregnant mice (Klauber *et al.*, 1997) must be balanced by recent evidence of systemic administration of the angiogenesis inhibitors angiostatin (O'Reilly *et al.*, 1994) and of endostatin (O'Reilly *et al.*, 1997) which did not inhibit wound healing in mice and may suggest systemic administration is safe.

Antiangiogenic therapy has been shown to be cytostatic rather than cytotoxic. Experimental data suggests that effective therapy requires the continuous presence of the inhibitor in the blood. Furthermore a disadvantage of local or regional administration is that it cannot treat the difficult clinical situations of disseminated and occult metastases. Tumor deposits may also be inaccessible for intramural injection.

Using muscle based gene therapy large amounts of recombinant proteins can be secreted *in vivo* for systemic therapy on a continuous basis. If the therapy is specific for tumor angiogenesis it could form a novel therapy. Published data suggests that angiogenic gene therapy will not be a single line treatment but will a useful addition to the modalities of surgery, chemotherapy and radiotherapy.

REFERENCES

Ayling, C., Moreland, B.H., Zanelli, J.M., and Schulster, D., (1989), Human growth hormone treatment of hypophysectomized rats increases the proportion of type-1 fibres in skeletal muscle. J. Endocrinol. **123**:429–435.

Baker, J., Liu, J.P., Robertson, E.J., and Efstratiadis, A., (1993), Role of Insulin-like growth factors in embryonic and postnatal growth. Cell **75**:73–82.

Baserga, R., Hongo, A., Rubini, M., Prisco, M., and Valentis, B., (1997), The IGF-1 receptors in cell growth, transformation and apoptosis. Biochim. Biophy. Acta. **1332**:105–126.

Bergsma, D.J., Grichnikm, J.M., Gossett, I.M., and Schwartz, R.J., (1986), Delimitation and characterization of cis-acting of DNA sequences required for the regulated expression and transcriptional control of the chicken skeletal a-actin gene. Mol. Cell Biol. **6**:2452–2475.

Bouvet, M., Ellis, L.M., Nishizaki, M., Fujiwara, T., Liu, W., Bucana, C.D., Fang, B., Lee, J.J., and Roth, J.A., (1998), Adenovirus-mediated wild type p53 gene transfer down-regulates vascular endothelial growth factor expression and inhibits angiogenesis in human colon cancer. Cancer Res. **58**:2288–2292.

Brahm, H., Piehl-Aulin, K., Saltin, B., and Ljunghall, S., (1997), Net fluxes over working thigh of hormones, growth factors and biomarkers of bone metabolism during lasting dynamic exercise. Calcified Tissue **60**:175–189.

Bittner, R.E., Shony, S., Streubel, B., Hubner, C., Volt, T., and Kress, W., (1995), Serum antibodies to the deleted dystrophin sequence after cardiac transplantation in a patient with Becker's muscular dystrophy. New Eng. J. Med. **333**:732–733.

Bucbinder, L., Talbott, R., Velasco-Miguel, S., Takenaka, I., Faha, B., Seizinger, B.R., and Kley, N., (1995), Induction of the growth inhibitor IGF-binding protein 3 by p53. Nature **377**:646–649.

Chen, H.H., Mack, L.M., Kelly, R., Ontell, M., Kochanek, S., and Clemens, P.R., (1997), Persistence in muscle of an adenoviral vector that lacks all viral genes. Proc. Natl. Acad. Sci. USA **94**:1645–1650.

Clemens, P.R., Kochanek, S., Sunada, Y., Chan, S., Chen, H.H., Campbell, K.P., and Caskey, C.T., (1996), In vivo muscle gene transfer of full-length dystrophin with an adenoviral vector that lacks all viral genes. Gene Therapy **3**:965–972.

Coleman, M.E., DeMayo, F., Yin, K.C., Lee, H.M., Geske, R., Montgomery, C., and Schwartz, R.J., (1995), Myofenic vector expression of insulin-like growth factor I stimulates muscle cell differentiation and myofiber hypertrophy in transgenic mice. J. Biol. Chem. **270**:12109–12116.

Corry, R.M., LoBuglio, A.F., Loechel, F., Moore, S.E., Sumerel, L.A., Barlow, D.L., and Curiel, D.T., (1995), A carcinoembryonic antigen polynucleotide vaccine has in vivo antitumor activity. Gene Therapy **2**:59–65.

Corry, R.M., White, S.A., Fultz, P.N., Khazaeli, M.B., Strong, T.V., Allen, K.O., Barlow, D.L., Moore, S.E., Coan, P.N., Davis, I., Curiel, D.T., and LoBuglio, A.F., (1998), Clinical Cancer Research **4**:2903–2912.

Corrado, K., Wills, P.L., and Chamberlain, J.S., (1994), Deletion analysis of the dystrophin-actin binding domain. FEBS Lett. **344**:255–260.

Cox, G.A., Cole, N.M., Matsumura, K., Phelps, S.F., Hauschka, S.D., Campbell, K.P., Faulkner, J.A., and Chamberlain, J.S., (1993), Overexpression of dystrophin in transgenic mdx mice eliminates dystrophic symptoms without toxicity. Nature **364**:725–729.

Dai Y., Roman, M., Naviaux, R.K., and Verma, I.M., (1992), Gene therapy via primary myoblasts, long-term expression of factor IX protein following transplantation in vivo. Proc. Natl. Acad. Sci. USA **89**:10892–10895.

Dai Y., Schwarz, E.M., Gu, D., Zhang, W.W., and Verma, I., (1995), Cellular and humoural immune responses to adenoviral vectors containing factor IX gene. Tolerization of Factor IX and vector antigens allows for long-term expression. Proc. Natl. Acad. Sci. USA **92**:1401–1405.

D'Ambrosio, C., Ferber, A., Resnicoff, M., and Baserga, R., (1996), A soluble insulin-like growth factor 1 receptor that induces apoptosis of tumor cells in vivo and inhibits tumorigenesis. Cancer Res. **56**:4013–4020.

Daughaday, W.H., and Rotwein, P., (1989), Insulin-like growth factors I and II. Peptide, messenger ribonucleic acid and gene structures, serum, and tissue concentrations. Endocrine Rev. **10**:68–91.

Davis, H.L., Demeneix, B.A., Quantin, B., Coulombe, J., and Whalen, R.G., (1993), Plasmid DNA is superior to viral vectors for direct gene transfer into adult mouse skeletal muscle. Hum. Gene. Ther. **4**:733–740.

D'Ercole, J.A., Stiles, A.D., and Underwood, L.E., (1984), Tissue concentrations of somatemedin C: further evidence for multiple sites of synthesis and paracrine or autocrine mechanisms of action. Proc. Natl. Acad. Sci. USA **81**:935–939.

Dhawan, J., Pan, L.C., Pavlath, G.K., Travis, M.A., Lanctot, A.M., and Blau, H.M., (1991), Systemic delivery of human growth hormone by injection of genetically engineered myoblasts. Science **254**:1509–1512.

Draghia-Akli, R., Li, X., and Schwartz, R.J., (1997), Enhanced growth by ectopic expression of growth hormone releasing hormone using an injectable myogenic vector. Nature Biotechnol. **15**:1285–1289.

Eisenbraun, M.D., Fuller, D.H., and Haynes, J.R., (1993), Examination of parameters affecting the elicitation of humoral immune responses by particle bombardment-mediated genetic immunisation. DNA Cell Biol. **12**:791–797.

England, S., Nicholson, L.V.B., Johnson, M.A., Forrest, S.M., Bundy, S., Zubrzycka-Gaarn, E., Love, D.R., and Davies, K.E., (1990), Very mild muscular dystrophy associated with the deletion of 46% of dystrophin. Nature **343**:180–182.

Fabbrizio, E., Pons, F., Robert, A., Hugon, G., Bonet-Kerrache, A., and Momet, D., (1994), The dystrophin superfamily, variability and complexity. J. Mucle Res. Cell Motil. **15**:595–606.

Fassati, A., Wells, D.J., Sgro Serpente, P.A., Walsh, F.S., Brown, S.C., Strong, P.N., and Dickson, G., (1997), Genetic correction of dystrophin deficiency and skeletal muscle remodeling in adult MDX mouse via transplantation of retroviral producer cells. J. Clin. Invest. **100**:620–628.

Fields, P., Herzog, R., Arruda, V., Hagstrom, N., Pasi, K.J., and High, K., (1998b), Immune response to a secretable protein following gene transfer into muscle by plasmid based, adenoviral or adeno-associated viral vectors. Blood **92**:147a.

Fields, P., Murdoch, P., Bayele, H., Perry, D., Wells, D., Watt, D., Goldspink, G., and Pasi, J., (1998a), Immune responses to direct plasmid injection in muscle, implications for transgene expression. Keystone Symposia, Molecular and cellular biology of gene therapy. SUPPLEMENT **25A**:P435.

Fisher, K., Jooss, K., Alston, J., Yang, Y., Haecker, S., High, K., Pathak, R., Raper, S., and Wilson, J.M., (1997), Recombinant Adeno-associated virus for muscle directed gene therapy. Nature Med. **3**:3227–3234.

Fisher, K.J., Choi, H., Burda, J., Chen, S., and Wilson, J.M., (1996), Recombinant adenoviruses deleted of all viral genes for gene therapy of cystic fibrosis. Virology **217**:11–22.

Floyd, S.S. Jr., Clemens, P.R., Ontell, M.R., Kochanek, S., Day, C.S., Yang, J., Hauschka, S.D., Balkir, L., Morgan, J., Moreland, M.S., Feero, G.W., Epperly, M., and Huard, J., (1998), Ex vivo gene transfer using adenovirus-mediated full-length dystrophin delivery to dystrophic muscles. Gene Therapy **5**:19–30.

Folkman, J., (1995), Angiogenesis in cancer, vascular, rheumatoid and other diseases. Nature Medicine **1**:27–31.

Goldspink, G., and Booth, F., (1992), General Remarks for Editorial Issue—Mechanical signals and gene expression in muscle. Am. J. Physiol. **262**:R327–R328.

Goldspink, G., Yang, S.Y., Skarli, M., and Vrbova, G., (1996), Local growth regulation is associated with an isoform of IGF-1 that is expressed in normal muscles but not in dystrophic muscles when subjected to stretch. J. Physiol. **496**:P10.

Gorecki, D.C., and Simons, J.P., (1999), The dangers of DNA vaccination (Letter). Nature Medicine **5**:126.

Hardy, S., Kitamura, Harris, T., Dai, Y., and Phipps, L., (1997), Construction of adenovirus through Cre Lox Recombination. J. Virol. **71**:1842–1849.

Griscelli, F., Li, H., Bennaceur-Griscelli, A., Soria, J., Opolon, P., Soria, C., Perricaudet, M., Yeh, P., and Lu, H., (1998), Angiostatin gene transfer: Inhibition of tumor growth in vivo by blockage of endothelial cell proliferation associated with mitosis arrest. Proc. Natl. Acad. Sci. USA **95**:6367–6372.

Hansen, E., Goldspink, G., Butterworth, P.W., and Chang, K-C., (1991), Strong expression of some mammalian gene constructs in fish muscle following direct gene transfer. FEBS Letts. **290**:73–76.

Herzog, R., Hagstrom, J., Kung, Z., Tai, S.J., Wilson, J.M., Fisher, K., and High, K.A., (1997a), Stable gene transfer and expression of human blood coagulation factor IX after intramuscular injection of recombinant adeno-associated virus. Proc. Natl. Acad. Sci. USA **94**:5804–5809.

Herzog, R., Hagstrom, J.N., Kung, S.Z., Yang, E.Y., Couto, L.B., Kurtzmann, G.J., and High, K.A., (1997b), Absence of antibodies against factor IX following IM injection of an AAV vector encoding a species specific transgene. Blood **90**:1957.

Hollstein, M., Sidransky, D., Vogelstein, B., and Harris, C.C., (1991), p53 mutations in human cancers. Science **253**:49–53.

Huard, J., Krisky, D., Oligino, T., Marconi, P., Day, C.S., Watkins, S.C., and Glorioso, J.C., (1997), Gene transfer to muscle using herpes simplex-based vectors. Neuromusc. Disord. **7**:299–313.

Kay, M.A., Holterman, A.X., Meuse, L., Gown, A., Linsley, P., and Wilson, C.B., (1995), Long term hepatic adenovirus mediated gene expression in mice following CTLAIg administration. Nat. Genet. **11**:191–197.

Keir, S.D., Mitchell, W.J., Feldman, L., and Martin, J.R., (1995), Targeting and gene expression in spinal cord motorneurones following intramuscular inoculation of an HSV-1 vector. J. Neurovirol. **1**:259267.

Kessler, P.D., Podsakoff, G.M., Chen, X., McQiuston, S.A., Colosi, P.C., Kurtzmann, G., and Byrne, B., (1996), Gene delivery to skeletal muscle results in sustained expression and systemic delivery of a therapeutic protein. Proc. Natl. Acad. Sci. USA **93**:14082–14087.

Klauber, N., Rohan, R., Flynn, E., and D'Amato, R.J., (1997), Critical components of the female reproductive pathway are suppressed by the angiogenesis inhibitor AGM-1470. Nature Medicine **3**:443–446.

Kochanek, S., Clemens, P.R., Mitani, K., Chen, H.H., Chan, S., and Caskey, C.T., (1996), A new adenoviral vector, Replacement of all viral coding sequences with 28 kb of DNA independently expressing both full-length dystrophin and Beta-galactosidase. Proc. Natl. Acad. Sci. USA **93**:5731–5736.

Kong, H., and Crystal, R.G., (1998), Gene therapy strategies for tumor antiangiogenesis. Journal of the National Cancer Institute **90**:273–286.

Li, Q., Kay, M.A., Finegold, M., Stratford-Perricaudet, L.D., and Woo, S.L., (1993), Assessment of recombinant adenoviral vectors for hepatic gene therapy. Hum. Gene Ther. **4**:403–409.

Kong, H., Hechet, D., Song, W., Kovesdi, I., Hackett, N.R., Yayon, A., and Crystal, R.G., (1998), Regional suppression of tumor growth in vivo transfer of a cDNA encoding a secreted form of the extracellular domain of the flt-1 vascular endothelial growth factor receptor. Human Gene Therapy **9**:823–833.

Lin, P., Buxton, J.A., Acheson, A., Radziejewski, C., Maisonpierre, P.C., Yancopoulos, G.D., Channon, K.M., Hale, L.P., Dewhirst, M.W., George, S.E., and Peters, K.G., (1998), Antiangiogenic therapy targeting the endothelium tyrosine kinase TIE2. Proc. Natl. Acad. Sci. USA **95**:8829–8834.

Lochmuller, H., Petrof, B.J., Pari, G., Larochelle, N., Dodelet, V., Wang, Q., Allen, C., Prescott, S., Massie, B., Nalbantoglu, J., and Karpati, G., (1996), Transient immunosuuression by FK506 permits a sustained high-level dystrophin expression after adenovirus-mediated dystrophin minigene transfer to skeletal muscles of adult dystrophic (mdx) mice. Gene Therapy **3**:706–716.

Lund, P.K., (1994), Insulin-like growth factor I: molecular biology and relevance to tissue-specific expression and action. Recent. Prog. Horm. Res. **49**:125–148.

Michou, A., Santoro, L., Christ, M., Jiullard, V., Pavirani, A., and Mehtali, M., (1997), Adenovirus-mediated gene transfer influence of transgene, mouse strain and type of immune response on persistence of transgene expression. Gene Therapy **4**:473–482.

Miller, G., Steinbrecher, R.A., Murdock, P.J., Tuddenham, E.G.D., Lee, C.A., Pasi, K.J., and Goldspink, G., (1995), Expression of factor VII by muscle cells in vitro and in vivo following direct gene transfer, modelling gene therapy for haemophilia, Gene Therapy **2**:736–742.

Miller, R.G., Sharma, K.R., Pavlath, G.K., Gussoni, E., Mynhier, M., Lanctot, A.M., Greco, C.M., Steinman, L., and Blau, H.M., (1997), Myoblast implantation in Duchenne muscular dystrophy, the San Francisco study. Muscle and Nerve. **20**:469–478.

Milner, R.E., Busaan, J.L., Wang, J.H., and Michalak, M., (1993), Phosphorylation of dystrophin. The carboxy-terminal region of dystrophin is a substrate for in vitro phosphorylation by p34cdc2 protein kinase. J. Biol. Chem. **268**:21901–21905.

Monahan, P.E., Samulski, R.J., Tazelaar, J., Xiao, X., Nichols, T.C., Bellinger, D.A., and Read, M.S., (1998), Direct intramuscular injection with recombinant AAV vectors results in sustained expression in a dog model of haemophilia. Gene Therapy **5**:40–49.

Murphy, L.J., Bell, G.I., and Friesen, H.G., (1987), Tissue distribution of insulin-like growth factor I and II messenger nucleic acid in the adult rat. Endocrinology **120**:1279–1282.

Nov F.J., Kruszewski, A., MacDermot, K.D., Goldspink, G., and Gorecki, D.C., (1995), Editing of human alpha-galactosidase RNA resulting in a pyrimidine to purine conversion. Nucleic. Acids. Res. 2636–2640.

Ohlsson, C., Kley, N., Werner, H., and LeRoith, D., (1998), p53 regulates insulin-like growth factor I receptor expression and IGF-1-induced tyrosine phosphorylation in an osteosarcoma cell line: interaction between p53 and Sp1. Endocrinology **139**:1101–1107.

O'Reilly, M.S., Holmgren, L., Shing, Y., Chan, C., Rosenthal, R.A., Moses, M., Lane, W.S., Cao, Y., Sage, E.H., and Folkman, J., (1994), Angiostatin: A novel angiogenesis inhibitor that mediates the suppression of metastases by a Lewis lung carcinoma. Cell **79**:315–328.

O'Reilly, M.S., Boeham, T., Shing, Y., Fukai, N., Vasios, G., Lane, W.S., Flynne, E., Birkhead, J.R., Olsen, B.R., and Folkman, J., (1997), Endostatin: An endogenous inhibitor of angiogenesis and tumor growth. Cell **88**:277–285.

Paillard, F., (1998), Suppression of angiogenesis: a means to fight cancer, Human Gene Therapy **9**:768–770.

Palmer, T.D., Thompson, A.R., and Miller, A.D., (1989), Production of human factor IX in animals by genetically modified skin fibroblasts, Potential therapy for Haemophilia B. Blood **73**:438–445.

Parrizas, M., Saltiel, A.R., and LeRoith, D., (1997), Insulin-like growth factor 1 inhibits apoptosis using the phosphaditylinositol 3′-kinase and the mitogen-activated protein kinase pathways. J. Biol. Chem. **272**:154–161.

Powell-Braxton, L., Hollingshead, P., Warburton, C., Dowd, M., Pitts-Meek, S., Dalton, D., Gillet, N., and Stewart, T.A., (1993), IGF-1 is required for normal embryonic growth in mice. Genes Dev. **7**:2609–2617.

Ranco, T.A., and Blau, H.M., (1994), Primary mouse myoblast purification, characterization, and transplantation for cell-mediated gene therapy. J. Cell Biol. **125**:1275–1287.

Ranco, T.A., Pavlath, G.K., and Blau, H.M., (1995), The fate of myoblasts following transplantation into mature muscle. Exp. Cell Res. **220**:383–389.

Reiss, K., D'Ambrosio, C., Tu, X., Tu, C., and Baserga, R., (1998), Inhibition of tumor growth by a dominant negative mutant of the insulin-like growth factor 1 receptor with a bystander effect (1998). Clinical Cancer Res. **4**:2647–2655.

Resnicoff, M., Abraham, D., Yutanawiboonchai, W., Rotman, H.L., Kajstura, J., Rubin, R., Zoltick, P., and Baserga, R., (1995), The insulin-like growth factor 1 receptor protects tumor cells from apoptosis in vivo. Cancer Res. **55**:2463–2469.

Ribotta, M.G.Y., Revah, F., Pradier, L., Loquet, I., Mallet, J., and Privat, A., (1997), Prevention of motoneuron death by adenovirus-mediated neurotrophic factors. J. Neurosc Res. **48**:281–285.

Rudman, D.M., Kutner, M.H., Rogers, C.M., Lubin, M.F., Fleming, G.A., and Brain, R.P., (1981), Impaired growth hormone secretin in the adult population. J. Clin. Inv. **67**:1361–1369.

Sato, Y., Roman, M., Tighe, H., Lee, D., Corr, M., Nguyen, M., Silverman, G., Lotz, M., Carson, D., and Raz, E., (1996), Immunostimulatory DNA sequences necessary for effective intradermal gene immunization. Science **273**:352–354.

Schelechter, N.L., Russell, S.M., Spencer, E.M., and Nicoll, C.S., (1986), Evidence suggesting that the direct growth promoting effect of growth hormone on cartilage in vivo is mediated by local production of somatomedin. Proc. Natl. Acad. Sci. USA **83**:7932–7934.

Shemanko, C.S., Sanghera, J.S., Milner, R.E., Pelech, S., and Michalak, M., (1995), Phosphorylation of the carboxyl terminal region of dystrophin by mitogen-activated protein (MAP) kinase. Mol. Cell. Biochem. **152**:63–70.

Shively, J., and Beatty, J., (1985), Molecular biological and clinical significance. CRC Crit. Rev. Oncol. Hematol. **2**:355–399.

Skarli, M., Kiri, A., Vrbova, G., Lee, C.A., and Goldspink, G., (1998), Myosin regulatory elements as vectors for gene transfer by intramuscular injection. Gene Therapy **5**:514–520.

Snyder, R., Spratt, S., Lgarde, C., Bohl, D., Kasper, B., Sloan, B., Cohen, L.K., and Danos, O., (1997), Efficient and stable adeno-associated virus mediated transduction in the skeletal muscle of adult immunocompetent mice. Human Gene Therapy **8**:1891–1900.

Stratford-Perricaudet, L.D., Levrero, M., Chasse, J.F., Perricaudet, M., and Briand, P., (1990), Evaluation of the transfer and expression in mice of an enzyme-encoding gene using a human adenovirus vector. Hum. Gene Ther. **1**:241–256.

Sunada, Y., Bernier, S.M., Kozak, C.A., Yamada, Y., and Campbell, K.P., (1994), Deficiency of merosin in dystrophic mice and genetic linkage of laminin M chain gene to dy locus. J. Biol. Chem. **269**:13729–13732.

Svenson, E., Black, H., Dugger, D., Tripathy, S., Goldwasser, E., Hao, Z., Chu, L., and Leiden, J., (1997), Long term erythropoietin expression in rodents and non human primates following intramuscular injection of a replication defective adenoviral vector. Hum. Gene Ther. **8**:797–1806.

Tanaka, T., Manome, Y., Wen, P., Kuffe, D.W., and Fine, H.A., (1997), Viral vector-mediated transduction of a modified platelet factor 4. Cancer Research **58**:3362–3369.

Thompson, J., Grunert, F., and Zimmerman, W., (1991), CEA gene family: molecular biology and clinical perspectives. J. Clin. Lab. Anal. **5**:344–366.

Tighe, H., Corr, M., Roman, M., and Raz, E., (1997), Gene vaccination, plasmid DNA is more than just a blueprint. Immunol. Today **19**:89–97.

Tinsley, J.M., Potter, A.C., Phelps, S.R., Fisher, R., Tricket, J.I., and Davies, K.E., (1996), Amelioration of the dystrophic phenotype of mdx mice using a truncated utrophin transgene. Nature **384**:349–353.

Ullrich, A., Bell, J.R., Chen, E.Y., Herrera, R., Petruzzelli, L.M., Dull, T.J., Gray, A., Coussens, L., Liao, Y.C., Tsubokawa, M., Mason, A., Seeburg, P., Grunfield, C., Rosen, O.M., and Ramachandran, J., (1985), Human insulin receptor and its relationship to the tyrosine kinase family of oncogenes. Nature **313**:756–761.

Ullrich, A., Gray, A., Tam, A.W., Yang-Feng, T., Tsubokawa, M., Collins, C., Henzel, W., Le Bon, T., Kathuria, S., Chen, E., Jacobs, S., Francke, U., Ramachandran, J., and Fujita-Yamaguchi, Y., (1986), insulin-like growth factor I receptor primary structure: comparison with insulin receptor suggests determinants that define functional specificity. EMBO J. **5**:2503–2512.

Verma, I.M., and Dai, Y., (1994), Progress in gene therapy. Proc XXI Int Congr world federation of haemophilia, 15.

Wang, J.M., Zheng, H., Blaivas, M., and Kurachi, K., (1997), Persistent systemic production of human Factor IX in mice by skeletal myoblast-mediated gene transfer, feasibility of repeat application to obtain therapeutic levels. Blood **90**:1075–1082.

Wang, Y., and Becker, D., (1997), Antisense targeting of basic fibroblast growth factor receptor-1 in human melanomas blocks intratumoral angiogenesis and tumor growth. Nature Medicine **3**:887–893.

Wells, D.J., Maule, J., McMahon, J., Mitchell, R., Damien, E., Poole, A., and Wells, K.E., (1998), Evaluation of plasmid DNA for in vivo gene therapy, factors affecting the number of transfected fibers. J. Pharm. Sci. **87**:763–768.

Wells, D.J., and Goldspink, G., (1992), Age and sex influence expression of plasmid DNA directly injected into mouse skeletal muscle. FEBS Letts **306**:203–205.

Wells, K.E., Walsh, F.S., Goldspink, G., Love, D.R., Chan-Thomas, P., Dunckley, M., Piper, T., and Dickson, G., (1992), Human dystrophin expression corrects the myopathic phenotype in transgenic mdx mice. Hum. Mol. Genet. **1**:35–40.

Werner, H., Karnieli, E., Rauscher, III F.J., and LeRoith, D., (1996), Wild type and mutant p53 differentially regulate transcription of the insulin-like growth factor I receptor gene. Proc. Natl. Acad. Sci. USA **93**:8318–8323.

Wolff, J.A., Ludtke, J.L., Acsadi, G., Williams, P., and Jani, A., (1992), Long-term persistence of plasmid DNA and foreign gene expression in mouse muscle. Hum. Mol. Genet. **1**:363–369.

Wolff, J.A., Malone, R.W., Williams, P., Chong, W., Acasadi, G., Jani, A., and Felgner, P.L., (1990), Direct gene transfer into mouse muscle in vivo. Science **247**:1465–1468.

Wright, M.J., Rosenthal, E., Stewart, L., Withtman, L.M., Miller, A.D., Latchman, D.S., and Marber, M.S., (1998), beta-galactosidase staining following intracoronary infusion of cationic liposomes in the in vivo rabbit heart is produced by microinfarcion rather than effective gene transfer, a cautionary tale. Gene Therapy **5**:301–308.

Xiao, X., Li, J., and Samulski, R.J., (1996), Efficient long-term gene transfer into muscle tissue of immunecompetent mice by adeno-associated virus vector. J. Virol. **70**:8098–8108.

Xu, H., Christmas, P., Wu, X.P., Wewer, U.M., and Engvall, E., (1994), Defective muscle basement membrane and lack of M-laminin in the dystrophic dy/dy mouse. Proc. Natl. Acad. Sci. **91**:5572–5576.

Yang, S.Y., Alnaqeeb, M., Simpson, H., and Goldspink, G., (1996), Cloning and characterisation of an IGF-1 isoform expressed in skeletal muscle subjected to stretch. J. Muscle Res. Cell Motil. **17**:487–495.

Yang, Y., Ertl, H.C., and Wilson, J.M., (1994), MHC class I restricted cytotoxic T lymphocytes to viral antigens create barriers to lung directed gene therapy with recombinant adenoviruses. Immunity **1**:433–442.

Yang, Y., Nunes, F.A., Berensci, K., Furth, E.E., Gonzzol, E., and Wilson, J.M., (1994), Cellular immunity to viral antigens limits E1 deleted adenovirus for gene therapy. Proc. Natl. Acad. Sci. USA **91**:4407–4411.

Yao, S.N., and Kuracchi, K., (1992), Expression of human factor IX in mice after injection of genetically modified myoblasts. Proc. Natl. Acad. Sci. USA **89**:3357–3361.

Yao, S.N., Wilson, J.M., Nabel, E.G., Kurachi, K., Hachiya, H.L., and Kurachi, K., (1991), Expression of human factor IX in rat capillary endothelial cells, Toward somatic gene therapy for Haemophilia B. Proc. Natl. Acad. Sci. USA **88**:8101–8105.

Zdanowitz, M.M., Moyse, J., Wingert Zahn, M.A., O'Connor, M., Teichberg, S., and Slonim, A.E., (1995), Effect of insulin-like growth factor I in murine muscular dystrophy. Endocrinology **136**:4880–4886.

III

CELL CYCLE CONTROL

11

ADHESION MOLECULES IN CANCER BIOLOGY

Yaw Ohene-Abuakwa* and Massimo Pignatelli

Division of Investigative Science
Imperial College School of Medicine
Hammersmith Campus
Du Cane Road
London W12 ONN

INTRODUCTION

Cell-cell and cell-extracellular matrix interactions profoundly influence a variety of biological processes such as cell growth, differentiation, and migration. A large number of transmembrane glycoproteins receptors namely Cell Adhesion Molecules (CAMs) and Substratum Adhesion Molecules (SAMs) which mediate adhesion between cells and/or between cells and extracellular matrix protein respectively have been identified. Their functions in many physiopathological processes have been studied using monoclonal antibodies, cell-cell and cell-substratum adhesion assays. It has become apparent that these molecules are expressed early during development on most cells and are subsequently maintained in a number of adult tissues derived from the three germ layers (Edelman and Crossin, 1991). Evidence from many laboratories suggests that adhesion molecules play important role in signal transduction processes (Howe *et al.*, 1998), tumour progression and may promote tumour growth and organ-specific metastasis. Certain adhesion molecules may also function as tumour suppressor genes (Fawcett and Harris, 1992). It is now clear that many of the histopathological (e.g., loss of intercellular adhesion, derangement of cytological and architectural organisation) and behavioural features of neoplastic cells (e.g., invasion and metastasis) which constitute the basis of the classical morphological assessment may result from changes in adhesion molecule expression and/or function (Nigam *et al.*, 1994; Pignatelli and Bodmer, 1988).

Recent developments in cell and molecular biology techniques have led to the identification of five families of adhesion molecules which include the cadherins, integrins, immunoglobulin superfamily, selectins (Etter *et al.*, 1997) and CD44 (Tuszynski *et al.*, 1997). These receptors are genetically and biochemically distinct although, in some cases, are functionally related (Pignatelli, 1993b; Ruoslahti and Obrink, 1996).

---

* Tel/Fax: 0181 383 2432, e-mail: m.pignatelli@rpms.ac.uk

Cancer Gene Therapy: Past Achievements and Future Challenges, edited by Habib
Kluwer Academic/Plenum Publishers, New York, 2000.

In this chapter we describe recent data on the role of selected CAMs and SAMs in normal tissues and the evidence implicating them in key aspects of tumour biology including differentiation, invasion and metastasis.

INTEGRINS

Integrins are transmembrane glycoproteins comprising of 16 alpha and 9 beta subunits which form at least 22 different non-covalently αβ heterodimers involved in cell-cell and cell-matrix interactions. Originally, Hynes (1987) divided integrins into three main subfamilies (β1,β2,β3) capable of associating with specific groups of α subunits, but more recent work has shown that some α subunits can associate with more than one β subunit, making this classification more complex. Because of this complexity, it has proved useful to group the integrin families according to their ligand binding activity, which segregate them into three groups: those that function as cell-cell adhesion molecules (mainly found on leukocytes), those that bind primarily to the major constituents of the basement membrane (i.e. type IV collagen and laminin) and those that bind primarily to the extracellular matrix protein (ECM) found during early development, inflammation and wound healing (i.e. fibronectin, fibrinogen, vitronectin and thrombospodin) (Albelda, 1993; Charpin *et al.*, 1997). The majority of known integrins function as receptors for ECM proteins and remain the principal receptors that facilitate cellular interaction to the ECM through binding to specific sites in collagen, fibronectin and laminin (Hynes, 1992). Some of them have overlapping binding profiles, e.g. α6β4 binds laminin in colon carcinoma cells (Lotz *et al.*, 1990) whereas in keratinocytes it binds other basement membrane proteins (Marchisio *et al.*, 1993). Those mediating adhesion between cells, i.e. functioning as CAMs are α4β1 and lymphocyte function associated antigen 1 (LFA-1) which bind to members of the Ig superfamily, (vascular cell adhesion molecule, VCAM and intercellular adhesion molecule, ICAM respectively) expressed on other cell types (Elices *et al.*, 1990; Springer, 1990). β1 is the most promiscuous of all integrins subunits, participating in 10 different complexes. Integrins α1β1 and α2β1 are the major cell surface collagen receptors whose binding depends on the presence of Mg2+. α1β1 integrin may prefer type IV collagen over type I collagen, and α2β1 also binds laminin-1 and tenascin (For review see Heino, 1996). Several unique peptide epitopes have been identified as recognition sequences for integrin receptors, and the best characterised of these is the tripeptide arginine-glycine-asparate (RGD) (Gui *et al.*, 1997).

To investigate the role of integrin receptor molecules in neoplasia, knowledge of the in vivo distribution of these molecules is required both in normal and neoplastic tissues. Such studies will indicate those receptors that are more likely to be significant either because of under- or over-expression, and may reflect an important role in tumour progression. The availability of monoclonal antibodies against specific α and β subunits has made possible to determine the distribution of integrins of cultured cells and more importantly normal and neoplastic cells (Albelda, 1991; Albelda and Buck., 1990; Albelda *et al.* 1990b; Helmer, 1990). A number of investigators have compared integrin expression on normal and transformed cell lines to determine the importance of integrins in tumour invasion and metastasis. It has then been noted that transformed fibroblasts tend to express lower amount and altered distribution of integrins, especially fibronectin receptor (Akiyama *et al.*, 1990). The integrin repertoire of cultured cells does not always mimic that found in tumours in vivo therefore it is important to study their distribution in a variety of human malignancies at different stages in order to understand the role of integrins in tumour growth. A large number of studies have analysed the expression of integrins in breast (Pignatelli *et al.*, 1991; Zutter *et al.*, 1993), pancreas (Hall *et al.*, 1991), prostate (Bonkhoff *et al.*, 1993), stomach, colon and skin tumours (Koukoulis *et al.*, 1993; Nigam *et al.*, 1993; Pignatelli *et al.*, 1990a; Stamp and Pignatelli, 1991). For example, α3β1 is often expressed by carcinomas (Pignatelli *et al.*, 1990c; Pignatelli *et al.*, 1992), as well as the epithelial components of mixed epithelial-mesenchymal tumours (Miettinen *et al.*, 1993).

The expression of α2 subunits is progressively diminished or lost in a manner that correlates with loss of the differentiated epithelial phenotype in breast cancer. The contribution of altered α2β1 integrin expression to breast cancer progression was assessed by using a gain of function model where the re-expression of α2β1 integrin in poorly differentiated mammary carcinoma cell Mm5MT with no detectable α2 resulted in a dramatic conversion from a malignant phenotype to a differentiated epithelial phenotype (Hong *et al.*, 1998). Other studies have shown that the α6β4 integrin complex is highly expressed in mouse mammary tumour cells and in human carcinoma cell lines. This integrin is also found to be expressed in human and murine lung carcinomas and that its expression increases in highly metastatic murine tumour cells (Falconi *et al.*, 1988; Kennel *et al.*, 1989). However, it has been suggested that a complex of integrin receptors is required for the differentiated epithelial phenotype.

The alteration in the pattern of integrin expression is also well documented in colon cancers. A number of studies have reported a reduction of the integrin α3 and α5 subunits in adenomas and a diminished expression of α2, α6 and β4 in carcinomas, but with no changes observed for the β1 and αv subunits (Agrez and Bates, 1994; Basora *et al.*, 1998).

A relationship between the integrity of the basement membrane and the integrin expression has also been observed in some studies. It has been postulated that altered levels or profiles of integrins might be relevant in metastasis. For example, changes in cell-substratum and cell-cell adhesion are both required for cells to leave the main tumour mass and invade locally, enter endothelial lined channels, metastasise and establish themselves in distant sites. This hypothesis has been put to test using both *in vitro* and *in vivo* experiments. Transformation of fibroblast cells by Rous sarcoma virus is accompanied by a reduction in the expression of α5β1 integrin, a fibronectin receptor. It has been observed elsewhere that cells transfected with *ras* gene produced greater quantities of the type IV collagenase (Sleeg *et al.*, 1994) that can destroy the basement membrane complex. However, cells transfected with and overexpressing the fibronectin receptor α5β1 demonstrated reduced tumorigenicity in soft agar and in vivo when injected into nude mice (Giancotti and Ruoslahti, 1990; Varner and Cheresh, 1996), suggesting that this adhesion receptor may act as a tumour suppressor gene and its loss may be responsible for the acquisition of the malignant phenotype. (Akiyama *et al.*, 1990). Conversely, when human rhabdomyosarcoma cells were transfected with α2 cDNA, an increased in the number of metastatic colonies was observed in nude mice as compared with non -transfected controls which lacked functional α2β1 laminin/collagen receptor (Chan, 1991). These results suggest that the function of α2 integrin molecule may be cell type dependent and require the interaction with other regulatory molecules in mediating these multiple diverse biological effects.

Although most studies have described a reduction of integrin levels in many common human tumours, heterogeneous expression is often observed. Downregulation of the α6 integrin chain has also been correlated with melanoma progression, although in malignant melanoma the expression of α2β1 seems to be stronger than in melanoma *in situ*. Loss of or derangement of the normal co-localisation of α6β4 and collagen VII, a component of the hemidesmosomal anchoring complex, is seen in the majority of invasive bladder cancers, an alteration that may facilitate invasion and metastasis (Korman and Hrabovsky, 1993; Liebert *et al.*, 1994). Integrins are involved in the regulation of proliferation, differentiation and cell survival in a variety of cell types, events that are dependent upon tyrosine phosphorylation (Manie *et al.*, 1997). In colorectal cancer for instance, the $\alpha 2\beta 1$ receptor is down-regulated, especially in poorly differentiated tumours. This integrin is of fundamental importance in the formation of differentiated glandular structures by colon carcinoma cell lines grown in three dimensional collagen gels and its loss would theoretically explain progression to a less differentiated state (Liu *et al.*, 1994; Pignatelli and Bodmer, 1990).

Integrin-mediated cell adhesion impacts on another key aspect of growth, by regulating apoptosis, besides influencing the activity of the basal cell cycle machinery. It has recently become clear that loss of integrin-mediated cell-matrix contact can induce apoptosis (anoikis) in certain cell types (Howe *et al.*, 1998). Epithelial and endothelial cells have been shown to undergo apoptosis when displaced from their extracellular matrix (Ruoslahti and Reed, 1994), and this may prevent detached cells from re-attaching to new matrices and developing a malignant phenotype. Angiogenic endothelial cells which are particularly dependent on the $\alpha v\beta 3$ integrin *in vivo* can be induced to undergo apoptosis by treatment with anti-integrin molecules, on the other hand ligation of this integrin on the endothelial cells is said to suppress p53 activity, inhibit p21 expression and increase bcl-2:bax ratio to promote cell survival (Varner and Cheresh, 1996). It is still not completely understood the integrin-specific signalling machinery and its role in anoikis. What does seem clear, however, is that the ability of integrins to suppress or induce anoikis is cell type dependent (Frisch and Ruoslahti, 1997).

CADHERINS

Cadherin cell adhesion molecules are fundamental determinants of tissue organisation in developing and adult organisms. They are transmembrane glycoproteins mediating cell-cell adhesion via calcium dependent, homotypic (cadherin to cadherin) interactions. Classical cadherins families are E- (epithelial), N- (neural), P- (placental), and *Xenopus* C-cadherin (Takeichi, 1991; Yap *et al.*, 1998). These are structurally and functionally related molecules which require the presence of calcium for their binding, in the absence of which, they are degraded by protease action (Overduin *et al.*, 1995). E-cadherin, the most important member, also known as L-CAM, uvomorulin, and Arc-1, has been mapped on chromosome 16q22.1, and it encodes a 120 kDa transmembrane protein. N-cadherin is expressed in neuroectodermal and mesodermal tissues (Hatta and Takeichi, 1987), and has also been mapped on chromosome 18q, (Walsh *et al.*, 1990). P-cadherin was originally described in mouse placenta, but has also been found in epithelial tissues including lung epithelia, basal cells of the skin and myoepithelial cells of the mammary gland (Soler *et al.*, 1997).

Although cadherins are considered tumour suppressor genes and defects in their expression or function have been associated with tumour progression (Hedrick *et al.*, 1993), it has been found that the expression of cadherins in tumour cells can be used to trace the histogenetic origin of tumours, to serve as differential diagnostic marker between tumours of similar phenotype but different histogenesis (Soler *et al.*, 1995). *In vitro* the selective loss of E-cadherin expression or function is associated with changes in cellular phenotype and with the development of invasive behaviour of tumour cells, an effect which can be reversed by transfection of E-cadherin cDNA (Mayer *et al.*, 1993). Several efforts have been made to understand the mechanisms by which E-cadherin expression is modulated, and the possibilities include transcriptional down-regulation, methylation of the E-cadherin promoter region, and inactivation by mutation (Hariguri *et al.*, 1998). There is enough evidence that down-regulation of E-cadherin can occur during tumourigenesis. This has been shown in carcinogen-induced skin tumours in mice, and also results from immunohistochemical investigations indicate decreased E-cadherin expression in various human carcinomas *in vivo*.

Loss of heterozygosity on the E-cadherin locus, i.e. chromosome 16q, is frequently found in many human cancers including breast (Sato *et al.*, 1991), prostate (Carter *et al.*, 1990) and liver (Slagle *et al.*, 1993), suggesting that the E-cadherin gene has been inactivated by two hits (mutation and allele loss) similarly to the inactivation of tumour suppressor genes. Decreased levels of the E-cadherin protein expression have frequently been observed in a wide range of human tumours including squamous cell carcinoma of the head and neck (Schipper *et al.*, 1991), pancreas (Ansari *et al.*, 1994), bladder (Bringuier *et al.*, 1993), colon and rectum (Pignatelli *et al.*, 1993). In diffuse type human gastric carcinoma, mutation of the E-cadherin gene which may destroy the putative calcium binding site of the molecule has also been detected in about 50% cases. Reduced or loss of expression of E-cadherin protein has been found in a higher percentage of early gastric cancer than in advanced intestinal type gastric cancer, suggesting that the down-regulation of E-cadherin may be an early event, (Pignatelli *et al.*, unpublished). Transcriptional inactivation of E-cadherin expression was shown to occur frequently in the progression of some tumour types, (Hiroshashi, 1998; Paul *et al.*, 1998). In epithelial ovarian cancers for instance, the expression of E-cadherin was found to be reduced or absent in poorly differentiated tumours, but not in cystadenomas (Davies *et al.*, 1998). In cervical intraepithelial neoplasia (CIN) it has been shown that the distribution of elevated and abnormal localisation of E-cadherin also correlates with the grade of CIN lesion and may help to define their invasive potential (Vessey *et al.*, 1995). Besides its ability to suppress both proliferation and invasion in carcinomas, E-cadherin-mediated adhesion has also been found to be maintain cell survival and prevent apoptosis (Hermiston and Gordon, 1995; St. Croix *et al.*, 1998).

Cadherin function is partly regulated through the cytoplasmic domain by a group of cytoskeletal proteins termed the catenins (α-, β-, and γ-) (Gumbiner and McCrea, 1993). α-catenin is believed to be important in linking E-cadherin to the actin cytoskeleton by binding directly to either β- or γ-catenin. Analysis of different cancer cell lines which show loose cell-cell association has revealed the expression of aberrant mRNA of α-catenin (Oyama *et al.*, 1994) β-catenin is involved in two major functions: cell adhesion and the Wnt-1 signal transduction pathway in vertebrates. Activation of the Wnt-1 signal pathway results in the inhibition of glycogen synthase kinase-3β, and a lack of phosphorylation of β-catenin, leading accumulation of the protein in the cytoplasm and nucleus due to lack of degradation by the proteasome system (Wong *et al.*, 1998). β-catenin level is also controlled by the APC gene, therefore malfunctioning of this gene can lead to upregulation and nuclear localisation of β-catenin (De La Coste *et al.*, 1998; Valizadeh *et al.*, 1997). Increased cytoplasmic accumulation and nuclear localisation of γ-catenin has also been reported in many human malignancies including colorectal carcinoma (Valizadeh *et al.*, 1997).

IMMUNOGLOBULIN (Ig) SUPERFAMILY

This is a group of molecules which contain immunoglobulin-like domains and in some cases also contain membrane proximal fibronectin type III repeats that share evolutionary homology and function primarily in recognition and binding processes on cell surfaces. They include proteins that mediate endothelial cell interactions with lymphocytes, monocytes, neutrophils and tumour cells that result in adhesion and transendothelial migration. (Beckner *et al.*, 1995). Among this group is the oncofetal protein, carcinoembryonic antigen (CEA), a 180 kDa glycosylated cell surface protein. Transfection and blocking experiments with monoclonal antibodies showed that CEA mediate the calcium independent, homotypic aggregation of colon carcinoma cell lines, and may also act as an accessory adhesion molecule mediating cell-matrix binding (Landuzzi *et al.*, 1996; Pignatelli *et al.*, 1990a). It has been suggested that CEA may also function as an anti-adhesive molecule since colorectal cells at the edge of an invading tumour express higher levels, about 2.6 times of CEA mRNA, than found in the luminal surface (Jothy *et al.*, 1993) Other Ig-related molecules also include CD2 (sheep erythrocyte receptor) and its ligand CD58 (LFA-3), neural cell-adhesion molecule CD56 (NCAM), and vascular adhesion molecule VCAM-1, CD54 (ICAM-1), ICAM-2 and ICAM-3. Although a role in tumour invasion has been postulated for this group of adhesion molecules, the divergent nature of individual members means there is no clear pattern emerging. N-CAM has been shown to have homology with the gene product encoded by the DCC gene on chromosome 18q (Fearon *et al.*, 1990) which is frequently altered in colorectal and other cancers including prostate, breast, pancreatic and bladder (Wu *et al.*, 1991; Hohne *et al.*, 1992). It has been proposed that loss of this gene could lead to an abnormality of cell-cell adhesion predisposing to a malignant phenotype. Evidence for the possible mechanism of action of the N-CAM molecule has been suggested by report which showed that N-CAM was in fact involved in contact-dependent inhibition of growth of a mouse fibroblast cell line. Immunohistochemical evaluation of N-CAM expression showed varying degree of positivity in tumours of neural and mesenchymal origin (Edelman and Crossin, 1991; Garin-Chesa *et al.*, 1991) and be negligible in carcinomas.

Other members of the Ig superfamily which may play a part in tumour cell interactions are V-CAM (vascular cell adhesion molecule) and I-CAM-1 (intercellular adhesion molecule 1). V-CAM is a cytokine-inducible cell adhesion molecule primarily found on vascular endothelial cells and serves as a counter-receptor for $\alpha 4\beta 1$ integrin expressed on leukocytes as well as on malignant melanoma (Albelda *et al.*, 1990), suggesting that V-CAM might function as an endothelial tumour adhesion receptor for malignant melanomas and myeloid malignancies expressing $\alpha 4\beta 1$. Overexpression of I-CAM-1 has been shown to be correlated with melanoma progression and increased risk of metastasis (Johnson *et al.*, 1989).

SELECTINS

Selectins are transmembrane glycoproteins composed of a single Ca^{2+} dependent N-terminal C-type lectin domain and mediate heterotypic interactions between blood cells and endothelial cells during lymphocyte homing and leukocyte adhesion. The lectin domain binds to specific carbohydrate groups such as the sialylated Lewis$_x$ antigen, that are fucosylated and sialylated (Hynes and Lander, 1992). Three selectins are known; E-selectin, P-selectin and L-selectin. These are closely related proteins, but have distinct roles and patterns of expression. E-selectin (also known as ELAM-1), is found principally on endothelium after cytokine activation and mediates adhesion of a memory T cell subset and myeloid cells. Colorectal carcinoma cell lines which express the sialylated Lewis$_x$ antigen are able to bind to cytokine-activated endothelial cells (Matsushita *et al.*,

1990), a binding inhibited by E-selectin antibodies (Lauri *et al.*, 1991). P-selectin (also known as GMP-140 or PADGEM) is also expressed on endothelia and platelets, and may provide a binding site for cells expressing the appropriate carbohydrate ligand. L-selectin is expressed constitutively on all classes of white cells and is involved in the binding of lymphocytes to high endothelial venules (HEV) of lymph nodes. Two HEV-associated selectin ligands of approximately 50kDa and 90kDa have been identified (Imai *et al.*, 1992). The interaction of L-selectin with these ligands is also carbohydrate directed. Expression of these adhesion molecules and the interaction with their ligands may underlie the mechanism whereby tumour cells are localised to certain organs, a hypothesis known as the "seed and soil" (Paget, 1989)

CD44

CD44, also known as the Hermes antigen, H-CAM, pgp-1 antigen is a widely distributed integral membrane glycoprotein expressed by a variety of normal and neoplastic cell types (Albelda, 1993). It was first discovered in haematopoietic cells and was shown to facilitate lymphocyte homing and T-cell activation (Tuszynski *et al.*, 1997). Furthermore, CD44 plays a role in cell-cell and cell-matrix interaction, and it is the primary cell-surface receptor for hyaluronan, a proteoglycan found in abundance in the extracellular matrix (Aruffo *et al.*, 1990). CD44 exists as multiple isoforms encoded by a single gene found on the short arm of chromosome 11. Genomic and cDNA cloning data have demonstrated that the diversity of CD44 isoform is generated by alternative splicing. Splicing of CD44 gene occurs at two different regions, the membrane proximal extracellular domain and the cytoplasmic carboxyl tail region. At least 20 different CD44 isoforms have been described (Matsumura and Tarin, 1992). The 90kDa isoform, CD44H, is broadly distributed in haematopoietic cells, fibroblasts, and numerous tumours of both mesenchymal and neuroectodermal origin (Stamenkovic *et al.*, 1989). The expression of the 150kDa isoform, CD44E, appears restricted to subsets of epithelial cells, although keratinocytes express another isoform of 230kDa molecular weight (Stamenkovic *et al.*, 1991).

CD44 has been implicated in tumour cell invasion and in the metastatic process. The pattern of expression of this receptor in tumorigenic cell lines is variable. Breast and colonic carcinomas and their metastases showed amplified expression of at least nine alternative spliced products of the CD44 gene in contrast to non-neoplastic tissues (Matsumura and Tarin, 1992). Strong CD44H expression has been associated with poor outcome of breast ductal carcinoma, non-Hodgkin's lymphoma (Jalkanen *et al.*, 1991; Joensuu *et al.*, 1993) and ovarian cancer (Cannistra *et al.*, 1993). Recent data have also shown that variant isoforms of CD44, CD44v, are expressed by various malignant cells. CD44v expression has been described in human cancers such as melanomas, lung cancers and brain metastases (Hofmann *et al.*, 1991). The expression of CD44v in colorectal carcinoma is increased in advanced tumours and may correlate with increasing metastatic potential (Wielenga *et al.*, 1993).

For the metastasising tumour cells to move through the extracellular matrix means the tumour cell must be able to recognise components of ECM. Hyaluronan makes up a significant component of the basement membrane and attachment to this proteoglycan is augmented by CD44H. Degradation of hyaluronan has been implicated as a necessary step for angiogenesis to occur, permitting tumour cells to migrate into a vascularised bed (Bartolazzi *et al.*, 1994). As mentioned above, one function of CD44 receptor is related to lymphocyte homing. Therefore it has been speculated that metastatic cells might be mimicking the adhesive and transmigratory properties of lymphocytes in order to promote lymphatic metastases. Gunthert *et al.*, (1991) demonstrated that transfection of CD44v metastatic variant construct into a non- metastatic rat pancreatic carcinoma cell line confers a metastatic potential. This has recently been confirmed by other investigators. Li *et al.*, (1993) demonstrated that whilst CD44H is found in glioblastomas and glioma cell lines, cells with low metastatic potential, CD44v expression was not detected. However, brain metastases showed high incidence of CD44v isoform. The role of CD44 in metastasis has further been elaborated by the isolation of cell lines that overexpress CD44. Melanoma cell lines which are CD44 positive were found to metastasise to the lung in contract to CD44 negative or low expressor cell lines (Birch *et al.*, 1991). A human homologue of this metastasis-associated variant has been found to be expressed in all invasive colorectal carcinomas and their metastases. (REF)

Much debate exists concerning the importance of CD44H versus CD44v, in tumour metastasis, but since tumour transformation is a dynamic process, several CD44 isoforms may contribute to tumour cell extravasation, dissemination and growth. Furthermore, the ability of a tumour to metastasise by haematogenous or lymphatic spread may depend at least in part on which CD44 isoform is expressed on the tumour cell surface

CONCLUSION

Changes in the expression and/or function of SAMs and CAMs have now been identified in transformed cells in vitro as well as in vivo. These changes may explain at least in part the phenotype and behaviour of malignant tumours. Differences between expression of adhesion molecules in tumours have been noted in many studies and this may reflect the small size and mixed-case pathological profile and the antibody specificities. However, the overwhelming interest in investigating the molecular control of adhesion molecule function, localisation and level of cell surface expression hopefully will lead to a better understanding of their clinical relevance in the histological assessment of diagnosis and prognosis, and to the development of new therapeutic approaches based on their genetic and functional modulation.

REFERENCES

Agrez, M.V., and Bates, R.C., 1994, Colorectal cancer and the integrin family of cell adhesion receptors: current status and future directions, *Euro. J. Cancer*, **30A**, 2166–2170.

Akiyama, S.K., Larjava, H., and Yamada, K.M., 1990, Differences in the biosynthesis and differentiation of the fibronectin receptor in normal and transformed cultured human cells, *Cancer Res.*, **50**, 1601–1607.

Albelda, S.A., and Buck, C.A., 1990, Integrins and other cell adhesion molecules, *FASEB J.*, **4**, 2868–2880.

Albelda, S.M., 1991, Endothelial and epithelial cell adhesion molecules, *Am J Resp Cell Biol.*, **4**, 195–203.

Albelda, S.M., 1993, Role of Integrins and other cell adhesion molecules in tumour progression and metastasis, *Laboratory Investigation* **68**, 4–17.

Albelda, S.M., Mette, S.A., and Elder, D.E., 1990, Integrin distribution in malignant melanoma: Association of the β3 subunit with tumour progression, *Cancer Res.*, **50**, 6757–6764.

Ansari, T., Lemoine, N., Gunter, P., Liu, D., Hirano, S., Takeichi, M., Kloppel, G., and Pignatelli, M., 1994, Decreased E-cadherin expression correlates with lymph node metastasis, high grade and advanced stage in pancreatic cancer, *J. Pathol.*, **172**, A12.

Aruffo, A., Stamenkovic, I., Melnick, M., Underhill, C.B., and Seed, B., 1990, CD44 is the principal cell surface receptor for hyaluronate, *Cell*, **61**, 1303–1313.

Basora, N., Desloges, N., Chang, Q., Bouatrouss, Y., Gosselin, J., Poisson, J., Sheppard, D., and Beaulieu, J.-F., 1998, Expression of the α9β1 integrin in human colonic epithelial cells: Resurgence of the foetal phenotype in a subset of colonic cancers and adenocarcinoma cell lines, *Int. J. Cancer*, **75**, 738–743.

Bartolazzi, A., Peach, R., Aruffo, A., and Stamenkovic, I., 1994, Interaction between CD44 and hyaluronate is directly implicated in the regulation of tumour development, *J. Exp. Med.*, **180**, 53–66.

Beckner, M.E., Krutzsch, H.C., Stracke, M.L., Williams, S.T., Gallardo, J.A., and Liotta, L.A., 1995, Identification of a new immunoglobulin superfamily protein expression in blood vessels with a heparin-binding consensus sequence, *Cancer Res.*, **55**, 2140–2149.

Birch, M., Mitchell, S., and Hart, I.R., 1991, Isolation and characterisation of human melanoma cell variants expression high and low levels of CD44, *Cancer Res.*, **51**, 6660–6667.

Bonkhoff, H., Stein, U., and Remberger, K., 1993, Differential expression of α6 and α2 very late antigen integrins in the normal hyperplastic and neoplastic prostate, *Hum. Pathol.*, **24**, 243–248.

Bringuier, P.P., Umbas, R., Schaafsma, H.W., Karthaus, H.F.M., Debruyne, F.M.J., and Schaken, J.A., 1993, Decreased E-cadherin immunoreactivity correlates with poor survival in patients with bladder tumours, *Cancer Res.*, **53**, 3241–3245.

Cannistra, S.A., Kansas, G.S., Niloff, J., DeFranzo, B., Kim, Y., and Ottensmeier, C., 1993, Binding of ovarian cancer cells to peritoneal mesothelium in vitro is partly mediated by CD44H, *Cancer Res.*, **53**, 3830–3838.

Carter, B.S., Eing, C.M., Ward, W.S., Treiger, B.F., Aalders, T.W., Schalken, J.A., Epstein, J.I., and Isaacs, W.B., 1990, Allelic loss of chromosome 16q and 10q in human prostate cancer, *Prco. Natl. Acad. Sci.*, **87**, 8751–8755.

Chan, B.M., 1991, *In vitro* and *in vivo* consequences of VLA-2 expression on rhabdomyosarcoma cells, *Science*, **251**, 1600–1602.

Davies, B.R., Worsley, S.D., and Ponder, B.A., 1998, Expression of E-cadherin, α-catenin and β-catenin in normal ovarian surface epithelial and epithelial ovarian cancer, *Histopathology*, **32**, 69–80.

De La Coste, A., Romagnolo, B., Billuart, P., Renard, C.-A., Buendia, M.-A., Soubrane, O., Fabre, M., Chelly, J., Beldjord, C., Khan, A., and Perret, C., 1998, Somatic mutation of the β-catenin gene are frequent in mouse and human hepatocellular carcinomas, *Proc. Natl. Acad. Sci. USA*, **95**, 8847–8851.

Edelman, G.M., and Crossin, K.L., 1991, Cell adhesion molecules: implications for molecular histology, *Annu. Rev. Biochem.*, **60**, 155–190.

Elices, M.J., Osborn, L., Takada, Y., *et al.*, 1990, VCAM-1 on activated endothelium interacts with the leukocytes integrin VLA-4 at a site distinct from the VLA-4/fibronectin binding site, *Cell*, **60**, 577–584.

Falconi, R., Sacchi, A., Reseau, J., and Kennel, S.J., 1988, Monoclonal antibody to human carcinoma-associated protein complex: quantitation in normal and tumour tissue, *Cancer Res.*, **48**, 816–821.

Fearon, E.R., Cho, K.R., Nigro, J.M., Kern, S.E., Simmons, J.W., Ruppert, J.M., Hamilton, S.R., Preisinger, A.C., Thomas, G., Kinzler, K.W., and Volgestein, B., 1990, Identification of chromosome 18q gene that is altered in colorectal cancers, *Science*, **247**, 49–56.

Frisch, S.M., and Ruoslahti, E., 1997, Integrins and anoikis, *Curr. Opin. in Cell Biol.*, **9**, 701–706.

Garin-Chesa, P., Fellinger, E.J., Huvos, A.G., Beresford, H.R., Melamed, M.R., Triche, T.J., and Rettig, W.J., 1991, Immunohistochemical analysis of neural cell adhesion molecules, *Am. J. Pathol.*, **139**, 275–286.

Giancotti, F.G., and Ruoslahti, E., 1990, Elevated levels of the α5β1 fibronectin receptor suppress the transformed phenotype of Chinese hamster ovary cells, *Cell*, **60**, 849–859.

Gui, G.P.H., Puddefoot, J.R., Vinson, G.P., Wells, C.A., and Carpenter, R., 1997, Altered cell-matrix contact: a prerequisite for breast cancer metastasis? *Br. J. Cancer*, **75**, 623–633

Gumbiner, B.M., and McCrea, P.D., 1993, Catenins as mediators of the cytoplasmic functions of cadherins, *J. Cell Sci. Suppl*, **17**, 155–158.

Gunthert, U., Hofmann, M., Rudy, W., Reber, S., Zoller, M., Haubmann, I., Matzku, S., Wenbel, A., Ponta, H., and Herrlich, A., 1991, A new variant of glycoprotein CD44 confers metastatic potential to rat carcinoma cells, *Cell*, **65**, 13–24.

Hall, P.A., Coates, P., Lemoine, N.R., and Horton, M.A., 1991, Characterisation of integrin chains in normal and neoplastic human pancreas, *J. Pathol.*, **165**, 33–41.

Hatta, K., and Takeichi, M., 1987, Expression of N-cadherin adhesion molecules associated with early morphogenic events in chick development, *Nature*, **320**, 447–449.

Hedrick, L., Cho, K.R., and Volgelstein, B., 1995, Cell adhesion molecules as tumour suppressors, *Trends Cell Biol.*, **3**, 36–39.

Heino, J., 1996, Biology of tumour cell invasion: interplay of cell adhesion and matrix degradation, *Int. J. Cancer*, **65**, 717–722.

Helmer, M.E., 1990, VLA proteins in the integrin family: Structures, functions and their role on leukocytes, *Annu. Rev. Immunol.*, **8**, 549–555.

Hermiston, M.L., and Gordon, J.I., 1995, In vivo analysis of cadherin function in the mouse intestinal epithelium: essential roles in adhesion, maintenance of differentiation and regulation of programme cell death, *J. Cell Biol.*, **129**, 489–506.

Hiraguri, S., Godfrey, T., Nakamura, H., Graff, J., Collins, C., Shayesteh, L., Doggett, N., Johnson, K., Wheelock, M., Herman, J., Baylin, S., Pinkel, D., and Gray, J., 1998, Mecahinsm of inactivation of E-cadherin in breast cancer cell lines, *Cancer Res.*, **58**, 1972–1977.

Hiroshashi, S., 1998, Inactivation of the E-cadherin mediated cell adhesion system in human cancers, *Am. J. Pathol.*, **153**, 333–339.

Hofmann, M., Rudy, W., Zoller, M., Tolf, C., Ponta, H., Herrlich, P., and Gunthert, U., 1991, CD44 splice variant confer metastatic behaviour in rats: Homologous sequences are expressed in human tumour cell lines, *Cancer Res.*, **51**, 5292–5297.

Hohne, M.W., Halatsch, M.E., Kahl, G.F., and Weinel, R.J., 1992, Frequent loss of expression of potential tumour suppressor gene DCC in ductal pancreatic adenocarcinoma, *Cancer Res.*, **52**, 2616–2619.

Hong, S., Santoro, S.A., and Zutter, M.M., 1998, Downstream events in mammary gland morphogenesis mediated by re-expression of the α2β1 integrin: The role of α6 and β4 integrin subunits, *Cancer Res.*, **58**, 2224–2233.

Howe, A., Aplin, A.E., Alahari, S.K., and Juliano, R.L., 1998, Integrin signaling and cell growth control, *Curr. Opin. in Cell Biol.*, **10**, 220–231.

Hynes, R.O., 1992, Integrins: versatility, modulation and signalling in cell adhesion, *Cell*, **69**, 11–25.

Hynes, R.O., and Lander, A.D., 1992, Contact and adhesive specificity in the association, migration and targeting of cells and axons, *Cell*, **68**, 303–322.

Imai, Y., Lasky, L.A., and Rosen, S.D., 1992, Further characterisation of the interaction between L-selectin and its endothelial ligands, *Glycobiology*, **2**, 373–381.

Jalkenan, S., Joensuu, H., Soderstorm, K.O., and Klemi, P.J., 1991, Lymphocyte homing and clinical behaviour of non-Hodgkin's lymphoma, *J. Clin. Invest.*, **87**, 1835–1840.

Joensuu, H., Klemi, P.J., Toikkanen, S., and Jalkenan, S., 1993, Glycoprotein CD44 expression and its association with survival in breast cancer, *Am. J. Pathol.*, **143**, 867–874.

Johnson, J.P., Stade, B.G., Holzmann, B., Schwable, W., and Riethmuller, G., 1989, *De novo* expresion of intercellular-adhesion molecule in melanoma correlates with increased risk of metastasis, *Proc. Natl. Acad. Sci. USA*, **86**, 641–644.

Jones, J.L., Royall, J.E., Critchley, D.R., and Walker, R.A., 1997, Modulation of myoepithelial-associated α6β4 integrin in a breast cancer cell line alters invasive potential, *Expt. Cell Res.*, **235**, 325–333.

Jothy, S., Yuan, S.-Y., and Shirota, K., 1993, Transcription of carcinoembryonic antigen in normal colon and colon carcinoma: in situ hybridisation study and the implication for a new in vivo functional model, *Am. J. Pathol.*, **143**, 250–257.

Kennel, S.J., Foote, L.J., Falconi, R., Sonnenberg, A., Crouse, C., and Helmer, M.E., 1989, Analysis of the tumour associated antigen TSP-180: identity with α6β4 in the integrin superfamily, *J. Biol. Chem.*, **264**, 15515–15521.

Korman, N.J., and Hrabovsky, S.L., 1993, basal cell carcinoma display extensive abnormalities in the hemidesmosome anchoring fibril complex, *Exp. Dematol.*, **2**, 139–144.

Koukoulis, G.K., Virtanen, I., Moll, R., Quaranta, V., and Gould, V.E., 1993, Immunolocalisation of integrins in the normal and neoplastic colonic epithelium, *Virchows Archiv. B. Cell Pathol.*, **63**, 373–383.

Landuzzi, L., Frabetti, F., Rossi, I., Griffoni, C., De Giovanni, C., Nicoletti, G., Nanni, P., Miniero, R., Palmieri, G., Santoni, A., and Lollini, P.-L., 1996, Expression of transduced carcinoembryonic antigen gene in human rhabdomyosarcoma inhibits metastasis, *Cancer Res.*, **56**, 4503–4508.

Lauri, D., Needham, L., Martin-Padura, I., and Dejana, E., 1991, Tumour cell adhesion to endothelial cells: Endothelial leukocyte adhesion molecule-1 as an inducible adhesive receptor specific for colon carcinoma cells, *J. Natl. Cancer Inst.*, **83**, 1321–1324.

Li, H., Hamou, M., Tibolet, N., Jaufeerally, R., Hofmann, M., Diserens, A., and Van Meir, E.G., 1993, Variant CD44 adhesion molecules are expressed in human brain metastases but not in glioblastomas, *Cancer Res.*, **53**, 5345–5349.

Liebert, M., Washington, R., Wedemeyer, G., *et al.*, 1994, Loss of co-localisation of α6β4 integrin VII in bladder cancer, *Am. J. Pathol.*, **144**, 787–795.

Liu, D., Gagliardi, G., Nasim, M.M., *et al.*, 1994, Transforming growth factor alpha can act as morphogen and/or mitogen in colorectal carcinoma cell line, *Int. J. Cancer*, **56**, 603–608.

Lotz, M.L., Korzelius, C.A., and Mercurio, A.M., 1990, Human colon carcinoma cells use multiple receptors to adhere to laminin: involvement of α6β4 and α2β1 integrins, *Cell Regul.*, **1**, 249–257.

Manie, S.N., Astier, A., Haghayeghi, N., Canty, T., Druker, B.J., Hirai, H., and Freedman, A.S., 1997, Regulation of integrin-mediated $p130^{cas}$ Tyrosine phosphorylation in human B cells, *J. Biol. Chem.*, **272**, 15636–15641.

Marchisio, P.C., Cremona, O., Savoia, P., *et al.*, 1993, The basement membrane protein BM-600/nicein codistributes with kalinin and the integrin α6β4 in human cultured keratinocytes, *Exp. Cell Res.*, **205**, 205–212.

Matsamura, Y., and Tarin, D., 1992, Significance of CD44gene products for cancer diagnosis and disease evaluation, *Lancet*, **340**, 1053–1058.

Matsushita, Y., Cleary, K.R., Ota, D.A., Hoff, S.D., and Irimura, T., 1990, Sialyldomeric lewis-X antigen expressed on mucin-like glycoprotein in colorectal cancer matastases, *Lab. Invest.*, **63**, 780–791.

Mayer, B., Johnson, J.P., Leitl, F., Jauch, K.W., Heiss, M.M., Schilberg, F.W., Birchmeier, W., and Funke, I., 1993, E-cadherin expression in primary and metastatic gastric cancer: down-regulation correlates with cellular dedifferentiation and glandular disintegration, *Cancer Res.*, **53**, 1690–1695.

Nigam, A.K., Pignatelli, M., and Boulos, P.B., 1994, Current concepts in metastasis, *Gut*, **35**, 996–1000.

Nigam, A.K., Savage, F.J., Boulos, P.S., Stamp, G.W.H., Liu, D., and Pignatelli, M., 1993, Loss of cell-cell and cell-matrix adhesion molecules in colorectal cancer, *Br. J. Cancer*, **68**, 507–514.

Oyama, T., Kanai, Y., Ochiai, A., Akimoto, S., Oda, T., Yanagihara, K., Nagafuchi, A., Tsukita, S., Shibamoto, S., Ito, F., Takeichi, M., Matsuda, H., and Hirohashi, S., 1994, A truncated β-catenin disrupts the interaction between E-cadherin and α-catenin: Acause of loss of intercellular adhesiveness in human cancer cell lines, *Cancer Res.*, **54**, 6282–6287.

Paget, S., 1989, The distribution of secondary growths in cancer of the breast, *Lancet*, **I**, 571–573.

Paul, R., Ewing, C.M., Jarrard, D.F., and Isaacs, W.B., 1998, Mechanisms of inactivation of the cadherin-catenin pathway include LOH, gene deletion and gene promoter hypermethylation, *Br. J. Urol. Suppl*, **1**, 37–43.

Pignatelli, M., 1993, E-cadherin: a biological marker of tumour differentiation, *J. Pathol.*, **171**, 81–82.

Pignatelli, M., and Bodmer, W.F., 1988, Genetics and biochemistry of collagen binding triggered glandular differentiation in human colon carcinoma cell lines, *Proc. Natl. Acad. Sci. USA*, **85**, 5561–5565.

Pignatelli, M., and Bodmer, W.F., 1990, Integrin cell adhesion molecules in colorectal cancer, *J. Pathol.*, **162**, 95–97.

Pignatelli, M., Durbin, H., and Bodmer, W.F., 1990a, Carcinoembryonic antigen functions as an accessory adhesion molecule mediating colon epithelial cell-collagen interactions, *Proc. Natl. Acad. Sci. USA*, **87**, 1541–1545.

Pignatelli, M., Hanby, A.M., and Stamp, G.W.H., 1991, Low expression of β1, α2 and α3 subunits of VLA integrins in malignant mammary tumours, *J. Pathol.*, **165**, 25–32.

Pignatelli, M., Smith, M.E.F., and Bodmer, W.F., 1990b, Low expression of collagen receptors in moderate and poorly differentiated colorectal adenocarcinomas, *Br. J. Cancer*, **61**, 636–638.

Ruoslahti, E., and Reed, J., 1994, Anchorage independence, integrins and apoptosis, *Cell*, **77**, 477–478.

Sato, T., Akiyama, F., Sakamoto, G., Kasumi, F., and Nakamura, Y., 1991, Accumulation of genetic alterations and progression of primary breast cancer, *Cancer Res.*, **51**, 5794–5799.

Schipper, J.H., Frixen, U.H., Behrens, J., Unger, A., Jahnke, K., and Birchmeier, W., 1991, E-cadherin expression in squamous cell carcinomas of the head and neck: inverse correlation with tumour dedifferentiation and lymph node metastasis, *Cancer Res.*, **51**, 6328–6337.

Slagle, B.L., Zhou, Y.-Z., Birchmeier, W., Scorsone, K.A., 1993, Deletion of the E-cadherin gene in hepatitis virus-posirive Chinese hepatocellular carcinomas, *Hepatology*, **18**, 757–762.

Sleeg, P., Bevilacqua, G., Sobel, M., and Liotta, L., 1994, Invasion and metastasis in: molecular genetics in cancer diagnosis, pp 49–53, Ed., Jeffrey Cossman, *Elsevier*.

Soler, A.P., Harner, G.D., Knudsen, K.A., McBrearty, F.X., Grujic, E., Salazar, H., Han, A.C., and Keshgegian, A.A., 1997, Expression of P-cadherin identifies prostate specific-antigen-negative cells in epithelial tissues of male sexual accessory organs and in prostatic carcinomas, *Am. J. Path.*, **151**, 471–478.

Soler, A.P., Knudsen, K.A., Tecson-Miguel, A., McBrearty, F.X., Han, A.C., Salazar, H., 1995, The differential expression of N-cadherin and E-cadherin distinguishes pleural mesotheliomas from lung adenocarcinomas, *Hum. Path.*, **26**, 1363–1369.

Springer, T.A., 1990, Adhesion receptors of the immune system, *Nature*, **346**, 425–435.

St. Croix, B., Sheehan, C., Rak, J.W., Florense, V.A., Slingerland, J.M., and Kerbel, R.S., 1998, E-cadherin-dependent growth suppression is mediated by cyclin-dependent kinase inhibitor $p27^{KIP1}$, *J. Cell Biol.*, **142**, 557–571.

Stamenkovic, I., Amiot, M., Pesando, J.M., and Seed, B., 1989, A lymphocyte molecule implicated in lymph node homing is a member of the cartilage link protein family, *Cell*, **56**, 1057–1062.

Stamenkovic, I., Aruffo, A., Amiot, M., and Seed, B., 1991, The hematopoietic and epithelial forms of CD44 are distinct polypeptydes with different adhesion potentials for hyaluronate bearing cells, *EMBO J.*, **10**, 343–348.

Stamp, G.W.H., and Pignatelli, M., 1991, Distribution of β1, α1, α2 and α3 integrin chain in basal cell carcinomas, *J. Pathol.*, **163**, 307–313.

Tuszynski, G.P., Wang, T.N., and Berger, D., 1997, Adhesive protein and the hematogenous spread of cancer, *Acta Haematologica*, **97**, 29–39.

Valizadeh, A., Karayiannakis, A.J., El-Hariry, I., Kimot, W., and Pignatelli, M., 1997, Expression of E-cadherin-associated molecules (α-,β- and γ-catenins and p120) in colorectal polyps, *Am. J-Path.*, **150**, 1977–1984.

Varner, J.A., and Cheresh, D.A., 1996, Integrins and cancer, *Curr Opin in Cell Biol.*, **8**, 742–730.

Vessey, C.J., Wilding, J., Folarin, N., Hirano, S., Takeichi, M., Soutter, P., Stamp, G.W.H., and Pignatelli, M., 1995, Altered expression and function of E-cadherin in cervical intraepithelial neoplasia and invasive squamous cell carcinoma, *J. Pathol.*, **176**, 151–159

Wielenga, V.J.M., Heider, K.-H., Offerhaus, G.J.A., Adolf, G.R., van der Berg, F.M., Ponta, H., Herrlick, P., and Pals, S.T., 1993, Expression of CD44 variant protein in human colorectal cancer is related to tumour progression, *Cancer Res.*, **53**, 4754–4756.

Wu, S.Q., Storer, B.E., Bookland, E.A., *et al.*, 1991, Nonrandom chromosome losses in stepwise neoplastic transformation in vitro of human uroepithelial cells, *Cancer Res.*, **51**, 3323–3326.

Yap, A.S., Niessen, C.M., and Gumbiner, B.M., 1998, The juxtamembrane region of the cadherin cytoplasmic tail support lateral clustering, adhesive strengthening, and interaction with p120^{ctn}, *J. Cell Biol.*, **141**, 779–789.

Zutter, M.M., Krigman, H.R., and Santoro, S.A., 1993, Altered integrin expression in adenocarcinoma of the breast, *Am. J. Pathol.*, **142**, 1439–1448.

12

CELL CYCLE CONTROL

Vivien J. Tannoch[1], Phil W. Hinds[1], and Li-Huei Tsai[1,2]

[1]Harvard Medical School
Department of Pathology
200 Longwood Ave
Boston Massachusetts 02115
[2]Howard Hughes Medical Institute
200 Longwood Ave
Boston Massachusetts 02115

1. INTRODUCTION

Regulation of the cell cycle is maintained through a delicately balanced array of mechanisms. These regulatory mechanisms extend from controlling the actual transcription of genes encoded in the genome, to regulating the catalytic activity of enzymes in the cell. This chapter will highlight specific examples of these control-ling mechanisms, the cell cycle machinery, and discuss some of the genes that are deregulated in disease states, such as cancer, which involves the deregulation of normal cell proliferation.

Our understanding of the cell cycle was substantially increased when the cyclins and cyclin-dependent kinases (cdks) were discovered (Sherr, 1993; Sherr, 1996). These proteins have multiple family members, and together the cyclins and cdks form active holoenzymes capable of phosphorylating their specific substrates. Progression through each phase of the cell cycle is tightly correlated with the expression and rapid degradation of these cyclin-cdk complexes. This tight regulation of the cyclin-cdk complexes is critical since it is their catalytic activity that determines the phosphorylation status of other proteins important in the regulation of the cell cycle. One such protein is the retinoblastoma (pRb) protein (Mittnacht, 1998). The *Rb* gene was originally identified as a tumour suppressor that was genetically mutated in families with a predisposition to retinoblastoma and osteosarcomas (Friend *et al.*, 1986). pRb is now recognised to exert potent control over cell cycle progression from G1 to the S phase. This is in part due to the ability of pRb to bind, and modulate the activity of transcription factors such as E2F, which are important in the regulation of gene transcription (Dyson, 1998; Nevins, 1992). The phosphorylation of pRb controls its ability to bind these transcription factors (Knudsen and Wang, 1997; Mittnacht, 1998). Release of the transcription factors by phosphorylation of pRb permits the activation, or de-repression, of transcription of yet other genes encoding proteins of importance in cell proliferation.

Cancer Gene Therapy: Past Achievements and Future Challenges, edited by Habib
Kluwer Academic/Plenum Publishers, New York, 2000.

Thus we are beginning to understand how the cell cycle is controlled by a wide variety of proteins which are in turn regulated by a number of mechanisms including phosphorylation and proteolytic degradation, all of which are critical to the successful completion of the cell cycle, see Fig. 1. This multi-layered complexity in the regulation of the cell cycle also allows the cell to decide when not to proceed with cell division due to problems in the cell (Elledge, 1996; Paulovich, Toczyski, and Hartwell, 1997). For example, DNA damage will trigger the normal cell to either arrest in the cell cycle process to allow for repair mechanisms to be initiated, or else if the damage is perceived as irreparable, then the cell may enter into the self-destructive apoptotic pathway (Paulovich *et al.*, 1997). These checkpoints serve to keep the cell from replicating damaged DNA and thus preserve genome integrity. This checkpoint control in the cell is also often subject to deregulation in the cancerous cell since the replication of damaged DNA can also give rise to the fortuitous acquisition of mutations that are advantageous to the diseased cell. These mutations can lead to aberrant control of the cell cycle, and provide an inappropriate growth advantage.

2. CYCLINS AND CYCLIN-DEPENDENT KINASES

The cyclins bind to, and regulate the cyclin-dependent kinases (cdks). Together these proteins form a holoenzyme with specific serine/threonine kinase activity. From early G1, the D-type cyclins form complexes with cdks 4 or 6 (Bates *et al.*, 1994; Matsushime *et al.*, 1992; Meyerson and Harlow, 1994), which initiates the hyperphosphorylation of pRb (Dowdy *et al.*, 1993; Kitagawa *et al.*, 1996). Currently, there are three D-type cyclins identified: D1, D2, and D3 (Inaba *et al.*, 1992; Xiong, Menninger, Beach, and Ward, 1992). The different D-type cyclins exhibit tissue-specific expression but may be functionally redundant in some, but not all, cell types (Inaba *et al.*, 1992; Sicinski *et al.*, 1995). D-type cyclin expression is stimulated in response to mitogenic growth factors in conjunction with the appropriate external stimuli such as cell adhesion (Sherr, 1993; Zhu *et al.*, 1996). Until the cell has progressed past the so-called restriction point in the G1 phase of the cell cycle, the cell is dependent on the stimulation by mitogens for cell cycle progression (Pardee, 1989). Removal of mitogens from the cell before this point has been passed will allow for the rapid degradation of the D-type cyclins, prevent cyclin E-cdk2 activation, and cause the cell to withdraw from the cell cycle. However, removal of mitogens after the cell has passed this G1 restriction point will not cause the cell to exit the cell cycle. The cell is said to be mitogen-independent at this point (Pardee, 1989).

Progression from G1 to the S phase of the cell cycle is associated with high level expression of cyclin E and cdk2 which further hyperphosphorylates pRb (Lundberg and Weinberg, 1998). The sequential hyperphosphorylation of pRb is thought to be necessary for the appropriate inactivation of pRb, which releases the E2F transcription factor family members and initiates the subsequent activation of genes important in cell proliferation (Dyson, 1998; Kitagawa *et al.*, 1996; Lundberg *et al.*, 1998). Although notably, in a cell culture system which has pRb that cannot be phosphorylated, these cells can be still induced to enter the cell cycle through ectopic expression of cyclin E (Knudsen *et al.*, 1998; Lukas *et al.*, 1997). However, these cells arrest later in the cell cycle, suggesting that phosphorylated pRb is required for completion of the cell cycle.

Cyclin E-cdk2 has also been reported to be in complex with other proteins, including components of the pre-mRNA splicing machinery, NFκB, as well as another

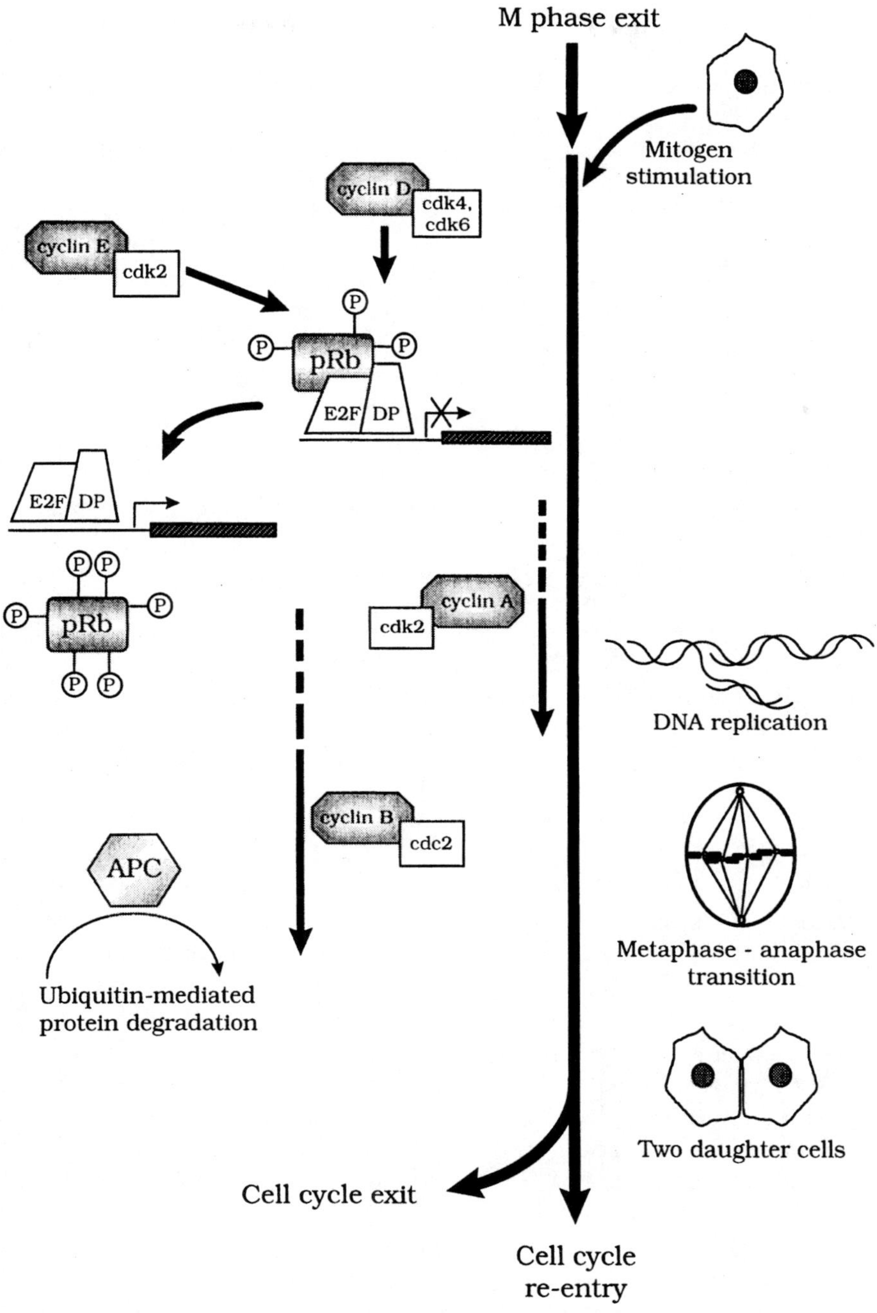

Figure 1. Schematic outline of events in the cell cycle. Entry into the G1 phase of the cell cycle follows mitogen stimulation of cells in Go, or following M phase exit in continuously proliferating cells. Cyclin D is expressed in response to mitogen stimulation, which complexes with either cdk4 or cdk6 and initiates the hyperphosphorylation of pRb. Subsequent hyperphosphorylation of pRb is mediated by cyclin E-cdk2, and this releases the transcription factors sequestered by pRb, such as E2F and DP, thus permitting E2F-mediated transcription. Cyclin A-cdk2 is expressed during DNA synthesis in the cell, and later in metaphase, cyclin B-cdc2 is active. The anaphase promoting complex (APC) targets specific proteins for ubiquitin-mediated degradation, which is necessary for the cell to complete mitosis and exit the cell cycle. Together, these proteins are regulated by many events, including phosphorylation, degradation and transcriptional activation.

putative substrate NPAT (Chen and Li, 1998; Seghezzi *et al.*, 1998; Zhao *et al.*, 1998). These results indicate that pRb is not likely to be the only important substrate for these cyclin-cdk complexes, and it will take many more studies to fully appreciate the contribution of these complexes to the regulation of the cell cycle.

Following the transition between G1 and S, the principle cyclin-cdk complex expressed in S phase is cyclin A-cdk2. During S phase, the cell is undergoing DNA duplication by initiating DNA synthesis at multiple origins of replication in the mammalian genome (Stillman, 1996). Entry into this phase of the cell cycle is tightly regulated, and it appears that one of the critical roles of cyclin A-cdk2 is in its participation with a large multiprotein complex that also contains SKP1, SKP2 and CUL-1 (Lisztwan *et al.*, 1998; Michel and Xiong, 1998; Yu, Gervais, and Zhang, 1998b). By analogy with the yeast complex containing these proteins, this multi-protein complex acts as an E3-like ubiquitin ligase, an enzyme important in substrate specificity in ubiquitin-mediated degradation of proteins (see below). In yeast, this complex mediates the specific degradation of the yeast cdk inhibitor Sic1, and this removal of Sic1 from the cell is necessary for initiating DNA replication (Feldman, Correll, Kaplan, and Deshaies, 1997; Verma *et al.*, 1997a; Verma, Feldman, and Deshaies, 1997b).

The next phases of the cell cycle are regulated primarily by cyclin B in complex with cdc2 (the M phase promoting factor, MPF). Entry into mitosis is dependent upon activation of MPF by cdc25-mediated dephosphorylation (Lammer *et al.*, 1998; Millar and Russell, 1992; Strausfeld *et al.*, 1991), and the physical re-localisation of cyclin B to the nucleus (Hagting *et al.*, 1998; Yang *et al.*, 1998). Finally, for the cell to complete mitosis, cyclin B must be subsequently destroyed and this is coordinated by the anaphase promoting complex (APC) (King *et al.*, 1995; Sudakin *et al.*, 1995).

3. OTHER FUNCTIONS FOR THE CYCLINS?

Evidence is now coming to light that the cyclins might not only function as the regulatory subunits of the cdks. Notably, cyclin D1 is often amplified and overexpressed in a large majority of human breast tumours that are oestrogen receptor positive. Transgenic mice that overexpress cyclin D1 are more susceptible to breast cancer (Wang *et al.*, 1994), and furthermore, the oestrogen-responsive breast epithelium tissue that normally proliferates during pregnancy, fails to do so in cyclin D1 knockout animals (Sicinski *et al.*, 1995). Early evidence suggests that cyclin D1 can potentiate oestrogen receptor-mediated transcription, independent of binding to cdks (Zwijsen *et al.*, 1997), and this effect is specific to cyclin D1. Furthermore, the D-type cyclins have been identified to bind and inhibit the transactivating ability of DMP1 and v-Myb (Ganter, Fu, and Lipsick, 1998; Inoue and Sherr, 1998). Unlike the co-activating function of cyclin D1 with the oestrogen receptor, the D-type cyclins antagonise DMP1 function by binding to the DNA binding domain of DMP1, thus directly inhibiting DMP1 binding and transactivation activity (Inoue *et al.*, 1998). Again, binding by the cyclins to DMP1 and v-Myb is independent of cdk binding and specific to the D-type cyclins. Transfection experiments demonstrated that expression of DMP1 in NIH 3T3 cells caused an inhibition of S phase in the cell cycle. The sequestering of DMP1 by the D-type cyclins therefore alleviates this inhibition and complements the activity of the D-type cyclins in complex with their cdks to promote cell cycle progression. Obviously it remains to be determined how much these other activities of the cyclins contribute to the cell cycle in cells with normal physiological levels of these proteins. However this does intriguingly present other potential roles for the cyclins in the cell and the further expanding the role of the cyclins in the control of the cell cycle.

4. CYCLIN-DEPENDENT KINASE ACTIVATION KINASES (CAKs)

Formation of a cyclin-cdk complex is not in itself sufficient for optimal substrate recognition and subsequent kinase activity. Full activation of cdk2 and cdc2 in complex with their cyclin subunits requires the phosphorylation of a specific threonine of the cdk located in a region known as the T loop. Crystallographic analysis has revealed the T loop structure of cdk2 to be in a position, which until bound by cyclin A, and phosphorylated by the cdk activation kinase (CAK), interferes with substrate binding to cdk2 by obstructing the substrate binding site (Jeffrey *et al.*, 1995; Russo *et al.*, 1996a). However, once in complex with cyclin A and phosphorylated at the T loop, the T loop is realigned to fully expose the substrate binding site (Russo, Jeffrey, and Pavletich, 1996b).

Mammalian CAK consists of yet another cyclin-cdk complex: cyclin H and cdk7. These proteins have been demonstrated *in vitro* to reconstitute CAK activity towards both cdk2 and cdc2 (Fisher and Morgan, 1994). Formation and enhanced activation of this complex appears to be regulated by either an assembly factor, MAT1, or by phosphorylation of cdk7 in its T loop region (Devault *et al.*, 1995; Fisher, Jin, Chamberlin, and Morgan, 1995). Interestingly, this cyclin H-cdk7 complex has also been shown to be a component of the basal transcriptional complex TFIIH (Feaver, Svejstrup, Henry, and Kornberg, 1994; Serizawa *et al.*, 1995; Shiekhattar *et al.*, 1995). It may be that the function of the CAK is to coordinately regulate not only the activities of the cyclin-cdks during the cell cycle, but also transcriptional activity in the cell. Of interest, another CAK protein has also been identified in yeast, Civ1/Cak1p (Kaldis, Sutton, and Solomon, 1996; Thuret, Valay, Faye, and Mann, 1996). This yeast CAK exhibits properties distinct from those described of mammalian CAK (Kaldis *et al.*, 1998) and is not a component of TFIIH, although both yeast and mammalian CAK phosphorylate the same residues on cdk6 (Kaldis *et al.*, 1998).

5. CYCLIN-DEPENDENT KINASE INHIBITORS

The cyclin-dependent kinase inhibitors can regulate the activity of the cyclins and the cyclin-cdk complexes through direct interaction of these proteins. To date, there are two defined classes of cyclin-dependent kinase inhibitors: the p16 Ink family (p15, p16, p18 and p19) and the Cip/Kip family (p21, p27 and p57). These families are defined through structural homology and exhibit distinct specificity in their inhibition.

The Ink family of inhibitors is refined in their activity against the G1 kinases, cdk4 and cdk6. These inhibitors bind to the cdks and prevent their association with their cyclin subunits, thus inhibiting kinase activity. It has been demonstrated by crystallographic studies of p16-bound cdk6 that the inhibitor binds to the kinase adjacent to the catalytic site, but opposite the cyclin binding region (Russo *et al.*, 1998). Thus it would seem that the inhibition of cyclin binding is mediated by a conformational change in the kinase that prevents association of the cyclins and distorts the ATP-binding site (Russo *et al.*, 1998). In contrast, the Cip/Kip family of inhibitors can exert their inhibitory activity throughout the cell cycle in a tertiary complex with the cyclin-cdks. These inhibitors are able to bind pre-formed complexes of cyclin-cdks, thereby directly inhibiting the kinases. The crystal structure of p27 bound to cyclin A/cdk2 has been resolved and illustrates how the N-terminal region of the inhibitor, which is conserved in the Cip/Kip family, binds to cdk2 (Russo *et al.*, 1996a). p27 destabilises the ATP binding site, and mimics ATP binding by occupying this region, and furthermore, the p27 molecule also binds to cyclin A at region which would normally permit substrate binding (Russo *et al.*, 1996a). In contrast to the Ink family of inhibitors which specifically inhibit the G1 cdks, the Cip/Kip family of inhibitors has been found in complex with nearly all cyclin-cdk complexes (Sherr and Roberts, 1995), and so it would seem that these inhibitors could act throughout the entire cell cycle. Together, the cdk inhibitors regulate the cell cycle by directly regulating the activity of the kinases.

In addition to binding the cdks, p21 and p57 have also been demonstrated to bind PCNA (proliferating cell nuclear antigen) (Flores-Rozas *et al.*, 1994; Gulbis *et al.*, 1996; Waga, Hannon, Beach, and Stillman, 1994; Watanabe *et al.*, 1998). PCNA is involved in DNA replication, and in overexpression studies, binding of p21 or p57 prevents PCNA-mediated DNA synthesis and arrests the cell cycle (Chen, Jackson, Kirschner, and Dutta, 1995; Luo, Hurwitz, and Massague, 1995; Watanabe *et al.*, 1998). Induction of p21 expression can be mediated by p53 (El-Deiry *et al.*, 1993; Kim, 1997). Therefore in a simple model, it could be expected that if the cell detected DNA damage, p53 expression would increase through protein stabilisation and this would lead to an induction of p21. Subsequently, p21 would bind to, and inhibit, both the cdks and PCNA, which together would efficiently mediate cell cycle arrest in response to DNA damage.

6. RETINOBLASTOMA AND FAMILY MEMBERS

The retinoblastoma (Rb) gene was discovered by virtue of its mutation in retinoblastoma cancers (Friend *et al.*, 1986). However, pRb and related family members, p107 and p130, are now recognised to be central players in the regulation of the cell cycle of all cell types, not just particularly in cells of the eye. Fundamental to the regulation of pRb function is its phosphorylation status (Mittnacht, 1998). This is critical for the ability of pRb to bind and inhibit the transcription factors E2F and DP (Dyson, 1998; Nevins, 1992). In a resting cell that is not actively dividing, pRb remains in a hypophosphorylated state. Here the transcription factors are bound by pRb and sequestered from activating transcription. During cell cycle progression, pRb becomes progressively hyperphosphorylated by the cyclin-cdks. This in turn leads to the release of the bound transcription factors E2F and DP. Cyclin D-cdk4 or 6, and cyclin-E-cdk2, have been demonstrated to specifically phosphorylate pRb (Dowdy *et al.*, 1993; Kitagawa *et al.*, 1996; Lundberg *et al.*, 1998). These events are critical in the G1/S progression of the cell through the cell cycle. pRb remain hyperphosphorylated throughout the remainder of the cell cycle. Related family members p130 and p107 also regulate E2F-dependent transcription, however, pRb is by far more widely implicated in the regulation of the cell cycle. This current vision of the pRb and related proteins may be modified when we begin to more fully appreciate the contribution of p107 and p130 to the regulation of the cell cycle (Mulligan and Jacks, 1998). Since the phosphorylation status of pRb is central to its function, the regulatory mechanisms governing the activity of the kinases that phosphorylate pRb will have direct consequences on the activity of pRb. Hence, for example, the cdk inhibitors and the regulated expression of the cyclins and cdks will directly impact pRb function. In light of this, pRb is generally regarded as the point of integration in the cell for the multitude of signals received by the cell, either to proliferate in response to mitogenic signals, or to withdraw from the cell cycle in response to cues to differentiate, quiesce or senesce. Not surprisingly, elements of the pathway involving pRb and its function are most frequently targets for mutation in disease states such as cancer (see below).

7. E2F/DP

The activity of the E2F transcription factors can be regulated by their interaction with pRb (Dyson, 1998; Nevins, 1992). E2F transcription factors are critical in the regulation of the G1/S phase of the cell cycle by regulating the transcription of genes. In mammalian cells, at least six members of the E2F family and two of the DP family of genes have been identified. These proteins share highly homologous dimerization and DNA binding domains, although only E2F-1-5 proteins have potent transactivation domains. The DP proteins, which heterodimerize with E2F, are thought to act by potentiating the activity of the E2F factors, and E2F-6, which does not possess a transactivating domain, is proposed to repress E2F-mediated transcription (Morkel *et al.*, 1997; Trimarchi *et al.*, 1998). The promoters of many genes have been found to contain consensus E2F binding sites, including cdc2, B-myb, cyclin E, dihydrofolate reductase (DHFR), E2F-1 and E2F-2 (Dalton, 1992; Geng *et al.*, 1996; Johnson, Ohtani, and Nevins, 1994; Lam and Watson, 1993; Sears, Ohtani, and Nevins, 1997; Wade *et al.*, 1992). Within the E2F transactivation domain lies a pRb-binding region. Therefore binding of pRb to the E2F proteins should directly interfere with E2F transactivation activity. Due to the large number of potential E2F heterodimer complexes with all of the E2F and DP family members, it can be anticipated that these different complexes will regulate the expression of a variety of genes in a specific manner. Furthermore, these E2F-responsive genes are differentially regulated by the different pRb family members (Hurford-Jr, Cobrinik, Lee, and Dyson, 1997). Release of the E2F transcription factors from pRb allows for both direct activation of E2F-responsive genes and the derepression of genes that were bound at their promoters by pRb-complexed E2F (Dyson, 1998). In addition to pRb regulation of E2F, the activity of these transcription factors is also affected by their phosphorylation status, their expression levels (both at the transcriptional level and protein degradation levels), and through cellular localisation (Helin, 1998). This complexity of E2F activity and its regulation belies its importance in the control of gene expression and the cell cycle.

8. PROTEOLYTIC DEGRADATION IN THE CELL CYCLE: UBIQUITIN-MEDIATED DEGRADATION

Many proteins are subject to rapid degradation during the cell cycle. Their degradation permits the transition of the cell to the next phase of the cell cycle. In this way, the cell maintains the precise orchestration of the cell cycle allowing the completion of one phase of the cell cycle before entry into the next phase. This specific degradation of the many proteins, including cyclins and cdk inhibitors, is achieved by targeted ubiquitination that signals their rapid degradation by the 26S proteosomes (King, Deshaies, Peters, and Kirschner, 1996; Krek, 1998; Murray, 1995). Transfer of ubiquitin to target proteins begins with the ATP-dependent thioester bond formation between an E1 enzyme and a ubiquitin molecule. This ubiquitin molecule is next transferred to a reactive cysteine residue of a second enzyme, E2. Finally, the ubiquitin is transferred, either directly from the E2 enzyme, or in concert with an E3 ubiquitin ligase (also referred to as an SCF: Skp1-Cullin-F-box) complex to the target protein. The E3 ubiquitin ligase is thought to contribute most to substrate specificity in ubiquitination by phosphorylation-dependent targeting. Ultimately, the target protein becomes multi-ubiquitinated and degraded to small peptides by the 26S proteosome complex which consists of a central 20S proteosome and two 19S subunits.

At present, we are beginning to recognise many proteins as targets of the ubiquitin-mediated degradation pathway, including cyclin D1, cyclin E and p27 (Clurman *et al.*, 1996; Diehl, Zindy, and Sherr, 1997; Pagano *et al.*, 1995; Sheaff *et al.*, 1997; Vlach, Hennecke, and Amati, 1997; Won and Reed, 1996). Serum withdrawal from the cell normally results in the rapid degradation of the D-type cyclins that are expressed in response to mitogens. This degradation has been demonstrated to be dependent on the phosphorylation of cyclin D1 at threonine 286 which stimulates its destruction by the ubiquitin degradation pathway (Diehl *et al.*, 1997). This rapid response to the withdrawal of serum from the cells ensures that the cells do not enter the cell cycle before they are fully activated and have passed the restriction point in G1 (Pardee, 1989). The periodic expression of cyclin E-cdk2 at the G1/S transition phase of the cell cycle is also tightly related to its destruction by ubiquitination-mediated degradation, thus keeping the expression of this cyclin-cdk complex in tight association with only the G1/S phase (Dulic, Lees, and Reed, 1992; Koff *et al.*, 1992). Not only are the cyclins/cdks subject to ubiquitin-mediated degradation, but so are the cdk inhibitors. The relative abundance of p27 in cycling and quiescent cells is regulated, at least in part, by its rate of degradation (Alessandrini, Chiaur, and Pagano, 1997; Pagano *et al.*, 1995; Vlach *et al.*, 1997). Clearly, a number of components of the cell cycle are regulated through ubiquitin-mediated degradation. The virtue of this mechanism for cell cycle regulation is that once the protein components of the cell cycle have fulfilled their function, they are irreversibly destroyed. In this manner, timely progression of the cell cycle is ensured without risk of the cell attempting to repeat any phase of the cell cycle inappropriately.

9. ANAPHASE PROMOTING COMPLEX (APC)

Critical for the initiation of mitosis and cell cycle exit is a specific E3-type ubiquitin ligase now recognised as the anaphase promoting complex (APC), or the cyclosome (King *et al.*, 1995; Peters, King, Hoog, and Kirschner, 1996; Yu *et al.*, 1998a). This large, multi-protein complex consists of at least eight subunits, and is known to specifically target cyclin B for ubiquitin-mediated degradation. APC selectively targets proteins for ubiquitination that contain the consensus nine amino acid "destruction box" motif, RXALGVIXN, which is found in the mitotic cyclins (Glotzer, Murray, and Kirschner, 1991). More proteins, other than the mitotic cyclins, are also being discovered which are regulated by APC-mediated degradation. The degradation of these proteins in mitosis, for example, Pds1p, Ase1p and geminin, is required for the successful completion of mitosis and exit from the cell cycle (Cohen-Fix, Peters, Kirschner, and Koshland, 1996; Juang *et al.*, 1997; McGarry and Kirschner, 1998). Mutation of the destruction boxes in these proteins renders them resistant to APC-mediated degradation and their expression in the cell is stabilised. Consequently the cell cycle is disrupted.

Just as the expression of proteins, and their activity during the cell cycle are necessary for normal cell cycle progression, so is the timely destruction of proteins during the cell cycle. This knowledge furthers our understanding of the cell cycle and brings to light the enormously intricate mechanisms involved in the regulation of the normal cell cycle.

10. CHECKPOINT CONTROLS

When damage is perceived in the cell, or a phase of the cell cycle has not been successfully completed, the cell can arrest and correct these events before progressing in the cell cycle. These regulatory mechanisms have been generally defined as checkpoint controls. Several events can trigger a "checkpoint arrest" including incomplete DNA replication, DNA damage induced by a variety of mechanisms including irradiation and chemical agents, incomplete or incorrect attachment of chromosomes to the mitotic spindle, and incorrect centrosome duplication events (Elledge, 1996; Paulovich *et al.*, 1997; Rudner and Murray, 1996; Winey, 1996).

The DNA replication and DNA damage checkpoints may be overlapping controlling mechanisms, as evidenced by yeast studies (Elledge, 1996; Zeng *et al.*, 1998). To prevent mitosis when there is DNA damage or incomplete DNA replication in the cell, the activation of cdc2 is prevented by inactivating cdc25 through its own phosphorylation and preferential binding to the 14-3-3 proteins (Peng *et al.*, 1997; Sanchez *et al.*, 1997). Cdc25 is a phosphatase that normally activates the cyclin B-cdc2 complex during mitosis. Phosphorylation, and therefore the inhibition of cdc25, is mediated by the yeast proteins Cds1 and Chk1 (Zeng *et al.*, 1998). By inhibiting the cyclin B-cdc2 kinase activity, mitosis is held in check until the DNA damage is repaired or DNA replication is completed, thus providing an elegant checkpoint in the cell cycle for maintenance of genome integrity. The study of DNA damage in mammalian cells has identified p53 and ATM (mutated in ataxia telangiectasia) as genes also important for this checkpoint (Livingstone *et al.*, 1992; Meyn, 1995; Painter and Young, 1980; Yin *et al.*, 1992). Mutation of p53 is possibly the most common event in human cancers (Sherr, 1996). Activation of p53 occurs in response to DNA damage, although how this mechanism of activation occurs is still unclear. Cells that possess a defective p53 gene fail to arrest normally in response to irradiation (Bouffler, Kemp, Balmain, and Cox, 1995; Kemp, Wheldon, and Balmain, 1994). As a transcription factor, p53 induces the expression of a number of genes including the cdk inhibitor, p21 (El-Deiry *et al.*, 1993; Kim, 1997). Conceivably, overexpression of p21 will inhibit the cell cycle by inhibiting the cyclins and thus allow for repair of the damaged DNA.

The correct segregation of the chromosomes during mitosis is also monitored in a checkpoint control mechanism (Rudner *et al.*, 1996). For anaphase to proceed in the mitotic cell, the bipolar spindle must be correctly assembled and attached through a kinetochore to the chromosomes aligned in metaphase (Rieder, Schultz, Cole, and Sluder, 1994; Sorger, Dobles, Tournebize, and Hyman, 1997). Each chromosome must be bound through its sister chromatids to opposite spindle poles. The spindle assembly checkpoint will prevent actual chromosome segregation if these events are not completed (Rieder *et al.*, 1994). Notably, microtubule dynamics and its control are important elements in this checkpoint (Sorger *et al.*, 1997).

11. CELL PROLIFERATION IN DISEASE

From a simplistic viewpoint, cancer and many other disease states are characterised by inappropriate cell proliferation (Sherr, 1996). Many of the genes that are mutated in cancer cause their inappropriate expression, and as a consequence, these mutations directly impact the regulation of the cell cycle. An apparently frequent target for inactivation in human tumours is the cdk inhibitor, p16 (Foulkes, Flanders, Pollock, and Hayward, 1997; Walker *et al.*, 1998). Conversely, cyclin D1 is often overexpressed through gene amplification in a variety of tumours (Dickson *et al.*, 1995). Both of these mutations can be predicted to have a net result on the activation status of pRb. By overexpressing cyclin D1, the cells initiate the hyperphosphorylation of pRb and through the inactivation of p16, inhibition of cdk4 and 6 is relieved, again potentially leading to the initiation of pRb hyperphosphorylation. Furthermore, we are now beginning to find examples of virally encoded cyclins which are resistant to the normal cdk inhibtiors (Swanton *et al.*, 1997). Knowledge of how the cell regulates its cell cycle is fundamental to finding therapeutic agents effective against disease states that deregulate normal cell proliferation. It can be readily envisioned that many other proteins involved in the control of cell proliferation may be taken and used to design targeted therapeutic applications. For example, although still in early studies, proteosome inhibitors are already being explored as potential therapeutic agents for targeting the ubiquitin-mediated degradation of proteins (Lee and Goldberg, 1998). This is proving to be an exciting time as we continue to dissect the cell cycle and understand its regulation.

REFERENCES

Alessandrini, A., Chiaur, D.S., and Pagano, M., 1997, Regulation of the cyclin-dependent kinase inhibitor p27 by degradation and phosphorylation, *Leukemia* **11**:342–345.

Bates, S., Bonetta, L., MacAllan, D., Parry, D., Holder, A., Dickson, C., and Peters, G., 1994, CDK6 (PLSTIRE) and CDK4 (PSK-J3) are a distinct subset of the cyclin-dependent kinases that associate with cyclin D1, *Oncogene* **9**:71–79.

Bouffler, S.D., Kemp, C.J., Balmain, A., and Cox, R., 1995, Spontaneous and ionizing radiation-induced chromosomal abnormalities in p53-deficient mice, *Cancer Res* **55**:3883–3889.

Chen, E., and Li, C.C., 1998, Association of Cdk2/cyclin E and NF-kappa B complexes at G1/S phase, *Biochem Biophys Res Comm* **249**:728–734.

Chen, J., Jackson, P.K., Kirschner, M.W., and Dutta, A., 1995, Separate domains of p21 involved in the inhibition of Cdk kinase and PCNA, *Nature* **374**:386–388.

Clurman, B.E., Sheaff, R.J., Thress, K., Groudine, M., and Roberts, J.M., 1996, Turnover of cyclin E by the ubiquitin-proteosome pathway is regulated by cdk2 binding and cyclin phosphorylation, *Genes Dev* **10**:1979–1990.

Cohen-Fix, O., Peters, J.-M., Kirschner, M.W., and Koshland, D., 1996, Anaphase initiation in Saccharomyces cerevisiae is controlled by the APC-dependent degradation of the anaphase inhibitor Pds1p, *Genes Dev* **10**:3081–3093.

Dalton, S., 1992, Cell cycle regulation of the human cdc2 gene, *EMBO J* **11**:1797–1804.

Devault, A., Martinez, A.-M., Fesquet, D., Labbe, J.-C., Morin, N., Tassan, J.-P., Nigg, E.A., Cavadore, J.-C., and Doree, M., 1995, MAT1 ("menage a trois") a new RING finger protein subunit stabilizing cyclin H-cdk7 complexes in starfish and Xenopus CAK, *EMBO J* **14**:5027–5036.

Dickson, C., Fantl, V., Gillett, C., Brookes, S., Bartek, J., Smith, R., Fisher, C., Barnes, D., and Peters, G., 1995, Amplificaton of chromosome band 11q13 and a role for cyclin D1 in human breast cancer, *Cancer Lett* **90**:43–50.

Diehl, J.A., Zindy, F., and Sherr, C.J., 1997, Inhibition of cyclin D1 phosphorylation on threonine-286 prevents its rapid degradation via the ubiquitin-proteosome pathway, *Genes Dev* **11**:957–972.

Dowdy, S.F., Hinds, P.W., Louie, K., Reed, S.I., Arnold, A., and Weinberg, R.A., 1993, Physical interaction of the retinoblastoma protein with human D cyclins, *Cell* **73**:499–511.

Dulic, V., Lees, E., and Reed, S.I., 1992, Association of human cyclin E with a periodic G1-S phase protein kinase, *Science* **257**:1958–1961.

Dyson, N., 1998, The regulation of E2F by pRB-family proteins, *Genes Dev* **12**:2245–2262.

El-Deiry, W.S., Tokino, T., Velculescu, V.E., Levy, D.B., Parsons, R., Trent, J.M., Lin, D., Mercer, E., Kinzler, K.W., and Vogelstein, B., 1993, WAF1, a potential mediator of p53 tumor suppression, *Cell* **75**:817–825.

Elledge, S.J., 1996, Cell cycle checkpoints: preventing an identity crisis, *Science* **274**:1664–1672.

Feaver, W.J., Svejstrup, J.Q., Henry, N.L., and Kornberg, R.D., 1994, Relationship of CDK-activating kinase and RNA polymerase II CTD kinase TFIIH/TFIIK, *Cell* **79**:1103–1109.

Feldman, R.M., Correll, C.C., Kaplan, K.B., and Deshaies, R.J., 1997, A complex of Cdc4p, Skp1p, and Cdc53p/cullin catalyzes ubiquitination of the phosphorylated CDK inhibitor Sic1p, *Cell* **91**:221–230.

Fisher, R.P., Jin, P., Chamberlin, H.M., and Morgan, D.O., 1995, Alternative mechanisms of CAK assembly require an assembly factor or an activating kinase, *Cell* **83**:47–57.

Fisher, R.P., and Morgan, D.O., 1994, A novel cyclin associates with MO15/cdk7 to form the CDK-activating complex, *Cell* **78**:713–724.

Flores-Rozas, H., Kelman, Z., Dean, F.B., Pan, Z.-Q., Harper, J.W., Elledge, S.J., O'Donnell, M., and Hurwitz, J., 1994, Cdk-interacting protein 1 directly binds with proliferating cell nuclear antigen and inhibits DNA replication catalyzed by the DNA polymerase δ holoenzyme, *Proc Natl Acad Sci USA* **91**:8655–8659.

Foulkes, W.D., Flanders, T.Y., Pollock, P.M., and Hayward, N.K., 1997, The CDKN2A (p16) gene and human cancer, *Mol Med* **3**:5–20.

Friend, S.H., Bernards, R., Rogelj, S., Weinberg, R.A., Rapaport, J.M., Albert, D.M., and Dryja, T.P., 1986, A human DNA segment with properties of the gene that predisposes to retinoblastoma and osteosarcoma, *Nature* **323**:643–646.

Ganter, B., Fu, S.I., and Lipsick, J.S., 1998, D-type cyclins repress transcriptional activation by the v-Myb but not the c-Myb DNA-binding domain, *EMBO J* **17**:255–268.

Geng, Y., Eaton, E.N., Picon, M., Roberts, J.M., Lundberg, A.S., Gifford, A., Sardet, C., and Weinberg, R.A., 1996, Regulation of cyclin E transcription by E2Fs and retinoblastoma protein, *Oncogene* **12**:1173–1180.

Glotzer, M., Murray, A.W., and Kirschner, M.W., 1991, Cyclin is degraded by the ubiquitin pathway, *Nature* **349**:132–138.

Gulbis, J.M., Kelman, Z., Hurwitz, J., O'Donnell, M., and Kuriyan, J., 1996, Structure of the C-terminal region of p21 (WAF1/CIP1) complexed with human PCNA, *Cell* **87**:297–306.

Hagting, A., Karlsson, C., Clute, P., Jackman, M., and Pines, J., 1998, MPF localization is controlled by nuclear export, *EMBO J* **17**:4127–4138.

Helin, K., 1998, Regulation of cell proliferation by the E2F transcription factors, *Current Opinion in Genetics & Development* **8**:28–35.

Hurford-Jr, R.K., Cobrinik, D., Lee, M.-H., and Dyson, N., 1997, pRb and p107/p130 are required for the regulated expression of different sets of E2F responsive genes, *Genes Dev* **11**:1447–1463.

Inaba, T., Matsushime, H., Valentine, M., Roussel, M.F., Sherr, C.J., and Look, A.T., 1992, Genomic organization, chromosomal localization, and independent expression of human cyclin D genes, *Genomics* **13**:565–574.

Inoue, K., and Sherr, C.J., 1998, Gene expression and cell cycle arrest mediated by transcription factor DMP1 is antagonized by D-type cyclins through a cyclin-dependent-kinase-independent mechanism, *Mol Cell Biol* **18**:1590–1600.

Jeffrey, P.D., Russo, A.A., Polyak, K., Gibbs, E., Hurwitz, J., Massague, J., and Pavletich, N.P., 1995, Mechanism of CDK activation revealed by the structure of a cyclin A-cdk2 complex, *Nature* **376**:313–320.

Johnson, D.G., Ohtani, K., and Nevins, J.R., 1994, Autoregulatory control of E2F-1 expression in response to positive and negative regulators of cell cycle expression, *Genes Dev* **8**:1514–1525.

Juang, Y.-L., Huang, J., McLaughlin, M.E., Tai, C.-Y., and Pellman, D., 1997, APC-mediated proteolysis of Ase1 and the morphogenesis of the mitotic spindle, *Science* **275**:1311–1314.

Kaldis, P., Russo, A.A., Chou, H.S., Pavletich, N.P., and Solomon, M.J., 1998, Human and yeast cdk-activating kinases (CAKs) display distinct substrate specificities, *Mol Biol Cell* **9**:2545–2560.

Kaldis, P., Sutton, A., and Solomon, M.J., 1996, The cdk-activating kinase (CAK) from budding yeast, *Cell* **86**:553–564.

Kemp, C.J., Wheldon, T., and Balmain, A., 1994, p53-deficient mice are extremely susceptible to radiation-induced tumorigenesis, *Nat Genet* **8**:66–69.

Kim, T.K., 1997, *In vitro* transcriptional activation of p21 promoter by p53, *Biochem Biophys Res Comm* **234**:300–302.

King, R.W., Deshaies, R.J., Peters, J.-M., and Kirschner, M.W., 1996, How proteolysis drives the cell cycle, *Science* **274**:1652–1659.

King, R.W., Peters, J.-M., Tugendreich, S., Rolfe, M., Hieter, P., and Kirschner, M.W., 1995, A 20S complex containing CDC27 and CDC16 catalyses the mitosis-specific conjugation of ubiquitin to cyclin B, *Cell* **81**:279–288.

Kitagawa, M., Higashi, H., Jung, H.K., Suzuki-Takahashi, I., Ikeda, M., Tamai, K., Kato, J.Y., Segawa, K., Yoshida, E., Nishimura, S., and Taya, Y., 1996, The consensus motif for phosphorylation by cyclin D1-Cdk4 is different from that for phosphorylation by cyclin A/E-Cdk2, *EMBO J* **15**:7060–7069.

Knudsen, E.S., Buckmaster, C., Chen, T.-T., Feramisco, J.R., and Wang, J.Y.J., 1998, Inhibition of DNA synthesis by pRb: effects on G1/S transition and S-phase progression, *Genes Dev* **12**:2278–2292.

Knudsen, E.S., and Wang, J.Y.J., 1997, Dual mechanisms for the inhibition of E2F binding to RB by cyclin-dependent kinase-mediated RB phosphorylation, *Mol Cell Biol* **17**:5771–5783.

Koff, A., Giordano, A., Desai, D., Yamashita, K., Harper, J.W., Elledge, S., Nishimoto, T., Morgan, D.O., Franza, B.R., and Roberts, J.M., 1992, Formation and activation of a cyclin E-cdk2 complex during the G1 phase of the human cell cycle, *Science* **257**:1689–1694.

Krek, W., 1998, Proteolysis and the G1-S transition: the SCF connection, *Current Opinion in Genetics & Development* **8**:36–42.

Lam, E.W.-F., and Watson, R.J., 1993, An E2F-binding site mediates cell-cycle regulated repression of mouse B-myb transcription, *EMBO J* **12**:2705–2713.

Lammer, C., Wagerer, S., Saffrich, R., Mertens, D., Ansorge, W., and Hoffmann, I., 1998, The cdc25B phosphatase is essential for the G2/M phase transition in human cells, *J Cell Sci* **111**:2445–2453.

Lee, D.H., and Goldberg, A.L., 1998, Proteosome inhibitors: valuable new tools for cell biologists, *Trends Cell Biol* **8**:397–403.

Lisztwan, J., Marti, A., Sutterluty, H., Gstaiger, M., Wirbelauer, C., and Krek, W., 1998, Association of human CUL-1 and ubiquitin-conjugating enzyme CDC34 with the F-box protein p45 (SKP2): evidence for evolutionary conservation in the subunit composition of the CDC34-SCF pathway, *EMBO J* **17**:368–383.

Livingstone, L.R., White, A., Sprouse, J., Livanos, E., Jacks, T., and Tlsty, T.D., 1992, Altered cell cycle arrest and gene amplification potential accompany loss of wild-type p53, *Cell* **70**:923–935.

Lukas, J., Herzinger, T., Hansen, K., Moroni, M.C., Resnitzky, D., Helin, K., Reed, S.I., and Bartek, J., 1997, Cyclin E-induced S phase without activation of the pRb/E2F pathway, *Genes Dev* **11**:1479–1492.

Lundberg, A.S., and Weinberg, R.A., 1998, Functional inactivation of the retinoblastoma protein requires sequential modification by at least two distinct cyclin-cdk complexes, *Mol Cell Biol* **18**:753–761.

Luo, Y., Hurwitz, J., and Massague, J., 1995, Cell-cycle inhibition by independent CDK and PCNA binding domains in p21Cip1, *Nature* **375**:159–161.

Matsushime, H., Ewen, M.E., Strom, D.K., Kato, J., Hanks, S.K., Roussel, M.F., and Sherr, C.J., 1992, Identification and properties of an atypical catalytic subunit (p34PSK-J3/cdk4) for mammalian D-type G1 cyclins, *Cell* **71**:323–334.

McGarry, T.J., and Kirschner, M.W., 1998, Geminin, an inhibitor of DNA replication is degraded during mitosis, *Cell* **93**:1043–1053.

Meyerson, M., and Harlow, E., 1994, Identification of a G1 kinase activity for cdk6, a novel cyclin D partner, *Mol Cell Biol* **14**:2077–2086.

Meyn, M.S., 1995, Ataxia-telangiectasia and cellular responses to DNA damage, *Cancer Res* **55**:5991–6001.

Michel, J.J., and Xiong, Y., 1998, Human CUL-1, but not other cullin family members, selectively interacts with SKP1 to form a complex with SKP2 and cyclin A, *Cell Growth Diff* **9**:435–449.

Millar, J.B.A., and Russell, P., 1992, The cdc25 M-phase inducer: an unconventional protein phosphatase, *Cell* **68**:407–410.

Mittnacht, S., 1998, Control of pRB phosphorylation, *Curr Opin Gen Dev* **8**:21–27.

Morkel, M., Wenkel, J., Bannister, A.J., Kouzarides, T., and Hagemeier, C., 1997, An E2F-like repressor of transcription, *Nature* **390**:567–568.

Mulligan, G., and Jacks, T., 1998, The retinoblastoma gene family: cousins with overlapping interests, *Trends Genet* **14**:223–229.

Murray, A., 1995, Cyclin Ubiquitination: the destructive end of mitosis, *Cell* **81**:149–152.

Nevins, J.R., 1992, E2F: A link between the Rb tumor suppressor protein and viral oncoproteins, *Science* **258**:424–429.

Pagano, M., Tam, S.W., Theodoras, A.M., Beer-Romero, P., Del-Sal, G., Chau, V., Yew, P.R., Draetta, G.F., and Rolfe, M., 1995, Role of the ubiquitin-proteosome pathway in regulating abundance of the cyclin-dependent kinase inhibitor p27, *Science* **269**:682–685.

Painter, R.B., and Young, B.R., 1980, Radiosensitivity in ataxia-telangiectasia: a new explanation, *Proc Natl Acad Sci USA* **77**:7315–7317.

Pardee, A.B., 1989, G1 events and regulation of cell proliferation, *Science* **246**:603–608.

Paulovich, A.G., Toczyski, D.P., and Hartwell, L.H., 1997, When checkpoints fail, *Cell* **88**:315–321.

Peng, C.Y., Graves, P.R., Thoma, R.S., Wu, Z., Shaw, A.S., and Piwnica-Worms, H., 1997, Mitotic and G2 checkpoint control: regulation of 14-3-3 protein binding by phosphorylation of Cdc25C on serine-216, *Science* **277**:1501–1505.

Peters, J.-M., King, R.W., Hoog, C., and Kirschner, M.W., 1996, Identification of BIME as a subunit of the anaphase-promoting complex, *Science* **274**:1199–1201.

Rieder, C.L., Schultz, A., Cole, R., and Sluder, G., 1994, Anaphase onset in vertebrate somatic cells is controlled by a checkpoint that monitors sister kinetochore attachment to the spindle, *J Cell Biol* **127**:1301–1310.

Rudner, A.D., and Murray, A.W., 1996, The spindle assembly checkpoint, *Curr Opin Cell Biol* **8**:773–780.

Russo, A.A., Jeffrey, P.D., Patten, A.K., Massague, J., and Pavletich, N.P., 1996a, Crystal structure of the p27 Kip1cyclin-dependent-kinase inhibitor bound to the cyclin A-Cdk2 complex, *Nature* **382**:325–331.

Russo, A.A., Jeffrey, P.D., and Pavletich, N.P., 1996b, Structural basis of cyclin-dependent kinase activation by phosphorylation, *Nature Struc Biol* **3**:696–700.

Russo, A.A., Tong, L., Lee, J.-O., Jeffrey, P.D., and Pavletich, N.P., 1998, Structural basis for inhibition of the cyclin-dependent kinase Cdk6 by the tumour suppressor p16INK4a, *Nature* **395**:237–243.

Sanchez, Y., Wong, C., Thoma, R.S., Richman, R., Wu, Z., Piwnica-Worms, H., and Elledge, S.J., 1997, Conservation of the Chk1 checkpoint pathway in mammals: linkage of DNA damage to Cdk regulation through Cdc25, *Science* **277**:1497–1501.

Sears, R., Ohtani, K., and Nevins, J.R., 1997, Identification of positively and negatively acting elements regulating expression of the E2F2 gene in response to cell growth signals, *Mol Cell Biol* **17**:5227–5235.

Seghezzi, W., Chua, K., Shanahan, F., Gozani, O., Reed, R., and Lees, E., 1998, Cyclin E associates with components of the pre-mRNA splicing machinery in mammalian cells, *Mol Cell Biol* **18**:4526–4536.

Serizawa, H., Makela, T.P., Conaway, J.W., Conaway, R.C., Weinberg, R.A., and Young, R.A., 1995, Association of cdk-activating kinase subunits with transcription factor TFIH, *Nature* **374**:280–282.

Sheaff, R.J., Groudine, M., Gordon, M., Roberts, J.M., and Clurman, B.E., 1997, Cyclin E-cdk2 is a regulator of p27Kip1, *Genes Dev* **11**:1464–1478.

Sherr, C.J., 1993, Mammalian G1 cyclins, *Cell* **73**:1059–1065.

Sherr, C.J., 1996, Cancer cell cycles, *Science* **274**:1672–1677.

Sherr, C.J., and Roberts, J.M., 1995, Inhibitors of mammalian G1 cyclin-dependent kinases, *Genes Dev* **9**:1149–1163.

Shiekhattar, R., Mermelstein, F., Fisher, R.P., Drapkin, R., Dynlacht, B., Wessling, H.C., Morgan, D.O., and Reinberg, D., 1995, Cdk-activating kinase complex is a component of human transcription factor TFIIH, *Nature* **374**:283–287.

Sicinski, P., Donaher, J.L., Parker, S.B., Li, T., Fazeli, A., Gardner, H., Haslam, S.Z., Bronson, R.T., Elledge, S.J., and Weinberg, R.A., 1995, Cyclin D1 provides a link between development and oncogenesis in the retina and breast, *Cell* **82**:621–630.

Sorger, P.K., Dobles, M., Tournebize, R., and Hyman, A.A., 1997, Coupling cell division and cell death to microtubule dynamics, *Curr Opin Cell Biol* **9**:807–814.

Stillman, B., 1996, Cell cycle control of DNA replication, *Science* **274**:1659–1664.

Strausfeld, U., Labbe, J.-C., Fesquet, D., Cavadore, J.-C., Picard, A., Sadhu, K., Russell, P., and Doree, M., 1991, Dephosphorylation and activation of a p34cdc2/cyclin B complex in vitro by human CDC25 protein, *Nature* **351**:242–245.

Sudakin, V., Ganoth, D., Dahan, A., Heller, H., Hershko, J., Luca, F.C., Ruderman, J.V., and Hershko, A., 1995, The cyclosome, a large complex containing cyclin-selective ubiquitin ligase activity, targets cyclins for destruction at the end of mitosis, *Mol Biol Cell* **6**:185–198.

Swanton, C., Mann, D.J., Fleckenstein, B., Neipel, F., Peters, G., and Jones, N., 1997, Herpes viral cyclin/cdk6 complexes evade inhibition by CDK inhibitor proteins, *Nature* **390**:184–187.

Thuret, J.Y., Valay, J.G., Faye, G., and Mann, C., 1996, Civ1 (CAK in vivo), a novel Cdk-activating kinase, *Cell* **86**:565–576.

Trimarchi, J.M., Fairchild, B., Verona, R., Moberg, K., Andon, N., and Lees, J.A., 1998, E2F-6, a member of the E2F family that can behave as a transcriptional repressor, *Proc Natl Acad Sci USA* **95**:2850–2855.

Verma, R., Annan, R.S., Huddleston, M.J., Carr, S.A., Reynard, G., and Deshaies, R.J., 1997a, Phosphorylation of Sic1p by G1 Cdk required for its degradation and entry into S phase, *Science* **278**:455–460.

Verma, R., Feldman, R.M., and Deshaies, R.J., 1997b, SIC1 is ubiquitinated in vitro by a pathway that requires CDC4, CDC34, and cyclin/CDK activities, *Mol Cell Biol* **8**:1427–1437.

Vlach, J., Hennecke, S., and Amati, B., 1997, Phosphorylation-dependent degradation of the cyclin-dependent kinase inhibitor p27, *EMBO J* **16**:5334–5344.

Wade, M., Kowalik, T.F., Mudryj, M., Huang, E.S., and Azizkhan, J.C., 1992, E2F mediates dihydrofolate reductase promoter activation and multiprotein complex formation in human cytomegalovirus infection, *Mol Cell Biol* **12**:4364–4374.

Waga, S., Hannon, G.J., Beach, D., and Stillman, B., 1994, The p21 inhibitor of cyclin-dependent kinases controls DNA replication by interaction with PCNA, *Nature* **369**:574–578.

Walker, G.J., Flores, J.F., Glendening, J.M., Lin, A.H., Markl, I.D., and Fountain, J.W., 1998, Virtually 100% of melanoma cell lines harbor alteration at the DNA level within CDKN2A, CDKN2B, or one of their downstream targets, *Genes Chromosomes Cancer* **22**:157–163.

Wang, Y., Cardiff, R.D., Zukerberg, L., Lees, E., Arnold, A., and Schmidt, E.V., 1994, Mammary hyperplasia and carcinoma in MMTV-cyclin D1 transgenic mice, *Nature* **369**:669–671.

Watanabe, H., Pan, Z.-Q., Schreiber-Agus, N., DePinho, R.A., Hurwitz, J., and Xiong, Y., 1998, Suppression of cell transformation by the cyclin-dependent kinase inhibitor p57KIP2 requires binding to proliferating cell nuclear antigen, *Proc Natl Acad Sci USA* **95**:1392–1397.

Winey, M., 1996, Keeping the centrosome on track. Genome stability, *Curr Biol* **6**:962–964.

Won, K.A., and Reed, S.I., 1996, Activation of cyclin E/cdk2 is coupled to site-specific autophosphorylation and ubiquitin-dependent degradation of cyclin E, *EMBO J* **15**:4182–4193.

Xiong, Y., Menninger, J., Beach, D., and Ward, D.C., 1992, Molecular cloning and chromosomal mapping of CCND genes encoding human D-type cyclins, *Genomics* **13**:575–584.

Yang, J., Bardes, E.S., Moore, J.D., Brennan, J., Powers, M.A., and Kornbluth, S., 1998, Control of cyclin B1 localization through regulated binding of the nuclear export factor CRM1, *Genes Dev* **12**:2131–2143.

Yin, Y., Tainsky, M.A., Bischoff, F.Z., Strong, L.C., and Wahl, G.M., 1992, Wild-type p53 restores cell cycle control and inhibits gene amplification in cells with mutant p53 alleles, *Cell* **70**:937–948.

Yu, H., Peters, J.-M., King, R.W., Page, A., Hieter, P., and Kirschner, M.W., 1998a, Identification of a cullin homology region in a subunit of the anaphase-promoting complex, *Science* **279**:1219–1222.

Yu, Z.K., Gervais, J.L.M., and Zhang, H., 1998b, Human CUL-1 associates with the SKP1/SKP2 complex and regulates p21 (CIP1/WAF1) and cyclin D proteins, *Proc Natl Acad Sci USA* **95**:11324–11329.

Zeng, Y., Forbes, K.C., Wu, Z., Moreno, S., Piwnica-Worms, H., and Enoch, T., 1998, Replication checkpoint requires phosphorylation of the phosphatase Cdc25 by Cds1 or Chk1, *Nature* **395**:507–510.

Zhao, J., Dynlacht, B., Imai, T., Hori, T., and Harlow, E., 1998, Expression of NPAT, a novel substrate of cyclin E-cdk2, promotes S-phase entry, *Genes Dev* **12**:456–461.

Zhu, X., Ohtsubo, M., Bohmer, R.M., Roberts, J.M., and Assoian, R.K., 1996, Adhesion-dependent cell cycle progression linked to the expression of cyclin D1, activation of cyclin E-cdk2, and phosphorylation of the retinoblastoma protein, *J Cell Biol* **133**:391–403.

Zwijsen, R.M.L., Wientjens, E., Klompmaker, R., Sman, J.V.D., Bernards, R., and Michalides, R.J.A.M., 1997, CDK-independent activation of estrogen receptor by cyclin D1, *Cell* **88**:405–415.

IV

APOPTOSIS

13

KILLER/DR5, A NOVEL DNA-DAMAGE INDUCIBLE DEATH RECEPTOR GENE, LINKS THE p53-TUMOR SUPPRESSOR TO CASPASE ACTIVATION AND APOPTOTIC DEATH

Gen Sheng Wu, Kunghong Kim, and Wafik S. El-Deiry*

Laboratory of Molecular Oncology and Cell Cycle Regulation
Howard Hughes Medical Institute
Departments of Medicine
Genetics, Cancer Center, and the Institute for Human Gene Therapy
University of Pennsylvania School of Medicine
415 Curie Blvd., Philadelphia
Pennsylvania 19104

1. INTRODUCTION

Programmed cell death, or apoptosis, is a genetically controlled mechanism that is essential for the maintenance of tissue homeostasis, development and the elimination of unwanted cells. This phenomenon has been observed since the early 1970's by Kerr and colleagues (1972), although only recently have the biochemical events begun to be understood. Much knowledge of the downstream events that signal apoptosis has been learned from studies of C. elegans (Yuan, Shaham, Ledoux, Ellis, and Horvitz, 1993). Since the TNF family was found to induce apoptosis in human cells in the early 1990's, there has been a great deal of effort to understand the role of cytokines in initiating apoptosis (Itoh, Yonehara, Ishii,Yonehara, Mizushima, Sameshima, Hase, Seto, and Nagata, 1991). The evidence thus far suggests that apoptosis induction involves the engagement of a set of ligands and their corresponding receptors and then transmission of the apoptotic signal in the cytoplasm by a number of proteases, known as caspases (cysteine proteases that cleave next to certain aspartic acid residues), which are activated in an ordered fashion (Nagata, 1997; Yuan, 1997; Ashkenazi and Dixit, 1998). There is also some evidence for caspase-independent apoptosis (Miller, Moulder, Knudson, Creedon, Deshmukh, Korsmeyer, and Johnson, 1997). The activated caspases cleave cellular proteins

* Corresponding Author

Cancer Gene Therapy: Past Achievements and Future Challenges, edited by Habib
Kluwer Academic / Plenum Publishers, New York, 2000.

which are essential for cell growth and viability (Nagata, 1997). The morphological hallmarks of apoptosis include cell shrinkage, membrane blebbing, nuclear condensation and fragmentation, and loss of mitochondrial function. There are a number of ligand-receptor families and target caspases involved in apoptosis. These death-inducing receptors belong to the TNF superfamily and include Fas, TNFR's and the TRAIL receptors (Nagata, 1997; Yuan, 1997). In this chapter, we will focus on TRAIL (TNF-related apoptosis inducing ligand), a recently identified member of the TNF family (Pitti, Marsters, Ruppert, Donahue, Moore, and Ashkenazi, 1996; Wiley, Schooley, Smolak, Din, Huang, Nicholl, Sutherland, Smith, Rauch, Smith, and Goodwin, 1995), and its receptors (Ashkenazi and Dixit, 1998; Golstein, 1997; Goodwin and Smith, 1998) in apoptosis and tumor suppression, as well as their potential application for cancer gene therapy.

2. TRAIL LIGAND

The TRAIL ligand was identified by two groups using the sequence LVVXXGLYYVYXQVXF, a highly conserved region in the TNF family (Wiley *et al.*, 1995) to search the EST database (Wiley *et al.*, 1995; Pitti *et al.*, 1996). A novel ligand named TRAIL or AP0-2L was identified and was found to be expressed in many human tissues, particularly in the spleen, prostate, and lung (Pitti *et al.*, 1996; Wiley *et al.*, 1995). The cDNA of TRAIL is 1769 base pairs in length and encodes a protein of 281 amino acids in human and 291 amino acids in mouse cells (Pitti *et al.*, 1996; Wiley *et al.*, 1995). The structure of TRAIL is that of a type II membrane protein with the C-terminal region exposed to the extracellular environment (Pitti *et al.*, 1996; Wiley *et al.*, 1995). TRAIL has been mapped to human chromosome 3q26. TRAIL induces apoptosis of cells in vitro (Wiley *et al.*, 1995). Interestingly, TRAIL selectively kills transformed cells and cancer cells but has little effect on normal cell viability (Pitti *et al.*, 1996; Wiley *et al.*, 1995). Although the TNF family members Fas and TNF are potent inducers of apoptosis in a variety of cell types, their application in cancer therapy has been limited due to toxicity (Ogasawara, Watanabe-Fukunaga, Adachi, Matsuzawa, Kasugai, Kitamura, Itoh, Suda, and Nagata, 1993). In contrast, the feature of selective killing of cancer cells by TRAIL and its receptors suggests a potential application for cancer therapy.

3. TRAIL RECEPTORS

3.1. DR4

Following the discovery of the TRAIL ligand, a conserved sequence derived from TNF receptors was used to search an EST database. The first TRAIL receptor, DR4 (death receptor-4), was identified by Dixit and his colleagues (Pan, O'Rourke, Chinnaiyan, Gentz, Ebner, Ni, and Dixit, 1997a). DR4 contains two extra-cellular cysteine-rich repeats and has a conserved cytoplasmic death-domain sequence near its C-terminus. Functional studies revealed that overexpression of DR4 rapidly induces apoptosis of mammalian cells (Pan *et al.*, 1997a). It was subsequently found that apoptosis signaling by TRAIL through DR4 is independent of FADD (Pan *et al.*, 1997a). Some studies showed that apoptosis signaling upon TRAIL binding may involve FADD and may also activate NFκ-B perhaps enhancing cell survival (Chaudhary, Eby, Jasmin, Bookwalter, Murray, and Hood, 1997; Schneider, Thome, Burns, Bodmer, Hofmann, Kataoka, Holler, and Tschopp, 1997b). However, FADD knockout cells were found to undergo TRAIL-mediated apoptosis, and also a dominant negative mutant of FADD had little effect on TRAIL-mediated apoptosis further confirming that TRAIL mediated-apoptosis can occur through a FADD-independent pathway (Marsters, Sheridan, Pitti, Donahue, Ruppert, Bauer, and Ashkenazi, 1996; Yeh, De, Pompa, McCurrach, Shu, Elia, Shahinian, Ng, Wakeham, Khoo, Mitchell, El-Deiry, Lowe, Goeddel, and Mak, 1998).

3.2. KILLER/DR5

In a search for p53-downstream targets involved in apoptosis, a subtractive hybridization technique was used to isolate adriamycin-inducible transcripts from chemosensitive but not chemoresistant ovarian cancer cells. This strategy led to the isolation of *KILLER/DR5* as a DNA damage-inducible *p53*-regulated death receptor gene (Wu, Burns, McDonald, Jiang, Meng, Krantz, Kao, Gan, Zhou, Muschel, Hamilton, Spinner, Markowitz, Wu, and El-Deiry, 1997). Sequence analysis revealed that the KILLER/DR5 gene encodes a 411 amino acid polypeptide with features characteristic of the TNF receptors (Wu *et al.*, 1997). KILLER/DR5 was identified independently as a TRAIL receptor by other groups using conserved TNF receptor sequences to search an EST database, and is also known as *TRAIL-R2* or *TRICK2* (hereafter referred to as *KILLER/DR5*) (Chaudhary *et al.*, 1997; MacFarlane, Ahmadi, Srinivasula, Fernandes-Alnemri, Cohen, and Alnemri, 1997; Pan, Ni, Wei, Yu, Gentz, and Dixit, 1997b; Schneider, Bodmer, Thome, Hofmann, Holler, and Tschopp, 1997a; Screaton, Mongkolsapaya, Xu, Cowper, McMichael, and Bell, 1997; Sheridan, Marsters, Pitti, Gurney, Skubatch, Baldwin, Ramakrishman, Gray, Baker, Wood, Goddard, Godowski, and Ashkenazi, 1997; Walczak, Degli-Espositi, Johnson, Smolak, Waugh, Boiani, Timour, Gerhart, Schooley, Smith, Goodwin, and Rauch, 1997). The death-domain containing *KILLER/DR5* protein is highly homologous to *DR4*, and is a potent inducer of apoptosis (Pan *et al.*, 1997a, Sheridan *et al.*, 1997; Wu *et al.*, 1997). Like DR4, KILLER/DR5 has two cysteine-rich extra-cellular repeats, whereas TNFR-1, Fas and DR3 have four, three, and four cysteine-rich repeats, respectively (Fig. 1). It is well established that upon binding of ligands to death receptors, the death-domain containing molecules FADD, TRADD, and RIP are recruited into the different receptor complexes through protein-protein interactions involving the death domain (Ashkenazi and Dixit, 1998; Nagata, 1997; Yuan, 1997). Thus, FADD is directly recruited into the activated Fas (Boldin, Varfolomeev, Pancer, Mett, Camonis, and Wallach, 1995; Chinnaiyan, O'Rourke, Tewari, and Dixit, 1995), while TRADD brings FADD into the activated TNF receptor complex (Hsu, Xiong, and Goeddel, 1995). Our studies suggest that like the case of DR4, FADD does not appear to be required for the KILLER/DR5-mediated apoptotic signal (Bedi, Ravi, Engstrom, Wu, Yeh, Mak, El-Deiry, and Fuchs, submitted).

3.3. Decoy Receptors of TRAIL

Studies aimed at understanding why TRAIL kills transformed but not normal cells led to the discovery of naturally occurring TRAIL-antagonist or "decoy" receptors. Several groups isolated these two other members of the TRAIL-receptor family known as TRID or DcR1 (Pan *et al.*, 1997a; Sheridan *et al.*, 1997) and decoy receptor 2 (DcR2) or TRUNDD (Marsters *et al.*, 1997; Pan, Ni, Yu, Wei, and Dixit, 1998). Within their extra-cellular domains, the decoy receptors share significant homology with the pro-apoptotic receptors DR4 and KILLER/DR5. However, unlike DR4 and KILLER/DR5, TRID lacks a cytoplasmic domain and TRUNDD contains a truncated death-domain. Thus, the decoy receptors act as negative regulators of TRAIL signaling by competing

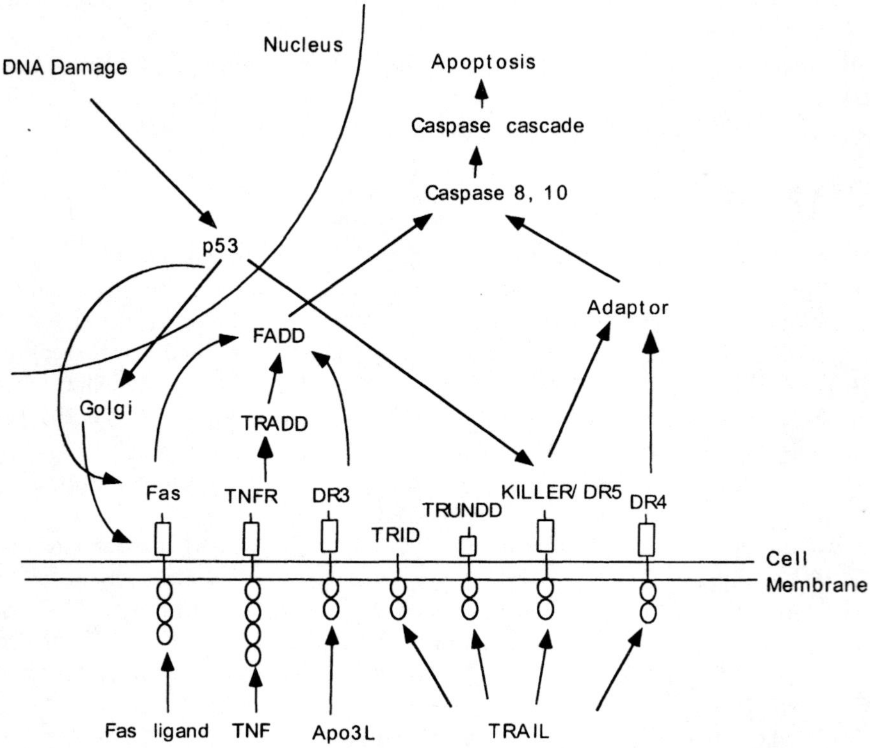

Figure 1. p53-Dependent regulation of death receptors. p53 upregulates both death receptors KILLER/DR5 and Fas/APO1 following DNA damage. The apoptosis signaling by Fas/APO1 leads to the recruitment of FADD into the activated receptor, while KILLER/DR5 signals to an as yet unidentified adaptor. It has been shown that both receptors can activate caspase 8 leading to cell death. Recent evidence suggested that Fas/APO1 receptor translocation from the Golgi to the cell surface as a potential mechanism of p53-mediated transcription-independent apoptosis (Bennett, Macdonald, Chan, Luzio, Simari, and Weissberg, 1998). However, neither the Fas receptor nor the adaptor FADD are required for p53-mediated apoptosis induced by anti-cancer drugs (Bedi *et al.*, submitted). The TRAIL lignad can bind to the pro-apoptotic receptors DR4 and KILLER/DR5 or the decoy receptors TRID or TRUNDD. Decoy receptors can antagonize TRAIL-mediated apoptosis by competing for ligand binding.

for ligand binding (Marsters *et al.*, 1997; Pan *et al.*, 1997b, Pan *et al.*, 1998; Sheridan *et al.*, 1997). It has also been reported that both decoy receptors preferentially expressed in normal cells but not tumor cells providing a potential explanation for the resistance of normal cells to TRAIL toxicity. Thus, the modulation of expression of decoy receptors may sensitize the TRAIL killing effect on tumor cells and could be useful for directing the application for cancer therapy.

4. ROLE OF TRAIL AND ITS RECEPTORS IN CANCER

Interestingly, all TRAIL receptors have been mapped to human chromosome 8p21(Marsters *et al.*, 1997; Walczak *et al.*, 1997; Wu *et al.*, 1997), a region where tumor suppressors are found in several human cancers including colon cancer, prostate cancer, bladder cancer, breast cancer, and head and neck cancer (Mitelman, Mertens and Johnsson, 1997). A mutation of KILLER/DR5 in a head and neck cancer was found encoding a truncated death domain that was defective in apoptosis induction (Pai, Wu, Ozoren, Wu, Jen, Sidransky and El-Deiry, 1998). Our more recent studies have identified a homozygous deletion of the DR4 gene in a nasopharyngeal cancer and this was associated with TRAIL resistance (Ozoren, Fisher, Kim, Liu, Lisitsyn, and El-Deiry, submitted).

5. ROLE OF DEATH RECEPTORS IN p53-DEPENDENT APOPTOSIS

p53 is the most commonly mutated gene in human cancer. p53 acts as a tumor suppressor by inducing cell cycle arrest and apoptosis (Levine, 1997). It is well known that p53-mediated cell cycle arrest involves in the activation of p21WAF1/CIP1, a cyclin-CDK inhibitor; whereas the mechanism of p53-mediated apoptosis appears to involve multiple downstream effector target genes (El-Deiry 1998). Several p53 targets involved in p53-mediated apoptosis have been identified (Fig. 2). Bax was the first gene to be implicated as a target in p53-dependent apoptosis (Miyashita, Krajewski, Krajewska, Wang, Lin, Liebermann, Hoffman, and Reed, 1994; Miyashita and Reed, 1995). It was subsequently found that bax knockout cells undergo p53-mediated apoptosis following irradiation (Knudson, Tung, Tourlellotte, Brown, and Korsmeyer, 1995; McCurrach, Corror,

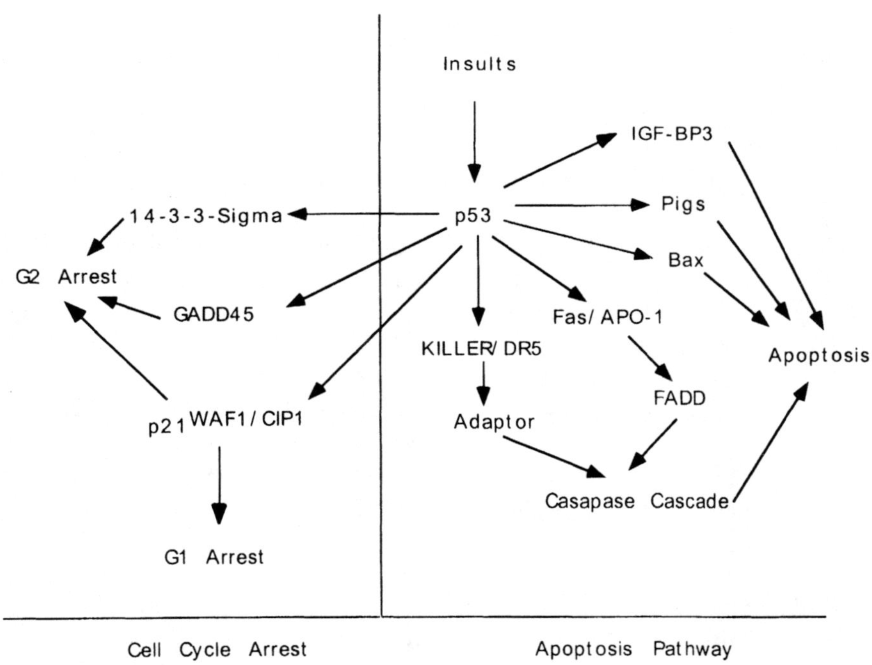

Figure 2. A model for p53-dependent apoptosis and cell cycle arrest. p53 is activated following insults leading to either cell cycle arrest by activating p21WAF1/CIP1 for G1 arrest and by activating GADD45 and 14-3-3-sigma for G2 arrest or apoptosis. Recent evidence has implicated $p21^{WAF1/CIP1}$ in the G2 check-point (Bunz, Dutriaux, Lengauer, Waldman, Zhou, Brown, Sedivy, Kinzler, and Vogelstein, 1998). p53-dependent-apoptosis is involved in the transcriptional activation of bax, KILLER/DR5, Pigs, Fas/APO1 and IGF-PB3. The activation of KILLER/DR5 and Fas/APO1 initiate a caspase-dependent apoptotit cascade, while the activation of bax leads to mitochondrial release of cytochrome c resulting in the activation of caspase 9. PIG genes are involved in generating toxic reactive oxygen species (Polyak *et al.*, 1997).

Knudson, Korsmeyer, and Lowe, 1997). Using Serial Analysis of Gene Expression (SAGE) technology, a group of genes named PIGs (p53-induced genes) involved in the generation of reactive oxygen species were linked to apoptosis upon p53 overexpression (Polyak, Xia, Zweier, Kinzler, and Vogelstein, 1997). Another gene implicated in p53-dependent apoptosis is Fas/APO1 (Owen-Schaub, Zhang, Cusack, Angelo, Santee, Fujiwara, Roth, Deisseroth, Zhang, and Kruzel, 1995), which can trigger the classic apoptosis pathway upon the activation. However, like bax, Fas/APO1 is not absolutely required for p53-dependent apoptosis (Bedi, Ravi, Engstrom, Wu, Yeh, Mak, El-Deiry, and Fuchs, submitted). KILLER/DR5 is induced by DNA damaging agents and appears to be regulated by p53 (Kastan, 1997; Wu *et al.*, 1997). It was shown that overexpression of KILLER/DR5 could efficiently eliminate colony formation of cancer cells (Hopkin, 1997; Wu *et al.*, 1997). Thus, KILLER/DR5 is a potent inducer of apoptosis, is induced by DNA damage and appears to be regulated by p53. KILLER/DR5 and Fas may be involved in activation of the death signal in response to p53. Our recent experiments suggest that the DNA damage-induced p53-dependent KILLER/DR5 upregulation occurs in cells undergoing apoptosis but not p53-dependent p21WAF1/CIP1-associated cell cycle arrest (Wu, Burns, Meng, and El-Deiry, submitted). Because death receptors have a potential application for human cancer therapy to bypass the malfunction of p53 caused by mutation, efforts are underway to screen for agents that can upregulate KILLER/DR5 expression by p53-dependent and -independent mechanisms. In this regard, recent studies have shown that the synthetic retinoid CD437 can upregulate KILLER/DR5 expression in lung cancer cells expressing wild-type (but not degraded) p53 and undergoing apoptosis (Sun, Yue, Wu, El-Deiry, Shroot, Hong, and Lotan, in press).

6. APPLICATION OF TRAIL IN CANCER THERAPY

Because TRAIL selectively kills transformed cells but not normal cells, the possibility that TRAIL may be a useful agent for treating cancers has been tested. Experiments carried out in vitro revealed that human glioma cell lines are susceptible to TRAIL-induced apoptosis (Rieger, Naumann, Glaser, Ashkenazi, and Weller, 1998). It has also been found that TRAIL kills melanoma cells which are resistant to other TNF family members including Fas (Thomas and Hersey, 1998; Griffith, Chin, Jackson, Lynch, and Kubin, 1998). It has been shown that sensitivity to TRAIL may not be reduced by the presence of a mutated p53 in some human cancer cells; however, expression of p53 enhanced the cytotoxic effects of TRAIL and Fas ligand (Rieger *et al.*, 1998). Our most recent experiments suggest that although the expression of decoy receptors is not the only factor that determines TRAIL sensitivity (Kim, and El-Deiry, submitted). It appears that there is a subgroup of TRAIL resistant cancer cells characterized by extremely low or absent DR4 expression, as well as increased expression of the caspase inhibitor FLIP (Kim and El-Deiry, submitted). Moreover, the targeted disruption of wild-type p53 expression through degradation led to TRAIL resistance in matched human colon, lung and ovarian cancer cell lines (Kim and El-Deiry, submitted). These results suggest that there are a number of potentially important determinants that contribute to the sensitivity of cancer cells to the cytotoxic effects of TRAIL. Because KILLER/DR5 is inducible by DNA damaging agents, we tested the hypothesis that there may be synergistic cancer cell killing through the use of a combination of TRAIL therapy and DNA damaging chemotherapy. We indeed found evidence for cancer-specific synergy in killing when chemotherapy was combined with TRAIL therapy (Kim and El-Deiry, submitted).

7. POTENTIAL ROLE OF DEATH RECEPTORS IN GENE THERAPY

The overexpression of death-domain containing receptors can induce apoptosis independent of ligand in vitro (Boldin, Mett, Vafolomeev, Chumakov, Shemer-Avni, Camonis, and Wallach, 1995). Moreover, even expression of death domain alone can cause cell death by apoptosis (Wu and El-Deiry, unpublished observations), suggesting that the initiation of apoptosis can bypass ligand binding and that the signal can be direcetly transduced to downstream molecules that activate caspases. Because overexpression of KILLER/DR5 eliminates colony formation of cancer cells (Hopkin, 1997; Wu *et al.*, 1997) and because both death receptors, Fas/APO1 and KILLER/DR5, appear to be regulated by the tumor suppressor gene p53, introduction of these two receptors into cancer cells would be expected to kill cancer cells. However, unlike the case of the TRAIL cytokine, there is no evidence at present that death receptor mediated killing is cancer-cell specific.

8. SUMMARY

TRAIL and its emerging receptors are the newest members of the TNF receptor super-family. The activation of TRAIL receptors by ligand binding leads to apoptosis through caspase activation through an as yet unclear signaling pathway that does not require the FADD adaptor. The TRAIL receptor KILLER/DR5, is induced by DNA damage and appears to be regulated by the tumor suppressor gene p53. Both the Fas receptor and KILLER/DR5 provide potential links between DNA damage-mediated activation of the p53 tumor suppressor and caspase activation. While further evaluation of the role of TRAIL receptors in human cancer is ongoing, initial studies suggest that both KILLER/DR5 and DR4 may be targets for inactivation and that these pro-apooptotic receptors may be tumro suppressor genes. Understanding the regulation of TRAIL and its receptors may thus be beneficial for the development of novel approaches for cancer treatment. TRAIL appears to be a cancer-specific cytotoxic agent and thus offers promise as a novel therapy for cancer either through replacement of the cytokine or potentially via gene replacement. Preliminary studies suggest the potential to combine TRAIL with classical cytotoxic chemotherapeutic drugs to achieve synergistic cell killing.

ACKNOWLEDGMENTS

W.S.E.-D. is an Assistant Investigator of the Howard Hughes Medical Institute.

REFERENCES

Ashkenazi, A., and Dixit, V.M., 1998, Death receptors: signaling and modulation, *Science* **281**:1305–1308.

Bedi, A., Ravi, R., Engstrom, L.W., Wu, G.S., Yeh, W.-C., Mak, T.W., El-Deiry, W.S., and Fuchs, E.J., 1998, Fas (CD95)-independent, p53-mediated apoptosis, submitted.

Bennett, M., Macdonald, K., Chan, S.W., Luzio, J.P., Simari, P., and Weissberg, P., 1998, Cell surface trafficking of Fas: A rapid mechanism of p53-mediated apoptosis, *Science* **282**:290–293.

Boldin, M.P., Mett, I.L., Varfolomeev, E.E., Chumakov, I., Shemer-Avni, Y., Camonis, J.H., and Wallach, D., 1995, Self-association of the "death-domains" of the p55 tumor necrosis factor (TNF) receptor and Fas/APO1 prompts signaling for TNF and Fas/APO1 effects, *J. Biol. Chem.* **270**:387–391.

Boldin, M.P., Varfolomeev, E.E., Pancer, Z., Mett, I.L., Camonis, J.H., and Wallach, D., 1995, A novel protein that interacts with the death domain of Fas/APO1 contains a sequence motif related to the death domain, *J. Biol. Chem.* **270**:7795–7798.

Bunz, F., Dutriaux, A., Lengauer, C., Waldman, T., Zhou, S., Brown, J.P., Sedivy, J.M., Kinzler, K.W., and Vogelstein, B., 1998, Requirement for p53 and p21 to sustain G2 arrest after DNA damage, *Science* **282**:149–150.

Chaudhary, P.M., Eby, M., Jasmin, A., Bookwalter, A., Murray, J., and Hood, L., 1997, Death receptor 5, a new member of the TNFR family, and DR4 induce FADD-dependent apoptosis and activate the NF-κB pathway, *Immunity* **7**:821–830.

Chinnaiyan, A.M., O'Rourke, K., Tewari, M., and Dixit, V.M., 1995, FADD, a novel death domain-containing protein, interacts with the death domain of Fas and initiates apoptosis, *Cell* **81**:505–512.

Clarke, A.R., Purdie, C.A., Harrison, D.J., Morris, R.G., Bird, C.C., Hooper, M.L., and Wyllie, A.H., 1993, Thymocyte apoptosis induced by p53-dependent and independent pathways, *Nature* **362**:849–852.

El-Deiry, W.S., 1998, The p53 pathway and cancer therapy, *The Cancer J.* **11**:229–236.

Golstein, P., 1997, Cell death: TRAIL and its receptors, *Curr. Biol.* **7**:R750–R753.

Goodwin, R.G., and Smith, C.A., 1998, The TRAIL of death, *Apoptosis* **3**:83–88.

Griffith, T.S., Chin, W.A., Jackson, G.C., Lynch, D.H., and Kubin, M.Z., 1998, Intracellular regulation of TRAIL-induced apoptosis in human melanoma cells, *J. Immunol.* **161**:2833–2840.

Hsu, H., Xiong, J., and Goeddel, D.V., 1995, The TNF receptor 1-associated protein TRADD signals cell death and NF-κB activation, *Cell* **81**:495–504.

Hopkin, K., 1997, For p53, there's more than one way to kill a cell, *J. NIH Res.* **9**:22–24.

Itoh, N., Yonehara, S., Ishii, A., Yonehara, M., Mizushima, S., Sameshima, M., Hase, A., Seto, Y., and Nagata, S., 1991, The polypeptide encoded by the cDNA for human cell surface antigen Fas can mediate apoptosis, *Cell* **66**:233–243.

Kastan, M., 1997, On the TRAIL from p53 to apoptosis?, *Nature Genet.* **17**:130–131.

Kerr, J. F.R., Wyllie, A.H., and Currie, A.R., 1972, Apoptosis: A Basic Biological Phenomenon with Wide-Ranging Implications in Tissue Kinetics, *Br. J. Cancer* **26**:239–257.

Kim, K.H., and El-Deiry, W.S., 1998, Molecular determinants of response to TRAIL combined with chemotherapy in killing of normal and cancer cells, submitted.

Knudson, C.M., Tung, K.S., Tourlellotte, W.G., Brown, G.A., and Korsmeyer, S.J., 1995, Bax-deficient mice with lymphoid hyperplasia and male germ cell death, *Science* **270**:96–99.

Levine, A.J., 1997, p53, the cellular gatekeeper for growth and division, *Cell* **88**:323–331.

MacFarlane, M., Ahmad, M., Srinivasula, S.M., Fernandes-Alnemri, T., Cohen, G.M., and Alnemri, E., 1997, Identification and molecular cloning of two novel receptors for cytotoxic ligand TRAIL, *J. Biol. Chem.* **272**:25417–25420.

Marsters, S.A., Pitti, R.M., Donahue, C.J., Ruppert, S., Bauer, K.D., and Ashkenazi, A., 1996, Activation of apoptosis by Apo-2 ligand is independent of FADD but blocked by CrmA, *Curr. Biol.* **6**:750–752.

Marsters, S.A., Sheridan, J.P., Pitti, R.M., Huang, A., Skubatch, M., Baldwin, D., Yuan, J., Gurney, A., Goddard, A.D., Godowski, P., and Ashkenazi, A., 1997, A novel receptor for Apo2L/TRAIL contains a truncated death domain, *Curr. Biol.* **7**:1003–1006.

McCurrach, M.E., Connor, T.M., Knudson, C.M., Korsmeyer, S.J., and Lowe, S.W., 1997, bax-deficiency promotes drug resistance and oncogenic transformation by attenuating p53-dependent apoptosis, *Proc. Natl. Acad. U.S.A.* **94**:2345–2349.

Miller, T.M., Moulder, K.L., Knudson, C.M., Creedon, D.J., Deshmukh, M., Korsmeyer, S.J., and Johnson, E.M. Jr., 1997, Bax deletion further orders the cell death pathway in cerebellar granule cells and suggests a caspase-independent pathway to cell death, *J. Cell Biol.* **139**:205–217.

Mitelman, F., Mertens, F., and Johansson, B., 1997, A breakpoint map of recurrent chromosomal rearrangements in human neoplasia, *Nature Genet.* **15**:417–474.

Miyashita, T., Krajewski, S., Krajewski, M., Wang, H.G., Lin, H.K., Liebermann, D., Hoffman, B., and Reed, J.C., 1994, Tumor suppressor p53 is a regulator of bcl2 and bax gene expression in vitro and in vivo, *Oncogene* **9**:1799–1805.

Miyashita, T., and Reed, J.C., 1995, Tumor suppressor p53 is a direct transcriptional activator of human bax gene, *Cell* **80**:293–299.

Nagata, S., 1997, Apoptosis by death factor, *Cell* **88**:355–365.

Ogasawara, J., Watanabe-Fukunaga, R., Adachi, M., Matsuzawa, A., Kasugai, T., Kitamura, Y., Itoh, N., Suda, T., and Nagata, S., 1993, Lethal effect of the anti-Fas antibody in mice, *Nature* **364**:806–809.

Owen-Schaub, L.B., Zhang, W., Cusack, J.C., Angelo, L.S., Santee, S.M., Fujiwara, T., Roth, J.A., Deisseroth, A.B., Zhang, W.W., and Kruzel, E., 1995, Wild-type human p53 and a temperature-sensitive mutant induce Fas/APO-1 expression, *Mol. Cell Biol.* **15**:3032–3040.

Ozoren, N., Fisher, M.J., Kim, K.H., Liu, C.X., Lisitsyn, N.A., and El-Deiry, W.S., 1998, Homozygous deletion of the death receptor DR4 gene in a nasopharyngeal cancer cell line is associated with TRAIL resistance, submitted.

Pai, S.I., Wu, G.S., Ozoren, N., Wu, L., Jen, J., Sidransky, D., and El-Deiry, W.S., 1998, Rare loss-of-function mutation of a death receptor gene in head and neck cancer, *Cancer Res.* **58**:3513–1318.

Pan, G., O'Rourke, K., Chinnaiyan, A.M., Gentz, R., Ebner, R., Ni, J., and Dixit, V.M., 1997a, The receptor for the cytotoxic ligand TRAIL, *Science* **276**:111–113.

Pan, G., Ni, J., Wei, Y.F., Yu, G., Gentz, R., and Dixit, V.M. An antagonist decoy receptor and a death domain-containing receptor for TRAIL., 1997b, *Science* **277**:815–818.

Pan, G., Ni, J., Yu, G., Wei, Y.F., and Dixit, V.M., 1998, TRUNDD, a new member of the TRAIL receptor family that antagonizes TRAIL signalling, *FEBS Letters* **424**:41–45.

Pitti, R.M., Marsters, S.A., Ruppert, S., Donahue, C.J., Moore, A., and Ashkenazi, A., 1996, Induction of apoptosis by Apo-2 ligand, a new member of the tumor necrosis factor cytokine family, *J. Biol. Chem.* **271**:12687–12690.

Polyak, K., Xia, Y., Zweier, J.L., Kinzler, K.W., and Vogelstein, B., 1997, A model for p53-induced apoptosis, *Nature* **389**:300–305.

Rieger, J., Naumann, U., Glaser, T., Ashkenazi, A., and Weller, M., 1998, APO2 ligand: a novel lethal weapon against malignant glioma?, *FEBS Letters* **427**:124–128.

Schneider, P., Bodmer, J.L., Thome, M., Hofmann, K., Holler, N., and Tschopp, J., 1997a, Characterization of two receptors for TRAIL, *FEBS Letters* **416**:329–334.

Schneider, P., Thome, M., Burns, K., Bodmer, J.L., Hofmann, K., Kataoka, T., Holler, N., and Tschopp, J., 1997b, TRAIL receptors 1 (DR4) and 2 (DR5) signal FADD-dependent apoptosis and activate NF-κB, *Immunity* **7**:831–836.

Screaton, G.R., Mongkolsapaya, J., Xu, X.N., Cowper, A.E., McMichael, A.J., and Bell, J., 1997, TRICK2, a new alternatively spliced receptor that transduces the cytotoxic signals from TRAIL, *Curr. Biol.* **7**:693–696.

Sheridan, J.P., Marsters, S.A., Pitti, R.M., Gurney, A., Skubatch, M., Baldwin, D., Ramakrishnan, L., Gray, C.L., Baker, K., Wood, W.I., Goddard, A.D., Godowski, P., and Ashkenazi, A., 1997, Control of TRAIL-induced apoptosis by a family of signaling and decoy receptors, *Science* **277**:818–821.

Sun, S.-Y., Yue, P., Wu, G.S., El-Deiry, W.S., Shroot, B., Hong, W.K., and Lotan, R., 1998, Mechanisms of apoptosis induced by the synthetic retinoid CD437 in human non-small cell lung carcinoma cells, *Oncogene*, in press.

Thomas, W.D., and Hersey, P., 1998, TNF-relating apoptosis-inducing ligand (TRAIL) induces apoptosis in Fas ligand-resistant melanoma cells and mediates CD4 T cell killing of target cells, *J. Immnol.* **161**-:2195–2200.

Walczak, H., Degli-Esposito, M.A., Johnson, R.S., Smolak, P.J., Waugh, J.Y., Boiani, N., Timour, M.S., Gerhart, M.J., Schooley, K.A., Smith, C.A., Goodwin, R.G., and Rauch, C.T., 1997, TRAIL-R2: a novel apoptosis-mediating receptor for TRAIL, *EMBO J.* **16**:5386–5397.

Wiley, S.R., Schooley, K., Smolak, P.J., Din, W.S., Huang, C.-P., Nicholl, J.K., Sutherland, G.R., Smith, T.D., Rauch, C., Smith, C.A., and Goodwin, R.G., 1995, Identification and characterization of a new member of the TNF family that induces apoptosis, *Immunity* **3**:673–682.

Wu, G.S., Burns, T.F., McDonald, E.R., III, Jiang, W., Meng, R., Krantz, I.D., Kao, G., Gan, D.-D., Zhou, J.-Y., Muschel, R., Hamilton, S.R., Spinner, N.B., Markowitz, S., Wu, G., and El-Deiry, W.S., 1997, *KILLER/DR5* is a DNA damage-inducible p53-regulated death receptor gene, *Nature Genet.* **17**: 141–143.

Wu, G.S., Burns, T.F., Meng, R.D., and El-Deiry, W.S., 1998, Involvement of the TRAIL receptor KILLER/DR5 in p53-dependent apoptosis but not growth arrest, submitted.

Yeh, W.C., Pompa, J.L., McCurrach, M.E., Shu, H.B., Elia, A.J., Shahinian, A., Ng, M., Wakeham, A., Khoo, W., Mitchell, K., El-Deiry, W.S., Lowe, S.W., Goeddel, D.V., and Mak, T.W., 1998, FADD: Essential for embryo development and signaling from some, but not all, inducers of apoptosis, *Science* **279**-:1954–1958.

Yuan, J., Shaham, S., Ledoux, S., Ellis, H.M., and Horvitz, H.R., 1993, The C. elegans cell death gene ced-3 encodes a protein similar to mammalian interleukin-1β-converting enzyme, *Cell* **75**:641–652.

Yuan, J., 1997, Transducing signals of life and death, *Curr. Opin. in Cell Biol.* **9**:247–251.

14

APOPTIN®

Alexandra Pietersen and Mathieu H. M. Noteborn

Leadd BV, and Apoptin Group
Department of Molecular Cell Biology
Leiden University Medical Center
P.O. Box 9503, 2300 RA Leiden
The Netherlands

1. APOPTOSIS

Apoptosis is a programmed physiological process for eliminating superfluous, or altered cells. Apoptosis can be induced by many stimuli through different pathways, all of which seem to converge into one evolutionary conserved process (Jacobson *et al.*, 1997; Chinnaiyan and Dixit, 1996). This process is characterized by distinct morphological events such as shrinkage of cells, segmentation of the nucleus, and condensation and cleavage of the DNA. Finally, the apoptotic remnants are enveloped by membranes and will undergo rapid phagocytosis by neighboring cells (Vaux and Strasser, 1996; Wyllie, 1995).

Apoptosis plays an essential role during embryogenesis and in adult organisms regulating e.g. the shape of organs and/or tissues or the function of the immune system. Apoptosis is also linked to various diseases, such as AIDS, Alzheimer disease, autoimmune diseases or cancer (Duke *et al.*, 1996).

1.1. The Involvement of Apoptosis in Tumor Development

Until recently, it was thought that the formation of tumors was caused solely by enhanced cell proliferation. Research data shows, however, that a decreased level of apoptosis also contributes to tumor formation (Duke *et al.*, 1996). Cells with mutations in their DNA have to repair this DNA damage before they will divide. If they are not able to repair the mutations, the genomic alterations may turn them into tumor cells. Alternatively, damaged cells may decide to undergo apoptosis and, by doing so, prevent tumor growth (Hunter, 1997; Paulovich, 1997).

For instance, the tumor-suppressor protein p53 plays an important role in the prevention of tumor growth. Individuals with a herited p53 defect have sarcomas, breast tumors or have adrenal carcinomas in more than 90% of the cases. It is generally assumed that tumor suppression is mainly due to p53-induced apoptosis and that, during tumor

Cancer Gene Therapy: Past Achievements and Future Challenges, edited by Habib
Kluwer Academic/Plenum Publishers, New York, 2000.

formation, a selection for non-functional p53 occurs (Smith and Fornace, 1995; Levine, 1997). The anti-apoptotic protein Bcl-2, and some related proteins, e.g. Bcl-x-l, BAG-1, Bak, Bax (Reed, 1997; Rao and White, 1997) are also involved in tumor development. As a result of chromosomal translocations, Bcl-2 is over-expressed in various tumors, e.g. leukemia, lymphoma, and breast cancer. Another example is chronic myeloid leukemia caused by a translocation involving the BCR and ABL genes. The proteins (protein-tyrosine kinases) mutated by this translocation inhibit apoptosis and cause

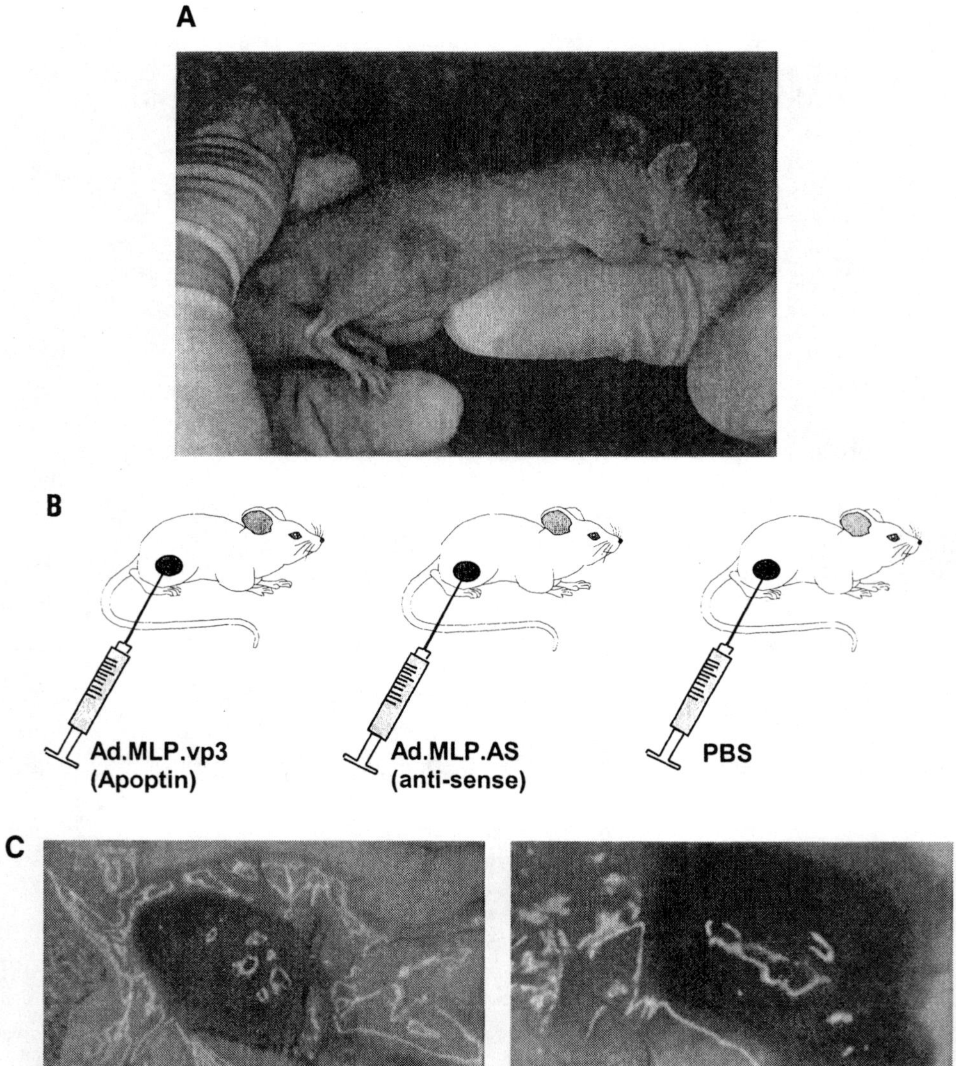

Figure 1. Anti-tumor effect of Ad-Apoptin. Panel A. Nude mice bearing an established tumor consisting of human hepatoma (HepG2) cells were treated with a single intratumoral injection. Panel B. The mice were distributed over three groups, receiving either a replication deficient adenovirus vector encoding Apoptin (AdMLP-vp3), an adenovirus vector containing the Apoptin gene in the antisense (AS) orientation, thus not encoding any proteins (AdMLP-AS) or saline. Panel C. The Apoptin treatment (left panel) resulted in a significant reduction of tumor growth, accompanied by drastic morphological changes: the tumors appeared pale, contained vesicle-like structures and showed a reduction in tumor vasculature. In contrast, these phenomena were not observed in the control-treated tumors (right panel).

the tumor to expand (Hunter, 1997). Several oncogenic DNA viruses encode proteins that disturb the function of p53 and, as a consequence, p53-induced apoptosis. Alternatively, DNA viruses have developed proteins that mimic and/or interfere with components of the apoptotic pathway, and in that way inhibit apoptosis (Teodoro and Branton, 1997).

1.2. Tumor Treatment Is Also Based on Apoptosis: a Paradox

Apoptosis is not only essential for tumor growth, but may also be exploited for the treatment of tumors. Many tumor cells have a disabling mutation in the decision machinery for apoptosis, but their execution system is still intact. This means that tumor cells will still die, if they are provided with an effective apoptotic signal (McDonell *et al.*, 1995). Chemo-therapeutic agents often induce apoptosis via activation of p53. However, more than 50% of the tumors, such as melanomas, lung cancer, or colon carcinomas do not contain functional p53. Patients with these types of tumors have a very low chance of responding to (chemo)-therapy (Levine, 1997; Lowe *et al.*, 1994). Likewise, over-expression of the anti-apoptosis proteins Bcl-2, or BCR-ABL has a negative influence on the chemo-therapeutic treatment of a large number of lymphomas (Rao and White, 1997; Martin and Green, 1994).

New anti-tumor therapies are underway, which are based on the induction of apoptosis and are not hampered by the lack of p53 and/or over-expression of anti-apoptosis genes. Recently, in our laboratory the apoptosis-inducing protein Apoptin was discovered. In this report, its interactions with apoptosis-mediating (e.g. p53) and apoptosis-inhibiting proteins (e.g. Bcl-2) will be discussed, as well as the development of a prototype anti-tumor therapy based on an adenovirus vector expressing Apoptin.

2. APOPTIN, A CHICKEN ANEMIA VIRUS (CAV) PROTEIN, INDUCES APOPTOSIS

2.1. CAV Induces Apoptosis

The chicken anemia virus (CAV) causes depletion of thymocytes and erythroblastoid cells in young chickens via apoptosis. In-vitro, these apoptotic processes were also observed in CAV-susceptible transformed chicken cells (Coombes and Crawford, 1996; Noteborn and Koch, 1995). DNA isolated from the CAV-infected cells shows the apoptosis-specific laddering pattern. Electron-microscopic analysis shows cells containing condensed chromatin, and apoptotic bodies in the cytoplasm of neighboring epithelial cells.

2.2. Apoptin®

The CAV genome, which was cloned in our laboratory, consists of a single-stranded circular DNA of 2319 nucleotides (Noteborn *et al.*, 1991). From the double-stranded replicative intermediate, a single polyadenylated polycistronic mRNA is transcribed (Noteborn and Koch, 1995) which encodes three distinct genes VP1 (51.6 kDa) being the capsid protein, VP2 (24 kDa) and VP3 (13,6 kDa). In transformed chicken cells, expression of VP3 alone was sufficient for the induction of apoptosis, leading to naming this protein Apoptin®, from **apopt**osis **in**duction (Noteborn *et al.*, 1994; Noteborn *et al.*, 1998a).

Apoptin is a rather small protein of 121-amino-acids long, which does not resemble any other sequenced animal or viral protein. It carries several known protein domains, including 2 nuclear-localization signals, and a nuclear-export signal. Furthermore, it has regions rich in proline or basic amino acids and overall Apoptin contains a high percentage of serine and threonine residues (Noteborn *et al.*, 1991).

3. APOPTIN INDUCES CELL DEATH IN HUMAN TRANSFORMED AN/OR TUMORIGENIC CELL LINES, BUT NOT IN NORMAL CELLS

3.1. Apoptin Induces Apoptosis in Various Human Tumor Cells

Interestingly, Apoptin can also induce apoptosis in various human transformed and or tumorigenic cell lines, derived from e.g. hepatomas, lymphomas, leukemias, melanomas, breast and lung tumors, neuroblastomas, cholangio-, colon- and squamous cell carcinomas (Noteborn *et al.*, 1998b; Noteborn, unpublished results). Apoptin-induced apoptosis is faster in one tumor cell line than in another, but it always reaches 90–100% of the Apoptin-positive cells 6 days after transfection. Soon after synthesis, the protein is dispersed throughout the nucleus and somewhat later the cell undergoes apoptosis. At this point in time, Apoptin aggregates and the cellular DNA condenses and/or is fragmented (Noteborn *et al.*, 1998a,b).

3.2. Apoptin Does Not Induce Apoptosis in Normal Human Cells

In vitro, Apoptin fails to induce apoptosis in normal lymphoid, dermal, fibroblastic, epidermal, endothelial and smooth-muscle cells. In addition, rodent embryo diploid fibroblasts and hepatocytes from adult rats were also shown not to undergo apoptosis upon expression of Apoptin. Long-term expression of Apoptin in normal human fibroblasts made clear that Apoptin has no toxic or transforming activity in these cells (Danenvan Oorschot *et al.*, 1997).

The possible cause for this phenomenon is that, in tumor cells, Apoptin is located in the nucleus, whereas in normal cells it was found to be present in the cytoplasm. Zhuang *et al.* (1995) have provided further evidence that its nuclear location is important for its apoptotic activity. Deletion of one of the nuclear location signals caused partial translocation of Apoptin in the perinuclear structures. Under these circumstance, the Apoptin induced apoptosis was significantly reduced. Therefore, it seems to be essential for Apoptin induction of apoptosis that Apoptin co-localizes with chromatin. The basic regions of Apoptin may allow interaction with nucleic acids. The presence of Apoptin in the chromatin structure, and its high proline content, may cause disturbance of the supercoil organization, resulting in apoptosis. Another possibility is that Apoptin acts as a transciptional and/or inhibitor of genes, which directly mediate apoptosis (Noteborn *et al.*, 1998a,b).

3.3. Transforming Agents Enable Apoptin-Induced Apoptosis

We have found that transient transfection of diploid cells with the SV40 transforming gene large T antigen will render these normal cells susceptible to Apoptin-induced apoptosis (Noteborn *et al.*, 1998c). The Apoptin protein becomes translocated from the cytoplasm into the nucleus. These results imply that expression of a transforming protein is sufficient to render normal cells susceptible to Apoptin, and the establishment of a stably transformed state is not required.

UV-irradiation or X-ray treatment causes an aberrant SOS-response in primary diploid cells from cancer-prone individuals (e.g. Li-Fraumeni and Lynch Type 2 syndrome; Abrahams *et al.*, 1996). This response can be described as a transient transformation and as expected, under these circumstances Apoptin can induce apoptosis (Zhang *et al.*, unpublished results).

All these features clearly indicate that Apoptin induces apoptosis in a transformed/tumorigenic cellular background.

4. APOPTIN DOES NOT ACT VIA p53 AND IS STIMULATED BY BCL-2

The tumor-specific activity of Apoptin makes it a highly potential candidate as an anti-tumor agent. To explore its potential further, we have analysed whether Apoptin can induce apoptosis in cases where conventional therapies fail, due to the absence of p53 or presence of over-expressed anti-apoptotic proteins.

The apoptotic decision process is a cascade of events, followed by the execution stage. At each decision step after an apoptotic trigger, a go/no-go decision can be made. A large panel of cytotoxic agents induces apoptosis via activation of the tumor-suppressor protein p53 (Levine, 1997; Guchelaar *et al.*, 1997). The next step in the cascade might be controlled by Bcl-2 (Reed, 1997; Rao and White, 1997) and, finally, activation of a cascade of specific cystein-proteases (caspases) is required, which is regarded as the point of no return after which the cells will rapidly undergo apoptosis (Nicholson, 1997).

4.1. Apoptin Induces p53-Independent Apoptosis

To examine whether Apoptin requires expression of wild-type p53 to induce apoptosis, the human osteosarcoma cell lines U2OS (wild-type p53), Saos-2/ala143 (mutant p53), and Saos-2 (no p53) were transiently transfected with plasmid DNA encoding Apoptin (Zhang *et al.*, 1995b). At various time points after transfection, the percentage of Apoptin-expressing p53-minus cells, in which the nuclei became apoptotic, was similar to that in U2OS cells, containing wild-type p53. In addition, the Apoptin-induced apoptosis does not seem to be slower in p53-minus Saos-2 cells than the one induced by p53, which indicates that the distinct Apoptin pathway has as much potential as the p53-regulated one. Apoptin can induce apoptosis in various other human or mammalian cells that do not express p53 (Noteborn *et al.*, 1998a). Besides p53 gene deletion and point mutations, p53 can be inactivated by proteins encoded by DNA-tumor viruses. We have obtained evidence that expression of the adenovirus E1B 55K (Zhuang *et al.*, unpublished results) or of the SV40 large T antigen (Danen-van Oorschot, 1997) do not interfere with apoptosis induced by Apoptin. All these data imply that Apoptin acts via a p53-independent apoptotic pathway (Table 1).

4.2. Bcl-2 Stimulates Apoptin-Induced Apoptosis

Next, we have analysed whether the proto-oncogene Bcl-2 can inhibit Apoptin-induced apoptosis. Bcl-2 and related proteins, are shown to inhibit both p53-dependent

Table 1. Characteristics of Apoptin®-induced apoptosis

Requirement for:	
P53	No
Inhibition by:	No
Ad5 E1B-55K	No
SV40 large T antigen	No
BCR-ABL	No
CrmA	No
ZVAD	No
Stimulation by:	
Bcl-2	Yes
BAG-1	Yes
Apoptin induced apoptosis:	
Normal (human) cells	No
Transformed/tumorigenic (human) cells	Yes
Pre-clinical results	
Toxicity in healthy animals	No
Anti-tumor activity	Yes

and p53-independent apoptosis (Rao and White, 1997). Danen-van Oorschot *et al.* (1997b) has shown that the Bcl-2-associating protein BAG-1 itself and in combination with Bcl-2 can inhibit p53-induced apoptosis in human transformed cells.

We have reported that Apoptin can induce apoptosis in human hematologic malignant DoHH-2 cells over-expressing Bcl-2 due to a t(14;18)(q31;21) chromosomal translocation. Apoptin induced apoptosis is faster in these Bcl-2-over-expressing cells than in ones with a normal level of Bcl-2 (Zhuang *et al.*, 1995b). Next, we examined in further detail whether Bcl-2 enhances Apoptin activity. To that end, human p53-minus Saos-2 cells were transiently transfected with plasmids encoding Apoptin and/or Bcl-2. Several days after transfection, human Saos-2 cells transfected with both Apoptin and Bcl-2 were shown to undergo apoptosis much faster than cells expressing Apoptin alone (Danen-Van Oorschot *et al.*, 1998). Co-expression of Apoptin with BAG-1 also resulted in an enhancement of Apoptin activity, although much less pronounced than Bcl-2 (Danen-Van Oorschot *et al.*, 1997b; Table 1).

The observation that over-expression of the proto-oncogene Bcl-2 does not force Apoptin to induce apoptosis in normal diploid human keratinocytes is interesting (Danen-Van Oorschot, 1998). Thus far, Bcl-2 is related to tumor formation due to its anti-apoptotic features and not due to its transforming ability (Rao and White, 1997). Apparently, non-transformed normal cells will not be (more) susceptible to Apoptin due to Bcl-2 over-expression, as is the case for transformed/tumorigenic ones (Noteborn *et al.*, 1998a).

Apoptin induces apoptosis in human K562 cells (derived from acute myeloid leukemias) which produce a chimeric BCR-ABL protein (Zhuang *et al.*, 1995b). From a therapeutic point of view these properties of Apoptin are important, because several tumors such as lymphomas and leukemias carry an (over)-expressed Bcl-2 gene or BCR-ABL gene resulting in their derailment and resistance to (chemo)-therapeutic agents (Table 1).

4.3. Apoptin Requires a Different Panel of Caspases Than p53

Preliminary data show that the cowpox-virus protein CrmA, which inhibits caspase 1 activity does not negatively influence Apoptin-induced apoptosis, but can inhibit p53-induced apoptosis (Danen-van Oorschot, unpublished results). Similar results were obtained with modified peptides, known to specifically inhibit caspases, also those more downstream in the caspase cascade (Table 1). These features illustrate that Apoptin acts via an unknown apoptotic pathway and/or very downstream in the apoptotic cascade.

5. DEVELOPMENT OF AN ANTI-TUMOR THERAPY

Specificity is an essential prerequisite for cancer gene therapy. Therefore, the fact that the Apoptin gene can specifically induce apoptosis in tumor cells, whereas other agents such as chemotherapeutic agents cannot, clearly make Apoptin a promising agent for anti-tumor therapy.

To investigate the potential of Apoptin for cancer gene therapy, the Apoptin gene has to be efficiently introduced into malignant cells *in vivo*. At present, the most efficient system to achieve this is based on replication-deficient adenovirus vectors. Recombinant adenoviral vectors can be grown to high titers. They are able to transduce non-mitotic cells, and do not integrate in the host genome. Many clinical gene-therapy trials are underway, which are based on (replication-deficient) adenovirus vectors.

We have developed a replication-deficient adenovirus vector AdMLP-VP3, expressing Apoptin under the regulation of the Ad5 major late promoter (Pietersen *et al.*, to be published). This Apoptin vector can be propagated in adenovirus-transformed cell lines such as 293, 911 and PER.C6 cells (Fallaux, 1996), but do not replicate in target cells. Pietersen *et al.* (unpublished results) have found that after adenovirus transfer of the Apoptin gene, human hepatoma HepG2 and Hep3B cells are susceptible to Apoptin, whereas primary hepatocytes are not. Similar results were found for other combinations of cells, e.g. keratinocytes versus squamous cell carcinoma. Infection of HepG2 cells with an adenoviral vector encoding Apoptin results in induction of apoptosis faster than transfection with a plasmid encoding Apoptin. Apparently, the adenoviral infection somehow stimulates Apoptin-induced apoptosis due to changes induced by the adenoviral infection, or, alternatively the effective intracellular level of Apoptin is higher after viral infection than after transfection. The effective anti-tumor activity of Adenovirus expression of Apoptin was demonstrated in vitro by inducing almost 100% cell death in a dish with human hepatoma or cholangiocarcinoma cells. In contrast, when these cells are infected with control adenovirus vectors, they continue to grow.

To evaluate the possible toxic effects of an adenovirus expressing Apoptin, single and/or multiple injections of virus were given to healthy rats. These toxicity studies prove that recombinant adenovirus expressing Apoptin does not cause obvious abnormalities, confirming and extending the results of the *in-vitro* studies. The animals show a normal increase in body weight and look healthy. Upon autopsy, the liver looks normal, although after intravenous injection about 90% of (recombinant) adenovirus accumulates in the liver. By means of BrdU labelling, it was shown that no increase in liver cell proliferation was found compared to control Ad-infection. Also, no Apoptin-related increase of apoptotic cells was observed in AdMLP-VP3 expressing Apoptin versus control adenoviruses (Dierendonck and Van der Eb, unpublished results). The fact that we were also successful in generating transgenic-Apoptin mice (expressing Apoptin in various tissues under the regulation of the broad MHC-1 promoter) strengthens the non-toxic character of Apoptin (Erkeland, unpublished results).

Further experiments showed that adenovirus-encoded Apoptin also has anti-tumor potential *in vivo*. Nude mice bearing subcutaneous human hepatomas were injected intratumorally with a single batch of recombinant-Apoptin adenovirus, control adenovirus or buffer. Only tumors treated with adenovirus expressing Apoptin showed a significant reduction of tumor growth. Remarkably, the Apoptin-treated tumors showed a reduced vascularization, in comparison with the control groups, which received an intratumoral injection with control adenovirus vectors. These results indicate that adenovirus vectors producing Apoptin may constitute a powerful tool for the treatment of solid tumors.

6. SUMMARY

Tumor cells are often characterized by the traithey are more resistant to apoptosis induced by e.g. cytotoxic agents than normal cells. Resistance to apoptosis induction can be a direct consequence of mutations in certain tumor-suppressor genes (p53) or of certain proto-oncogenes (Bcl-2). Therefore, new cancer therapies are under development to bypass the resistance to chemo- and radio-therapy of tumors.

Apoptin acts independently of p53, is stimulated by Bcl-2 and is insensitive to BCR-ABL, which means that Apoptin can induce apoptosis in cases where present (chemo)-therapeutic agents, unfortunately, will fail. The fact that Apoptin induces apoptosis in human tumorigenic cells but not in normal diploid cells, implies that side-effects of Apoptin treatment are expected to be minor.

In-vivo results with a first prototype of anti-tumor therapy based on expression of Apoptin indicate that Apoptin has low acute toxicity and is effective as an anti-tumor agent.

REFERENCES

Abrahams, P.J., Houweling, A., Cornelissen-Steijger, P.D.M., Arwert, F., Menko, F.H., Pinedo, H.M., Terleth, C., and Van der Eb, A.J. 1996, Inheritance of abnormal expression of SOS-like response in Xeroderma pigmentosum and hereditary cancer-prone syndromes. *Cancer Res.* **56**:2621–2625.

Chinnaiyan, A.M., and Dixit, V.M. 1996, The cell-death machine. *Curr. Biol.* **6**:555–562.

Coombes, A.L., and Crawford, G.R. 1996, Chicken anaemia virus: a short review. *World's Poultry Sci. J.* **52**:267–277.

Danen-van Oorschot, A.A.A.M., Fischer, D.F., Grimbergen, J.M., Klein, B., Zhuang, S.-M., Falkenburg, J.H.F., Backendorf, C., Quax, P.H.A., Van der Eb, A.J., and Noteborn, M.H.M. 1997, Apoptin induces apoptosis in human transformed and malignant cells but not in normal cells. *Proc. Natl. Acad. Sci. USA* **94**:5843–5847.

Danen-Van Oorschot, A.A.A.M., Den Hollander, A., Takayama, S., Reed, J., Van der Eb, A.J., and Noteborn, M.H.M. 1997, BAG-1 inhibits p53-induced but not apoptin-induced apoptosis. *Apoptosis* **2**:395–402.

Danen-Van Oorschot, A.A.A.M., Zhang, Y.-H., Erkeland, S., Fischer, D.F., Van der Eb, A.J., and Noteborn, M.H.M. 1998, The effect of Bcl-2 on Apoptin in normal versus transformed human cells. *Leukemia*, In press.

Duke, R.C., Ojcius D.M., and Young, J.D.-E. 1996, Cell suicide in health and disease. *Sci. American* **274**:48–55.

Fallaux, F. 1996, Gene Therapy for hemophilia A: Towards the use of adenoviral vectors? PhD Thesis, Leiden University, The Netherlands.

Guchelaar, H.J., Vermes, A., Vermes, I., and Haanen, C. 1997, Apoptosis: molecular mechanisms and implications for cancer chemotherapy. *Pharm. World Sci.* **19**:119–125.

Hunter, T. 1997, Oncoprotein Networks. *Cell* **88**:333–346.

Jacobson, M.D., Weil, M., and Raff, M. 1997, Programmed cell death in animal development. *Cell* **88**:347–354.

Levine, A.J. 1997, p53, the cellular gatekeeper for growth and division. *Cell* **88**:323–331.

Lowe, S., Bodis, S., McClathey A., Remington, L., Ruley, H.E., Fisher, D.E., Houseman, D.E., and Jacks, T. 1994, p53 status and the efficacy of cancer therapy *in vivo*. *Science* **266**:807–810.

Martin S.J., and Green, D.R. 1994, Apoptosis as a goal of cancer therapy. *Curr. Opin. Oncol.* **6**:616–621.

McDonnell, T.J., Meyn, R.E., and Robertson, L.E. 1995, Implications of apoptotic cell death regulation in cancer therapy. *Sem. Cancer Biol.* **6**:53–60.

Nicholson, D.W., and Thornberry, N.A. 1997, Caspases: killer proteases. *Tr. Biol. Sci.* **22**:299–306.

Noteborn, M.H.M., and Koch, G. 1995, Chicken anemia virus infection: molecular basis of pathogenicity. *Avian Pathol.* **24**:11–31.

Noteborn, M.H.M., Danen-Van Oorschot, A.A.A.M., and Van der Eb, A.J. 1998, The Apoptin® gene of chicken anemia virus in the induction of apoptosis in human tumorigenic cells and in gene therapy of cancer. *Gene Ther. Mol. Biol.* **1**:399–406.

Noteborn, M.H.M., Danen-Van Oorschot, A.A.A.M., and Van der Eb, A.J. 1998, Chicken anemia virus: Induction of apoptosis by a single protein of a single-stranded DNA virus. *Sem. Virol.* **8**:497–504.

Noteborn, M.H.M., De Boer, G.F., Van Roozelaar, D., Karreman, C., Kranenburg, O., Vos, J.G., Jeurissen, S.H.M., Hoeben, R.C., Zantema, A., Koch, G., Van Ormondt, H., and Van der Eb, A.J. 1991, Characterization of cloned chicken anemia virus DNA that contains all elements for the infectious replication cycle. *J. Virol.* **65**:3131–3139.

Noteborn, M.H.M., Zhang, Y.-H., and Van der Eb, A.J. 1998, Apoptin® specifically causes apoptosis in tumor cells and after UV-treatment in untransformed cells from cancer-prone individuals: A review. *Mutation Research* **400**:447–455.

Noteborn, M.H.M., Todd, D., Verschueren, C.A.J., De Gauw, H.W.F.M., Curran, W.L., Veldkamp, S., Douglas, A.J., McNulty, M.S., Van der Eb, A.J., and Koch, G. 1994, A single chicken anemia virus protein induces apoptosis. *J. Virol.* **68**:346–351.

Paulovich, A.G., Toczysцki, D.P., and Hartwell, L.H. 1997, When Checkpoints fail. *Cell* **88**:315–321.

Rao, L., and White, E. 1997, Bcl-2 and the ICE family of apoptotic regulators: making a connection. *Curr. Opin. Genet. Dev.* **7**:52–58.

Reed, J.C. 1997, Double identity for proteins of the Bcl-2 family. *Nature* **387**:773–776.

Smith, M.L., and Fornace, Jr. A.J. 1995, Genomic instability and the role of p53 mutations in cancer cells. *Curr. Opin. Oncol.* **7**:69–75.

Teodoro, J.G., and Branton, P.E. 1997, Regulation of Apoptosis by viral gene products. *J. Virol.* **71**:1739–1746.

Vaux, D.L., and Strasser, A. 1996, The molecular biology of apoptosis. *Proc. Natl. Acad. Sci. USA* **93**: 2239–2244.

Wyllie, A.H. 1995, The genetic regulation of apoptosis. *Curr. Opin. Gen. Dev.* **5**:97–104.

Zhuang, S.-M., Landegent, J.E., Verschueren, C.A.J, Falkenburg, J.H.F., Van Ormondt, H., Van der Eb, A.J., and Noteborn, M.H.M. 1995, Apoptin, a protein encoded by chicken anemia virus, induces cell death in various human hematologic malignant cells *in vitro*. *Leukemia* **9 S1**:118–120.

Zhuang, S.-M., Shvarts, A., Van Ormondt, H., Jochemsen, A.-G., Van der Eb, A.J., and Noteborn, M.H.M. 1995, Apoptin, a protein derived from chicken anemia virus, induces a p53-independent apoptosis in human osteosarcoma cells. *Cancer Res.* **55**:486–489.

ADENOVIRUS-MEDIATED HERPES SIMPLEX THYMIDINE KINASE GENE THERAPY FOR BRAIN TUMORS

Anu-Maaria Sandmair[1*,2†] Matti Vapalahti[2,4],
and Seppo Ylä-Herttuala[1*,3,4]

[1]A.I. Virtanen Institute
University of Kuopio
P.O.B. 1627, FIN-70211 Kuopio
Finland
[2]Department of Neurosurgery
[3]Department of Medicine and
[4]Gene Therapy Unit
University Hospital of Kuopio
P.O.B. 1777, FIN-70110 Kuopio
Finland

1. INTRODUCTION

Intracranial neoplasms continue to challage physicians and scientists, since the poor prognosis specially of malignant glioma has shown only minor improvements despite all attemps to improve the outcome of these patients. Gene therapy is one of the most interesting novel treatment modalities in attempting to change the survival of malignant glioma patients. While holding the promise for an effective treatment of these patients, the transfer of gene therapy from cell culture to animal models and finally to clinical applications presents a great challenge.

2. MALIGNANT BRAIN TUMORS

Intracranial neoplasms are divided into primary tumors derived from the central nervous system (CNS) or other structures in cranial cavity, and secondary or metastatic tumors, which spread to CNS without CNS origin. Classification of brain tumors are

* Fax: +358-17-163030, † Fax: +358-17-163030

Cancer Gene Therapy: Past Achievements and Future Challenges, edited by Habib
Kluwer Academic/Plenum Publishers, New York, 2000.

based on either origin, histological growth pattern and to a lesser extent on radiological and operative appearance. The most frequent primary brain tumors are neuroglial, i.e., astrocytomas, anablastic astrocytomas, glioblastoma multiforme, oligodendrogliomas and ependymomas (Fuller *et al.*, 1996). Etiology of CNS tumors is largely unknown, and the causes can be multiple: maldevelopment, genetic factors, chromosomal aberrations, trauma, viruses, chemical compounds and immunological dysfunctions. However, it has not been possible to draw any final conclusions regarding any single causative mechanism.

The overall prognosis of most benign neoplasms have shown improvements, in which easier diagnosis and reduced surgical morbidity have changed the outlook. On the other hand, malignant tumors, especially glioblastoma multiforme, still have a poor outlook, even though prompt diagnosis, aggressive treatment with combined surgery, radiation and chemotherapy have achived some positive effects.

3. MALIGNANT GLIOMA

3.1. Definition and Classification

All gliomas occur from well differentiated to anablastic tumors. The term malignant glioma is usually used for those gliomas which show evidence of histological and biological malignancy with anablasia, invasiveness, tumor recurrency and metastatic spread. These tumors include glioblastoma, somewhat less aggressive anablastic astrocytoma, malignant oligodendroglioma and malignant ependymoma (Fuller *et al.*, 1996).

Gliomas have been classified by several different features, sometimes in a confusing way. The World Health Organization (WHO) classification is most widely accepted (Kleilhues *et al.*, 1993), but Kernohan classification is also commonly used (Kernohan *et al.*, 1949). Both recognize three or four grades of increasing malignancy according to anaplasia and invasiveness. However, the the WHO classification have been given almost an authorial position. The WHO classification recognizes four grades of neoplasms (Grades I, II, III, IV) but in effect divides diffuse neoplasms into three grades, since Grade I is assigned to the pilocytic astrocytoma.

Glioblastoma multiforme is widely recognized as an extreme manifestation of dedifferentation of glioma cells, mostly astrocytes. Already macroscopically, during brain surgery, the diagnosis of glioblastoma can be quite evident when necrosis, fistulous blood vessels containing arterial blood and hemorrages of various ages are present. Microscopically most glioblastomas have marked pleomorphism, high mitotic rate, necrosis borded by glial cells with an appearance of pseudopalisanding and microvasculature with hyperplasia and endothelial proliferation often associated with brain oedema (Fuller *et al.*, 1996).

Anablastic astrocytoma (WHO Gr III) is distinguished from glioblastoma multiforme (WHO Gr IV) particularly by the absence of necrosis. Anaplastic astrocytoma patients are significantly different in age, duration of symptoms and particulary in outcome showing 50–100% better survival than those with glioblastoma (Salcman, 1980). Anablastic oligodendroglioma is a densely cellular tumor with pleomorphic and prominent nuclei, with typical features of calsification and microcysts.

Clinical features of gliomas consist of headache, epilepsy, mental changes or hemiparesis, in this order of frequency. Laboratory findings are usually not diagnostic at all. Focal clinical features such as visual failure, changes of personality, ataxia, dysphasia or sensory or motor changes affecting limbs can provide evidence of the probable site of the tumor. Vomiting and headache are non-specific symptoms but associate with brain tumor occuring often in the early morning with raised intracranial pressure. Visual failure due to papilloedema is a symptom not infrequently found by ophtalmologist.

3.2. Epidemiology and Prognosis

Malignant gliomas are the most common primary brain tumors in adults and the leading course of cancer related deaths. The average annual incidence of primary brain tumors is 9/100,000, gliomas being the most common, representing more than half of the cases (58%) and glioblastoma alone 20%. Intracranial neoplasms (ICN) account 5–10% of all tumors in all age groups except 20–40% of all malignancies in childhood (Salcman, 1980). There are two age peaks for the insidence of central nervous system tumors: the first occures in childhood at the age of 5–9 years and the second at the age of 50–55 years (Salcman, 1980). In childhood medulloblastoma is the most common tumor, malignant gliomas are rare, and the tumors occure more frequently in the posterior fossa. Cerebellar hemispheres are the most common sites of adult brain tumors (Salcman, 1996).

Median survival of malignant glioma patients is 14 months and the enjoyable, good quality of life is even shorter, 45–50 weeks (Salcman, 1980) On an average, malignant glioma recures usually 35 weeks after operation which is a disappointingly short time period. The prognosis of glioblastoma patients is very poor with an average survival of 10 months after an active treatment (Salcman,1980).

Age is the most important prognostic factor for malignant glioma patients. Other significant factors are the duration of symptoms, histological features of the tumor and performance status at the time of the diagnosis. Also, the extent of tumor resection has an effect on the prognosis.

3.3. Cellular and Molecular Alterations

The loss of growth control in malignant glioma involves activation of growth promoting genes (proto-oncogenes) or inactivation of genes which control cell growth (tumor suppressor genes). Actively growing tumors usually show evidence for both activation of several proto-oncogenes and inactivation of several tumor suppressor genes, which is called a multi-step theory of tumorigenesis.

One of the common features in malignant gliomas is an increased amplification of epidermal growth factor reseptor (EGFR) (Watkins and Rouleau, 1994). Several studies have demonstrated a correlation between EGFR overexpression and biological behavior of higher grade tumors. Also, an overexpression of platlet-derived growth factor and vascular endothelial growth factor have been reported in malignant gliomas.

Immunologically, normal central nervous system constitutes a relatively priviledged site because of the blood brain barrier and lack of lymphatic system. However, microglia and astrocytes are able to function as antigen presenting cells and glioma cells are capable of secreting some cytokines, such as IL-1, IL-3 and IL-6 (Dietrich *et al.*, 1997). In spite of these features, host-immune responses in malignant gliomas are week compared to other cancers, even though the blood brain barrier is partly disturbed in malignant glioma patients (Dietrich *et al.*, 1997).

3.4. Current Treatment Strategies

Brain tumors rarely spread via the bloodstream and therapeutic approach can focus on the local problem. However, this target is extreamly difficult to achieve with current treatment strategies. Generally, surgery is the principal treatment method for malignant glioma and is indicated to obtain biopsy samples for histological diagnosis of the tumor. When possible, tumor mass reduction should be done for the relief of symptoms from intracranial pressure and brain shifts. Also, the surgical excision may increase the efficacy of adjuvant treatments. However, macroscopically radical removal is often not possible due to the tumor cell seeding in nearby tissue, often extending along anatomical tracts of the brain. In case of elongate area tumors, where resection would cause unacceptable risks for the patient, tumors biopsy is usually performed to get a histological diagnosis, often stereotactically or under image quidance.

Radiation therapy (RT) is the most important treatment modality following surgery of brain tumors. Radiation therapy at doses 45–60 Gy significantly prolonges survival (Walker *et al.*, 1978, 1980; Davies *et al.*, 1996). However, the radiation sensitivity of the whole brain and neural tissue sets a limit to radiotherapy, with brain stem, spinal cord and infantile brain beeing particulary sensitive to radiation. Radiation necrosis raises at the dose of 60 Gy to 4% and is even more frequent with higher doses (Liebel *et al.*, 1989). In most brain tumors, the results obtained with RT alone are not satisfactory. Postoperative RT is generally advocated for the following tumors: astrocytoma gr III-IV, ependymoma, medulloblastoma, pinealoma, sarcoma, hemangioblastoma and malignant lymphomas. In the following tumors, postoperative RT is sometimes indicated, especially when the tumor removal has been incomplete: astrocytoma gr II, oligodendroglioma, meningeoma, pituitary adenoma, craniopharyngeoma, chordoma and neurinoma. Lymphomas and medulloblastomas are highly sensitive to radiation. In contrast, glioblastoma multiforme is one of the most malignant and the most treatment-resistant human malignancies. Doses higher than 60–70 Gy have shown no effect on gr IV tumors. Instead, the incidense of late radiation necrosis has increased.

Satisfactory results in the treatment of malignant glioma have not been achieved with surgery and raditation. One is therefore faced with the choice of chemotherapy, although, the treatment response has not been much better (Hildebrandt *et al.*, 1997). Malignant glioma, still, even with prompt diagnosis and an aggressive treatment with combined surgery, radiation and chemotherapy has a poor prognosis.

4. GENE THERAPY FOR MALIGNANT GLIOMA

Malignant glioma was one of the first diseases treated with gene therapy. Some features, such as a relatively localized growth, nondividing surrounding tissues and the feasibility of local administration during tumor resection, together with a poor prognosis with current treatment strategies have made malignant glioma one of the most favorable targets for gene therapy.

4.1. Herpes Simplex Virus Thymidine Kinase (HSVtk)

Thymidine kinase gene transfer combined with a treatment with guanosine analoque, ganciclovir (GCV) is the most widely used method for gene therapy of malignant gliomas (Moolten *et al.*, 1994). The HSVtk gene has been cloned from herpes simplex virus. HSVtk leads to the phosphorylation of GCV to a monophosphate form which cellular kinases further phosphorylate to a triphosphate form. Phosphorylated GCV acts as a toxic nucleotide analoque and inhibits the function of DNA-polymerase leading to the death of the target cell (Moolten *et al.*, 1994). This method seems to be very attractive for the treatment of malignant brain tumors, since the inhibition of DNA-polymerase targets the treatment effect to the actively proliferating malignant cells, instead of non-dividing neurons and glia cells. HSVtk is as a relatively nontoxic product and only forms highly toxic substance in the presence of GCV. The treatment effect is further enhanced by a phenomenon called a bystander effect (Bi *et al.*, 1993). The mechanism of this effect is generally thought to be related to the release of toxic nucleotide analoques to adjacent tumor cells via gap junctions leading to the death of nearby cells that lack the HSVtk gene.

4.2. Adenoviruses

Adenoviruses as gene therapy vectors have been described elsewhere in this book (pp ••–••). Shortly, adenoviruses have several features which make them very suitable for gene therapy of brain tumors: Adenoviruses are efficient gene transfer vectors and they also infect non-dividing cells (Puumalainen *et al.*, 1999; Boviatsis *et al.*, 1994; Laitinen *et al.*, 1998). With HSVtk gene transfer this is not a serious limitation since the inhibition of DNA-polymerase targets the treatment effect to the actively proliferating malignant cells, instead of non-dividing neurons and glia cells. Adenoviruses lead to an efficient short term gene expression which is useful for the destruction of tumor cells. In addition, adenoviruses usually raise an immune response (Kajiwara *et al.*, 1997) which limits the usefulness of repeated gene transfers. However, immunostimulation may also be advantageous by causing a response againts the glioma cells (Wilson, 1995). Finally, in animal experiments adenoviral HSVtk gene therapy has been successfully used for the treatment of malignant glioma (Culver *et al.*, 1997; Chen *et al.*, 1994).

4.3. Adenovirus-Mediated Gene Therapy for the Treatment of Malignant Glioma

4.3.1. Animal Studies. There are a number of studies showing positive results for adenovirus HSVtk gene therapy in animal models of malignant glioma. Chen et al showed >500-fold brain tumor volume regression in nude mice compared to control group after injection of 3×10^8 virus particles into 8 days old C6-tumors (Chen *et al.*, 1994). According to Maron et al there was a 28 -fold brain tumor volume regression and four-fold longer survival time in Wistar rats after treatment of 5 days old C6 tumors with 2.5×10^8 pfu of ADVtk (Maron *et al.*, 1997). Colak et al demonstrated the treatment effect of 5.6×10^7 pfu of ADVtk in 7 days old 9 L spinal cord tumors with a prolonged symptom-free time and a reduced tumor size (Colak *et al.*, 1995). Significant prolongation of survival and symptom-free period of Fisher 344 rats with 3 days old 9L brain tumors were detected by Vincent et al after 10^{6-8} pfu ADVtk injection (Vincent *et al.*, 1997, 1996). Our own studies are in line with these findings with prolonged survival (Tyynelä *et al.*, 1998).

4.3.2. Clinical Studies. Results from ADVtk gene therapy trials for the treatment of malignant glioma patients have not yet been published. Ram et al used retrovirus-mediated HSVtk gene therapy for clinical cases and showed safety and limited efficacy

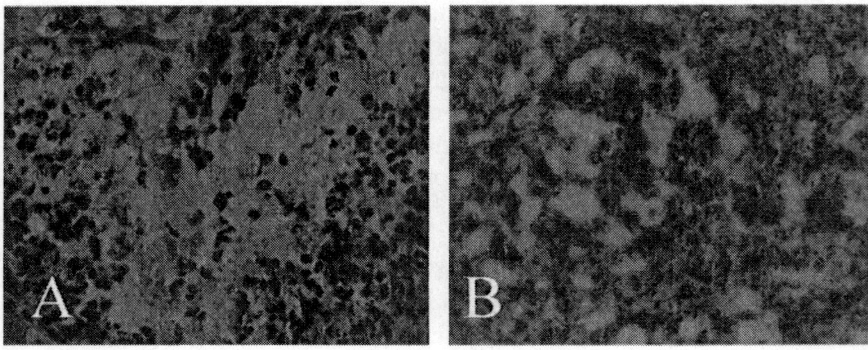

Figure 1. A. Adv-HSVtk treatment effect in BT4C rat glioma model shows a necrotic area at the injection site of the vector with some surrounding viable glioma cells. Hematoxylin-eosin stainig. **B.** Injection of Adv-vector causes a microglia reaction at the site of injection as seen with a Ferritin immunostaining (Kaneko *et al.*, 1989). Magnifications: A: 10×; B: 20×.

of the treatment (Ram *et al.*, 1997). Adv-mediated LacZ gene transfers have been used to evaluate the safety and efficacy of gene transfers before therapeutical trials with Adv-HSVtk (Puumalainen *et al.*, 1998). According to these studies gene transfers have been safe and well tolerated. Gene transfer efficiency varied between <0.01 and 11% (Puumalainen *et al.*, 1998). Adv-HSVtk phase I and II clinical trials are ongoing and preliminary results should be available shortly.

5. CONCLUSIONS

Adenovirus-mediated HSVtk gene therapy has been shown to be eefective for the treatment of malignant glioma when the gene transfer efficiency is high. Adenoviruses are currently the most promising vectors for gene therapy of malignant glioma especially because of their high efficiency of gene transfer. Experiences from preclinical in vitro and in vivo studies are promising and in the near future the first reports of clinical phase I and II studies should be available. These results will significantly advance our understanding about the efficacy of adenovirus-mediated gene transfers in the treament of malignant glioma in man.

ACKNOWLEDGMENTS

This study was supported by grants from Finnish Academy and Kuopio University Hospital (EVO grants 5021 and 5118).

REFERENCES

Ammirati, M., Galicich, J.H., Arbit, E., and Liao, Y. 1987, Reoperation in the treatment of recurrent intracranial malignant gliomas. *Neurosurgery* **21**:607–614.

Bi, W.L., Parysek, L.M., Warnick, R., and Stambrook, P.J., 1993, In vitro evidence that metabolic cooperation is responsible for the bystander effect observed with HSVtk retroviral gene therapy. *Hum.Gene Ther.* **4**:725–731.

Boviatsis, E.J., Chase, M., Wei, M.X., Tamiya, T., Hurford, R.K.Jr., Kowall, N.W., Tepper, R.I., Breakefield, X.O., and Chiocca, E.A. 1994, Gene transfer into experimental brain tumors mediated by adenovirus, Herpes simplex virus, and retrovirus vectors. *Hum Gene Ther* **5**:183–191.

Chen, S.-H., Shine, H.D., Goodman, J.C., Grossman, R.G., and Woo, S.L.C. 1994, Gene therapy for brain tumors: Regression of experimental gliomas by adenovirus-mediated gene transfer *in vivo*. *Proc Natl Acad Sci USA* **91**:3054–3057.

Colak, A., Goodman, J.C., Chen, S.H., Woo, S.L.C., Grossman, R.G., and Shine, H.D., 1995, Adenovirus-mediated gene therapy for experimental spinal cord tumors: tumoricidal efficacy and functional outcome. *Brain Research*. **691**:76–82.

Davies, E., Clarke, C., and Hopkins, A., 1996, Malignant cerebral glioma-1: Survival, disability, and morbidity after radiotherapy. *BMJ*, **313**:1507–1512.

Dietrich, P.Y., Walker, P.R., Saas, P., and de Tribolet, N., 1997, Immunobiology of Gliomas: New Perspectives for Therapy, *Ann NY Acad Sci* **824**:124–140.

Fuller, G.N., and Burger, P.C., 1996, Gliomas: Pathology, in: *Neurosurgery*, Volume 1 (Wilkins, R.H., and Rengachary, S.S., eds), pp. 735–747, McGrawn-Hill, Health professions Division.

Hildebrandt, J., Dewitte, O., Dietrich, P.Y., and de Tribolet, N., 1997, *Eur Neurol*, **38:**238–253.

Kajiwara, K., Byrnes, A.P., Charlton, H.M., Wood, M.J.A., and Wood, K.J., 1997, Immune Responses to Adenoviral Vectors During Gene Transfer in the Brain. *Hum. Gene Ther*. **8**:253–265.

Kaneko, Y., Kitamoto, T., Tateishi, J., and Yamaguchi, K. 1989, Ferritin immunohistochemistry as a marker for microglia. *Acta Neuropathol*, **79**:129–136.

Kernohan, J.W., Mabon, R.F., Svien, H.J., and Adson, A.W., 1949, A simplified classification of the gliomas. *Proc Staff Mtg Mayo Clin*, **24**:71–75.

Kleihues, P., Burger, P.C., and Scheithauer, B.W., 1993, The new WHO classification of brain tumours. *Brain Pathol*, **3**:255–268.

Laitinen, M., Mäkinen, K., Manninen, H., Matsi, P., Kossila, M., Agrawal, R.S., Pakkanen, T., Luoma, J.S., Viita, H., Hartikainen, J., Alhava, E., Laakso, M., and Ylä-Herttuala, S., 1998, Adenovirus-mediated gene transfer to lower limb artery of patients with chronic critical leg ischaemia. *Hum Gene Ther*, **9**:1481–1486.

Le Gal La Salle, G., Robert, J.J., Berrard, V., Ridoux, L.D., Stratford-Perricaudet, L.D., Perricaudet, M., and Mallet, J., 1993, An adenovirus vector for gene transfer into neurons and glia in the brain. *Science* **259**:988–990.

Leibel, S.A., Gutin, P.H., Wara, W.M., Silver, P.S., Larson, D.A., Edwards, M.S., Lamb, S.A., Ham, B., Weaver, K.A., and Barnett, C., 1989, Survival and quality of life after interstitial implantation of removable high-activity iodine-125 sources for the treatment of patients with recurrent malignant gliomas. *Int J Radiat Oncol Biol Phys*, **17**:1129–1139.

Levy, M.L., and Apuzzo, M.L.J., Immunobiolgy of Brain Tumors, in: *Neurosurgery*, Volume 1 (Wilkins, R.H., and Rengachary, S.S., eds), pp. 707–713, McGrawn-Hill, Health professions Division.

Maron, A., Havaux, N., Le Roux, A., Knoops, B., Perricaudet, M., and Octave, J.N., 1997, Differential toxicity of ganciclovir for rat neurons and astrocytes in primary culture following adenovirus-mediated transfer of the HSVtk gene. *Gene Ther.* **4**:25–31.

Maron, A., Le Roux, A.L., Mottet, I., Dedieu, J.F., Brion, J.P., Demeure, R., Perricaudet, M., and Octave, J.N., 1996, Gene therapy of rat C6 glioma using adenovirus-mediated transfer of the herpes simplex virus thymidine kinase gene: long-term follow-up by magnetic resonance imaging. *Gene Therapy*. **3**:315–322.

Moolten, F.L., 1994, Drug sensitivity ("suicide") genes for selective cancer chemotherapy. *Cancer Gene Ther*, **1**:279–287.

Perez-Cruet, M.J., Trask, T.W., Chen, S.H., Goodman, J.C., Woo, S.L.C., Grossman, R.G., and Shine, H.D., 1994, Adenovirus-Mediated Gene Therapy of Experimental Gliomas. *Journal of Neuroscience Research*. **39**:506–511.

Puumalainen, A.M., Puranen, P., Loimas, S., Puranen, M., Hurskainen, H., Kossila, M., Tyynelä, K., Turunen, M., Vanninen, R., Lehtolainen, P., Paljärvi, L., Johansson, R., Vapalahti, M., and Ylä-Herttuala, S. Thymidine kinase gene therapy for human malignant glioma using replication-deficient retroviruses or adenoviruses. Phase I safety and efficacy trial. Submitted.

Puumalainen, A.M., Turunen, M., Loimas, S., Vainio, P., Tyynelä, K., Vanninen, R., Vapalahti, M., Bjerkvig, R., Jänne, J., and Ylä-Herttuala, S., Herpes Simplex virus thymidine kinase gene therapy in experimental rat BT4C glioma model. Submitted.

Puumalainen, A.M., Vapalahti, M., Agrawal, R., Kossila, M., Laukkanen, M., Lehtolainen, P., Viita, H., Paljärvi, L., Vanninen, R., and Ylä-Herttuala, S., 1998, B-galactosidase gene transfer to human malignant glioma *in vivo* using replication-deficient retroviruses and adenoviruses. *Hum. Gene Ther.* **9**:1769–1774.

Ram, Z., Culver, K.W., Oshiro, E.M., Viola, J.J., DeVroom, H.L., Otto, E., Long, Z., Chiang, Y., McCarrity, G.J., Muul, L.M., Katz, D., Blaese, M., and Oldfield, E.H., 1997, Therapy of malignant brain tumors by intratumoral implantation of retroviral vector-producing cells. *Nature Medicine* **3**:1354–1361.

Salcman, M., 1980, Survival in glioblastoma. Historical prespective. *Neurosurgery*, **7**:435–439.

Salcman, M., 1996, Supratentorial Gliomas: Clinical Features and Surgical Therapy, in: *Neurosurgery,* Volume 1 (Wilkins, R.H., and Rengachary, S.S., eds), pp. 777–788, McGrawn-Hill, Health professions Division.

Tyynelä, K., Puumalainen, A.M., Turunen, M., Hakumäki, J., Kauppinen, R., Johansson, R., and Ylä-Herttuala, S., 1998, Adenovirus-mediated Herpes simplex virus-thymidine kinase gene therapy in BT4C rat glioma *in vivo*. Submitted.

Vincent, A.J.P.E., Esandi, M.C., Avezaat, C.J.J., Vecht, C., Smitt, P.S., Bekkum, D.W., Valerio, D., Hoogerbrugge, P.M., and Bout, A., 1997, Preclinical Testing of Recombinant Adenoviral Herpes Simplex Virus-Thymidine Kinase Gene Therapy for Central Nervous System Malignancies. *Neurosurgery*. **41**:442–452.

Vincent, A.J.P.E., Esandi, M.C., Someren, G., Notebloom, J.L., Avezaat, C.J.J., Vecht, C., Smitt, P.A.E.S., Bekkum, D.W., Valerio, D., Hoogerbrugge, P.M., and Bout, A., 1996, Treatment of leptomeningeal metastases in a rat model using a recombinant adenovirus containing the HSV-tk gene. *J Neurosurg*. **85**:648–654.

Vincent, A.J.P.E., Vogels, R., Someren, G.V., Esandi, M.C., Notebloom, J.L., Avezaat, C.J.J., Vecht, C., Bekkum, D.W.V., Valerio, D., Bout, A., and Hoogerbrugge, P.M., 1996, Herpes Simplex Virus Thymidine Kinase Gene Therapy for Rat Malignant Brain Tumors. *Human Gene Therapy*. **7**:197–205.

Viola, J.J., Ram, Z., Walbridge, S., Oshiro, E.M., Trapnell, B., Tao-Cheng, J.H., and Oldfield, E.H., Adenovirally mediated gene transfer into experimental solid brain tumors and leptomeningeal cancer cells. *J Neurosurg* **82**:70–76.

Walker, M.D., Alexander, E., Hunt, W.E., MacCarty, C.S., Mahaley, J., and Norell, H.A., *et al.*, 1978, *J Neurosurg*, **49**:333–343.

Walker, M.D., Green, S.B., Byar, D.P., Alexander, E., Batzdorf, U., Brooks, W.H., Hunt, W.E., MacCarty, C.S., Mahaley, M.S., Mealey, J., Owens, G., Ransohoff, J., Robertson, J.T., Saphiro, W.R., Smith, K.R. Jr., Wilson, C.B., and Strike, T.A., 1980, Randomized comparisons of radiotherapy and nitrosoureas for the treatment of malignant glioma after surgery. *N Engl J Med*, **303**:1323–1329.

Watkins, D., and Rouleau, G.A., 1994, Genetics, Prognosis and Therapy of Central Nervous System Tumors. *Cancer Detection and Prevention*, **18**:139–144.

Wilson, J., 1995, Adenoviruses as Gene-Delivery Vehicles. *N Engl J Med* **334**:1185–1187.

Ylä-Herttuala, S., 1998, Gene therapy for cancer. In: *Tumor Invasion: Biological, Clinical and Therapeutical Considerations.* (Mikkelsen, Bjerkvig, Laerum, Rosenblum eds), pp. 435–445, John Wiley & Sons.

DEVELOPMENT OF CLINICAL TRIAL OF *E1A* GENE THERAPY TARGETING HER-2/*neu*-OVEREXPRESSING BREAST AND OVARIAN CANCER

M. C. Hung[1]*, G. N. Hortobagyi[2], and N. T. Ueno[3]

[1]Section of Molecular Cell Biology
Department of Cancer Biology and
Department of Surgical Oncology
The University of Texas M. D. Anderson Cancer Center
[2]Department of Breast Medical Oncology
The University of Texas M. D. Anderson Cancer Center
[3]Department of Blood and Marrow Transplantation
and Section of Molecular Cell Biology
Department of Cancer Biology
The University of Texas M. D. Anderson Cancer Center

1. INTRODUCTION

HER-2/*neu* overexpression among patients with breast and ovarian cancers are considered unfavorable prognosis with a shorter overall survival and a lower response rate to chemotherapeutic agents. Therefore, downregulation of HER-2/*neu* overexpression in cancer cells by repressing their HER-2/*neu* promoter activity of cancer cells is an attractive strategy to reverse both the transformation phenotype and the chemoresistance induced by HER-2/*neu* overexpression. In this chapter, we will review the rationale and our recent experience of *E1A* gene therapy for HER-2/*neu*-overexpressing breast and ovarian cancer. We have previously identified a viral transcriptional regulator the adenovirus type 5 *E1A*, which can repress the HER-2/*neu* overexpression. Expression of *E1A* in HER-2/*neu*-overexpressing cancer cells resulted in downregulation of the HER-2/*neu*

* To whom correspondence should be addressed. University of Texas M. D. Anderson Cancer Center, Department of Cancer Biology, Section of Molecular Cellular Biology, Box 79, 1515 Holcombe Boulevard, Houston, Texas 77030. Phone: (713) 792-3668, Fax: (713) 794-4784, E-mail: mchung@odin.mdacc.tmc.edu, nueno@notes.mdacc.tmc.edu

Cancer Gene Therapy: Past Achievements and Future Challenges, edited by Habib
Kluwer Academic/Plenum Publishers, New York, 2000.

overexpression which reversed the malignant phenotype and restoring the sensitivity to chemotherapeutic agents *in vitro*. These findings were followed by a series of *in vivo* studies to demonstrate that *E1A* gene can act therapeutically as a tumor suppressor gene for HER-2/*neu*-overexpressing cancers in ovarian and breast cancer xenograft model. Based on these preclinical studies, phase I trial of *E1A* gene therapy was recently conducted at The University of Texas M. D. Anderson Cancer Center. This trial showed that the *E1A* gene could be delivered to cancer cells by cationic liposome delivery system. Furthermore, HER-2/*neu* overexpression was downregulated and apoptosis was induced by the *E1A* gene.

2. WHY TARGET HER-2/*neu* OVEREXPRESSION?

The HER-2/*neu* (*c-erbB2*) gene encodes a receptor tyrosine kinase (p185) with significant structural and functional homology to the epidermal growth factor receptor (EGFR) (Bargmann, Hung, and Weinberg, 1986b; Schechter, Hung, Vaidyanathan, Weinberg, Yang, Francke, Ullrich, and Coussens, 1985; Yamamoto, Ikawa, Akiyama, Semba, Nomura, Miyajima, Saito, and Toyoshima, 1986). The HER-2/*neu* receptor contains an extracellular domain, a transmembrane domain, and an intracellular domain with intrinsic tyrosine kinase. The ligand for the HER-2/*neu* receptor has not been well defined. Identification of the oncogenic HER-2/*neu* oncogene and the oncogenic property of the HER-2/*neu* were originally demonstrated in the rat *neu* oncogene that was isolated from rat *neu*roblastoma (Bargmann, Hung, and Weinberg, 1986a; Hung, Schechter, Chevray, Stern, and Weinberg, 1986; Shih, Padhy, Murray, and Weinberg, 1981). The activated rat *neu* oncogene contains a point mutation in the transmembrane domain of its protein, which results in a constitutively active tyrosine kinase (Bargmann *et al.*, 1986a; Bargmann and Weinberg, 1988; Hung, Yan, and Zhao, 1989; Stern, Kamps, and Cao, 1988) and escapes a negative regulation of tyrosine kinase activity in metaphase (Kiyokawa, Yan, Brown, and Hung, 1995). This increased tyrosine kinase activity of HER-2/*neu*-encoded p185 by mutation resembles overexpression of the normal HER-2/*neu* gene. In human, the normal HER-2/*neu* gene is frequently overexpressed or amplified without mutation in many types of cancer: breast cancer (Slamon, Clark, Wong, Levin, Ullrich, and McGuire, 1987; Slamon, Godolphin, Jones, Holt, Wong, Keith, Levin, Stuart, Udove, Ullrich, and Press, 1989), ovarian cancer (Berchuck, Kamel, Whitaker, Kerns, Olt, Kinney, Soper, Dodge, Clarke-Pearson, Marks *et al.*, 1990; Slamon *et al.*, 1989), lung cancer (Kern, Schwartz, Nordberg, Weiner, Greene, Torney, and Robinson, 1990; Schneider, Hung, Chiocca, Manning, Zhao, Fang, and Roth, 1989; Shi, He, Cao, Pan, Zhang, Yu, and Hung, 1992; Weiner, Nordberg, Robinson, Nowell, Gazdar, Greene, Williams, Cohen, and Kern, 1990), gastric cancer (Kameda, Yasui, Yoshida, Tsujino, Nakayama, Ito, Ito, and Tahara, 1990; Yokota, Yamamoto, Toyoshima, Terada, Sugimura, Battifora, and Cline, 1986), and oral cancer (Xia, Lau, Zhang, Liu, Li, Kiyokawa, Clayman, Katz, and Hung, 1997). The overall survival of these patients who have HER-2/*neu*-overexpressing cancer is significantly shorter than patients who have low HER-2/*neu*-expressing cancer. Additionally, overexpression of the normal human HER-2/*neu* gene enhanced metastatic potential in human breast, ovarian, and non-small-cell lung carcinoma (NSCLC) cells (Benz, Scott, Sarup, Johnson, Tripathy, Coronado, Shepard, and Osborne, 1993; Chazin, Kaleko, Miller, and Slamon, 1992; Tan, Yao, and Yu, 1997; Yu, Hamada, Zhang, Nicolson, and Hung, 1992; Yu, Wolf, Scanlon, Price, and Hung, 1993b; Yu and Hung, 1991a) and resulted in enhanced resistance to a panel of chemotherapeutic agents (Tsai, Perng, Chen, Jan, Hung, Ku, and Chang, 1994; Tsai, Yu, Chang, Wu, Perng, Ibrahim, and Hung, 1995; Ueno, Yu, and Hung, 1997; Yu, Jing, Liu, Yao, Tan, McDonnell, and Hung, 1998b; Yu, Liu, Jing, Sun, Price, Singletary, Ibrahim, Hortobagyi, and Hung, 1998c; Yu, Liu, Tan, Li, Wang, and Hung, 1996). Especially, HER-2/*neu* overexpression in breast and ovarian cancer cells induced chemoresistance to paclitaxel (Taxol) (Ueno *et al.*, 1997, Yu, 1998 #3906; Yu *et al.*, 1998b; Yu *et al.*, 1996). In contrast, Pegram *et al.* suggested that HER-2/*neu* overexpression does not confer resistance to chemotherapeutic agents including paclitaxel (Pegram, Finn, Arzoo, Beryt, Pietras, and Slamon, 1997). However, further studies of HER-2/*neu*-induced paclitaxel resistance was recently shown to be distinct from the multiple drug resistance (MDR)-mediated pathway (Yu *et al.*, 1998c) and a molecular mechanism was proposed that HER-2/*neu* overexpression causes resistance to paclitaxel-induced apoptosis by induction of $p21^{Cip1}$ inhibiting $p34^{cdc2}$-kinase (Yu *et al.*, 1998b). In summary, HER-2/*neu* overexpression has a major contribution to tumorigenicity, metastasis, and drug resistance in human HER-2/*neu*-overexpressing cancer cells; therefore, molecular therapies that target HER-2/*neu* overexpression have become highly attractive approaches.

3. THE ADENOVIRUS TYPE 5 *E1A* GENE CAN BE A TUMOR SUPPRESSOR GENE

The adenovirus type 5 *E1A* gene associated with tumor suppression activates through at least two independent pathways; one targeting HER-2/*neu*-overexpressing and the other low-HER-2/*neu*-expressing cancer cells. The adenovirus type 5 *E1A* gene, which encodes two major proteins (289R and 243R), is the first viral gene expressed in the cells after adenovirus infection. The *E1A* gene product is a transcription factor that activates viral gene transcription and reprogramming of the host's cellular gene expression (Nevins, 1990; Nevins, Raychaudhuri, Yee, Rooney, Kovesdi, and Reichel, 1988). It is noteworthy that for the purpose of this chapter that *E1A* refers to the type 5 or type 2, which is a non-transforming *E1A,* because other sera type of *E1A* such as type 12 is capable of inducing transforming activities by itself. The type 5 and type 2 *E1A* was initially classified as an "immortalization oncogene" due to its ability to cooperate with the transforming *ras* or *E1B* genes to transform primary embryo cells (Byrd, Grand, and Gallimore, 1988; Montell, Courtois, Eng, and Berk, 1984; Ruley, 1983). However, "immortalization phenotypes" per se are not equal to "transforming phenotypes," as many established cell lines have been immortalized but do not produce transforming activities such as anchorage-independent growth and tumorigenicity in immunodeficient mice. Furthermore, expression of the type 5 *E1A* gene itself does not induce transforming phenotypes in normal fibroblast cell lines (Yu, Scorsone, and Hung, 1991b) and there are no epidemiological studies to suggest that adenoviruses can induce tumors in human beings.

We first demonstrated that the *E1A* gene products inhibit HER-2/*neu* expression in both rodent and human cancer cells. The *E1A*-mediated HER-2/*neu* repression occurs through transcriptional repression at the level of the HER-2/*neu* promoter (Yan, Chang, and Hung, 1991; Yu, Suen, Yan, Chang, and Hung, 1990). Transfection of the *E1A* gene into genomic HER-2/*neu* oncogene-transformed mouse embryo fibroblast cell lines virtually abolishes the tumorigenicity and metastatic potential induced by the HER-2/*neu* oncogene through repression of HER-2/*neu* expression (Yu *et al.*, 1992; Yu *et al.*, 1991b).

Re-expression of the HER-2/*neu*-encoded p185 protein in these *E1A* transfectants by transfection of an HER-2/*neu* cDNA construct driven by a promoter that cannot be inhibited by *E1A* restored virtually all the transforming phenotypes including the tumorigenicity when injected subcutaneously, the ability to grow in soft agarose, and the *in vitro* growth rate. However metastatic potential was only partially recovered, suggesting that EIA may also have anti-metastatic activity independent of downregulation of HER-2/*neu* (Yu, Shi, Scanlon, and Hung, 1993a). We have also demonstrated that *E1A* can function as a tumor suppressor in HER-2/*neu*-overexpressing human ovarian cancer cell line by repressing HER-2/*neu* p185 expression (Yu *et al.*, 1993b). Thus, for HER-2/*neu*-overexpressing transforming cells, including mouse fibroblasts and human cancer cells, *E1A* functions as a tumor suppressor and transcriptional repression of the HER-2/*neu* oncogene plays a critical role in tumor suppression activities.

Recently, *E1A* was reviewed and recognized as a tumor suppressor gene (Chinnadurai, 1992; Mymryk, 1996; Yu and Hung, 1998a), and several studies have indicated that *E1A* is associated with metastasis- or tumor-suppression activities by variety of mechanisms in addition to HER-2/*neu* overexpression (Deng, Xia, and Hung, 1998; Frisch, 1991; Frisch and Dolter, 1995; Pozzatti, McCormick, Thompson, and Khoury, 1988). The *E1A* gene was reported to reduce the metastatic potential of *ras*-transformed rat embryo fibroblast cells (Pozzatti *et al.*, 1988). In addition, in low-HER-2/*neu*-expressing human tumor cell lines, stable expression of the *E1A* gene reduced anchorage-independent growth and tumorigenic potential, induced flat morphology, and restored contact inhibition (Frisch, 1991; Frisch *et al.*, 1995). Thus, *E1A* may function as a tumor suppressor, depending on the individual oncogenesis backgrounds of the cancer cells. In addition, *E1A* has also been shown to induce apoptosis under some conditions in tissue culture (Lowe, Jacks, Housman, and Ruley, 1994; Lowe and Ruley, 1993a; Lowe, Ruley, Jacks, and Housman, 1993b; Shao, Karunagaran, Zhou, Li, Lo, Deng, Chiao, and Hung, 1997) and in the animal model (Deng *et al.*, 1998). This property is similar to the ability of the well-known tumor suppression gene *p53* to induce apoptosis.

To further investigate whether the *E1A* gene can be a therapeutic gene for HER-2/*neu*-overexpressing human breast and ovarian cancers, a tumor-bearing mouse model was established and the *E1A* gene was delivered by either DC-Chol cationic liposome or recombinant replication-deficient adenoviral vector. DC-Chol is a novel cationic liposome, which allows transfection of both *in vitro* and *in vivo* mammalian cells without major toxicity. Weekly injection of *E1A* complexed with DC-Chol cationic liposome into nude mice bearing human HER-2/*neu*-overexpressing ovarian cancer achieved 60–80% of long-term disease free survival rate among tumor-bearing mice. In addition, the number of mice with distant metastases of breast cancer was significantly reduced even though *E1A* was delivered to mammary fat pad where human HER-2/*neu*-overexpressing cancer was induced in the orthotopic breast cancer model (Chang, Xia, Shao, Sorgi, Hortobagyi, Huang, and Hung, 1997; Yu, Matin, Xia, Sorgi, Huang, and Hung, 1995; Zhang, Yu, Xia, and Hung, 1995). In addition to the breast and ovarian cancer animal models, we have shown similar therapeutic effect of *E1A* in a HER-2/*neu*-overexpressing lung cancer animal model (Chang, Xia, Shao, and Hung, 1996). In this study, the tumor-bearing mice were established through intratracheal inoculation of lung cancer cells and then the *E1A* gene was delivered intravenously by an adenovirus vector. In summary, the preferred tumor suppression effect of *E1A* on HER-2/*neu*-overexpressing cancer cells can be demonstrated through two independent gene delivery systems and three different animal models.

4. PHASE I TRIAL OF *E1A* GENE THERAPY IN HER-2/*neu*-OVEREXPRESSING BREAST AND OVARIAN CANCER

Based on the above preclinical results, we designed and conducted phase I trial of *E1A* gene therapy in patients with metastatic or recurrent breast and ovarian cancers. Our goal was to evaluate the safety and gene transfer efficiency of *E1A*- DC/Chol cationic liposome complex in 12 patients with metastatic or recurrent breast and ovarian cancers. Furthermore, we examined the biological activity of *E1A* in these patients by measuring the level of HER-2/*neu* downregulation by *E1A*. To achieve this objective, we initially required our patients to have pleural effusions or ascites to analyze the samples for HER-2/*neu* expression level in a time course manner.

For the gene delivery system, we chose the DC-Chol cationic liposome used in the preclinical animal model because it was known to be biodegradable, non-mutagenic and low in immunogenicity (Nabel, Gordon, Yang, Xu, San, Plautz, Wu, Gao, Huang, and

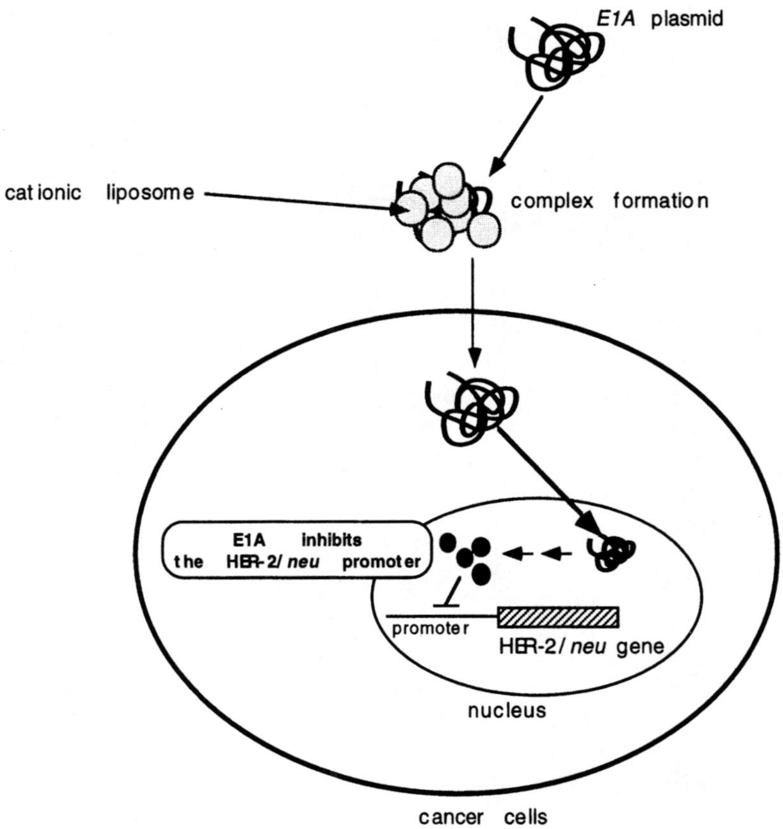

Figure 1. Downregulation of HER-2/*neu* by *E1A* gene in cancer cell A small portion of *E1A* plasmid DNA enters the nucleus and produces E1A protein through traditional cellular machinery (i.e.: transcription and translational processes). The E1A protein, then interacts with other nuclear proteins to inhibit promoter activity of HER-2/*neu* gene (Chen, Yu, Chinnadurai, Karunagaran, and Hung, 1997; Yu *et al.*, 1990). This process then results in downregulation of HER-2/*neu* gene product.

Nabel, 1992; Nabel, Nabel, Yang, Fox, Plautz, Gao, Huang, Shu, Gordon, and Chang, 1993; Xing, Liu, Xia, Stephens, Huang, Lopez-Berestein, and Hung, 1997; Xing, Zhang, Chang, Tucker, Chen, Huang, and Hung, 1998; Yu *et al.*, 1995). Prior to initiating a phase I clinical trial, we conducted studies to evaluate the safety of intraperitoneal injection of *E1A*-DC/Chol complex in normal mice. The cumulative doses up to 40 times of the *E1A*-DC/Chol complex that was proposed for the phase I clinical trial had no adverse effects on renal, hepatic and hematological parameters. No major organ pathologic changes were observed. Furthermore, *E1A* DNA was detected in lung and kidney 1 1/2 years after the last injection of *E1A*; however, *E1A* DNA was not detected in liver, heart, spleen, brain, uterus, or ovary. Therefore, we expected that DC-Chol cationic liposome gene delivery system would allow us to inject the E1A/DC-Chol complex multiple times into our patients without any major toxicity (Xing *et al.*, 1997; Xing *et al.*, 1998).

For the phase I trial, patients with a tumor that overexpressed HER-2/*neu* were preferred. A tumor was considered positive for HER-2/*neu* overexpression if more than 10% of tumor cells gave the HER-2/*neu* signal and this signal intensity is stronger than 1+ (Xia *et al.*, 1997). Patients received weekly injections of *E1A*/DC-Chol cationic liposome complex through a Tenckhoff catheter placed either in the pleural or cavity (in breast cancer patients) peritoneal cavity (in ovarian cancer patients). After three consecutive weekly injections, patients were given a week off (one cycle). The starting dose of *E1A* complexed to the lipid carrier DC-Chol/DOPE was 1.8 mg/m^2 that was converted from the effective dose in the animal studies (Chang *et al.*, 1997; Yu *et al.*, 1995). The *E1A* dose was escalated from 1.8, 3.6, to 7.2 mg/m^2/injection; the median number of injections was 6 (1–8); the most common toxicity was fever for 2–3 days after each injections regardless of the dose levels. At 7.2 mg/m^2, there was grade 3–4 nausea/vomiting or pain which led us to reduce the dose to 3.6 mg/m^2. There were 3 patients with catheter-related infection requiring discontinuation of the treatment. There was no drug-related fatality. Several patients experienced stable disease with transient decrease in their tumor markers (CA-215, CEA). Immunohistochemical staining for HER-2/*neu* expression has been completed in 6 patients. All 6 patients showed downregulation of HER-2/*neu*. We also performed RT-PCR in 1 pt and detected *E1A* expression in multiple organs (lung, tumor, liver, kidney, etc.), but not in brain or ovary. In addition, *E1A* expression by immunohistochemical staining and induction of apoptosis was confirmed after *E1A*/DC-Chol complex treatment. Thus, we concluded that *E1A* gene delivery is feasible, and *E1A* gene expression can be detected in cancer cells.

5. CONCLUSION AND FUTURE DIRECTION

In summary, we have shown that molecular approach to target the HER-2/*neu* promoter by the *E1A* gene can be translated from the basic sciences to the clinical practice. In our preclinical study, downregulating HER-2/*neu* overexpression resulted in reversing the malignant phenotype and retarding tumor growth in animals. In the phase I trial, we showed that the *E1A* gene can be delivered to the cancer cells with HER-2/*neu* downregulation and induction of apoptosis as expected from our preclinical experiments. We are currently planning a phase II trial in which we will target patient populations with HER-2/*neu*-overexpressing ovarian cancer but with minimal residual disease that possibly allows us to obtain maximum transfection efficiency of the *E1A* gene. Theoretically, high transfection efficiency will further enhance the anti-tumor activity of *E1A* and translate into prolonged progression-free survival. As described in the background, *E1A* is known to have anti-tumor activity by a variety of mechanisms in addition to HER-2/*neu* downregulation. Therefore, in the future phase II trial we plan to monitor other biological endpoints such as suppression of metastatic markers, inducibility of apoptosis and induction of immunological reaction so as to determine the precise mechanism of the *E1A* anti-tumor activity. Furthermore, anti-tumor activity of *E1A* gene in low HER-2/*neu*-expressing cancer cells needs to be further addressed in preclinical animal model. It is possible that higher doses of *E1A* gene may be required to observe therapeutic efficacy for low HER-2/*neu*-expressing cancer cells than HER-2/*neu*-overexpressing cancer cells in the preclinical animal models which were done in immunodeficient mice. If the mechanism of *E1A* anti-tumor activity is dependent on tumor type rather than simple downregulation of HER-2/*HER-2/neu*. For instance the *E1A*-induced immune response and anti-metastatic activity may contribute to anti-tumor activities in clinical trials, this allows us to expand the implication of *E1A* in variety of disease setting.

Recently, based on the fact that HER-2/*neu* overexpression induced chemoresistance to paclitaxel, we demonstrated that *E1A*-mediated HER-2/*neu* downregulation sensitize response to HER-2/*neu*-overexpressing cancer cells to paclitaxel in preclinical animal study (data unpublished), which suggests that combination therapy of paclitaxel and *E1A* gene therapy may be a feasible approach as a phase II clinical trial. Finally, so far all clinical trials using *E1A*/DC-Chol complex have been limited to local delivery systems; e.g., intra-cavitary injection for breast and ovarian cancer as described in this chapter or intratumoral injections of *E1A*/DC-Chol complex in head & neck cancer and breast cancer (Murray, Yoo, Lopez-Berestein, Hung, Booser, Rahman, Valero, Hortobagyi, Ensley, Jacobs, Brewer, Carey, and Daifuku, 1998; Yoo, Ensley, Jacobs, Sakr, Johnson, Wei, Carey, and Daifuku, 1998). This local delivery system does not allow delivering the therapeutic gene in a high efficient systemic manner; therefore, there is an urgent need for development of systemic intravenous gene delivery system.

ACKNOWLEDGMENT

The authors are supported by National Institutes of Health Grants CA76450-1 (to N.T.U.), CA58880, CA78623 and CA77858 (to M.C.H.) and Nelly Connelly Breast Cancer Research Fund (to G.N.H.) The clinical trial was supported by Targeted Genetics, Co., Seattle, WA, 98101. We finally thank the many scientists and physicians involved in the development of the project and all the patients for their courage, support and corporation for this trial.

REFERENCES

Bargmann, C.I., Hung, M.C., and Weinberg, R.A. (1986a). Multiple independent activations of the neu oncogene by a point mutation altering the transmembrane domain of p185. *Cell*, **45**, 649–657.

Bargmann, C.I., Hung, M.C., and Weinberg, R.A. (1986b). The neu oncogene encodes an epidermal growth factor receptor-related protein. *Nature*, **319**, 226–230.

Bargmann, C.I., and Weinberg, R.A. (1988). Increased tyrosine kinase activity associated with the protein encoded by the activated neu oncogene. *Proc Natl Acad Sci USA*, **85**, 5394–5398.

Benz, C.C., Scott, G.K., Sarup, J.C., Johnson, R.M., Tripathy, D., Coronado, E., Shepard, H.M., and Osborne, C.K. (1993). Estrogen-dependent, tamoxifen-resistant tumorigenic growth of MCF-7 cells transfected with HER2/neu. *Breast Cancer Res Treat*, **24**, 85–95.

Berchuck, A., Kamel, A., Whitaker, R., Kerns, B., Olt, G., Kinney, R., Soper, J.T., Dodge, R., Clarke-Pearson, D.L., Marks, P., *et al.* (1990). Overexpression of Her-2/Neu Is Associated With Poor Survival in Advanced Epithelial Ovarian Cancer. *Cancer Research*, **50**, 4087–4091.

Byrd, P.J., Grand, R.J., and Gallimore, P.H. (1988). Differential transformation of primary human embryo retinal cells by adenovirus E1 regions and combinations of E1A + Ras. *Oncogene*, **2**, 477–484.

Chang, J.Y., Xia, W., Shao, R., Sorgi, F., Hortobagyi, G.N., Huang, L., and Hung, M.C. (1997). The tumor suppression activity of E1A in HER-2/neu-overexpressing breast cancer. *Oncogene*, **14**, 561–568.

Chang, J.Y., Xia, W.Y., Shao, R.P., and Hung, M.C. (1996). Inhibition of intratracheal lung cancer development by systemic delivery of E1A. *Oncogene*, **13**, 1405–1412.

Chazin, V.R., Kaleko, M., Miller, A.D., and Slamon, D.J. (1992). Transformation Mediated By the Human Her-2 Gene Independent of the Epidermal Growth Factor Receptor. *Oncogene*, **7**, 1859–1866.

Chen, H., Yu, D., Chinnadurai, G., Karunagaran, D., and Hung, M.C. (1997). Mapping of adenovirus 5 E1A domains responsible for suppression of neu-mediated transformation via transcriptional repression of neu. *Oncogene*, **14**, 1965–1971.

Chinnadurai, G. (1992). Adenovirus E1a as a tumor-suppressor gene. *Oncogene*, **7**, 1255–1258.

Deng, J., Xia, W., and Hung, M.C. (1998). Adenovirus 5 E1A-mediated tumor suppression associated with E1A-mediated apoptosis in vivo. *Oncogene*, **17**, 2167–2175.

Frisch, S.M. (1991). Antioncogenic effect of adenovirus e1a in human tumor cells. *Proc Natl Acad Sci USA*, **88**, 9077–9081.

Frisch, S.M., and Dolter, K.E. (1995). Adenovirus E1a-mediated tumor suppression by a c-erbB-2/neu-independent mechanism. *Cancer Research*, **55**, 5551–5555.

Hung, M.C., Schechter, A.L., Chevray, P.Y., Stern, D.F., and Weinberg, R.A. (1986). Molecular cloning of the neu gene: absence of gross structural alteration in oncogenic alleles. *Proc Natl Acad Sci USA*, **83**, 261–264.

Hung, M.C., Yan, D.H., and Zhao, X.Y. (1989). Amplification of the proto-neu oncogene facilitates oncogenic activation by a single point mutation. *Proc Natl Acad Sci USA*, **86**, 2545–2548.

Kameda, T., Yasui, W., Yoshida, K., Tsujino, T., Nakayama, H., Ito, M., Ito, H., and Tahara, E. (1990). Expression of erbb2 in human gastric carcinomas: relationship between p185erbb2 expression and the gene amplification. *Cancer Research*, **50**, 8002–8009.

Kern, J.A., Schwartz, D.A., Nordberg, J.E., Weiner, D.B., Greene, M.I., Torney, L., and Robinson, R.A. (1990). p185neu expression in human lung adenocarcinomas predicts shortened survival. *Cancer Research*, **50**, 5184–5187.

Kiyokawa, N., Yan, D.H., Brown, M.E., and Hung, M.C. (1995). Cell cycle-dependent regulation of p185neu: a relationship between disruption of this regulation and transformation. *Proc Natl Acad Sci USA*, **92**, 1092–1096.

Lowe, S.W., Jacks, T., Housman, D.E., and Ruley, H.E. (1994). Abrogation of oncogene-associated apoptosis allows transformation of p53-deficient cells. *Proc Natl Acad Sci USA*, **91**, 2026–2030.

Lowe, S.W., and Ruley, H.E. (1993a). Stabilization of the p53 tumor suppressor is induced by adenovirus 5 E1A and accompanies apoptosis. *Genes Dev*, **7**, 535–545.

Lowe, S.W., Ruley, H.E., Jacks, T., and Housman, D.E. (1993b). p53-dependent apoptosis modulates the cytotoxicity of anticancer agents. *Cell*, **74**, 957–967.

Montell, C., Courtois, G., Eng, C., and Berk, A. (1984). Complete transformation by adenovirus 2 requires both E1A proteins. *Cell*, **36**, 951–961.

Murray, J.L., Yoo, G.H., Lopez-Berestein, G., Hung, M.-C., Booser, D., Rahman, Z., Valero, V., Hortobagyi, J., Ensley, J., Jacobs, J., Brewer, H., Carey, R., and Daifuku, R. (1998). Phase I trial of intratumoral liposomal-E1A gene therapy in patients with recurrent/refractory breast (BC) and head and neck (H&N) cancer. In *Proc Annu Meet Am Soc Clin Oncol*, Vol. 17. pp. 431a.

Mymryk, J.S. (1996). Tumor suppressive properties of the adenovirus 5 E1A oncogene. *Oncogene*, **13**, 1581–1589.

Nabel, E.G., Gordon, D., Yang, Z.Y., Xu, L., San, H., Plautz, G.E., Wu, B.Y., Gao, X., Huang, L., and Nabel, G.J. (1992). Gene transfer in vivo with DNA-liposome complex: lack of autoimmunity and gonadal localization. *Hum Gene Ther*, **3**, 649–656.

Nabel, G.J., Nabel, E.G., Yang, Z.Y., Fox, B.A., Plautz, G.E., Gao, X., Huang, L., Shu, S., Gordon, D., and Chang, A.E. (1993). Direct gene transfer with DNA-liposome complexes in melanoma: expression, biologic activity, and lack of toxicity in humans. *Proc Natl Acad Sci USA*, **90**, 11307–11311.

Nevins, J.R. (1990). Adenovirus E1A-dependent trans-activation of transcription. *Semin Cancer Biol*, **1**, 59–68.

Nevins, J.R., Raychaudhuri, P., Yee, A.S., Rooney, R.J., Kovesdi, I., and Reichel, R. (1988). Transactivation by the adenovirus E1A gene. *Biochem Cell Biol*, **66**, 578–583.

Pegram, M.D., Finn, R.S., Arzoo, K., Beryt, M., Pietras, R.J., and Slamon, D.J. (1997). The effect of HER-2/neu overexpression on chemotherapeutic drug sensitivity in human breast and ovarian cancer cells. *Oncogene*, **15**, 537–547.

Pozzatti, R., McCormick, M., Thompson, M.A., and Khoury, G. (1988). The *E1A* gene of adenovirus type 2 reduces the metastatic potential of ras-transformed rat embryo cells. *Mol Cell Biol*, **8**, 2984–2988.

Ruley, H.E. (1983). Adenovirus early region 1A enables viral and cellular transforming genes to transform primary cells in culture. *Nature*, **304**, 602–606.

Schechter, A.L., Hung, M.C., Vaidyanathan, L., Weinberg, R.A., Yang, F.T., Francke, U., Ullrich, A., and Coussens, L. (1985). The neu gene: an erbB-homologous gene distinct from and unlinked to the gene encoding the EGF receptor. *Science*, **229**, 976–978.

Schneider, P.M., Hung, M.C., Chiocca, S.M., Manning, J., Zhao, X.Y., Fang, K., and Roth, J.A. (1989). Differential expression of the c-erbB-2 gene in human small cell and non-small cell lung cancer. *Cancer Res*, **49**, 4968–4971.

Shao, R., Karunagaran, D., Zhou, B.P., Li, K., Lo, S.S., Deng, J., Chiao, P., and Hung, M.C. (1997). Inhibition of nuclear factor-kappaB activity is involved in E1A- mediated sensitization of radiation-induced apoptosis. *J Biol Chem*, **272**, 32739–32742.

Shi, D., He, G., Cao, S., Pan, W., Zhang, H.Z., Yu, D., and Hung, M.C. (1992). Overexpression of the c-erbB-2/neu-encoded p185 protein in primary lung cancer. *Mol Carcinog*, **5**, 213–218.

Shih, C., Padhy, L.C., Murray, M., and Weinberg, R.A. (1981). Transforming genes of carcinomas and neuroblastomas introduced into mouse fibroblasts. *Nature*, **290**, 261–264.

Slamon, D.J., Clark, G.M., Wong, S.G., Levin, W.J., Ullrich, A., and McGuire, W.L. (1987). Human breast cancer: correlation of relapse and survival with amplification of the HER-2/neu oncogene. *Science*, **235**, 177–182.

Slamon, D.J., Godolphin, W., Jones, L.A., Holt, J.A., Wong, S.G., Keith, D.E., Levin, W.J., Stuart, S.G., Udove, J., Ullrich, A., and Press, M.F. (1989). Studies of the HER-2/neu proto-oncogene in human breast and ovarian cancer. *Science*, **244**, 707–712.

Stern, D.F., Kamps, M.P., and Cao, H. (1988). Oncogenic activation of p185neu stimulates tyrosine phosphorylation in vivo. *Mol Cell Biol*, **8**, 3969–3973.

Tan, M., Yao, J., and Yu, D. (1997). Overexpression of the c-erbB-2 gene enhanced intrinsic metastasis potential in human breast cancer cells without increasing their transformation abilities. *Cancer Research*, **57**, 1199–1205.

Tsai, C.M., Perng, R.P., Chen, M.H., Jan, Y.H., Hung, M.C., Ku, T.Y., and Chang, K.T. (1994). Greater enhancement of chemosensitivity by caffeine in high-p185neu-expressing human non-small-cell lung cancer cell lines. *J Natl Cancer Inst*, **86**, 1018–1020.

Tsai, C.M., Yu, D., Chang, K.T., Wu, L.H., Perng, R.P., Ibrahim, N.K., and Hung, M.C. (1995). Enhanced chemoresistance by elevation of p185neu levels in HER-2/neu-transfected human lung cancer cells. *J Natl Cancer Inst*, **87**, 682–684.

Ueno, N.T., Yu, D., and Hung, M.C. (1997). Chemosensitization of HER-2/*neu*-overexpressing human breast cancer cells to paclitaxel (Taxol) by adenovirus type 5 *E1A*. *Oncogene*, **15**, 953–960.

Weiner, D.B., Nordberg, J., Robinson, R., Nowell, P.C., Gazdar, A., Greene, M.I., Williams, W.V., Cohen, J.A., and Kern, J.A. (1990). Expression of the neu gene-encoded protein (p185neu) in human non-small cell carcinomas of the lung. *Cancer Research*, **50**, 421–425.

Xia, W.Y., Lau, Y.K., Zhang, H.Z., Liu, A.R., Li, L., Kiyokawa, N., Clayman, G.L., Katz, R.L., and Hung, M.C. (1997). Strong correlation between c-Erbb-2 overexpression and overall survival of patients with oral squamous cell carcinoma. *Clin Cancer Res*, **3**, 3–9.

Xing, X., Liu, V., Xia, W., Stephens, L.C., Huang, L., Lopez-Berestein, G., and Hung, M.C. (1997). Safety studies of the intraperitoneal injection of E1A-liposome complex in mice. *Gene Ther*, **4**, 238–243.

Xing, X., Zhang, S., Chang, J.Y., Tucker, S.D., Chen, H., Huang, L., and Hung, M.C. (1998). Safety study and characterization of E1A-liposome complex gene delivery in an ovarian cancer model. *Gene Ther*, **5**, 1538–1544.

Yamamoto, T., Ikawa, S., Akiyama, T., Semba, K., Nomura, N., Miyajima, N., Saito, T., and Toyoshima, K. (1986). Similarity of protein encoded by the human c-erb-B-2 gene to epidermal growth factor receptor. *Nature*, **319**, 230–234.

Yan, D.H., Chang, L.S., and Hung, M.C. (1991). Repressed expression of the HER-2/c-erbB-2 proto-oncogene by the adenovirus E1a gene products. *Oncogene*, **6**, 343–345.

Yokota, J., Yamamoto, T., Toyoshima, K., Terada, M., Sugimura, T., Battifora, H., and Cline, M.J. (1986). Amplification of c-erbB-2 oncogene in human adenocarcinomas in vivo. *Lancet*, **1**, 765–767.

Yoo, G.H., Ensley, J., Jacobs, J., Sakr, W., Johnson, R., Wei, W.Z., Carey, M., and Daifuku, R. (1998). Intratumoral E1A gene therapy for patients with unresectable and recurren head and neck cancer. In *Proc Annu Meet Am Assoc Cancer Res*, Vol. 39. p. 322.

Yu, D., Hamada, J., Zhang, H., Nicolson, G.L., and Hung, M.C. (1992). Mechanisms of c-erbB2/neu oncogene-induced metastasis and repression of metastatic properties by adenovirus 5 E1A gene products. *Oncogene*, **7**, 2263–2270.

Yu, D., and Hung, M.C. (1998a). The erbB2 gene as a cancer therapeutic target and the tumor- and metastasis-suppressing function of E1A. *Cancer Metastasis Rev*, **17**, 195–202.

Yu, D., Jing, T., Liu, B., Yao, J., Tan, M., McDonnell, T.J., and Hung, M.C. (1998b). Overexpression of ErbB2 blocks Taxol-induced apoptosis by upregulation of p21Cip1, which inhibits p34Cdc2 kinase. *Mol Cell*, **2**, 581–591.

Yu, D., Liu, B., Jing, T., Sun, D., Price, J.E., Singletary, S.E., Ibrahim, N., Hortobagyi, G.N., and Hung, M.C. (1998c). Overexpression of both p185c-erbB2 and p170mdr-1 renders breast cancer cells highly resistant to taxol. *Oncogene*, **16**, 2087–2094.

Yu, D., Matin, A., Xia, W., Sorgi, F., Huang, L., and Hung, M.C. (1995). Liposome-mediated in vivo E1A gene transfer suppressed dissemination of ovarian cancer cells that overexpress HER-2/neu. *Oncogene*, **11**, 1383–1388.

Yu, D., Shi, D., Scanlon, M., and Hung, M.C. (1993a). Reexpression of neu-encoded oncoprotein counteracts the tumor-suppressing but not the metastasis-suppressing function of E1A. *Cancer Research*, **53**, 5784–5790.

Yu, D., Suen, T.C., Yan, D.H., Chang, L.S., and Hung, M.C. (1990). Transcriptional repression of the neu protooncogene by the adenovirus 5 E1A gene products. *Proc Natl Acad Sci USA*, **87**, 4499–4503.

Yu, D., Wolf, J.K., Scanlon, M., Price, J.E., and Hung, M.C. (1993b). Enhanced c-erbB-2/neu expression in human ovarian cancer cells correlates with more severe malignancy that can be suppressed by E1A. *Cancer Res*, **53**, 891–898.

Yu, D.H., and Hung, M.C. (1991a). Expression of activated rat neu oncogene is sufficient to induce experimental metastasis in 3T3 cells. *Oncogene*, **6**, 1991–1996.

Yu, D.H., Liu, B.L., Tan, M., Li, J.Z., Wang, S.S., and Hung, M.C. (1996). Overexpression of c-erbb-2/neu in breast cancer cells confers increased resistance to taxol via MDR-1-independent mechanisms. *Oncogene*, **13**, 1359–1365.

Yu, D.H., Scorsone, K., and Hung, M.C. (1991b). Adenovirus type 5 E1A gene products act as transformation suppressors of the neu oncogene. *Mol Cell Biol*, **11**, 1745–1750.

Zhang, Y., Yu, D., Xia, W., and Hung, M.C. (1995). HER-2/neu-targeting cancer therapy via adenovirus-mediated E1A delivery in an animal model. *Oncogene*, **10**, 1947–1954.

V

TUMOUR SUPPRESSOR GENES

17

PRE-CLINICAL STUDIES WITH TUMOR SUPPRESSOR GENES

Prem Seth

Medical Breast Cancer Section
Medicine Branch
Building 10, Room 12 N226
National Cancer Institute,
National Institutes of Health,
Bethesda, Maryland 20892

Cancer progression is considered to be a multi-stage process. Several genetic and epi-genetic factors have been suggested to participate in this process. One molecular event which has been implicated in the development of many cancers is the loss of a tumor suppressor function of many genes. Therefore, it is a reasonable idea to target these defects in cancer cells, in an effort to correct these abnormalities in the cancer cells. Several tumor suppressor genes have been isolated in recent years. Although a significant amount of work has been conducted using *p53* gene, in theory each one of these tumor suppressor genes can be potentially useful for cancer gene therapy. The purpose of the following section is to review our current status of the use of tumor suppressor genes and some of the future challenges we are likely to face for the successful therapeutic applications of these tumor suppressor genes.

1. p53 GENE

p53 is a tumor suppressor gene that is frequently mutated in many cancers (Nigro *et al.*, 1989). It is estimated that at least half of the human malignancies including breast, ovary, lung, colorectal carcinomas are associated with mutations in p53. It has been postulated that wild type p53 can regulate the cell growth either by controlling the cell cycle progression at G1 to S, and G2 to M phase, or by inducing apoptosis (programmed cell death) (Ko and Prives, 1996). Thus during DNA damage by a variety of genotoxic stress situations, p53 can induce cell cycle arrest, thus giving an opportunity to repair the DNA. If cells fail to repair the DNA, p53 can induce apoptosis of the potentially damaged cell. It is believed that cancer cells defective in p53 have lost this ability to undergo cell growth by cell cycle progression and apoptosis, and hence acquire the transformed phenotype. It is therefore only logical to investigate if replacing the mutant p53 by a wild type copy

Cancer Gene Therapy: Past Achievements and Future Challenges, edited by Habib
Kluwer Academic/Plenum Publishers, New York, 2000.

of p53 or targeting the mutant form of p53 will reverse the phenotype. Many laboratories have evaluated the effect of over-expressing p53 protein on cancer cells (Seth, 1998). This has been studied by ecotopic expression of wild type p53 using non-viral vectors, and viral vectors as described below.

Initial experiments were conducted using plasmid-mediated p53 expression. Cancer cells tranfected with a plasmid DNA and a selectable marker neomycin, exhibited a reduction in the colony number when treated with G418. Interestingly this colony number was dramatically reduced in cells with endogenous mutant p53 (Baker *et al.*, 1990; Casey *et al.*, 1991). In another study, a plasmid containing p53 driven by a metallothionein promoter was shown to induce apoptosis in human colon cancer cell lines when simultaneously exposed to the metallic inducers (Shaw *et al.*, 1992). In these studies it was further shown that over-expression of wild type p53 resulted in the inhibition of the tumor growth when the cells were subsequently injected into nude mice. Recent studies from other laboratories have also shown that when a plasmid containing p53 cDNA was mixed with liposomes and injected into the human breast or glioblastoma xenografts grown in nude mice, it prevented the tumor growth (Hsiao *et al.*, 1997; Xu *et al.*, 1997). These pre-clinical studies have led to a clinical trial under the direction of Dr. N. Habib (Habib *et al.*, 1996). In this protocol patients with hepatic cellular carcinoma are administered with a plasmid expressing p53 directly into the hepatic artery.

Retroviruses containing p53 cDNA have also been tested for the potential benefits for cancer gene therapy. When a retrovirus containing p53 was used to infect osteosarcoma and rhabdomyosarcoma cells, and the cells injected into nude mice, there was a significant inhibition of tumor growth (Chen *et al.*, 1990). Moreover, when a retrovirus containing p53 gene were injected directly into pre-established human lung xenografts, it inhibited the growth of tumors in nude mice (Cai *et al.*, 1993). Using a multicellular tumor spheroid model this laboratory further demonstrated that retroviruses can penetrate into these three dimensional structures. Moreover, the cells which received the virus expressing p53 appear to undergo apoptosis. These *in vitro* studies have led to a clinical trials in which retrovirus expressing p53 is administered in to the lungs of patients with small cell carcinoma. Preliminary results of this trail have been published recently (Roth *et al.*, 1996).

In an effort to increase the efficiency of gene transfer many laboratories have developed recombinant adenoviruses containing p53 cDNA. Infection of cancer cells with adenoviral vectors carrying wild-type *p53* (AdWTp53) has been demonstrated to produce high levels of p53 protein in a number of epithelial cancer cell lines. These include breast, prostate, ovary, pancreatic and neuroblastoma cell lines, just to mention a few (Katayose *et al.*, 1995a). While all cell lines expressing adenoviral-mediated p53 show growth inhibition, these vectors are much more cytotoxic to cancer cells expressing a mutant form of endogenous *p53*. Normal human cells are generally resistant to the cytotoxic effects of AdWTp53 (Katayose *et al.*, 1995a). We have further shown that p53 overexpression results in growth arrest in all the cell lines examined. However, breast cancer cells such as MDA-MB-231 which endogenously express mutant form of p53 undergo apoptosis, while cells such as a breast cancer cell line MCF-7 which endogenously express a wild type p53 is resistant to apoptosis (Katayose *et al.*, 1995a, b). Since p53-mediated apoptosis appears to be a key determinant for cancer gene therapy, we are interested in understanding the mechanisms of "sensitivity" and "resistance" to AdWTp53. It is known that p53 is a transcription factor and can either upregulate or down regulate the expression of many genes potentially involved in apoptosis. It has also been suggested that perhaps some non-transcriptional mechanisms such as protein-protein interaction could be also important for apoptosis (Seth *et al.*, 1997a; Seth, 1998; Seth *et al.*, 1998). In any event, the idea is the p53 will either activate the pathway leading to apoptosis, or down regulate the pathways which favor the cell survival. It has been shown that the expression of several genes are upregulated during p53-mediated apoptosis. These include *Bax, Fas (CD95/APO-1)* and *IGFBP3* (Insulin like growth factor binding protein 3). Similarly, p53 can also down down-regulate the activity of many genes, some of which are survival factors for the cells. Examples of these genes include the *fibroblast growth factor receptor-1, thyroid receptor* β*-1* and *insulin like growth factor-1* (Seth *et al.*, 1997a; Seth, 1998; Seth *et al.*, 1998; Rakkar *et al.*, 1999). Thus, p53-mediated apoptosis is an active area of research in which adenoviral vectors expressing wild-type p53 protein can potentially be very useful.

In recent years, the effects of recombinant adenoviral vectors expressing wild-type p53 on inhibiting the tumor growth in animal models have been extensively reported. Human cancer cells (lung cancer cells, head and neck cancer cells, prostate cancer cells, breast cancer cells among others) when pre-infected with these vectors and subsequently injected into the nude mice show an inhibition of tumor formation. In fact, direct injection of these vectors into pre-established tumors in nude mice, lead to a significant inhibition of tumor growth (Seth, 1998). Cells from these AdWT53-injected tumors *in vivo* have been shown to exhibit high levels of p53 expression and to undergo apoptosis. Notably, however, direct injection of adenoviral vectors into the tumor mass can infect only a limited number of cells (generally less than 5% of the total cell population). Hence it has been postulated that adenoviral-mediated p53 expression and cell death may lead to a "bystander effect"; i.e., inhibition of growth of cells which are not infected with the recombinant virus and hence do not express exogenous wild type p53 protein. Several mechanisms of this putative bystander effect have been proposed. Down regulation of the angiogenesis factors such as vascular derived endothelial growth factors or up-regulation of angiogenesis inhibiting factors like thrombospondin 1 could be important components of this bystander effect (Seth, 1998; Li *et al.*, 1998; Rizk *et al.*, 1998). It is also conceivable, though not yet proven, that the adenovirus backbone and the p53 protein can activate an immune response directed against the cancer cells. Adenovirally delivered human wild type *p53* has also been shown to enhance sensitivity of cancer cells to chemotherapeutic agents as well as external radiation. This observation was initially made in lung cancer cells treated, *in vitro* and *in vivo,* with a combination of cis-platin and the adenoviral vector. We have observed similar results even in multi-drug resistant breast cancer cells *in vitro* (Seth *et al.*, 1997 a, b; Song *et al.*, 1997; Song *et al.*, 1998; Ifiyinwa *et al.*, 1997). Based on this pre-clinical research, clinical protocols using adenovirus expressing p53, alone or in combination with chemotherapeutic agents, are currently accruing patients (Roth *et al.*, 1998; Clayman *et al.*, 1998).

Since nude mice lack a functional immune system, to address the critical issues such as the p53-mediated bystander effect, and the effects of AdWTp53 on the immune system, it is desirable to use an immune competent animal model. For this purpose our laboratory has recently shown that an adenovirus expressing p53 can be also successfully used to induce apoptosis in a mouse Renca cell lines (Li *et al.*, 1998). Moreover, when Renca cell were injected into Balb/C mice and the tumor formed were administered with AdWTp53 complete tumor regressions were observed in 30% of the animals (Li *et al.*, 1998). We further showed that these cured animals were resistant to the tumor formation when subsequently challenged with Renca cells. Using this animal model it will be now feasible to address some of the issues relevant to the clinical application of the adenoviruses in the patients.

While it is worthwhile to pursue the clinical trials using AdWTp53 an interesting idea of AdWTp53 application is its possible utility in bone marrow purging. Since adenoviruses poorly infect hematopoeitic cells due the lack of adenovirus receptors on these bone marrow cells, an idea arose that if a mixed population of cancer cells and bone marrow cells was simultaneously exposed to AdWTp53, the later will specifically kill the tumor cells while protecting the bone marrow cells (Seth *et al.*, 1996). In fact when a mixed population of breast cancer cells and CD34+ cell were exposed to AdWTp53, while there was a complete inhibition of the colony formation of the tumor cells, there was no significant inhibition of CFUGM colonies (Seth *et al.*, 1996). These results indicated a potential of using AdWTp53 in purging the breast cancer cells contaminating the bone marrow cells during the autologous bone marrow transplantation.

Recently our laboratory has also used an AAV vector expressing p53 (Qazilbash *et al.*, 1997). Like Adp53 this vector also induced apoptosis in cancer cells expressing null p53 or devoid of p53. In an animal model (H-358) this vector also appears to be effective in inhibiting the tumor growth (Qazilbash *et al.*, 1997). However no clinical trial using AAVp53 has yet been initiated.

2. p16/INK4A AND OTHER CYCLIN-DEPENDENT KINASE INHIBITORS

Another tumor suppressor gene frequently mutated or absent in some cancers is $p16^{INK4A}$. The protein product of this gene is a cyclin-dependent kinase inhibitor of INK4 family, the over expression of which is expected to inhibit the phosphorylation of Rb and other Rb like proteins resulting in the cell cycle arrest and hence inhibition of cell growth (Stone *et al.*, 1995).

Much of the initial experiments were conducted using transfection based assays in which p16 was overexpressed in the cancer cell using plasmid DNAs. It was established that p16 can induce cell cycle arrest particularly in cells containing Rb protein (Koh *et al.*, 1995). For gene therapy purposes, a recombinant adenovirus expressing $\mathrm{p16^{INK4A}}$ has been constructed in our laboratory, and evaluated for its effects on cell growth (Craig *et al.*, 1998). Human cancer cells infected with Adp16 produce high level of $\mathrm{p16^{INK4A}}$ protein, and show growth inhibition. These inhibitory effects are much more pronounced in cells devoid of endogenous $\mathrm{p16^{INK4}}$, but containing wild type retinoblastoma protein. However, cells containing wild type p16 and defective Rb proteins were resistant to the cytotoxic effects of Adp16 (Craig *et al.*, 1998). Cytotoxic effects of Adp16 were due to the inhibition of cell cycle progression and possibly apoptosis if the cells. Thus, cancer cells defective in $\mathrm{p16^{INK4}}$, are a good target for gene therapy by Adp16. Recent studies have shown that direct injection of Adp16 into human xenografts including head and neck cancer, lung cancer and esophageal cancers can inhibit the growth of these tumors when grown in nude mice (Rocco *et al.*, 1998).

In addition to $\mathrm{p16^{INK4}}$, adenoviral vectors expressing other cyclin kinase inhibitors $\mathrm{p21^{WAF1/Cip1}}$, and $\mathrm{p27^{Kip}}$ have also been reported. We have shown that over expression of both $\mathrm{p21^{WAF1/Cip1}}$and $\mathrm{p27^{Kip1}}$ inhibits cell growth and induces G1/S arrest in the infected cells (Katayose *et al.*, 1995b; Craig *et al.*, 1997). Interestingly unlike Adp16 the cytotoxic effects of Adp21 and Adp27 were observed on all the cancer cell lines examined (Craig *et al.*, 1997). However between the two vectors Adp27 was in general much more cytotoxic than Adp21. Similarly the effects of Adp27 on the inhibition of cell cycle progression was also more proven with Adp27 (Craig *et al.*, 1997). Another difference was that Adp27 induced apoptosis in all the cancer cell examined while Adp21 failed to induce apoptosis (Katayose *et al.*, 1997). Moreover, in the breast tumor models examined in our laboratory we found that while direct administration of Adp27 into the tumors was very effective in inhibiting the tumor growth and increasing the median animal survival, Adp21 failed to exhibit such effects (Rakkar *et al.*, 1998). So far, no Phase 1 clinical trial evaluating adenoviral vectors transducing cyclin dependent kinase inhibitors has been initiated.

Some recent reports have focused on combinations of adenoviral vectors delivering tumor suppressor genes to cancer cells. Whereas adenoviral delivery of $p16^{INK4}$ or *p53* on their own did not lead to apoptosis, simultaneous exposure of cells to adenoviral vectors expressing of $p16^{INK4}$ and *p53* induced apoptosis and led to inhibition of tumor growth in human hepatocellular carcinoma xenografts in nude mice (Sandig *et al.*, 1997).

3. RB GENE

The retinoblastoma gene (*RB*) is a prototypic tumor suppressor gene. Rb can control the entry of cells into S-phase by interacting with the transcription factor E2F (Weinberg, 1992). When Rb is present in the phosphorylated state, it interacts with E2F and this results in the movement of the cells into DNA synthesis phase. However, if Rb phosphorylation is inhibited for example by cyclin-dependent kinase inhibitors, Rb protein can no longer control E2F transcription and the cells remain arrested in G1 phase. Thus overexpression of Rb by vectors can also result in the presence of excess of Rb protein not complexed with E2F, and hence resulting in G1/S arrest.

Studies have demonstrated that tumorigenicity of the non-small cell lung carcinoma and bladder carcinoma cells in nude mice is suppressed following treatment by a recombinant adenovirus vector expressing the N-terminal truncated retinoblastoma (RB) protein. Notably, gene therapy of established human RB– and RB+ bladder xenograft cancers in nude mice by this vector resulted in regression of the treated tumors (Xu *et al.*, 1996). In another recent study, decreased tumor cell proliferation and inhibition of human breast tumor growth was shown with a recombinant adenovirus carrying *RB* cDNA (Demers *et al.*, 1998).

4. OTHER TUMOR SUPPRESSOR GENES

In recent years several tumor suppressor genes have been isolated. These include —APC, PTEN, BRCA1, VHL genes among others. Preliminary work using these tumor suppressor genes suggests that they are all potentially useful for cancer gene therapy.

4.1. Adenomatous polyposis coli (APC) is a tumor suppressor gene, mutations of which have been shown to initiate sporadic and inherited forms of the human colorectal cancer. APC gene product has been shown to associate with catenins, suggesting a link between the tumor initiation and the cell adhesion (Su *et al.*, 1993). More importantly using a inducible expression system it has been shown that overexpression of APC alleles resulted in the significant inhibition of cell growth. It was also documented that these effects of APC are due to the induction of apoptosis of the cancer cells (Morin *et al.*, 1996). Therefore, it is likely that overexpression of APC gene could be a potential gene therapy approach.

4.2. PTEN is a potential tumor suppressor genes that appears to be mutated in several human cancers including glioblastomas, prostate and breast cancers. The protein product of PTEN gene has a sequence homology with tyrosine kinase and the cytoskeletal protein tensin (Li *et al.*, 1997). Over expression of PTEN using adenoviral vectors caused cell cycle arrest and inhibition of glioblastoma tumor cells' growth *in vitro* (Cheney *et al.*, 1998). Moreover anchorage-independent growth in soft agarose was also inhibited by the adenovirus expressing PTEN gene. These investigators also showed that PTEN overexpressing cancer cells were completely non-tumorigenic (Cheney *et al.*, 1998). Thus *in vivo* tumor suppression activity of PTEN suggests that adenovirus expressing PTEN has a potential application in cancer gene therapy.

4.3. Another tumor suppressor gene, BRCA1 influences the susceptibility to breast and ovarian cancer (Miki *et al.*, 1994). Recently BRCA1 protein has been implicated in controlling the cell cycle progression and induction of apoptosis (Shao *et al.*, 1996; Somasundaram *et al.*, 1997). A retrovirus expressing BRCA1 was shown to inhibit the growth of ovarian cancer cells *in vitro* and caused the suppression of tumor growth *in vivo* in a nude mouse xenograft model (Obermiller *et al.*, 1997). This has resulted in testing this retrovirus containing BRCA1 cDNA in a Phase 1 clinical trial (Obermiller *et al.*, 1997).

4.4. Inactivation of Von Hippel-Lindau (VHL) a tumor suppressor gene results in a number of abnormalities including renal cell carcinoma, cerebral hemanglioblastomas, pheochomocytomas and retinal angiomata (Kaelin and Maher, 1998). *In vitro* studies have shown that wild type, but not mutant, VHL protein can inhibit elongation of transcription by inhibiting the interaction of elongin A subunit with elongin B and elongin C. Still other studies have implicated VHL protein expression in the down-regulation of various proteins including vascular endothelial growth factor, platelet derived growth factor, and glucose transporter (Kaelin and Maher, 1998). Renal cancer cells transduced with a plasmid expressing wild type VHL and neomycin genes inhibited the colony formation in the presence of G418 (Chen *et al.*, 1995). Other studies suggested that renal cancer cells which expressed a wild type VHL undergo cell cycle exit when serum starved, thereby suggesting the role of VHL in controlling cell cycle progression (Pause *et al.*, 1998). Transduction of a renal cancer cell line 786–0, which contains mutant VHL gene, with a vector expressing wild type VHL, resulted in diminished tumor formation in nude mice (Ilioupolos *et al.*, 1995). Recently, our laboratory has constructed an adenovirus expressing VHL gene and showed that infection of cancer cells with this virus results in the inhibition of cell growth, and caused cell cycle arrest. Interestingly, VHL was shown to upregulate the expression of p27/Kip1 protein previously known to induce apoptosis in cancer cells (Kim *et al.*, 1998; Katayose *et al.*, 1997). These studies suggest a possible application of such viral vectors in cancer gene therapy.

5. CONCLUSION AND FUTURE CHALLENGES

In recent years tremendous progress has been made in the pre-clinical research using tumor suppressor genes. These studies have demonstrated that it is possible to deliver the normal alleles of these tumor suppressor genes to the cancer cells expressing a defective form of these genes. In many cases it has been shown that this results in either cell cycle arrest and/or apoptosis of the target cells. Moreover, using various animals models it has been also successfully shown that vector-mediated expression of the tumor suppressor genes in the tumor cells also prevents their ability to form tumors. Many studies have indicated that direct injection of vectors expressing tumor suppressor genes can inhibit the tumor growth in rodent models. All these studies have led to several Phase 1 clinical trials.

While the pre-clinical results obtained are potentially interesting, several challenges remain in this field. One problem is that most vector-driven cancer gene therapy using tumor suppressor genes has been conducted by directly administrating the vectors into the tumor mass or by regional delivery into the tumors. While this prevents the growth of this tumor locally, very little research has been done to understand how this approach will kill the cancer cells at the distant metastatic sites. So, research in the area should be a priority.

Another problem with the currently used vectors is that most vectors when given directly into the tumors can only infect only a certain fraction (5–10%) of cells in the tumor mass; hence a large portion of tumor cells are not accessible to the vector. So, perhaps one needs to devise delivery methods which can increase the distribution of the vectors to the maximum number of cells in the tumor. However, it is worth pointing again that in spite of the poor distribution of vectors into the tumors by the existing techniques, many studies have shown complete tumor regressions post vector-administrations into the tumor. So, therefore there must be certain bystander effect which influence the outcome of this therapy. Hence it is very important to understand the mechanism of this bystander effect.

Another potential area of research is the development of vectors which will infect or at-least express the therapeutic genes only in the cancer cells while completely protecting the normal cells of the body. This could be accomplished by several ways including the use of cell and tissue specific promoters, transcriptional targeting, pseudotyping the vectors among other approaches.

Finally, in spite of the various problems, one must continue conducting the clinical trials using the available vectors expressing tumor suppressor genes. If one selects patients who have exhausted all the classical therapeutic approaches such as surgery, radiation and chemotherapy, then perhaps one need not to explain the possible failure of cancer gene therapy in patients-after all other therapies have already failed on these patients! Who knows these clinical trials might actually yield a therapeutic approach which would never be available unless such clinical trial are actively pursued.

REFERENCES

Baker, S.J., Markowitz, S., Fearon, E.R., Willson, J.K., and Vogelstein, B., 1990, Suppression of human colorectal carcinoma cell growth by wild-type p53, *Science* **249**:912–915.

Cai, D.W., Mukhopadhyay, T., Liu, Y., Fujiwara, T., and Roth, J.A., 1993, Stable expression of the wild-type p53 gene in human lung cancer cells after retrovirus-mediated gene transfer, *Hum Gene Ther.* **4**:617–624.

Casey, G., Lo-Hsueh, M., Lopez, M.E., Vogelstein, B., and Stanbridge, E.J. (1991) Growth suppression of human breast cancer cells by the introduction of a wild-type p53 gene, *Oncogene* **6**:1791–1797.

Chen, F., Kishida, T., Duh, F.M., Renbaum, P., Orcutt, M.L., Schmidt, L., and Zbar, B., 1995, Suppression of growth of renal carcinoma cells by the von Hippel-Lindau tumor suppressor gene, *Cancer Res.* **58**: 4804–4807.

Chen, P., Chen, Y., Bookstein, R., and Lee, W., 1990, Genetic mechanisms of tumor suppression by the human p53 gene. *Sceince (Washington, DC)* **250**:1576–1579.

Cheney, I.W., Johnson, D.E., Vaillancourt, M.T., Avanzini, J., Morimoto, A., Demers, G.W., Wills, K.N., Shabram, P.W., Bolen, J.B., Tavtigian, S.V., and Bookstein, R., 1998, Suppression of tumorigenicity of glioblastoma cells by adenovirus-mediated MMAC1/PTEN gene transfer, *Cancer Res.* **58**:2331–2334.

Clayman, G.L., el-Naggar, A.K., Lippman, S.M., Henderson, Y.C., Frederick, M., Merritt, J.A., Zumstein, L.A., Timmons, T.M., Liu, T.J., Ginsberg, L., Roth, J.A., Hong, W.K., Bruso, P., and Goepfert, H.,

1998, Adenovirus-mediated p53 gene transfer in patients with advanced recurrent head and neck squamous cell carcinoma, *J. Clin. Oncol.* **16**:2221–2232.

Craig, C., Min, K., Ohri, E., Wersto, R., Li, Z., Katayose, D., and Seth, P., 1998, Consequences of adenoviral-mediated p16 protein expression on cell cycle arrest are determined by the endogenous Rb and p16 status in human breast cancer cells, *Oncogene* **16**:265–272.

Craig, C., Wersto, R., Kim, M., Ohri, E., Li, Z., Katayose, D., Lee, S.J., Trepel, J., Cowan, K., and Seth, P., 1997, recombinant adenovirus expressing p27Kip1 induces cell cycle arrest and loss of cyclin-Cdk activity in human breast cancer cells, *Oncogene* **14**:2283–2289.

Demers, G.W., Harris, M.P., Wen, S.F., Engler, H., Nielsen, L.L., and Maneval, D.C., 1998, A recombinant adenoviral vector expressing full-length human retinoblastoma susceptibility gene inhibits human tumor cell growth, *Cancer Gene Ther.* **5**:207–214.

Habib, N.A., Ding, S.F., el-Masry, R., Mitry, R.R., Honda, K., Michail, N.E., Dalla Serra, G., Izzi, G., Greco, L., Bassyouni , M., el-Toukhy, M., and Abdel-Gaffar, Y., 1996, Preliminary report: the short-term effects of direct p53 DNA injection in primary hepatocellular carcinomas, *Cancer Detect Prev.* **20**:103–107.

Hsiao, M., Tse, V., Carmel, J., Tsai, Y., Felgner, P.L., Haas, M., and Silverberg, G.D., 1997, Intracavitary liposome-mediated p53 gene transfer into glioblastoma with endogenous wild-type p53 in vivo results in tumor suppression and long-term survival, *Biochem. Biophys. Res. Commun.* **233**:359–364.

Ifeyinwa, A., Seth, P., Cowan, K.H., and Sinha, B.K., 1998, Role of wild-type-53 in the enhancement of camptothecin cytotoxicity against human prostate tumor cells, *Anticancer Res.* **18**:1845–1850.

Iliopoulos, O., Kibel, A., Gray, S., and Kaelin, W.G. Jr., 1995, Tumour suppression by the human von Hippel-Lindau gene product, *Nat. Med.* **1**:822–826.

Kaelin, W.G. Jr., and Maher, E.R., 1998, The VHL tumour-suppressor gene paradigm, *Trends Genet.* **14**: 423–426.

Katayose, D., Gudas, J., Nguyen, H., Srivastava, S., Cowan, K., and Seth, P., 1995a, Cytotoxic effects of a recombinant adenovirus expressing human wild type p53 on tumor and normal mammary epithelial cells, *Clinical Cancer Res.* **8**:889–898.

Katayose, D., Wersto, R., Cowan, K.H., and Seth, P., 1995b, Effects of a recombinant adenovirus expressing WAF1/Cip1 on cell growth, cell cycle, and apoptosis, *Cell Growth Differ.* **6**:1207–1212.

Katayose, Y., Kim, M., Rakkar, A., Li, Z., Cowan, K., and Seth, P., 1997, Promoting apoptosis: a novel activity associated with cyclin-dependent kinase inhibitor p27/Kip1, *Cancer Res.* **57**:5441–5447.

Kim, M., Katayose, Y., Rakkar, A., Li, Z., Hwang, S., Katayose, D., Cowan, K., and Seth, P., 1998, A recombinant adenovirus expressing tumor suppressor gene VHL induces cell cycle arrest by upregulating cyclin-dependent kinase inhibitor p27Kip1, *Biochem. Biophys. Res. Commm.* **253**:672–677.

Ko, L.J., and Prives, C., 1996, p53: puzzle and paraqdigm, *Genes Dev.* **10**:1054–1072.

Koh, J., Enders, G.H., Dynlacht, B.D., and Harlow, E., 1995, Tumour-derived p16 alleles encoding proteins defective in cell-cycle inhibition, *Nature* **375**:506–510.

Li, J., Yen, C., Liaw, D., Podsypanina, K., Bose, S., Wang, S.I., Puc, J., Miliaresis, C., Rodgers, L., McCombie, R., Bigner, S.H., Giovanella, B.C., Ittmann, M., Tycko, B., Hibshoosh, H., Wigler, M.H., and Parsons, R., 1997, PTEN, a putative protein tyrosine phosphatase gene mutated in human brain, breast, and prostate cancer, *Science* **275**:1943–1947.

Li, Z., Rakkar, A., Katayose, Y., Kim, M., Shanmugam, N., Moul, J.W., McLeod, D.G., Cowan, K.H., and Seth, P., 1998, Efficacy of multiple administrations of a recombinant adenovirus expressing wild type p53 in an immune-competent mouse tumor model, *Gene Ther.* **5**:605–613.

Miki, Y., Swensen, J., Shattuck-Eidens, D., Futreal, P.A., Harshman, K., Tavtigian, S., Liu, Q., Cochran, C., Bennett, L.M., and Ding, W., 1994, A strong candidate for the breast and ovarian cancer susceptibility gene BRCA1, *Science* **266**:66–71.

Morin, P.J., Vogelstein, B., and Kinzler, K.W., 1996, Apoptosis and APC in colorectal tumorigenesis, *Proc. Natl. Acad. Sci. USA.* **93**:7950–7954.

Nigro, J.M., Baker, S.J., Preisinger, A.C., Jessup, J.M., Hostetter, R., Cleary, K., Bigner, S.H., Davidson, N., Baylin, S., Devilee, P., Glover, T., Collins, F.C., Weston, A., Modali, R., Harris, C.C., and Vogelstein, B., 1989, Mutations in the p53 gene occur in diverse human tumour types, *Nature* **342**:705–708.

Obermiller, P.S., Redlin-Frazier, S., Jensen, R.A., Welcsh, P., Dann, J., King, M.C., Johnson, D.H., Holt, J.T., and Tait, D.L., 1997, A Phase I Trial of Retroviral BRCA1sv Gene Therapy in Ovarian Cancer, *Clin Cancer Res.* **3**:1959–1968.

Pause, A., Lee, S., Lonergan, K.M., and Klausner, R.D., 1998, The von Hippel-Lindau tumor suppressor gene is required for cell cycle exit upon serum withdrawal, *Proc, Natl. Acad. Sci. USA.* **95**:993–998.

Qazilbash, M.H., Xiao, X., Seth, P., Cowan, K.H., and Walsh, C.E., 1997, Cancer gene therapy using a novel adeno-associated virus vector expressing human wild-type p53, *Gene Therapy* **4**:675–682.

Rakkar, A., Li, Z., Katayose, Y., Kim, M., Cowan, K.H., and Seth, P., 1998, Adenoviral expression of the cyclin-dependent kinase inhibitor p27/Kip1: A stategy for breast cancer gene therapy, *J. Natl. Cancer Institute* **90**:1836–1838.

Rakkar, A., Katayose, Y., Kim, M., Craig, C., Choi, Y., Lee, S., Ohri, E., Li, Z., Cowan, K.H., and Seth, P., 1999, p53 mediated apoptosis is associated with up-regulation of Fas/CD95/Apo-1, *Cell Death and Differentiation* (in press).

Rizk, R., Kaiser, L., Seth, P., and Albelda, S., 1999, Bystander effect in the context of Ad.p53 virus, *Cancer Gene Ther.* (in press).

Rocco, J.W., Li, D., Liggett, W.H. Jr., Duan, L., Saunders, J.K.Jr., Sidransky, D., and O'Malley, B.W. Jr., 1998, p16INK4A adenovirus-mediated gene therapy for human head and neck squamous cell cancer, *Clin Cancer Res.* 4:1697–1704.

Roth, J.A., Nguyen, D., Lawrence, D.D., Kemp, B.L., Carrasco, C.H., Ferson, D.Z., Hong, W.K., Komaki, R., Lee, J.J., Nesbitt, J.C., Pisters, K.M., Putnam, J.B., Schea, R., Shin, D.M., Walsh, G.L., Dolormente, M.M., Han, C.I., Martin, F.D., Yen, N., Xu, K., Stephens, L.C., McDonnell, T.J., Mukhopadhyay, T., and Cai, D., 1996, Retrovirus-mediated wild-type p53 gene transfer to tumors of patients with lung cancer, *Nat. Med.* **2**:985–991.

Roth, J.A., Swisher, S.G., Merritt, J.A., Lawrence, D.D., Kemp, B.L., Carrasco, C.H., El-Naggar, A.K., Fossella, F.V., Glisson, B.S., Hong, W.K., Khurl, F.R., Kurie, J.M., Nesbitt, J.C., Pisters, K., Putnam, J.B., Schrump, D.S., Shin, D.M., and Walsh, G.L., 1998, Gene therapy for non-small cell lung cancer: a preliminary report of a phase I trial of adenoviral p53 gene replacement, *Semin Oncol.* **25**:33–37.

Sandig, V., Brand, K., Herwig, S., Lukas, J., Bartek, J., and Strauss, M., 1997, Adenovirally transferred p16INK4/CDKN2 and p53 genes cooperate to induce apoptotic tumor cell death, *Nat. Med.* **3**:313–319.

Seth P., Katayose, Y., Rakkar, A., and Li, Z., 199a, Recombinant adenoviral vector expressing p53 for cancer gene therapy. In Cancer Gene Therapy (ed H. Mazarakisa, and S. Swart) IBC Library Series pp 171–179, Southborough, MA, USA.

Seth, P., Katayose, D., Li, Z., Kim, M., Wersto, R., Craig, C., Shanmugam, N., Ohri, E., Mudasar, B., Rakkar, A.N.S., Kodali, P., and Cowan, K., 1997b, A recombinant adenovirus expressing wild type p53 induces apoptosis in drug resistant breast cancer cells: a gene therapy approach for drug reistant cancers, *Cancer Gene Ther.* **4**:383–390.

Seth, P., Rakkar, A., Kim, M., and Katayose, Y., 1998, Cancer gene therapy with adenoviral vectors expressing tumor suppressor genes, in: *IBC Library Series* (N. Habib, ed) Southborough, MA, USA (in press).

Seth, P. (1998) Vectors for cancer gene therapy, in: *Gene therapy for the treatment of cancer: Progress and Prospects* (B. Huber, and I. Magrath, eds) pp 41–77, Cambridge University Press, Cambridge, UK.

Seth, P., Brinkmann, U., Schwartz, G.N., Katayose, D., Gress, R., Pastan, I., and Cowan, K., 1996, Adenovirus-mediated gene transfer to human breast tumor cells: an approach for cancer gene therapy and bone marrow purging, *Cancer Res.* **56**:1346–1351.

Shao, N., Chai, Y.L., Shyam, E., Reddy, P., and Rao, V.N., 1996, Induction of apoptosis by the tumor suppressor protein BRCA1, *Oncogene* **13**:1–7.

Shaw, P., Bovey, R., Tardy, S., Sahli, R., Sordat, B., and Costa, J., 1992, Induction of apoptosis by wild-type p53 in a human colon tumor-derived cell line. *Proc. Natl. Acad. Sci. USA.* **89**:4495–4499.

Somasundaram, K., Zhang, H., Zeng, Y.X., Houvras, Y., Peng, Y., Zhang, H., Wu, G.S., Licht, J.D., Weber, B.L., and El-Deiry, W.S., 1997, Arrest of the cell cycle by the tumour-suppressor BRCA1 requires the CDK-inhibitor p21WAF1/CiP1, *Nature* **389**:187–190.

Song K.M., Fukushima, P., Seth, P., and Sinha, B.K., 1998, Role of p53 and apoptosis in sensitization of cis-diaminedoichloroplatinum antitumor activity by interleukin-1α in ovarian carcinoma cells, *Int. J. Oncol.* **12**:299–304.

Song, K., Li, Z., Seth, P., Cowan, K., and Sinha, B.K., 1997, Chemosensitization of ovarian carcinomas by wild type p53 gene transfer by adenovirus vector, *Oncology Res.* **9**:603–609.

Spitz, F.R., Nguyen, D., Skibber, J.M., Cusack, J., Roth, J.A., and Cristiano, R.J., 1996, In vivo adenovirus-mediated p53 tumor suppressor gene therapy for colorectal cancer, *Anticancer Res.* **16**:3415–3422.

Stone, S., Jiang, P., Dayananth, P., Tavtigian, S.V., Katcher, H., Parry, D., Peters, G., and Kamb, A., 1995, Complex structure and regulation of the P16 (MTS1) locus, *Cancer Res.* **55**:2988–2994.

Su, L.K., Vogelstein, B., and Kinzler, K.W., 1993, Association of the APC tumor suppressor protein with catenins, *Science* **262**:1734–1737.

Weinberg, R.A., 1992, The retinoblastoma gene and gene product, *Cancer Surv.* **12**:43–57.

Xu, H.J., Zhou, Y., Seigne, J., Perng, G.S., Mixon, M., Zhang, C., Li, J., Benedict, W.F., and Hu, S.X., 1996, Enhanced tumor suppressor gene therapy via replication-deficient adenovirus vectors expressing an N-terminal truncated retinoblastoma protein, *Cancer Res.* **56**:2245–2249.

Xu, M., Kumar, D., Srinivas, S., Detolla, L.J., Yu, S.F., Stass, S.A., and Mixson, A.J., 1997, Parenteral gene therapy with p53 inhibits human breast tumors in vivo through a bystander mechanism without evidence of toxicity, *Hum Gene Ther.* **8**:177–185.

18

GENE THERAPY FOR LIVER TUMOURS

Ragai R. Mitry, Marc R. Mansour, Roman Havlík, and Nagy A. Habib

Liver Surgery Section
Division of Surgery
Anaesthetics and Intensive Care
Imperial College School of Medicine
Hammersmith Hospital Campus
Du Cane Road, London W12 ONN
United Kingdom

1. EPIDEMIOLOGY

1.1. Hepatocellular Carcinoma

Hepatocellular carcinoma (HCC) is one of the most common tumours worldwide with an estimated 1 million new cases diagnosed per year (Carr *et al.*, 1997). In the Western world HCC is relatively rare representing less than 2% of all cancers. It is the most frequent primary liver malignancy, the remainder being made up of cholangiocarcinoma and fibrolamellar variant. There is a male to female ratio of 4:1. It is estimated that 60–90% of HCC cases occur in cirrhotic livers, depending on whether the tumour occurs in a high- or low-incidence areas (Wands and Blum, 1991).

Geographically HCC clusters in areas such as Africa, South East Asia and the Middle East where hepatitis infection is endemic. Chronic hepatitis B infection, hepatitis B carrier status (Caselmann, 1995) and to a lesser degree chronic hepatitis C infection (Gerber, 1995) all confer an increased risk to HCC, primarily as a result of integration of viral DNA into the genome (which can be isolated in tumour specimens) with consequent mutations of tumour suppressor genes such as *p53* (Wang *et al.*, 1995). Aflotoxin B_1 (AFB_1) exposure also induces mutations in *p53* gene, primarily through the formation of AFB_1–DNA adducts (Gerbes and Caselmann, 1993; Kobertz *et al.*, 1997). Cirrhosis as related to alcohol (reviewed in Seitz *et al.*, 1998) and haemochromatosis (Fellows *et al.*, 1988; reviewed in Barisani *et al.*, 1996) are also important factors, as is exposure to chemical hepatocarcinogens such as nitrosamine (Olajos, 1977), organic halides (Weisburger, 1977), and the contrast agent Thorotrast (Kato, 1987). Long-term use of the oral contraceptive pill (reviewed by Dourakis *et al.*, 1998) and steroids (Eagon *et al.*, 1996) are implicated in the development and progression of several liver diseases including HCC.

Cancer Gene Therapy: Past Achievements and Future Challenges, edited by Habib
Kluwer Academic/Plenum Publishers, New York, 2000.

1.2. Colorectal Liver Metastases

In the UK, tumour involvement of the liver is more often due to metastatic deposits, particularly from abdominal, pelvic, lung or breast malignancies, compared to primary liver tumours. Colorectal cancer is the most frequent primary source acounting for over 90% of liver secondaries, accountable partly by its high prevalence as it is one of the most common tumours and partly by the venous drainage via the portal vein. Synchronous liver metastases are found in 15–25% of all cases at initial presentation and 20–50% will develop metachronous liver recurrence (Sheiner *et al.*, 1994). If untreated 5 year survival after diagnosis of metastatic liver lesions from primary colorectal cancer is extremely rare and median survival is 5–6 months (Blumgart and Fong, 1995).

2. CURRENT TREATMENT OPTIONS

2.1. Hepatocellular Carcinoma

The prognosis of HCC is poor with only few patients surviving beyond 1 year. The only curative option is liver resection or transplantation (Philosophe *et al.*, 1998). Surgical resection can achieve up to 50% 5 year survival rate (Nagashima *et al.*, 1996). However, in most patients the tumours are unresectable because of multicentricity or poor hepatic function. Transplantation is indicated in patients with poor liver function that have no extra-hepatic spread. Five year survival rates for small cancers post-transplantation is nearly the same as after surgical resection (Gundlach *et al.*, 1998). Systemic chemotherapy and/or radiotherapy do not offer a survival advantage. For patients who are not candidates of potentially curative surgery, percutaneous ethanol injection and transarterial chemotherapy, with or without embolisation, could be benefitial and slow progression of the tumours (Allgaier *et al.*, 1998). Hyperthermia resulting from laser beam or radiofrequency using cooled needle tip have been used to provide palliation in these patients (Jiao *et al.*, 1999; Marone *et al.*, 1998). Another option is targeted therapy such as radiotherapy, chemotherapy and immunotherapy. Locoregional immunotherapy using interleukin-2 and interferon-gamma emulsified in a lipiodol-urografin mixture combined with locoregional chemotherapy is a promising therapeutic approach in patients with advanced HCC (Lygidakis *et al.*, 1995).

2.2. Colorectal Liver Metastases

Liver resection is still the only potentially curative treatment for metastatic colorectal cancer to the liver and offers between 25%–35% 5 year survival (Sheiner *et al.*, 1994; Fong and Blumgart, 1998). However, only 20% of patients with colorectal liver metastases are suitable for curative resection. Surgery should be attempted in those with absence of extrahepatic disease where all detectable tumour can be completely removed (Blumgart and Fong, 1995). Systemic chemotherapy and regional chemotherapy via the hepatic artery using a surgically implanted pump have significant response rates in comparison to no therapy, while regional chemotherapy provides roughly double response as compared to systemic chemotherapy (Rougier 1998; Vaughn *et al.*, 1993; Steele 1993). Isolated liver perfusion allows the use of higher doses of chemotherapy but needs further development and randomised trials (Rougier, 1998). Heat ablation, especially radiofrequency ablation and cryoablation, can achieve long-term survival in patients with colorectal metastases in liver only, but neither have proven to be curative (Fong and Blumgart, 1998, Curley and Vecchio, 1998). Ethanol injection and chemoembolization are other palliative options (Livraghi *et al.*, 1993).

3. GENETICS OF HCC AND CLM

Cancer is a distinct type of genetic disease characterised by failure to maintain the fidelity of DNA due to germ-line and/or somatic gene changes (Bishop, 1986, 1987; Solomon *et al.*, 1991). It follows what is known as the "multi-hit" concept, i.e. it arises from the accumulation of multiple rather than a single genetic mutation, with some 3 to 6 mutations required for initiation and development of cancer. Each genetic mutation leads to an increase in cellular proliferation resulting in gradual increase in tumour size, disorganisation and malignancy (Vogelstein and Kinzler, 1993).

Of all the aetiological agents involved in HCC, the hepatatis virus is the most important. HBV DNA integrated into the human genome has been identified in HCC tumour specimens and is thought to play a role in the genetic anomalies (Kew, 1996).

The genetic abnormalities seen in HCC tumours mainly include inactivation of tumour suppressor genes (Weinberg, 1991), and activation of oncogenes (Bishop, 1983). The tumour suppresor gene of most interest is *p53*, which is located on the short arm of chromosome 17(17p13). Deletion or mutation of wild-type *p53* (wt-*p53*) gene has been identified in approximately 50% of all tumour types and its presence in HCC carries a worse prognosis (Honda *et al.*, 1998). Activation of oncogenes from proto-oncogenes results in overexpression of proteins responsible for growth regulation and differentiation (Bishop, 1983). In HCC there has been interest in the *ras* proto-oncogene family (Weinberg, 1989), myc family and in telomerase (Tahara *et al.*, 1995).

4. GENE THERAPY

Non-surgical treatments such as chemotherapy are limited not only by their efficacy at the tumour site but also by its dose-limiting cytotoxicity of other tissue, particularly bone marrow. Gene therapy is an exciting approach because it offers the potential of correcting the underlying genetic defect, as well as specifically targeting the tumour preferentially to other tissue. The use of therapeutic strategies that discriminate effectively between tumour and non-tumour will revolutionise the approach of oncology in the future.

The use of gene therapy as a definitive treatment in its own right is limited at present, its initial role likely to be as an adjuvant therapy in combination with radiotherapy, chemotherapy and surgery. Surgically non-resectable tumours could be induced to regress to resectable, and post-operatively adjuvant gene therapy could improve upon present recurrence rates. Drug-sensitising genes could be introduced to tumour to help enhance the toxicity of chemotherapy, or conversely the dose-limiting cytotoxic effects of chemotherapy could be diminished by introducing drug-resistance genes to tissue such as bone marrow.

Gene therapy follows straightforward concepts. For instance, wt-*p53* has been used in gene therapy by introducing it into tumour cells lacking normal p53 protein activities. Overactive oncogenes have been inactivated using an anti-sense approach. Moreover, "killer" genes such as thymidine kinase (TK) and/or cytosine deaminase (CD) have been used in combination with pro-drugs, and immunomodulator genes such as cytokines with/without an anti-cancer vaccine approach, are all being studied for clinical application, with clinical trials already underway (Habib *et al.*, 1996; Crystal, 1997).

4.1. Tumour-Suppressor Genes (TSGs)

In liver tumours, mutation(s), deletion(s) and inactivation of TGSs are common. Bressac and colleagues studied the structure and expression of the *p53* gene in seven human HCC cell lines. Three of the seven cell lines Hep3B, PLC/PRF/5 and FOCUS showed abnormal structure and expression of *p53* (Bressac *et al.*, 1990; Bressac *et al.*, 1991). Both Hep3B and PLC/PRF/5 cell lines express the HBV surgface antigen (HBsAg), which is believed to induce malignant transformation in HCC. Evidence continues to accumulate implicating HBV infection, both directly and indirectly, in HCC carcinogenesis (Kew, 1996). Several research groups working on *p53* mutation analysis in HCC tumours have shown a higher frequency of mutations occuring in areas where chronic hepatitis infection or aflatoxin B1 exposure is common (Scorsone *et al.*, 1992; Lunn *et al.*, 1997; Yotsuyanagi *et al.*, 1995; Tabor, 1994). Therefore, it seems that the *p53* gene is a good candidate as a therapeutic agent for gene therapy in HCC. Moreover, metastatic cancers to the liver also have a high frequency of liver mutations and can also be considered for *p53* gene therapy.

4.1.1. p53 *Tumour Suppressor Gene.* Wild-type *p53* (wt-*p53*) gene is the principal tumour suppressor gene (TSG) in humans (Hollstein *et al.*, 1991) and is commonly mutated in cancers, and is known to be frequently mutated in HCC (Sheu *et al.*, 1992; Oda *et al.*, 1994; Honda *et al.*, 1997). It encodes a 53 kD protein, which is known as the "guardian of the genome" and acts as a "molecular policeman" monitoring the integrity of the genome (Lane, 1992). While wt-p53 (wt-p53) has anti-oncogenic functions, if mutated it can act as an oncogene (Finlay *et al.*, 1989). DNA damage leads to induction of p53 and its accumulation leads to either cell growth arrest or apoptosis, protecting the genome from accumulating excess mutations, i.e. p53 acts as an "emergency brake" (Ko and Prives, 1996).

4.1.1.1. *p53—In Vitro & In Vitro* Studies. Several research teams studied the effects of wt-*p53* on HCC cell lines, in order to show how effective wt-p53 would be in killing HCC cells. The *in vitro* HCC models included animal and human HCC cell lines with normal or mutant *p53*, with or without integrated hepatitis B viral (HBV) sequence. In fact the best models involved the use of cell lines with mutant *p53* and integrated HBV sequences, as these two features are closely associated. Various vectors encoding wt-*p53* were used in these studies, including plasmids, adenoviruses (Ad) and retroviruses (Puisieux *et al.*, 1993).

The cell killing is achieved through programmed cell death or apoptosis (Mitry *et al.*, 1997). Adenoviruses encoding *p53* (Ad-*p53*) vector are more effective than plasmids in killing human HCC cells, such as Hep3B, HepG2, and PLC/PRF/5. 50% kiling or apotosis could be achieved at about 15 virus particles per cell. This suggest that *p53* is an ideal gene for *in vitro* treatment of HCC, which was tested in animal models. Habib and colleagues, induced subcutaneous HCC tumours in nude mice, by injecting PLC/PRF/5 (an HCC cell line with mutant *p53* and integrated HBV sequence) cells in the flanks. The mice were injected intratumourally with a total of 1×10^9 particles of Ad-*p53*, while the control goup received a blank Ad vector (i.e. does not encode *p53*) injections at same dose. The results showed that Ad-*p53* inhibited the HCC tumour growth with some mice were "tumour-free" at the end of treatment period. Moreover, the control tumours continued grew till the end of treatment period (unpublished data). A recent study by Anderson and colleagues (1998) supported the findings of Habib and colleagues, where they developed an HCC synergic rat model, in which the rats have developed multifocal HCC nodules in the liver. Their results showed that the treatment with a recombinant adenovirus encoding *p53* resulted in effcient gene expression, growth suppression, and apoptosis (Anderson *et al.*, 1998).

4.1.1.2. *p53*—Clinical Studies. To our knowledge, few gene therapy clinical trials for HCC have been reported. The only published trial was a Phase I study which involved the use of wt-*p53* gene. In this study, Habib and colleagues (1996) investigated the effects of percutaneous intratumoural injection of plasmid encoding human wt-*p53* gene in eight cirrhotic patients with histologically proven HCC due to viral hepatitis. Each patient was injected with 2 mg of "naked" wt-*p53* plasmid DNA intratumourally under CT scan guidance. The tumour size, serum alpha-fetoprotein (AFP) tumour marker level and liver function tests were monitored every month prior to and during the period of study. Of the eight patients, one had complete response for 19 months, and three had significant reduction in tumour volume (90%, 75%, 50%). The results suggest that the clinical benefit of wt-*p53* injection in HCC is promising. Moreover, such investigations should be extended to include a larger number of patients in a prospective randomised clinical trial in order to prove its benefits and understand the mechanism of action. Side effects of the treatment were minimal. Another clinical trial by Venook and his team was mentioned in a review by Roth and Cristiano (1997). It was initiated to investigate the effects of adenovirus encoding *p53* in primary and metastatic liver cancers.

4.2. Suicide Gene Therapy

Viral and bacterial genes that encode enzymes able to convert non-toxic pro-drugs into toxic metabolites are currently being researched for the treatment of many cancers including colorectal, prostatic, thyroid, brain, breast, melanoma and HCC. They have the potential to offer tumour selectivity by:

i) using tumour-specific vectors.
ii) controling suicide gene transcription using a tumour-specific promoter, thus limiting suicide gene expression to target cells only. Examples include PSA promoters for prostatic carcinoma, CEA promoters for colorectal and lung carcinoma, erbB2 for breast and pancreatic carcinomas, and AFP for HCC (reviewed by Kanai *et al.*, 1997).

4.2.1. Thymidine Kinase. The nucleoside analogues, acyclovir and ganciclovir are used extensively in clinical practice for the treatment of herpes simplex viral (HSV) infections. They are essentially pro-drugs that are phosphorylated to the active form by thymidine kinase (TK), an enzyme not normally expressed in non-foetal human cells. The HSV1-TK gene is expressed after herpetic viral integration into the genome, leading to activation of the drug and cell death.

One of the most investigated suicide gene systems used to date involves HSV1-TK gene transfer (typically using an adenovirus vector) followed by ganciclovir administration. In order to limit expression of HSV1-TK to HCC cells only, an alpha-fetoprotein (AFP) promoter has been constructed such that only cells expressing AFP will be killed.

4.2.1.1. TK—*In Vitro* Studies. Moolten was the pioneer of the HSV-TK suicide gene system (1986) and was the first to show its application in vitro (Moolten *et al.*, 1990). An adenoviral vector has been used to introduce the HSV1-TK gene, under the control of an AFP promoter, into various HCC cell lines (Kanai *et al.*, 1996). Cell death after ganciclovir administration occurred in hepatoma cells producing AFP, namely HuH-7 and HepG2 cell lines, while HLF and HLE cell lines that do not produce AFP were unaffected. When the AFP promoter is used, ganciclovir sensitivity increases in parallel with the cells' ability to produce AFP (Su *et al.*, 1996).

Phosphorylated ganciclovir is thought to cause cell death by inhibition of DNA polymerase and chain termination of DNA synthesis in dividing cells (Moolten *et al.*, 1990). Furthermore, a "bystander effect" has been observed both *in vitro* and *in vivo* whereby neighbouring HSV1-TK-negative cells have died when in contact with HSV1-TK-positive cells after ganciclovir treatment (Freeman *et al.*, 1993).

Cell death by apoptosis occurs via a *p53*-independent pathway in non-proliferating thyrocytes expressing HSV1-TK after ganciclovir administration (Wallace *et al.*, 1996) and is supported by the effective killing of hepatoma cell lines expressing mutant *p53* such as Hep3B. This is of relevence to the therapeutic viability of this suicide model given the high frequency of *p53* mutations in HCC.

A glucocorticoid responsive element exists upstream of the AFP gene in hepatoma cells, explaining the increase in AFP observed after addition of dexamethasone into the culture medium of such cells (Nagabayashi *et al.*, 1989). The potency of the TK/ganciclovir suicide system has been enhanced by the addition of dexamethasone to a HepG2 cell line (Ido *et al.*, 1995).

4.2.1.2. TK—*In Vivo* Studies. The *in vitro* work has translated well to *in vivo* studies. Most have used subcutaneously grown human HCC tumours followed by transfection with an AFP-HSV1-TK gene via an adenoviral vector. Significant regression has been induced in HuH-7 established tumours in athymic mice, while normal tissue remains unaffected, highlighting tumour selectivity (Ueki *et al.*, 1998).

While other methods of gene therapy are seriously limited by the low transfection rate of tumour cells by the various vectors used, the bystander effect seen in this suicide model enables significant tumour regression to occur when only 10% of the tumour mass expresses HSV1-TK (Freeman *et al.*, 1993). An immunostimulatory mechanism is thought to play a role: upregulation of costimulatory molecules such as ICAM, B7.1 and B7.2 have been seen in ovarian tumour cells (Freeman, 1998).

Furthermore, a similar therapeutic response has been seen in rats with CLM. The established colorectal metastases were directly injected with HSV-TK-producing packaging cells (M11) leading to a 60 fold reduction in tumour mass compared to controls after treatment with GCV (Caruso *et al.*, 1993). This was confirmed using an adenoviral vector and direct intratumoural injection, and enhanced cytotoxicity by coadministration of an IL-2 gene (Chen *et al.*, 1995).

4.2.1.3. TK—Clinical Trials. A phase I trial of patients with relapsed ovarian carcinoma has shown complete remission in 3 of 18 patients after TK/ganciclovir administration (Freeman, 1998). The trial has been modified by immunising patients with the HER-2/Neu protein prior to treatment and presently in phase I/II.

4.2.2. Cytosine Deaminase Gene. Cytosine deaminase (CD) is a non-toxic gene found in some fungi and bacteria where it has a role in the conversion of cytosine to uracil. Organisms containing the gene are extremely efficient at metabolising 5-Fluorocytosine (5-FC) to the toxic 5-Fluorouracil (5-FU), an agent used in chemotherapy. Thus the E.coli CD gene is being used as a suicide gene such that genetically modified cells "commit suicide" in the presence of 5-FC.

4.2.2.1. CD—*In Vitro* Studies. The CD gene has been used with an AFP promoter to kill only AFP-positive HCC cell lines in the presence of 5-FC (Kanai *et al.*, 1997). A bystander effect also occurs whereby neighbouring cells not transduced with CD are also killed, though unlike the TK/ganciclovir system, this is irrespective of cell-to-cell contact with transduced cells. 5-FU converted from 5-FC in transduced cells is released into the medium on cell lysis and is likely to be responsible for the bystander effect, with 5-FU levels in the medium correlating well with degree of cytotoxicity (Kuriyama *et al.*, 1998).

It has been shown that the cytotoxic effect of the CD/5-FC system can be enhanced by the concomitant addition of the gene uracil phosphoribosyltransferase, that converts 5-FU to 5-fluoroUMP. This may work in part by increasing the diffusion gradient across the cell membrane and was shown in an *E. coli* model only and not yet in HCC cell lines (Tiraby *et al.*, 1998).

4.2.2.2. CD—*In Vivo.* AFP-positive human HCC tumours established subcutaneously *in vivo* have been shown to regress significantly after adenoviral-mediated insertion of the CD gene (with an AFP promoter) and subsequent 5-FC administration. Non-tumour tissue was unaffected (Kanai *et al.*, 1997).

4.2.2.3. CD—Clinical Trial. A phase 1 trial has been initiated with 18 patients with proven CLM. They are being treated with direct intratumoural injection of CD gene using an adenoviral vector, in combination with oral 5-FC (Crystal *et al.*, 1997).

4.3. Antisense Gene Therapy

This is an exciting approach to genetic modulation that can be used to diminish gene expression by introducing a molecule that identifies and inactivates or cleves a specifically targeted mRNA sequence thus blocking translation. Thus it is particularly useful for overactive oncogenes. Such antisense molecules include complementary oligonucleotides and ribozymes. The latter are RNA enzymes capable of cleaving specific RNA sequences. They have the advantage that their role as catalytic agents allows them to digest large numbers of mRNA. This is particularly important as cells tend to upregulate mRNA expression in response to such a challenge (Scanlon, 1997).

4.3.1. Antisense—In Vitro & In Vivo *Studies.* There has been considerable interest in the *ras* onogene family—37% of preneoplastic liver tissues and 50% of HCC tumours exhibit Ha-*ras* mutations. The *ras* proto-oncogene is most frequently converted to the oncogene by a single base mutation at codon 12 (GGC to GUC), 13 or 61 (CAA to CTA). Significant inhibition of HCC formation has been seen in rats treated with Ha-*ras* antisense during exposure to the hepatocarcinogen N-nitrosomorpholine (Baba *et al.*, 1997). An antisense H-*ras* oligonucleotide has been effective at inducing apoptosis in BEL-7402 human HCC cells in vitro, and also retards tumour growth in vivo (Liao *et al.*, 1997).

A ribozyme cleaving at GUC has been designed, i.e. only the oncogene mRNA is inactivated, while the normal mRNA GGC sequence is unaffected (Ohta *et al.*, 1996). This has been effective in melanomas grown subcutaneously on athymic mice though not yet applied to HCC cell lines.

4.4. Immunogene Therapy

The immunocompromised, such as those on long-term immunosuppressive therapy following transplantation or those suffering with AIDS, are susceptable to a variety of tumour types due to lack of the normal T-cell response that identifies and destroys abnormal cells.

In the immunocompetent, cancer cells survive by evading the host immune system. This occurs for two reasons: firstly tumour cells are poorly immunogenic. They not only have a poor ability at presenting antigen to T-cells via MHC class I and II molecules, they also lack the co-stimulatory molecules required for T-cell activation such as B7.1, B7.2 and 41BB. Secondly, T-cells in many cancer patients have been identified as missing the ζ chain involved in T-cell activation even if this process has taken place. The aim of immunogene therapy is to address one or both of these abnormalities, enabling immunostimulation against the tumour and thus tumour regression. It also offers the possibility of generating vaccines for both the treatment and prevention of various cancers.

4.4.1. Costimulators—IGF1, B7.1, and B7.2

4.4.1.1. Costimulators—*In Vitro* & *In Vivo* Studies. Insulin-like growth factor, IGF1 is an important growth factor expressed in both normal and tumour cells, its level being particularly high in HCC cells. An antisense IGF1 vector can not only completely inhibit IGF1 production in several different HCC cell lines in vitro but can also lead to tumour regression of HCC *in vivo* (Guo, Trojan, and Anthony, 1997).

The inhibition of tumour growth is not as straightforward as it sounds. They found that tumour regression was in fact occuring by an immune reaction, and identified an increase in MHCI expression on the HCC cell surface. However, HCC cells transfected with antisense IGF1 were still able to form tumours in 60% of syngenic animals when injected subcutaneously. Whilst these cells had higher MHCI antigens on their cell surface, they lacked B7.1 and B7.2, which are costimulatory antigens important to T-cell activation. They have thus used a dual approach, transfecting cells with both antisense IGF1 and a gene encoding the B7.1 antigen. These double-transduced cells were unable to form tumour in syngenic animals when injected subcutaneously, though they were still able to in immunocompromised animals such as SCID and nude mice.

4.4.1.2. Costimulators—Clinical Trials. Cotransfecting established tumours with these genes is a form of *in vivo* cancer vaccine. A phase I/II trial using the IGF1/B7.1 cotransfection approach has been initiated in 33 patients with HCC (Guo, 1997). The IGF1 antisense gene and the B7.1 antigen gene were mixed with lipedol and injected intratumourally. Tumour samples were shown to contain double-transfected cells by growth in a double-selection medium. 12 of the patients had a significant drop in AFP levels, though in 7 this later returned to pre-treatment levels. 4 patients with inoperable tumours had a greater than 50% decrease in tumour size allowing operative intervention. Tumour samples from these 4 patients were shown to have a CD3+ CD8+

and CD3+ CD4+ predominant lymphocytic infiltrate. No delayed hypersensitivity was observed.

4.4. Telomerase

Telomerase is a ribonucleoprotein polymerase that is absent from most normal cells. After cell division there is loss of several hundred terminal nucleotides resulting in a shorter telomere, the progressive loss of chromosomal length thought to play a role in cell ageing and initiation of apoptosis. Telomerase maintains chromosomal length and is thus implicated in immortalising cells. Immortal tumour cells do not lose telomere length after cell division (Kanazawa, 1997).

4.4.1. Telomerase—In Vitro *Studies.* Telomerase activity has been detected in the majority of HCC samples analysed. In 1 series 15 of 18 HCC samples were telomerase positive, while only 1 of 22 non-tumour liver samples was positive (Tahara *et al.*, 1995). Interestingly, 19 of 38 hepatitis tissues and 6 of 8 cirrhotic tissues sampled were also positive suggesting telomerase may play a role in carcinogenesis. Furthermore, telomerase activity in HCC has been correlated to tumour differentiation and patient survival (Kishimoto *et al.*, 1998).

A hammerhead ribozyme that cleaves the RNA component of telomerase has been used to inhibit telomerase activity in HepG2 and Huh-7 cell extracts (Kanazawa *et al.*, 1996), though it has not yet been shown to work effectively in living cell lines.

5. CHOICE OF GENE DELIVERY SYSTEM

Therapeutic DNA-based therapy or gene therapy is used to correct genetic defects. It requires the introduction of the recombinant DNA into the target cells/tissue and the subsequent expression of the gene(s), (Anderson, 1984). The initial step in gene therapy would be transfection or transduction of target cells. The gene delivery systems are either non-viral or viral. The non-viral systems or physio-chemical methods, include gene gun, direct injection of "naked" DNA, liposomes/DNA complexes, and electroporation. They have a number of features that make them attractive alternatives to viral systems. These features include safety, ease in manipulation, and stability. However, their efficiency in delivering the genes is very low compared to that of the viruses.

Viral vectors mainly include adenoviruses (Ad), adeno-associated viruses (AAV) and retroviruses, plus other viral vectors such as lentiviruses and herpes virsuses. The viral gene delivery systems are well reviewed by Jolly (1994). To date no single delivery system can provide a flexible approach to the many applications of gene therapy.

6. ROUTE OF ADMINISTRATION

Following the choice of the gene(s) of interest, the second important step is route of administration of the vector carrying the gene(s) in order to transfect/transduce as many target cells as possible. In the case of liver tumours, several routes have been tried in animal models and human (Anderson *et al.*, 1998) including direct intratumoural injection of "naked" plasmid DNA (Habib *et al.*, 1996) or replication deficient adenovirus (Crystal, 1997), intra-hepatic artery and intra-peritoneal.

REFERENCES

Allgaier, H.P., Deibert, P., Becker, G., and Blum, H.E., 1998, Hepatocellular carcinoma: non-surgical treatment. *Schweiz Rundsch Med Prax* **87**:1466–1470.

Anderson, W.F., 1984, Prospects for human gene therapy. *Science* **226**:401–409.

Anderson, S.C, Johnson, D.E., Harris M.P., Engler, H., Hancock, W., Huang, W.-M., Wills, K.N., Gregory, R.J., Sutjipto, S., Wen, S.F., lofgren, S., Shepard, H.M., and Maneval, D.C., 1998, p53 Gene therapy in a rat model of hepatocellular carcinoma: intra-arterial delivery of recombinant adenovirus. *Clin. Cancer Res.* **4**:1649–1659.

Baba, M., Yamamoto, R., Iishi, H., and Tatsuta, M., 1997, Ha-*ras* mutations in N-nitrosomorpholine-induced lesions and inhibition of hepatocarcinogenesis by antisense sequences in rat liver. *Int. J. Cancer* **72**:815–820.

Barisani, D., Green. R.M., and Gollan, J.L., 1999, Genetic hemochromatosis: pathogenesis, diagnosis, and therapy. *Dig. Dis.* **14**(15):304–315.

Bishop, J.M., 1987, The molecular genetics of cancer. *Science* **235**:305–311.

Bishop, J.M., 1986, Oncogenes and proto-oncogenes. *J. Cell Physiol*, **suppl 4**:1–5.

Blumgart, L.H., and Fong, Y. 1995, Surgical options in the treament of hepatic metastases from colorectal cancer. *Curr. Probl. Surg.* **32**:333–428.

Bressac, B., Galvin, K.M., Liang, T.J., Isselbacher, K.J., Wands, J.R., and Ozturk, M., 1990, Abnormal strusture and expression of p53 gene in human hepatocellular carcinoma. *Proc. Natl. Acad. Sci. USA* **87**:1973–1977.

Bressac, B., Kew, M., Wands, J., and Ozturk, M., 1991, Selective G to T mutations of p53 gene in hepatocellular carcinoma from southern Africa. *Nature* **350**:429–431.

Carr, B.I., Flickinger, J.C., and Lotze, M.T. Hepatobiliary Cancers. In: DeVita VT, Jr. Hellman S, Rosenberg SA. eds. *Cancer: Principles and Practice of Oncology*. 5th ed. Philadelphia: J.B. Lippincott Co., **1997**:1087–1114.

Caruso, M., Panis, Y., Gagandeep, S., Houssin, D., Salzmann J.L., and Klatzmann, D., 1993, Regression of established macroscopic liver metastases after in situ transduction of a suicide gene. *Proc. Natl. Acad. Sci.* USA **90**:7024–7028.

Caselmann, WH., 1995, Transactivation of cellular gene expression by hepatitis B viral proteins: a possible molecular mechanism of hepatocarcinogenesis. *J. Hepatol.* **21**:34–37.

Crystal, R.G., 1997, Phase I study of direct administration of a replication deficient adenovirus vector containing *E. coli* cytosine deaminase gene to metastatic colon carcinoma of the liver in association with the oral administration of the pro-drug 5-fluorocytosine. *Hum. Gene Theapy* **8**:985–1001.

Curley, S.A., and Vecchio, R., 1998, New trends in the surgical treatment of colorectal cancer liver metastases. *Tumori* **84**:281–288.

Dourakis, S.P., and Tolis, G., 1998, Sex hormonal preparations and the liver. *Eur. J. Contracept. Reprod. Health Care* **3**(1):7–16.

Eagon, P.K., Elm, M.S., Epley, M.J., Shinozuka, H., and Rao, K.N., 1996, Sex steroid metabolism and receptor status in hepatic hyperplasia and cancer in rats. *Gastroentrol* **110**(4): 1199–1207.

Fellows, I.W., Stewart, M., Jeffcoate, W.J., Smith, P.G., and Toghill, P.J., 1988, Hepatocellular carcinoma in primary hemochromatosis in the abscence of cirrhosis. *Gut*, **29**, 1603–1606.

Finlay, C.A., Hinds, P.W., and Levine, A.J., 1989, The p53 proto-oncogene can act as a suppressor of transformation. *Cell* **57**(7):1083–1093.

Fong, Y., and Blumgart, L.H., 1998, Hepatic colorectal metastasis: current status of surgical therapy. *Oncology* **12**:1489–98.

Freeman, S.M. Therapeutic Applications of HSV-TK Suicide Gene. 3rd International Cancer Gene Therapy Meeting, Imperial College School of Medicine. May **1998**.

Freeman, S., Abboud, C., Whartenby, K., Packman, C., Moolten, F., and Abraham, G., 1993, The "bystander effect"; tumour regression when a fraction of the tumour mass is genetically modified. *Cancer Research* **53**:5274–5283.

Gerber, M.A., 1995, Pathobiological effects of hepatitis C. *J. Hepatol.* **22**:83–86.

Gerbes, A.L., and Caselmann, W.H., 1993, Point mutations of the p53 gene, human hepatocellular carcinoma and aflatoxins. *J. Hepatol.* **19**:312–315.

Gou, Y. Cytokines and Immunotherapy. In: Mazarakis H, Swart SJ., eds. Cancer Gene Therapy. MA: IBC, **1997**:155–168.

Gundlach, M., Brunken, C., Broelsch, C.E., 1998, Hepatocellular carcinoma: surgical treatment. *Schweiz Rundsch Med Prax* **87**:1471–1474.

Habib, N.A., Ding, S.-F., El-Masry, R., Mitry, R.R., Honda, K., Michail, N.E., Dalla Serra, G., Izzi, G., Greco, L., Bassyouni, M., El-Toukhy M., and Abdel-Gaffar, Y., 1996, Preliminary Report: The short-term effects of direct p53 DNA injection in primary hepatocellular carcinomas. *Cancer Detection and Prevention* **20**:103–107.

Harris, M.P., Sutjipto, S., Wills, K.N., Hancock, W., Cornell, D., Johnson, D.E., Gregory, R.J., Shepard, H.M., and Maneval, D.C., 1996, Adenovirus-mediated p53 gene transfer inhibits growth of human tumor cells expressing mutant p53 protein. *Cancer Gene Ther.* **3**(2):121–30.

Hollstein, M., Sidransky, D., Vogelstein, B., and Harris, C.C., 1991, p53 Mutations in Human Cancers. *Science* **253**:49–53.

Honda, K., Sbisa, E., Tullo, A., Papeo, P.A., Saccone, C., Poole, S., Pignatelli, M., Mitry, R.R., Ding, S., Isla, A., Davies, A., and Habib, N.A., 1998, p53 mutation is a poor prognostic indicator for survival in patients with hepatocellular carcinoma undergoing surgical tumour ablation. *Br. J. Cancer* 1998, **77**(5):776–782.

Ido, A., Nakata, K., Kato, Y., Nakao, K., Murata, K., and Nagataki, S., 1995, Gene therapy for hepatoma cells using a retrovirus vector carrying herpes simplex virus thymidine kinase gene under the control of human alpha-fetoprotein gene promoter. *Cancer Res.* **55**:3105–3109.

Jiao, R.L., Hansen, P.D., Havlik, R., Mitry, R.R., Pignatelli, M., and Habib, N.A., 1999, Clinical short-term results of radiofrequency ablation in primary and secondary liver tumours. *Am. J. Surg.* [Accepted].

Jolly, D., 1994, Viral vector systems for gene therapy. *Cancer Gene Therapy* **1**:51–64.

Kanai, F., Hamada, H., Shiratori, Y., and Omata, M., 1997, Adenovirus-mediated enzyme-prodrug therapy for cancer. *The Cancer Journal* **10**(6) [Novermber/December].

Kanai, F., Lan, K.H., Shiratori, Y., Tanaka, T., Ohashi, M., and Omata, M., 1997, In vivo gene therapy for alpha-fetoprotein—producing hepatocellular carcinoma by adenovirus-mediatd transfer of cytosine deaminase gene. *Cancer Res.* **57**(3):461–465.

Kanai, F., Shiratori, Y., Yoshida, Y., Kanegae, Y., Nakabayashi, H., and Omata, M., 1996, Gene therapy for alpha-fetoprotein-producing human hepatoma cells by adenovirus-mediated transfer of the herpes simplex virus thymidine kinase gene. *Hepatology* **23**:1359–1367.

Kanazawa, Y. Antisense Approach. In: Mazarakis H, Swart SJ., eds. Cancer Gene Therapy. MA: IBC, **1997**:223–231.

Kanazawa, Y., Ohkawa, K., Ueda, K., Mita, E., Takehara, T., and Hayashi, N., 1996, Hammerhead ribozyme-mediated inhibition of telomerase activity in extracts of human hepatocellular carcinoma cells. *Biochem. Biophys. Res. Commun.* **225**(2):570–576.

Kato, I., and Kido, C., 1987, Increased risk of death in thorotrast-exposed patients during the late follow-up period. *Jpn. J. Cancer Res.* **78**(11):1187–1192.

Kew, M.C., 1996, The hepatitis B virus and the genesis of hepatocellular carcinoma. *G. I. Cancer* **1**:143–148.

Kishimoto, K., Fujimoto, J., Takeuchi, M., Yamamoto, H., Ueki, T., and Okamoto, E., 1998, Telomerase activity in hepatocellular carcinoma and adjacent liver tissues. *J. Surg. Oncol.* **69**(3):119–24.

Ko, L.J., Prives, C., 1996, p53: puzzle and paradigm. *Genes Dev.* **10**(9):1054–1072.

Kobertz, W.R., Wang, D., Wogan, G.N., and Essigmann, J.M., 1997, An intercalation inhibitor altering the target selectivity of DNA damaging agents: synthesis of site-specific alflatoxin B1 adducts in a p53 mutational hotspot. *Proc. Natl. Acad. Sci. USA* **94**(18):9579–9584.

Kuriyama, S., Masui, K., Sakamoto, T., Nakatani, T., Kikukawa, M., and Tsujii, T., 1998, Bystander effect caused bu cytosine deaminase gene and 5-fluorocytosine in vitro is substantially mediated by generated 5-fluorouracil. *Anticancer Research* **18**(5A):3399–406.

Lane D.P., 1992, p53, guardian of the genome. *Nature* **358**, 15–16.

Liao, Y., Tang, Z.Y., Liu K.D., Ye, S.L., and Huang, Z., 1997, Apoptosis of human BEL-7402 hepatocellular carcinoma cells released by antisense H-ras DNA—in vitro and in vivo studies. *J. Cancer. Res. Clin. Oncol.* **123**(1):25–33.

Livraghi, T., Vettori, C., and Lazzaroni, S., 1991, Liver metastases: results of percutaneous ethanol injection in 14 patients. *Radiology* **179**:709–712.

Lunn, R.M., Zhang, Y.-J., Wang, L.-Y., Chen, C.-J., Lee, P.-H., Lee, C.-S., Tsai, W.Y., and Santella, R.M., 1997, p53 mutations, chronic Hepatitis B virus infection, and aflatoxin exposure in hepatocellular carcinoma in Taiwan. *Cancer Res.* **57**:3471–3477.

Lygidakis, N.J., Kosmidis, P., Yiras, N., Parissis, J., Kyparidou, E., 1995, Combined transarterial targeting locoregional immunotherapy-chemotherapy for patients with unresectable hepatocellular carcinoma: a new alternative for an old problem. *J Interferon Cytokine Res* **15**:467–472.

Marone, G., Francica, G., D'Angelo, V., Iodice, G., Pastore, P., and Siani, A., 1998, Echo-guided radiofrequency percutaneous ablation of hepatocellular carcinoma in cirrhosis using a cooled needle. *Radiol Med* **95**:624–629.

Mitry, R.R., Sarraf, C.E., Wu, C.G., Pignatelli, M., and Habib, N.A., 1997, Wild-type p53 induces apoptosis in Hep3B through up-regulation of bax expression. *Lab. Invest.* 1997, **77**(4):369–378.

Moolten, F.L., and Wells, J.M., 1990, Curability of tumours bearing herpes thymidine kinase genes transferred by retroviral vectors. *J. Natl. Cancer Inst.* **82**:297–300.

Moolten, F.L., 1986, Tumour chemosensitivity conferred by inserted herpes thymidine kinase genes: Paradigm for a prospective cancer control strategy. *Cancer Res.*; **46**:5276–5281.

Nagashima, I., Hamada, C., and Naruse, K., 1996, Surgical resection for small hepatocellular carcinoma. *Surgery* **119**:40–45.

Nakabayashi, H., Watanabe, K., Saito, A., Otsuru, A., Sawadaishi, K., and Tamaoki, T., 1989, Transcriptional regulation of alpha-fetoprotein expression by dexamethasone in human hepatoma cells. *J. Biol. Chem.* **264**:266–271.

Oda. T., Tsuda. H., Sakamoto. M., and Hirohashi., S., 1994, Different mutations of the p53 gene in nodule-in-nodule hepatocellular carcinoma as a evidence for multistage progression. *Cancer Lett.* **83**(1–2):197–200.

Ohta, Y., Kijima, H., Ohkawa, T., Kashani-Sabet, M., and Scanlon, K.J., 1996, Tissue specific expression of an anti-ras ribozyme inhibits proliferation of human malignant cells. *Nuc. Acids Res.* **24**(5): 938–942.

Olajos, E.J., 1977, Biological interactions of N-nitroso compounds: a review. *Ecotoxicol. Environ. Safety* **1**:175–196.

Philosophe, B., Greig, P.D., Hemming, A.W., Cattral, M.S., and Langer, B., 1998, surgical management of hepatocellular carcinoma: resection or transplantation? *J Gastrointestinal Surg* **2**:21–27.

Puisieux, A., Galvin, K., Troalen, F., Bressac, B., Marcais, C., Galun, E., Ponchel, F., Yakicier, C., Ji, J., and Ozturk, M., 1993a, retinoblastoma and p53 tumor suppressor genes in human hepatoma cell lines. *FASEB J.* **7**:1407–1413.

Puisieux, A., Ponchel, F., and Ozturk, M., 1993b, p53 as a growth suppressor gene in HBV-related hepatocellular carcinoma cells. *Oncogene* **8**:487–490.

Roth, J.A., and Cristiano, R.J., 1997, Gene therapy for cancer: what have we done and where are we going? *J. Natl. Cancer Inst.* **89**:21–39.

Rougier, P., 1998, Are there indications for intraarterial hepatic chemotherapy or isolated liver perfusion? The case of liver metastases from colorectal cancer. *Recent Results Cancer Res* **147**:3–12.

Scanlon, K. Antisense Approach. In: Mazarakis H, Swart SJ., eds. Cancer Gene Therapy. MA: IBC, **1997**:197–212.

Scorsone, K.A., Zhou, Y.-Z., Butel, J.S., and Slagle, B.L., 1992, *p53* Mutations Cluster at Codon 249 in Hepatitis B Virus-positive Hepatocellular Carcinomas from China. *Cancer Res.* **52**:1635–1638.

Seitz, H.K., Poschl, G., and Simanowski, U.A., 1998, Alcohol and cancer. *Recent. Dev. Alcohol* **14**:67–95.

Sheiner, P.A., Brower, S.T., Treatment of metastatic cancer to the liver. Sem Liver Dis 1994, **14**:169–177.

Sheu, J.C., Huang, G.T., Lee, P.H., Chung, J.C., Chou, H.C., Lai, M.Y., Wang, J.T., Lee, H.S., Shih, L.N., Yang, P.M., *et al.*, 1992, Mutation of p53 gene in hepatocellular carcinoma in Taiwan. *Cancer Res.* **52**(21):6098–6100.

Solomon. E., Borrow. J., and Goddard, A.D., 1991, Chromosome aberrations and cancer. *Science* **254**:1153–1160.

Spitz, F.R., Nguyen, D., Skibber, J.M., Cusack, J., Roth, J.A., and Cristiano, R.J., 1996, In vivo adenovirus-mediated p53 tumor suppressor gene therapy for colorectal cancer. *Anticancer Res.* **16**:3415–22.

Steele, G., Standard postoperative monitoring of patients after resection of colon and rectum cancer. Cancer, 1993; 71;1993.

Su, H., Chang, J., Mei Xu, S., and Wai Kan, Y., 1996, Selective killing of AFP-positive hepatocellular carcinoma cells by adeno-associated virus transfer of the herpes simplex virus thymidine kinase gene. *Hum. Gene Therapy* **7**:463–470.

Tabor, E., 1994, Tumor suppressor genes, growth factor genes, and oncogenes in Hepatitis B virus-associated hepatocellular carcinoma. *J. Med. Virol.* **42**:357–365.

Tahara, H., Nakanishi, T., Kitamoto, M., Nakashio, R., Shay, J.W., and Ide, T., 1995, Telomerase activity in human liver tissues: comparison between chronic liver disease and hepatocellular carcinomas. *Cancer Res.* **55**(13):2734–6.

Tiraby, M., Cazaux, C., Baron, M., Drocourt, D., Reynes, J.P., and Tiraby, G., 1998, Concomitant expression of E.coli cytosine deaminase and uracil phosphoribosyltransferase improves the cytotxicity of 5-fluorocytosine. *FEMS Microbiol. Lett.* **167**(1):41–49.

Ueki, T., Nakata, K., Mawatara, F., Tsuruta, S., Ido, A., and Eguchi, K., 1998, Retrovirus-mediated gene therapy for human hepatocellular carcinoma transplanted in athymic mice. *Int. J. Mol. Med.* **1**(4):671–675.

Vaughn, D.J., Haller, D.G., 1993 Nonsurgical management of recurrent colorectal cancer. Cancer, **71**: 4278–4292.

Vogelstein, B., and Kinzler, K.W., 1993, The multistep nature of cancer. *TIG* **9**(4):138–141.

Wallace, H., Clarke, A.R., Harrison, D.J., Hooper, M.L., and Bishop, J.O., 1996, Ganciclovir-induced ablation non-proliferating thyrocytes expressing herpesvirus thymidine kinase occurs by p53-independent apoptosis. *Oncogene* **13**(1):55–61.

Wands, I.R., and Blum, H.E., 1991, Primary hepatocellular carcinoma. *New Engl. J. Med.* **325**:729–731.

Wang, X.W., Gibson, M.K., Vermeulen, W., Yeh, H., Forrester, K., Stürzbecher, H.-W., Hoeijmakers, J.H.J., and Harris, C.C., 1995, Abrogation of p53-induced apoptosis by the hepatitis B vris X gene. *Cancer Res.* **55**:6012–6016.

Weinberg, R.A., 1991, Tumour suppressor genes. *Science* **254**:1138–1146.

Weisburger, E., 1977, Carcinogenicity studies on halogenated hydrocarbons. *Environ. Health Perspect* **21**:25–30.

Xu GW, Sun ZT, Forrester K, Wang XW, Coursen J, Harris CC., 1996, Tissue-specific growth suppression and chemosensitivity promotion in human hepatocellular carcinoma cells by retroviral-mediated transfer of the wild-type p53 gene. *Hepatology* **24**(5):1264–1268.

Yotsuyanagi, H., Koike, K., Yasuda, K., Moriya, K., Hino, K., Kurokawa, K., and Iino, S., 1995, Hepatitis C Virus Genotypes and Development of Hepatocellular Carcinoma. *Cancer* **76**:1352–1355.

19

ADENOVIRUS-MEDIATED TRANSFER OF A p53 GENE IN OVARIAN CANCER

Junzo Kigawa and Naoki Terakawa

Departments of Obstetrics and Gynecology
Tottori University School of Medicine
36-1 Nishimachi, Yonago 6838504
Japan

1. INTRODUCTION

Ovarian cancer, one of the major causes of death from cancer in women, is commonly diagnosed at advanced stage. (Boente *et al.*, 1993; Perez *et al.*, 1993; Thigpen *et al.*, 1993) Cytoreductive surgery followed by platinum-based combination chemotherapy is currently the standard treatment for patients with ovarian cancer. However, the success of treatment is limited by the development of resistance to platinum-based chemotherapy. (Kigawa *et al.*, 1993; Maggie, 1991) As a result, the survival rate for patients with ovarian cancer has been not improved through the past 20 years. (Thigpen *et al.*, 1993) Resistance to anti-cancer drugs remains a serious obstacle for improving the prognosis of patients with ovarian cancer. A new approach for the treatment is necessary to improve the prognosis of patients with chemo-resistant ovarian cancer.

Several mechanisms of drug resistance have been proposed, including a decrease in the accumulation of the drug, an increase in cellular detoxification of the drug, and an increase in DNA repair activity. (Andrew *et al.*, 1988; Fujiwara *et al.*, 1990; Kasahara *et al.*, 1991; Masuda *et al.*, 1988; Minagawa *et al.*, 1994) Recent findings have shown that several genes are involved in control of chemo-resistance including the p53 gene. (Bichat *et al.*, 1998; Ding *et al.*, 1998; 1998; Kigawa *et al.*, 1998; Lowe *et al.*, 1994; Wang *et al.*, 1995) and that gene therapy may be a potential strategy for the therapy of chemo-resistant ovarian cancer. (Barnes *et al.*, 1997; Kanamori *et al.*, 1998)

Ovarian cancer frequently develops widespread intraperitoneal dissemination and multi-organ involvement within the peritoneal cavity. This disease may be an appropriate target for gene therapy, because of easy access to cancer cells in abdominal cavity. Several types of gene therapy for ovarian cancer such as suicide gene and tumor suppressor gene have been reported. (Kanamori *et al.*, 1998; Kaye, 1996; Tong *et al.*, 1997) In this chapter, gene therapy with adenovirus-vector of a p53 gene for ovarian cancer is discussed.

Cancer Gene Therapy: Past Achievements and Future Challenges, edited by Habib
Kluwer Academic/Plenum Publishers, New York, 2000.

2. OVARIAN CANCER AND p53 GENE

The p53 gene encodes a cell-cycle checkpoint protein that functions in the G1 phase of the cell cycle and has a pivotal role in inducing apoptosis. (Kellen, 1994) The p53 protein is a transcriptional regulator of cellular response to DNA damage. (Williams *et al.*, 1993) Several studies have shown the critical role of the p53 gene in executing cell death in response to cytotoxic drugs, and that the mutations in the p53 gene have been associated with the lack of response to these agents. (Kastan *et al.*, 1995; Man *et al.*, 1995) Changes in Bcl-2-Bax expression balance regulates p53-dependent apoptosis. p53 and Bax expressions increased and Bcl-2 decreased after exposure to DNA-damaging agents. (Perego *et al.*, 1996) Additionally, Bax expression significantly increased in tumors with the wild-type p53 gene after chemotherapy, but did not increase with mutation of the p53 gene in ovarian cancer. (Jonathan *et al.*, 1996) p53-dependent apoptosis in tumors is strongly related to sensitivity to chemotherapy.

p53 gene mutation has been commonly found in ovarian cancer. The frequency of p53 gene mutation is 50–72% of ovarian cancer. (Kohler *et al.*, 1993; Okada *et al.*, 1998) Ovarian cancers with mutation of the p53 gene have been significantly less sensitive to chemotherapy and more aggressive than those with the wild-type p53 gene. Interestingly, mutations of p53 gene have increased after chemotherapy. (Okada *et al.*, 1998) The result can be explained by two mechanisms. First, the clone with p53 gene mutation which was resistant to chemotherapy could remain after chemotherapy. Second, p53 gene mutation was induced by chemotherapy. It is not easy to determine which mechanism enrolled in each case, however, chemotherapy might develop the aggressiveness of cancer in either case. Therefore, gene therapy with p53 gene is considered to be a potential strategy for the therapy of chemo-resistant ovarian cancer.

3. ADENOVIRUS-VECTOR

3.1. Vector

It is crucial for any gene therapy to deliver the therapeutic gene to the target cells and express the gene at appropriate levels. To accomplish gene delivery, a variety of viral and non-viral systems such as liposome vector have been developed. (Barnes *et al.*, 1997) Because adenovirus-vector shows the highest efficiency of gene transduction (more than 10-fold compared to the other vectors), a recombinant adenovirus is most suitable for cancer gene therapy.

3.2. Development of Adenovector

The procedure to construct a recombinant adenovirus carrying the wild-type p53 gene (AxCAp53) is summarized in Fig. 1. Briefly, a 1.95-kb, BamHI-EcoRI fragment containing the full-length cDNA for human wild-type p53 was isolated from a plasmid pProSp53 (Health Science Research Resources Bank, Tokyo, Japan). The p53 coding sequence was inserted into a cassette cosmid pAxCAwt. This cassette cosmid contains human adenovirus type 5 (Ad5) genomic DNA lacking the E1A, E1B and E3 regions and the CAG promoter. The promoter consists of the cytomegalovirus immediate early enhancer, the chicken β-actin promoter, and the rabbit β-globin polyadenylation signal. According to the method described by Miyake *et al.* (1996), the p53-bearing cassette

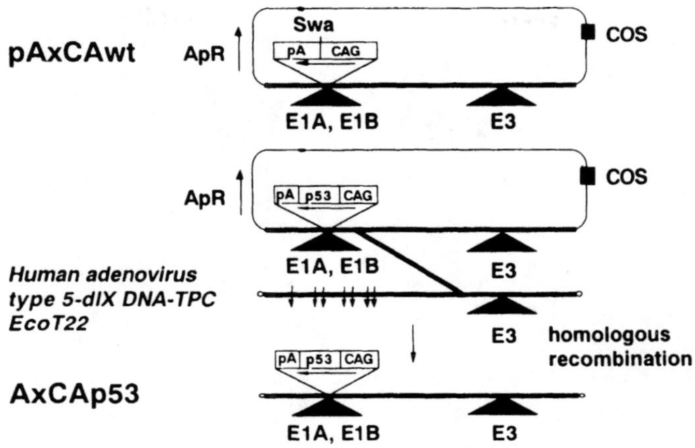

Figure 1. The procedure to construct a recombinant adenovirus carrying the wild-type p53 gene (AxCAp53). CAG, CAG promoter; pa, polyadenylation signal; ▲, deletion; ApR, ampicillin-resistance gene; COS, cos site of λ phage.

cosmid and EcoT22 I-digested DNA-terminal protein complex (TPC) of Ad5-dlX were co-transfected into the 293 cell line (the American Type Culture Collection, Rockville, MD) which was a human embryonal kidney cell line transformed by Ad5 E1A and E1B genes. AxCAp53 was generated by homologous recombination between the cosmid and EcoT22 I-digested Ad5-dlX DNA-TPC in the 293 cells.

4. EFFICACY OF ADENOVIRUS-MEDIATED TRANSFER OF A p53 GENE

4.1. Transfection of Adenovirus-Vector

It is known that promoter controls the transduction of adenovirus vector. To determine the efficiency of the recombinant adenovirus to transduce cells, the cells were infected with AxCALacZ, encoding for the bacterial LacZ gene under the control of the CAG promoter, and the transduced cells was detected by β-galactosidase staining. The magnitude of staining was quantitated by counting the percentage of blue cells. The efficiency of the recombinant adenovirus to transduce SK-OV-3 cells, that had homozygous deletion of the p53 gene, was 89% for 6.25 multiplicities of infection (MOI), 94% for 12.5 MOI, and 100% for over 25 MOI. In another study, 100 MOI of the ACN53, a different type of recombinant adenovirus carrying the wild-type p53 gene, is necessary to achieve more than 98% of transduction efficiency to SK-OV-3 cells. (Mujoo *et al.*, 1996)

The CAG promoter was used as the expresser of our adenovirus vector. This promoter seems to be a high expresser, because an expression plasmid vector containing the CAG promoter has shown the highest levels of a reporter gene expression among 9 vectors containing different promoters such as the cytomegalovirus enhancer/promoter and the SO40 enhancer/promoter. (Fukuchi *et al.*, 1994) It is, therefore, likely to consider that the CAG promoter induced the higher efficiency of gene transduction for this adenovirus vector.

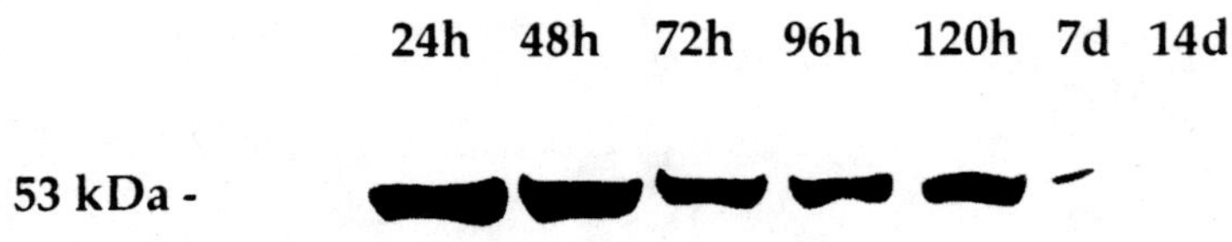

Figure 2. The expression of the p53 protein was monitored by Western blot analysis at various time points up to 14 days after infection. The p53 protein expressed at 24 h and disappeared 14 days after infection with AxCAp53.

4.2. Expression of p53 Protein

The p53 protein expressed at 24 h and disappeared 14 days after infection with AxCAp53. The peak of the expression was 48 h after infection in each MOI (Fig. 2). The p53 protein coordinately expressed with infectious units of virus per cell of AxCAp53. (Kanamori *et al.*, 1998)

5. EFFICACY OF ADENOVIRUS-MEDIATED TRANSFER OF A p53 GENE IN OVARIAN CANCER

5.1. Effect on the Cell Growth

The growth of SK-OV-3 cells after infection with AxCAp53 or AxCAp53R, that is a recombinant adenovirus inserted the p53 gene and did not express the p53 protein, is shown in Fig. 3. AxCAp53 significantly suppressed cell growth in each MOI. The

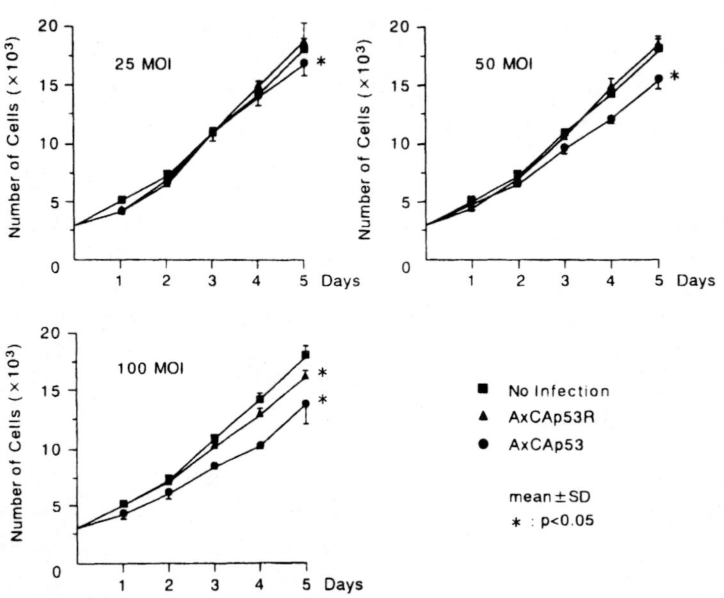

Figure 3. Changes in the cell growth curve of SK-OV-3 cells after infection with AxCAp53 or AxCAp53R. The upper left shows growth curve at 25 MOI, the upper right at 50 MOI, and the lower left at 100 MOI. AxCAp53 significantly suppressed cell growth in each MOI. Cell growth was not affected by less than 100 MOI of AxCAp53R.

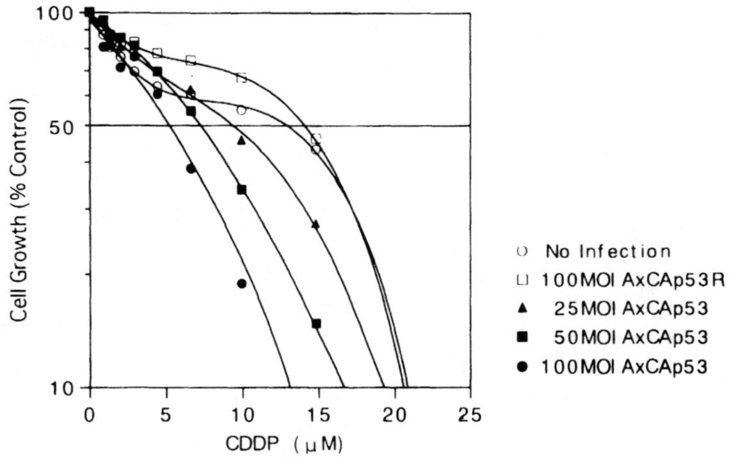

Figure 4. Dose-response curves of CDDP in combination with AxCAp53. Infection with AxCAp53 increased sensitivity to CDDP in the SK-OV-3 cell.

suppression rate 5 days after treatment was 6.8% for 25 MOI, 14.6% for 50 MOI, and 23.4% for 100 MOI. Although AxCAp53 significantly suppressed cell growth in each MOI, the effect of introducing wild type p53 into a p53 negative cells was relatively poor. Whereas, cell growth was not affected by less than 100 MOI of AxCAp53R. Those recombinant adenoviruses have less cytotoxity, and the p53 protein expressed by infection with AxCAp53 can just suppress the growth of SK-OV-3 cells.

5.2. Combination Effect of AxCAp53 and CDDP

The efficacy of AxCAp53 to suppress the cell growth was not sufficient. Polyak *et al.* (1996) showed that p53 gene transfer to p53 gene delete cells did not always induce apoptosis. The combination of p53 gene therapy and CDDP, a key drug for chemotherapy for ovarian cancer, is evaluated in this part.

Dose-response curves of CDDP in combination with AxCAp53 are shown in Fig. 4. IC50 to CDDP was 12.9 μM for non-infected SK-OV-3 cells, 9.3 μM for AxCAp53 infected cells at 25 MOI, 7.3 μM for 50 MOI, and 5.2 μM for 100 MOI. Interestingly, AxCAp53 increased sensitivity to CDDP in ovarian cancer cells with deletion of the p53 gene. The synergistic effect in combination of p53 gene transfection and administration of CDDP in ovarian cancer cell line. (Kanamori *et al.*, 1998) Therefore, the combination of p53 gene transfection and CDDP may be an effective strategy for the therapy of ovarian cancer with mutant p53 gene which frequently shows chemoresistance.

5.3. p53-Dependent Apoptosis in Ovarian Cancer with Transfection of p53 Gene

The induction of p53 protein expression by DNA-damaging agents including cisplatin (Fritsche *et al.*, 1993; Landesman *et al.*, 1993; Williams *et al.*, 1993). The expression of p53 protein increased in nonresponders after chemotherapy (Okada *et al.*, 1998). Accumulation of p53 protein by cytotoxic agents delays the cell growth to allow

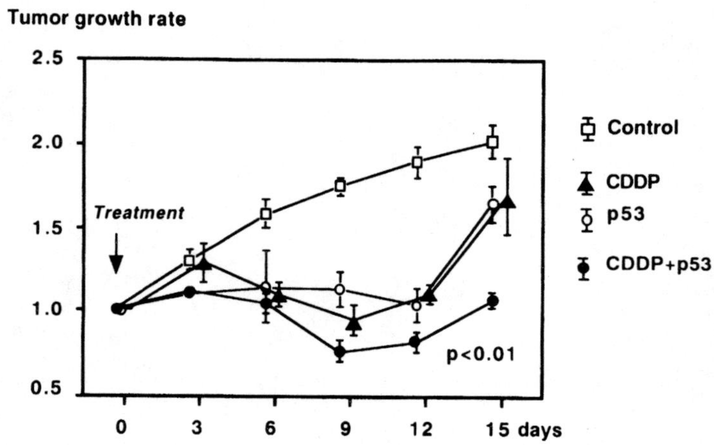

Figure 5. Tumor growth rate is determined by the following formula: tumor growth = tumor size / tumor size on day when treatment has been started. The tumor suppression rate was significantly greater in combination treatment of AxCAp53 and CDDP compared to single treatment of AxCAp53 or CDDP.

repair processes (Maestro *et al.*, 1992). These findings suggest that p53 protein can function not only in induction of apoptosis but also in DNA repair. It is important to clarify whether transfection of p53 gene induces apoptosis or enhances DNA repair.

Growth suppression rate of implantation tumor of SK-OV-3 in scid mice is shown in Fig. 5. The mice were evenly divided into 4 groups according to the treatment; no treatment, AxCAp53 alone, CDDP alone, combination treatment of AxCAp53 and CDDP. The suppression rate was significantly greater in combination treatment of AxCAp53 and CDDP compared to single treatment of AxCAp53 or CDDP. In addition, apoptotic index was significantly greater in combination treatment. In contrast, PCNA did not differ between tumor with transfection of a wild-type p53 gene and tumor without the transfection. Bax expression significantly increased after CDDP administration in tumors with p53 gene transfection. These findings suggest that transfection of a wild-type p53 gene induces apoptosis and does not enhance the repair system in ovarian cancer with deletion of p53 gene.

6. CDDP ADMINISTRATION AND p53 GENE TRANSFER

Fujiwara *et al.* (1994) indicated that the timing of CDDP administration and p53 gene transfer was critical: cisplatin administration simultaneous with or subsequent to p53 gene transfer was less effective than cisplatin-first sequential treatment in lung cancer cells. Although the suppression rate of cell growth was highest for p53 gene prior to CDDP, followed by simultaneous exposure and then CDDP-first sequential treatment in SK-OV-3 cells, an in vivo study did not show significant difference in tumor suppression rate among the timing of CDDP administration and p53 gene transfer. The further study is necessary to clarify the more effective timing of CDDP administration and p53 gene transfer.

7. SUMMARY

Given the lack of effective conventional therapy, those patients with recurrent or refractory ovarian cancer should be considered for currently approved investigational gene therapy protocols. Several studies have shown a potential modality of p53 gene transfer in cancer gene therapy. We also developed a new recombinant adenovirus carrying a wild-type p53 gene (AxCAp53). Although the efficacy of AxCAp53 to suppression of cell growth was not sufficient, AxCAp53 increased sensitivity to CDDP in ovarian cancer cells with deletion of the p53 gene. The combination of CDDP and AxCAp53 may be a potential strategy for the therapy of CDDP resistant ovarian cancer.

REFERENCES

Andrews, P.A., and Howell, S.B., 1990, Cellular pharmacology of cisplatin: Perspectives on mechanisms of acquired resistance. Cancer Cell. 2:35–43.

Andrews, P.A., Velury, S., Mann, S.C., and Howell, S.B., 1988, Cis-Diamminedichloroplatinum (II) accumulation in sensitive and resistant human ovarian carcinoma cells. Cancer Res. 48:68–73.

Barnes, M.N., Deshane, J.S., Rosenfeld, M., Siegal, G.P., Curiel, D.T., and Alvarez, R.D., 1997, Gene therapy and ovarian cancer: a review. Obstet Gynecol. 89:145–155.

Bichat, F., Solis, R.G., Poullain, M.G., Poupon, M.F., Khayat, D., and Bastian, G., 1998, S9788 modulation of P-glycoprotein and Multidrug-related protein-mediated multidrug resistance by servier 9788 in doxorubicin-resistant MCF7 cells. Biochem Pharmacol. 56:497–502.

Boente, M.P., Hurteau, J., Rodriguez, G.C., Bast, R.C. Jr., and Berchuck, A., 1993, The biology of ovarian cancer. Cur Opin Oncol. 5:900–907.

Deb, S., Jackson, C.T., Subler, M.A., and Martin, D.W., 1992, Modulation of cellular and viral promoters by mutant human p53 proteins found in tumor cells. J Virol 66:6164–6170.

Ding, W., Li, Y.P., Nobile, L.M., Grills, G., Carrera, I., Paietta, E., Tallman, M.S., Wiernik, P.H., and Gallagher, R.E., 1998, Leukemic cellular retinoic acid resistance and missense mutations in the PML-RARalpha fusion gene after relapse of acute promyelocytic leukemia from treatment with all-trans retinoic acid and intensive chemotherapy. Blood. 92:1172–1183.

Fujiwara, T., Grimm, E.A., Mukhopadhyay, T., Zhang, W.W., Owen, S.L.B., and Roth, J.A., 1994, Induction of chemosensitivity in human lung cancer cells in vivo by adenovirus-mediated transfer of the wild-type p53 gene. Cancer Res. 54:2287–2291.

Fujiwara, Y., Sugimoto, Y., Kasahara, K., Bungo, M., Yamakido, M., Tew, K.D., and Saijo, N., 1990, Determinants of drug response in a cisplatin-resistant human lung cancer cell line. Jpn J Cancer Res. 81:527–535.

Fukuchi, K., Hearn, M.G., Deeb, S.S., Smith, A.C., Dang, N., Miyazaki, J., Bothwell, M., and Martin, G.M., 1994, Activity assays of nine heterogeneous promoters in neural and other cultured cells. In Vitro Cell Dev. Biol. Anim. 30:300–305.

Fritsche, M., Haessler, C., and Brandner, G., 1993, Induction of nuclear accumulation of the tumor-suppressor protein p53 by DNA-damaging agents. Oncogene. 8:307–318.

Jonathan, J., Herod, O., Eliopoulos, A.G., Warwick, J., Niedobitek, G., Young, L.S., and Kerr, D.J., 1996, The prognostic significance of Bcl-2 and p53 expression in ovarian cancer. Cancer Res. 56:2178–2184.

Kanamori, Y., Kigawa, J., Minagawa, Y., Irie, T., Oishi, T., Shimada, M., Takahashi, M., Nakamura, T., Sato K., and Terakawa, N. 1998, A newly developed adenovirus-mediated transfer of a wild-type p53 gene increases sensitivity to cis-diamminedichloroplatinum (II) in p53-deleted ovarian cancer cells. Eur J Cancer. 34:1802–1806.

Kasahara, K., Fujiwara, Y., Nishio, K., Ohmori, T., Sugimoto, Y., Komiya, K., Matsuda, T., and Saijo, N., 1991, Metallothionein content correlates with the sensitivity of human small cell lung cancer cell lines to cisplatin. Cancer Res. 51:3237–3242.

Kastan, M.B., Canman, C.E., and Leonard, C.J., 1995, p53 cell cycle control and apoptosis: implications for cancer. Cancer Metastasis Rev. 14:3–15.

Kaye, S.B., 1996, Ovarian cancer from the laboratory to the clinic: challenges for the future. Ann Oncol. 7:9–13.

Kellen, J.A., 1994, Molecular interrelationships in multidrug resistance. Anticancer Res. 14:433–435.

Kigawa, J., Minagawa, Y., Ishihara, H., Kanamori, Y., and Terakawa, N., 1993 Tumor DNA ploidy and prognosis of patients with serous cystadenocarcinoma of the ovary. Cancer. 72:804–808.

Kigawa, J., Minagawa, Y., Cheng X., and Terakawa, N., 1998, Gamma-glutamyl cysteine synthetase upregulates glutathione andmultidrug resistance-associated protein in patients with chemoresistant epithelial ovarian cancer. Clin Cancer Res. 4:1737–1741.

Kohler, M.F., Marks, J.R., Wiseman, R.W., Jacobs, I.J., Davidoff, A.M., Clarke, P., Soper, J.T., Bast, R.C. Jr., and Berchuck, A., 1993, Spectrum of mutation and frequency of allelic deletion of the p53 gene in ovarian cancer. J Natl Cancer Inst. 85:1513–1519.

Landesman, Y, Bringold, F., and Kimchi, A., 1994, p53 undergoes eitopic changes in vitro by sodium-vanadate. Oncogene. 9:1241–1245.

Lowe, S.W., Bodis, S., McClatchey, A., Remington, L., Ruley, H.E., Fisher, D.E., Housman, D.E., and Jacks, T., 1994, p53 status and the efficacy of cancer therapy in vivo. Science. 266:807–810.

Maestro, R., Dolcetti, R., Gasparotto, D., Doglioni, C., Pelucchi, S., Barzan, L., Grandi, E., and Boiocchi, M., 1992, High frequency of p53 gene alterations associated with protein overexpression in human squamous cell carcinoma of the larynx. Oncogene. 7:1159–1166.

Maggie, F.M., 1991, Cisplatin update. Semin Oncol. 8:1–4.

Man, S., Bani, M.R., Adachi, D., Hawley, R.G., Kerbel, R.S., Ben, D.Y., and Lu, S., 1995, Retroviral insertional mutagenesis as a strategy for the identification of genes associated with cis-diamminedichloroplatinum (II) resistance. Cancer Res. 55:1139–1145.

Masuda, H., Ozols, R.F., Lai, G.M., Fojo, A., Rothenberg, M., and Hamilton, T.C., 1988, Increased DNA repair as a mechanism of acquired resistance to cis-Diammine-dichloroplatinum (II) in human ovarian cancer cell lines. Cancer Res. 48:5713–5716.

Minagawa, Y., Kigawa, J., Ishihara, H., Itamochi, H., and Terakawa, N., 1994, Synergistic enhancement of cisplatin cytotoxicity by SN-38, an active metabolite of CPT-11, for cisplatin-resistant Hela cells. Jpn J Cancer Res. 85:966–971.

Miyake, S., Makimura, M., Kanegae, Y., Harada, S., Sato, Y., Takamori, K., Tokuda, C., and Saito, I., 1996, Efficient generation of recombinant adenoviruses using adenovirus DNA-terminal protein complex and a cosmid bearing the full-length virus genome. Proc. Natl. Acad. Sci. USA. 93:1320–1324.

Mujoo, K., Maneval, D.C., Anderson, S.C., and Gutterman, J.U., 1996, Adenoviral-mediated p53 tumor suppressor gene therapy of human ovarian carcinoma. Oncogene. 12:1617–1623.

Okada, M., Kigawa, J., Minagawa Y., Kanamori, Y., Itamochi, H., Cheng, X., Ohishi, T., and Terakawa, N., 1998, Alterations of the p53 gene and p53 protein expression in epithelialovarian cancer before and after chemotherapy. Int J Clin Oncol 3:240–246.

Perego, P., Giarola, M., Righetti, S.C., Supino, R., Caserini, C., Delia, D., Pierotti, M.A., Miyashita, T., Reed, J.C., and Zunino, F., 1996, Association between cisplaton resistance and mutation of p53 gene and reduced bax expression in ovarian carcinoma cell systems. Cancer Res. 56:556–562.

Perez, R.P., Hamilton, T.C., Ozols, R.F., and Young, R.C., 1993, Mechanisms and modulation of resistance to chemotherapy in ovarian cancer. Cancer. 71:1571–1580.

Polyak, K., Waldman, T., He, T.C., Kinzler, K.W., and Volgelstein, B., 1996, Genetic determinations of p53-induced apoptosis and growth arrest. Genes Dev. 10:1945–1952.

Thigpen, J.T., Bertelsen, K., Eisenhauer, E.A., Hacker, N.F., Lund, B., and Sessa, C., 1993, Long-term follow-up of patients with advanced ovarian carcinoma treated with chemotherapy. Ann Oncol. 4:35–40.

Tong, X.W., Agoulnik, I., Blankenburg, K., Contant, C.F., Hasenburg, A., Runnebaum, L.B., Stickeler, E., Kaplan, A.L., Woo, S.L., and Kieback, D.G., 1997, Human epithelial ovarian cancer xenotransplants into nude mice can be cured by adenovirus-mediated thymidine kinase gene therapy Anticancer Res. 17:811–813.

Wang, X.W., Yeh, H., Scheaffer, L., Roy, R., Monocollin, V., Egly, J.M., Wang, Z., Friedberg, E.C., Evans, M.K., Taffe, B.G., Bohr, V.A., Weeda, G., Hoeijmakers, J.H.J., Forrester, and K., Harris, C.C., 1995, p53 modulation of TFIIH-associated nucleoide excision repair activity. Nat Genet. 10:188–195.

Williams, G.T., and Smith, C.A., 1993, Molecular regulation of apoptosis: genetic controls on cell death. Cell. 74:777–779.

VI

OTHER SYSTEMS

ELICITING HYPERACUTE REJECTION AS A TUMOR KILLING STRATEGY

Herpes Amplicon Vector Transfer of the α(1,3)Galactosyltransferase Gene

Charles J. Link*, Daniel J. Hellrung, Tatiana Seregina, and Suming Wang

Human Gene Therapy Research Institute
1415 Woodland Avenue, Suite 218
Des Moines, Iowa 50309

INTRODUCTION

In 1997 an estimated 560,000 people died from cancer despite all currently available therapies (Parker *et al.*, 1997). New approaches are greatly needed to combat this deadly disease. Gene therapy is among the current experimental strategies for patients with cancer who have failed standard therapy. Despite more than 100 gene therapy trials evidence of success is very limited (Roth *et al.*, 1997). Therefore, new candidate genes and delivery methods must be pursued.

Great interest has developed in the idea of using farm animals as a source of organs or tissues for humans. Unfortunately, strong immunological barriers to xenotransplants can destroy a transplanted solid organ within minutes, a process termed hyperacute rejection. The hyperacute rejection model of xenograft survival is typically a vascularized xenograft directly exposed to blood serum (Pruitt, 1994). Research has demonstrated that hyperacute rejection with porcine xenografts transplanted to baboons occurs secondary to porcine α(1,3)GT gene expression and αgal presentation (Pruitt *et al.*, 1994), (Platt *et al.*, 1990). The enzyme α(1,3)GT is expressed in all mammalian species including *Mus musculus*, but not in Old World primates, apes, or humans (Galili *et al.*, 1988). The α(1,3)GT gene is not active in humans due to the presence of two base pair frameshift mutations (Larsen *et al.*, 1990). α(1,3)GT catalyzes the transfer of galactose from uridine diphosphate galactose (UDP-Gal) to the N-acetyl-lactosamine acceptors on carbohydrate side chains in a specific α(1,3)linkage on glycoproteins or glycolipids (Galβ1 → 4GlcNAc-R).

* Phone: 515-241-8787, Fax: 515-241-8788

Cancer Gene Therapy: Past Achievements and Future Challenges, edited by Habib
Kluwer Academic / Plenum Publishers, New York, 2000.

Galβ1 → 4GlcNAc-R + UDP-Gal → Galα1 → 3Galβ1 → 4GlcNAc-R

Anti-αgal Ab present in the human serum can recognize this epitope (Galili *et al.*, 1988). In fact, preexisting human Ab against αgal represent almost 1% of total human Ab (Galili, 1993) and are the basis for complement-mediated hyperacute rejection (Sandrin *et al.*, 1993). Murine vector producing cells (VPC) implanted into humans is another type of xenograft. Our laboratory and others have demonstrated that murine retroviral VPC and the viral vectors they produce express αgal and therefore are lysed by Ab and complement within 30 min after being exposed to human serum (Welsh *et al.*, 1975), (Rother *et al.*, 1995a), (Link *et al.*, 1997). The effect of this serum inactivation on VPC and retroviruses is due to αgal expression on the cells (Rother *et al.*, 1995a), (Link *et al.*, 1997), (Rother *et al.*, 1995b). We are also conducting a trial of intraperitoneal infusion of murine VPC into women with ovarian cancer. Two patients showed some tumor regression despite minimal transgene transfer suggesting that hyperacute rejection of the murine VPC might inhibit the cancer. Increases in anti-αgal titer were observed in women treated with higher VPC doses.

Viral vectors can efficiently transduce human tumor cells with anti-tumor therapeutic genes. As part of a study of blocking retroviral destruction by using human packaging cells, the transfer of the porcine α(1,3)GT to human fibroblasts was shown to result in sensitivity to Ab and complement destruction (Collins *et al.*, 1994). Human tumors are not sensitive to lysis by autologous human serum. We have found that the retroviral transfer of the α(1,3)GT gene into human tumor cells results in the expression of αgal. The murine α(1,3)GT enzyme was cloned into the LNCX retroviral vector backbone (Larsen *et al.*, 1989), (Miller and Buttimore, 1986). These human cells displayed αgal and became sensitive to human serum (Link *et al.*, 1998). The sensitive tumor cell clones expressed αgal by FACS analysis. These initial data suggested that the presentation of αgal on human tumor cells induces human serum sensitivity. In this study, we employed a Herpes amplicon vector to transfer and express the α(1,3)GT gene. *In vitro* assays demonstrated that human serum could destroy Herpes vector transduced human cancer cells by a hyperacute rejection mechanism. This approach may allow us to generate direct anti-tumor responses by changing the tumor's glycosylation pattern to include αgal. This presents an innovative approach to cancer gene therapy that exploits a naturally occurring physiologic process in humans. We propose to use the murine α(1,3)GT gene as a therapeutic transgene to direct hyperacute rejection to a new target: tumor tissue.

METHODS AND RESULTS

Phase I Clinical Trial of LTKOSN.1 VPC for Women with Recurrent Ovarian Cancer

The implantation of VPC to deliver genes into tumor cells was first demonstrated in a brain tumor model by engrafting LacZ VPC into rodents with C6 glioma tumors (Short *et al.*, 1990). In these animal tumor models all of the rodents are αgal positive. Obviously, in these αgal positive models hyperacute rejection is not contributing to the observed responses with the HStk system. However, the human is a very different animal in this respect. We are conducting a dose escalation Phase I trial using the intraperitoneal infusion (xenotransplant) of murine VPC to attempt to transfer the HStk gene (Link *et al.*, 1996). Results from this clinical trial of LTKOSN.1 VPC for women with recurrent ovarian cancer suggest a role for hyperacute rejection in anti-tumor responses in humans. Five patients have been enrolled thus far. Two patients had evidence of anti-tumor activity, but this did not correlate with any significant gene transfer (<1%) into tumors.

Titers of Anti-αgal Antibody in Patients Treated with Murine Vector Producer Cells

We expected that patients who received murine cell xenografts would generate increased anti-αgal titers. Thus, the increased anti-αgal titer was the result of a specific immune response against αgal epitopes on the xenograft (Fig. 1). In the first three patients treated with 1×10^6 VPC/kg (range 54 to 57×10^6 VPC), no increases in anti-αgal titers were seen. However, patients 4 and 5 treated at the 10×10^6 VPC/kg dose had 8-fold and 16-fold increases, respectively, in anti-αgal titers by ELISA. Thus a strong and significant response to the αgal antigen was observed at the higher doses. This actually raises an interesting possibility that patients could be boosted to increase their anti-αgal titers before receiving gene transfer of the α(1,3)GT to their solid tumors. This might be another way to enhance the immune effects against the tumor and break tolerance.

Gene Delivery with the pHE700 Herpes Simplex Amplicon Vector

The next fundamental problem in trying to develop an effective cancer therapeutic is gene delivery. Several gene delivery methods are under development for *in vivo*

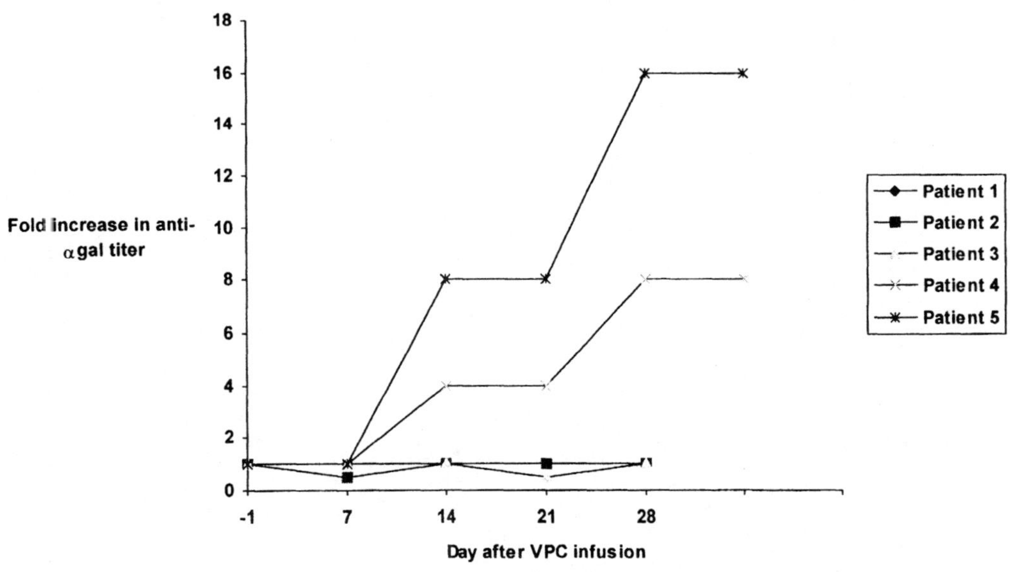

Figure 1. Patients serial blood samples were drawn before VPC infusions and then weekly thereafter and evaluated by ELISA assay for anti-αgal Ab titers. We expected that patients who received murine cell xenografts would generate increases in anti-αgal titers. In the first three patients treated with 1×10^6 VPC/kg (range 54–57 $\times 10^5$ VPC) no increases in anti-αgal titers were seen. However, patients 4 and 5 treated at the 10×10^6 VPC/kg dose had 8-fold and 16-fold increases, respectively, in anti-αgal titers by ELISA. Thus a strong and significant response to the αgal antigen was observed at the higher doses.

transfer of genes into tumor cells, including two methods with which our group has extensive experience. The first method considered was the engraftment of murine retroviral VPC as performed in the Phase I trial discussed above (Link *et al.*, 1996), (Galili *et al.*, 1995). Our preliminary data suggest that this method if attempted outside of the peritoneal cavity will have substantial barriers since VPC are quickly lysed by serum Ab and complement (Link *et al.*, 1997). Our laboratory has developed an improved gene transfer method based on the HSV viral amplicon system (Wang *et al.*, 1997). This project used HSV amplicons to transfer and test the effectiveness of α(1,3)GT gene therapy against solid tumors. One of two broad categories of HSV-based vectors is amplicons (Spaete and Frenkel, 1982). Plasmids containing the HSV-1 lytic replication origin (ori S) and the HSV-1 terminal packaging signal sequences, can be amplified and packaged into infectious HSV-1 virions in the presence of transactivating helper virus (Spaete and Frenkel, 1982), (Kwong and Frenkel, 1984), (Geller and Breakefield, 1988), (Geller and Freese, 1990). We have constructed a novel HSV vector, pHE700 that also contains Epstein-Barr virus (EBV) sequences that maintain the plasmid as an episome in the transfected cell nucleus (Wang *et al.*, 1997), (Wang and Vos, 1996). EBV contains a latent replication origin (ori P) which directs viral self-replication and maintenance in cells without entering the lytic cycle (Reisman *et al.*, 1985), (Sugden *et al.*, 1985), (Yates *et al.*, 1985). The Epstein-Barr virus nuclear antigen 1 (EBNA-1) encodes a DNA binding transactivator for ori P (Reisman *et al.*, 1985), (Sugden *et al.*, 1985), (Yates *et al.*, 1985). The EBV ori P and EBNA-1 gene provide more stability to eukaryotic expression vectors. The combination of the HSV amplicon with the EBV sequences improves the ease of use of the HSV amplicon system (Figs. 2 and 3). Our replication incompetent pHE700 vectors maintained wide tropism for delivering transgene(s) into both dividing and quiescent cells with high efficiency both *in vitro* and *in vivo* (Wang *et al.*, 1997). The vector could be produced at high titer and packages a stable vector with a 21 kb DNA insert (Wang *et al.*, 1997).

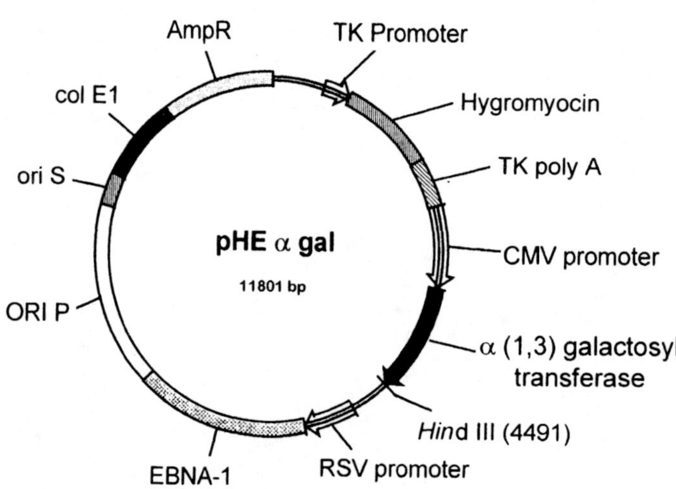

Figure 2. Plasmid map of pHEαgal1 vector. Amp^R, ampicillin resistant; "a", HSV-1 packaging signal; HSV-tk promoter, HSV-1 thymidine kinase promoter; hyg+, hygromycin resistance; MCS, multi-cloning site; ΔEBNA-1, modified EBV nuclear antigen gene; ori P, EBV unique latent replication origin; ori S, HSV-1 replication origin.

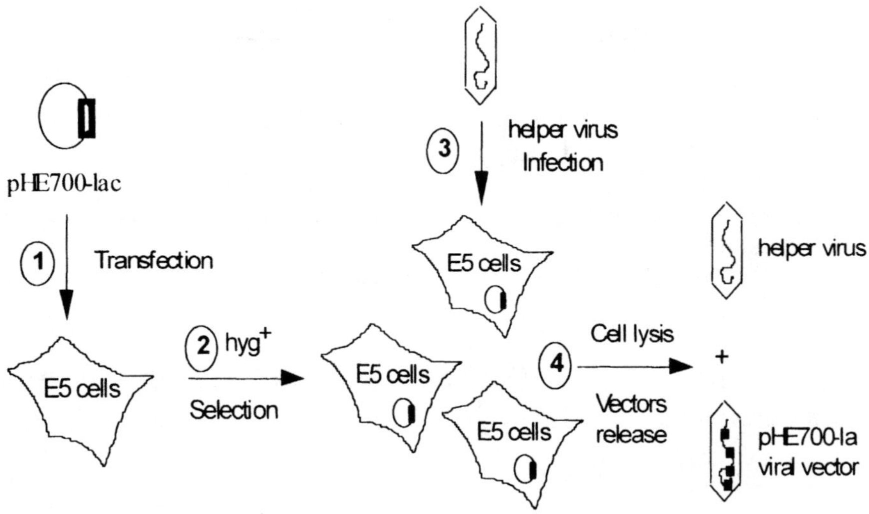

Figure 3. HSV amplicon packaging system. 1) pHE700-lac plasmid is transfected into E5 helper cells that express the IE3 gene in trans. 2) Hygromycin selection of transfected E5 cells to obtain population of cells containing pHE vector. 3) Infection with helper virus that provides replication and packaging functions. 4) pHE700-lac vector and helper virus replicate and induce lytic release.

Amplicon Vector Packaging and Transduction

The maintenance of pHE700 vector as an episome was demonstrated by transfection of pHE700-lac into E5 cells and selection with hygromycin. The pHE700-lac vector contains the bacterial LacZ gene cloned into the multi-cloning site of pHE700 (Wang *et al.*, 1997). By day 16 of drug selection, almost all cells expressed β-galactosidase. To generate viral stocks, selected E5 cells containing pHE700-lac plasmid were infected with d120 helper virus (kindly provided by N. DeLuca, University of Pittsburgh). The resulting supernatants contain both the pHE700-lac vector and helper virus. The multiplicity of infection (MOI) of the helper virus added was between 0.01 to 0.1 to induce viral vector production within 24–36 hr. The average titer obtained was 10^6 bfu/ml with a ratio of amplicon to d120 helper virus from 1 : 1 to 1 : 10. The pHE700-lac containing supernatants were used to transduce human target cells *in vitro*. The β-galactosidase gene expression was evaluated after infection with pHE700-lac vector (3–10 MOI) in various cultured human tumor cells and normal fibroblasts. All cells were fixed and stained with X-gal after infection. The expression continued for approximately 2 weeks with peak expression occurring 48–72 hr after transduction. At MOI of 10, >95% of human tumor cells could be transduced (Wang *et al.*, 1997). This compares favorably with higher MOI often required for adenoviral vectors (up to 3000 MOI *in vitro*) (Chen *et al.*, 1995).

Cloning and Production of the α(1,3)GT Gene in pHE Amplicon Vector

Cloning of the pHE7αgal1 vector was accomplished by introducing the complete murine α(1,3)gene into the pHE700 multicloning site under control of the CMV immediate early promoter (Fig. 2). The cloned pHE7αgal plasmid was transfected into E5 cells expressing the replication permissive IE3 gene (Wang *et al.*, 1997). This amplicon was packaged into virions using E5 cells infected with CgalΔ3 helper virus (d120 virus that expresses β-galactosidase marker) as described above. The pHE7αgal1 amplicon undergoes rolling circle replication in E5 cells that are transcomplemented by CgalΔ3 (kindly provided by P. Johnson, University of California, San Diego), an IE3 gene defective helper virus. The E5 cells became αgal positive as determined by lectin binding after selection of the cells in hygromycin. In the presence of helper virus, the amplicon replicates until a viral DNA of 152 kb long is obtained and then packaged. Thus, for the pHE7αGT plasmid of approximately 13 kb, at least 11 expression cassettes will be present as tandem repeats in the resulting viral vector. Hygromycin resistant colonies were infected with 0.1 MOI CgalΔ3. HE7αgal1 supernates from these cells yielded a titer of 1.52×10^6 αgal unit/ml vector (positive cells by lectin staining).

α(1,3)GT Gene Transfer Renders Human Tumor Cells Sensitive to Cytolysis by Human Serums

Investigators have demonstrated that certain lectins bind specifically to αgal on glycoproteins or glycolipids (Wood *et al.*, 1979). This method was used to quantify the amount of αgal present in transduced cell clones. RPMI media (500 µl) containing 1×10^6 cells (Gibco BRL, Gaithersburg, MD) was stained with 10 µl of 1 mg/ml solution of *Griffonia simplicifolia* isolectin B4 (Vector Laboratories, Burlingame, CA). In a preliminary experiment, these supernates were then used to transduce two human cell lines, IGROV1 ovarian carcinoma and HCT15 colon carcinoma cells. Cells were incubated with vector for 24 hr and then stained with IB4 lectin and observed for αgal presence. Intense green fluorescence was detected on the outer rim of the cells (data not shown). This represents the surface location of modified glycoproteins and glycolipids on transduced human tumor cells expressing the αgal epitope. IGROV1 ovarian carcinoma cells and HCT15 colon cancer cells were trypsinized, counted and plated in 24-well plates. When the cells were nearly confluent, 3 MOI of HE7αgal1 vector was added for 2 hr at 37 °C. Forty-eight hr later, cells were tested for serum sensitivity by exposure to 50% human serum for 1 hr at 37 °C. PA317 mouse fibroblasts were a positive control for serum killing. The results are shown in Fig. 4. Complement and Ab mediated sensitivity occurred in

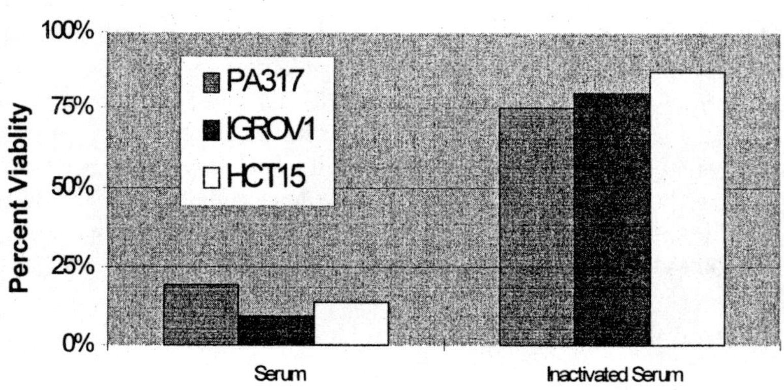

Figure 4. Human serum sensitivity after transduction of human tumor cells with HEαgal vector. Human IGROV1 ovarian carcinoma cells and HCT15 colon cancer cells were transduced with 3 MOI of HE7αgal1 vector and 48 hr later were tested for serum sensitivity by exposing the cells to 50% human serum for 1 hr at 37 °C. PA317 mouse fibroblasts were a positive control for serum killing.

the HE7αgal1 transduced tumor cells. At 48 hr, no cytotoxicity is observed from pHE-tk control vector infection of human tumor cells.

DISCUSSION

Clinical Results Suggest A Role for Hyperacute Rejection in Anti-Tumor Responses

Results from the only published clinical trial using murine VPC xenografts were recently reported (Ram *et al.*, 1997). This trial used multiple stereotaxic injections to introduce murine HStk VPC into the enhancing portion of brain tumors. Anti-tumor activity was observed in selected tumor deposits in some patients (Ram *et al.*, 1997). It is important to note that limited (minimal) gene transfer was observed. Of note, the authors did report an increase in VPC binding Ab after VPC injection that may represent anti-αgal Ab (Ram *et al.*, 1997). Other groups have shown that anti-pig Ab present in human serum are predominantly anti-αgal IgM Ab (Vaughan *et al.*, 1995). Previously, diabetic patients grafted with fetal porcine islet cells had been noted to develop an increase in anti-αgal titers (8 to 64-fold), while total immunoglobulin concentration was unchanged (Galili *et al.*, 1995). The time frame of Ab titer increase noted in our study with murine VPC intraperitoneal infusions (Fig. 1) is similar to that reported for diabetic patients transplanted with porcine islet xenografts (Galili *et al.*, 1995). The demonstration of tumor regression in our initial results without significant gene transfer suggests that hyperacute rejection of the murine cells might inhibit the cancer by an innocent bystander effect as described for red blood cells (RBC) in penicillin antigen-antibody complex mediated hemolytic anemia (Schwartz *et al.*, 1995). The penicillin antigen-antibody complexes adhere to the RBC membrane and activate complement against the RBC. The αgal and anti-αgal antigen-antibody may adhere to tumor cells and possibly activate complement and injure the tumor cells.

HSV Amplicon Gene Transfer

Anti-tumor gene therapy requires highly efficient gene transfer and expression of therapeutic genes (Roth *et al.*, 1997). HSV vectors offer both of these properties and will be the gene transfer vehicle for our anti-tumor approach with α(1,3)GT therapy. These vectors are extremely attractive since they efficiently transduce nondividing cells (G_0 stage). Most human epithelial neoplasms have a significant percentage of their cells in the G_0 stage of the cell cycle. Breast cancers, for example, have only approximately 9% of their cells in S phase (Witzig *et al.*, 1993). Murine retroviral vectors are a poor choice in this case, since mitosis is required to transduce tumor cells (Miller *et al.*, 1990). Another advantage of our Herpes amplicon vector is their large carrying capacity (Wang *et al.*, 1997). The Goldie-Coldman hypothesis predicts that spontaneous mutations in cancer cells provide resistance to chemotherapy and that therapeutic failures are directly related to tumor burden (Goldie and Coldman, 1979). This tenet argues that multiple, independent therapeutic targets need to be attacked for success. The predicted tumor resistance will likely extend to gene therapy as well; for example, our group has demonstrated that resistance to the Herpes simplex thymidine kinase (HStk) gene and ganciclovir (GCV) killing is common in tumor cells (Seregina *et al.*, 1995). In the future, gene delivery methods that transfer multiple therapeutic genes in concert or genes with multiple mechanisms of action will dominate approaches for cancer treatment. The pHE700 plasmid system permits easier cloning and functions as a shuttle vector carrying genomic information between prokaryotic and eukaryotic cells (Kwong and Frenkel, 1985). The amplicon systems retain the merits of HSV-1 vectors, but tend to have lower titers (Spaete and Frenkel, 1982). Previously reported amplicon systems were unable to package large DNA fragments (>15 kb) efficiently (Kwong and Frenkel, 1984). Several groups are working intensively to develop HSV vectors as anti-cancer therapeutics, since these vectors effectively transfer genes into tumors *in vivo* (Palella *et al.*, 1989), (Chiocca *et al.*, 1990), (Andersen *et al.*, 1992), (Ho *et al.*, 1993), (During *et al.*, 1994), (Glorioso *et al.*, 1995), (Martuza *et al.*, 1991), (Takamiya *et al.*, 1992), (Boviatsis *et al.*, 1994), (Mineta *et al.*, 1994), (Chambers *et al.*, 1995). For these reasons, we chose this vector to delivery the α(1,3)GT gene into tumor cells.

α(1,3)GT Gene Expression from the HE7agal Vector Induces Human Tumor Cell Killing

The results in Fig. 4 clearly demonstrate the rapid destruction of ovarian or colon carcinoma cells by human serum after amplicon vector transfer and expression of the α(1,3)GT gene. The killing effect mediated by anti-αgal Ab present in human serum is rapid, extensive and occurs without the need to add other exogenous co-factors, prodrugs or cells. The degree of tumor cell death is similar to that of xenogeneic murine VPC. Therefore, the α(1,3)GT gene in fact represents a unique "suicide" gene to induce destruction of the tumor since the mediators of cell death (anti-αgal Ab and complement) are present in the human immune system before transgene delivery. One potential benefit of our gene therapy approach is that the α(1,3)GT expression will induce multiple tumor-associated glycoproteins to be modified simultaneously and hopefully, improve tumor antigen presentation to these multiple targets.

Other Potentially Important Effects of α(1,3)GT Gene Expression on the Immune System

Since the αgal modifications affect multiple glycolipids and glycoproteins on the cell surface, the immune system should have an increased opportunity to detect and process tumor specific antigens. Previous studies have established that multiple membrane proteins in pig cells present αgal antigens and are targets for human anti-αgal Ab binding (Platt *et al.*, 1990). In addition, many glycoproteins possess multiple carbohydrate side chains and can express multiple αgal epitopes. Since both opsonization and complement fixation are dependent on epitope density, α(1,3)GT gene therapy offers clear advantages over other candidate transgenes. Opsonization of tumor cells can occur by anti-αgal Ab that should significantly enhance tumor antigen presentation to macrophages by αgal-antibody complex binding to their Fc receptors. The uptake of human tumor cells into macrophages could be significantly enhanced by cell membrane modification with enzymes in solution using a recombinant, truncated form of an α(1,3)GT (Galili and Latemple, 1997). These results using enzymatic modifications of human cell membranes should be transferable to our gene therapy approach. Furthermore, α(1,3)GT gene expression could permit a large number of membrane glycoproteins to be induced to express these epitopes simultaneously. Previous attempts of immunotherapy have mainly employed single cytokine molecules (e.g., IL-2, GM-CSF) (Golumbek *et al.*, 1991),

(Dranoff *et al.*, 1993), (Fearon *et al.*, 1990) or single rejection antigens (e.g., HLA B7, melanoma specific tumor antigen) (Nabel *et al.*, 1993), (Reeves *et al.*, 1996). These results suggest that further development of α(1,3)GT gene therapy as an anti-tumor strategy is warranted.

ACKNOWLEDGMENTS

This grant was in part supported by the American Cancer Society, grant number RPG-98-091-01-MBC.

REFERENCES

Andersen, J.K., Garber, D.A., Meaney, C.A., and Breakefield, X.O. (1992). "Gene transfer into mammalian central nervous system using herpes virus vectors: extended expression of bacterial lacZ in neurons using the neuron-specific enolase promoter." *Hum. Gene. Ther.*, 3(5), 487–499.

Boviatsis, E.J., Scharf, J.M., Chase, M., Harrington, K., Kowall, N.W., Breakefield, X.O., and Chiocca, E.A. (1994). "Antitumor activity and reporter gene transfer into rat brain neoplasms inoculated with herpes simplex virus vectors defective in thymidine kinase or ribonucleotide reductase." *Gene. Ther.*, 1(5), 323–331.

Chambers, R., Gillespie, G.Y., Soroceanu, L., Andreansky, S., Chatterjee, S., Chou, J., Roizman, B., and Whitley, R.J. (1995). "Comparison of genetically engineered herpes simplex viruses for the treatment of brain tumors in a scid mouse model of human malignant glioma." *Proc. Natl. Acad. Sci. U.S.A.*, 92(5), 1411–1415.

Chen, S., Li Chen, X.H., Wang, Y., Kosai, K., Finegold, M.J., Rich, S.S., and Woo, S.L. (1995). "Combination gene therapy for liver metastasis of colon carcinoma in vivo." *Proc. Natl. Acad. Sci.*, 92, 2577–2581.

Chiocca, E.A., Choi, B.B., Cai, W.Z., DeLuca, N.A., Schaffer, P.A., DiFiglia, M., Breakefield, X.O., and Martuza, R.L. (1990). "Transfer and expression of the lacZ gene in rat brain neurons by herpes simplex virus mutants." *New Biol.*, 2, 739–746.

Collins, M.K., Takeuchi, Y., Cosset, F.L., Tailor, C., and Weiss, R.A. (1994). "Development of recombinant retroviruses suitable for in vivo gene delivery." *Cold Spring Harbor Meeting: Gene Therapy*, Sept. 1994, 97.

Dranoff, G., Jaffee, E., Lazenby, A., Golumbek, P., Levitsky, H., Brose, K., Jackson, V., Hamada, H., Pardoll, D., and Mulligan, R. (1993). "Vaccination with irradiated tumor cells engineered to secrete murine GM-CSF stimulates potent, specific and long lasting anti-tumor immunity." *Proc. Natl. Acad. Sci. (USA)*, 90, 3539–3543.

During, M.J., Naegele, J.R., O'Malley, K.L., and Geller, A.I. (1994). "Long-term behavioral recovery in Parkinsonian rats by an HSV vector expressing tyrosine hydroxylase." *Science*, 266, 1399–1403.

Fearon, E.R., Pardoll, D.M., Itaya, T., Golumbek, P., Levitsky, H.I., Simons, J.W., Karasuyama, H., Vogelstein, B., and Frost, P. (1990). "Interleukin-2 production by tumor cells bypasses T helper function in the generation of an antitumor response." *Cell*, 60, 397–403.

Galili, U. (1993). "Evolution and pathophysiology of the human natural anti-alpha-galactosyl IgG (anti-Gal) antibody." *Springer Semin Immunopathol*, 15, 155.

Galili, U., and Latemple, D.C. (1997). "Natural anti-gal antibody as a universal augmentor of autologous tumor vaccine immunogenicity." *Immunol. Today*, 18, 281–285.

Galili, U., Shohet, S.B., Kobrin, E., Stults, C.L., and Macher, B.A. (1988). "Man, apes, and Old World monkeys differ from other mammals in the expression of alpha-galactosyl epitopes on nucleated cells." *J. Biol. Chem.*, 263(33), 17755–17762.

Galili, U., Tibell, A., Samuelsson, B., Rydberg, L., and Groth, C.G. (1995). "Increased anti-Gal activity in diabetic patients transplanted with fetal porcine islet cell clusters." *Transplantation*, 59(11), 1549–1556.

Geller, A.I., and Breakefield, X.O. (1988). "A defective HSV-1 vector expresses Escherichia coli β-galactosidase in cultured peripheral neurons." *Science*, 241, 1667–1669.

Geller, A.I., and Freese, A. (1990). "Infection of cultured central nervous system neurons with a defective herpes simplex virus 1 vector results in stable expression of Escherichia coli beta-galactosidase." *Proc. Natl. Acad. Sci. U.S.A.*, 87, 1149–1153.

Glorioso, J.C., DeLuca, N.A., and Fink, D.J. (1995). "Development and application of herpes simplex virus vectors for human gene therapy." *Annu. Rev. Microbiol.*, 49, 675–710.

Goldie, J.H., and Coldman, A.J. (1979). "A mathematic model for relating the drug sensitivity of tumors to their spontaneous mutation rate." *Cancer Treat. Rep.*, 63(11–12), 1727–1733.

Golumbek, P.T., Lazenby, A.J., Levitsky, H.I., Jaffee, L.M., Karasuyama, H., Baker, M., and Pardoll, D.M. (1991). "Treatment of established renal cancer by tumor cells engineered to secrete interleukin-4." *Science*, 254, 713–716.

Ho, D., Mocarski, E., and Sapoloski, R. (1993). "Altering central nervous system physiology with a defective herpes simplex virus vector expressing the glucose transporter gene." *Proc. Natl. Acad. Sci. U.S.A.*, 90, 3655–3659.

Kwong, A.D., and Frenkel, N. (1984). "Herpes simplex virus amplicon: Effect of size on replication of constructed defective genomes containing eukaryotic DNA sequences." *J. Virol.*, 51, 595–603.

Kwong, A.D., and Frenkel, N. (1985). "The Herpes simplex virus amplicon. Efficient expression of a chimeric chicken ovalbumin gene amplified within defective virus genomes." *Virology*, 142, 421–425.

Larsen, R.D., Rajan, V.P., Ruff, M.M., Kukowska-Latallo, J., Cummings, R.D., and Lowe, J.B. (1989). "Isolation of a cDNA encoding a murine UDPgalactose: beta-D-galactosyl- 1,4-N-acetyl-D-glucosaminide alpha-1,3-galactosyltransferase: expression cloning by gene transfer." *Proc. Natl. Acad. Sci. U.S.A.*, 86(21), 8227–8231.

Larsen, R.D., Rivera-Marrero, C.A., Ernst, L.K., Cummings, R.D., and Lowe, J.B. (1990). "Frameshift and nonsense mutations in a human genomic sequence homologous to a murine UDP-Gal:beta-D-Gal(1,4)-D-GlcNAc alpha(1,3)- galactosyltransferase cDNA." *J. Biol. Chem.*, 265(12), 7055–7061.

Link, C.J., Levy, J.P., Seregina, T., Atchinson, R., and Moorman, D. (1997). "Protection of complement mediated lysis of murine vector producer cells by sCR1 and low molecular weight heparins." Cancer Gene Therapy, H. Mazarakis and S.J. Swart, eds., IBC Library Series, London, United Kingdom, 135–152.

Link, C.J., Jr., Moorman, D., Seregina, T., Levy, J.P., and Schabold, K.J. (1996). "A phase I trial of in vivo gene therapy with the herpes simplex thymidine kinase/ganciclovir system for the treatment of refractory or recurrent ovarian cancer." *Hum. Gene. Ther.*, 7(9), 1161–1179.

Link, C.J., Seregina, T.S., Atchison, R., Hall, A., Muldoon, R., and Levy, J.P. (1998). "Eliciting hyperacute xenograft response to treat human cancer: α(1,3)galactosyltransferase gene therapy." *Anticancer Res.*, 18, 2301–2308.

Martuza, R.L., Malick, A., Markert, J.M., Ruffner, K.L., and Coen, D.M. (1991). "Experimental therapy of human glioma by means of a genetically engineered virus mutant." *Science*, 252, 854–856.

Miller, A.D., and Buttimore, C. (1986). "Redesign of retrovirus packaging cell lines to avoid recombination leading to helper virus production." *Molec. Cell Biol.*, 6, 2895–2902.

Miller, D.G., Adam, M.A., and Miller, A.D. (1990). "Gene transfer by retrovirus vectors occurs in cells that are actively replicating at the time of infection." *Mol. Cell Biol.*, 10, 4139–4242.

Mineta, T., Markert, J.M., Takamiya, Y., Coen, D.M., Rabkin, S.D., and Martuza, R.L. (1994). "CNS tumor therapy by attenuated herpes simplex viruses." *Gene. Ther.*, 1(1).

Nabel, G.J., Nabel, E., Yang, Z., Fox, B.A., Plautz, G.E., Gao, X., Huang, L., Shu, S., Gordon, D., and Chang, A.E. (1993). "Direct gene transfer with DNA-liposome complexes in melanoma: Expression, biologic activity, and lack of toxicity in humans." *Proc. Natl. Acad. Sci. (USA)*, 90, 11307–11311.

Palella, T.D., Hidaka, Y., Silverman, L.J., Levine, M., Glorioso, J., and Kelley, W.N. (1989). "Expression of human HPRT mRNA in brains of mice infected with a recombinant herpes simplex type 1 vector." *Gene.*, 80, 137–144.

Parker, S.L., Tong, T., Bolden, S., and Wingo, P.A. (1997). "Cancer statistics, 1997." *CA, Cancer J. for Clinicians*, 47, 5–27.

Platt, J.L., Vercellotti, G.M., Dalmasso, A.P., Matas, A.J., Bolman, R.M., Najarian, J.S., and Bach, F.H. (1990). "Transplantation of discordant xenografts: a review of progress." *Immunol. Today*, 17, 450–457.

Pruitt, S.K., Kirk, A.D., Bollinger, R.R., Marsh, J., Henry C., Collins, B.H., Levin, J.L., Mault, J.R., Heinle, J.S., Ibrahim, S., Rudolph, A.R., Baldwin, I., William M., and Sanfilippo, F. (1994). "The effect of soluble complement receptor type 1 on hyperacute rejection of porcine xenografts." *Transplantation*, 57, 363–370.

Ram, Z., Culver, K.W., Oshiro, E.M., Viola, J.J., DeVroom, H.L., Otto, E., Long, Z., Chiang, Y., McGarrity, G.J., Muul, L.M., Katz, D., Blaese, R.M., and Oldfield, E.H. (1997). "Therapy of malignant brain tumors by intratumoral implantation of retroviral vector-producing cells [see comments]." *Nat. Med.*, 3(12), 1354–1361.

Reeves, M.E., Royal, R.E., Lam, J.S., Rosenberg, S.A., and Hwu, P. (1996). "Retroviral transduction of human dendritic cells with a tumor-associated antigen gene." *Cancer Res.*, 56(24), 5672–5677.

Reisman, D., Yates, J., and Sugden, B. (1985). "A putative origin of replication of plasmids derived from Epstein-Barr virus is composed of two cis-acting components." *Mol. Cell. Biol.*, 5, 1822–1832.

Roth, J.A., Cristiano, R.J., Roth, J.A., and Cristiano, R.J. (1997). "Gene therapy for cancer: what have we done and where are we going?" *J. Natl. Cancer Inst.*, 89(1), 21–39.

Rother, R.P., Fodor, W.L., Springhorn, J.P., Birks, C.W., Setter, E., Sandrin, M.S., Squinto, S.P., and Rollins, S.A. (1995a). "A novel mechanism of retrovirus inactivation in human serum mediated by anti-alpha-galactosyl natural antibody." *J. Exp. Med.*, 182(5), 1345–1355.

Rother, R.P., Squinto, S.P., Mason, J.M., and Rollins, S.A. (1995b). "Protection of retroviral vector particles in human blood through complement inhibition." *Hum. Gene. Ther.*, 6(4), 429–435.

Sandrin, M.S., Vaughan, H.A., Dabkowski, P.L., and McKenzie, I.F. (1993). "Anti-pig IgM antibodies in human serum react predominantly with Gal(alpha 1–3)Gal epitopes." *Proc. Natl. Acad. Sci. U.S.A.*, 90(23), 11391–11395.

Schwartz, R.S., Silberstein, L.E., and Berkman, E.M. (1995). "Autoimmune hemolytic anemia." Hematology Basic Principles and Practice, R. Hoffman, E.J. Benz, S.J. Shattil, B. Furie, H.J. Cohen, and L.E. Silberstein, eds., Churchill Livingstone, New York, 723–724.

Seregina, T., Levy, J., and Link, C. "Development of ganciclovir resistance in cells transduced with the HStk gene." *Fourth International Conference on the Gene Therapy of Cancer*, San Diego, CA, A332.

Short, M.P., Choi, B.C., Lee, J.K., Malick, A., Breakefield, X.O., and Martuza, R.L. (1990). "Gene delivery to glioma cells in rat brain by grafting of a retrovirus packaging cell line." *J. Neurosci. Res.*, 27, 427–439.

Spaete, R.R., and Frenkel, N. (1982). "The herpes simplex virus amplicon: A new eukaryotic defective-virus cloning-amplifying vector." *Cell*, 30, 295–304.

Sugden, B., Marsh, K., and Yates, J. (1985). "A vector that replicates as a plasmid and can be efficiently selected in B-lymphoblasts transformed by Epstein-Barr virus." *Molec Cell Biol*, 5(2), 410–413.

Takamiya, Y., Short, M.P., Ezzeddine, Z.D., Moolten, F.L., Breakefield, X.O., and Martuza, R.L. (1992). "Gene therapy of malignant brain tumors: a rat glioma line bearing the herpes simplex virus type 1-thymidine kinase gene and wild type retrovirus kills other tumor cells." *J. Neurosci Res.*, 33(3), 493–503.

Vaughan, H.A., McKenzie, I.F., and Sandrin, M.S. (1995). "Biochemical studies of pig xenoantigens detected by naturally occurring human antibodies and the galactose alpha(1–3)galactose reactive lectin." *Transplantation*, 59(1), 102–109.

Wang, S., and Vos, J. (1996). "A hybrid infectious vector based on Epstein-Barr virus and Herpes simplex virus type I for gene transfer into Human Gene." *J Virol*, 70, 8422–8430.

Wang, S., Young, W.-B., Jacobson, C., and Link, C.J. (1997). "A novel Herpesvirus amplicon system for in vivo gene therapy." *Gene. Ther.*, 4, 1132–1141.

Welsh, R.M., Cooper, N.R., Jensen, F.C., and Oldstone, M.B. (1975). "Human serum lyses RNA tumor viruses." *Nature*, 257, 612–614.

Witzig, T.E., Ingle, J.N., Schaid, D.J., Wold, L.E., Barlow, J.F., Gonchoroff, N.J., Gerstner, J.B., Krook, J.E., Grant, C.S., and Katzmann, J.A. (1993). "DNA ploidy and percent S-phase as prognostic factors in node-positive breast cancer: results from patients enrolled in two prospective randomized trials." *J. Clin. Oncol.*, 11(2), 351–359.

Wood, C., Kabat, E.A., Murphy, L.A., and Goldstein, I.J. (1979). "Immunochemical studies on the cpmbining site of two isolectins A4 and B4 isolated from Bandeirea simpliciflia." *Arch. Biochem. Biophys.*, 198, 1–11.

Yates, J.L., Warren, N., and Sugden, B. (1985). "Stable replication of plasmids derived from Epstein-Barr virus in various mammalian cells." *Nature*, 313, 812–815.

21

INNATE IMMUNE THERAPY FOR CANCER

Screen for Molecules Capable of Activating the Innate Immune System

Tsukasa Seya[1,3], Nasim A. Begum[1], Midori Nomura[1,3], Shoutaro Tsuji[1,3], Misako Matsumoto[1,3], Akira Hayashi[3], Ichiro Azuma[2], and Kumao Toyoshima[1]

[1]Department of Immunology
Osaka Medical Center for Cancer
and Cardiovascular Diseases
Higashinari-ku, Osaka 537
[2]Institute of Immunological Science
Hokkaido University
Kita-ku, Sapporo 060, and
[3]Organization for Pharmaceutical Safety and Research (OPSR)
Tokyo 113 Japan

I. SUMMARY

Tumor cells usually express antigens which are distinguishable from normal "self" antigens and are thereby recognized by the host immune system. However, the host immune system barely responds to tumors in patients. Supplementation with adjuvant (such as BCG-CWS) in patients with cancer contributes to regression of intrinsically growing cancer. The adjuvant targets antigen-presenting cells, i.e. innate immunity, but not lymphocytes, and promotes up-regulation of MHC, co-stimulators and initial cytokines in antigen-presenting cells. We hypothesized that the role of the adjuvant is to provide conditions suitable for antigen-presentation where antigens are available and the lack of adjuvant-induced priming of antigen-presenting cells results in unresponsiveness to tumor antigens. Here, we report innate immune therapy applicable to cancer patients by supplementation with adjuvants for induction of potent immune responses against tumors.

Cancer Gene Therapy: Past Achievements and Future Challenges, edited by Habib *et al.*
Kluwer Academic/Plenum Publishers, New York, 2000.

II. INTRODUCTION

In the human immune system, lymphocytes are major effectors involved in destruction of foreign material. Each of the lymphocytes exresses a specific antigen receptor which is generated by somatic gene rearrangement and is clonally distributed on T and B cells. A tremendous number of clones each expressing unique antigen receptors can cover a huge variety of nonself antigens. To eliminate foreign cells, CD8-positive T lymphocytes mature into CTL equipped with perforin, granzymes and Fas ligand, and B lymphocytes mature into plasma cells to produce antibodies (Abs). Antigen (Ag) is absolutely required to trigger activation of the lymphocyte immune system, a system designated as acquired immunity (Janeway and Travers, 1997).

However, *in vitro* animal experiments, minimal or ineffective Ab production or CTL induction has been induced by stimulation of pure Ag without adjuvant (Janeway, 1989). Animals effectively produce Ab against foreign material when they are co-sensitized with adjuvant plus purified Ag. Adjuvants consisting of mineral oil and bacterial contituents do not directly activate lymphocytes, but potentially induce immune responses involving lymphocytes. Recent findings have indicated that the targets of the adjuvant are not lymphocytes but antigen-presenting cells (APC), and it has been proposed that the immune system involving APC, namely innate immunity, must be elicited prior to priming of the lymphocyte system (Fearon and Locksley, 1996; Fearon, 1997). If this is the case, lymphocyte activation is strictly controlled by the more basic immune system which is regulated by adjuvants. However, the role of adjuvant in the innate immune system has not been clarified, since the innate immune system *per se* has not been well-characterized compared to the acquired immune system (Fearon and Locksley, 1996).

Indeed, all of the known microbial products that have immunostimulatory activity represent invariant structures shared by large groups of microorganisms (Medzhitov and Janeway, 1997), and these are absolutely essential for the microbe's survival; lipoteichoic acids and LPS are common components of Gram-positive and Gram-negative bacteria, respectively. Double-stranded RNA is a structural signature of several groups of RNA viruses. Specific CpG repeats are also present in bacteria. Mannans and galactans are conserved as characteristic components of yeasts and mycobacteria. Bacteria cannot survive without certain peptideglycans, which are not shared with mammals. None of these compounds are produced by the host organism, and all are essential for the physiology and survival of the respective microbes. Notably, these compounds exhibit potent immunomodulatory activity to the host immune system similarly to adjuvants, and are designated as pathogen-associated molecular patterns (PAMPs) (Medzhitov and Janeway, 1997). Based on these findings, we hypothesized the importance of immune stimulation with both adjuvants and Ag in induction of strong host immune responses.

Tumor cells share the same MHC as host cells and barely express PAMPs. Tumor-associated Ags have been identified particularly in melanomas (Rosenberg *et al.*, 1998). Thus, we hypothesized that the inability of tumors to provoke host immune responses is due to the lack of PAMPs in patients with cancer. Supplementation with PAMPs as adjuvants may result in an increase in efficiency of immune responses against tumor Ags. This was supported by the observation that severe infection sometimes improves the prognosis of patients with potentially lethal cancer. We have challenged cancer patients with BCG-cell wall skeleton (BCG-CWS) for this purpose and have obtained good results (Yamamura *et al.*, 1976; Hayashi *et al.*, 1994, 1998). We call this PAMP-supplementation therapy "innate immune therapy".

III. INNATE IMMUNE RECEPTORS AND INTRACELLULAR EVENTS

The roles of PAMPs in immune modulation and cancer healing will be clarified if one can identify the receptors for each PAMP. Unlike Ag receptors, these PAMP receptors are generated from the germline-encoded genomic sequences. The possible receptors for PAMPs are shown in Table 1. CD14 (Wright, 1995), lectin receptors including macrophage mannose receptor (MMR) (Prigozy *et al.*, 1997) and DEC205 (Jiang *et al.*, 1995), and mannose binding lectin (MBL)/C1q (Law and Reid, 1989) have been identified as PAMP-recognition molecules on macrophages/dendritic cells (Mf/DC). More recently, Toll-related proteins (TRP) were discovered to participate in recognition of PAMP or mediators produced secondary to responses to PAMP (Medzhitov and Janeway, 1997; Rock *et al.*, 1998). The common features of these molecules are as follows: 1. they recognize characteristic components of microorganisms; 2. they are distributed on APC particularly Mf and DC; 3. they share protein motifs which confer a consensus structure with variable amino acids corresponding to a variety of PAMPs; and 4. they signal the APCs to mediate immune responses. These stimuli are known to be essential for the conversion of immature APCs to the mature form. For example, LPS activates monocytes into activated Mϕ and causes maturation of immature DC (Fig. 1). During maturation, the cells show altered phagocytic capacity and express co-stimulatory molecules, MHC and secretory cytokines such as proinflammatory cytokines, IL-12 and IL-18 (Fearon, 1997). Thus, a number of messages could be up- and down-regulated during maturation of Mf/DC, i.e. by PAMP stimulation.

Although it is still unclear what is required for immune modulation in the human immune system, we have identified the messages affected by BCG-CWS (Begum *et al.*, 1999), a potent PAMP for immune activator, by differential display and/or subtraction

Table 1. Putative adjuvant receptors in the innate immune system

Protein family	Site of expression	Example	Ligands
C-type lectins			
Humoral	Plasma protein	Collectins (MBL)	Bacterila and viral carbohydrates
Cellular	Macrophage, dendritic cell	Macrophage C-type lectin	GalNAc receptor
	Macrophage, dendritic cell	Macrophage mannose receptor	Terminal mannose
	Macrophage, dendritic cell	DEC 205	Terminal mannose
	NK cells	NKR-P1	Unkown sugars
	NK cells	Ly 49	MHC class I, sugars?
Leucine-rich proteins	Macrophage, epithelial cells	CD14	LPS
	Monocytes, others	TRP	Unknown ligand
	B cells	RP105	Unkown
Scavenger receptors	Macrophage	Macrophage scavenger receptor	Bacterial cell walls
	Macrophage subset	MARCO	Bacterial cell walls
Pentraxins	Plasma protein	C-reactive protein	Phosphatidyl choline
	Plasma protein	Serum amyloid P	Bacterial cell walls
Lipid transferases	Plasma protein	LBP	LPS, other LS
	Leukocytes	BPI	LPS, other LS
Integrins	Macrophage, dendritic cell, NK, T cells	CD11b, c; CD18	LPS
SCR proteins	Macrophage, dendritic cell, NK, T cells	CD21, CD35, CD46, CD55	Components of virus, bacteria and complement

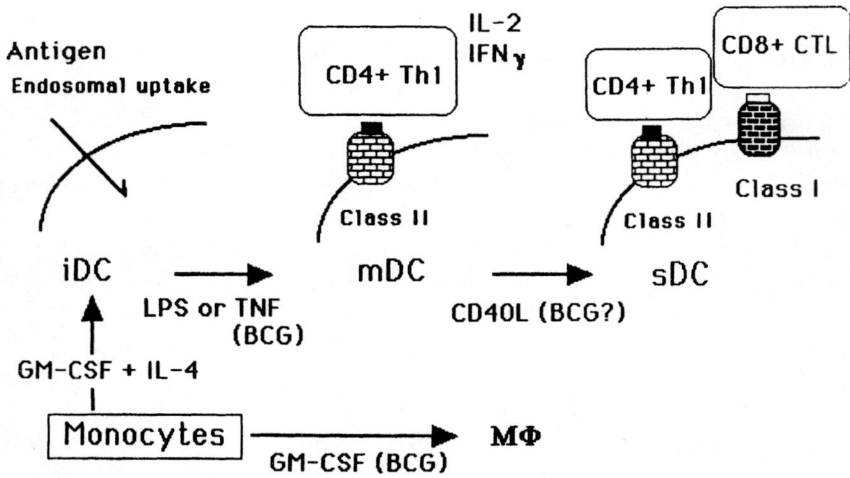

Figure 1. Signals for BCG-CWS-mediated up-regulation in human monocytes identified by cDNA subtraction screening and Northern analysis. Left lanes, RNA (2μg) from BCG-CWS-stimulated human monocytes. Right lanes, RNA(2μg) from BCG-CWS-unstimulated human monocytes. The conditions employed for BCG-CWS stimulation and subtraction screening are described by Begum *et al.* (1999).

screening using BCG-CWS-stimulated and -unstimulated human monocytes. mRNA levels of the cytokines, IL-1β, IL-8, MCP-JE and PBEF, and signaling molecules including EF1b, TNF-stimulated gene, SOS1, EMT tyrosine kinase, redox factor 1 and clathrin assembly lymphoid myeloid leukemia gene, are up-regulated in monocytes activated by BCG-CWS. mRNA levels of enzymes such as disintegrin protease, MnSOD, phospholipase A2 and G6PD are also up-regulated, while MD-1, a human homologue of chicken macrophage differentiation factor-1, is down-regulated at both mRNA and protein levels by stimulation with BCG-CWS (Begum *et al.*, 1999). A number of unknown clones have been harvested in repetitive subtraction trials suggesting the presence of a number of BCG-CWS-affected transcripts (Begum *et al.*, unpublished). These results were confirmed by Northern analysis (Fig. 2). We are currently planning to identify the full sequences of these clones and to successively characterize both structural and functional properties of the deduced proteins. Although these are not the only genes that are affected by BCG-CWS, one can predict the factors important for immune stimulation from the step-by-step maturation schema of APCs shown in Fig. 1.

IV. TOLL FAMILY PROTEINS AND THEIR LIGANDS

During this study, we confirmed that BCG is a PAMP, although it is not a representative of PAMPs, and found that Toll-related proteins (TRPs) and its family serve as receptors for IL-1β (Parnet *et al.*, 1996) and IL-18 (Torigoe *et al.*, 1997), and presumably also for other PAMPs (Yang *et al.*, 1998). In humans, 5 TRPs have already been cloned and their numbers and families are expanding rapidly. RP-105 was recently identified as a Toll family protein expressed on B cells (Miyake *et al.*, 1995). More recently, MD-1 and/or MD-2 were reported to bind RP-105 to sustain its stable expression (Miyake *et*

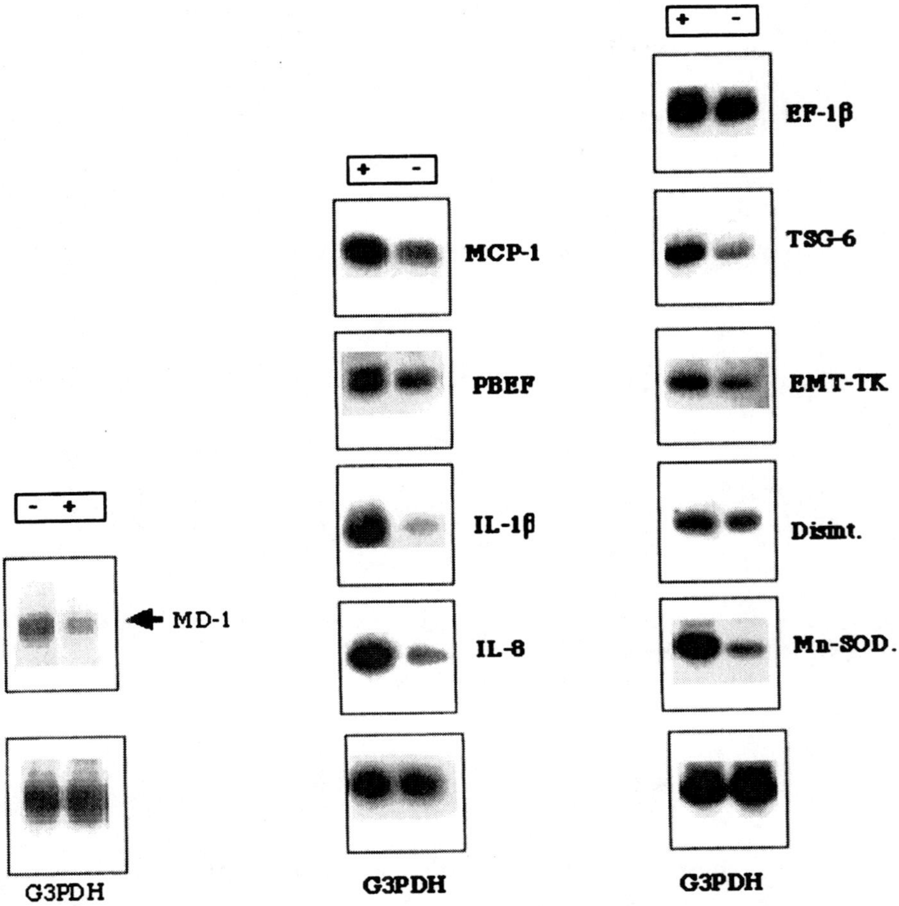

Figure 2. Maturation of APCs from human peripheral blood monocytes. The scheme shown in this figure are referred to Immunobiology (Janeway and Travers, 1997). The acting points of BCG-CWS in the maturation process of APCs are indicated.

al., 1998). Thus, to our knowledge, TRPs bind ligands belonging to three different cathegories, IL-1-like structured cytokines, MD-1 family proteins and PAMPs.

This raises the question of what happens on Mf/DC via Toll-mediated ligand binding. Most TRP are believed to form a complex with Myd88 inside the cells, and ligand stimulation induces activation of NFkB (Burns *et al.*, 1998; Gerard, 1998). Fig. 3 shows a schematic diagram for the pathway of TRP-mediated NFkB activation. It is generally accepted that IRAK and TRAF 6 are involved in TRP-mediated NFkB activation (Muzio *et al.*, 1998; Stancovski and Baltimore, 1998). This diagram is supported by the finding that IL-1β and IL-18 share the IL-1 receptor family which also possess Toll receptor-like motifs in the cytoplasmic tails (Medzhitov *et al.*, 1997; Adachi *et al.*, 1998). It is likely that PAMP-mediated signaling in monocytes triggers an initial event occurring in infectious hosts through activation of the innate immune system. However, PAMPs and the cytokine ligands induce completely different cellular responses which are unexplainable merely by the NFkB activation

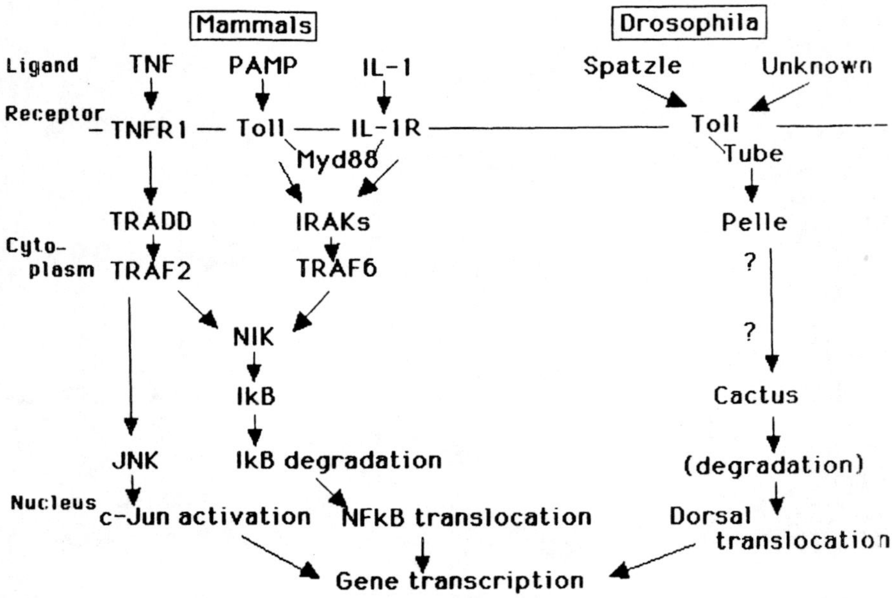

Figure 3. A possible signaling pathway through TRPs. Possible NFkB activation pathways in mammalians are compared with those found in dropsophila.

theory. Alternative intracellular responses will need to be identified for the divergent functional properties of these cytokines. The roles of MD-1/MD-2 and identity of what kind of TRPs associating with these molecules remain unknown. These points should be tested to verify the control system of the initial immune responses induced by TRPs.

BCG-CWS was found to down-regulate MD-1 (Begum *et al.*, 1999). These observations, together with the above findings regarding murine MD-1 (Miyake *et al.*, 1998), suggest two possibilities in terms of innate immune modulation. First, MD-1 down-regulation may destabilize the surface-expression of Toll family proteins on BCG-activated monocytes. If TRPs including RP-105 are important for recognition and phagocytosis of foreign material, reducing surface levels of TRP may lead to activation-induced suppression of phagocytosis. In fact, this is true in the dendritic cells, which lose phagocytic activity during maturation. Second, down-regulation of RP-105 may cancel the RP-105-mediated apoptosis pathway (Yamashita *et al.*, 1996). RP-105 renders murine B cells susceptible to anti-IgM-mediated apoptosis and probably functions as a receptor for second signaling in Ag-receptor-mediated B cell apoptosis. RP-105 is also present on the monocyte lineage (Miura *et al.*, 1998). BCG-CWS-activated macrophages could be protected from apoptosis secondary to stimulation by foreign material by down-regulating MD-1 and then TRP. This is reminiscent of the connection of TRP with the adaptor molecule Myd88 which possesses the death domain closely associated with apoptosis (Bonnert *et al.*, 1997; Burns *et al.*, 1998; Adachi *et al.*, 1998).

V. PERSPECTIVE OF INNATE IMMUNE THERAPY

Adjuvants have a long history in which they have been utilized as immune potentiators, the mechanism of which remained unidentified (Azuma *et al.*, 1974; Zabar *et al.*, 1974). The most commonly used of these molecules is Freund's complete adjuvant, consisting of mineral oil, detergent (Arlacel A) and dead bacteria (*Mycobacterium smegmatis*). This adjuvant mixture has been frequently used for efficient induction of Ab production and T cell-mediated cytolysis by *in vivo* application to animals. BCG-CWS is a constituent of the Mycobacterium, *M. bovis* . This adjuvant has been applied for cancer treatment without knowledge concerning its role in immune potentiation (Hayashi *et al.*, 1994; 1998). Current technology will provide a breakthrough to facilitate comprehensive understanding of the modes of action of therapeutic adjuvants and development of efficient adjuvants for cancer treatment.

We speculated that BCG-CWS is responsible for provoking innate immune responses and receptors for BCG-CWS which must be PAMP-recognition proteins play a key role in the adjuvant effects. Although we have not yet identified the receptor for BCG-CWS, our preliminary findings suggest that some of the TRPs may be related to the final responses of monocytes stimulated by BCG-CWS. BCG-CWS therapy for cancer patients will provide important markers for predicting the effectiveness of innate immune therapy. Increased INFγ level is a good marker for prognosis of cancer patients undergoing BCG-CWS therapy (Hayashi *et al.*, 1998). However, the levels of IFNγ may not be a "specific" indicator for innate immune activation. We provided more accurate and activation step-specific markers reflecting the advantage of innate immune therapy by clarifying the steps directed by each PAMP as shown in Fig. 1.

By choosing appropriate PAMP, it may be possible to control the potency of innate immune activation. Mf/DC come to express three major effectors, primary cytokines (IL-1β, IL-12, IL-18 etc.), co-stimulators (CD80, CD86, CD40 etc.) and Ag-presenting units (CD1 family, MHC class I, class II etc.) during their maturation in a situation-dependent manner (Janeway and Travers, 1997). The three major responses of Mf/DC in the innate immune system can be separately controlled by combinations of TRP ligands to fully instruct the lymphocyte acquired immunity. How adjuvants act on the immune system has long been obscure, but recent findings on innate immunity have provided clues for elucidation of the detailed roles of adjuvants. We believe that development of more effective PAMPs for innate immune therapy than BCG-CWS and their application to cancer patients will improve the prognoses of these patients.

ACKNOWLEDGMENTS

We are grateful to Drs. T. Masaoka and Y. Suzuki (Osaka Medical Center for Cancer) for invaluable discussions.

REFERENCES

Azuma, I., Ribi, E.E., Meyer, T.J., and Zbar, B. 1974. Biologically active components from mycobacterial cell walls. I. Isolation and composition of cell wall sekleton and component P3. J. Natl. Cancer Inst. 52:95–101.

Begum, N.A., Tsuji, S., Nomura, M., Matsumoto, M., Seya, T., Azuma, I., Hayashi, A., and Toyoshima, K. 1999. A BCG-CWS-regulated gene in human monocytes by differential display: Identification of a human homologue of chicken MD-1. Biochem. Biophys. Res. Commun. (in press).

Bonnert, T.P., Garka, K.E., Parnet, P., Sonoda, G., Testa, J.R., and Sims, J.E. 1997. The cloning and characterization of human MyD88: a member of an IL-1 receptor related family. FEBS Lett. 402: 81–84.

Burns, K., Martinon, F., Esslinger, C., Pahl, H., Schneider, P., Bodmer, J.L., Di Marco, F., French, L., and Tschopp, J. 1998. MyD88, an adapter protein involved in interleukin-1 signaling. J. Biol. Chem. 273:12203–12209.

Fearon, D.T., and Locksley, R.M. 1996. The instructive role of innate immunity in the acquired immune response. Science 272:50–54.

Fearon, D.T. 1997. Seeking wisedom in innate immunity. Nature 388:323–324.

Gerard, C. 1998. For whom the bell tolls. Nature 395:217–219.

Hayashi, A. 1994. Interferon-gamma as a marker for the effective cancer immnotherapy with BCG-Cell wall skeleton. Proc. Jpn. Acad. 70:Ser. B, 205–209.

Hayashi, A., Doi, O., Azuma, I., and Toyoshima, K. 1998. Immuno-friendly use of BCG-cell wall skeleton remarkably improves the survival rate of various cancer patients. Proc. Jpn. Acad. 74:Ser. 50–55.

Janeway, C.A. 1989. Approaching the asymptote? Evolution and revolution in immunology. Cold Spring Harbor Symp. Quant. Biol. 54:1.

Janeway, C.A., and Travers, P. 1997. The recognition of antigen. Immunobiology, 3rd ed., Garland Publishing Inc. New York, USA.

Jiang, W., Swiggard, W.J., Heufler, C., Peng, M., Mirza, A., Steinman, R.M., and Nussenzweig, M.C. 1995. The receptor DEC-205 expressed by dendritic cells and thymic epithelial cells is involved in antigen processing. Nature 375:151–155.

Law, S.K.A., and Reid, K.B.M. 1989. Complement. IRL press. Oxford, UK.

Medzhitov, R., and Janeway, C.A. 1997. Innate immunity: the virtues of a nonclonal system of recognition. Cell 91:295–298.

Medzhitov, R., Preston-Hurlburt P., and Janeway, C.A. Jr. 1997. A human homologue of the Drosophila Toll protein signals activation of adaptive immunity. Nature 388:394–397.

Miura, Y., Shimazu, R., Miyake, K., Akashi, S., Ogata, H., Yamashita, Y., Narisawa, Y., and Kimoto, M. 1998. RP105 is associated with MD-1 and transmits an activation signal in human B cells. Blood 92: 2815–2822.

Miyake, K., Shimazu, R., Kondo, J., Niki, T., Akashi, S., Ogata, H., Yamashita, Y., Miura, Y., and Kimoto, M. 1998. Mouse MD-1, a molecule that is physically associated with RP105 and positively regulates its expression. J. Immunol. 161:1348–1353.

Muzio, M., Natoli, G., Saccani, S., Levrero, M., and Mantovani, A. 1998. The human toll signaling pathway: divergence of nuclear factor kappaB and JNK/SAPK activation upstream of tumor necrosis factor receptor-associated factor 6 (TRAF6). J. Exp. Med. 187:2097–2101.

Adachi, O., Kawai, T., Takeda, K., Matsumoto, M., Tsutsui, H., Sakagami, M., Nakanishi, K., and Akira, S. 1998. Targeted Disruption of the MyD88 Gene Results in Loss of IL-1- and IL-18-Mediated Function. Immunity 9:143–150.

Parnet, P., Garka, K.E., Bonnert, T.P., Dower, S.K., and Sims, J.E. 1996. IL-1Rrp is a novel receptor-like molecule similar to the type I interleukin-1 receptor and its homologues T1/ST2 and IL-1R AcP. J. Biol. Chem. 271:3967–3970.

Prigozy, T.I., Sieling, P.A., Clemens, D., Stewart, P.L., Behar, S.M., Porcelli, S.A., Brenner, M.B., Modlin, R.L., and Kronenberg, M. 1997. The mannose receptor delivers lipoglycan antigens to endosomes for presentation to T cells by CD1b molecules. Immunity 6:187–197.

Rock, F.L., Hardiman, G., Timans, J.C., Kastelein, R.A., and Bazan, J.F. 1998. A family of human receptors structurally related to Drosophila Toll. Proc Natl Acad Sci U S A 95:588–593.

Rosenberg, S.A., Yang, J.C., Schwartzentruber, D.J., Hwu, P., Marincola, F.M., Topalian, S.L., Restifo, N.P., Dudley, M.E., Schwarz, S.L., Spiess, P.J., Wunderlich, J.R., Parkhurst, M.R., Kawakami, Y., Seipp, C.A., Einhorn, J.H., and White, D.E. 1998. Immunologic and therapeutic evaluation of a synthetic peptide vaccine for the treatment of patients with metastatic melanoma. Nat. Med. 4:321–327.

Stancovski, I., and Baltimore, D. 1998. NF-kB activation: the IkB kinase revealed? Cell 91:299–302.

Torigoe, K., Ushio, S., Okura, T., Kobayashi, S., Taniai, M., Kunikata, T., Murakami, T., Sanou, O., Kojima, H., Fujii, M., Ohta, T., Ikeda, M., Ikegami, H., and Kurimoto, M. 1997. Purification and characterization of the human interleukin-18 receptor. J. Biol. Chem. 272:25737–25742.

Wright, S.D. 1995. CD14 and innate recognition of bacteria. J. Immunol. 6–8.

Yamamura, Y., Azuma, I., Taniyama, T., Sugimura, K., Hirano, F., Tokuzen, R., Okabe, M., Nakahara, W., Yasumoto, K., and Ohta, M. 1976. Immunotherapy of cancer with cell wall skeleton of Mycobacterium bovis-Bacillus Calmette-Guerin: experimental and clinical results. Ann. N. Y. Acad. Sci. 277:209–227.

Yamashita, Y., Miyake, K., Miura, Y., Kaneko, Y., Yagiya, H., Suda, T., Nagata, S., Nomura, J., Sakaguchi, N., and Kimoto, M. 1996. Activation mediated by RP105 but not CD40 makes normal B cells susceptible to anti-IgM-induced apoptosis: a role for Fc receptor coligation. J. Exp. Med. 184:113–120.

Yang, R-B., Mark, M.R., Gray, A., Huang, A., Xie, M.H., Goddard, A., Wood, W.I., Gurney, A.L., and Godowski, P.J. 1998. Toll-like receptor-2 mediates lipopolysaccharide-induced cellular signalling. Nature 395:284–288.

Zbar, B., Ribi, E.E., Meyer, T.J., Azuma, I., and Rapp, H.J. 1974. Immunotherapy of cancer: regression of established intradermal tumors after intralesional injection of mycobacterial cell walls attached to oil droplets. J. Natl. Cancer Inst. 52:1571–1577.

22

Mda-7, A NOVEL MELANOMA DIFFERENTIATION ASSOCIATED GENE WITH PROMISE FOR CANCER GENE THERAPY

Malavi T. Madireddi[1], Zao-Zhong Su[1], Charles S. H. Young[4], Neil I. Goldstein[5], and Paul B. Fisher[1,2,3]

Departments of Urology[1]
Pathology[2], Neurosurgery[3] and Microbiology[4]
Herbert Irving Comprehensive Cancer Center
Columbia University
College of Physicians and Surgeons
New York, New York, 10032
and GenQuest Incorporated[5]
New York, New York, 10032

1. GENERAL INTRODUCTION

The carcinogenic process is multistep in terms of its etiology and multifactor with respect to its development (Fisher, 1984; Bishop, 1991; Vogelstein and Kinzler, 1991; Knudson, 1993; Du *et al.*, 1998; Kieff, 1998). Tumor cells often contain multiple genetic abnormalities, including inactivation of tumor suppressor genes, activation of oncogenes and changes in the expression of cell cycle regulatory genes, DNA damage and repair genes and genes that can directly promote cancer aggressiveness (Fisher, 1984; Bishop, 1991; Vogelstein and Kinzler, 1991; Knudson, 1993; Levine, 1993; Hartwell and Kastan, 1994; Su *et al.*, 1997). The ability of a single tumor suppressor gene, such as p53, to reverse the cancerous properties of human tumors with multiple genetic abnormalities *in vitro* and *in vivo* in human tumor xenograft models has provided the impetus for defining strategies for the gene-based therapy of human cancer (Crystal, 1995; Xu *et al.*, 1996; Roth and Cristiano, 1997; Anderson *et al.*, 1998; Chen and Mixson, 1998; Hwang *et al.*, 1998; Roth *et al.*, 1998; Takeda *et al.*, 1998).

The observation that a number of genetic mutations can result in neoplastic transformation has led many investigators to speculate that gene therapy could become a useful modality to treat cancer (Fisher, 1984; Levine, 1993; Crystal, 1995; Takeda *et al.*, 1998). Conceptually, gene replacement therapy appears to be a reasonable and viable approach to treat neoplastic diseases; by restoring more normal gene expression patterns with regards to tumor suppressor gene function within the tumor cell it may be possible

Cancer Gene Therapy: Past Achievements and Future Challenges, edited by Habib
Kluwer Academic/Plenum Publishers, New York, 2000.

to revert the transformed phenotype to a more normal state. However, for this approach to be realized it will be necessary to develop improved delivery systems, both viral and non-viral, and to identify suppressor genes that selectively target cancer cells for growth arrest and apoptosis while sparing normal cells and tissues from damage.

Since tumor cells undergo a number of genetic changes in addition to the loss of function of tumor suppressors, the ultimate objective of cancer gene-therapy should not simply be restoration of function of the specific lost tumor suppressor. Instead, it is more important to reestablish growth control to an extent that would singly or in combination with additional tumor suppressors or chemotherapeutic agents lead to selective tumor cell growth suppression and apoptosis.

To date, a number of gene therapy strategies have been explored. These include, cytokine gene transfer and tumor cell vaccination (Rosenberg, 1992), inhibition of activated oncogenes by antisense mechanisms (Zhang *et al.*, 1993; Su *et al.*, 1998a), transfer of tumor suppressor genes (Xu *et al.*, 1996; Prabhu *et al.*, 1998; Rocco *et al.*, 1998), transfer of viruses specifically active in cancer cells with mutated p53 genes (Bischoff *et al.*, 1996; Heise *et al.*, 1997), transfer of viruses in which a tissue specific promoter is driving adenovirus replicative functions and cytotoxicity (Dematteo *et al.*, 1997; Gotoh *et al.*, 1997; Rodriguez *et al.*, 1997) and selective prodrug activation by genes referred to as "suicide genes" (Borrelli *et al.*, 1988; Moolten and Wells, 1990; Culver *et al.*, 1992; Smythe *et al.*, 1994). A more recent approach employed a combinatorial tumor suppressor gene expression strategy using the cell cycle regulator gene p16 and the apoptosis inducing wild-type p53 gene (Sandig *et al.*, 1997). This dual gene approach (cell cycle arrest and activation of apoptosis) resulted in specific inhibition of tumor growth in nude mice (Sandig *et al.*, 1997) suggesting that combination tumor or cell cycle gene delivery may prove of value for cancer gene therapy.

Treatment of human melanoma cells with the combination of IFN-β + MEZ results in a rapid and irreversible loss of proliferative ability, an inhibition of tumorigenic potential in nude mice and terminal differentiation (Fisher *et al.*, 1985; Jiang and Fisher, 1993; Jiang *et al.*, 1993a, 1994). Subtraction hybridization of temporally spaced cDNAs constructed from actively growing human melanoma cells from temporally spaced cDNAs produced from human melanoma cells treated with IFN-β + MEZ permitted the cloning of a novel melanoma differentiation associated gene, mda-7 (Jiang and Fisher, 1993; Jiang *et al.*, 1994, 1995a). When ectopically expressed by means of a replication defective adenovirus in diverse human cancer cell types with single and multiple genetic defects, Ad.*mda*-7 S induces growth suppression and apoptosis (Jiang *et al.*, 1996; Su *et al.*, 1998b, 1998c). In contrast, infection of normal human epithelial or fibroblast cells with Ad.*mda*-7 S elicits no deleterious effect on growth or cellular physiology (Jiang *et al.*, 1996; Su *et al.*, 1998b, 1998c). The selective growth inhibition and targeted killing of cancer cells by Ad.*mda*-7 S places this novel tumor growth suppressor gene in a unique position for potential gene therapy applications and distinguishes it from the majority of currently identified tumor suppressor genes. The present review describes the cloning and characterization of *mda*-7 and the pre-clinical studies of *mda*-7 that provide support for its potential use as a novel gene-based therapeutic for human cancer.

2. BACKGROUND

The specific molecular and biochemical alterations associated with and controlling carcinogenesis and tumor progression in the majority of human malignancies remain to be elucidated. A consistent theme in cancer is an aberrant pattern of differentiation, and in more progressed cancers a further suppression of differentiation-associated traits (Jiang *et al.*, 1994; Waxman, 1995). A well-known example of one such cancer subtype is malignant melanoma which displays defined changes in differentiated, biological and biochemical properties as it evolves from melanocyte to nevus to radial growth phase (RGP) primary melanoma to vertical growth phase (VGP) primary melanoma to metastatic melanoma (Herlyn, 1990; Clark, 1991). Of the various forms of cancers plaguing the North American population, melanoma is increasing at the fastest rate (Clark, 1991). With time, melanoma cells can progress from a readily curable state (RGP and VGP primary melanomas, when ≤0.4 mm in depth) to an extremely aggressive form, metastatic melanoma, that is highly refractive to conventional therapies and invariably fatal in advanced stages (Herlyn, 1990; Kerbel, 1990; Lu and Kerbel, 1994). These findings underscore the importance of identifying new and more effective therapeutic strategies to treat patients with metastatic melanoma.

A potentially less toxic approach to cancer therapy involves a strategy termed differentiation therapy (Sachs, 1978; Jimenez and Yunis, 1987; Jiang *et al.*, 1994; Waxman, 1995). Several assumptions serve as the basis for this therapeutic modality. (i) Neoplastic cells often display aberrant patterns of differentiation resulting in unrestrained growth. (ii) In many contexts, failure to display normal programs of differentiation results from the inability of the cancer cell to express the appropriate genes necessary for normal growth control and differentiation. (iii) The genes necessary for inducing controlled growth and terminal differentiation are functional in cancer cells but their expression is reduced below a threshold level necessary for biological activity. (iv) Treatment of cancer cells with the appropriate agent(s) can result in reprogramming of tumor cell gene expression thereby inducing a loss of proliferative capacity and terminal differentiation. Pivotal to this hypothesis is the assumption that the genes mediating normal differentiation in many tumor cells are not genetically defective, and failure to differentiate normally is a consequence of uncoordinated and/or inappropriate gene expression.

The successful application of differentiation therapy in specific instances, both *in vitro* and *in vivo* may be initiated by subtle, yet specific gene expression alterations that induce terminal differentiation (rev. in Roberts *et al.*, 1998). This could involve the induction or elevated expression of silent or active genes, respectively, the suppression of specific target genes or a combination of these changes in the differentiating cell. During the process of differentiation, modifications in gene expression would be expected to alter the existing message and protein pools of differentiation regulating molecules, by posttranscriptional and posttanslational mechanisms that affect message stability and/or protein modification states, respectively. Metastatic human melanoma has been used to test the differentiation/modified gene expression hypothesis (Fisher *et al.*, 1985; Jiang and Fisher, 1993; Jiang *et al.*, 1993a, 1994). Treatment of human melanoma cells with the combination of IFN-β plus MEZ results in a rapid cessation of growth, induction of profound morphological changes (such as the appearance of dendritic processes), significant alterations in gene expression, modifications in antigenic phenotype, an increase in melanin synthesis, the appearance of new E2F transcription factor complexes and an irreversible loss in proliferative capacity and terminal cell differentiation (Fisher *et al.*, 1985; Guarini *et al.*, 1989, 1992; Jiang and Fisher, 1993; Jiang *et al.*, 1993a, 1993b, 1994, 1995a, 1995b, 1995c, 1996; Lin *et al.*, 1996, 1998; Kang *et al.*, 1988; Su *et al.*, 1998b, 1998c). In contrast, treatment of the same melanoma cells with equivalent doses of either IFN-β or MEZ individually results in specific differentiation-related and immunologically-related changes and growth suppression, but terminal differentiation does not ensue (Fisher *et al.*, 1985; Jiang and Fisher, 1993; Jiang *et al.*, 1993a, 1994).

2.1. Differentiation Induction Subtraction Hybridization (DISH) Approach for Identification of Differentiation Associated Genes

To define the spectrum of gene expression changes occurring as a consequence of induction of terminal differentiation in human melanoma cells, a modified subtraction hybridization approach was used (Jiang and Fisher, 1993). By combining differentiation induction with the molecular procedure of subtraction hybridization, the DISH protocol, it is possible to identify and clone genes associated with and potentially mediating growth arrest and terminal differentiation (Jiang and Fisher, 1993; Huang *et al.*, 1998). The DISH cDNA library was prepared from temporally spaced mRNAs isolated from actively proliferating human HO-1 melanoma cells that were subtracted from a cDNA library produced from temporally spaced mRNA populations harvested after HO-1 cells were treated with IFN-β + MEZ for 2, 4, 6, 8, 12 and 24 hr (Jiang and Fisher, 1993).

In a primary screening of the DISH library, 70 random cDNAs were isolated and 23 of these were found to display differential expression in differentiated versus undifferentiated HO-1 human melanoma cells (Jiang and Fisher, 1993; Jiang *et al.*, 1994). Of the 23 melanoma differentiation associated (*mda*) genes originally identified, 14 represented known sequences and 9 were classified as novel, without sequence homology to gene sequences reported in DNA data bases (Jiang and Fisher, 1993). One of the known genes is the key cell cycle regulator p21 (*mda*-6), a general inhibitor of cyclin-dependent kinases (El Deiry *et al.*, 1993; Harper *et al.*, 1993; Jiang and Fisher, 1993; Noda *et al.*, 1994; Jiang *et al.*, 1995b; 1996a; Chellappan *et al.*, 1998; Soos *et al.*, 1998; Gervais *et al.*, 1998). Another novel gene identified by DISH is the *mda*-9 gene that is transiently induced during terminal differentiation (Lin *et al.*, 1996). *Mda*-9 is also induced by gamma-interferon (Lin *et al.*, 1998). Subsequently, *mda*-2 was found to be homologous to the germ-cell-specific transcriptional repressor Tctex-1 (Roux *et al.*, 1994; O'Neill and Artzt, 1995) and *mda*-4 was shown to be a novel member of the human interferon-inducible gene family associated with control of tumorigenicity in a model of human melanoma (DeYoung *et al.*, 1997). One of the novel *mda* genes with potential clinical relevance is the *mda*-7 gene, a ubiquitous cancer growth suppressor gene that exerts its growth suppression activity specifically in cancer cells and not in normal cells (Jiang *et al.*, 1995a, 1996).

2.2. Cloning and Characterization of *mda*-7

The level of *mda*-7 mRNA is elevated in human melanoma cells induced to growth arrest irreversibly and undergo terminal differentiation (Jiang *et al.*, 1995a). The *mda*-7 cDNA encodes a novel polypeptide, with an apparent molecular mass of 23.8 kDa (Fig. 1), which shows no homology to any known protein in the Genebank or EMBL databases. The primary amino acid sequence of *MDA*-7 between amino acid 105–120 displays a signature sequence consensus common to the IL-10 gene family (Mosmann, 1994). Other potential consensus domains present in the *MDA*-7 amino acid sequence are phosphorylation sites for cyclin dependent kinase 2 (Cdk2) and protein kinase C (PKC). The significance and relevance of these motifs/domains in determining *MDA*-7 action are not understood at the present time.

A basic premise of the aberrant differentiation/modified gene expression model of melanoma development is that specific genes regulating melanoma growth and differentiation are expressed in normal melanocytes, whereas their expression is decreased or

```
CTTGCCTGCAAACCTTTACTTCTGAAATGACTTCCACGGCTGGGACGGGAACCTTCCACC 60
CACAGCTATGCCTCTGATTGGTGAATGGTGAAGGTGCCTGTCTAACTTTTCTGTAAAAAG 120
AACCAGCTGCCTCCAGGCAGCCAGCCCTCAAGCATCACTTACAGGACCAGAGGGACAAGA 180
CATGACTGTGATGAGGAGCTGCTTTCGCCAATTTAACACCAAGAAGAATTGAGGCTGCTT 240
GGGAGGAAGGCCAGGAGGAACACGAGACTGAGAGATGAATTTTCAACAGAGGCTGCAAAG 300
                                   M  N  F  Q  Q  R  L  Q  S

CCTGTGGACTTTAGCCAGACCCTTCTGCCCTCCTTTGCTGGCGACAGCCTCTCAAATGCA 360
 L  W  T  L  A  R  P  F  C  P  P  L  L  A  T  A  S  Q  M  Q

GATGGTTGTGCTCCCTTGCCTGGGTTTTACCCTGCTTCTCTGGAGCCAGGTATCAGGGGC 420
 M  V  V  L  P  C  L  G  F  T  L  L  L  W  S  Q  V  S  G  A

CCAGGGCCAAGAATTCCACTTTGGGCCCTGCCAAGTGAAGGGGGTTGTTCCCCAGAAACT 480
 Q  G  Q  E  F  H  F  G  P  C  Q  V  K  G  V  V  P  Q  K  L

GTGGGAAGCCTTCTGGGCTGTGAAAGACACTATGCAAGCTCAGGATAACATCACGAGTGC 540
 W  E  A  F  W  A  V  K  D  T  M  Q  A  Q  D  N  I  T  S  A

CCGGCTGCTGCAGCAGGAGGTTCTGCAGAACGTCTCGGATGCTGAGAGCTGTTACCTTGT 600
 R  L  L  Q  Q  E  V  L  Q  N  V  S  D  A  E  S  C  Y  L  V

CCACACCCTGCTGGAGTTCTACTTGAAAACTGTTTTCAAAAACTACCACAATAGAACAGT 660
 H  T  L  L  E  F  Y  L  K  T  V  F  K  N  Y  H  N  R  T  V

TGAAGTCAGGACTCTGAAGTCATTCTCTACTCTGGCCAACAACTTTGTTCTCATCGTGTC 720
 E  V  R  T  L  K  S  F  S  T  L  A  N  N  F  V  L  I  V  S

ACAACTGCAACCCAGTCAAGAAAATGAGATGTTTTCCATCAGAGACAGTGCACACAGGCG 780
 Q  L  Q  P  S  Q  E  N  E  M  F  S  I  R  D  S  A  H  R  R

GTTTCTGCTATTCCGGAGAGCATTCAAACAGTTGGACGTAGAAGCAGCTCTGACCAAAGC 840
 F  L  L  F  R  R  A  F  K  Q  L  D  V  E  A  A  L  T  K  A

CCTTGGGGAAGTGGACATTCTTCTGACCTGGATGCAGAAATTCTACAAGCTCTGAATGTC 900
 L  G  E  V  D  I  L  L  T  W  M  Q  K  F  Y  K  L  *

TAGACCAGGACCTCCCTCCCCCTGGCACTGGTTTGTTCCCTGTGTCATTTCAAACAGTCT 960
CCCTTCCTATGCTGTTCACTGGACACTTCACGCCCTTGGCCATGGGTCCCATTCTTGGCC 1020
CAGGATTATTGTCAAAGAAGTCATTCTTTAAGCAGCGCCAGTGACAGTCAGGGAAGGTGC 1080
CTCTGGATGCTGTGAAGAGTCTACAGAGAAGATTCTTGTATTTATTACAACTCTATTTAA 1140
TTAATGTCAGTATTTCAACTGAAGTTCTATTTATTTGTGAGACTGTAAGTTACATGAAGG 1200
CAGCAGAATATTGTGCCCCATGCTTCTTTACCCCTCACAATCCTTGCCACAGTGTGGGGC 1260
AGTGGATGGGTGCTTAGTAAGTACTTAATAAACTGTGGTGCTTTTTTTGGCCTGTCTTTG 1320
GATTGTTAAAAAACAGAGAGGGATGCTTGGATGTAAAACTGAACTTCAGAGCATGAAAAT 1380
CACACTGTCTGCTGATATCTGCAGGGACAGAGCATTGGGGTGGGGGTAAGGTGCATCTGT 1440
TTGAAAAGTAAACGATAAAATGTGGATTAAAGTGCCCAGCACAAAGCAGATCCTCAATAA 1500
ACATTTCATTTCCCACCCACACTCGCCAGCTCACCCCATCATCCCTTTCCCTTGGTGCCC 1560
TCCTTTTTTTTTTATCCTAGTCATTCTTCCCTAATCTTCCACTTGAGTGTCAAGCTGACC 1620
TTGCTGATGGTGACATTGCACCTGGATGTACTATCCAATCTGTGATGACATTCCCTGCTA 1680
ATAAAAGACAACATAACTCAAAAAAAAAAAAAAAAAAA 1718
```

Figure 1. Nucleotide sequence and deduced amino acid sequence of the *mda*-7 cDNA. The hydrophobic domain is underlined; the 3′ untranslated sequence shows three putative instability motifs ATTTA (underlined) and two consensus signals AATAAA for poly(A) addition (underlined). Genbank Accession number U16261. Data from Jiang *et al.* (1995a).

absent in metastatic melanoma (Jiang and Fisher, 1993; Jiang *et al.*, 1994). Analysis of *mda*-7 expression during natural melanoma progression, i.e., melanocyte to metastatic melanoma (Jiang *et al.*, 1995a) and using the Matrigel-assisted melanoma progression model (Kobayashi *et al.*, 1994) demonstrate an inverse correlation between *mda*-7 expression and melanoma progression. The levels of *mda*-7 expression in normal melanocytes and nevi are statistically higher than in late VGP and metastatic human melanoma cell lines (Fig. 2) (Jiang *et al.*, 1995a). Similarly, in RGP and early VGP primary melanomas selected for autonomous growth in nude mice by coinjection with Matrigel, the Matrigel-assisted melanoma progression model, the levels of *mda*-7 mRNA are lower in the progressed variants than in the parental cells (Jiang *et al.*, 1995a). These

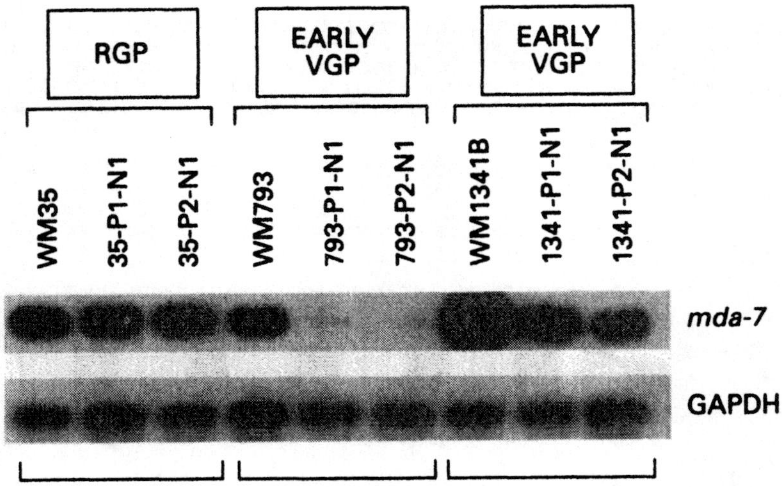

Figure 2. Expression of *mda*-7 as a function of "Matrigel-mediated" progression in primary human melanomas. The levels of *mda*-7 and GAPDH were determined by RT-PCR from the indicated actively proliferating cell lines. P1 and P2 refer to the first and second passage, respectively, through nude mice of RGP or early VGP human melanoma cells in combination with Matrigel. Data from Jiang *et al.* (1995a).

data provide support for the aberrant differentiation/modified gene expression model of melanoma development and demonstrate that *mda*-7 expression levels decrease with melanoma progression.

If one assumes that treatment of human melanoma cells with agents inducing loss of growth potential and terminal differentiation results from the activation of specific target genes that serve as negative regulators of these processes, then the levels of defined genes should be increased or induced following such a treatment protocol (Jiang and Fisher, 1993; Jiang *et al.*, 1994). In contrast, in normal melanocytes the levels of the negative regulators of melanoma progression should be constitutively expressed and treatment with the same differentiation inducing agents should not dramatically alter expression of these negative regulators of growth and differentiaton. Growth of HO-1 human melanoma cells in medium containing a combination of IFN-β and MEZ, and to a lesser extent with only MEZ, causes growth arrest and induction of *mda*-7 expression (Fig. 3). Elevated *mda*-7 mRNA levels are apparent within 24 hr of treatment with the differentiation inducing agents and high levels of expression persist over the next 96 hr, at which time the cells are terminally differentiated (Fig. 3). In contrast, in SV40-transformed human melanocytes (Melber *et al.*, 1989), *mda*-7 is constitutively expressed and expression is not significantly altered following treatment with IFN-β + MEZ (Jiang *et al.*, 1995a). Transfection of *mda*-7 expression constructs into HO-1 and C8161 human melanoma cells reduces growth and inhibits colony formation (Jiang *et al.*, 1995a). These data demonstrate that an increase in *mda*-7 mRNA expression correlates with a loss of growth potential and terminal differentiation in human melanoma cells, an effect not seen in immortalized normal human melanocytes. Moreover, evidence is presented that ectopic expression of *mda*-7 in human melanoma cells directly inhibits growth. In this context, *mda*-7 may contribute to the process of terminal differentiation in human melanoma cells treated with IFN-β + MEZ.

Immunoprecipitation analyses using peptide-derived rabbit polyclonal and mouse monoclonal antibodies detect increased *MDA*-7 protein of ~23.8 kDa in MEZ and

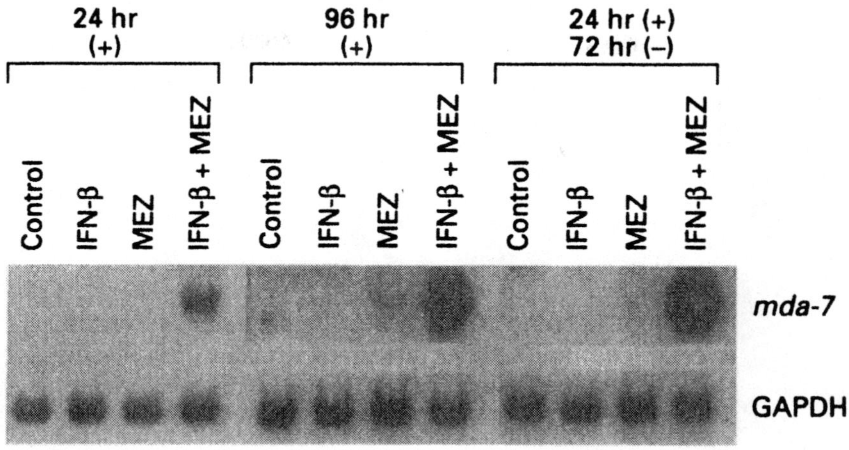

Figure 3. Induction of *mda*-7 expression in HO-1 cells as a function of differentiation. HO-1 cells were untreated (control) or treated for 24 hr, 96 hr, or 24 hr followed by growth for 72 hr in the absence of inducer with IFN-β (2000 U/ml), MEZ (10 ng/ml) or IFN-β + MEZ (2000 U/ml + 10 ng/ml). RNA isolation, Northern blotting and hybridization with *mda*-7 and GAPDH was performed as described (Jiang and Fisher, 1993; Jiang *et al.*, 1993a). Data from Jiang *et al.* (1995a).

IFN-β + MEZ treated HO-1 cells (Jiang *et al.*, 1995a). Interestingly, a second high molecular weight (HMC), *mda*-7-associated protein co-precipitates in this immune complex. The relative molecular mass of the HMC protein is about 90–110 kDa (Fig. 4). Since *mda*-7 lacks a nuclear localization signal, yet immunofluorescence analysis indicates that *MDA*-7 stains chromatin in cells undergoing mitosis (evident during prophase, metaphase, anaphase and telophase) (Madireddi and Fisher, unpublished data), it is possible that the HMC protein functions as a chaperone allowing *MDA*-7 to enter the nucleus. This possibility and the functional relevance of the interaction between *MDA*-7 and the HMC are currently under investigation.

3. PRE-CLINICAL GENE THERAPY STUDIES WITH *Mda*-7 AND Ad.*mda*-7 S

On the basis of the growth suppressive properties of *mda*-7 in human melanoma cells, it was considered meaningful to determine if this growth inhibitory action was specific to this target cell type or more ubiquitous in its action (Jiang *et al.*, 1996; Su *et al.*, 1998b, 1998c). To address these questions the effects of *mda*-7 on growth and colony formation of normal and tumor cells of various origins and tumor cells with a range of genetic defects (p53 WT/Mut, Rb WT/Mut) were studied *in vitro* using transfection approaches with a recombinant *mda*-7 cDNA cloned into a pREP-4 vector (Jiang *et al.*, 1996). These experiments indicated that *mda*-7 was growth inhibitory toward a wide spectrum of cancer cell types, while sparing normal cells from any overt deleterious effects (Jiang *et al.*, 1996). Since transfection of plasmid DNA into cells is an inherently inefficient process, a replication incompetent adenovirus was constructed that expresses *mda*-7, Ad.*mda*-7 S (Su *et al.*, 1998b, 1998c). With this type of genetic reagent it was possible to more accurately determine the biological effects of *mda*-7 on the growth and physiology of normal and cancer cells (Su *et al.*, 1998b, 1998c). As observed with DNA

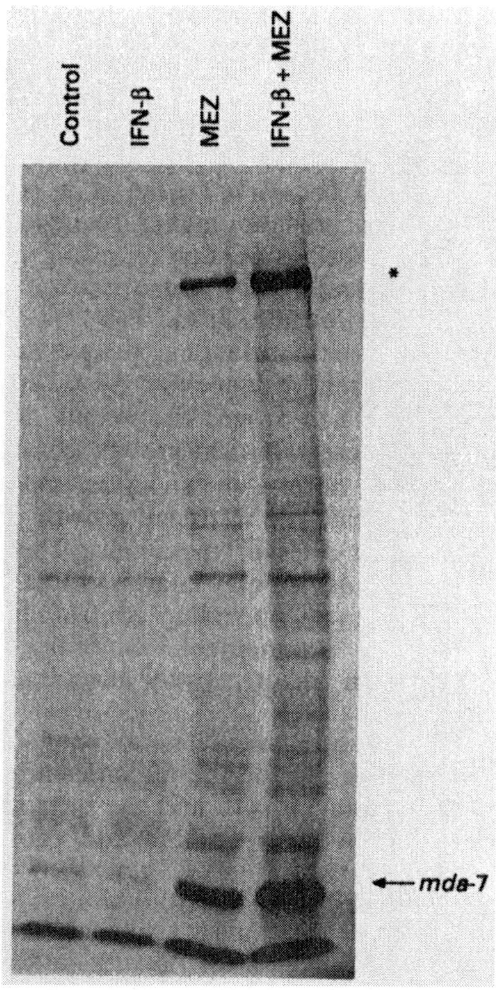

Figure 4. Effect of IFN-β, MEZ and IFN-β + MEZ on *MDA*-7 protein and high molecular weight-*MDA*-7 complexing (HMC) protein in HO-1 cells. Lysates from untreated (Control) and HO-1 cells treated with IFN-β (2000 U/ml), MEZ 10 ng/ml or IFN-β + MEZ (2000 U/ml + 10 ng/ml) for 96 hr were immunoprecipitated with *MDA*-7 anti-peptide rabbit polyclonal antibodies. * Designates the HMC protein. Data from Jiang *et al.* (1995a).

transfection, Ad.*mda*-7 S displayed selective antiproliferative activity toward human cancer versus normal human epithelial and fibroblast cells (Su *et al.*, 1998b, 1998c).

3.1. *In vitro* Induction of Differential Growth Suppression in Cancer and Transformed Cells by *mda*-7

Human cancers of diverse origins and with multiple genetic alterations including, breast, cervical, colorectal, nasopharyngeal and prostate carcinomas and glioblastoma multiforme and osteosarcoma, were transfected with an *mda*-7 expression plasmid and assayed for colony formation (Jiang *et al.*, 1996). Significant reductions in colony formation were observed in the different human tumor lines that were analyzed (Table 1). The ability of transformation to elicit sensitivity to *mda*-7-induced growth suppression has been demonstrated using cloned rat embryo fibroblast (CREF) cells and the Ha-*ras* oncogene. CREF cells exhibited only a minor reduction (≤1.7-fold in comparison with vector transfected cultures, average of six independent assays) in colony formation after transfection with *mda*-7, whereas CREF cells stably transformed with the Ha-*ras*

Table 1. Effect of *mda*-7 on monolayer colony formation of human cancer, normal rat embryo fibroblast (CREF) and Ha-ras-transformed CREF cells

Cell Type	RSV-vector*	RSV-*mda*-7 (S)†	RSV-*mda*-7 (AS)
Human cancer cell lines‡			
MCF-7 (breast-Ca)	118 ± 24	42 ± 16 (3.5)	146 ± 20
T47D (breast-Ca)	172 ± 9	44 ± 7 (4.2)	186 ± 28
HeLa (cervix-Ca)	1,571 ± 446	117 ± 107 (15.2)	1,771 ± 385
LS174T (colorectal-Ca)	130 ± 14	30 ± 3 (5.4)	160 ± 15
HONE-1 (nasopharyngeal-Ca)	219 ± 19	71 ± 8 (3.5)	250 ± 19
DU-145 (prostate-Ca)	174 ± 18	54 ± 8 (3.1)	166 ± 12
T98G (glioblastoma)	99 ± 9	32 ± 4 (3.6)	115 ± 14
Saos-2 (osteosarcoma)	126 ± 22	35 ± 6 (3.9)	138 ± 14
Rat embryo fibroblast			
CREF (normal rat embryo)	60 ± 10	35 ± 5 (1.7)	66 ± 7
CREF-*ras* (transformed)	147 ± 16	25 ± 4 (6.0)	151 ± 16

*Logarithmically growing cells were seeded at 1×10^6 cells per 100-mm plate and transfected with 10 μg of vector [pREP4 (RSV)] containing no insert, *mda*-7 (S), or *mda*-7 (AS). After 24 hr, cells were replated at $\approx 2 \times 10^5$ cells per 100-mm plate in medium containing 100 μg of hygromycin per ml. Medium was changed every 3 or 4 days and plates were fixed in formaldehyde and stained with Giemsa at day 14 or 21. Colonies containing 50 or more cells were enumerated. Values shown are the average Hyg^R colonies formed in four to five replicate plates ± SD.
†Values in parentheses indicate fold-decrease in colony formation versus RSV-*mda*-7 (AS) transfected cells.
‡MCF-7, T47D, HeLa, LS174T, DU-145, and HONE-1 are human carcinoma (Ca) cell lines isolated from the indicated anatomical site. T98G is a human glioblastoma multiforme cell line. CREF-*ras* is a Ha-*ras* (T24) oncogene transformed CREF clone.

oncogene exhibited a 6-fold reduction in colony formation when ectopically expressing *mda*-7 (Table 1) (Jiang *et al.*, 1996; unpublished data).

Supporting evidence for a direct effect of *mda*-7 on human cancer cell growth has been provided using human cervical (HeLa) and prostate (DU-145) cancer cells stably expressing elevated levels of *mda*-7 (Jiang *et al.*, 1996). Transfection of HeLa cells with *mda*-7 dramatically reduces monolayer colony formation, both size and number (Table 1 and Fig. 5) (Jiang *et al.*, 1996). To determine if the ability of colonies to form after transfection of HeLa cells with *mda*-7 resulted from failure of the transfected cells to continue expressing the transfected *mda*-7 gene 10 random colonies were isolated, RNA was

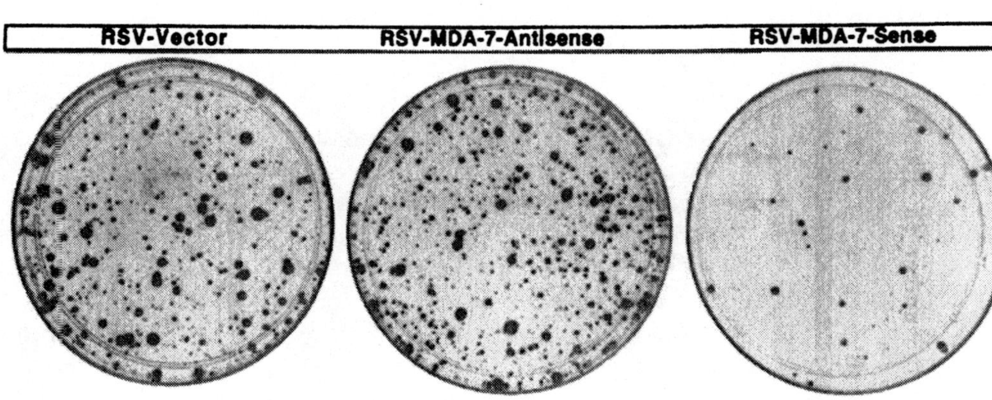

Figure 5. Effect of *mda*-7 expression on Hyg^R colony formation in HeLa cells. HeLa cells were transfected with 10 μg of pREP4 vector (RSV-vector), *mda*-7 cloned in and antisense orientation in the pREP4 vector (RSV-*mda*-7 antisense), or *mda*-7 cloned in the sense orientation in the pREP4 vector (RSV-*mda*-7 sense) and selected in media containing 100 μg/ml hygromycin. Data from Jiang *et al.* (1996).

extracted and then analyzed by Northern blotting for *mda*-7 expression. Seven of the 10 Hela clones did not express detectable levels of *mda*-7, whereas 2 clones produced low levels of *mda*-7 mRNA and 1 clone (HeLa clone 2) expressed moderate levels of *mda*-7 mRNA (Jiang *et al.*, 1996). These findings support the hypothesis that continued *mda*-7 expression in HeLa cells may be necessary to prevent colony formation and reduce colony size. When compared to parental HeLa or vector transformed HeLa (HeLa clone 1) cells, the *mda*-7 expressing HeLa clone 2 cells grew at a slower rate in monolayer culture and formed fewer colonies when grown in agar. HeLa clone 2 cells also expressed the *MDA*-7 protein and the HMC protein. Infection with an adenovirus expressing antisense *mda*-7, Ad.*mda*-7 AS, resulted in an elimination of the *MDA*-7 protein and the HMC protein and a return in growth properties to that of parental HeLa and vector transformed HeLa clone 1 cells (Jiang *et al.*, 1996). Similarly, *mda*-7 expression in DU-145 cells, by means of an expression cassette containing a dexamethasone (DEX) inducible MMTV-promoter controlling *mda*-7 expression, resulting from the addition of DEX produced a reduced growth rate and a reduction in agar colony formation (data not shown). These results document a direct correlation between elevated *mda*-7 expression and reduction of growth and suppression of transformation related properties in human cancer cells. Moreover, since DU-145 cells produce a defective Rb protein and HeLa cells do not contain functional p53 or Rb proteins, the ability of *mda*-7 to modify the phenotype of human cancer cells occurs independent of the action of these two widely studied tumor suppressor genes.

In contrast to the dramatic effect of ectopic expression of *mda*-7 in human cancer cells, overexpression of *mda*-7 in normal cells, including human mammary epithelial, human skin fibroblasts and CREF cells, did not alter significantly growth or colony formation (Jiang *et al.*, 1996; Su *et al.*, 1998b, 1998c). The absence of homology of *mda*-7 to any of the previously identified growth suppressor genes and its ability to differentially regulate growth in cancer cells versus normal cells is intriguing. It is tempting to speculate that *mda*-7's mode of action may be unique, in the sense that its "activity" may be regulated by the microenvironment (normal versus cancer cells) in which it is expressed. Some of the potential differences in *mda*-7 induced effects in cancer versus normal cells may be due to post-translational modification by phosphorylation, appropriate/inappropriate localization of the protein to its subcellular location and/or half-life of the message and protein. Current data suggest that *mda*-7 may represent a novel class of cancer cell growth suppressor genes with selective anti-tumor activity.

3.2. *Mda*-7, a Ubiquitous Inducer of Apoptosis in Cancer Cells

To better define the mechanism of action of *mda*-7 induced growth suppression of cancer cells and its potential application as a cancer gene therapeutic, a "replication deficient" recombinant adenovirus (Crystal, 1995; Smith, 1995) expressing *mda*-7 (Ad.*mda*-7 S) was constructed (Su *et al.*, 1998b, 1998c). The advantages of using viral vectors over nonviral vectors are discussed elsewhere (Roth and Cristiano, 1997; Kong and Crystal, 1998). For the effective utilization of viral vectors as gene transfer vehicles, sufficient viral genes are removed to permit insertion of the expression cassette and to render the virus replication deficient and nontoxic. The viral vector used was comprised of the adenovirus backbone minus the Ad E1A and E1B gene sequences that have been replaced by the CMV promoter and appropriate RNA processing signals (also called the viral cassette). The open-reading frame of the *mda*-7 cDNA was inserted downstream of the CMV promoter. Viral vectors capitalize on the ability of the viruses to efficiently enter cells through specific receptors, which are present on most human cells. The mechanisms inherent in the virus structure facilitate effective transfer of their genome to a site in the host cell where the host cell machinery will transcribe and translate the expression cassette integrated in the viral genome. Adenovirus vectors are currently the "gold standard" vectors for *in vivo* gene transfer in that, they are easy to produce in large amounts with high-purity, transfer genes to almost all types of cells (quiescent or proliferating and regardless of the state of differentiation of the cell), and the resulting expression of the transgene is robust (Crystal, 1995; Smith, 1995). The only disadvantage of the adenoviral vector is that, they do not integrate into the target cell genome and their expression is transient, lasting from days to weeks. Transient expression of the *mda*-7 transgene in the target cells is sufficient to induce growth suppression and integration of *mda*-7 into the genome is not required for its action.

Infection of human tumor cells, including metastatic melanoma (C8161), glioblastoma multiforme (T98G), osteosarcoma (Saos 2) and carcinoma of the breast (MCF-7, T47D), cervix (HeLa), colon (SW480, WiDr and LS174T), nasopharynx (HONE-1) and prostate (DU-145 and LNCaP), with 100 pfu/cell of Ad.*mda*-7 S reduced colony formation by >50% and inhibited growth in diverse cancers in comparison with untreated cells or those infected with the virus vector lacking the *mda*-7 gene (Ad.vec) (Figs. 6 and 7) (Su *et al.*, 1998b). To assess the infection efficiency of the virus, a parallel set of cultures were infected with 100 pfu/cell of Ad.β-Gal virus (Su *et al.*, 1998b, 1998c). The viral infection efficiency in Ad.β-Gal cultures ranged between 92–100% (unpublished data). Therefore, the growth inhibition and reduction in colony formation in cancer cells is a direct consequence of expression of the *mda*-7 transgene. In stark contrast, no discernable effect was evident in normal human mammary epithelial, human skin fibroblast or an immortalized (SV40 containing) human breast epithelial cell line (HBL-100) infected with 100 pfu/cell of Ad.*mda*-7 S, even though replicate cultures infected with

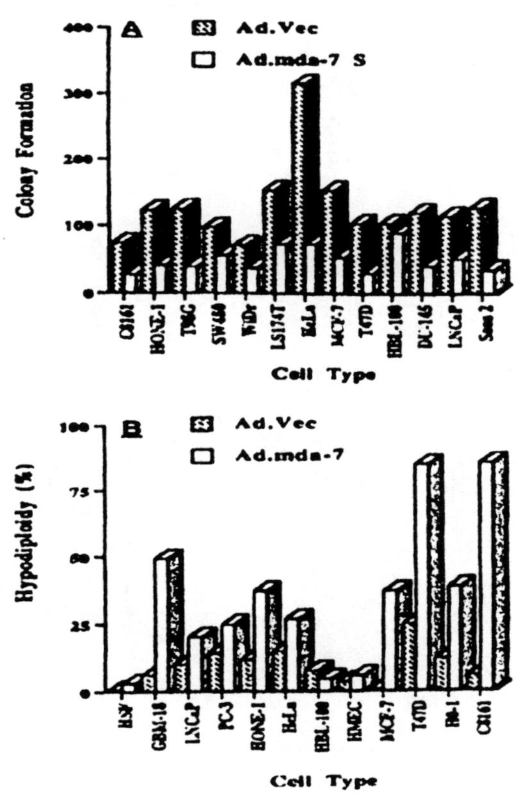

Figure 6. Effect of Ad.*mda*-7 on colony formation and induction of hypodiploidy in normal and cancer cells. Cells were infected with 100 pfu/cell of Ad.*Vec* or Ad.*mda-7 S* and analyzed for monolayer colony formation (A) or for hypodiploidy (A_o) by FACS analysis (B). Colony forming assays were scored after 2–3 weeks. Hypodiploidy assays were performed 96 hr post-infection. Replicate samples varied by <10%. Similar results (±15%) were obtained in an additional replicate study. Human cell types: C8161 and HO-1: metastatic melanoma; HONE-1: nasopharyngal carcinoma; T98G and GBM-18: glioblastoma multiform; SW480, WiDr and LS174T: colon carcinoma; HeLa: cervical carcinoma; MCF-7 and T47D: breast carcinoma; DU-145, LNCaP and PC-3: prostate carcinoma; Saos-2: osteosarcoma; HBL-100: normal breast epithelial cell line; HSF: early passage skin fibroblasts; HMEC: early passage normal mammary epithelial cells. Data from Su *et al.* (1998c).

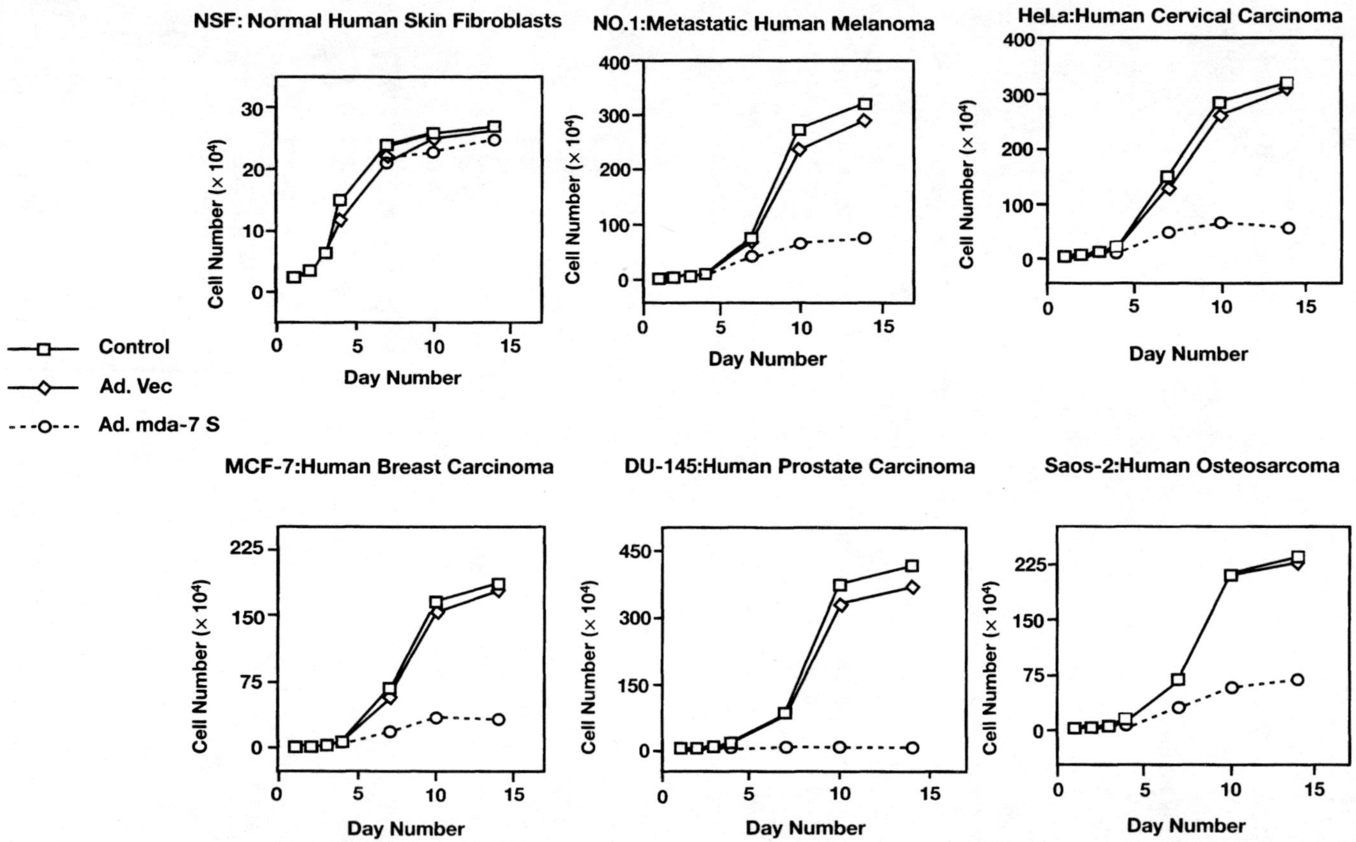

Figure 7. Effect of Ad.*Vec*, Ad.β-gal and Ad.*mda-7* S on the growth of normal human skin fibroblasts and various human cancer cell lines. The indicated cell types were infected with 100 pfu/cell of Ad.*Vec*, Ad.β-gal and Ad.*mda-7* S and cell growth was determined over a 14-day period. Triplicate samples varied by <10%. Similar results (±15%) were obtained in an additional replicate study. Data from Su *et al.* (1998c).

100 pfu/cell of Ad.β-gal demonstrated high levels of infectivity and transgene expression (Su *et al.*, 1998b).

An additional parameter analyzed upon ectopic expression of Ad.*mda*-7 S was cell cycle progression, as assayed by fluorescence activated cell sorter (FACS). FACS data indicated a marked increase in hypodiploid cell populations in cancer cells infected with Ad.*mda*-7 S, an effect that was not apparent after infection with Ad.vec or Ad.β-Gal (Su *et al.*, 1998b). No induction or changes in hypodiploidy were apparent in normal human mammary epithelial, human skin fibroblast or HBL-100 cells infected with Ad.*mda*-7 S. These data strongly suggested that ectopic expression of *mda*-7 in cancer cells may be inducing programmed cell death (apoptosis) (Su *et al.*, 1998b, 1998c). To confirm this possibility, cell samples were analyzed after infection with Ad.*mda*-7 S for, inter-nucleosomal DNA degradation by DNA ladder assay, DNA cleavage by TUNEL assay, and changes in phospholipid localization in the cell membrane (an early indicator of apoptosis) by annexin V assay (Fig. 8 and data not shown) (Su *et al.*, 1998b). All of these assays confirmed that Ad.*mda*-7 S was selectively inducing apoptosis in cancer versus normal cells (Su *et al.*, 1998b, 1998c).

To scrutinize the molecular events involved in *mda*-7 initiated apoptosis some of the better characterized regulators of apoptosis were analyzed, including the Bcl-2 gene family members (review, Adams and Cory, 1998; Green and Reed, 1998; Schendel *et al.*, 1998). Of the various Bcl-2 gene family proteins analyzed by Western blotting of whole cell proteins only one of the pro-apoptotic factors, BAX (Ottavi *et al.*, 1993) was consistently upregulated and its induction coincided with expression of *mda*-7 and development of apoptosis (Fig. 9). No evidence for a consistent increase in BAX protein levels

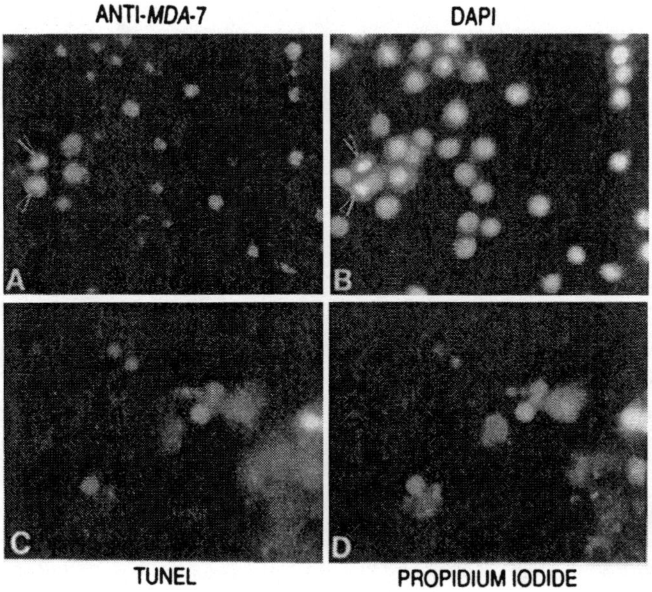

Figure 8. Nuclear localization of *mda*-7 and induction of a positive TUNEL reaction in MCF-7 cells infected with Ad.*mda*-7 S. MCF-7 cells were doubly stained with Anti-*MDA*-7 antibody (A) and DAPI (B) 2 days after infection with 100 pfu/cell of Ad.*mda*-7 S. The position of two mitotic cells stained with Anti-*mda*-7 antibody are shown in (A) and corresponding DAPI counter stain is indicated in (B, arrows label metaphase chromosomes). MCF-7 cells 4 days after infection with 100 pfu/cell of Ad.*mda*-7 S were doubly stained by the TUNEL method (C) and propidium iodide (D). Data from Su *et al.* (1998b).

A

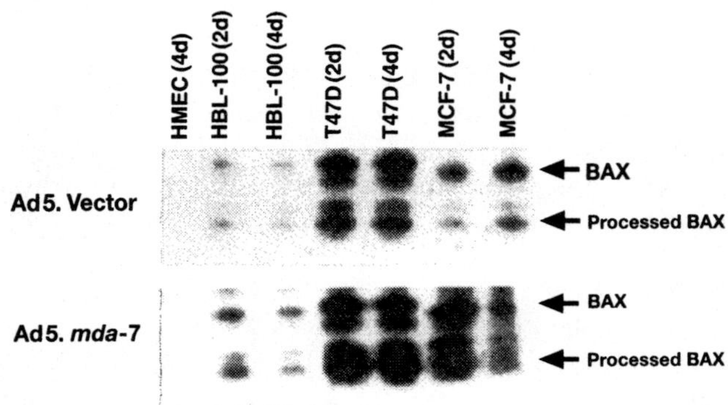

B

Protein	Expti. Condition*	HMEC+	HBL-100	MCF-7	T47D
BCL-2	Ad. *Vec* (4 D)	+	+++	-	-
BCL-2	Ad. *mda*-7 S (4 D)	+	++++	-	+/-
BAK	Ad. *Vec* (4 D)	+	++	+	+++
BAK	Ad. *mda*-7 S (4 D)	++	+++	+	+++

*Cells were infected with 100 pfu/cell of Ad. *Vec* or Ad. *mda*-7 S and cell lysates were prepared after 4 days and analyzed by Western blotting using antibodies to the indicated proteins.

+Relative expression levels indicated by + and - designations, with 4+ maximum and +/- minimum expression. A - designation indicates no detectable protein.

Figure 9. Expression of BCL-2, BAX and BAK proteins in normal breast epithelial and breast carcinoma cells infected with Ad.*mda*-7 S or Ad.*Vec*. (A) Immunoblot analyses of BAX protein in normal mammary epithelial and cancer cells. Equal amounts of whole cell protein from 2 and 4 day cell cultures infected with 100 pfu/cell of Ad.*Vec* or Ad.*mda*-7 S were resolved by SDS-PAGE (4–20%), immunoblotted and probed with Anti-BAX MAb. Note that both intact BAX and processed BAX proteins are visible in the breast cancer cell lines 2 days after Ad.*mda*-7 infection. Low levels of BAX in 4 day MCF-7 is due to proteolytic degradation because the cells in this study were 70% apoptotic by 4 days post-infection with Ad.*mda*-7 S. (B) Tabular compilation of protein levels of the Bcl-2 gene family members, BCL-2 and BAK, 4 days after infection with 100 pfu/cell of Ad.*mda*-7 S or Ad.*Vec* in normal breast epithelial cells and breast carcinoma cell lines. Data from Su *et al.* (1998b).

was found in normal control cells infected with Ad.*mda*-7 S. In contrast to BAX, Bcl-2, Bcl-X_L and Ad E1B, the viral homologue of Bcl-2, are anti-apoptotic and promote cell survival by dimerizing with free BAX (Reed, 1995, 1997; White, 1996; Adams and Cory, 1998; Green and Reed, 1998). A comparison of BAX : Bcl-2 ratios revealed significantly higher ratios in cancer cells infected with Ad.*mda*-7 S in comparison to normal control cells, implicating BAX as a significant contributor to *mda*-7 induced cell death (Su *et al.*, 1998b).

A proposed mechanism for the anti-apoptotic effect of Bcl-2 involves the formation of a stable hetero-dimeric complex between BAX and Bcl-2 or BAX and Ad E1B, thereby nullifying the apoptotic inducing effect of BAX (Yin *et al.*, 1994; Han *et al.*, 1996; Huang *et al.*, 1997). Since high levels of Bcl-2 or Ad E1B can counter the pro-apoptotic signal from BAX and BAX appears to be a major contributor to mda-7 induced cell death (Su *et al.*, 1998b), studies were performed to determine if free BAX is essential for *mda*-7 induced cell death. Bcl-2 or Ad E1B were transiently expressed in two human breast carcinoma cell lines (MCF-7 and T47D) in the presence or absence of *mda*-7 (regulated by a DEX inducible MMTV promoter) to determine effects on cell growth and apoptosis (Su *et al.*, 1998b). In both cases, expression of *mda*-7 failed to induce growth suppression and the cells were resistant to apoptosis when cells were coexpressing Bcl-2 or Ad E1B (Su *et al.*, 1998b). These studies confirm the selective growth inhibitory effect of *mda*-7 toward cancer cells and they provide direct functional evidence for a role of the BAX mediated apoptotic pathway in regulating *mda*-7 action in tumor cells.

3.3. *In vivo* Inhibition of Human Tumor Growth in Nude Mice by Ad.*mda*-7 S

The *in vitro* growth suppression and apoptosis inducing activity of *mda*-7 specifically toward cancer cells prompted us to evaluate this gene for its potential *in vivo* potency toward human tumors grown in nude mice. These types of studies are prerequisites for the ultimate use of *mda*-7 as a gene-based therapeutic for human cancer. For these experiments the Ad.*mda*-7 S was used as a gene delivery system (Su *et al.*, 1998b, 1998c). Two experimental protocols were employed to evaluate the potential antitumor activity of *mda*-7. (i) A "Tumorigenicity Model", where human breast carcinoma cells (MCF-7) which were infected with 100 pfu/cell of Ad.mda-7 S or Ad.vec 96 hr prior to implantation in nude mice (Xu *et al.*, 1996; Su *et al.*, 1998b, 1998c). In the control animal group inoculated with 1×10^6 MCF-7 cells infected with Ad.vec, tumors developed in all groups within 7 days and grew progressively during the course of the experiment (4 weeks). In distinct contrast, the experimental group inoculated with 1×10^6 MCF-7 cells infected with 100 pfu/cell of Ad.*mda*-7 S exhibited statistically ($p < 0.01$) significant suppression of tumor development, as defined by tumor volume and weight (Su *et al.*, 1998b, 1998c) (Fig. 9). (ii) A more stringent test for potential anti-tumor activity is the "*in vivo* Therapy Model" (Su *et al.*, 1998c). Here, human cervical cancer cells (HeLa) were first implanted in nude mice and tumors were allowed to grow until they reached an average volume of 150–200 mm^3. The treatment regimen consisted of 3 weekly intratumoral injections of Ad.*mda*-7 S or Ad.vec (1×10^8 pfu/injections) over 4 sites for 3 weeks followed by an additional one week observation. The experiment was terminated due to the extreme size of the tumors in the control group (Ad.vec). However, statistically significant ($p < 0.05$) inhibition of tumor growth was observed in the experimental group (Ad.*mda*-7 S) compared to the control group (Fig. 10) (Su *et al.*, 1998c). These *in vivo* results provide direct experimental evidence suggesting that *mda*-7 may have significant potential as a cancer gene therapeutic.

4. PROPOSED MECHANISM OF APOPTOSIS INDUCTION IN CANCER CELLS BY Ad.*mda*-7 S

Experimental evidence suggest a relationship between the induction of elevated levels of BAX protein and an increase in the BAX to BCL-2 protein ratio as major factors in determining *mda*-7 induction of apoptosis in human cancer cells (Su *et al.*, 1998b, 1998c). In this context, any model of *mda*-7 action must take into account our current status of understanding of the mechanism of action of the BAX and BCL-2 proteins. A hypothetical model of *mda*-7 action in cancer cells, which incorporates a conceptualization of BAX protein action, is shown in Fig. 11. Cotransfection of *mda*-7 with an expression vector containing the Bax promoter linked to a luciferase reporter gene does not

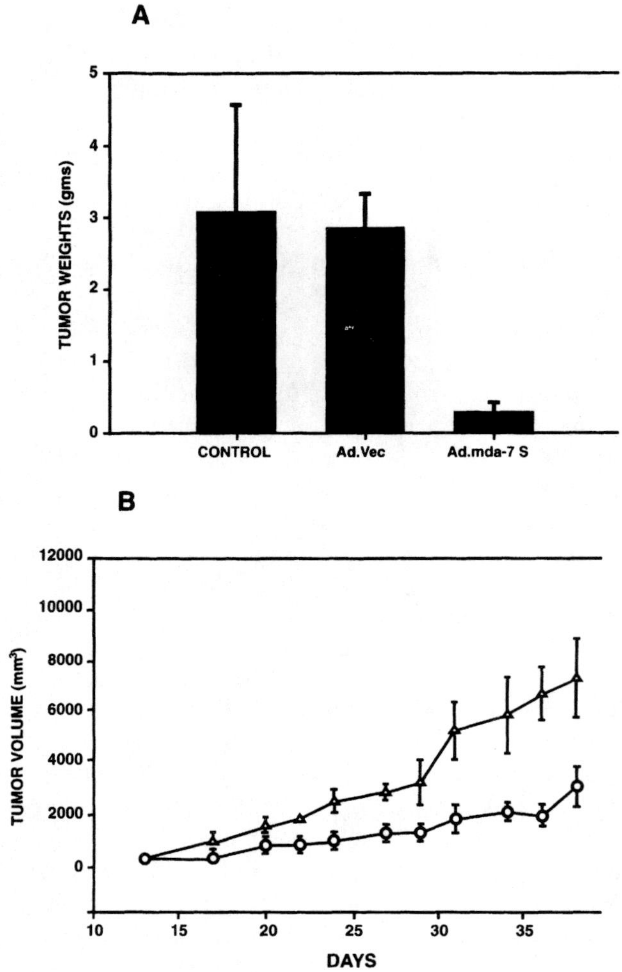

Figure 10. Effect of *mda*-7 on human tumor formation and progression in nude mice. MCF-7 tumorigenicity studies (A). MCF-7 cells were uninfected (CONTROL) or infected with 100 pfu/cell of Ad.*Vec* or Ad.*mda*-7 S and incubated for 4 days at 37 °C. Cells were removed with trypsin, mixed with Matrigel (1 : 1) and injected into nude mice (10^6 cells/animal). Results are shown as tumor weight in grams ± the standard deviation. Statistical significance was determined with a Student's t test using the computer program Sigma Stat (Jandel, San Rafael, CA) and a p value of <0.05 is indicated by an asterisk. Two additional tumorigenicity studies have been performed with qualitatively similar results. HeLa therapy study (B). HeLa cells were mixed with Matrigel (1 : 1) and injected into nude mice. When tumors reached a volume of 150–200 mm3 the animals were divided into groups and treated with either 10^8 pfu of *Ad.Vec* (Δ) or *Ad.mda*-7 S (□) 3X a week for three weeks. After an additional week without treatment, animals were sacrificed and their tumors were removed, snap frozen, and weighed. Results are shown as tumor progression (tumor volume over time) ± standard deviation. Statistical significance was determined by ANOVA. A second therapy experimental model using the same protocol with HeLa xenografts has been completed yielding identical results. Data from Su *et al.* (1998c).

result in promoter activation suggesting that the effect of *mda*-7 on Bax expression is not mediated by a direct activation of the Bax promoter (Madireddi and Fisher, unpublished data). As depicted in Fig. 11, Ad.*mda*-7 S may induce BAX protein in cancer cells by an indirect mechanism that involves additional molecules (?) that remain to be identified. The ability of elevated BAX protein to induce apoptosis in breast cancer cells can be inhibited by the simultaneous overexpression of BCL-2 or Ad E1B proteins (Su *et al.*,

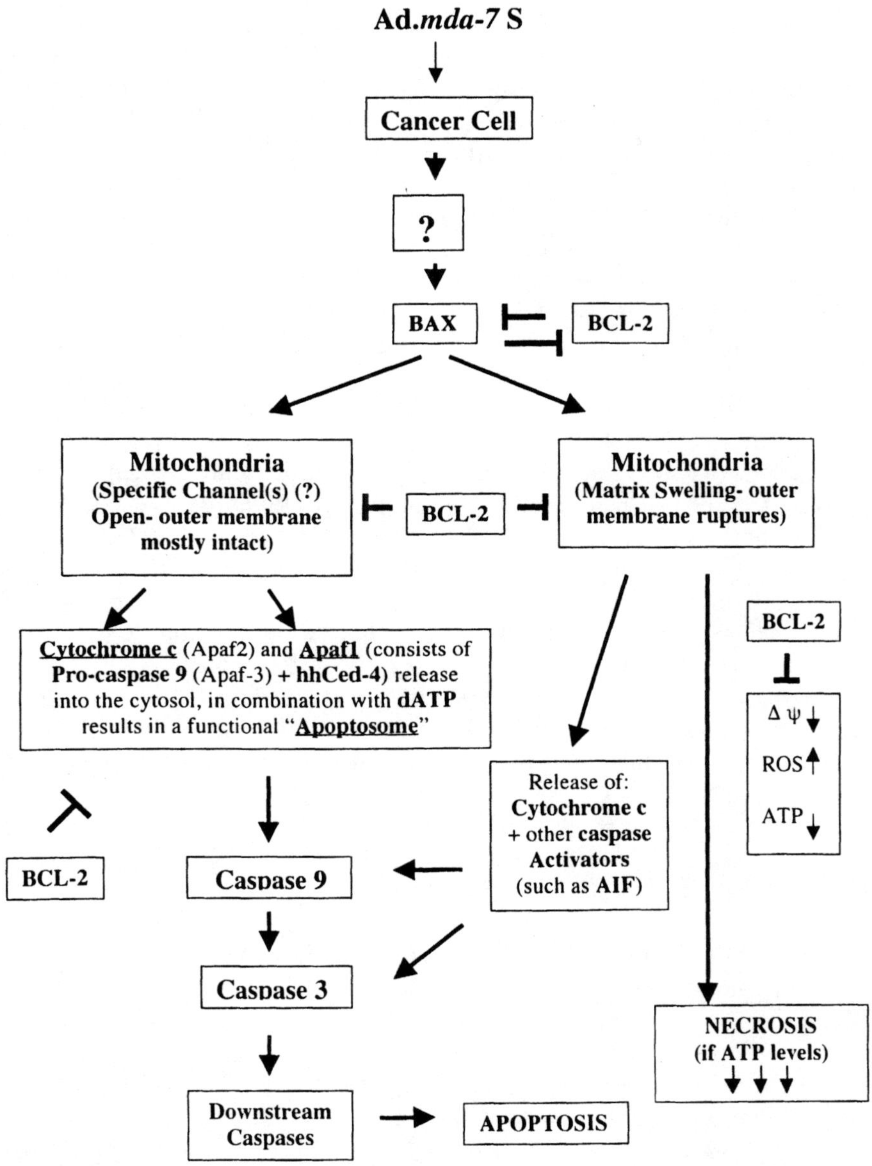

Figure 11. Hypothetical model for *mda*-7 induction of BAX resulting in apoptosis and necrosis, specifically in cancer cells.

1998b), supporting the importance of the ratio of pro- to anti-apoptotic molecules in defining cell fate. If overexpression of BAX protein is not nullified in the cancer cell two possible scenarios can occur, both of which initially involve the mitochondria and ultimately result in cell death (Green and Reed, 1998; Schendel *et al.*, 1998; Wilson, 1998). In one model, BAX protein is proposed to induce channel-formation in the mitochondria, a process inhibited by BCL-2 protein, resulting in the release of cytochrome c (cyt c) (referred to as Apaf2) and Apaf1 (which consists of two active components, procaspase 9 (referred to as Apaf3) and hhCed-4) (Li *et al.*, 1997b; Zuo *et al.*, 1997; Green and Reed, 1998; Schendel *et al.*, 1998; Wilson, 1998). In combination with ATP, cyt c, pro-caspase 9 and hhCed-4 a tetramolecular complex (a functional "Apoptosome") (Li *et al.*, 1997a) that activates caspase 9 which then mediates the activation of caspase 3 and downstream caspases culminating in apoptosis (Li *et al.*, 1997a; Wilson, 1998). A limitation of this model is the fact that induction of ion channels by BAX protein has only been observed in synthetic membranes (Antonsson *et al.*, 1997). A second model proposes that BAX protein, either by overexpression or forced dimerization (Gross *et al.*, 1998), may directly modify mitochondrial integrity thereby mediating caspase-independent death (Adams and Cory, 1998; Green and Reed, 1998; Schendel *et al.*, 1998; Wilson, 1998). In this scheme, the mitochondria are damaged by the action of BAX protein resulting in matrix swelling and rupturing of the outer membrane (Adams and Cory, 1998; Green and Reed, 1998; Schendel *et al.*, 1998; Wilson, 1998). These changes in mitochondrial architecture may result from induction of mitochondrial permeability transition (PT) and loss of mitochondrial transmembrane potential ($\Delta\Phi_M$) or may involve changes in the phosphorylation of BCL-2 (located in mitochrondrial membranes) that is required to protect cells from apoptosis (Marchetti *et al.*, 1996; Ito *et al.*, 1997; Kroemer, 1997; Adams and Cory, 1998; Green and Reed, 1998; Schendel *et al.*, 1998; Wilson, 1998). This model predicts disruption of mitochondrial organization resulting in cell killing (necrosis) by alterations in electron transport, oxidative phosphorylation and ATP production and changes in cellular reduction-oxidation (redox) potential (Green and Reed, 1998; Schendel *et al.*, 1998; Wilson, 1998). BAX protein-induced PT alterations may also mediate release of cyt c and other caspase activators (such as AIF) contributing to cell death (Adams and Cory, 1998; Green and Reed, 1998; Schendel *et al.*, 1998; Wilson, 1998). In this context, cell death induced by this latter pathway may not be completely caspase-independent.

5. CONCLUSIONS

By appropriate chemical intervention human melanoma cells can be induced to revert to a more normal cellular state and terminally differentiate, with a concomitant irreversible loss in proliferative ability (Fisher *et al.*, 1985; Jiang and Fisher, 1993; Jiang *et al.*, 1993a, 1994). This experimental finding provides additional support for a potentially less toxic form of cancer therapy, differentiation therapy (Sachs, 1978; Fisher *et al.*, 1985; Jimenez and Yunis, 1987; Jiang *et al.*, 1993a, 1994; Waxman, 1995). The combination of this approach with the molecular strategy of subtraction hybridization, the DISH protocol, is permitting an identification and cloning of the complete repertoire of cDNAs involved in the processes of growth control and differentiation in human cancer cells (Jiang and Fisher, 1993; Jiang *et al.*, 1994; Huang *et al.*, 1998). This information will result in a molecular definition of cancer cell growth and terminal differentiation, including the identification of pertinent genes associated with and potentially mediating these processes.

The legitimacy of the DISH approach as a means of cloning relevant genes involved in growth control and differentiation is supported by the identification and isolation of *mda*-7 (Jiang *et al.*, 1994, 1995, 1996; Su *et al.*, 1998b, 1998c). *Mda*-7 is upregulated in human melanoma cells as a consequence of induction of terminal differentiation and ectopic expression of *mda*-7 in human melanoma cells evokes growth suppression. Moreover, the selective ability of *mda*-7 to induce programmed cell death in cancer cells of diverse origins with multiple genetic alterations but not in normal cells (by a BAX-mediated pathway) places *mda*-7 in a new class of tumor growth suppressor genes. The direct demonstration of *in vivo* anti-tumor activity of *mda*-7, when used in a replication incompetent adenovirus (Ad.*mda*-7 S), toward injected human tumor cells and human tumor xenografts in nude mice suggests that this cancer growth suppressing gene may provide benefit for the therapy of human cancer. Future studies will determine if this promise is realized.

6. FUTURE PERSPECTIVES

Experiments are currently in progress to define *mda*-7 genomic structure, to elucidate the mechanism by which *mda*-7 selectively induces growth arrest and apoptosis in cancer cells and to determine the role of *mda*-7 in normal cellular physiology and development. Analysis of *mda*-7 genomic structure will provide insights into the possible role of mutations in this gene in cancer development and progression. Identification and characterization of the promoter region of *mda*-7 will shed light on potentially important transcriptional processes and assist in the identification and elucidation of the roles of positive and negative transcriptional regulatory elements and interacting proteins in controlling *mda*-7 gene expression. Based on the nucleotide sequence of *mda*-7, the role of message stability in regulating *mda*-7 gene expression is also being evaluated. Studies are underway to define approaches for enhancing the efficacy of Ad.*mda*-7 S for inducing cancer growth suppression and apoptosis. These include combining Ad.*mda*-7 S with physical and chemical therapeutic agents and using Ad.*mda*-7 S with adenoviruses expressing the cell cycle inhibitory gene p21 (WAF1/cip1*mda*-6) (Jiang and Fisher, 1993; El Deiry *et al.*, 1993; Harper *et al.*, 1993; Noda *et al.*, 1994), or other growth inhibitory and/or apoptosis inducing genes such as p53, p16 and p27. Because of the perceived importance of BAX in *mda*-7-induced apoptosis in transformed cells, experiments are planned to determine if absence of BAX in transformed mouse embryo cells, i.e., BAX −/− and BAX +/− cultures, renders these cells resistant to the biological effects of *mda*-7. Since the *MDA*-7 protein has no nuclear localization sequence, yet it appears to localize in the nucleus during mitosis and specifically complex with chromatin during prophase, metaphase, anaphase and telophase, studies are focusing on identifying and cloning genes encoding interacting proteins, such as the HMC, that may serve as chaperones in translocating *MDA*-7 into the nucleus. Experiments are also planned to study the *MDA*-7 protein in detail, including posttranslational modification, subcellular localization, targeting and protein stability. To define the role of *mda*-7 in normal mouse development and the processes of aging and cancer, transgenic mice and homozygous deleted knockout mice will be generated and studied. On the basis of the results obtained to date with *mda*-7, the studies briefly outlined above are well justified and they will provide important insights into a novel gene with potential relevance to cancer development and progression. Moreover, the direct cancer growth suppressing and apoptosis inducing properties of *mda*-7 suggest that this gene may prove amenable for developing new and improved cancer therapeutics.

ACKNOWLEDGMENTS

The contributions of our colleagues in specific studies described in this review are acknowledged and appreciated, including Drs. Hongping Jiang, Shinichi Kitada, Jian

Lin and John C. Reed and Mr. Jiao Jiao Lin. The research described in this review was supported in part by National Institutes of Health grants CA35675 and GM 31452, an award from the Army Department of Defense Initiative on Breast Cancer (DAMD17-98-1-8053), the Samuel Waxman Cancer Foundation and the Chernow Endowment. PBF is the Chernow Research Scientist in the Departments of Neurosurgery, Pathology and Urology.

REFERENCES

Adams, J.M., and Cory, S., 1998, The bcl-2 protein family: arbiters of cell survival, *Science* **281**:1322–1326.

Anderson, S.C., Johnson, D.E., Harris, M.P., Engler, H., Hancock, W., Huang, W.M., Wills, K.N., Gregory, R.J., Sutjipto, S., Wen, S.F., Lofgren, S., Shepard, H.M., and Maneval, D.C., 1998, p53 gene therapy in a rat model of hepatocellular carcinoma: intra-arterial delivery of a recombinant adenovirus, *Clin. Cancer Res.* **4**:1649–1659.

Antonsson, B., Conti, F., Ciavatta, A., Montessuit, S., Lewis, S., Martinou, I., Bernasconi, L., Bernard, A., Mermod, J.J., Mazzei, G., Maundrell, K., Gambale, F., Sadoul, R., and Martinou, J.C., 1997, Inhibition of Bax channel-forming activity by Bcl-2, *Science* **277**:370–372.

Bischoff, J.R., Kirn, D.H., Williams, A., Heise, C., Horn, S., Muna, M., Ng, L., Nye, J.A., Sampson-Johannes, A., Fattaey, A., and McCormick, F., 1996, An adenovirus mutant that replicates selectively in p53-deficient human tumor cells. *Science* **274**:373–376.

Bishop, J.M., 1991, Molecular themes in oncogenesis, *Cell* **64**:235–248.

Borrelli, E., Heyman, R., His, M., and Evans, R.M., 1988, Targeting of an inducible toxic phenotype in animal cells. *Proc. Natl. Acad. Sci. USA* **85**:7572–7576.

Chellappan, S.P., Giordano, A., and Fisher, P.B., 1998, Role of cyclin-dependent kinases and their inhibitors in cellular differentiation and development, *Curr. Topics Microbiol. Immunol.* **227**:57–103.

Chen, Q.R., and Mixson, J.A., 1998, Systemic gene therapy with p53 inhibits breast cancer: recent advances and therapeutic implications, *Frontiers Bioscience* **3**:D997–D1004.

Clark, W., 1991, Tumor progression and the nature of cancer, *Br. J. Cancer* **64**:631–644.

Crystal, R.G., 1995, Transfer of genes to humans: early lessons and obstacles to success, *Science* **270**:404–410.

Culver, K.W., Ran, Z., Wallbridge, S., Ishii, H., Oldfield, E.H., and Blaese, R.M., 1992. In vivo gene transfer with retroviral vector-producer cells for treatment of experimental brain tumors, *Science* **256**:1550–1552.

Dematteo, R.P., McClane, S.J., Fisher, K., Yeh, H., Chu, G., Burke, C., and Raper, S.E., 1997, Engineering tissue-specific expression of a recombinant adenovirus: selective transgene transcription in the pancreas using the amylase gene, *J. Surgical Res.* **72**:155–161.

DeYoung, K.L., Ray, M.E., Su, Y.A., Anzick, S.L., Johnstone, R.W., Trapani, J.A., Meltzer, P.S., and Trent, J.M., 1997, Cloning a novel member of the human interferon-inducible gene family associated with control of tumorigenicity in a model of human melanoma, *Oncogene* **15**:453–457.

Du, Y.X., Zhou, B.S., and Wu, J.M., 1998, Lifestyle factors and human lung cancer: an overview of recent advances (review), *Intl. J. Oncology* **13**:471–479.

El-Deiry, W.S., Tokino, T., Velculescu, V.E., Levy, D.B., Parsons, R., Trent, J.M., Lin, D., Mercer, W.E., Kinzler, K.W., and Vogelstein, B., 1993, WAF1, a potential mediator of p53 tumor suppression, *Cell* **75**:817–825.

Fisher, P.B., 1984, Enhancement of viral transformation and expression of the transformed phenotype by tumor promoters, in: *Tumor Promotion and Carcinogenesis In Vitro, Mechanisms of Tumor Promotion*, (T.J. Slaga, ed.), pp. 57–123, CRC Press, Boca Raton, Florida.

Fisher, P.B., Prignoli, D.R., Hermo, H., Jr., Weinstein, I.B., and Pestka, S., 1985, Effects of combined treatment with interferon and mezerein on melanogenesis and growth in human melanoma cells, *J. Interferon Res.* **5**:11–22.

Gervais, J.L., Seth, P., and Zhang, H., 1998, Cleavage of CDK inhibitor p21(Cip1/Waf1) by caspases is an early event during DNA damage-induced apoptosis, *J. Biol. Chem.* **273**:19207–19212.

Gotoh, A., Kamidono, S., and Chung, L.W., 1997, Clinical application for gene therapy in prostate cancer, *Acta Urologica Japonica* **43**:829–833.

Green, D.R., and Reed, J.C., 1998, Mitochondria and apoptosis, *Science* **281**:1309–1312.

Gross, A., Jockel, J., Wei, M.C., and Korsmeyer, S.J., 1998, Enforced dimerization of BAX results in its translocation, mitochondrial dysfunction and apoptosis, *EMBO J.* **17**:3878–3885.

Guarini, L., Temponi, M., Edwalds, G.M., Vita, J.R., Fisher, P.B., and Ferrone, S., 1989, In vitro differentiation and antigenic changes in human melanoma cell lines, *Cancer Immunol. Immunother.* **30**:262–268.

Guarini, L., Graham, G.M., Jiang, H., Ferrone, S., Zucker, S., and Fisher, P.B., 1992, Modulation of the antigenic phenotype of human melanoma cells by differentiation-inducing and growth-suppressing agents, *Pigment Cell Res. Suppl.* **2**:123–131.

Han, Z., Malik, N., Carter, T., Reeves, W.H., Wyche, J.H., and Hendrickson, E.A., 1996, DNA-dependent protein kinase is a target for a CPP32-like apoptotic protease, *J. Biol. Chem.* **271**:25035–25040.

Harper, J.W., Adami, G.R., Wei, N., Keyomarsi, K., and Elledge, S.J., 1993, The p21 Cdk-interacting protein Cip1 is a potent inhibitor of G1 cyclin-dependent kinases, *Cell* **75**:805–816.

Hartwell, L.H., and Kastan, M.B., 1994, Cell cycle control of cancer, *Science* **266**:1821–1828.

Heise, C., Sampson-Johannes, A., Williams, A., McCormick, F., von Hoff, D.D., and Kirn, D.H., 1997, ONYX-015, an E1B gene-attenuated adenovirus, causes tumor-specific cytolysis and antitumoral efficacy that can be augmented by standard chemotherapeutic agents. *Nature Med.* **3**:639–645.

Herlyn, M., 1990, Human melanoma: development and progression, *Cancer Metastasis Rev.* **9**:101–112.

Huang, F., Adelman, J., Jiang, H., Goldstein, N.I., and Fisher, P.B., 1998, Differentiation induction subtraction hybridization (DISH): an approach for defining the spectrum of genes differentially expressed during induction of irreversible growth arrest and terminal differentiation in human melanoma cells, *Oncogene* in press.

Huang, D.C., Cory, S., and Strasser, A., 1997, Bcl-2, Bcl-XL and adenovirus protein E1B19kD are functionally equivalent in their ability to inhibit cell death, *Oncogene* **14**:405–414.

Hwang, R.F., Gordon, E.M., Anderson, W.F., and Parekh, D., 1998, Gene therapy for primary and metastatic pancreatic cancer with intraperitoneal retroviral vector bearing the wild-type p53 gene, *Surgery* **124**:143–150.

Ito, T., Deng, X.M., Carr, B., and May, W.S., 1997, Bcl-2 phosphorylation required for anti-apoptosis function, *J. Biol. Chem.* **272**:11671–11673.

Jiang, H., and Fisher, P.B., 1993, Use of a sensitive and efficient subtraction hybridization protocol for the identification of genes differentially regulated during the induction of differentiation in human melanoma cells, *Mol. Cell Different.* **2**:221–239.

Jiang, H., Su, Z.-Z., Boyd, J., and Fisher, P.B., 1993a, Gene expression changes associated with reversible growth suppression and the induction of terminal differentiation in human melanoma cells, *Mol. Cell Different.* **1**:41–66.

Jiang, H., Waxman, S., and Fisher, P.B., 1993b, Regulation of c-*fos*, c-*jun* and *jun*-B expression in human melanoma cells induced to terminally differentiate, *Mol. Cell Different.* **1**:197–214.

Jiang, H., Lin, J., and Fisher, P.B., 1994, A molecular definition of terminal differentiation in human melanoma cells, *Mol. Cell Different.* **2**:221–239.

Jiang, H., Lin, J.J., Su, Z.-Z., Goldstein, N.I., and Fisher, P.B., 1995a, Subtraction hybridization identifies a novel melanoma differentiation associated gene, *mda*-7, modulated during human melanoma differentiation, growth and progression, *Oncogene* **11**:2477–2486.

Jiang, H., Lin, J., Su, Z.-Z., Herlyn, M., Kerbel, R.S., Weissman, B.E., Welch, D.R., and Fisher, P.B., 1995b, The melanoma differentiation-associated gene *mda*-6, which encodes the cyclin-dependent kinase inhibitor p21, is differentially expressed during growth, differentiation and progression in human melanoma cells, *Oncogene* **10**:1855–1864.

Jiang, H., Lin, J., Young, S.-M., Goldstein, N.I., Waxman, S., Davila, V., Chellappan, S.P., and Fisher, P.B., 1995c, Cell cycle gene expression and E2F transcription factor complexes in human melanoma cells induced to terminally differentiate, *Oncogene* **11**:1179–1189.

Jiang H., Su, Z.-Z., Lin, J.J., Goldstein, N.I., Young, C.S.H., and Fisher, P.B., 1996, The melanoma differentiation associated gene *mda*-7 suppresses cancer cell growth, *Proc. Natl. Acad. Sci. USA* **93**:9160–9165.

Jimenez, J.J., and Yunis, A.A., 1987, Tumor cell rejection through terminal cell differentiation, *Science* **238**:1278–1280.

Kang, D.-C., Motwani, M., and Fisher, P.B., 1998. Role of the transcription factor AP1 in melanoma differentiation (review), *Intl. J. Oncology*, in press.

Kerbel, R.S., 1990, Growth dominance and the metastatic cancer cell: cellular and molecular aspects, *Adv. Cancer Res.* **55**:87–132.

Kieff, E., 1998, Current perspectives on the molecular pathogenesis of virus-induced cancers in human immunodeficiency virus infection and acquired immunodeficiency syndrome. [Review], *J. Nat. Cancer Inst. Monographs* **23**:7–14.

Kobayashi, H., Man, S., MacDougall, J.R., Graham, C.H., Lu, C., and Kerbel, R.S., 1994, Variant sublines of early-stage human melanomas selected for tumorigenicity in nude mice express a multicytokine-resistant phenotype, *Amer. J. Pathol.* **144**:776–786.

Kong, H.L., and Crystal, R.G., 1998, Gene therapy strategies for tumor antiangiogenesis, *J. Natl. Cancer Inst.* **90**:270–271.

Knudson, A.G., 1993, Antioncogenes and human cancer, *Proc. Natl. Acad. Sci. USA* **90**:10914–10921.

Kroemer, G., 1997, The proto-oncogene Bcl-2 and its role in regulating apoptosis, *Nature Med.* **6**:614–620.

Levine, A.J., 1993, The tumor suppressor genes, *Annu. Rev. Biochem.* **62**:623–651.

Li, P., Deepak, N., Budihardjo, I., Srinivasula, S.M., Ahmad, M., Alnemri, E.S., and Wang X., 1997a, Cytochrome c and dATP-dependent formation of Apaf-1/caspase-9 complex initiates an apoptotic cascade, *Cell* **91**:479–489.

Li, F., Srinivasan, A., Wang, Y., Armstrong, R.C., Tomaselli, K.J., and Fritz, L.C., 1997b, Cell-specific induction of apoptosis by microinjection of cytochrome c-Bcl-K_L has activity independent of cytochrome c release, *J. Biol. Chem.* **272**:30299–30305.

Lin, J.J., Jiang, H., and Fisher, P.B., 1996, Characterization of a novel melanoma differentiation associated gene, *mda*-9, that is down-regulated during terminal cell differentiation, *Mol. Cell Different.* **4**:317–333.

Lin, J.J., Jiang, H., and Fisher, P.B., 1998, Melanoma differentiation associated gene-9, mda-9, is a human gamma interferon responsive gene, *Gene* **207**:105–110.

Lu, C., and Kerbel, R.S., 1994, Cytokines, growth factors and the loss of negative growth controls in the progression of human cutaneous malignant melanoma, *Curr. Opin. Oncol.* **6**:213–234.

Marchetti, P., Castedo, M., Susin, S.A., Zamzami, N., Hirsch, T., Macho, A., Haeffner, A., Hirsch, F., Geuskens, M., and Kroemer, G., 1996, Mitochondrial permeability transition is the central coordinating event of apoptosis. *J. Exptl. Med.* **184**:1153–1160.

Melber, K., Zhu, G., and Diamond, L., 1989, SV40-transformed human melanocyte sensitivity to growth inhibition by the phorbol ester 12-0-tetradecanoylphorbol-13-acetate, *Cancer Res.* **49**:3650–3655.

Moolten, F.L., and Wells, J.M., 1990, Curability of tumors bearing herpes thymidine kinase genes transferred by retroviral vectors. *J. Natl. Cancer Inst.* **82**:297–300.

Mosmann, T.R., 1994, Properties and functions of interleukin-10. [Review], *Advan. Immunol.* **56**:1–26.

Noda, A., Ning, Y., Venable, S.F., Pereira-Smith, O.M., and Smith, J.R., 1994, Cloning of senescent cell-derived inhibitors of DNA synthesis using an expression screen, *Exptl. Cell Res.* **211**:90–98.

O'Neill, M.J., and Artzt, K., 1995, Identification of a germ-cell-specific transcriptional repressor in the promoter of Tctex-1. *Development* **12**:561–568.

Ottavi, Z.N., Miliman, C.L., and Korsmeyer, S.J., 1993, Bcl-2 heterodimerizes *in vivo* with a conserved homolog, Bax, that accelerates programmed cell death, *Cell* **74**:609–619.

Prabhu, N.S., Somasundaram, K., Satyamoorthy, K., Herlyn, M., and El-Deiry, W.S., 1998, p73beta, unlike p53, suppresses growth and induces apoptosis of human papillomavirus E6-expressing cancer cells, *Intl. J. Oncology* **13**:5–9.

Reed, J.C., 1995, Regulation of apoptosis by bcl-2 family proteins and its role in cancer and chemoresistance, *Curr. Opin. Oncol.* **7**:541–546.

Reed, J.C., 1997, Bcl-2 family proteins: regulators of apoptosis and chemoresistance in hematologic malignancies, *Seminars Hematol.* **34(4 Suppl 5)**:9–19.

Roberts, T., Vrana, J.A., Grant S., and Fisher, P.B., 1998, Differentiation therapy of cancer: basic science and clinical applications, *Pharmacol. Therapeut.* in press.

Rocco, J.W., Li, D., Liggett, W.H., Jr., Duan, L., Saunders, J.K., Jr., Sidransky, D., and O'Malley, B.W., Jr., 1998, p16INK4A adenovirus-mediated gene therapy for human head and neck squamous cell cancer, *Clinical Cancer Res.* **4**:1697–1704.

Rodriguez, R., Schuur, E.R., Lim, H.Y., Henderson, G.A., Simons, J.W., and Henderson, D.R., 1997, Prostate attenuated replication competent adenovirus (ARCA) CN706: a selective cytotoxic for prostate-specific antigen-positive prostate cancer cells. *Cancer Res.* **57**:2559–2563.

Rosenberg, S.A., 1992, Gene therapy for cancer. *J. Amer. Med. Assoc.* **268**:2416–2419.

Roth, J.A., and Cristiano, R.J., 1997, Gene therapy for cancer: what have we done and where are we going?, *J. Natl. Cancer Inst.* **89**:21–39.

Roth, J.A., Swisher, S.G., Merritt, J.A., Lawrence, D.D., Kemp, B.L., Carrasco, C.H., El-Naggar, A.K., Fossella, F.V., Glisson, B.S., Hong, W.K., Khurl, F.R., Kurie, J.M., Nesbitt, J.C., Pisters, K., Putnam, J.B., Schrump, D.S., Shin, D.M., and Walsh, G.L., 1998, Gene therapy for non-small cell lung cancer: a preliminary report of a phase I trial of adenoviral p53 gene replacement, *Sem. in Oncology* **25**:33–37.

Roux, A.F., Rommens, J., McDowell, C., Anson-Cartwright, L., Bell, S., Schappert, K., Fishman, G.A., and Musarella, M., 1994, Identification of a gene from Xp21 with similarity to the tctex-1 gene of the murine t complex. *Hum. Mol. Genet.* **3**:257–263.

Sachs L., 1978. Control of normal cell differentiation and the phenotypic reversion of malignancy in myeloid cells, *Nature* **274**:535–539.

Sandig, V., Brand, K., Herwig, S., Lukas, J., Bartek, J., and Strauss, M., 1997, Adenovirally transferred p16INK4/CDKN2 and p53 genes cooperate to induce apoptotic tumor cell death, *Nature Med.* **3**:313–319.

Schendel, S.L., Montal, M., and Reed, J.C., 1998, Bcl-2 family proteins as ion-channels. *Cell Growth & Different.* **5**:372–380.

Smythe, W.R., Hwang, H.C., Amin, K.M., Eck, S.L., Davidson, B.L., Wilson, J.M., Kaiser, L.R., and Albelda, S.M., 1994, Use of recombinant adenovirus to transfer the herpes simplex virus thymidine kinase (HSV-TK) gene to thoracic neoplasms: an effective *in vitro* drug sensitization system. *Cancer Res.* **54**:2055–2059.

Soos, T.J., Park, M., Kiyokawa, H., and Koff, A., 1998, Regulation of the cell cycle by CDK inhibitors. Results and problems. [Review], *Cell Differ.* **22**:111–131.

Smith, A.E., 1995, Viral vectors in gene therapy, *Annu. Rev. Microbiol.* **49**:807–838.

Su, Z.-Z., Shi, Y., and Fisher, P.B., 1997, Subtraction hybridization identifies a transformation progression-associated gene PEG-3 with sequence homology to a growth arrest and DNA damage inducible gene, *Proc. Natl. Acad. Sci. USA* **94**:9125–9130.

Su, Z.-Z., Goldstein, N.I., and Fisher, P.B., 1998a, Antisense inhibition of the PTI-1 oncogene reverses cancer phenotypes, *Proc. Natl. Acad. Sci. USA* **95**:1764–1769.

Su, Z.-Z., Madireddi, M.T., Lin, J.J., Young, C.S.H., Kitada, S., Reed, J.C., Goldstein, N.I., and Fisher, P.B., 1998b, The cancer growth suppressor gene mda-7 selectively induces apoptosis in human cancer cells and inhibits tumor growth in nude mice, *Proc. Natl. Acad. Sci. USA* in press.

Su, Z.-Z., Madireddi, M.T., Young, C.S.H., Jiang, H., Goldstein, N.I., and Fisher, P.B., 1998c, Melanoma differentiation associated gene-7 (*Mda*-7): a ubiquitous cancer growth suppressor gene, in *Cancer Gene Therapy*, (N. Habib, ed.), IBC Series, in press.

Takeda, S., Nakao, A., Miyoshi, K., and Takagi, H., 1998, Gene therapy for pancreatic cancer, *Semi. Surg. Oncol.* **15**:57–61.

Vogelstein, B., and Kinzler, K.W., 1991, The multistep nature of cancer, *Trends Genet.* **9**:138–141.

Waxman, S. (ed.), 1995, *Differentiation Therapy*, Vol. 10, pp. 1–531, Ares Serono Symposia, Rome, Italy.

White, E., 1996, Life, death, and the pursuit of apoptosis, Genes & Dev. **10**:1–15.

Wilson, M.R., 1998, Apoptosis: unmasking the executioner, *Cell Death & Different.* **5**:646–652.

Xu, H.J., Zhou, Y., Seigne, J., Perng, G.S., Mixon, M., Zhang, C., Li, J., Benedict, W.F., and Hu, S.X., 1996, Enhanced tumor suppressor gene therapy via replication-deficient adenovirus vectors expressing an N-terminal truncated retinoblastoma protein, *Cancer Res.* **56**:2245–2249.

Yin, X.M., Ottavi, Z.N., and Korsmeyer, S.J., 1994, BH1 and BH2 domains of Bcl-2 are required for inhibition of apoptosis and heterodimerization with Bax, *Nature* **26**:272–273.

Zhang, Y., Mukhopadhyay, T., Donehower, L.A., Georges, R.N., and Roth, J.A., 1993, Retroviral vector-mediated transduction of K-ras antisense RNA into human lung cancer cells inhibits expression of the malignant phenotype, *Human Gene Therapy* **4**:451–460.

Zuo, H., Henzel, W.J., Liu, X.S., Lutschg, A., and Wang, X.D., 1997, Apaf-1, a human protein homologous to *C. elegans* Ced-4, participates in cytochrome c-dependent activation of caspase-3, *Cell* **91**:405–413.

VII

ANTISENSE AND RIBOZYMES

23

ANTISENSE IGF AND ANTISENSE IGF-IR THERAPY OF MALIGNANCY

Hao Wang[1,2], Yanjun Liu[1,2], Lixin Wei[1], and Yajun Guo[1,2]

[1]Shanghai International Joint Cancer Institute
The Second Military Medical University
Shanghai, P. R. China
[2]Sidney Kimmel Cancer Center
San Diego, USA

Antisense strategy includes applying of RNA molecules or oligodeoxynulceotides (ODNs) binding to a complementary sequence (target sequence) of the gene of interest. Binding of the antisense RNA or ODNs disrupt the function of the target sequence either by blocking its function (e.g., by binding to a splice junction), promoting its degradation, or by altering the structure of the target sequence. The inhibition of gene expression by introducing a source of nucleotide sequences complementary to a given endogenous mRNA species has been successfully applied to a growing number of genes in cultured cells. Transfection of antisense to the myc gene (Negroni *et al.*, 1991; Whitesell *et al.*, 1991) or to basic fibroblast growth factor gene (Morrison, 1991) completely inhibits proliferation and growth rate of human neuroblastoma and transformed human astrocytes in vitro. Although it has been showed that antibody blockage of growth factors, for example, IGF-1, IGF-2 and TGF etc., is also effective in inhibiting cell growth, this approach has limitation and is unable to block intracellular autocrine loops of these growth factors (Heldin and Westermake, 1989). Recently, other investigators and we have demonstrated a tumorigenic role of IGF-1 and IGF-II. The function of IGF-1 and IGF-II is apparently distinct from its role in certain cellular systems as growth factor, and instead seems to relate to their capacities to regulate cell's differentiated state (immunogenicity). Hence, these findings highlight the fact that tumorigenic phenotypes may arise through unexpected molecular mechanisms relating to IGF-1 and IGF-II. This, in turn, raises the prospect of novel anti-cancer therapeutic approaches by regulating a tumorigenic phenotype and by directly inhibiting growth and development of tumor cells.

1. IGF-I, -II AND THEIR RECEPTOR IN MALIGNANCY

Insulin like growth factor-I (IGF-I) and IGF-II are critical regulators of cell proliferation. The growth-promoting function of both ligands is mediated by the type I IGF

Cancer Gene Therapy: Past Achievements and Future Challenges, edited by Habib
Kluwer Academic/Plenum Publishers, New York, 2000.

receptor (IGF-IR). There is no correlation between expression IGF-I, IGF-II and their receptors on normal tissue and tumor cells.

The IGF-I receptor belongs to the tyrosine kinase growth factor receptor family, sharing 70% homology with the insulin receptor (Ullrich *et al.*, 1986). It has two α- subunits and two β- subunits. The α- subunit locates extracellularly on cell membrane and contains the ligand-binding domain. The β- subunit has a transmembrane domain and an intracellular domain, which contains the ATP binding site and the tyrosine kinase catalytic domain. When the IGF-IR combines its ligands (IGF-I, IGF-II, or insulin at supraphysiological concentrations), the IGF-IR regulates cell proliferation. (See reviews by Sara and Hall, 1990; Baserga and Rubin, 1993; Resnicoff and Baserga, 1998).

IGF-I and IGF-II and their receptor were found to be associated with the normal differentiation of mesenchymal tissues and the growth of animal and human tumors (Ulrich *et al.*, 1986; Han *et al.*, 1987; Rubin *et al.*, 1995). Many animal and human malignancies, such as nervous system-derived tumors, colon and liver cancer, have been reported to express high level of IGF-I, and/or IGF-I receptor (Leroith *et al.*, 1995). Since one tumor cell could express IGF and IGF receptor together, an autocrine loop may exist and drive the tumorigenic phenotype (Kiess *et al.*, 1989). The autocrine hypothesis of malignant transformation has derived experimental support primarily from antibody blocking (Heldin *et al.*, 1989) and sense gene transfer studies. However, antibody-blocking is ineffective for the analysis of intracellular autocrine loops (Heldin *et al.*, 1989). The use of antisense RNA (or antisense ODNs) could generate phenocopies of null mutation of IGF or IGF-IR in tumor cells, providing a direct approach for accessing its autocrine contribution to the tumorigenic phenotype. According to the data reported previously by *Baserga in 1995* and our group, the importance of the IGF-I receptor in transformation has a broad spectrum at least including these tumor models described as follows:

Rat glioblastoma C6;
Rat prostate adenocarcinoma PA-III;
Mouse melanoma B16-F10;
Mouse embryonic carcinoma (EC) PCC3;
Human glioblastoma T98G;
Human breast carcinoma MCF-7;
Human small cell lung carcinoma CALA6;
Human melanoma FO-1;
Ewing's sarcoma cell line SK-ES-1, RD-ES and SK-N-MC;
Prostatic cancer cell lines, PC-3, DU-145.

IGF-I and IGF-I receptor also play an important role in tumor metastasis. Prostate cancer (the most commonly diagnosed cancer in men and the second leading cause of death from malignancy in men in the United State) preferentially metastasizes to bone, where IGFs are abundantly accumulated (Mohan S. and Baylink D. J., 1991; Canalis *et al.*, 1991). Prostate cancer cells express IGF-I receptor, which could facilitate the development of bone metastases because IGFs are required for the tumor cell growth. Antisense IGF-IR transfected murine H-59 carcinoma cells had a markedly reduced expression and lost the ability to metastases in vivo (Long *et al.*, 1995). IGF-R is commonly over-expressed in breast cancer (Peyrat and Bonneterre, 1992), while the serum IGF-I levels are also reported to be 25% higher in breast cancer patients compared to normal age-matched controls (Peyrat *et al.*, 1993). Recently study on MDA-MB-435 metastatic breast cancer cells showed that inhibition of IGF-R resulted in suppression of adhesion, invasion, and metastasis, providing a mechanistic rationale for targeting IGF-IR in the treatment of metastatic breast cancer (Dunn *et al.*, 1998).

Recently, results from a long-term prospective study correlates plasma levels of IGF-I with a risk of prostate cancer. This research shows that men with the highest natural levels of circulating IGF-I—but levels still within the normal range of 100–500 ng/ml had more than four times the risk of prostate cancer compared with those in a reference group. The IGF-I level was a more sensitive early-stage indicator of the risk of getting prostate cancer than the presence of prostate specific antigen (PSA) in serum, which is routinely used for clinical diagnosis of the disease (Chan *et al.*, 1998).

2. ANTISENSE VECTORS

Trojan *et al.* (1992, 1993) have first demonstrated that inhibition of IGF-I expression by antisense RNA could inhibit the tumorigenicity of C6 glio blastoma cells. While in the latter reports, most investigations were designed to block IGF-I receptor. According to the autocrine loop, blockade of either IGF-I or its receptor could disrupt the signaling provided by interaction between IGF-1 and its counter receptor (IGF-1 receptor). However, the key element within the loop should be the receptor. The antisense sequences had been designated aiming either the coden or the 3′, 5, non-translated region of the target gene, according to the cDNA sequence of these genes.

Ilan's and Trojan's group (Case Western Reserve University, Cleveland) (Trojan *et al.*, 1992) constructed an episomal based antisense vector that is able to express high level antisense IGF-1 mRNA in transfected tumor cells in the presence of Zn^{2+} in culture medium as promoter inducer. Diagram of the MT-1 antisense IGF-1 expression vector is described in Fig. 1.

The episome-based antisense IGF-1 vector containing metalothionein 1 promoter so that the expression of antisense mRNA can be *in vitro* induced by Zn^{2+} (50 μM). The MT-1 promoter is known to be active *in vivo* (Palmiter *et al.*, 1983). The antisense sequence was designed as a 1 Kb fragment inserted in antisense orientation.

The antisense IGF-2 vector was constructed similarly. A synthetic double-stranded oligonucleotide of 99 bp in length, spanning the 1st through the 33rd amino acid of the processed mouse IGF-II peptide, was inserted into the vector described above, replacing the IGF-I sequence, in the antisense orientation (Trojan *et al.*, 1994).

In the report of Baserga's group (Thomas Jefferson University, Philadelphia), the antisense IGF-I receptor sequence (300 bp) was under the control of *Drosophila* HSP70 promoter. The expression of these (including the sense sequence, as control) transcripts was strongly increased at 39.6 °C and it was only faintly detectable at 37 °C. There was not transcripts to be observed in 34 °C (Fig. 2) (Resnicoff *et al.*, 1994a, 1994b).

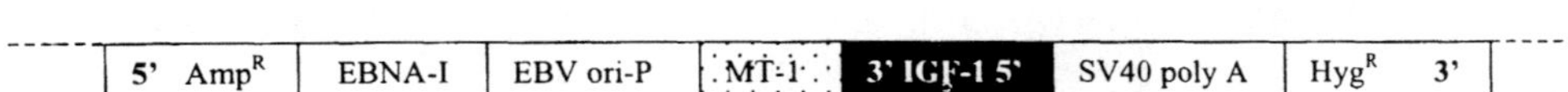

Figure 1. Diagram of the antisense IGF-I expression vector pAntiIGF-I (Trojan *et al.*, 1992). MT-1, metalothionein 1 promotor; EBNA-I, EBV-encoded nuclear antigen I; SV40, simian virus 40; Amp^R, ampicillin resistance gene; Hyg^R, hygromycin resistance gene.

Figure 2. Diagram of plasmid HSP-IGF-IR antisense. HSP-70: *Drosophila* hot-shook protein 70 promotor; Poly A: Hepatitis B polyadenylation signal sequence; PSV NEO: neomycin resistance gene under the SV40 promoter.

3. ANTISENSE OLIGODEOXYNUCLEOTIDES (ODNs)

Approach using nuclease-resistant phosphorothinate antisense ODNs is an alternative to antisense IGF protocol that is also effective in inhibiting the function of IGF-I and IGF receptors in different tumor cell types. The specificity of the antisense ODNs can be confirmed by measuring level of the targeted mRNA and proteins, respectively. To inhibit proliferation of tumor cells by ODNs, antisense and oligodeoxynucleotides need to be prepared corresponding to codons of the signal sequence of the subunit of IGF-I or IGF receptors preceding the proreceptor sequence and then, transfected into cultured target cells by a standard lipofectin approach. It has been reported that the application of the antisense ODN reduce potently the cell growth of unstimulated NIH: OVCAR3 cells *in vitro*. This inhibiting ability of ODNs can be greatly enhanced when delivered by the cationic lipid formulation lipofectin because antisense ODNs are unable to cross cellular membrane efficiently (Resnicoff *et al.*, 1994b; Markus Muller *et al.*, 1998). In comparison with antisense IGF mRNA approach, antisense S-ODNs approach is simple and cost effective. However, high concentration of ODNs may result a high and toxic accumulation in fraction of the cells because of a non-specific uptaking induced by lipid mediated delivery. For these reasons, local administration and low concentration of ODNs when applied together with cationic lipids might minimize toxic reaction and have a future therapeutic potential in cancer treatment.

4. *IN VITRO* INVESTIGATIONS

Transforming of the tumor cells with the antisense vectors were performed *in vitro* by conventional methods (i.e., Lepofectin, calcium phosphate precipitation, or electroporation) and then screened for positive clones with hygromycin or neomycin. It should be pointed out here that establishment of antisense IGF-1 transfected tumor cell line is very difficult in normal culture condition without supplement of a resource of IGF-1 protein. All most of all IGF-1 producing tumor cells after transfection with antisense IGF-1 and blocking IGF-1 production died within three days in the absence of exogenous IGF-1 in culture medium. Addition of exogenous IGF-1 or condition medium from IGF-1 producing tumor cell culture is essential to maintain the transfected tumor cell survival. This is even more critical for establishing transfectants from primary human tumor cells.

Expression of the antisense IGF-I RNA by Zn^{2+} activation of the MT-1 promoter in stable transfected tumor cells block the accumulation of endogenous IGF-I transcripts, detected by in situ hybridization and immunocytochemistry, respectively. Addition of $ZnSO_4$ in the culture medium resulted in high levels of antisense transcript accumulation and dramatically decreased levels of endogenous IGF-I mRNA and IGF-I protein

(Trojan *et al.*, 1992). Similar *in vitro* results were obtained in other antisense vector transformed tumor systems (Resnicoff *et al.*, 1994b; Rininsland *et al.*, 1997).

The growth of tumor cells in monolayer culture was not affected by the expression of the antisense RNA. However, decreased growth rates and clone formation rates in soft agar assay were observed. More impotently, it has been demonstrated that the cultured transfected tumor cells lines in culture, rat C-6 glioma, human hepatocellular carcinoma, human glioma and mouse teratocarcinoma, expressed an significantly increased level of MHC class I, intracellular adhesion molecule 1 (ICAM-1) and, in some circumstance, B7 costimulatory molecule. This increased expression of MHC class I and costimulatory molecules on transfected tumor cell surfaces, indicated by ^{51}Cr-release complement dependent cytotoxicity assay and immunofluorescent staining flow cytometery analysis, is directly associated with the final immune recognition of tumor immunogenicity (Trojan J., 1996).

5. *IN VIVO* INVESTIGATIONS

In comparison with the sense gene transfection control, none immunogenic and low immunogenic tumor cells transfected with antisense RNA vector become more immunogenic and lost their tumorigenicity in syngenetic animals. Subcutaneous injection of IGF-I (Trojan *et al.*, 1993) or IGF-IR (Resnicoff *et al.*, 1994b) antisense transfected C6 glioblastoma cells into syngenetic rats could not only protect animals from challenge with wild type tumor cells, but also caused regression of established brain glioblastomas. An antitumor immune response elicited by immunization with the antisense IGF-1 transfectants is antigen specific and mediated T lymphocytes. Tumor tissue necrosis and local T lymphocyte infiltration can be observed in the treated animals. Majority of T lymphocytes infiltrated in tumor tissue is CD8$^+$ lymphocytes demonstrated by immunofluorescent staining with anti CD8 monoclonal antibody. As described above, the enhanced immunogenicity of rat C6 glioma, human primary glioma and mouse teratocarcinoma after transfected with antisense IGF-1 mRNA is correlated to the expression of MHC class I and costimulation molecule B7 on the cell surfaces that provide functional signalings for T cell activation (Trojan *et al.*, 1996a,b).

The antisense IGF-I R transfected tumor also showed reduced ability of tumorigenicity and ability to metastasis in nude mice. In *Burfeind's* report (1996), mice injected with IGF-IR antisense-transfected PA-III cells (rat prostate adenocarcinoma) either developed tumors 90% smaller than controls or remained tumor-free for more than 60 days after tumor cell injection. When control-transfected PA-III cells were inoculated over the abraded calvaria of nude mice, large tumors formed with invasion of tumor cells into the brain parenchyma. In contrast, IGF-IR antisense transfectants formed significantly smaller tumors with no infiltration into brain. These results clearly show that signaling mediated by intersection between IGF-1 and its counter receptor play an important role in metastasis and provide a basis for targeting IGF-IR as a potential treatment for prostate cancer. Similar results were achieved in different models (Resnicoff, 1994b; Long *et al.*, 1995). Recent studies have shown that some rat and mouse hepatoma cells expressed increased level of MHC class I molecules on their cell surfaces after transfected with antisense IGF-1 expressing vector, which is closely correlated with the poor tumorigenicity of the transfected tumor cells in syngeneic animals. However, transfection of tumor cells with antisense IGF-1 is not uniformly effective in induction of immune responses (Guo *et al.*, 1997). Several IGF-1 producing tumor cell lines, with complete block of IGF-1 production and with enhanced expression of MHC class I molecules on their cell surfaces after transfection with antisense IGF-1 vector, still remained their tumorigenicity in syngeneic animals. According to the concepts for T cell mediated immune responses, both antigen specific and costimulatory signals are essential for full activation of anti-tumor immune responses. The interaction of MHC-peptide complex on tumor cells and T cell receptor on T cells deliver antigen specific signal. The costimulatory signals and amplification signals can be mediated by a number of cytokines and adhesion molecules. One well-characterized costimulatory signal is provided by the interaction of CD28 on T cells with its primary ligands B7.1 (CD80) and B7.2 (CD86) on antigen presenting cells (APCs). Absence of either MHC or costimulatory molecules on tumor cells result in a failure to stimulate T cell mediated antitumor immunity. Therefore, we reasoned that the unresponsiveness of some tumor cells to antisense IGF-1 gene transfer may be due to absence of costimulatory signals on these transfected tumor Cells. To support this hypothesis, we use hepa 1–6 hepatoma and SMCC-1 colon cancer model systems for further evaluating the efficacy of co-transfection with genes encoding antisense IGF-1 and mouse B7.1 molecules in eliciting antitumor immune responses. Both hepa 1–6 and SMCC-1 cell lines are chemically induced tumors originating in C57BL/6 mice and grow rapidly forming subcutaneous tumors. Because of lacking both MHC class I and B7 molecules on their cell surfaces and failing to induce host immune response even when transfected with genes encoding antisense IGF-1 or B7.1 molecules respectively, the two cell lines are considered as unique *in vivo* tumor model systems and are well-suited for experimental immunotherapy studies. Using double gene transfection approach, we successfully converted hepa 1–6 and SMCC-1 cells from poorly immunogenic to immunogenic ones. The transfetants were effective in priming naive spleen T cells from syngeneic mouse with generation of tumor specific CTLs *in vitro* and eliciting antitumor immune responses *in vivo*.

The data presented here suggest that co-transfection of tumor cells with B7.1 expression vector synergies with antisense IGF-1 gene transfer approach to generate immunogenic tumor cells *in vitro.* The resulting tumor cells could be used as cellular tumor vaccines to elicit anti-tumor specific immune responses that are effective in eradicating large established tumors. Modification of autologous tumor cells by this double gene transfer approach may provide a valuable strategy for generating therapeutic cellular vaccines for the treatment of cancers.

6. CONCLUDING REMARKS

The evidence from the studies conducted by different methods provide further support for the role of IGF in cell growth, differentiation, tumorigenicity and metastasis. Results from the studies using antisense strategy also help us realizing further function of IGF in regulating cell circle and cell program death. However, It still needs to be extensively investigated if this approach can be used as a novel immunogene therapy strategy for the treatment of human cancer. The IGF-I-IGF-IR autocrine loop certainly should be a target of antisense therapy strategies that directly inhibit tumor cell growth after elimination of IGF-1 or IGF-1 receptor expression (Lee *et al.*, 1996). However, it will be our goal to generate strongly immunogenic tumor cells by this approach that could be used as tumor vaccine to elicit antitumor specific immunity. Tumor cell may escape host immune response through a complicate mechanism. It is impossible to correct all of defects that contribute to tumorigenicity. In addition, condition of host immune system is another key issue that should be considered when conduction of immunotherapy. It may be helpful to generate therapeutic tumor vaccine cells using a combination of different approaches that have potential to enhance immunogenicity of tumor cells. Generation of immunogenic tumor cells by antisense approaches may provide a novel immunotherapeutic strategy for treatment of human cancer, especially for human hepatocellular carcinoma, an IGF high producing cancer.

REFERENCES

Baserga, R., 1995, The insulin-like growth factor I receptor: A key to tumor growth? *Cancer Res* **55**:249–252.

Baserga, R., and Rubin, R., 1993, Cell cycle and growth control. *Crit Rev Eukaryot Gene Expr* **3**:47–61.

Burfeind, P., Chernicky, C.L., Rininsland, F., Ilan, J., and Ilan, J., 1996, Antisense RNA to the type I insulin-like growth factor receptor suppresses tumor growth and prevents invasion by rat prostate cancer cells *in vivo*. *Proc Natl Acad Sci USA* **93**:7263–7268.

Canalis, E., McCarthy, T.L., and Centrella, M., 1991, Growth factors and cytokines in bone cell metabolism. *Annu Rev Med* **42**:17–24.

Chan, J.M., Stampfer, M.J., Giovannucci, E., Gann, P.H., Ma, J., Wilkinson, P., Hennekens, C.H., and Pollak, M., 1998, Plasma insulin-like growth factor-I and prostate cancer risk: a prospective study. *Science* **279**:563–566.

Dunn, S.E., Ehrlich, M., Sharp, N.J., Reiss, K., Solomon, G., Hawkins, R., Baserga, R., and Barrett, J.C., 1998, A dominant negative mutant of the insulin-like growth factor-I receptor inhibits the adhesion, invasion, and metastasis of breast cancer. *Cancer Res* 58:3353–3361.

Guo, Y., 1997, Immunogene therapy: Both *in vitro* and *in vivo* gene transfer approaches, in: *Cancer Gene Therapy* (H. Mazarakis, and S.J. Swart, eds.), pp. 155–168, International Business Communicatins, Inc., Southborough.

Han, V.K., D'Ercole, A.J., and Lund, P.K., 1987, Cellular localization of somatomedin (insulin-like growth factor) messenger RNA in the human fetus. *Science* **236**:193–197.

Heldin, C.H., and Westermark, B., 1989, Growth factors as transforming proteins. *Eur J Biochem* **184**:487–496.

Kiess, W., Lee, L., Graham, D.E., Greenstein, L., Tseng, L.Y., Rechler, M.M., and Nissley, S.P., 1989, Rat C6 glial cells synthesize insulin-like growth factor I (IGF-I) and express IGF-I receptors and IGF-II/mannose 6-phosphate receptors. *Endocrinology* **124**:1727–1736.

Lee, C.T., Wu, S., Gabrilovich, D., Chen, H., Nadaf-Rahrov, S., Ciernik, I.F., and Carbone, D.P., 1996., Antitumor effects of an adenovirus expressing antisense insulin-like growth factor I receptor on human lung cancer cell lines. *Cancer Res* **56**:3038–3041.

LeRoith, D., Werner, H., Beitner-Johnson, D., and Roberts, C.T. Jr., 1995, Molecular and cellular aspects of the insulin-like growth factor I receptor. *Endocr Rev* **16**:143–163.

Long, L., Rubin, R., Baserga, R., and Brodt, P., 1995, Loss of the metastatic phenotype in murine carcinoma cells expressing an antisense RNA to the insulin-like growth factor receptor. *Cancer Res* **55**:1006–1009.

Mohan, S., and Baylink, D.J., 1991, Bone growth factors. *Clin Orthop* **263**:30–48.

Morrison, R.S., 1991, Suppression of basic fibroblast growth factor expression by antisense oligodeoxynucleotides inhibits the growth of transformed human astrocytes. *J Biol Chem* **266**:728–734.

Müller, M., Dietel, M., Turzynski, A., and Wiechen, K., 1998, Antisense phosphorothioate oligodeoxynucleotide down-regulation of the insulin-like growth factor I receptor in ovarian cancer cells. *Int J Cancer* **77**:567–571.

Negroni, A., Scarpa, S., Romeo, A., Ferrari, S., Modesti, A., and Raschella, G., 1991, Decrease of proliferation rate and induction of differentiation by a MYCN antisense DNA oligomer in a human neuroblastoma cell line. *Cell Growth Differ* **2**:511–518.

Palmiter, R.D., Norstedt, G., Gelinas, R.E., Hammer, R.E., and Brinster, R.L., 1983, Metallothionein-human GH fusion genes stimulate growth of mice. *Science* **222**:809–814.

Peyrat, J.P., and Bonneterre, J., 1992, Type 1 IGF receptor in human breast diseases. *Breast Cancer Res Treat* **22**:59–67.

Peyrat, J.P., Bonneterre, J., Hecquet, B., Vennin, P., Louchez, M.M., Fournier, C., Lefebvre, J., and Demaille, A., 1993, Plasma insulin-like growth factor-1 (IGF-1) concentrations in human breast cancer. *Eur J Cancer* **29A**:492–497.

Resnicoff, M., and Baserga, R., 1998, The role of the insulin-like growth factor I receptor in transformation and apoptosis. *Ann N Y Acad Sci* **842**:76–81.

Resnicoff, M., Coppola, D., Sell, C., Rubin, R., Ferrone, S., and Baserga, R., 1994a, Growth inhibition of human melanoma cells in nude mice by antisense strategies to the type 1 insulin-like growth factor receptor. *Cancer Res* **54**:4848–4850.

Resnicoff, M., Sell, C., Rubini, M., Coppola, D., Ambrose, D., Baserga, R., and Rubin, R., 1994b, Rat glioblastoma cells expressing an antisense RNA to the insulin-like growth factor-1 (IGF-I) receptor are nontumorigenic and induce regression of wild-type tumors. *Cancer Res* **54**:2218–2222.

Rininsland, F., Johnson, T.R., Chernicky, C.L., Schulze, E., Burfeind, P., and Ilan, J., 1997, Suppression of insulin-like growth factor type I receptor by a triple-helix strategy inhibits IGF-I transcription and tumorigenic potential of rat C6 glioblastoma cells. *Proc Natl Acad Sci USA* **94**:5854–5859.

Rubin, R., and Baserga, R., 1995, Insulin-like growth factor-I receptor. Its role in cell proliferation, apoptosis, and tumorigenicity. *Lab Invest* **73**:311–331.

Sara, V.R., and Hall, K., 1990, Insulin-like growth factors and their binding proteins. *Physiol Rev* **70**:591–614.

Trojan, J., Blossey, B.K., Johnson, T.R., Rudin, S.D., Tykocinski, M., Ilan, J., and Ilan, J., 1992, Loss of tumorigenicity of rat glioblastoma directed by episome-based antisense cDNA transcription of insulin-like growth factor I. *Proc Natl Acad Sci USA* **89**:4874–4878.

Trojan, J., Johnson, T.R., Rudin, S.D., Ilan, J., Tykocinski, M.L., and Ilan, J., 1993, Treatment and prevention of rat glioblastoma by immunogenic C6 cells expressing antisense insulin-like growth factor I RNA. *Science* **259**:5091 94–97.

Trojan, J., Johnson, T.R., Rudin, S.D., Blossey, B.K., Kelley, K.M., Shevelev, A., and Abdul-Karim, F.W., Anthony D.D., Tykocinski M.L., and Ilan J., 1994, Gene therapy of murine teratocarcinoma: separate functions for insulin-like growth factors I and II in immunogenicity and differentiation. *Proc Natl Acad Sci USA* **91**:6088–6092.

Trojan, J., Duc, H.T., Lafarge-Frayssinet, C., Upegui-Gonzales, L.C., Swiercz, B., Bismuth, H., Hor, F., Anthony, D., Guo, Y., and Ilan, J., 1996a, Immunotherapy of tumors expressing IGF-I. *C R Seances Soc Biol Fil* **190**:165–169.

Trojan, J., Duc, H.T., Upegui-Gonzalez, L.C., Hor, F., Guo, Y., Anthony, D., and Ilan, J., 1996b, Presence of MHC-I and B-7 molecules in rat and human glioma cells expressing antisense IGF-I mRNA. *Neurosci Lett* **212**:9–12.

Ullrich, A., Gray, A., Tam, A.W., Yang-Feng, T., Tsubokawa, M., Collins, C., Henzel, W., Le Bon, T., Kathuria, S., and Chen, E., 1986, Insulin-like growth factor I receptor primary structure: comparison with insulin receptor suggests structural determinants that define functional specificity. *EMBO J* **5**:2503–2512.

Whitesell, L., Rosolen, A., and Neckers, L.M., 1991, Episome-generated N-myc antisense RNA restricts the differentiation potential of primitive neuroectodermal cell lines. *Mol Cell Biol* **11**:1360–1371.

24

SENSITIZATION OF TUMORS TO CHEMOTHERAPY THROUGH GENE THERAPY

Ruth A. Gjerset and Dan Mercola

Sidney Kimmel Cancer Center
10835 Altman Row
San Diego, California 92121

1. INTRODUCTION

The cellular response to DNA damage plays a critical role not only in tumor progression but also in the process of acquired drug resistance, a problem that affects about half of all cancer cases overall and remains one of the major obstacles to successful therapy of cancer. Modulation of these DNA damage response pathways may therefore provide us with a means to reverse acquired drug resistance and improve the outcome of therapy for a large fraction of cancer patients. In this article we will focus on two major pathways involved in the cellular response to DNA damage: The Jun kinase stress activated pathway, and the p53-mediated DNA damage response pathway leading to apoptosis. Through independent mechanisms, each of these pathways modulates the cellular response to DNA damaging chemotherapies and radiation. Therapeutic approaches based on inhibiting the Jun kinase pathway and/or restoring the p53 pathway may, therefore, provide us with new biological strategies for reversing acquired drug resistance, thus improving the outcome of therapy for most cancers.

2. THE p53 TUMOR SUPPRESSOR AND THE DNA DAMAGE RESPONSE

Over half of all cancers suffer loss of function of the p53 tumor suppressor (Levine, 1993), a key player in the induction of apoptosis in response to DNA damage (Clarke *et al*, 1993; Gjerset *et al.*, 1995; Lotem *et al.*, 1993; Lowe *et al.*, 1993), in addition to its roles in cell cycle regulation and DNA repair (Levine, 1997). Its involvement in DNA damage-induced apoptosis may derive from its ability to bind, alone or possibly in combination with other DNA damage recognition proteins, to sites of damaged DNA, including single stranded ends and insertion-deletion loops (Bakalkin *et al.*, 1995; Lee *et al.*, 1995; Levine 1997 (review)). p53 might also bind to DNA adducts and strand breaks induced by various therapies. The frequent loss of p53 function in

Cancer Gene Therapy: Past Achievements and Future Challenges, edited by Habib
Kluwer Academic/Plenum Publishers, New York, 2000.

cancer, often associated with disease progression and increased genomic instability, may reflect at least in part the role of p53 in DNA damage recognition and apoptosis. Most genome destabilizing events, including gene amplification, gene deletion and gene translocation, involve DNA strand breaks (Stark, 1993). These breaks could serve as triggers for p53-mediated apoptosis and provide the driving force for loss of p53.

The same process that underlies the progression of cancer, that is, genomic instability accompanied by loss of p53-mediated apoptosis, can also lead to therapy resistance. Support for the idea that loss of p53 could desensitize a cell to the damaging effects of drugs and radiation comes from studies of p53-null transgenic mice. In these studies it was observed that normal transgenic hematopoietic cells (Lotem and Sachs, 1993), E1A-expressing trangenic fibroblasts (Lowe *et al.*, 1993), and transformed transgenic fibroblasts (Lowe *et al.*, 1994) were all more resistant to apoptosis following treatment with any of a wide variety of anti-cancer agents including radiation, than were the comparable cells from the parental strain of mice that express wild-type p53. Cell killing was therefore enhanced in cells that expressed wild-type p53 and were able to trigger their own cell death program.

The possibility that p53 status is a factor in therapy responsiveness finds further support from a recent anti-cancer drug screening of human tumor cell lines expressing mutant or wild-type p53 (O'Connor *et al.*, 1997). In that study the growth inhibitory properties of some 123 anti-cancer agents were tested in 60 human tumor lines of known p53 status. It was found that cells expressing mutant p53 showed less growth inhibition than did wild-type p53-expressing cells lines following treatment with the majority of clinically used anti-cancer agents, including DNA cross-linking agents such as cisplatin, antimetabolites such as 5-fluorouracil, and topoisomerase I and II inhibitors. Since these agents are known to cause DNA damage, either directly or indirectly, these results are consistent with a role of p53 in mediating the cellular apoptotic response to DNA damage. One class of agents, the antimitotic agents, appeared to suppress growth in a p53-independent manner, consistent with the primary target for these agents being the mitotic apparatus rather than DNA.

Taken together, these studies suggest that p53 gene transfer could have clinical application in suppressing cancer and enhancing the responsiveness of tumors to a wide variety of DNA damaging therapies, a possibility that greatly expands the clinical application of p53-based approaches. Numerous *in vitro* studies in a variety of tumor cell systems support the use of p53 gene transfer to sensitize tumors to such therapy, including, 5-fluorouracil, cisplatin, topoisomerase I inhibitors and gamma radiation (see below).

2.1. *In Vitro* Sensitization of Tumor Cells to Chemotherapeutic Drugs by p53

In Figs. 1 and 2 we have summarized the results of screening tumor cells for sensitivity to various chemotherapeutic agents following exposure of cells to a replication-defective adenovirus encoding wild-type p53 (Ad-p53, Ad-βgal, and Ad-Luc (luciferase) vectors provided by Dr. Deborah R. Wilson, Introgen Therapeutics, Inc.). The cells were treated with virus under conditions which achieve about 70–80% infection efficiency, as judged by infection of parallel cultures with a β-galactosidase adenovirus. Following infection, cells were replated at low density in 96-well plates and treated with varying levels of chemotherapeutic agent, followed by an additional 5–7 days of incubation and measurement of cell viability.

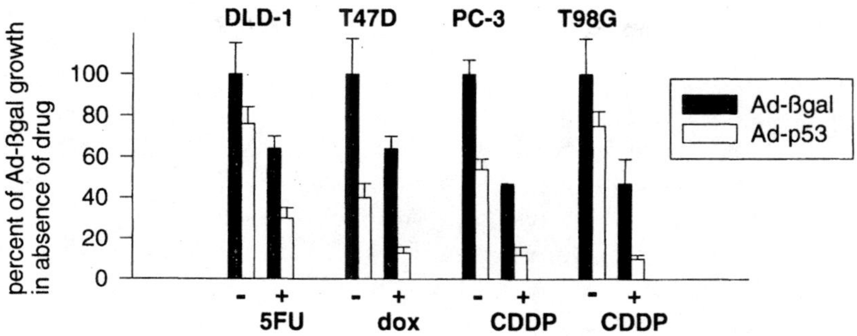

Figure 1. Viability assay showing that Ad-p53 suppresses growth and enhances sensitivity to DNA damaging chemotherapeutic drugs in p53-mutant-expressing cells (DLD-1 colon cancer, T47D breast cancer, PC-3 prostate cancer, and T98G glioblastoma). Infection efficiencies were 60–70% and drug treatments 1 day post-infection were as follows: 5-fluorouracil (5FU) 10 μM 1 hour; doxorubicin (dox) 3.7 μM 1 hour; cisplatin (CDDP) 30 μM 1 hour for PC-3 and 20 μM 1 hour for T98G. Viability was assayed 6 days post drug treatment in all cases except PC-3 which was 4 days post drug treatment, and expressed as a percent of control cells treated with Ad-βgal.

In numerous examples from a variety of cancer types, we have observed suppression of mutant-p53-expressing tumor cells *in vitro* following treatment with Ad-p53. As shown in Fig. 1, restoration of wild-type p53 in mutant p53-expressing cells (DLD-1, T47D, PC-3, T98G) results in marked enhancement of sensitivity to a variety of DNA damaging treatments (5-fluorouracil, doxorubicin, cisplatin), consistent with possibility that we have restored the p53-mediated pathway of DNA damage recognition and apoptosis. Apoptosis was confirmed by (a) propidium iodide staining of fixed cells followed by FACS analysis to reveal a sub G1 peak of apoptotic cells, and (b) an ELISA assay (Boehringer Mannheim Corp, Indianapolis, IN)) to detect oligonucleosomal fragments released from the nuclei of cells in the early phases of apoptosis (see Gjerset *et al.*, 1995).

In contrast, this enhancement of sensitivity is not observed in two wild-type p53-expressing cell lines, MCF7 and LS174T (Fig. 2). This suggests that wild-type p53 gene transfer may be effective in therapy sensitization only in the case of tumors that have lost wild-type p53 function.

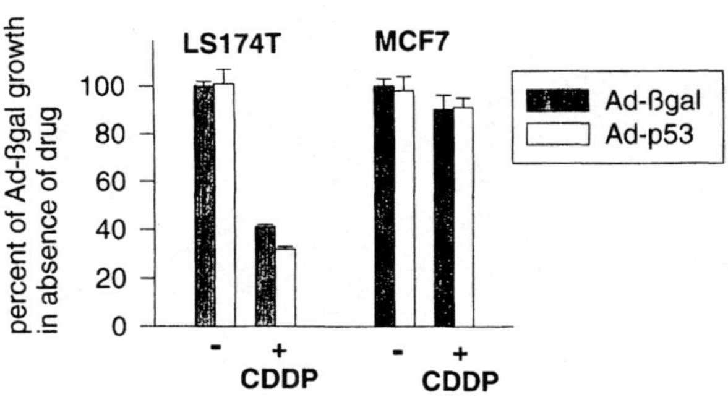

Figure 2. Viability assay showing that Ad-p53 neither suppresses growth nor enhances sensitivity to DNA damaging chemotherapeutic drugs in wild-type p53-expressing cells (LS174T colon cancer cells and MCF7 breast cancer cells). Infection efficiencies were 60–70%, and drug treatments 1 day post-infection were 30 μM cisplatin (CDDP) 1 hour for LS174T and 20 μM 1 hour for MCF7. Viability was assayed 6 days post drug treatment.

2.2. *In Vivo* Sensitization of Tumors to Chemotherapeutic Drugs by p53

We have extended our *in vitro* observations on tumor suppression and therapy sensitization to *in vivo* models for human head and neck cancer and colon cancer (Gjerset *et al.*, 1997) and prostate cancer (Fig. 3) in nude mice, where established tumors were treated *in vivo* with replication-defective p53-adenovirus, and chemotherapy. We have also studied chemosensitization by p53 using *ex vivo* modified cells in an orthotopic model of glioblastoma in Fisher rats (Dorigo *et al.*, 1998). The results are consistent with other *in vivo* studies in animal models showing a combined benefit of p53 and chemotherapy (Badie *et al.*, 1998; Fujiwara *et al.*, 1994; Miyake *et al.*, 1998; Nielsen *et al.*, 1998; Nguyen *et al.*, 1996). Together with the *in vitro* data, these results support the clinical application of adenovirus p53 combination approaches to tumors expressing mutant p53. In the experiment shown in Fig. 3, PC-3 prostate cancer cells which express mutant p53, were implanted in 20 nude mice (5×10^6 cells/animal) where they form rapidly growing subcutaneous tumors. At 5 days post-implantation, when tumor sizes had obtained a volume of about 50 mm^3, animals were randomized by tumor size and treatment was initiated. The first day of treatment was then designated day 1 and consisted of intraperitoneal injection of cisplatin (Platinol^{TM}, Bristol Laboratories, obtained through local pharmacies), at 4 mg/kg (LD10) on days 1 and 8. Vector (Ad-p53 or Ad-luc control) was administered intratumorally (10^8 pfu per tumor) of days 1, 3, 5, 8, and 10. In both cases, some suppression was observed with Ad-p53 alone, with a significant enhancement of suppression when Ad-p53 was combined with chemotherapy. The vector doses used here

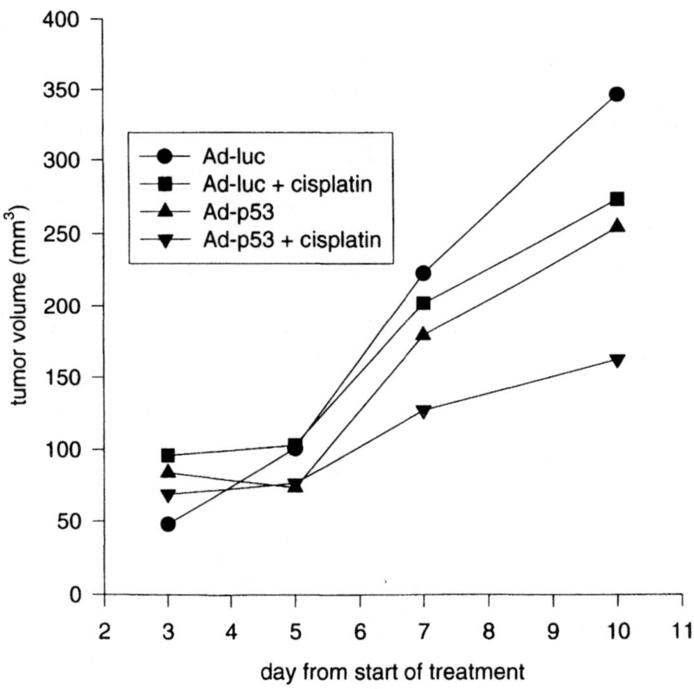

Figure 3. Suppression of PC-3 prostate tumor growth in nude mice by Ad-p53 plus cisplatin. Vector (10^8 pfu per tumor) was administered on days 1,3,5,8,10, and cisplatin was administered on days 1, 8.

were relatively low (10–30 pfu per tumor cell) and completely without side effects (as judged by periodic weight measurements on the animals and histological examination. We observe little or no toxicity in nude mice of vector doses as high as 10^9 pfu per animal (Gjerset *et al.*, 1997). Phase I clinical trials employing p53 adenovirus also demonstrate that this vector is well tolerated in patients (Roth *et al.*, 1998). Thus restoration of wild-type p53 function through gene therapy may provide a significant benefit for patients with advanced cancers, when used in combination with conventional therapies.

3. THE JUN KINASE/STRESS-ACTIVATED PROTEIN KINASE PATHWAY AND THE DNA DAMAGE RESPONSE

Another important cellular pathway is also triggered in response to DNA damage: the Jun Kinase/Stress-activated protein kinase pathway (JNK/SAPK), one of several distinct Mitogen-Activated Protein Kinase (MAPK) Pathways involved in signal transduction. Besides its role in the DNA damage response (discussed below), the JNK/SAPK pathway is also induced by growth factors such as EGF-1(Bost *et al.*, 1999), by oncogene expression (Binétruy *et al.*, 1991), and is essential for transformation of rat embryo fibroblasts (Smeal *et al.*, 1991). As we show below, inhibition of this pathway enhances sensitivity to DNA damaging therapies in both p53 mutant and p53 wild-type tumors, as T98G glioblastoma cells, PC-3 prostate cancer cells, and U87 glioblastoma cells which express mutant p53, and MCF7 breast cancer cells which express wild-type p53, are all sensitized to various DNA damaging treatments by expression of a dominant-negative inhibitor of the pathway. Thus it may be possible to modulate this pathway by itself or in combination with p53 gene replacement to enhance tumor cell responsiveness to DNA damaging chemotherapies.

The Mitogen Activated Protein Kinase (MAPK) pathways play roles in several cellular processes, including cellular transformation, proliferation, differentiation, and the DNA damage response. Through these pathways extracellular growth and stress stimuli transmit signals through a cascade of kinases and phosphorylation events that result in the phosphorylation and activation of transcription factors such as c-Jun (a heterodimeric component of the AP-1 complex and related transcription complexes), ATF-2 and Elk-1 (Bost *et al.*, 1997; Cavigelli *et al.*, 1995; Dérijard *et al.*, 1994; Gupta *et al.*, 1995; Hibi *et al.*, 1993; Livingstone *et al.*, 1995; Potapova *et al.*, 1997; Smeal *et al.*, 1991, 1992). This phosphorylation activates the transcriptional transactivation properties of AP-1 and related factors and leads to the subsequent induction of specific response genes. The cascade of events begins with the activation of the Mitogen Activated Protein Kinase Kinase Kinases (MAPKKK), followed by activation of the Mitogen Activated Protein Kinase Kinases (MAPKK), and finally activation of the Mitogen Activated Protein Kinases (MAPK), of which the Jun Kinase family of enzymes (including JNK 1,2,3) is one example.

3.1. *In Vitro* Sensitization to Chemotherapeutic Drugs Through Inhibition of the JNK/SAPK Pathway

T98G glioblastoma cells activate the JNK/SAPK pathway in response to treatment with the chemotherapeutic agent, cisplatin (Potapova *et al.*, 1997), which forms bifunctional DNA cross links between adjacent guanines or adenine-guanine dinucleotides. This is consistent with numerous other reports in which activation of the JNK/SAPK pathway has been observed in response to genotoxic DNA treatments, including UV irradiation (Dérijard *et al.*, 1994, Adler *et al.*, 1995a, 1995b, 1996), ionizing radiation (Kharbanda *et al.*, 1995), the mutagens methylmethane sulfate (MMS), N-nitro-N′-nitroso-guanidine (MNNG) and numerous genotoxic chemotherapeutic agents such as Ara-C (1-β-D-Arabinofuranosylcytosine) (van Dam *et al.*, 1995; Kharbanda *et al.*, 1995). For several of these agents such a UV-irradiation, the activation of the JNK/SAPK pathway is directly proportional to the number of DNA damaging events (Adler *et al.*, 1995a). Activation is often rapid, detectable within a few minutes. These characteristics have given rise to the hypothesis that the role of the JNK/SAPK pathway may be to mediate DNA repair (Potapova *et al.*, 1997).

To evaluate the role of JNK/SAPK activation in the cellular DNA damage response we selected clones of T98G cells modified to express a non-phosphorylatable mutant of c-Jun (T98G mJun), in which serines 63 and 73, the targets of Jun Kinase-mediated phosphorylation, have been replaced by alanines (Smeal *et al.*, 1991, 1992). These cells are therefore suppressed in JNK-mediated functions. We found that unlike control vector-modified cells (T98GLHCX) or wild-type c-Jun-modified cells (T98GcJun), both of which are highly resistant to cisplatin, the modified clones expressing mJun (T98GdnJun) showed a significant loss of viability following treatment with cisplatin (Fig. 4). Furthermore, the increase in cisplatin sensitivity correlated with the level of expression of mJun in several different clones (Fig. 5). We observed a similar increase in sensitivity to cisplatin in MCF7 breast cancer cells, in PC-3 prostate cancer cells, and in U87

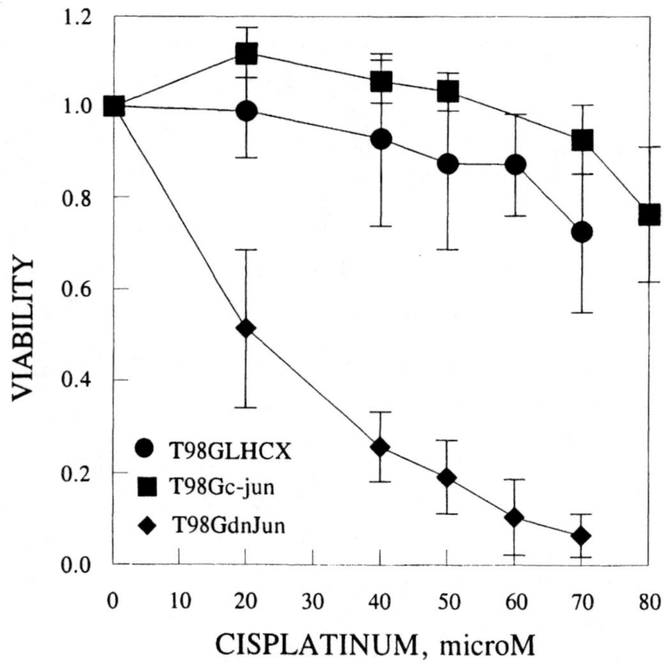

Figure 4. Viability assay showing that mutant Jun sensitizes T98G cells to cisplatin. Viability was assayed 5 days following a 1 hour treatment with cisplatin. Empty vector control cells (●); wild-type c-Jun-expressing cells (■); and mutant Jun-expressing cells (◆).

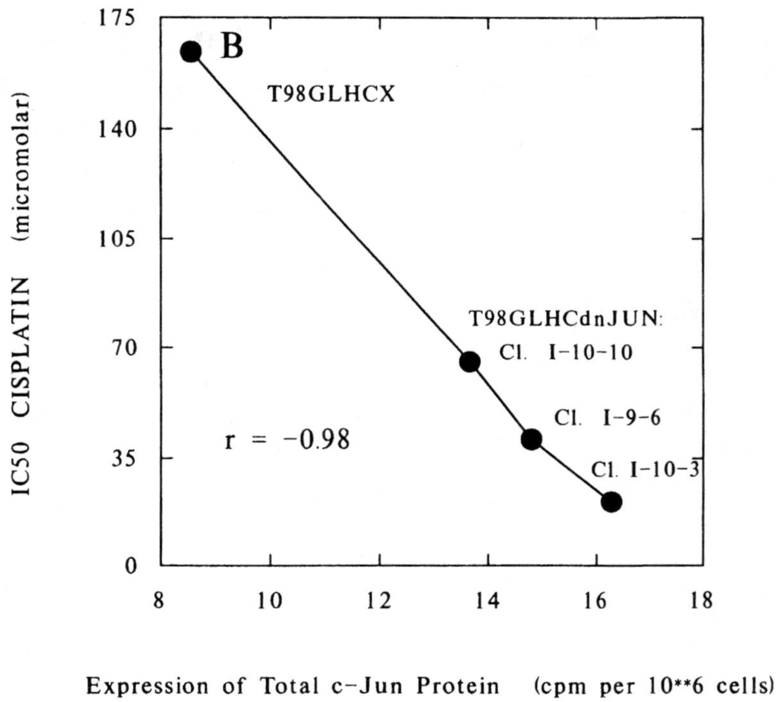

Figure 5. Dose response curve of the IC_{50} (cisplatin) versus total immunoreactive Jun (c-Jun + mutant Jun). Total immunoreactive Jun was determined by sequential immunoprecipitation of ^{35}S-labeled cells using specific Jun B, Jun D, followed by pan-Jun antiserum.

glioblastoma cells that were modified with the non-phosphorylatable mJun (not shown). We also observed enhanced sensitivity to cisplatin in cells modified with the TAM67 mutant of c-Jun (not shown). Tam67 has a truncated N-terminal domain and is known to be a dominant negative inhibitor of c-Jun phosphorylation (Grant *et al.*, 1996). However, we did not observe in mJun modified T98G cells nor in mJun modified PC-3 cells an increase in sensitivity to the anti-mitotic agent, taxotere™ (kindly provided by Dr. Pierre Potier, Centre Nationale de la Recherche Scientifique, Paris), whose primary target is the mitotic spindle rather than DNA (Fig. 6). Taken together these results argue that the JNK/SAPK pathway plays a role in resistance to DNA damaging agents but not to agents that do not damage DNA, and that inhibition of this pathway could be a means to reverse resistance to DNA damaging therapies used in cancer treatment.

3.1.1. Inhibition of DNA Repair by Inhibition of the JNK/SAPK Pathway. T98G clones modified to express mutant Jun are compromised in repair of cisplatin-DNA adducts (Potapova *et al.*, 1997), an observation that further supports the rationale for enhancing sensitivity to DNA damage through inhibition of the JNK/SAPK pathway. DNA repair is known to play a role in acquired resistance to DNA damaging chemotherapies (Barret and Hill, 1998; Crul *et al.*, 1997; Reed, 1998; Fink, Aebi and Howell, 1998; Saves and Masson, 1998; Scanlon, 1989). Furthermore, as summarized in Table 1, a number of the genes involved in DNA synthesis and repair are potentially regulated by the AP-1 transcription factor and related transcription factors targeted by the

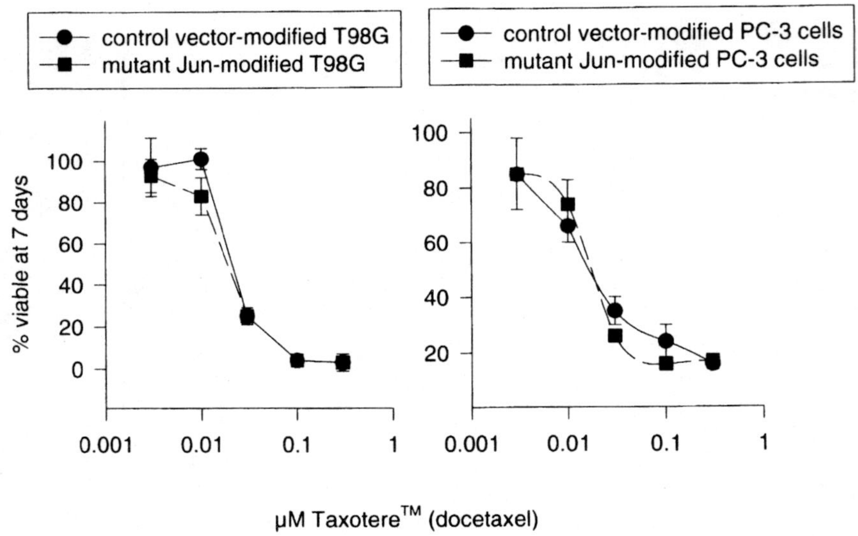

Figure 6. Viability assay showing that mutant Jun does not sensitize T98G cells to Taxotere™. Viability was assayed 7 days following a 1 hour treatment with Taxotere™.

JNK/SAPK pathway (i.e., c-Jun and ATF2, components of AP-1 and c-Jun-ATF2 heterodimeric transcription activating complexes, respectively). Furthermore, down regulation of AP-1 by treatment of cells with a c-fos antisense ribozyme, has previously been shown to correlate with down regulation of expression of certain of these enzymes (thymidylate synthetase, DNA polymerase β) and enhanced sensitivity to cisplatin (Scanlon *et al.*, 1991).

We analyzed DNA repair using a PCR-based assay developed by Jennerwein and Eastman (1991) based on their observations that DNA-cisplatin adducts block the progression of the Taq polymerase and lead to a decrease the yield of PCR product obtained from any given PCR amplicon in proportion to the extent of platination. The authors demonstrated that the relationship between the relative PCR signal strength (P) for a given amplicon from damaged versus undamaged templates, and overall platination level fits a Poisson distribution predicted from a random platination, so that $P = e^{-n}$, where n = adducts per amplicon-sized fragment.

We have used primers described by Oshita and Saijo (1994) to amplify a 2.7 Kb fragment of the human hypoxanthine phosphoribosyl transferase (HPRT) gene, a fragment large enough to sustain readily detectable levels of damage following cisplatin treatment of cells. As an internal control for the efficiency of the PCR reaction, we used a nested 5′ primer which amplified a 150 base fragment of the same gene. At levels of cisplatin used to treat cells, damage to the smaller fragment was undetectable. PCR reactions were performed under quantitative conditions such that the extent of reaction remained directly proportional to amount of template. Figure 7 shows an example of PCR amplification of the 2.7 Kb fragment and the 150 base fragment from DNA prepared from T98G glioblastoma cells treated with 0, 100, and 200 μM cisplatin for 1 hour. Quantitation of band intensities was accomplished using Kodak digital camera and analysis software, and relative band intensity measurements were then used to calculate adducts per Kb based on the Poisson relationship described above.

Table 1. DNA REPAIR ASSOCIATED GENES WHICH CONTAIN POTENTIAL JNK-REGULATED SEQUENCES[1]

Repair-associated gene	Element	Sequence (consensus sequence)[2]	Position[3]	LLG score	Reference for function
DNA Polymerase β	AP-1/TRE	CTGACTCA (- t g a c t c a)	337	2.0	Known to be functional and TPA-activated classic TRE (1).
	ATF/CREB	TTACGTAA (t t a c g t c a)	282	2.0	Known to be genotoxic-activated (1, 9).
DNA Polymerase α (26)	ATF/CREB	CACGTCA (t/g a c g t c a)	−82		Function of ATF/CREB (26) unknown.
		GGGGTCA (t g a g t c a)	−149		AP-1 sites thought to be significant in DNA repair: DPα is over-expressed in cisplatin resistent cells and anti-Fos ribozyme sensitizes (27).
Topoisomerase I	AP-1	GGGGGCGG (t g a g t c a)	753	2.0	(2–3)
	AP-1	TGACCCA (t g a c t c a)	217	2.0	(2–3)
	ATF/CREB	TGACGTCA (t g a c g t c a)	792	2.0	known to be functional & stress-activated (2–3).
Topoisomerase IIα	ATF/CREB	TGACGCCG (t g a c g t c a)	286	2.0	(10–11); Topo II is UV-inducible and functions early in UV-induced DNA damage repair.
	AP-1	TGATTGG (t g a g t c a)	337	2.0	(10)
Topoisomerase IIβ				2.0	(12)
	AP-1	TGACTCA (t g a c t c a)	3	1.6	(12)
	AP-1	AGAGTCA (t g a g t c a)	65		
ATF-3	AP-1	TGAGTAA (t g a c g t c a)	−1,600		(13) ATF-3 is stress-induced, anisomycin (JNK activator) induced, and induced by ATF-2/c-Jun coexpression suggesting a functional role for the ATF/CREB site. ATF-3/c-Jun heterodimers bind ATF/CREB sites and activate transcription (14–15) and ATF-3/c-Jun and ATF-3/JunD heterodimers have been shown to bind TTAGTTAC, a ATF/CREB sequence, which mediates EGF/ras/raf-stimulated transcription (28), however, a role in induction of DNA repair genes is not known.
	AP-1	ATAGTCA (t g a c g t c a)	−1,353		
	AP-1	AGACTAA (t g a c g t c a)	−605		
	AP-1	GAGTCA (t g a c g t c a)	−380		
	ATF/CRE	TTACGTCA–92 (t t a c g t c a)			

(*continued*)

Table 1. *Continued*

Repair-associated gene	Element	Sequence (consensus sequence)[2]	Position[3]	LLG score	Reference for function
c-Jun	ATF/CREB	TTACCTCA (t t a c g t c a)		2.0	(4–5); the "functional" association with DNA repair is strong induction of c-*Jun* by genotoxins known to activate JNK/SAPK
Uracil Glycosylase	AP-1	TGGGTCA (t g a g t c a)	141	2.0	not known
PCNA	AP-1	TGACTCA (t g a c t c a)	489	2.0	DNA polymerase-•• accessory protein function;
Proliferating Cell Nuclear Antigen	ATF/CREB	TGAGGTCAGGG (t g a c g t c a - - -)	209	1.64	(7–8); Il-2, a potent JNK/SAPK activator, induces PNCA expression via ATF/CREB promoter sites which is blocked by rapamycin.
	ATF/CREB	GTGACGTCAC (- t t a c g t c a - -)	1,253	1.60	
GADD153 (19)	ATF/CREB	ACTCCTGACCTT (t g/t a c g t c a - - - -)	207	1.63	Induction requires phosphorylation-dependent event that is *not* PKA, PKC (16), or p38 (25) mediated consistent with a role for JNK/SAPK (16). Moreover GADD153 is induced by MMS (16) & cisplatin (17–18). The role of the ATF/CREB site is unknown. GADD153 is phoshorylated and activated by p38 in response to stress (25).
Growth Arrest and DNA-damage" Inducible Gene	AP-1	TGACTCA (t g a c t c a)	710	2.0	
XRCC1 (20)	ATF/CREB	ACGTCA (a c g t c a)	1,815	2.0	The site at 466 is consistent with c-Jun/ATF-3 *vs.* ATF-2 (TESS).
X-ray Damage Repair Cross Completementing Gene Product.	ATF/CREB	GGACGTCAA (t g a c g t c a)	1,814	2.0	Functional roles of these ATF/CREB and AP-1 sites are not known zxzx.
	ATF/CREB	CCTGACCTCA (- - t g a c g t c a)	2,029	1.64	
	ATF/CREB	GCTGACGTCAG (- - t g a c g t c a -)	466	1.60	
		CCAATCA (t g/t a c/g t c a)	93	2.0	

MGMT (22)	ATF/CREB	TGCGTCA (t g a c g t c a)	1,661	2.0	MGMT is induced by genotoxic agents (21). The site at 1,674 is consistent with c-Jun/ATF-3 (TESS). The functional significance of these sites is unknown.
O6-Methylguanine-DNA-Methyl-transferase	ATF/CREB	GTGACATCAT (- t g a c t c a -)	1,195		
	AP-1	TGAGTCA (t g a g t c a)	734	2.0	
	AP-1	TTACTCA (t t a c t c a)	285	1.73	
MSH2 (23)	ATF/CREB	TGGCGTCA (t g a c t c a)	108	1.62	TESS does not recognize c-Jun participation at 108 site. Role in cisplatin induced repair unknown. MSH2 has been reported to selectively bind to cisplatin-DNA adducts
	AP-1	TGAATCA (t g a c/g t c a)	569	2.0	
	AP-1	TGATGAAA (t g a c/g t c a)	884	1.62	
Metallothionein IIA	AP-1	GAGCCGCAAGT (g a g t c a - - - - t GACTTCTAGCG g a c t c a a g t c CGGGGCGTG a - - - - - - t g)	188	2.0	(6); TPA and UV-light activated

[1]Repair associated protein for which only partial promoter sequences are known (i.e. in Genbank) without recognizable AP-1 regulated sites include ADP ribose poly merase, tif2/ref1, SSRP, Ercc1 and thymidylate synthetase.

[2]AP-1 consensus: TG/TAC/GTCA; CREB/ATF-2 consensus: TG/TACGTCA.

[3]Positions are based on TESS numbering of promoter sequences unless preceeded by (–).

References: 1. Srivastava *et al.* J. Biol. Chem., 1995;270:16408. 2. Baumgärtner *et al.*, Biochem. Biophys. Acta, 1994;1218:123. 3. Heiland *et al.* Eur. J. Biochem. 1993;217:813. 4. Kharbanda *et al.* Canc. Res., 1991;51:6636. 5. Kharbanda *et al.*, J. Clin. Inves., 1990;86:1517. 6. Lee *et al.* Nature, 1987;325:368. 7. Feuerstein N. *et al.* J. Biol. Chem., 1995;270:9454. 8. Huang D. *et al.* Molec. Cell. Biol., 1994;14:4233; 9. Kedar *et al.* PNAS USA, 88:3729. 10. Hochhauser D. *et al.* J. Biol. Chem., 1992;26:187961–187965. 11. Popanda O. Thielmann H.W. Carcinogenesis, 1992;13:2321. 12. Unpublished: Genebank Acc. #T29334, Inst. Genomics Res., 1995. 13. Liang *et al.* J. Biol. Chem., 1996;271:1695. 14. Hsu, J.-C. *et al.*, Mol. Cell. Biol., 1992;12:4654. 15. Chu, H.-M. *et al.* Mol. Endocrinol. 1994;8:59. 16. Luethy, J. and Holbrook, N. Cancer Res. 1994;54:1902S. 17. Gately, D. *et al.* Br. J. Cancer, 1994;70:1102. 18. Delmastro, D. *et al.* Canc. Chemoth. Pharmacol., 1997;39:245. 19. Park, J. *et al.*, Gene, 1992;116:259 (Acc. #S40707). 20. Lamerdin, J., *et al.*, Genomics, 1995;25:547 (Acc. #L34079). 21. Lefebvre, P. *et al.*, *Dna Cell Biol.*, 1993;12:233. 22. Iwakuma, T. *et al.*, DNA Cell Biol., 1996;15:863. 23. Seib, T. *et al.* Gene Bank direct submission Acc. #U232824. 24. Leach, F. *et al.* Cell, 1993;75:1215. 24. Kolodner, R. *et al.* Genomics, 1994;24:515. 24. Mello, J. *et al.* Chem. Biol. 1996;3:579. 25. XiaoZhong, W. and Ron, D. Science, 1997;272;1347–1349. 26. Pearson, B. *et al.* Molec. Cell. Biol., 1991;11:2081. 27. Scanlon, K. *et al.* PNAS USA, 1994;91:11123. 28. Nilsson *et al.* *Cell Growth & Diff.* 1997;8:913.

Abbreviations: LLG, Log likeli-hood score which is 2 for a perfect match of the candidate response element with with the consensus sequence (TESS criteria) and all ambiguous matches yield a score of 0; AP-1, activator protein-1 complex, a Jun and a Fos family member; CREB/ATF, cAMP response element binding proteins, N-terminal phosphorylated c-Jun and N-terminal phosphorylated ATF-2 for a cAMP-independent complex which bind the CRE-like sequence TTACGTCA. XRCC., MMS, methyl methanesulfonate.

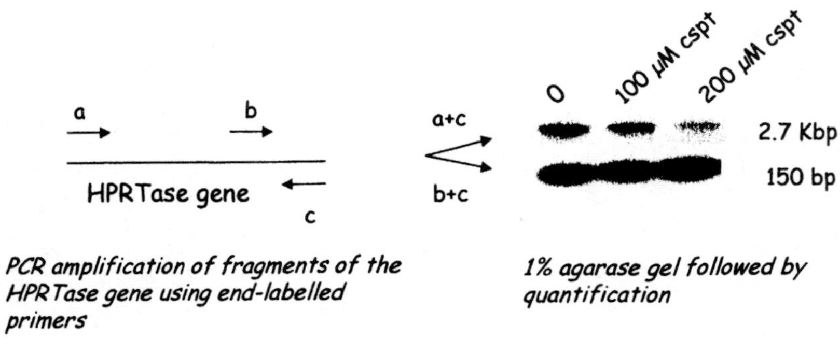

Figure 7. PCR amplification of DNA from T98G cells that had been incubated 1 hour with 0, 100 μM, and 200 μM cisplatin. Products were analyzed on an agarose gel and stained with ethidium bromide, followed by quantitation of band intensities using the Electrophoresis Documentation and analysis System 120 (Kodak Digital Science™).

Table 2 summarizes the cisplatin adducts per Kb observed on DNA from T98G glioblastoma cells treated for 1 hour with 200 μM cisplatin and then harvested either immediately or after a 6 hour recovery period. As shown in the table, cisplatin adducts occur at greater frequency in DNA from T98G mutant Jun expressing cells than in DNA from parental T98G cells. Furthermore, parental T98G cells repair more than half of the adducts during the 6 hour recovery period, whereas little repair occurs in T98G mJun cells during that period of time. Thus the increase in cisplatin sensitivity observed in T98G mJun cells can be accounted for at least in part by a defect in their ability to repair damaged DNA.

3.2. Mechanism of Action of JNK/SAPK in DNA Repair

How might the activation of JNK by genotoxic stress affect DNA repair? The answer to this question is not known in any detail, but a number of observations point to several testable hypotheses. It was noted that the promoters of certain genes known to be involved in carrying out cisplatin-DNA adduct repair contain AP-1-like or ATF2/CREB regulatory elements (section 3.1.1). The repair of DNA-cisplatin adducts is believed to require the nucleotide-excision repair process (Reed, 1998). In addition, a number of enzymes utilized for DNA synthesis are reported to be involved (reviewed in Zamble and Lippard, 1995). These included DNA polymerase β, topoisomerase I, topoisomerase II, uracyl glycosylase, PCNA, metallothionine and others (Table 1, and references therein). In each case the promoter regions contains one or more AP-1

Table 2. Cisplatin-DNA adducts per 2.7 KBASE

Treatment	T98G parental cells	T98G mutant jun cells
200 μM cisplatin 1 hour	0.48 ± 0.08	0.63 ± 0.2
200 μM cisplatin 1 hour, 6 hour recovery	0.20 ± 0.05	0.55 ± 0.2

or ATF2/CREB regulatory sequences. In several cases such as DNA polymerase beta, genotoxic stress in the form of UV-irradiation is know to induce the gene and that the ATF2/CREB sites are required for this event (Table 1 and references therein). Moreover, the N-terminal phosphorylation of c-Jun and ATF2 by the action of Jun Kinase leads to increased DNA binding and transactivation potential of heterodimeric complex formed by phosphorylated c-Jun and ATF2. Indeed, the c-Jun promoter itself is one of the better studied example of a gene that is up-regulated upon binding of the c-Jun-ATF2 heterodimer (van Dam *et al.*, 1995; Wilhelm *et al.*, 1995). N-terminal phosphorylation of both c-Jun and ATF2 is required for the activation of the c-jun gene (van Dam *et al.*, 1995; Wilhelm *et al.*, 1995). Other examples of gene that are up-regulated upon formation of c-Jun-ATF2 heterodimers include the ELAN promoter (De Luca, *et al.*, 1994) and the TNF-alpha promoter (Newell *et al.*, 1994). Thus, there is circumstantial evidence supporting the hypothesis that activation of the Jun Kinase pathway by genotoxic stress leading to the phosphorylation of both ATF2 and c-Jun and thereby causing preferential formation of ATF2-c-Jun heterodimers. These complexes preferentially bind promoters containing octomeric ATF2/CREB regulatory elements. This hypothesis predicts that several of the genes known to be involved in cisplatin-DNA adduct repair may be coordinately upregulated upon activation of the JNK pathway following damage to DNA by cisplatin. If verified, such an explanation suggests that the JNK pathway is a potentially useful target for intervention and that agents that inhibit JNK activity or formation may be useful in achieving enhanced sensitivity of tumor cells to cisplatin.

3.3. JNK/SAPK Pathway with Antisense Approaches

The above observations argue that targeting the JNK/SAPK pathway could have potential as a sensitizer to a variety of DNA damaging chemotherapies, a possibility that we have investigated directly using antisense oligonucleotides to either JNK 1 or 2 family of JNK isoforms.

Antisense mechanisms result most likely from oligonucleotide-target mRNA hybrid formation in the nucleus which stimulates the cleavage of the mRNA at one or more sites near the terminus of the hybrid complex by ribonuclease H (Crooke, 1992, 1998; Dean *et al.*, 1996), as well as from cytoplasmic complexes which may lead to translation arrest (Crooke, 1992, 1998; Dean *et al.*, 1996). Because steady state mRNA levels do not always predict steady state gene product levels, we monitor the effects antisense by the fractional reduction of the gene product. An enzymatic assay that discriminates among the various isoforms of Jun Kinase can be used (Hibi *et al.*, 1993), as well as Western analysis or PAGE analysis following immunoprecipitation.

The appropriate oligonucleotide controls for antisense efficacy have been described in detail (Stein and Kreig, 1994). and include a sense sequence, scrambled sequence or random control oligonucleotides, and antisense sequences bearing one or a small number of mismatched bases, i.e. mismatched with respect to the complementary target sequence.

For our studies we have used antisense oligonucleotides complementary to the JNK1 or JNK 2 families of JNK isoforms by selecting target sequences common to the major isoforms of human JNK1 and JNK2. Phosphorothioate backbone chemistry has been used as this formulation is nuclease resistant, readily available and a common element of second generation compounds (Crooke, 1992, 1998; Dean *et al.*, 1996; McKay *et al.*, 1999).

In collaboration with ISIS Pharmaceuticals, Inc. (Carlsbad, CA) we have identified effective antisense oligonucleotides capable of entering and inhibiting specifically JNK1 or JNK2 mRNA expression in A549 lung cancer cells (Bost *et al.*, 1997, 1999). As an example of a screening for JNK1 and JNK2, Fig. 8 is included. As shown, most candidate oligonucleotides had only a small effect on the steady state level of JNK1 or JNK2 mRNA (Fig. 8). However, their degree of activity greatly varied depending on the targeted region (Fig. 8). Among the most potent oligonucleotides of the array complementary to JNK1 is ISIS 12539 (JNK1ASISIS12539) (5′-CTCTCTGTAGGCC-CGCTTGG-3′) located in the 3′ end of the coding region. Treatment with this oligonucleotide leads to a reduction of steady state mRNA level by 95% (Fig. 8A) as observed 24 hours after a 4 hour exposure of the cells to 0.4 μM antisense oligonucleotides. ISIS 12560 (JNK2ASISIS12560) (5′-GTCCGGGCCAGGCCAAAGTC-3′), located in the 5′ end of the coding region of JNK2 genes, was the most efficient oligonucleotide in reducing JNK2 steady state mRNA levels. Lipofection with this oligonucleotide lead to a decrease in the steady-state JNK2 message level by 92% compared to the control JNK2 steady state mRNA level. The oligonucleotides targeted to the regions flanking either side of the sequences for JNK1ASISIS12539 and JNK2ASISIS12560 also inhibited JNK1 and JNK2 steady state mRNA but in a manner that decreased with increasing distance from the optimum sequence (Fig. 8A and B) suggesting that discrete regions of the target mRNA are accessible and sensitive to the antisense-mediated elimination of steady-state mRNA as previously described (Crooke, 1992, 1998).

Cross inhibition tests demonstrated isoform class specificity. As shown in Fig. 9, JNK1ASISIS12539 eliminates JNK1 mRNA and protein but does not affect JNK2 mRNA and protein and, conversely, JNK2ASISIS12560 has no effect on JNK1 mRNA and protein but abolishes JNK2 mRNA and protein. In the northern analysis we describe an example where two other antisense oligonucleotides from the initial gene screening are less efficient in reducing mRNA steady-state levels of their respective target gene (Fig. 9). Similarly, western analysis shows that lipofections with the antisense oligonucleotides leads to complete elimination of their respective target steady state protein levels, whereas the scrambled sequence versions of the candidate oligonucleotides, JNK1Scr or JNK2Scr have no effect (Fig. 10). Thus effective and specific reagents can be obtained.

These reagents have been used to examine the growth promoting roles of the Jun Kinase pathway in human T98G glioblastoma cells (Potapova *et al.*, 1998), human A549 NSCLC cells (Bost *et al.*, 1997, 1999), and human PC3 prostate carcinoma cells (Yang *et al.*, 1998) and shown to specifically inhibit growth and, in the case of A549 cells, anchorage independent growth. Moreover, these reagents have been used in systemic treatment of established PC3 xenografts and shown to block growth and promote regression of the established tumors in high frequency (Bost *et al.*, 1998). In this case, growth inhibition by antisense treatment of 78% was superior to that of cisplatin treatment alone (47%), however the effects of combined antisense JNK-cisplatin of 89% inhibition was greater than either alone suggesting that, indeed, it may be possible to sensitize solid tumors to cisplatin by elimination of Jun Kinase. It will be of interest, therefore, to determine whether JNK in involved in a general mechanism utilized in other tumor cell lines and/or mediates DNA damage repair of other DNA-damaging agents.

4. CONCLUSIONS

These results have clinical implications with regard to the application of therapies designed to alleviate drug resistance. As tumors progress, particularly if they have been exposed to DNA damaging therapies, they may upregulate their DNA repair response

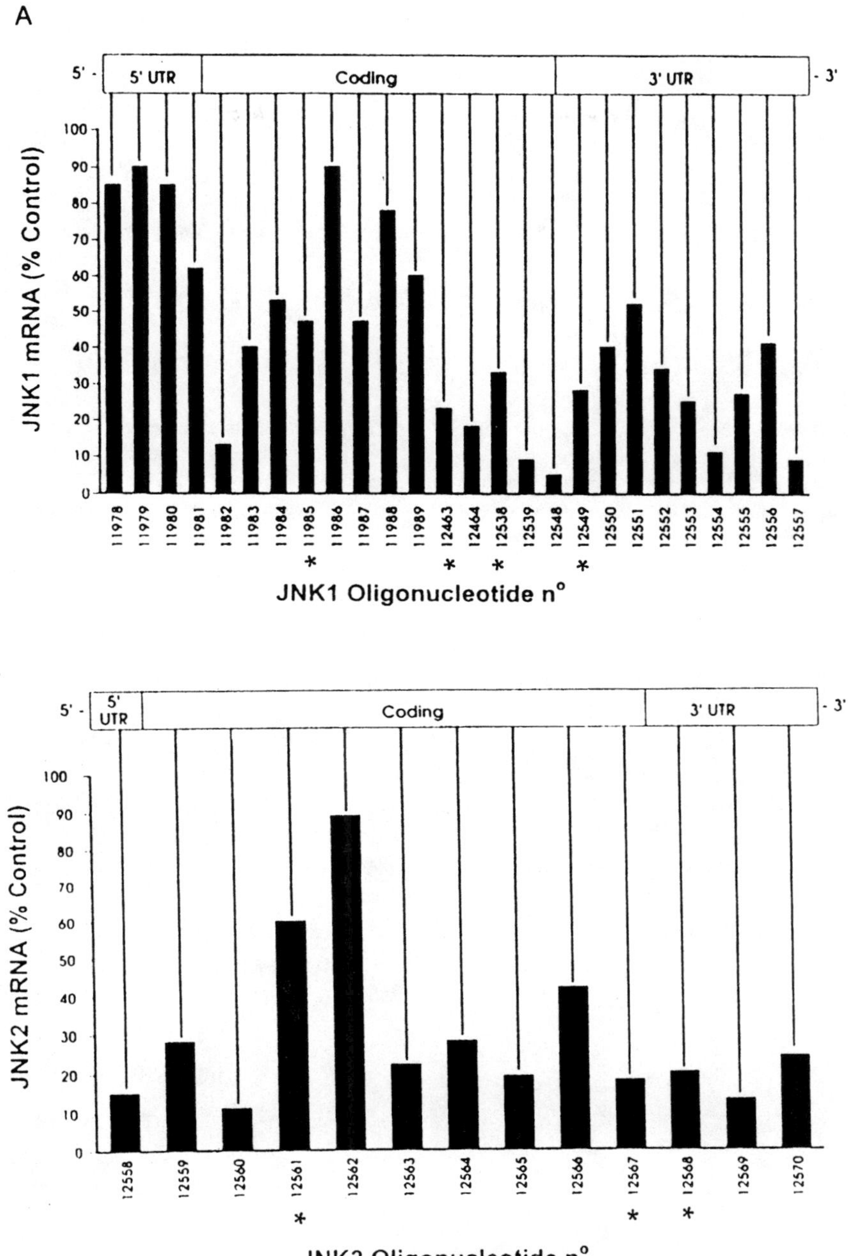

Figure 8. "Messenger walk" survey for the elimination of JNK1 and JNK2 mRNA levels following treatment with phosphorothioate antisense oligonucleotides targeted to JNK1 or JNK2 mRNA. A, JNK1 mRNA steady state levels in A549 cells following treatment with 26 different phosphorothioate antisense oligonucleotides targeted to JNK1 mRNA. JNK1AS^{ISIS12539} gave the most consistent results for the elimination of JNK1 mRNA steady state levels. B, Similar experiment to the previous one (Fig. 1A) was performed with 13 phosphorothioate antisense oligonucleotides complementary to the indicated regions of the JNK2 mRNA. JNK2AS^{ISIS12560} yielded the most consistent results for the elimination of JNK2 mRNA steady state levels. All oligonucleotides are arrayed relative to their complementary sequence along the JNK transcript. The asterisks indicate the oligonucleotides having a very similar nucleotide composition (2–4 bases) to the leading compound. These screenings have been repeated two times with similar results.

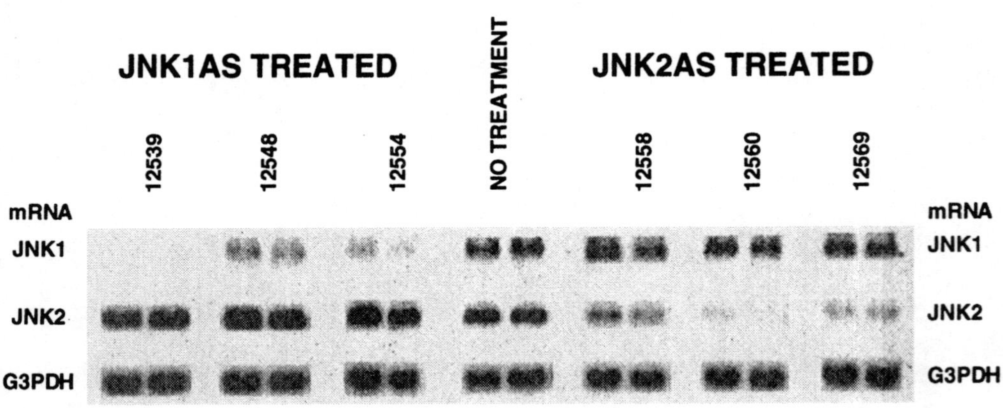

Figure 9. A549 cells were treated with 3 different antisense oligonucleotides complementary to JNK1 mRNA (including the active antisense oligonucleotide JNK1AS^{ISIS12539} and 3 different antisense oligonucleotides complementary to JNK2 mRNA (including the active antisense oligonucleotide JNK2AS^{ISIS12560}). 24 hours after a 4 hours transfection with 0.4 μM antisense oligonucleotide, mRNA was prepared and examined by northern analysis, the same membrane was hybridized successively with JNK1 probe, JNK2 probe and the G3PDH probe.

(see Reed, 1998 for review), which may in part involve AP-1 regulated DNA synthesis and repair genes. Elevated c-fos expression, one component of the AP-1 transcription factor, correlates with cisplatin resistance (Scanlon *et al.*, 1991; Funato *et al.*, 1992) and may directly affect the synthesis of AP-1 regulated DNA synthesis and repair genes, such as polymerase β, topoisomerase I and thymidylate synthetase (Scanlon and Kashini-Sabet, 1998). Loss of DNA damage-induced apoptosis through loss of wild-type p53 expression also correlates with drug resistance (O'Connor *et al.*, 1997) and represents an independent mechanism through which tumors acquire resistance to therapy. Because the success or failure of DNA repair could be critical in determining how a cancer cell responds to expression of p53, it may be necessary, in optimizing the benefits of

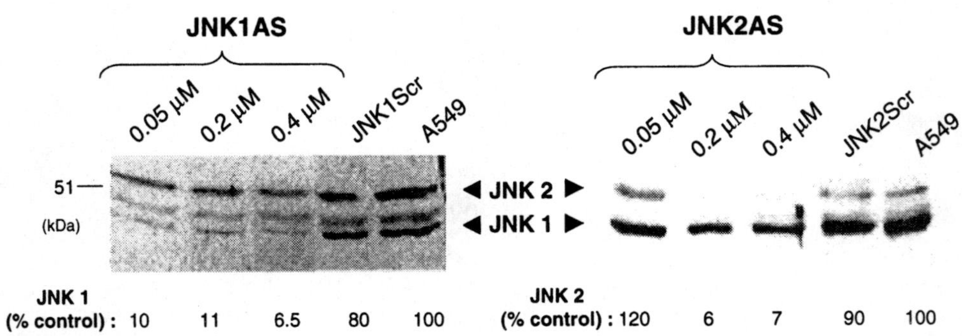

Figure 10. A549 cells were treated with the indicated concentration of phosphorothioate 2′-O-methoxyethyl-modified antisense oligonucleotides JNK1AS^{ISIS15346} or JNK2AS^{ISIS15353} and 0.4 μM of their respective control oligonucleotides JNK1Scr^{ISIS18076} and JNK2Scr^{ISIS18078}. Cell extracts were prepared 36 hours after the transfection and examined by western analysis using anti-JNK1 antibodies (SC-571). Protein level (below) was determined by comparison to respective protein level in untreated cells using the Electrophoresis Documentation and analysis System 120 (Kodak Digital Science™).

p53-based approaches, to concurrently down-regulate the cellular stress and DNA damage response pathways such as the Jun Kinase pathway. Combination approaches targeting independently these two cellular responses to DNA damage can be foreseen today, and may provide an optimal strategy for reversing or alleviating drug resistance in advanced cancers.

ACKNOWLEDGMENTS

The authors would like to acknowledge financial support form the following: grants CA69546 (RAG), CA56834 (DM), CA76173 (DM) from the National Cancer Institute, grant DAMD17-96-1-6038 from the U.S. Army Breast Cancer Research Program (RAG), grant 1PF0140 from the California Cancer Research Program (RAG), grant 83401 from the California Breast Cancer Research Program (DM), and Introgen Therapeutics, Inc. (RAG), The authors would also like to thank Drs. Fred Bost and Olga Potapova for making available to us manuscripts prior to publication. We thank Deborah Wilson, Ph.D. of Introgen Therapeutics, Inc. for providing us with replication recombinant adenoviruses (Ad-p53, Ad-βgal, Ad-Luc). We also thank R. McKay, Ph.D. and N. Dean, Ph.D. of ISIS Pharmaceuticals for their support and materials for the studies reviewed here.

REFERENCES

Badie, B., Kramar, M.H., Lau, R., Boothman, D.A., Economou, J.S., and Black, K.L., 1998, Adenovirus-mediated p53 gene delivery potentiates the radiation-induced growth inhibition of experimental brain tumors, *J. Neurooncol.* **37**:217–222.

Bakalkin, G., Selivanova, G., Yakovleva, T., Kiseleva, E., Kashuba, E., Magnusson, K.P., Szekely, L., Klein, G., Terenus, L., and Wiman, K.G., 1995, p53 binds single stranded DNA ends through the C-terminal domain and internal DNA segments via the middle domain, *Nucl. Acids Res.* **23**:362–269.

Barret, J.M., and Hill, B.T., 1998, DNA repair mechanisms associated with cellular resistance to antitumor drugs: potential novel targets, *Anticancer Drugs* **9**:105–123.

Binetruy, B., Smeal, T., and Karin, M., 1991, Ha-ras augments c-Jun activity and stimulates phosphorylation of its activation domain, *Nature* **351**:122–127.

Cavigelli, M., Dolfi, F., Claret, F.X., and Karin, M., 1995, Induction of c-fos expression through JNK-mediated TCF/Elk-1 phosphorylation, *EMBO J.* **14**:5957–5964.

Chelliah, J., and Jarvis, W.D., 1996, Effect of 1-beta-D-arabinofuranosylcytosine on apoptosis and differentiation in human monocytic leukemia cells (U937) expressing a c-Jun dominant-negative mutant protein (TAM67), *Cell Growth Differ.* **7**:603–613.

Clarke, A.R., Purdie, C.A., and Harrison, D.J., *et al.*, 1993, Thymocyte apoptosis induced by p53-dependent and independent pathways, *Nature* **362**:849–852.

Crooke, S.T., 1992, Therapeutic applications of oligonucleotides, *Annu. Rev. Pharmacol. Toxicol.* **32**:329–376.

Crooke, S.T., 1998, Molecular mechanisms of antisense drugs: Rnase H, *Antisense Nucleic Acid Drug Dev.* **8**:133–134.

Crul, M., Schellens, J.H., Beijnen, J.H., and Maliepaard, M., 1997, Cisplatin resistance and DNA repair, *Cancer Treat Rev.* **23**:341–366.

Dean, N.M., McKay, R., Miraglia, L., Geiger, T., Muller, M., Fabbro, D., and Bennett, C.F., 1996, Antisense oligonucleotides as inhibitors of signal transduction: development from research tools to therapeutic agents, *Biochem. Soc. Trans.* **24**:623–629.

De Luca, L.G., Johnson, D.R., Whitley, M.Z., Collins, T., and Pober, J.S., 1994, cAMP and tumor necrosis factor competitively regulated transcriptional activation through a nuclear factor binding to the cAMP-responsive element/activating transcription fact element of the endothelial leukocyte adhesion molecule-1 (E-selectin) promoter, *J. Biol. Chem.* **269**:19193–19196.

Derijard, B., Hibi, M., Wu, I.H., Barrett, T., Su, B., Deng, T., Karin, M., and Davis, R.J., 1994, JNK1: a protein kinase stimulated by UV light and Ha-Ras that binds and phosphorylates the c-Jun activation domain, *Cell* **76**:1025–1037.

Dorigo, O., Turla, S.T., Lebedeva, S., and Gjerset, R.A., 1998, Sensitization of rat glioblastoma multiforme to cisplatin *in vivo* following restoration of wild-type p53 function, *J. Neurosurg.* **88**:535–540.

Fink, D., Aebi, S., Howell, S.B., 1998, The role of DNA mismatch repair in drug resistance, *Clin. Cancer Res.* **4**:1–6.

Fujiwara, T., Grimm, E.A., Mukhopadhyay, T., Zhang, W.W., Owen-Schaub, L.B., and Roth, J.A., 1994, Induction of chemosensitivity in human lung cancer cells *in vivo* by adenovirus-mediated transfer of the wild-type p53 gene, *Cancer Res.* **54**:2287–2291.

Funato, T., Yoshida, E., Jiao, L., Tone, T., Kashani-Sabet, M., and Scanlon, K.J., 1992, The utility of an anti-fos ribozyme in reversing cisplatin resistance in human carcinomas, *Adv. Enzyme Regul.* **32**:195–209.

Gjerset, R.A., Turla, S.T., Sobol, R.E., Scalise, J.J., Mercola, D., Collins, H., Hopkins, and P.J., 1995, Use of wild-type p53 to achieve complete treatment sensitization of tumor cells expressing endogenous mutant p53. *Molec. Carcinog.* **14**:275–285.

Gjerset, R.A., Dorigo, O., and Maung, V., *et al.*, 1997, Tumor regression *in vivo* following p53 combination therapy, *Cancer Gene Therapy* **4**:0–2.

Grant, S., Freemerman, A.J., Birrer, M.J., Martin, H.A., Turner, A.J., Szabo, E., and Reed, E., 1998, Platinum-DNA adduct, nucelotide excision repair and platinum based anti-cancer chemotherapy, *Cancer Treat Rev.* **24**:331–344.

Gupta, S., Campbell, D., Derijard, B., and Davis, R.J., 1995, Transcription factor ATF2 regulation by the JNK signal transduction pathway, *Science* **267**:389–393.

Hibi, M., Lin, A., Smeal, T., Minden, A., and Karin, M., 1993, Identification of an oncoprotein- and UV-responsive protein kinase that binds and potentiates the c-Jun activation domain, *Genes Dev.* **7**:2135–2148.

Jennerwein, M.M., and Eastman, A., 1991, A polymerase chain reaction-based method to detect cisplatin adducts in specific genes, *Nucleic Acids Res.* **19**:6209–6214.

Lee, S., Elenbas, B., Levine, A., and Griffith, J., 1995, p53 and its 14kDa C-terminal domain recognize primary DNA damage in the form of insertion/deletion mismatches, *Cell* **81**:1013–1021.

Levine, A.J., 1993, The tumor suppressor genes, *Annu. Rev. Biochem.* **62**:623–651.

Levine, A.J., 1997, p53, the cellular gatekeeper for growth and division, *Cell* **88**:323–331.

Livingstone, C., Patel, G., and Jones, N., 1995, ATF-2 contains a phosphorylation-dependent transcriptional activation domain, *EMBO J.* **14**:1785–1797.

Lotem, J., and Sachs, L., 1993, Hematopoietic cells from mice deficient in wild-type p53 are more resistant to induction of apoptosis by some agents, *Blood* **82**:1092–1096.

Lowe, S.W., Ruley, H.E., and Jacks, T., and Housman, D.E., 1993, p53-mediated apoptosis modulates the cytotoxicity of anti-cancer agents, *Cell* **74**:957–967.

Lowe, S.W., Bodis, S., McClatchy A., Remington, L., Ruley, H.E., Fisher, D.E., Housman, D.E., and Jacks, T., 1994, p53 status and the efficacy of cancer therapy *in vivo*, *Science* **266**:807–810.

McKay, R.A., Miraglia, L.J., Cummins, L.L., Owens, S.R., Sasmor, H., and Dean, N.M., Characterization of a potent and specific class of antisense oligonucleotide inhibitor of human protein kinase C-alpha expression, *J. Biol. Chem.* **274**:1715–1722.

Miyake, H., Hara, I., Gohji, K., Yamanaka, K., Arakawa, S., and Kamidono, S., 1998, Enhancement of chemosensitivity in human bladder cancer cells by adenoviral-mediated p53 gene transfer, *Anticancer Res.* **18**:3087–3092.

Newell, C.L., Deisseroth, A.B., and Lopez-Berestein, G., 1994, Interaction of nuclear proteins with an AP-1/CRE-like promoter sequence in the human TNF-alpha gene, *J. Leukocyte Biol.* **56**:27–35.

Nguyen, D.M., Spitz, F.R., Yen, N., Cristiano, R.J., and Roth, J.A., 1996, Gene therapy for lung cancer: enhancement of tumor suppression by a combination of sequential systemic cisplatin and adenovirus-mediated p53 gene transfer, *J. Thorac. Cardiovasc. Surg.* **112**:1372–1376, discussion 1376–1377.

Nielsen, L.L., Lipari, P., Dell, J., Gurnani, M., and Hajian, G., 1998, Adenovirus-mediated p53 gene therapy and paclitaxel have synergistic efficacy in models of human head and neck, ovarian, prostate, and breast cancer, *Clin. Cancer Res.* **4**:835–846.

O'Connor, P.M., Jackman, J., Bae, I., Myers, T.G., Fan, S., Mutoh, M., Scuderio, D.A., Monks, A., Sausville, E.A., Weinstein, J.N., Friend, S., Fornace, A.J., Jr., and Kohn, K.W., 1997, Characterization of the p53 tumor suppressor pathway in cell lines of the National Cancer Institute anticancer drug screen and correlations with the growth-inhibitory potency of 123 anticancer agents, *Cancer Res.*, **57**:4285–4300.

Oshita, F., and Saijo, N., 1994, Rapid polymerase chain reaction assay to detect variation in the extent of gene-specific damage between cisplatin- or VP-16-resistant and sensitive lung cancer cell lines, *Jpn J. Cancer Res.* **85**:669–673.

Potapova, O., Haghighi, A., Bost, F., Liu, C., Birrer, M.J., Gjerset, R., and Mercola, D., 1997, The JNK/stress-activated protein kinase pathway functions to regulate DNA repair and inhibition of the pathway sensitizes tumor cells to cisplatin, *J. Biol. Chem.* **272**:14041–14044.

Potapova, O., Bost, F., Dean, N., and Mercola, D., 1998, Inhibition of the JNK pathway blocks cell cycle progression in human tumor cells, *Proc. AACR*, 89th Annual Meeting, abstract #1720, p. 251.

Reed, E., 1998, Platinum-DNA adduct, nucleotide excision repair and platinum based anti-cancer chemotherapy, *Cancer Treatment Reviews* **24:** 331–344.

Roth, J.A., Swisher, S.G., Merritt, J.A., Lawrence, D.D., Kemp, B.L., Carrasco, C.H., El-Naggar, A.K., Fossella, F.V., Glisson, B.S., Hong, W.K., Khurl, F.R., Kurie, J.M., Nesbitt, J.C., Pisters, K., Putnam, J.B., Schrump, D.S., Shin, D.M., and Walsh, G.L., 1998, Gene therapy for non-small cell lung cancer: a preliminary report of a phase I trial of adenoviral p53 gene replacement, *Semin. Oncol.* **25**:33–37.

Saves, I., and Masson, J.M., 1998, Mechanisms of resistance to xenobiotics in human therapy, *Cell Mol Life Sci.* **54**:405–426.

Scanlon, K.J., Kashani-Sabet, M., 1988, Elevated expression of thymidylate synthase cycle genes in cisplatin-resistant human ovarian carcinoma A2780 cells, *Proc. Natl. Acad. Sci. USA.* **85**:650–653.

Scanlon, K.J., Kashani-Sabet, M., Miyachi, H., Sowers, L.C., and Rossi, J., 1989, Molecular basis of cisplatin resistance in human carcinomas: model systems and patients, *Anticancer Res.* **9**:1301–1312.

Scanlon, K.J., Jiao, L., Funato, T., Wang, W., Tone, T., Rossi, J.J., and Kashani-Sabet, M., 1991, Ribozyme-mediated cleavage of c-fos mRNA reduces gene expression of DNA synthesis enzymes and metallothionein, *Proc. Natl. Acad. Sci. USA.* **88**:10591–10595.

Smeal, T., Binetruy, B., Mercola, D., Birrer, M., and Karin, M., 1991, Oncogenic and transcriptional cooperation with Ha-ras requires phosphorylation of c-Jun on serines 63 and 73, *Nature* **354**:494–496.

Smeal, T., Binetruy, B., Mercola, D., Grover-Bardwick, A., Heidecker, G., Rapp, U.R., and Karin, M., 1992, Oncoprotein-mediated signalling cascade stimulates c-Jun activity by phosphorylation of serines 63 and 73, *Mol. Cell. Biol.* **12**:3507–3513.

Stark, G.R., 1993, Regulation and mechanisms of mammalian gene amplification, *Adv. Cancer Res.* **61**:87–113.

Stein, C.A., and Kreig, A.M., 1994, Problems in interpretation of data derived from *in vitro* and *in vivo* use of antisence oligodeoxynucleotides, *Antisense Res. Dev.* **4**:67–69.

Van Dam, H., Wilhelm, D., Herr, I., Steffen, A., Herrlich, P., and Angel, P., 1995, ATF-2 is preferentially activated by stress-activated protein kinases to mediate c-jun induction in response to genotoxic agents, *EMBO J.* **14**:1798–1811.

Wilhelm, D., van Dam, H., Herr, I., Baumann, B., Herrlich, P., and Angel, P., 1995, Both ATP-2 and c-Jun are phosphorylated by stress-activated protein kinases in response to UV irradiation, *Immunobiology* **193**:143–148.

25

DESIGN AND TESTING OF RIBOZYMES FOR CANCER GENE THERAPY

James S. Norris, Brian Hoel, Dale Voeks, Frideriki Maggouta, Michael Dahm[1], Weihua Pan[2], and Gary Clawson[2]

Department of Microbiology and Immunology
Medical University of South Carolina
171 Ashley Avenue
Charleston, SC 29425
[1]HEXAL Gentech Forschungs GmbH
Industriestr, 25
D-83607 Holzkirchen
Germany
[2]Department of Pathology C7768
Penn State University
500 University Drive
Hershey, Pennsylvania 17033

INTRODUCTION

The following chapter will cover the subject of the development of a series of ribozymes that target the AC40 subunit of RNA polymerase I (RNA pol I), an essential cellular gene that, when inactivated, leads to cell death. The ribozymes are designed to be delivered in a virus or a liposome as a therapy for prostate cancer. In the course of development of these ribozymes it became clear that target selection in any given mRNA was subject to rules that were not easily defined. Therefore, the following chapter will describe the efforts of this laboratory to develop ribozymes against one of the essential subunits of RNA pol I first using an *in vitro* target selection system coupled with computer modeling followed by functional analyses of release and target cleavage activities.

RNA POLYMERASE I

Yeast and higher eukaryotes contain three nuclear RNA polymerases. RNA pol I is used to transcribe rDNA genes, while the other two are involved in transcription of

Cancer Gene Therapy: Past Achievements and Future Challenges, edited by Habib
Kluwer Academic/Plenum Publishers, New York, 2000.

messenger RNA, 5S, tRNAs, U6 and 7SK RNAs and certain viral RNAs. RNA pol I consists of at least 14 subunits of which ten appear to be essential for the function of the enzyme. Two subunits AC40 and AC19 are thought to be essential and likely play a role in subunit assembly. The AC40 subunit is also shared by RNA pol III (Lanzendorfer *et al.*, 1997). The mouse and human homologue of AC40 are cloned (Song *et al.*, 1994; unpublished).

Where studied, ribosomal RNA genes have been found to be tandemly arrayed with 100–1,000 copies spanning several million base pairs (Wallace and Birnstiel, 1966). RNA pol I transcribes these genes and the primary transcripts are processed to form 18S, 5.8S and 28S ribosomal RNA which form the core of the ribozyme. Ribosomal genes account for 40% of the total cellular transcription and 80% of the RNA content of the cell while constituting only 1% of the genome (Moss and Stefanovsky, 1995). Mature ribosomal RNA and ribosomal proteins form the ribosome complex. We hypothesized that any disruptions of ribosome complex formation would likely result in dysregulation of translation and/or death of the cell. The retinoblastoma gene product has been shown to disrupt RNA pol I mediated transcription by sequestering upstream binding factor UBF, thus down-regulating cellular proliferation via disruption of ribosome assembly (Cavanaugh *et al.*, 1995; Voit, Schafer, and Grummit, 1997).

PROSTATE CANCER

Prostate cancer is the now the most frequently diagnosed cancer in US males and the second leading cause of death (Proctor *et al.*, 1997). Already a major health problem, increases in incidence and mortality are predicted in the next 15 years as the population ages. The molecular mechanisms behind disease initiation and progression still remain largely a mystery. The overall value of screening and treatment also remains controversial. A sensitive marker for screening and monitoring prostate cancer is prostate-specific antigen (PSA). Concerns have been raised about the diagnostic accuracy of this molecule and its predictive ability although newer methodology measuring bound and free PSA have been developed and may be more predictive.

Therapies for prostate cancer are largely defined by stage of disease and include radical prostatectomy, radiation therapy, and cryotherapy. If the disease is not detected until it is more advanced, distant metastases have usually occurred and watchful waiting, radiation therapy, hormonal ablation therapy and limited chemotherapy have been employed, none with any great success. The occurrence rates and mortality of non-localized disease is very high (Cersosimo and Carr, 1996). Further, about one-third of patients at diagnosis are in an advanced stage of disease, thus, additional therapeutic options are clearly required before successful treatment of prostate cancer can be achieved.

There are a number of gene therapeutic approaches being tested for prostate cancer, including trials that involve stimulation of the immune system and trials where delivery of herpes simplex virus thymidine kinase gene is used to cause a bystander effect following application of gancyclovir. Other clinical trials are in progress. Presently the most serious deficiency in the application of gene therapy is the lack of delivery vehicles. This continues in 1998 and until such delivery systems have been developed, successful treatment of prostate cancer will be problematic, unless immunotherapy or successful bystander therapy are developed.

RIBOZYMES

The first encounter with catalytic RNA molecules was made independently by Cech (1986), Altman (1989) and Forster and Symons (1987). The hammerhead ribozyme, used in our research, was described by Buzayan (1986) and Forster and Symons (1987) and subsequent work by Uhlenbeck (1987) demonstrated it could function in trans against almost any RNA molecule containing the 3 base target recognition sequence NUX (N = any base and X = A, C or U). The basic structure of a hammerhead ribozyme is presented in Fig. 1, and is composed of a catalytic core of 24 conserved bases containing helix II and two self-associating helices, I and III which are antisense to the substrate and position the ribozyme to catalyze cleavage. Hammerhead ribozymes favor the recognition sequence GUC in naturally occurring self-cleavage reactions. The ribozyme cleaves 3′ of this ribonucleotide triplet by attacking the 2′-hydroxyl resulting in a 2′, 3′ cyclic phosphate and a 5′ hydroxyl on the 3′ fragment.

Hammerhead ribozymes are believed to function in complex with divalent metal cations such as magnesium or maganese (Orita *et al.*, 1996). For catalytic function there are three events that must occur. The first event is the interaction of helix I and III, with the substrate; the second event is the trans-esterification reaction. The third event is dissociation of helices I and III from the RNA target. The latter activity is critical for enzymatic function to be maintained. Thus the length and base composition of the helix I and III arms, (generally be 6–8 bases) is a crucial design feature.

SCREENING TECHNOLOGY

The screening technology we use is adapted from Lieber and Strauss (1995) and depends on the construction of a library of oligonucleotides with the structure shown in Fig. 2. Anti-sense regions are synthesized randomly to give a complexity approaching 4×10^{15}. Two bases of the NUX target sequence are included to drive specificity for the target. Primers for PCR synthesis are included on either side of the random targeting

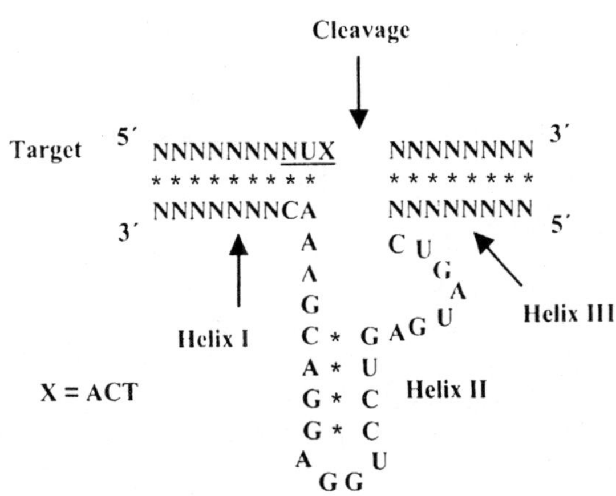

Figure 1. Schematic diagram of a typical hammerhead ribozyme. The figure illustrates the two antisense helices I (A). III (C) and helix II (B) in the hammerhead domain.

Figure 2. Cartoon of the oligonucleotide library used to prepare the antisense RNA molecules for identifying binding on the mRNA target.

sequence. A sample of the oligonucleotide library is PCR-amplified followed by synthesis of anti-sense RNA molecules using T7 polymerase and ^{32}P-CTP. The target mRNA is also cloned into an appropriate vector and similarly run off using T7 polymerase. A limitation of this technique is the maximum length of the target RNA must not exceed 1,500 bp and 1,000–1,200 bp is optimal. Following incubation of the unlabeled target mRNA with the radioactively labeled RNA transcript, free and bound products are separated by native gel electrophoresis. The band on the gel that represents the full-length target RNA will have some fraction of bound ^{32}P-RNA transcripts that base pair with accessible sites on the RNA. This fraction is excised from the gel and following extraction is subjected to RT PCR. The above procedure is repeated five—six times. Subsequently the enriched RNAs are reverse transcribed, cloned and sequenced. Sequence information is compared to the sequence of target RNA and likely target sites are chosen as depicted in Fig. 3 for the RNA pol I AC40 mRNA. Site selection is also based on the predicted secondary structure of the RNA. Figure 3 depicts five possible sites obtained by such a selection procedure. Clearly, at this point the MFOLD program has proven to be useful.

There are certain design rules employed for ribozyme construction (Zhao and Lemke) but generally the most important consideration for subsequent development of the ribozymes is a preliminary *in vitro* testing against the transcribed mRNA target. If cleavage is observed, ribozymes are then tested *in vivo* by transfection procedures and results are analyzed by Northern, Western, or phenotypic response. This laboratory and its collaborators have now developed over sixty ribozymes to different targets and tested many of them functionally both for their ability to cleave the target RNA *in vitro* and *in vivo* (Voeks, Norris, and Clawson, 1998; Schalles *et al.*, 1998).

TRIPLE RIBOZYME DESIGN

A second feature of our ribozyme design program was the development of triple ribozyme vectors that release an internally cloned ribozyme allowing for the development of very specific targeting sequences with few extraneous bases associated with the helix I and helix III domains. This provides for increased selectivity in its function (Voeks, Norris, and Clawson, 1998). This is necessary because the longer the anti-sense strands are, the more mismatches are likely to occur with the undesired consequence of nonspecific interaction and loss of function. For human gene therapy purposes this an unwanted outcome.

Figure 4 depicts a typical structure for a triple ribozyme in which the cis-acting domains (ribozyme A and C) have been engineered to provide more effective release of

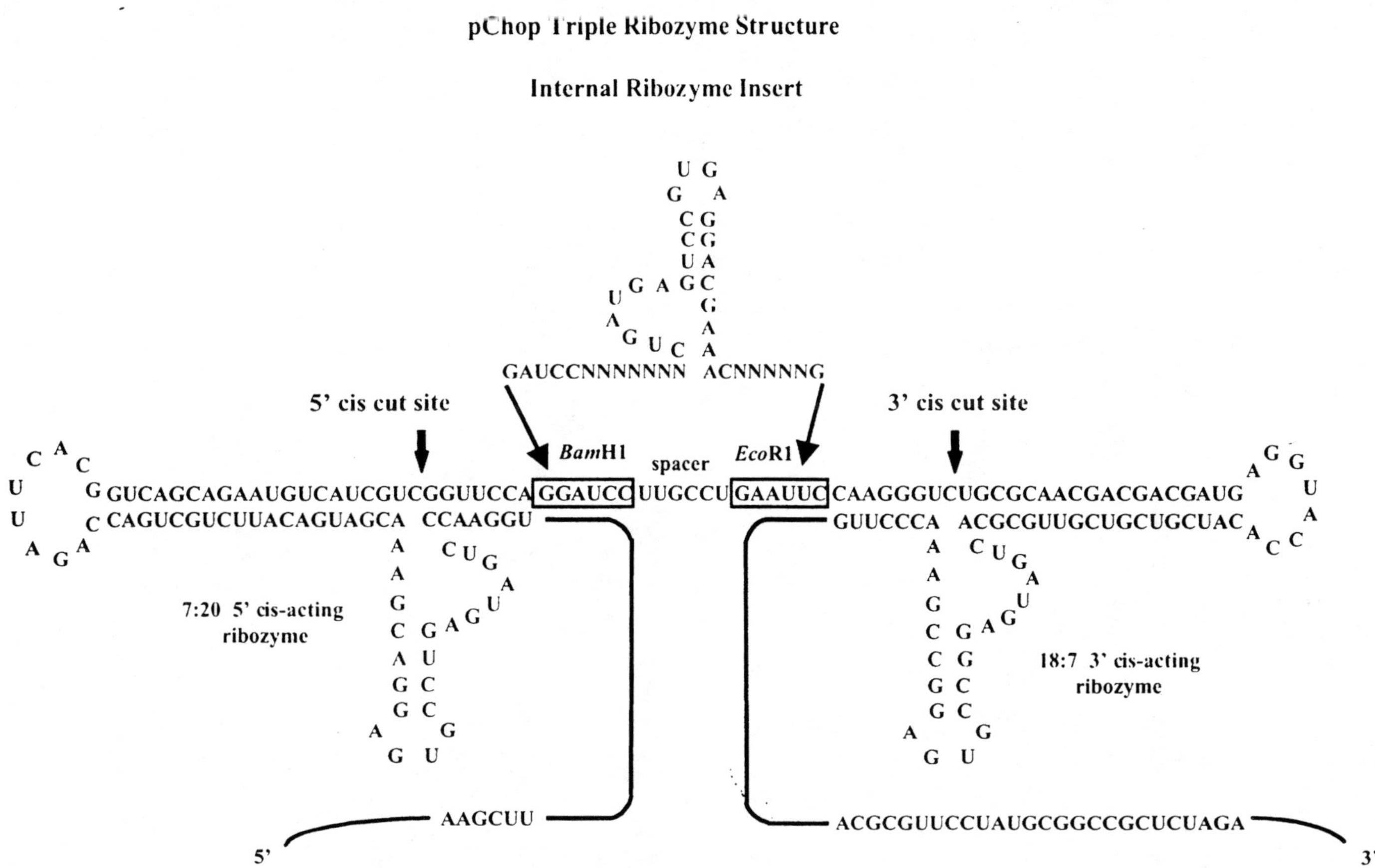

Figure 3. The AC40 mRNA target was folded using MFOLD (Zucker, 1989). Library sequences that bound the mRNA were cloned and any targets that contain the NUX motif were identified. Ribozymes with a ΔG of > -22 kcal/Mol are designed from this information and using the rules of Zhao and Lemke (1998).

Entire AC40 Substrate

Mfold predicted secondary structure

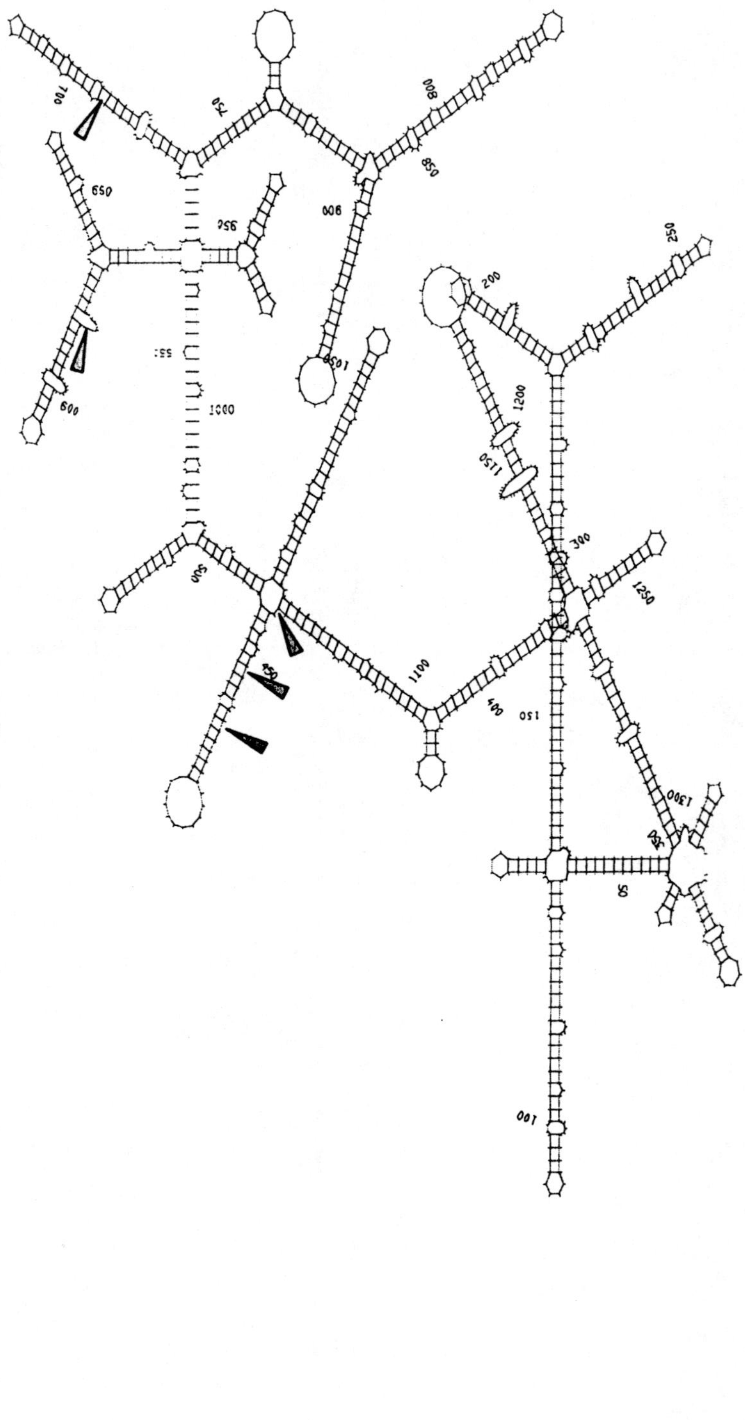

Figure 4. The triple ribozyme cloning vector pChop is depicted in this figure. Modification of the cis-acting ribozyme arms to achieve more efficient and complete release of internal ribozyme are designed into this vector and previously validated by experimentation (unpublished). The restriction sites *Eco*R1 and *Bam*H1 are used to clone subject ribozymes.

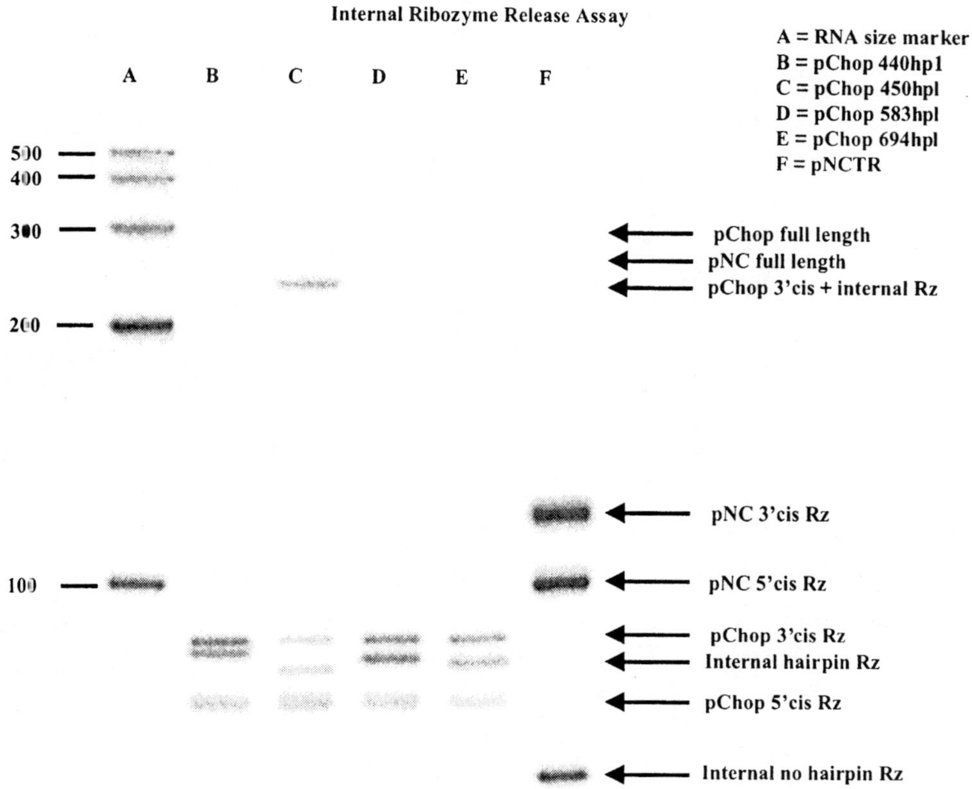

Figure 5. Depicts a self-cleavage reaction to determine if the internal ribozyme is efficiently released. Different ribozymes are identified at the top of the gel. The right lane are markers. Cleavage fragments are identified on the gel. Differences in fragment size are based on the presence or absence of hairpin structures being tested for their ability to stabilize the ribozymes *in vivo*. Reaction conditions are 37 °C in 40 mM TRIS, 2 mM spermidine, 50 μM $NaCl_2$ and 8 mM $MgCl_2$ pH 8.0 for 1 hour.

the internal ribozyme. RNA pol I ribozymes derived from the screening procedure are cloned into the *Eco*R1 and *Bam*H1 restriction site. So far eight ribozymes have been constructed to RNA pol I. Figure 5 depicts the self-cleavage reaction of five of these ribozymes. All clones release the internal ribozyme rapidly and with similar efficiency. This reaction is necessary to detect any interference, with respect to correct folding and cleavage, of the internal ribozyme (B) with flanking cis-acting motifs.

Figure 6 is an example of *in vitro* cleavage of radioactively labeled full-length pol I target mRNA by ribozymes depicted in Fig. 5. Ribozyme function is observed by formation of appropriate length cleavage products on the gel. Clearly not all ribozymes cleaved the target RNA under these conditions. This is generally, but not invariantly, predictive of *in vivo* function. The NCTR ribozyme is most efficient in this reaction and is also very efficient *in vivo* (Voeks, Norris, and Clawson, 1998).

SUMMARY

This chapter describes procedural aspects for development of ribozymes in general, and specifically, that cleave mRNA to an essential cellular gene, the AC40 subunit of

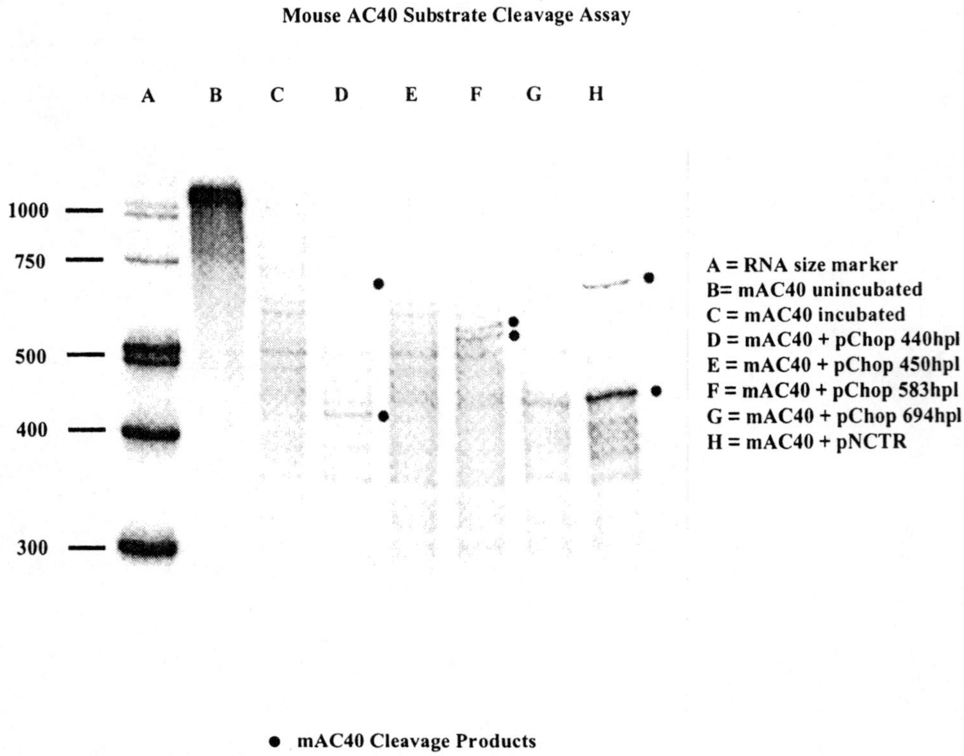

Figure 6. Cleavage of radioactively labeled full length gel electrophoresis. In this particular reaction there was some contamination by a nuclease that apparently came with the SP6 polymerase transcribed ribozymes (all but NCTR which was transcribed by T7 polymerase). Two of the SP6 transcribed ribozymes gave evidence of the correct length cleavage products on the gel (●) as does NCTR. Reaction conditions are 37 °C, 10 mM TRIS, 10 mM $MgCl_2$ at pH 8.0 for 4 hours.

RNA pol I. Ribozyme design includes functional selection of binding sites followed by computer modeling. These ribozymes are being used in vectors that target expression to the prostate via tissue specific promoters (Voeks, Norris, and Clawson, 1998) and have demonstrated efficacy.

ACKNOWLEDGMENTS

Special thanks to Ms. Janie Nelson for typing and graphic layout. This work was supported by NIH CA 69596 and by a research contract with Hexal, Pharmaceuticals, Inc.

REFERENCES

Altman, S., 1989, Ribonuclease P: an enzyme with a catalytic RNA subunit, *Advances in Enzymology & Related Areas of Molecular Biology* **62**:1–36.

Buzayan, J.M., Gerlach, W.L., and Bruening, G., 1986, non-enzymatic cleavage and ligation of RNAs complementary to a plant virus satellite RNA. *Nature* **323**:349–352.

Cavanaugh, A.H., Hempel, W.M., Taylor, L.J., Rogalsky, V., Todorov, G., and Rothblum, L.I., 1995, Activity of RNA polymerase I transcription factor UBF blocked by Rb gene product, *Nature* **374**:177–180.

Cech, T.R., and Bass, B.L., 1986, Biological catalysis by RNA, *Annual Review of Biochemistry* **55**:599–629.

Cersosimo, R.J., and Carr, D., 1996, Prostate cancer: current and evolving strategies, *American Journal of Health-System Pharmacy* **53**:381–396.

Forster, A.C., and Symons, R.H., 1987, Self-cleavage of plus and minus RNAs of a virusoid and a structural model for the active sites, *Cell* **49**:211–220.

Lanzendorfer, M., Smid, A., Klinger, C., Schultz, P., Sentenac, A., Carles, C., *et al.*, 1997, A shared subunit belongs to the eukaryotic core RNA polymerase, *Genes & Development* **11**:1037–1047.

Lieber, A., and Strauss, M., 1995, Selection of efficient cleavage sites in target RNAs by using a ribozyme expression library, *Molecular & Cellular Biology* **15**:540–551.

Moss, T., and Stefanovsky, V.Y., 1995, Promotion and regulation of ribosomal transcription in eukaryotes by RNA polymerase I, *Progress in Nucleic Acid Research & Molecular Biology* **50**:25–66.

Orita, M., Vinayak, R., Andrus, A., Warashina, M., Chiba, A., Kaniwa, H., *et al.*, 1996, Magnesium-mediated conversion of an inactive form of a hammerhead ribozyme to an active complex with its substrate. An investigation by NMR spectroscopy, *Journal of Biological Chemistry* **271**:9447–9454.

Parker, S.L., Tong, T., Bolden, S., and Wingo, P.A., 1997, Cancer statistics, 1997 [published erratum appears in CA Cancer J Clin 1997 Mar-Apr; 47(2):68], *Ca: a Cancer Journal for Clinicians* **47**:5–27.

Ren, L., Schalles, S.L., Pan, W., Isom, C.E., Loy, S.E., Lee, J.-H., Benedict, C.M., Pickering, M.T., Norris, J.S., and Clawson, G.A., 1998, Construction and deployment of triple ribozymes targeted to multicatalytic proteinase subunits C3 and C9. *Gene Therapy and Molecular Biology* (In Press).

Song, C.Z., Hanada, K., Yano, K., Maeda, Y., Yamamoto, K., and Muramatsu, M., 1994, High conservation of subunit composition of RNA polymerase I(A) between yeast and mouse and the molecular cloning of mouse RNA polymerase I 40-kDa subunit RPA40, *Journal of Biological Chemistry* **269**:26976–26981.

Uhlenbeck, O.C., 1987, A small catalytic oligoribonucleotide, *Nature* **328**:596–600.

Voit R., Schafer, K., and Grummt, I., 1997, Mechanism of repression of RNA polymerase I transcription by the retinoblastoma protein, *Molecular & Cellular Biology* **17**:4230–4237.

Wallace, H., and Birnstiel, M.L., 1966, Ribosomal cistrons and the nucleolar organizer, *Biochimica et Biophysica Acta* **114**:296–310.

Zuker, M., 1989, On finding all suboptimal foldings of an RNA molecule, Science **244**:48–52.

Voeks, D.J., Norris, J.S., and Clawson, G.A., 1998, Ribozymes in gene therapy of prostate cancer, in: *Ribozymes in the Gene Therapy of Cancer*, (K.J. Scanlon and M. Kashani-Sabet, eds.), chapter 13, pp. 165–173, R. G. Landes Company, Texas.

Zhao, J.J., and Lemke, G., 1998, Rules for ribozymes, *Molecular & Cellular Neurosciences* **11**:92–97.

26

RIBOZYMES

Their Design and Use in Cancer

Philip C. Turner

School of Biological Sciences
University of Liverpool
Crown Street
Liverpool
L69 7ZB
UK

1. INTRODUCTION

Although ribozymes (catalytic RNA molecules) have not yet been used in human patients, their promising use in tissue culture cells, as well as transgenic and other animal models, suggests that their application to treat human disorders, including cancers, is imminent. The purpose of this chapter is to outline the structure of the most common and promising ribozymes and consider how they are designed and delivered before reviewing some of their most recent uses in the field of cancer. More detailed reviews of particular aspects of the various classes of ribozymes will be referred to below.

2. RIBOZYME TYPES

The first ribozymes to be discovered were self-cleaving intons (Kruger *et al.*, 1982) and the RNA component of the enzyme RNase P which cleaves tRNA precursor molecules to generate their mature 5′ end (Guerrier-Takada *et al.*, 1983). Hammerhead ribozymes were discovered as parts of small pathogenic plant RNAs that undergo a self-cleaving reaction to generate monomeric genomes (Forster and Symons, 1987). Their small size, and the ease with which their reaction could be replicated *in vitro* (at neutral pH) enabled trans-cleaving hammerhead ribozymes to be rapidly developed (Uhlenbeck, 1987). Another self-cleaving ribozyme discovered in a plant virus pathogen was the somewhat larger hairpin ribozyme (Hampel and Triz, 1989) which also functions in viral replication to generate monomeric genomes. The human subviral pathogen hepatitis delta virus (HDV), which has around 1,700 nucleotides (nt) of single stranded, circular RNA

Cancer Gene Therapy: Past Achievements and Future Challenges, edited by Habib
Kluwer Academic/Plenum Publishers, New York, 2000.

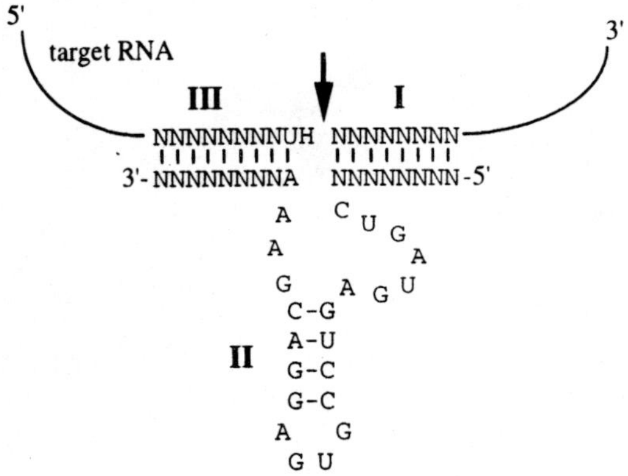

Figure 1. The hammerhead ribozyme showing the base pairing interaction with target RNA via helices I and III. The cleavage site is shown by an arrow.

in its genome is also replicated by producing linear genomic multimers that self cleave to monomers (Wu *et al.*, 1989). Both of these ribozyme activities have also been studied and simplified for use in trans-cleaving applications (Branch and Robertson, 1991; Hampel *et al.*, 1990). Other naturally occuring ribozymes are known, such as the *Neurospora* mitochondrial VS ribozymes (Saville and Collins, 1990), the newt ribozyme (Marusic *et al.*, 1997) and *in vitro* selection experiments have created novel ribozymes with unusual enzyme activities (Suga *et al.*, 1998). At the present time these particular ribozymes are either not appropriate or have not yet been developed for use in trans-cleaving applications such as specific cleavage of target RNA molecules to down regulate gene expression and will not be considered further here.

2.1. Hammerhead Ribozymes

Comparison of sequences of the hammerhead catalytic motif from many sources, together with extensive mutagenesis studies have enabled the conserved and important features of the hammerhead ribozyme to be deduced (Vaish *et al.*, 1998). These are shown in Fig. 1. Helices I and III base pair with the target RNA, either side of the cleavage site which is a 5′-UH-3′ sequence, where H is any nucleotide except G. Different 5′-UH-3′ sequences are cleaved with different efficiencies and most naturally occurring sites are 5′-GUC-3′, which is probably why this sequence features in most scientific reports. The catalytic core of the hammerhead ribozyme is formed mainly by the unpaired nucleotides between helix II and the nucleotides pairing with the target and inactive control ribozymes can be made by single nucleotide changes to several of these residues (Birikh *et al.*, 1997). Due to its small size, which makes it relatively easy to chemically synthesize and manipulate, the catalytic properties of the hammerhead ribozyme have been extensively studied. The parameters include symmetry and lengths of helices I and III, target sequence effects, ribozyme modifications, etc. The reader is referred to Birikh *et al.* (1997) and Vaish *et al.* (1998) for further information.

2.2. Hairpin Ribozymes

Deletion and mutagenesis studies on the hairpin ribozyme (reviewed in Walter and Burke, 1998) from the minus strand of tobacco ringspot virus satellite RNA showed that the essential features were a 50 nt ribozyme that could cleave a 14 nt target (substrate)

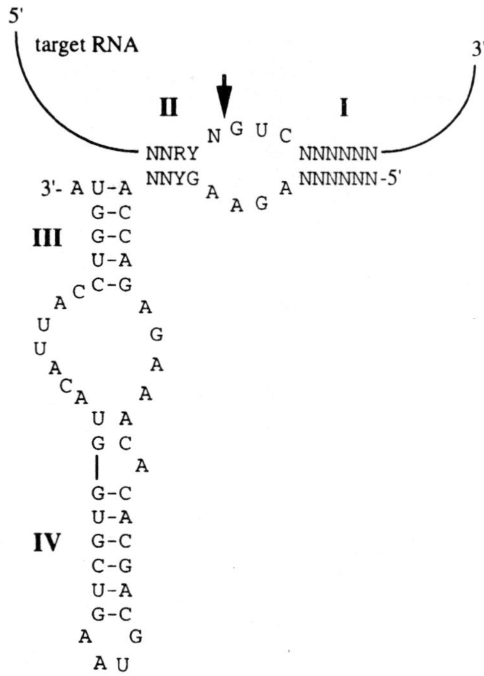

Figure 2. The hairpin ribozyme. Helices are labeled I–IV and the cleavage site is shown by an arrow. R = A or G, Y = C or T.

RNA (Fig. 2). The structure consists of 4 helices (labelled 1 to 4 from the 5′ end of the ribozyme, ie base no. 1) separated by 2 bulges and a terminal loop. Helices 1 and 2 form by base pairing with the target RNA sequence creating a bulge in between, containing the site of cleavage which is on the 5′ side of a conserved 5′-GUC-3′ sequence. The length of helix 1 can be varied depending on the target, but is usually between 4 and 12 bp, with 6–8 bp being a good starting point. Best cleavage of the target is observed if the cleavage site has the sequence 5′-RYNGUC-3′, but this can be relaxed to 5′-NBNGUC-3′ (where B is any nt except A). The length of helix 2 need only be 4 bp, but the residues in each strand must be complementary. Helix length variations and other mutations generally decrease ribozyme activity but C instead of U39 improves activity as does extending helix 4 from 3 to 6 bp and having GUAA as the terminator loop (Yu and Burke, 1997).

2.3. RNase P Ribozymes

Ribonuclease P, an RNA protein complex, is an RNA processing enzyme found in all cells that cleaves the 5′ leader sequence off precursor tRNA molecules (Altman *et al.*, 1993). Although the RNA component is catalytically active on its own, the protein component significantly enhances the reaction. In cells, RNase P processes over 50 different pre-tRNA species that have no conserved sequence around the cleavage site because the enzyme recognizes common higher order structural features of the pre-tRNAs. To get RNase P to cleave a selected target RNA, it is necessary to anneal an external guide sequence (EGS) to the target so that structural features of a pre-tRNA are produced which can be recognised and cleaved. Figure 3 shows the essential features of the EGS-target interaction. The stem loop is equivalent to the pre-tRNA's T-arm, whereas the

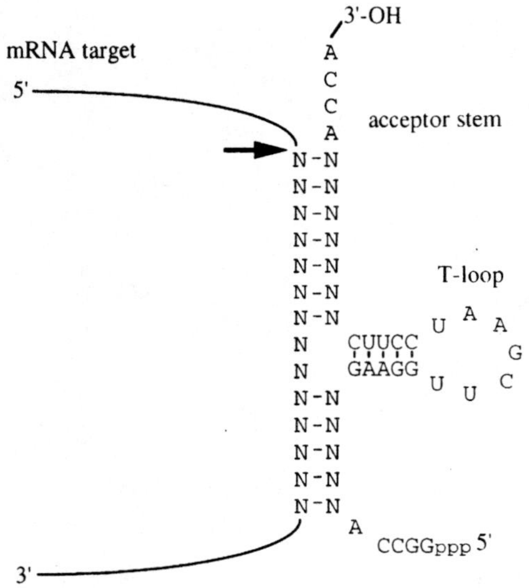

Figure 3. A trans-cleaving RNase P based ribozyme showing the interaction of the external guide sequence with the target RNA. Arrow indicates the cleavage site.

longer helix mimics the amino acid acceptor stem having a terminal 5′-CCA-3′ sequence. The shorter helix mimics the D-stem of pre-tRNA (Yuan and Liu, 1998). Although some target RNAs have been cleaved by RNase P when complexed with EGS RNAs, at significant rates, others were cleaved much more slowly. This was probably due to a failure of the EGS RNA and the target RNA to fold in a suitable way to be recognised by RNase P. Some enhancement of cleavage activity can be achieved by changing the sequence in the T-loop and an *in vitro* selection procedure has been developed for this purpose (Yuan, 1998).

2.4. Hepatitis Delta Virus Ribozymes

HDV replicates by a rolling circle mechanism and this infectious human pathogen actually has ribozyme activity in both genomic and anti-genomic RNA strands (Lai, 1995). The essential features of the genomic HDV ribozyme are shown in Fig. 4 where it is seen that there is only one nucleotide 5′ to the cleavage site and 84 nt 3′ to it. The ribozyme folds to form 4 helices of which II and III are most critical for activity (Ferré-d'Amaré *et al.*, 1998). Helix IV can be shortened and the sequence of helix I altered to some degree as long as base pairing is maintained. The lack of extensive sequence requirements 5′ of the cleavage site has meant that HDV ribozymes have been exploited to cleave in cis to release RNA molecules with precise 3′ ends (Chowrira *et al.*, 1994). To create trans-cleaving ribozymes that specifically cleave target RNA molecules has proved more difficult than for the other classes of ribozyme described above, and they are not yet as efficient as these other types. Figure 4B shows a circularly permuted and minimized HDV ribozyme that combines features of both genomic and anti-genomic versions (Been *et al.*, 1992). The sequence of helix I is varied to target to the desired substrate, however, residues either side of the cleavage site should be maintained to direct cleavage at a 5′-CG-3′ sequence in the target.

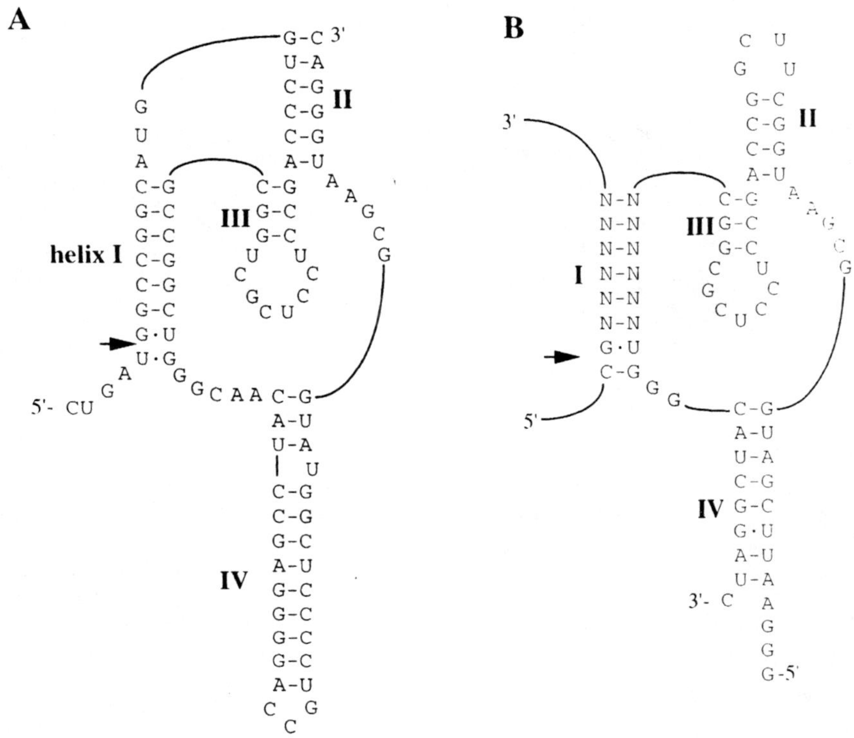

Figure 4. (A) The genomic HDV cis-cleaving ribozyme. (B) A trans-cleaving version combining genomic and anti-genomic features. Helices are labelled I–IV and the cleavage sites are indicated by arrows.

2.5. Tetrahymena Group I Intron

The natural reaction of the intron in the *Tetrahymena thermophila* 26S rRNA is to excise itself by two transesterification reactions, the second of which involves ligating the two exons together (Cech, 1990). A shortened version of this intron (L-21) can carry out the ligation (trans-splicing) step *in vitro* and this ribozyme activity has been harnessed to carry out *in vivo* repairs to defective mRNAs (Lan *et al.*, 1998). The kinds of defects that could be repaired by trans-splicing include those generated by nonsense or frameshift mutations or chromosomal rearrangements. The principle is shown in Fig. 5. The 5′ end of the ribozyme is designed to base pair to the target RNA upstream of a U residue and the 3′ end of the ribozyme carries the new 3′ sequence to be ligated downstream of the U residue in the target.

3. DELIVERY AND DESIGN OF RIBOZYMES

To down-regulate gene expression, ribozymes can be delivered to cells in two different ways. If pre-synthesised ribozymes are administred to cells or animals as one would deliver drugs, this is referred to as exogenous delivery. If, however, a gene encoding a ribozyme is introduced into cells so that the ribozyme is produced within the cell itself,

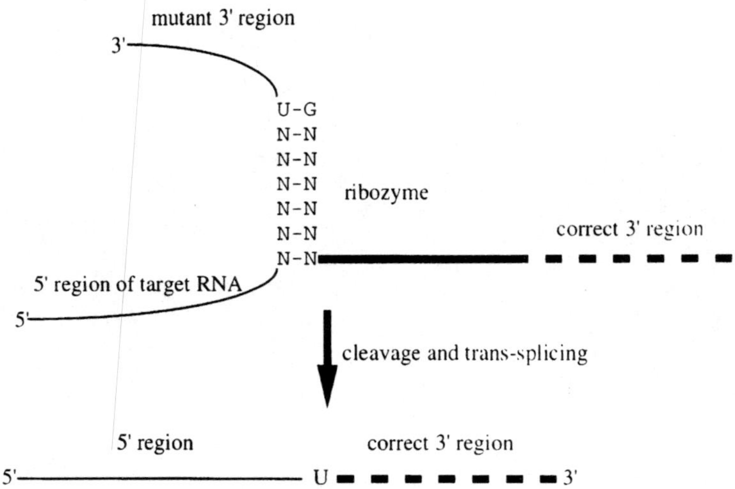

Figure 5. The process of trans-splicing whereby a ribozyme can replace a defective 3′ portion of a transcript with a functional segment of RNA.

this is known as endogenous delivery. Each method has its advantages and disadvantages which are presented below, followed by a discussion of the important points regarding target site selection.

3.1. Exogenous Delivery

The method of exogenous delivery is best suited to small ribozymes such as the hammerhead (Birikh *et al.*, 1997), since these molecules must be chemically synthesised and there are practical limits to the length of oligoribonucleotides that can currently be made. The major problem associated with exogenous delivery is that of stability of the ribozyme. RNA molecules are very sensitive to RNases that occur in serum and in culture media and their half-life can be very short. Substantial work has been done to explore what chemical modifications can be carried out to stabilise ribozymes against RNase attack without reducing their catalytic activity. Most commonly, allyl, amino, fluoro or o-methyl modifications are introduced at the 2′ positions of the ribose backbone together with phosphorothioate termini. These modified ribozymes can have much improved half lives in serum, media and cell extracts. A further problem of exogenous delivery is that of toxicity. Steps need to be taken to provide low continuous doses, as oligonucleotides have been shown to have toxic effects, and depending on the disorder, this could be for very long periods. The ribozyme must also be able to gain entry into the cell and be active therein. This may require packaging or encapsulation of the ribozyme to aid uptake by cells, for example in liposomes, however, even with this delivery method the ribozyme can end up in endosomes and fail to interact with its target RNA (Gregoriadis, 1995; Zelphati and Szoka, 1996).

3.2. Endogenous Delivery

There are two major parameters to consider when developing ribozymes for endogenous delivery, the design of the gene expression cassette and the method of administering the DNA to cells, for example choice of vector. While the latter can include direct microinjection, which is suitable for single cells such as when creating transgenic animals, it is not very useful when the objective is to introduce a ribozyme DNA construct into an organ in a whole animal. In such cases it is usual to make use of a viral vector system into which the ribozyme DNA cassette is inserted. Among the viral vector systems that have been employed, retroviral vectors have advantages because they are efficiently introduced (transduction) and can stably integrate into the genome of the host (Bertrand and Rossi, 1996; Sun *et al.*, 1996; Welch *et al.*, 1996). DNA viruses, such as adenovirus, have been used, but suffer from the drawback that expression is only transient and the host mounts an immune response which makes repeated application futile (Ali *et al.*, 1994). The non-pathogenic adeno associated virus (AAV) is being used as an alternative as it can integrate into a specific site on chromosome 19 and can also do so in non-dividing cells (Bertrand *et al.*, 1997).

There are many considerations to be taken into account when choosing/designing an expression cassette. Not only should the promoter be efficient, but the ribozyme transcript must be stable and localise with the target RNA inside the cell. RNA polymerase III (pol III) promoters (e.g. tRNA and 5S) are very efficient, but often have internal promoter elements (i.e. within the transcribed region) which constrains ribozyme design. The U6 snRNA promoter is efficiently transcribed by pol III and has no internal promoter elements, however, its transcripts are localised in the nucleus and this may prevent ribozyme-target interaction in the cytoplasm even though the ribozyme should be able to interact with the pre-mRNA target in the nucleus. Pol II-based expression cassettes have also been used. These include U1 and U7 snRNA genes as well as other human or viral promoters coupled to the SV40 poly(A) site. The former produce nuclear-localised transcripts while the latter are exported to the cytoplasm. It has been observed that lower levels of ribozyme transcript can be more effective if they co-localise with the target (Bertrand *et al.*, 1997). If the ribozyme transcript is not produced with a poly(A) tail, which should aid stability, it is wise to include a 3′ stem-loop structure on the transcript for this purpose. Usually the 5′ cap of pol II transcripts is sufficient to protect the 5′ end from 5′ exonucleases.

3.3. Target Site Selection

Natural RNA molecules that may be chosen for cleavage by ribozymes are sufficiently long that they can fold into complex secondary and tertiary structures. Hence not all of the potential cleavage sequences in the target RNA will be equally accessible to ribozyme attack. Attempts to predict accessible sites using computer programs (James and Cowe, 1997; Sczakiel and Tabler, 1997) demonstrated some success, however, the method does not guarantee that any particular sequence will be efficiently cleavable *in vitro*, let alone *in vivo*. It is possible to determine experimentally which parts of a target RNA molecule are able to bind short antisense oligodeoxynucleotides and thus be the sites of sensitivity to nuclease (RNase H) (Frank and Goodchild, 1997). Usually these sites at which oligodeoxynucleotides can anneal are also accessible to ribozymes, at least *in vitro*. Sites chosen in this way might well be cleavable *in vivo*, but may not be the best possible site for *in vivo* cleavage. Lieber and Strauss (1995) have developed a selection system in which a library of hammerhead ribozymes, containing a constant ribozyme core sequence, but random hybridizing arm sequences, is mixed with target RNA in cellular extracts. The various ribozymes will cleave the target RNA in different positions and to different degrees, but the most efficient will produce cleavage products that are the most abundant. The cleavage products are cloned by making cDNA and using PCR and the sequences of the cleavage sites determined by sequencing the clones. With this knowledge of the cleavage site, efficient ribozymes can be designed that work well *in vivo* (Lieber and Kay, 1996).

Other *in vivo* methods of selecting efficient ribozyme target sites are being developed, but do not yet permit exhaustive screening of all possible sites. Luciferase has been used as a reporter in which target sequences have been linked upstream of the luciferase gene (Kawasaki *et al.*, 1996). Ribozymes introduced, or co-expressed, in cells expressing the construct should cause a reduction in luciferase activity if they are active. An *E. coli* based system utilising positive selection with trimethoprim for hammerhead ribozymes cleaving in cis and a yeast system that requires cell-cycle arrest may also be worth considering if they are further developed (Fujita *et al.*, 1997).

4. APPLICATIONS OF THE USE OF RIBOZYMES IN CANCER

4.1. Multidrug Resistance

A major problem in many cancer treatment programs involving chemotherapy is the multidrug resistance phenotype in which cells become simultaneously resistant to a broad range of different drugs. A key feature of this phenotype is the induction of the MDR-1 gene, which encodes a 170 kDa glycoprotein called P-glycoprotein (P-gp) (Gros *et al.*, 1986). P-gp becomes associated with the plasma membrane and its 12 transmembrane domains form a pore through which drugs are actively pumped out of the cell using ATP. The MDR-1 gene is regulated by the transcription factor AP-1 (Madden *et al.*, 1993) and hence activation of members of the Fos or Jun families (which form AP-1) stimulate MDR-1 gene expression. Reversal of the MDR phenotype has been achieved in cell culture using ribozymes against MDR-1 (Scanlon *et al.*, 1994) and c-fos (Scanlon *et al.*, 1991). the ribozyme against MDR-1 (anti-MDR-1) was a hammerhead ribozyme with symmetrical hybridising arms of 10 nt (i.e. a 10:10 hammerhead ribozyme) to the GUC at position 2641 which was expressed from the pHAPr-1 vector which has a constitutive -actin promoter. The anti-*c-fos* ribozyme was a 12:12 hammerhead against the GUC at position 926, cloned into the pMAMneo vector, which has a murine moloney tumor virus (MMTV), dexamethasone-inducible promoter (Scanlon *et al.*, 1991). Northern blotting showed a reduction in the target RNA levels in the stably transfected, drug resistant ovarian carcinoma cell lines which then became more sensitive to chemotherapeutic agents. This work has been extended using a ribozyme against codon 196 of MDR-1 which was delivered to human breast cancer cells in culture as a liposomal complex. Ribozyme induced decreases in MDR-1 mRNA levels and P-gp production correlated with cells having high endocytotic activity, suggesting that this parameter would be a good predictor of success in liposome-mediated gene transfer (Masuda *et al.*, 1998). Other workers have used a retroviral vector with a MDR-1 ribozyme coupled to a promoter (CEA) expressed only in lung adenocarcinoma cells. Both in cell culture and in nude mice, drug sensitivity was restored (Gao *et al.*, 1997). Resistance of human colon cancer cells to cisplatin has also been reversed using a hammerhead ribozyme to *c-fos in vitro* and in implanted tumors (Funato *et al.*, 1997).

4.2. *Ras* Oncogenes

There are 3 functional members of the *ras* family of oncogenes in humans, H-*ras*, K-*ras* and N-*ras* and in a large number of human cancers one of them is activated by mutation. Mutations in H-*ras* are present in around 65% of human bladder cancers, usually as a single base change at codon 12 (GGC to GTC, gly to val) and the same mutation is seen in many malignant melanomas. This particular change actually creates one of the preferred cleavage sequences for the hammerhead ribozyme and permits these ribozymes to be designed to cleave only mutant H-*ras* (Koizumi *et al.*, 1989). To examine the effectiveness of such a ribozyme against bladder cancer, an 8:6 hammerhead ribozyme was cloned between the CMV promoter and the SV40 poly(A) site in the pAC-CMVpLpA vector. This was then co-transfected with pJM17 into 293 cells to produce recombinant adenovirus. Bladder cancer cells were then transduced with the recombinant adenovirus and inhibition of H-*ras* expression monitored by Western blotting for p21 or Northern blotting. *In vitro* studies examining the effects on tumor size can be carried out following administration of the recombinant adenovirus to tumor-bearing mice (Feng *et al.*, 1995). Similar tumor regression was observed in mice when DNA encoding an anti-H-*ras* ribozyme was directly injected into the tumors (Chang *et al.*, 1997). In the FEM human malignant melanoma cell line, the above 8:6 anti-H-*ras* hammerhead ribozyme targeted to codon 12 was transfected as either a human-actin, an MMTV or a CMV promoter construct and in all cases suppressed H-*ras* gene expression and inhibited cell growth (Ohta *et al.*, 1994). To direct ribozyme expression only to melanocytes, the tyrosinase promoter was used and again showed significant reversal of malignant phenotype (Ohta *et al.*, 1996).

The K-*ras* gene is mutated in codon 12 from GGU to GUU in over 80% of human pancreatic ductal adenocarcinomas and similar methodology has been used to reverse the phenotype of the human Capan-1 pancreatic carcinoma cell line (Kijima *et al.*, 1996). The ribozyme was a 6:6 hammerhead expressed in pHAPr-1 and was stably transfected into Capan-1 cells where it slowed growth and changed cell morphology. A similar K-*ras* ribozyme has been developed for testing against a common lung adenocarcinoma with the same codon 12 mutation. The 6:6 hammerhead ribozyme was cloned into pACCMVpLpARS$^+$ utilising the CMV promoter and SV40 poly(A) sequences. Co-transfection with pJM17 gave recombinant adenovirus for *in vitro* and *in vivo* use (Tong *et al.*, 1998).

Mutant N-*ras* genes have been found in various cancers, the most common point mutations being in codons 12, 13, 59 or 61. A series of exogenously delivered and chemically modified hammerhead ribozymes were tested against a N-*ras*-luciferase fusion gene stably transfected into Hela cells (Scherr *et al.*, 1997). The most efficient at inhibiting luciferase activity were 7:7 armed ribozymes against codon 13. In more recent work the ribozyme was transduced into Hela cells using a retroviral vector and it reduced luciferase activity by over 95%. When the IL-3-dependent cell line TF-1 was transformed by N-*ras* it lost its dependence on IL-3 and showed a proliferative response to erythropoietin. Both these effects were reversed when the ribozyme was introduced (Scherr *et al.*, 1998).

4.3. The *c-erb*B-2 (*HER2/neu*) Oncogene

In a variety of human cancers including breast and ovarian cancer, the c-erbB-2 gene is overexpressed. Although there is no difference in the sequence of the transcripts in these tumors, a feature which was exploited in the mutant *ras* genes above, ribozymes can still be utilised to down-regulate the overexpressed gene restoring it back to normal levels. In BT-474 breast cancer cells, transfection with a -actin driven 7:6 hammerhead ribozyme designed against the GUC at position 831 of *c-erb*B-2, considerably reduced cell growth and reduced *c-erb*B-2 levles on Northern blots (Suzuki *et al.*, 1998). Ribozymes have also been used to show that overexpression of *c-erb*B-2 occurs in certain aggressive human ovarian cancers and its reduction can reduce tumor growth (Juhl *et al.*, 1997). This was dramatically shown by the shrinkage of established tumors in nude mice when the tetracycline-regulated ribozyme gene was activated by antibiotic withdrawal. Others have shown that a hammerhead ribozyme to the GUC sequence at position 631 expressed as a pol III transcript inhibited production of green fluorescent protein (GFP) when co-transfected into NIH: OVCAR3 ovarian cancer cells with a *c-erb*B-2-GFP fusion gene (Wiechen *et al.*, 1998).

4.4. Prostate Cancer

During treatment of prostate cancer it is common to observe androgen receptor (AR) dependent or hormone refractory tumors. The oncogene *bcl*-2 suppresses apoptosis and its overexpression in prostate cells makes them resistant to hormone ablation. Hammerhead ribozymes designed against *bcl*-2 were transfected into *bcl*-2 overexpressing prostate carcinoma cells (LNCaP/*bcl*-2) and by both RNase protection and a cell viability assay there was a reduction in *bcl*-2 mRNA and increased cell death in transfected cells (Dorai *et al.*, 1997). In further work, a dimeric hammerhead ribozyme was shown to induce apoptosis in a low *bcl*-2 expressing cell line variant of LNCaP, but only restored the ability to respond to a second apoptotic agent in a high *bcl*-2 expressing LNCaP cell line (Chen *et al.*, 1998). In AR-dependent prostate tumors a hammerhead ribozyme co-transfected into PC3 cells with an AR-dependent CAT reporter gene showed a reduction in AR inducible CAT activity offering a route to treating androgen resistant, but AR-dependent prostate tumors (Chen *et al.*, 1998).

4.5. Glioma

Often malignant gliomas possess amplified and rearranged epidermal growth factor (*EGFR*) genes and one particular *EGFR* mutation in primary gliomas creates an aberrant mRNA lacking exons II to VI. A hammerhead riboozyme was designed to the GUA sequence created by the exon I–VII fusion which was then tested *in vitro* and *in vivo* (Yamazaki *et al.*, 1998). When co-transfected into NIH3T3 cells that had been made to express the aberrant *EGFR* there was a reduction in aberrant *EGFR* mRNA and the tumor forming capacity of these cells in nude mice was also reduced. A chemically synthesised hammerhead ribozyme with 2′-amino modifications to all pyrimidines was designed against protein kinase C alpha (Sioud and Sorensen, 1998). The ribozyme inhibited gliomal cell growth in culture by reducing protein kinase C alpha gene expression and a single injection of ribozyme as a cationic liposome complex also inhibited gliomal tumor growth.

4.6. Leukaemia

Many haematological malignancies arise from chromosomal translocation events and several workers have explored the possibility of using ribozymes to cleave the abnormal fusion gene transcript specifically, leaving transcripts from the normal alleles uncleaved. In the case of chronic myelogenous leukaemia (CML) patients, over 95% have a fusion of the *bcr* and *abl* genes, though the precise fusion point can vary. It is thought that the uncontrolled kinase activity in the fusion protein is critical to the malignancy. A hammerhead ribozyme targeted to a b3a2 *bcr-abl* fusion transcript was endogenously expressed in the CML cell line K562 and analysis showed a 53% decrease in fusion transcript level, a 63% decrease in colony forming activity and significant inhibition of cell proliferation (Wright *et al.*, 1998). Similar results have been obtained using multiunit ribozymes (Leopold *et al.*, 1995).

About 10% of acute myelogenous leukaemia (AML) patients have a fusion between the AML-1 gene and the MTG8 gene, both of which are transcription factors. The fusion protein is thought to act as an aberrant transcription factor in the development of AML. 8:9 and 9:8 hammerhead ribozymes were designed against CUC and AUC sequences near the fusion point and transfected into Kasumi-1 leukaemia cells using cationic liposomes (Matsushita *et al.*, 1995; 1998). Both ribozymes significantly inhibited cell proliferation.

In other work studying leukaemia viruses, hammerhead ribozymes have been designed against the *rex/tax* transcript and shown to be more effective than antisense RNAs at inhibiting viral replication in stably transfected cells (Cantor, 1998; Cantor *et al.*, 1996). Similarly, using hammerhead ribozymes against human T-cell leukaemia virus type-1 (HTLV-1), the *rex/tax* transcript was targeted in synovial cells from patients with HTLV-1 associated arthropathy. The ribozymes cleaved the transcripts and significantly inhibited synovial cell growth and induced apoptosis (Kitajima *et al.*, 1997).

4.7. Metastasis

The ability of tumor cells to spread and induce secondary tumors at other sites in the body is poorly understood but may involve inappropriate expression of cell adhesion molecules, proteases or their inhibitors, angiogenic factors or motility factors. The loss of expression of anti-metastatic genes could also be important (de la Rosa *et al.*, 1995; Dong *et al.*, 1995). An overexpressed gene that may promote metastasis has been isolated from mouse, mts-1 (Ebralidze *et al.*, 1989), rat, p9Ka (Barraclough *et al.*, 1987) and human, CAPL (Schäfer *et al.*, 1995). An 11:9 hammerhead ribozyme was designed against the GUC at position 39 in exon 2 of CAPL, cloned into pHAPr-1-neo and transfected into osteosarcoma cells that expressed high levels of CAPL mRNA and protein (Mælandsmo *et al.*, 1996, 1998). Although the tumorigenicity of clonally expanded cellls was unchanged following subcutaneous injection, the metastic potential was significantly reduced after intracardial injection.

Matrix metalloproteinases (MMPs) have also been linked to metastasis and to evaluate the role of MMP-9 (gelatinase B) in this process, an 8:7 hammerhead ribozyme against rat MMP-9 was cloned into pRC/CMV between the promoter and the SV40 poly(A) site. When transfected into metastatic rat embryo cells (2.10.10), these cells no longer were metastatic in a lung colonisation assay, even though as above, they retained tumorigenicity (Hua and Muschel, 1996).

The growth factor pleiotrophin may play a role in angiogenesis and metastasis and has also been a target for down-regulation using hammerhead ribozymes (Czubayko *et al.*, 1996, 1998). An 11:11 hammerhead ribozyme designed against the UUC of codon 22 was cloned into pRC/CMV and stably transfected into human adenocarcinoma cells (SW-13). Pleiotrophin levels in culture supernatant dropped to background levels and tumorigenicity was reduced. In human melanoma cells stably transfected with the same ribozyme construct, their ability to produce subcutaneous tumor growth in mice was also inhibited as was metastatic spread.

4.8. Other Examples

Many ovarian epithelial carcinomas co-express macrophage colony-stimulating factor (M-CSF) and its receptor (c-fms) creating an autocrine loop of self-stimulated growth. Codon 18 of the *c-fms* transcript was targeted by a hammerhead ribozyme by transfection using a vector for endogenous expression. Transfectants had a significant reduction in growth potential which correlated with reduced *c-fms* transcript and protein levels (Yokoyama *et al.*, 1997).

A triple ribozyme in which 2 cis-acting ribozymes released a central trans-acting ribozyme against the retinoblastoma (Rb) gene transcript was stably introduced into cells. The tetracycline-regulatable cells showed abnormal morphology, loss of contact inhibition and other features characteristic of the transformed phenotype (Benedict *et al.*, 1998). This system should permit *in vivo* models to be developed to study the function of the Rb tumor suppressor gene. Another use of ribozymes to study tumor suppressors involves the p16(INK4a) gene. A hammerhead ribozyme expressed in murine erythroleukaemia (MEL) cells reduced endogenous p16 protein by 70% and caused accelerated cell-cycle progression (Nylandsted *et al.*, 1998).

Since some experimental tumors are rejected when anti-CTLA-4 antibodies are administered, developing ribozymes that inhibit the production of the CTLA-4 receptor may have therapeutic use. A hairpin ribozyme against CTLA-4 was transfected into murine T-cells and showed over 90% reduction in CTLA-4 mRNA and no detectable protein (Cepero *et al.*, 1998). This ribozyme also shows promise as a tool to study the role of CTLA-4 in cancer.

5. FUTURE CHALLENGES

There are now several examples of ribozymes acting to decrease mRNA and protein levels in cells growing in culture. Some of these ribozymes act to inhibit growth of tumors in animal models and a few examples exist of transgenic animals that express ribozymes endogenously with no obvious side effects. So the stage is now set for ribozymes to play a role in human therapeutics, particularly in cancer traetment as the above work testifies. It is anticipated that a growing number of gene therapy trials will be approved in the near future and only then will it become clear whether the hopeful examples given in this chapter can be developed into useful, or even preferred, methods of treatment.

Although most of the promising results described in Section 4 made use of hammerhead ribozymes and only a few utilised the hairpin ribozyme, this mainly reflects the more recent discovery and analysis of the other ribozyme types described in Section 2. Given time for further research, the other types may prove more efficient either generally, or in particular circumstances. This is particularly true for the trans-splicing ribozyme (Section 2.5) which has been used successfully to correct mutant beta-globin transcripts in human erythrocyte precursors (Lan *et al.*, 1998), but could also be used to down-regulate aberrant transcripts where the "correction" approach does not work.

REFERENCES

Ali, M., Lemoine, N.R., and Ring, C.J.A., 1994, The use of DNA viruses as vectors for gene therapy, *Gene Ther.* **6**:367–384.

Altman, S., Kirsebom, L., and Talbot, S., 1993, Recent studies of ribonuclease P, *FASEB J.* **7**:7–14.

Barraclough, B.R., Savin, J., Dube, S.K., and Rudland, P.S., 1987, Molecular cloning and sequence of the gene for *p9Ka*. A cultured myoepithelial cell protein with strong homology to S-100, a calcium binding protein. *J. Mol. Biol.* **198**:13–20.

Been, M.D., Perrotta, A.T., and Rosenstein, S.P., 1992, Secondary structure of the self-cleaving RNA of hepatitis delta virus: application to catalytic RNA design, *Biochemistry* **31**:11843–11852.

Benedict, C.M., Pan, W.H., Loy, S.E., and Clawson, G.A., 1998, Triple ribozyme-mediated down-regulation of the retinoblastoma gene, *Carcinogenesis* **19**:1223–1230.

Bertrand, E., and Rossi, J.J., 1996, Anti-HIV therapeutic hammerhead ribozymes: targeting strategies and optimization of intracellular function, *Nucleic Acids. Mol. Biol.* **10**:301–314.

Bertrand, E., Castanotto, D., Zhou, C., Carbonnelle, C., Lee, N.S., Good, P., Chattergee, S., Grange, T., Pictet, R., Kohn, D., Engelke, D., and Rossi, J.J., 1997, The expression cassette determines the functional activity of ribozymes in mammalian cells by controlling their intracellular localization, *RNA* **3**:75–88.

Birikh, K.R., Heaton, P.A., and Eckstein, F., 1997, The structure, function and application of the hammerhead ribozyme, *Eur. J. Biochem.* **245**:1–16.

Branch, A.D., and Robertson, H.D., 1991, Efficient trans cleavage and a common structural motif for the ribozymes of the human hepatitis δ agent, *Proc. Natl. Acad. Sci. USA* **88**:10163–10167.

Cantor, G.H., 1998, Bovine leukaemia virus ribozymes, in: *Therapeutic Applications of Ribozymes*, (K.J. Scanlon, ed.), pp. 119–129, Humana Press, Totowa, NJ.

Cantor, G.H., Stone, D.M., McElwain, T.F., and Palmer, G.H., 1996, Comparison of the antiviral efficacy of ribozymes and antisense RNA directed against bovine leukemia virus rex/tax, *Antisense Nucleic Acid. Drug. Devel.* **6**:301–304.

Cech, T.R., Self-splicing of group I introns, *Annu. Rev. Biochem.* **59**:543–568.

Cepero, E., Hnatyszyn, H.J., Kraus, G., and Lichtenheld, M.G., 1998, Potent inhibition of CTLA-4 expression by an anti-CTLA-4 ribozyme, *Biochem. Biophys. Res. Comm.* **247**:838–843.

Chang, M.Y., Won, S.J., and Liu, H.S., 1997, A ribozyme specifically suppresses transformation and tumorigenicity of Ha-ras-oncogene-transformed NIH/3T3 cell lines, *J. Cancer Res. Clin. Oncol.* **123**:91–99.

Chen, S., Song, C.S., Lavrovsky, Y., Bi, B.Y., Vellanoweth, R., Chatterjee, B., and Roy, A.K., 1998, Catalytic cleavage of the androgen receptor messenger RNA and functional inhibition of androgen receptor activity by a hammerhead ribozyme, *Molec. Endocrinol.* **12**:1558–1566.

Chowrira, B.M., Pavco, P.A., and McSwiggen, J.A., 1994, In vitro and in vivo comparison of hammerhead, hairpin and hepatitis delta virus self-processing ribozyme cassettes, *J. Biol. Chem.* **269**:25856–25864.

Czubayko, F., Schulte, A.M., Berchem, G.J., and Wellstein, A., 1996, Melanoma angiogenesis and metastasis modulated by riboenzyme targeting of the secreted growth factor pleiotrophin, *Proc. Natl. Acad. Sci. USA* **93**:14753–14758.

Czubayko, F., Schulte, A.M., and Wellstein, A., 1998, Ribozyme targeting of the growth factor pleiotrophin, in: *Therapeutic Applications of Ribozymes*, (K.J. Scanlon, ed.), pp. 311–328, Humana Press, Totowa, NJ.

de la Rosa, A., Williams, R.L., and Steeg, P.S., 1995, *Nm23*/nucleoside diphosphate kinase: toward a structural and biochemical understanding of its biological functions, *Bioessays* **17**:53–62.

Dong, J.-T., Lamb, P.W., Rinker-Schaeffer, C.W., Vukanovic, J., Ichikawa, T., Isaacs, J.T., and Barrett, J.C., 1995, *KAI*1, a metastasis suppressor gene for prostate cancer on human chromosome 11p11.2, *Science* **268**:884–886.

Dorai, T., Goluboff, E.T., Olsson, C.A., and Buttyan, R., 1997, Development of a hammerhead ribozyme against bcl-2.2. Ribozyme treatment sensitizes hormone-resistant prostate cancer cells to apoptotic agents, *Anticancer Res.* **17**:3307–3312.

Dorai, T., Olsson, C.A., Katz, A.E., and Buttyan, R., 1997, Development of a hammerhead ribozyme against bcl-2.1. Preliminary evaluation of a potential gene therapeutic agent for hormone-refractory human prostate cancer, *Prostate* **32**:246–258.

Ebralidze, A., Tulchinsky, E., Grigorian, M., Afanasyeva, A., Senin, V., Revazova, E., and Lukanidin, E., 1989, Isolation and characterisation of a gene specifically expressed in different metastatic cells and whose deduced gene product has a high degree of homology to a Ca^{2+}-binding protein family, *Genes Dev.* **3**:1086–1093.

Feng, M., Cabrera, G., Deshane, J., Scanlon, K.J., and Curiel, D.T., 1995, Neoplastic reversion accomplished by high efficiency adenoviral-mediated delivery of an anti-*ras* ribozyme, *Cancer Res.* **55**:2024–2028.

Ferré-d'Amaré, A.R., Zhou, K.H., and Doudna, J.A., 1998, Crystal structure of a hepatitis delta virus ribozyme, *Nature* **395**:567–574.

Forster, A.C., and Symons, R.H., 1987, Self-cleavage of plus and minus RNAs of a virusoid and a structural model for the active sites, *Cell* **49**:211–220.

Frank, B.L., and Goodchild, J., 1997, Selection of accessible sites for ribozymes on large RNA transcripts, in: *Ribozyme Protocols*, (P.C. Turner, ed.), pp. 37–43, Humana Press Inc., Totowa, NJ.

Fujita, S., Koguma, T., and Ohkawa, J., 1997, Discrimination of a single base change in a ribozyme using the gene for dihydrofolate reductase as a selective marker in *Escherichia coli*, *Proc. Natl. Acad. Sci. USA* **94**:391–396.

Funato, T., Ishii, T., Kanbe, M., Scanlon, K.J., and Sasaki, T., 1997, Reversal of cisplatin resistance in vivo by an anti-fos ribozyme, *In Vivo* **11**:217–220.

Gao, Z.Q., Gao, Z.P., Liu, X.F., and Zhang, T., 1997, Selective reversal of drug resistance in drug-resistant lung adenocarcinoma cells by tumor-specific expression of MDR1 ribozyme gene mediated by retrovirus, Science In China Series C-Life Sciences **40**:122–127.

Gregoriadis, G., 1995, Engineering Liposomes for drug delivery: Progress and problems, *Trends Biotechnol.* **13**:527–537.

Gros, P., Croop, J., Roninson, I., Varshavsky, A., and Housman, D.E., 1986, Isolation and characterisation of DNA sequences amplified in multidrug resistant hamster cells, *Proc. Natl. Acad. Sci. USA* **83**:337–341.

Guerrier-Takada, C., Gardiner, K., Marsh, T., Pace, N., and Altman, S., 1983. The RNA moiety of ribonuclease P is the catalytic subunit of the enzyme, *Cell* **35**:849–857.

Hampel, A., and Tritz, R., 1989, RNA catalytic properties of the minimum (–) sTRSV sequence, *Biochemistry* **28**:4929–4933.

Hampel, A., Tritz, R., Hicks, M., and Cruz, P., 1990, Hairpin catalytic RNA model: evidence for helices and sequence requirement for substrate RNA, *Nucleic Acids Res.* **18**:299–304.

Hua, J., and Muschel, R.J., 1996, Inhibition of matrix metalloproteinase 9 expression by a ribozyme blocks metastasis in a rat sarcoma model system, *Cancer Res.* **56**:5279–5282.

James, W., and Cowe, E., 1997, Computational approaches to the identification of ribozyme target sites, in: *Ribozyme Protocols*, (P.C. Turner, ed.), pp. 17–26, Humana Press Inc., Totowa, NJ.

Jones, J.T., and Sullenger, B.A., 1997, Evaluating and enhancing ribozyme reaction efficiency in mammalian cells, *Nature Biotechnology* **15**:902–905.

Juhl, H., Downing, S.G., Wellstein, A., and Czubayko, F., 1997, HER-2/neu is rate-limiting for ovarian cancer growth—Conditional depletion of HER-2/neu by ribozyme targeting, *J. Biol. Chem.* **272**:29482–29486.

Kawasaki, H., Ohkawa, J., Tanishige, N., Yoshinari, K., Murata, T., Yokoyama, K., and Taira, K., 1996, Selection of the best target site for ribozyme-mediated cleavage within a fusion gene for adenovirus E1A-associated 300 kDA protein (p300) and luciferase, *Nucleic Acids Res.* **24**:3010–3016.

Kijima, H., Bouffard, D.Y., and Scanlon, K.J., 1996, Ribozyme-mediated reversal of human pancreatic carcinoma phenotype, in: *Bone Marrow Transplantation-Basic and Clinical Studies*, (Ikehara, S., Takaku, F., and Good, R.A., eds), pp. 153–163, Springer-Verlag, Tokyo.

Kitajima, I., Hanyu, N., Kawahara, K., Soejima, Y., Kubo, T., Yamada, R., Kaneda, Y., and Maruyama, I., 1998, Ribozyme-based gene cleavage approach to chronic arthritis associated with human T cell leukemia virus type I—Induction of apoptosis in synoviocytes by ablation of HTLV-I tax protein, *Arthritis And Rheumatism* **40**:2118–2127.

Koizumi, M., Hayase, Y., Iwai, S., Kamiya, H., Inoue, H., and Ohtsuka, E., 1989, Design of RNA enzymes distinguishing a single base mutation in RNA, *Nucleic Acids Res.* **17**:7059–7071.

Kruger, K., Grabowski, P.J., Zaug, A.J., Sands, J., Gottschling, D.E., and Cech, T.R., 1982, Self-splicing RNA: autoexcision and autocyclization of the ribosomal RNA intervening sequence of *Tetrahymena*, *Cell* **31**:147–157.

Lai, M.M.C., 1995, The molecular biology of hepatitis delta virus, *Annu. Rev. Biochem.* **64**:259–286.

Lan, N., Howrey, R.P., Lee, S.W., Smith, C.A., and Sullenger, B.A., 1998, Ribozyme-mediated repair of sickle beta-globin mRNAs in erythrocyte precursors, *Science* **280**:1593–1596.

Leopold, L.H., Shore, S.K., Newkirk, T.A., Reddy, R.M.V., and Reddy, E.P., 1995, Multi-unit ribozyme-mediated cleavage of *bcr-abl* mRNA in myeloid leukaemias, *Blood* **85**:2162–2170.

Lieber, A., and Kay, M.A., 1996, Adenovirus mediated expression of ribozymes in mice, *J. Virol.* **70**:3153–3158.

Lieber, A., and Strauss, M., 1995, Selection of efficient cleavage sites in target RNAs by using a ribozyme expression library, *Mol. Cell. Biol.* **15**:540–551.

Madden, M.J., Morrow, C.S., Nakagawa, M., Goldsmith, M.E., Fairchild, C.R., and Cowan, K.H., 1993, Identification of 5′ and 3′ sequences involved in the transcription of the human mdr-1 gene in vivo, *J. Biol. Chem.* **268**:8290–8297.

Masuda, Y., Kobayashi, H., Holland, J.F., and Ohnuma, T., 1998, Reversal of multidrug resistance by a liposome-MDR1 ribozyme complex, *Cancer Chemother. Pharmacol.* **42**:9–16.

Mælandsmo, G.M., Hovig, E., Skrede, M., Engebraaten, O., Fløренes, V.A., Myklebost, O., Grigorian, M., Lukanidin, E., Scanlon, K.J., and Fodstad, Ø., 1996, Reversal of the in vivo metastatic phenotype of human tumor cells by an anti-CAPL (mts-1) ribozyme, *Cancer Res.* **56**:5490–5498.

Mælandsmo, G.M., Kjønniksen, I., Hovig, E., Scanlon, K.J., and Fodstad, Ø., 1998, Ribozymes—applications in cancer metastasis research, in: *Therapeutic Applications of Ribozymes*, (K.J. Scanlon, ed.), pp. 241–260, Humana Press, Totowa, NJ.

Marusic, L., Luzi, E., Barsacchi, G., and Eckstein, F., 1997, Intermolecular cleavage by the newt ribozyme, *Eur. J. Biochem.* **247**:396–401.

Matsushita, H., Kobayashi, H., Mori, S., Kizaki, M., and Ikeda, Y., 1995, Ribozymes cleave the AML-1/MTG8 fusion transcript and inhibit proliferation of leukaemic cells with t(8;21), *Biochem. Biophys. Res. Commun.* **215**:431–437.

Matsushita, H., Kobayashi, H., Kizaki, M., and Ikeda, Y., 1998, ribozymes targeted against AML-1/MTG8, in: *Therapeutic Applications of Ribozymes*, (K.J. Scanlon, ed.), pp. 293–310, Humana Press, Totowa, NJ.

Nylandsted, J., Rohde, M., Bartek, J., and Strauss, M., 1998, Expression of a p16(INK4a)-specific ribozyme down-modulates p16(INK4a) abundance and accelerates cell proliferation, *FEBS Lett.* **436**:41–45.

Ohta, Y., Tone, T., Shitara, T., Funato, T., Jiao, L., Kashfian, B.I., Yoshida, E., Horng, M., Tsai, P., Lauterbach, K., Kashani-Sabet, M., Florenes, V.A., Fodstad, Ø.Y., and Scanlon, K.J., 1994, H-*ras* ribozyme mediated alteration of the human melanoma phenotype, *Ann. NY Acad. Sci.* **716**:242–253.

Ohta, Y., Kijima, H., Kashani-Sabet, M., and Scanlon, K.J., 1996, Suppression of the malignant phenotype of melanoma cells by anti-oncogene ribozymes, *J. Invest. Dermatol.* **106**:275–280.

Saville, B.J., and Collins, R.A., 1990, A site-specific self-cleavage reaction performed by a novel RNA in *Neurospora* mitochondria, *Cell* **61**:685–696.

Scanlon, K.J., Ishida, H., and Kashani-Sabet, M., 1994, Ribozyme mediated reversal of the multidrug-resistant phenotype, *Proc. Natl. Acad. Sci. USA* **91**:11123–11127.

Scanlon, K.J., Jiao, L., Funato, T., Wang, W., Tone, T., Rossi, J.J., and Kashnani-Sabet, M., 1991, Ribozyme-mediated cleavage of c-fos mRNA reduces gene expression of DNA synthesis enzymes and metallothionine, *Proc. Natl. Acad. Sci. USA* **88**:10591–10595.

Scherr, M., Grez, R., Ganser, A., and Engels, J.W., 1997, Specific hammerhead ribozyme-mediated cleavage of mutant N-ras mRNA in vitro and ex vivo—Oligoribonucleotides as therapeutic agents, *J. Biol. Chem.* **272**:14304–14313.

Scherr, M., Maurer, A.B., Klein, S., Ganser, A., Engels, J.W., and Grez, M., 1998, Effective reversal of a transformed phenotype by retrovirus-mediated transfer of a ribozyme directed against mutant N-ras, *Gene Therapy* **5**:1227–1234.

Sczakiel, G., and Tabler, M., 1997, Computer-aided calculation of the local folding potential of target RNA and its use for ribozyme design, in: *Ribozyme Protocols*, (P.C. Turner, ed.), pp. 11–15, Humana Press Inc., Totowa, NJ.

Sioud, M., and Sorensen, D.R., 1998, A nuclease-resistant protein kinase C alpha ribozyme blocks glioma cell growth, *Nature Biotechnol.* **16**:556–561.

Suga, H., Lohse, P.A., and Szostak, J.W., 1998, Structural and kinetic characterisation of an acyl transferase ribozyme, *J. Am. Chem. Soc.* **120**:1151–1156.

Sun, L.-Q., Gerlach, W.L., and Symonds, G., 1996, The use of ribozymes to inhibit HIV replication, *Nucleic Acids. Mol. Biol.* **10**:329–342.

Suzuki, T., Curcio, L.D., Tsai, J., and Kashani-Sabet, M., 1998, Anti-*c-erb*-B-2 ribozyme for breast cancer, in: *Therapeutic Applications of Ribozymes*, (K.J. Scanlon, ed.), pp. 223–239, Humana Press, Totowa, NJ.

Tong, A.W., Zhang, Y., Nemunaitis, J., and Mues, G., 1998, K-ras ribozyme for lung cancer, in: *Therapeutic Applications of Ribozymes*, (K.J. Scanlon, ed.), pp. 209–222, Humana Press, Totowa, NJ.

Uhlenbeck, O.C., 1987, A small catalytic oligoribonucleotide, *Nature* **328**:596–600.

Vaish, N.K., Kore, A.R., and Eckstein, F., 1998, Recent developments in the hammerhead ribozyme field, *Nucleic Acids. Res.* **26**:5237–5242.

Walter, N.G., and Burke, J.M., 1998, The hairpin ribozyme: structure, assembly and catalysis, *Curr. Opin. Chem. Biol.* **2**:24–30.

Welch, P.J., Hampel, A., Barber, J., Wong-Staal, F., and Yu, M., 1996, Inhibition of HIV replication by the hairpin ribozyme, *Nucleic Acids Mol. Biol.* **10**:315–327.

Wiechen, K., Zimmer, C., and Dietel, M., 1998, Selection of a high activity c-erbB-2 ribozyme using a fusion gene of c-erbB-2 and the enhanced green fluorescent protein, *Cancer Gene Therapy* **5**:45–51.

Wright, L.A., Milliken, S., Biggs, J.C., and Kearney, P., 1998, Ex vivo effects associated with the expression of a bcr-abl-specific ribozyme in a CML cell line, *Antisense Nucleic Acid. Drug Devel.* **8**:15–23.

Wu, H.-N., Lin, Y.-J., Lin, F.-P., Makino, S., Chang, M.-F., and Lai, M.M.C., 1989, Human hepatitis δ virus RNA subfragments contain an autocleavage activity, *Proc. Natl. Acad. Sci. USA* **86**:1831–1835.

Yamazaki, H., Kijima, H., Ohnishi, Y., Abe, Y., Oshika, Y., Tsuchida, T., Tokunaga, T., Tsugu, A., Ueyama, Y., Tamaoki, N., and Nakamura, M., 1998, Inhibition of tumor growth by ribozyme-mediated suppression of aberrant epidermal growth factor receptor gene expression, *J. Natl. Cancer Inst.* **90**:581–587.

Yokoyama, Y., Morishita, S., Takahashi, Y., Hashimoto, M., and Tamaya, T., 1997, Modulation of *c-fms* proto-oncogene in an ovarian carcinoma cell line by a hammerhead ribozyme, *Brit. J. Cancer* **76**:977–982.

Yu, Q., and Burke, J.M., 1997, Design of hairpin ribozymes for *in vitro* and cellular applications, in: *Ribozyme Protocols*, (P.C. Turner, ed.), pp. 161–169, Humana Press Inc., Totowa, NJ.

Yuan, Y., 1998, Application of in vitro selection in design of external guide sequences that direct efficient cleavage of RNA by RNase P, in: *Therapeutic Applications of Ribozymes*, (K.J. Scanlon, ed.), pp. 415–425, Humana Press, Totowa, NJ.

Yuan, Y., and Liu, F., 1998, Targeted cleavage of RNA using external guide sequences and eukaryotic RNase P, in: *Therapeutic Applications of Ribozymes*, (K.J. Scanlon, ed.), pp. 397–413, Humana Press, Totowa, NJ.

Zelphati, O., and Szoka, F.C., 1996, Mechanism of oligonucleotide release from cationic liposomes, *Proc Natl. Acad. Sci. USA* **93**:11493–11498.

Zhang, Y., Mukhopadhyay, T., Donehower, L.A., Georges, R.N., and Roth, J.A., 1993, Retroviral vector-mediated transduction of K-*ras* antisense RNA into human lung cancer cells inhibits expression of the malignant phenotype, *Human Gene Therapy* **4**:451–460.

27

THE APPROACH OF TRIPLE HELIX FORMATION IN CONTROL OF GENE EXPRESSION AND THE TREATMENT OF TUMORS EXPRESSING IGF-I

Lia C. Upegui-Gonzalez[1], Jean-Christophe François[2], Adama Ly[1], and Jerzy Trojan[1]

[1]Laboratoire de Neurologie du Développement
INSERM-CRI 9701
48 bd Sérurier 75019 Paris
France
[2]Laboratoire de Biophysique
INSERM U201, CNRS UA481
Muséum National d'Histoire Naturelle
43 rue Cuvier 75231 Paris Cedex 05
France

1. INTRODUCTION

Gene therapy and gene immunotherapy provide new approaches for clinical trials (Anderson, 1992; Dodet, 1993; Guo *et al.*, 1994; Ido *et al.*, 1995; Miller, 1992; Sokol and Gewirtz, 1996). Among these are strategies using antisense and triple helix technologies which lead to activation of the host immune system (Shevelev *et al.*, 1997; Trojan *et al.*, 1992). Antisense nucleic acids are designed to bind to RNA resulting in the formation of RNA-DNA or RNA-RNA hybrids with an arrest of translation. *In vitro* experiments have shown the potentiality of the antisense techniques to block expression of tumor-associated genes. Antisense oligonucleotides have been tested on animal models for different cancer types as Burkitt's lymphoma (Huang *et al.*, 1995), Philadelphia leukemia (Okabe *et al.*, 1993; Skorski *et al.*, 1995) and melamona (Leonetti *et al.*, 1996). The antisense oligonucleotides have reduced tumorigenicity of transfected cells and inhibited tumor progression.

Another kind of oligonucleotides that block gene expression are the triple-helix forming oligonucleotides (TFOs). They block RNA polymerases transit by forming a triple-helical structure on DNA. These TFOs promise to be a new class of sequence-specific DNA-binding drugs which will target malignancies at the transcriptionnal level

Cancer Gene Therapy: Past Achievements and Future Challenges, edited by Habib
Kluwer Academic/Plenum Publishers, New York, 2000.

(Chan and Glazer, 1997; Hélène *et al.*, 1998; Maher III, 1996; Vasquez and Wilson, 1998). TFOs have been used to form an intermolecular triplex structure blocking the binding of transcription factors and repressing transcription of genes such as epidermal growth factor receptor (Durland *et al.*, 1991), oncogene myc (Kim *et al.*, 1998; Postel *et al.*, 1991; Thomas *et al.*, 1995), HER2 (Ebbinghaus *et al.*, 1993; Porumb *et al.*, 1996), Ha-ras (Mayfield *et al.*, 1994) or Ki-ras (Alunni-Fabbroni *et al.*, 1996). TFOs may prove to be the basis of effective chemotherapy drugs for concerned cancers.

A protein commonly associated with tumors is Insulin-like growth factor I (IGF-I). IGF-I is expressed in many fetal tissues and is associated with cellular normal and neoplastic growth and differentiation (Baserga, 1995; Han and Hill, 1992; Kiess *et al.*, 1989). IGF-I antisense RNA and IGF-I triple helix RNA-DNA strategies have been shown to induce the arrest and rejection of following tumors *in vivo*: glioma (Shevelev *et al.*, 1997; Trojan *et al.*, 1992; Trojan *et al.*, 1993), teratocarcinoma (Trojan *et al.*, 1994) and hepatoma (Lafarge-Frayssinet *et al.*, 1997; Upegui-Gonzalez *et al.*, 1998a; Upegui-Gonzalez *et al.*, 1998b). These results have formed the basis for a gene therapy clinical trial for human primary glioblastoma and hepatocellular cancer.

2. ANTI-GENE STRATEGIES

Gene therapy aims to introduce genes into cells for therapeutic gain (Sokol and Gewirtz, 1996). Many strategies to modify or to suppress gene expression in tumor or infected cells have been proposed for reverse gene therapy including strategies employing short oligonucleotides (Hélène, 1994). Synthetic DNA or RNA oligonucleotides may be used to interdict individual gene expression in a sequence-specific manner. These short nucleic acids are designed to bind to their complementary targets via double or triple helix formation. Briefly, in the antisense strategy, the oligonucleotides are targeted to complementary sequence on messenger RNAs. In the antigene strategy, the oligonucleotides are targeted to double stranded DNA containing polypurine-polypyrimidine sequences via triple helix formation. These oligonucleotides may control gene expression at different levels (Fig. 1). Inhibition of the initiation of gene transcription could be achieved with triplex-forming oligonucleotides when they are targeted upstream to the transcriptional start sites in the region recognized by transcriptional activators (Fig. 1a). Oligonucleotides could also hybridize to the locally opened loop created by RNA polymerase during the transcription process via duplex or even triplex formation (b). Nascent RNAs synthesized during the elongation step could be also targeted in the nucleus by antisense or triplex-forming oligonucleotides (c). Inhibition of mRNA splicing or interference with the turnover of spliced mRNA could be achieved by targeting intron-exon junctions (d,e). The translation of the mRNA into the coded protein could be inhibited by blocking of the binding of initiation factors, interfering with the assembly of ribosomal subunits or with the reading of the messenger RNA (f,g,h). The antisense oligonucleotide when hybridized to the 5′ untranslated (f), the initiation codon (g) or the 3′ untranslated region (h) of the mRNA could induce the specific degradation of the mRNA via the specific ribonuclease H that recognized DNA-RNA hybrids and cleaved RNA strands of such hybrids. This RNAse H could theoritically cleave mRNA/DNA hybrids at each levels described in Fig. 1, except when oligonucleotide is targeted to DNA via triple-helix formation (a). Triplex-forming oligonucleotide could also be targeted to single-stranded nucleic acids via triple helix formation. A part of this "clamp" oligonucleotide bound to the mRNA and the second part fold to form a triple helical

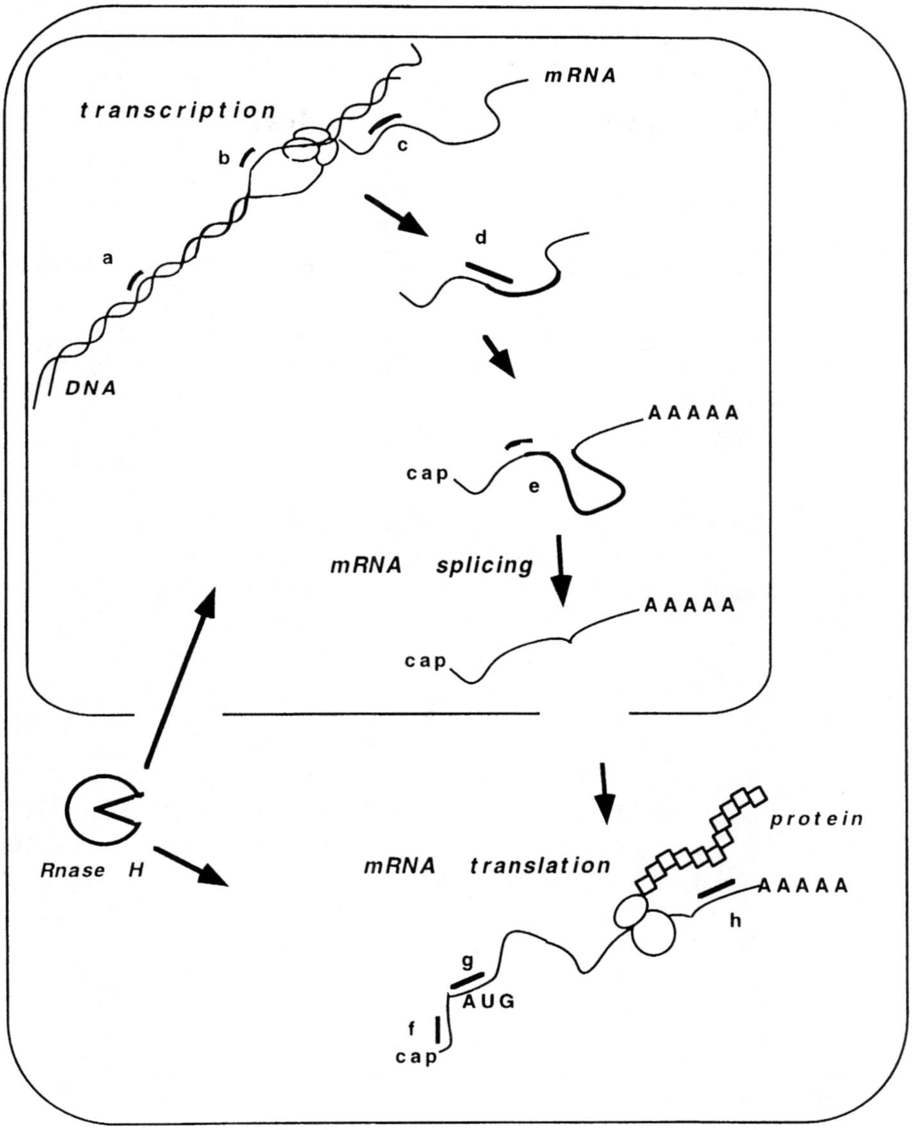

Figure 1. A schematic representation of potential binding sites for antisense and triplex-forming oligonucleotides.

structure (Hélène, 1994). These clamp oligonucleotides are not susceptible to cleavage via RNAse H.

Through the binding on their complementary sequence, anti-gene oligonucleotides may block specifically gene expression at transcriptional or translational levels. These anti-genes could be classified in three categories, the antisense molecules targeted to the complementary sequence in mRNA (including antisense RNA, antisense oligonucleotides and ribozymes (Couture and Stinchcomb, 1996)), the triplex-forming oligomers targeted to the double-stranded DNA gene, and the sense oligonucleotides designed to act as decoys to trap regulatory proteins (Morishita *et al.*, 1998). Here, we presented only some examples of the recent utilization of the triplex-involving strategy to control gene expression.

3. TRIPLEX-FORMING OLIGONUCLEOTIDES

The formation of a triple helical structure on a specific gene via oligonucleotide binding could inhibit transcriptional and/or DNA replication leading to specific inhibition of the gene expression. Short oligonucleotides are able to bind to duplex DNA containing polypurine-polypyrimidine sequences via triplex formation. The oligonucleotide bound into the major groove of the duplex through the formation of specific hydrogen bonds between bases of the third strand and the polypurine strand of the duplex DNA. The DNA remained double stranded upon binding of a triplex-forming oligonucleotide. The third strand could be either pyrimidine or purine-rich and synthesized as an RNA or an DNA molecule. Oligopyrimidines bound to duplex DNA by inducing Hoogsteen hydrogen bonds with the polypurine strand of the duplex target (Chan and Glazer, 1997; Giovannangeli and Hélène, 1997; Vasquez and Wilson, 1998). The orientation of the pyrimidine-rich third strand is parallel to this purine sequence. Cytosine and thymine of the CT-containing oligonucleotide recognized the CG and the TA base pairs to form the base triplets $CG^{*}C^{+}$ (where C+ indicates protonated cytosine) and TA*T, respectively. The formation of these base triads are in Hoogsteen configuration and required a slightly acidic pH to protonate cytosine. Substitution of cytosine with 5 methylcytosine greatly alleviates this pH dependence and allows triplex formation at physiological pH (Hausheer *et al.*, 1992). Purine-containing oligonucleotides could also be designed to bind duplex DNA via triplex formation (Chan and Glazer, 1997; Giovannangeli and Hélène, 1997; Vasquez and Wilson, 1998). The orientation of the TFO binding is antiparallel to the purine-containing strand of the target and reverse Hoogsteen bonds are formed. This purine-rich third strand formed triplex at physiological pH, but required the presence of magnesium ions.

The selectivity of binding of triplex-forming oligonucleotide within duplex DNA is at least as high as the Watson-Crick base pairing involved in double helix formation. Mismatches in a triple helical complex decreased greatly binding affinities depending on the location of the base triplet mismatch. As observed with double stranded complex, mismatch at the ends of the polypurine-polypyrimidine target is less destabilizing for the binding of TFO than a central mismatch (Mergny *et al.*, 1991). A lot of base analogs was used to increase the triplex-binding code which is summarized in (Chan and Glazer, 1997).

3.1. Replication Control with Triplex-Forming Oligonucleotides

A lot of studies showed that gene transcription could be blocked by oligonucleotide-directed triple helix formation (see below). Nevertheless, only several studies have been performed with intermolecular triple helical complexes concerning the inhibition of DNA replication. It was mainly shown that primer extension by DNA polymerases could be blocked by TFOs (Guieysse *et al.*, 1995; Hacia *et al.*, 1994; Samadashwily and Mirkin, 1994). Intramolecular triplex structures (H-DNA) were suggested to arrest DNA polymerization ((Krasilnikov *et al.*, 1997; Samadashwily *et al.*, 1993); and for a review on H-

DNA structures (Frank-Kamenetskii and Mirkin, 1995)). To our knowledge, no experiments were described concerning interference of DNA replication in cells by intermolecular triple helices. The initiation of DNA replication by the bacteriophage T7 polymerase could be efficiently and specifically inhibited depending on the primer position on the single-stranded template (Guieysse *et al.*, 1995). If the distance between the primer and the triplex-forming oligonucleotide is greater than 3 nucleotides, the inhibition became ineffective (Guieysse *et al.*, 1995). Primer extension by Klenow fragment of the E. Coli DNA polymerase I could be blocked by triplex-forming oligonucleotides (Hacia *et al.*, 1994). The inhibition is highly dependant on the length of the TFO used.

3.2. Control of Gene Transcription Using Triplex-Forming Oligonucleotides

Purine-rich or pyrimidine-rich triplex-forming oligonucleotides could inhibit gene transcription by occluding the binding of transcription factors or RNA polymerase to their targets. Other mechanisms have been proposed for the TFO-mediated transcriptional repression involving modification in DNA conformation and/or flexibility (reviewed in (Maher III, 1996)). Transcription initiation by bacteriophage RNA polymerase can be strongly inhibited by DNA and RNA triplex-forming oligonucleotides with binding sites overlapping homopurine operator sequence (Maher III, 1992; Skoog and Maher III, 1993b). Nevertheless, these triplex structures are greatly destabilized by elongating RNA polymerases (Maher III, 1992). The efficiency of triplex-forming oligonucleotides as artificial repressors may be limited by transcriptional activity in the gene control region. Antagonistic effects of other promoters acting at distance from the TFO binding site have been observed (Skoog and Maher III, 1993a). In fact, transcription elongation cannot be efficiently inhibited by TFOs targeted to double-stranded transcribed sequences (Duval-Valentin *et al.*, 1992; Skoog and Maher III, 1993a; Young *et al.*, 1991). Nonetheless, a more efficient inhibition of transcription elongation could be achieved involving stable triplex structures such as those formed with TFOs containing phosphoramidate linkages (Giovannangeli *et al.*, 1996) or such as covalent triplex induced by TFOs covalently linked to crosslinking molecules (Duval-Valentin *et al.*, 1992; Young *et al.*, 1991).

Most of triplex-forming oligonucleotides have been targeted to gene regulatory promoter regions such as binding sites recognized by transcriptional activators or RNA polymerase complexes. Triplex-mediated inhibition of gene transcription have been demonstrated in eukaryotic (Ing *et al.*, 1993; Maher III *et al.*, 1992; Mayfield *et al.*, 1994; Postel *et al.*, 1991) and prokaryotic (Duval-Valentin *et al.*, 1992) cell-free transcription systems. Interference with cellular or transfected gene expression by triplex-forming oligonucleotides have also been reported using either direct cellular transfection of TFOs (Aggarwal *et al.*, 1996; Imagawa *et al.*, 1994; Porumb *et al.*, 1996; Scaggiante *et al.*, 1994; Tu *et al.*, 1995), or cotransfection of TFOs with reporter genes containing target sequences (Alunni-Fabbroni *et al.*, 1996; Grigoriev *et al.*, 1992; Hobbs and Yoon, 1994; Ing *et al.*, 1993; Joseph *et al.*, 1997; Kim and Miller, 1998; Kim *et al.*, 1998; Kovacs *et al.*, 1996; Roy, 1994; Svinarchuk *et al.*, 1996).

Triplex-forming oligonucleotides targeted to a positive regulatory element of the human erythropoietin promoter (Epo) decreased the production of Epo mRNA in the human hepatoma cell line, Hep3B (Imagawa *et al.*, 1994). A purine-rich TFO targeted to the interferon-inducible responsive element (IRE) of the 6–16 gene was found to be able to inhibit specifically the reporter gene expression controled by IRE in HeLa cells. Interferon treatment could not induce gene transcription in presence of the specific TFO

(Roy, 1994). Similarly, the induction by IFN-gamma of HLA-DR molecules and other IFN gamma-inducible molecules such as the adhesion molecule ICAM-1 and the Fc receptor for IgG is downregulated by a TFO targeted to the regulatory region of the human MHC class II (DRA) gene (Fedoseyeva *et al.*, 1994). It has been shown that the expression of rat androgen receptor was decreased in transfected COS-1 cells by triplex-forming oligonucleotide directed to the activator domain of the gene (Song *et al.*, 1995).

Synthesis of the alpha1(I) collagen seemed to play an important role in various disease where abnormal collagen accumulated in interstitial spaces. A polypurine-polypyrimidine sequence found in the rat alpha1(I) collagen promoter which is a positive regulatory element of the collagen gene has been targeted by triplex-forming oligonucleotides. *In vitro* transcription assays and transient transfection experiments demonstrated that transcriptional activity of this promoter was downregulated by specific TFO likely by interfering with the binding of regulatory factors to the polypurine-polypyrimidine target (Kovacs *et al.*, 1996).

It is often difficult to provide clear demonstration of the involvment of triplex formation when inhibition of gene transcription is observed with triplex-forming oligonucleotides targeted to endogeneous genes. The synthesis of control oligonucleotides designed in such a way that they can not bind *in vitro* to the target sequence contributed partially to the answer. Nevertheless, scrambled sequences of oligonucleotides used as controls could not completely rule out non-specific effects of oligonucleotides on gene expression. For example, few unexpected effects of antisense oligonucleotides that are unrelated to their intended antisense activity but due to their peculiar structures have been described (for review see (Stein, 1996)).

Sequence specificity of transcription inhibition by oligomers in cells could be demonstrated using plasmid carrying the gene of interest and the target sequence for the TFO. This target site could be mutated by *in vitro* mutagenesis to provide evidences of triple helix formation in the observed inhibition. For example, it was shown that a triplex-forming oligonucleotide targeted to the promoter of the alpha subunit of the interleukin-2 receptor was able to inhibit specifically the transcription of the reporter gene (Grigoriev *et al.*, 1992). No inhibition was observed with the specific TFO and a plasmid containing a mutated target sequence, suggesting strongly the involvement of triplex formation in the biological effect (Grigoriev *et al.*, 1992).

3.3. Inhibition of Translation by Triplex-Forming Oligonucleotides

Two oligonucleotides could recognize a polypurine sequence on single-stranded nucleic acid by forming a triple helix. The first oligonucleotide bound to the single-stranded target forming Watson-Crick base pairs and the second one bound to the duplex via triple helix formation involving Hoogsteen or reverse Hoogsteen hydrogen bonds. These two oligonucleotides could be linked together to form either a clamp oligonucleotide or a circular oligonucleotide. This circular oligonucleotide formed tight triplex with its DNA single-stranded target likely because of entropic reasons (Kool, 1996). Clamp oligomers formed less stable complex than circular oligonucleotides but more stable complex than two separate oligonucleotides (Hélène *et al.*, 1998). These two types of triplex-forming oligonucleotides targeted to single-stranded polypurine sequence represent an interesting alternative for antisense oligonucleotides to inhibit translation steps. They can interfere with ribosomal machinery by physical blockade. To our knowledge, no study concerning circular and clamp oligonucleotides in cellular systems were provided demonstrating that the inhibition of translation is due to triplex-formation and not to RNAse H induced cleavage of the Watson-Crick double-helix formed by the oligomer and the messenger RNA. Nevertheless, it has been shown that peptide-nucleic acids (PNAs) could inhibit mRNA translation by triplex-formation. These PNAs are a novel DNA mimic in which the phosphodiester backbone of the oligonucleotide has been replaced with a polyamide backbone. These PNAs which are highly resistant to nucleases and to proteases represented promising tools to inhibit gene expression (for review see (Good and Nielsen, 1997)). They could bind to polypurine sequence forming a 2PNA:1RNA complex with the mRNA target and inactivate the RNA by steric blocking of translation (Good and Nielsen, 1997).

4. DNA ALTERATION BY TRIPLEX-FORMING OLIGONUCLEOTIDES

Triplex-mediated site directed mutagenesis could be achieved using triplex-forming oligonucleotides covalently linked to psoralen derivatives. This psoralen which is a bifunctional photoreactive mutagen, intercalates in the DNA double helix and covalently cross-link thymines on both strands upon UVA irradiation. DNA lesions obtained with these psoralen-TFOs were specifically repaired in mammalian cells leading to transversions, point mutation or microdeletion (Chan and Glazer, 1997; Havre and Glazer, 1993; Sandor and Bredberg, 1994; Wang *et al.*, 1995).

These psoralen-containing triplex-forming oligonucleotides could be used to form irreversible triple helical structure on double-stranded targets, and thereby increasing the inhibitory effect of TFOs on gene transcription. A triplex-forming oligonucleotide linked to psoralen and targeted to the interleukin-2 receptor promoter modify specifically the DNA target and inhibit transcription of the reporter gene (Grigoriev *et al.*, 1993). Psoralen-TFO conjugates could also serve as tools to demonstrate that a TFO recognized and bound specifically its double-stranded target located on nuclear DNA in intact chromatin structures (Giovannangeli *et al.*, 1997).

DNA cleaving agents such as Fe-EDTA (Moser and Dervan, 1987), Cu-phenanthroline (François *et al.*, 1989), were linked to triplex-forming oligonu-cleotides. These oligomer-conjugates which induced specific cleavage of the double-stranded target after chemical activation (addition of reducing agent) could be used as molecular scissors to cleave genomic DNA (for review see (Hélène, 1993)). These artificial nucleases that recognized longer sequences than restriction enzymes and cut genomic DNA less frequently could cleave specifically at one site a yeast chromosome (Strobel and Dervan, 1990). Single-site enzymatic cleavage of genomic DNA could be obtained by oligonucleotide-directed triple-helix formation (Dervan, 1992; Strobel *et al.*, 1991).

5. TRIPLEX STRATEGY IN CANCER GENE THERAPY

The goal of conventional anticancer chemotherapy is to block the abnormal proliferation of malignant cells. Unfortunately, some tumoral cells can escape to chemotherapy by developping multidrug resistance via the over-expression and sometimes amplification of a transmembrane glycoprotein (mdr1 gene). This protein is likely responsible for the rapid efflux of many anti-tumor chemicals. The human mdr1 gene was chosen as target for triplex-forming oligonucleotides. It was shown that this TFO down-regulated specifically the mdr1 expression in a human drug resistant cell line (Scaggiante *et al.*, 1994).

Triplex strategy was applied to the ras oncogenes which are the most frequently activated oncogenes in human cancer. *In vitro* transcription of human Ha-ras was inhibited by triplex-forming oligonucleotides targeted to sequences recognized by the Sp1 trancription factor (Mayfield *et al.*, 1994). Using transient transfection assays, it was demonstrated that a purine-rich TFO could also inhibit the transcription of murine c-Kiras gene in NIH 3T3 cells (Alunni-Fabbroni *et al.*, 1996). At present, no data is available concerning the effect of these TFOs on the expression of the endogeneous ras oncogenes in cell lines.

As overexpression of c-myc protein appears in a variety of malignant cells, c-myc protooncogene was chosen as target for triplex-forming oligonucleotides. Endogeneous c-myc gene expression was inhibited by TFO treatment in a human cervical carcinoma cell line (Hela) (Postel *et al.*, 1991) and in a breast carcinoma cell line (MCF-7) (Thomas *et al.*, 1995).

Triplex-forming oligonucleotides were shown to bind *in vitro* to human EGF-receptor promoter (Durland *et al.*, 1991), and to inhibit *in vitro* transcription of HER2/neu gene (Ebbinghaus *et al.*, 1993). The transcription of endogeneous human HER-2/neu oncogene, which is overexpressed in breast cancer and other human malignancies, has been inhibited by TFO treatment of a breast carcinoma cell line (MCF-7) (Porumb *et al.*, 1996).

A lot of growth factors are known to play a role in tumorigenesis, and thereby represent relevant targets for antigene therapies. The synthesis of human tumor necrosis factor (TNF), which acts as an autocrine growth factor in various tumor cell lines including neuroblastoma and glioblastoma, has been blocked by triplex-forming oligonucleotide treatment (Aggarwal *et al.*, 1996). Growth of the human glioblastoma tumor cells decreased in a dose-dependant manner with the TFO treatment. These results demonstrate the potential therapeutic utility of triplex-mediated strategy in modulating the expression of oncogenes or growth factors.

6. GENE THERAPY OF TUMORS EXPRESSING IGF-I

Insulin-like growth factor I (IGF-I) and its receptor (IGF-I-R) represent key regulators of cell growth in normal cells and also seem to play a major role in transformation and tumorigenesis of various cells such as glioblastoma, prostatic carcinoma and hepatocarcinoma cells. Glioblastoma is the most frequent brain tumor in man and is usually fatal. Both human and rat glioma cells express high amounts of IGF-I. It has been previously reported that rat C6 glioblastoma cells express high levels of IGF-I, which is enhanced when the cells are grown in serum-free medium (Kiess *et al.*, 1989). Futhermore, when C6 cells were transfected with a vector containing an antisense IGF-I cDNA transcriptional cassette, they lost their tumorigenicity (Trojan *et al.*, 1993). Moreover, when antisense-transfected cells were injected into syngeneic animals with existing glioblastoma tumors, the established tumors regressed (Trojan *et al.*, 1993). IGF-I seems to act by enabling cancer cells to evade immune recogntion (Baserga, 1995; Trojan *et al.*, 1993). Down-regulation in the expression of IGF-I is coincident with a reapparence of B-7 and MHC class I antigens at the surface of transfected cells (Trojan *et al.*, 1996). When injected subcutaneously, the transfected cancer cells initiate an immune reaction involving CD8+ lymphocytes which results in regression of tumor (Lafarge-Frayssinet *et al.*, 1997; Trojan *et al.*, 1994; Trojan *et al.*, 1993).

The Insulin-like growth factor-I (IGF-I) expression was also inhibited following stable transfection of C6 rat glioblastoma cells with a DNA plasmid from which a triplex-forming RNA is transcribed that is targeted to the IGF-I promoter via triple helix formation. The ability of this expression vector to inhibit IGF-I gene expression *in vitro* as well as tumorigenesis in an animal was tested (Shevelev *et al.*, 1997). A dramatic reduction of IGF-I RNA and protein levels in cultured cells occurred following transfection of rat C6 cells with an eukaryotic expression plasmid encoding the oligopurine sequence of the triple helix-forming RNA but not the oligopyrimidine or a control sequence (Shevelev *et al.*, 1997). The cells transfected with the oligopurine sequence displayed morphological changes, upregulation of MHC class I antigens, and increased expression of protease nexin I. Dramatic inhibition of tumor growth occurred in nude mice following injection of transfected C6 cells (Shevelev *et al.*, 1997).

Hepatocellular carcinoma (HCC) causes more than 1 million deaths per year in the world. Prognosis is dimal and depended on the possibility of resection of tumor. Surgery constitutes the cure for only few HCC patients. Cellular gene therapy offers new possibilities for treatment. Inhibition of cellular IGF-I expression by transfection with the vectors that continuously synthesize either antisense RNA to IGF-I cDNA or RNA strands forming a triple helix with IGF-I cDNA, causes tumor arrest when the transfected cells are injected into mice bearing hepatoma (Upegui-Gonzalez *et al.*, 1998a; Upegui-Gonzalez *et al.*, 1998b). Transgenic mice developping genetically programmed hepatoma (mice die between the ages of 6 and 7 months) and the derived hepatoma cell line (mhAT1F1) were used as models. Transfected cells changed their phenotype, stopped producing IGF-I, and recovered MHC-I expression (Upegui-Gonzalez *et al.*, 1998a; Upegui-Gonzalez *et al.*, 1998b). These cells expressed IL-6 but were down-regulated in production of IL-10 and TNF-alpha. The injection of transfected cells into mice with terminal-phase significantly prolonged their survival. The results suggest that the injection of transfected hepatoma cells using antisense or triple helix approaches could constitute a vaccine against murine hepatoma. Experiments designed to test these strategies for developments of immunogene vaccines towards HCC are in progress.

Another utilization of triplex-forming oligonucleotide delivered inside the cells via transcription of a plasmid vector was recently described with the receptor of the IGF-1 gene in glioblastoma cells (Rininsland *et al.*, 1997). This RNA TFO was targeted to the IGF-1 receptor gene (IGF-I-R). It should be noted that both triplex-forming transcripts targeted to IGF-I and IGF-I receptor contained an homopurine effector sequence that is designed to bind to DNA double helix via triplex formation. The pyrimidine motif for the triplex-forming RNA was used as control and did not give any inhibitory effect on IGF-1 or IGF-1 R expression (Rininsland *et al.*, 1997; Shevelev *et al.*, 1997). Previous *in vitro* studies have shown that under physiological conditions where a purine-rich oligodeoxynucleotide bound to DNA, its RNA counterpart does not bind (Semerad and Maher, 1994). To our knowledge, only one example of RNA third strand in purine motif was recently described (Ririe and Guntaka, 1998). Nevertheless, a pyrimidine-containing RNA TFO formed more stable triplex than the DNA isosequence (Escudé *et al.*, 1993). No evidence has been provided yet demonstrating the involvement of triplex formation in the case of triplex-forming RNA transcripts targeted to IGF-1 and IGF-1 R. However, several control sequences were used where purine-rich RNA effector sequence was replaced and did not give any inhibitory effect on IGF-1 or IGF-1 R expression (Rininsland *et al.*, 1997; Shevelev *et al.*, 1997). In addition, it remains possible that inside mammalian cells a protein recognized and stabilized triple helical structures formed with

RNA purine third strand (for a recent example of triplex-binding proteins, see (Musso *et al.*, 1998)).

7. CONCLUSION

The triplex-based strategy as the well-known antisense strategy seems to be a powerful approach to control gene expression in malignant cells. Few examples of the inhibitory activity of triplex-forming oligonucleotides on target genes involved in tumorigenesis are now available (for reviews see (Chan and Glazer, 1997; Giovannangeli and Hélène, 1997; Maher III, 1996; Vasquez and Wilson, 1998)). Most of the TFOs are targeted to polypurine-polypyrimidine sequences located in control regions of the gene of interest and are cell delivered via transfection with various chemical carriers.

An alternative way to introduce TFOs in cells is to use a plasmid vector that can drive the synthesis of an RNA triplex-forming oligonucleotide inside the cells. This TFO generated *in situ* is therefore protected from degradation by nucleases and could reach its DNA target without being trapped in lysosomal vesicles. Obviously, it could be transfected in cells via either standard cell transfection procedures or via ways similarly used in virus-based gene therapy. An application of this triplex-based approach has been used for the inhibition of the Insulin-like growth factor I protein which plays a major role in tumorigenesis of glioblastoma and hepatocarcinoma (Shevelev *et al.*, 1997; Upegui-Gonzalez *et al.*, 1998b). The inhibition of IGF-I using antisense or triple-helix technologies in these cells induced an immune response of syngenic animals (Lafarge-Frayssinet *et al.*, 1997; Shevelev *et al.*, 1997; Trojan *et al.*, 1993; Upegui-Gonzalez *et al.*, 1998a; Upegui-Gonzalez *et al.*, 1998b). These results obtained with IGF-I familly may ultimately lead to the use of the antisense and triplex-based approaches in clinical trials for anticancer gene therapy.

REFERENCES

Aggarwal, B., Schwarz, L., Hogan, M., and Rando, R., 1996, Triple helix-forming oligodeoxyribonucleotides targeted to the human tumor necrosis factor (TNF) gene inhibit TNF production and block the TNF-dependent growth of human glioblastoma tumor cells, *Cancer Res*, **56**:5156–5164.

Alunni-Fabbroni, M., Pirulli, D., Manzini, G., and Xodo, L., 1996, (A,G)-oligonucleotides form extraordinary stable triple helices with a critical R.Y sequence of the murine c-Ki-ras promoter and inhibit transcription in transfected NIH 3T3 cells, *Biochemistry*, **35**:16361–16369.

Anderson, W.F., 1992, Human gene therapy, *Science*, **256**:808–813.

Baserga, R., 1995, The insulin-like growth factor I receptor: a key to tumor growth?, *Cancer Res*, **55**:249–252.

Chan, P.P., and Glazer, P.M., 1997, Triplex DNA: Fundamentals, advances and potential applications for gene therapy, *J Mol Med*, **75**:267–282.

Couture, L.A., and Stinchcomb, D.T., 1996, Anti-gene therapy: the use of ribozymes to inhibit gene function, *Trends In Genet*, **12**:510–515.

Dervan, P., 1992, Reagents for the site-specific cleavage of megabase DNA, *Nature*, **359**:87–88.

Dodet, B., 1993, Commercial prospects for gene therapy—a company survey, *Trends Biotechnol*, **11**:182–189.

Durland, R., Kessler, D., Gunnell, S., Duvic, M., Pettitt, B., and Hogan, M., 1991, Binding of triple helix forming oligonucleotides to sites in gene promoters, *Biochemistry*, **30**:9246–9255.

Duval-Valentin, G., Thuong, N., and Hélène, C., 1992, Specific inhibition of transcription by triple helix-forming oligonucleotides, *Proc Natl Acad Sci U S A*, **89**:504–508.

Ebbinghaus, S., Gee, J., Rodu, B., Mayfield, C., Sanders, G., and Miller, D., 1993, Triplex formation inhibits HER-2/neu transcription *in vitro*, *J Clin Invest*, **92**:2433–2439.

Escudé, C., François, J.C., Sun, J.S., Ott, G., Sprinzl, M., Garestier, T., and Hélène, C., 1993, Stability of triple helices containing RNA and DNA strands: experimental and molecular modeling studies, *Nucleic Acids Res*, **21**:5547–5553.

Fedoseyeva, E., Li, Y., Huey, B., Tam, S., Hunt, C., Benichou, G., and Garovoy, M., 1994, Inhibition of interferon-gamma-mediated immune functions by oligonucleotides. Suppression of human T cell proliferation by downregulation of IFN-gamma-induced ICAM-1 and Fc-receptor on accessory cells, *Transplantation*, **57**:606–612.

François, J.C., Saison-Behmoaras, T., Barbier, C., Chassignol, M., Thuong, N., and Hélène, C., 1989, Sequence-specific recognition and cleavage of duplex DNA via triple-helix formation by oligonucleotides covalently linked to a phenanthroline-copper chelate, *Proc Natl Acad Sci U S A*, **86**: 9702–9706.

Frank-Kamenetskii, M., and Mirkin, S., 1995, Triplex DNA structures, *Annu Rev Biochem*, **64**:65–95.

Giovannangeli, C., Diviacco, S., Labrousse, V., Gryaznov, S., Charneau, P., and Hélène, C., 1997, Accessibility of nuclear DNA to triplex-forming oligonucleotides: the integrated HIV-1 provirus as a target, *Proc Natl Acad Sci U S A*, **94**:79–84.

Giovannangeli, C., and Hélène, C., 1997, Progress in developments of triplex-based strategies, *Antisense Nucleic Acid Drug Dev*, **7**:413–421.

Giovannangeli, C., Perrouault, L., Escude, C., Gryaznov, S., and Hélène, C., 1996, Efficient inhibition of transcription elongation *in vitro* by oligonucleotide phosphoramidates targeted to proviral HIV DNA, *J Mol Biol*, **261**:386–398.

Good, L., and Nielsen, P.E., 1997, Progress in developping PNA as a gene-targeted drug, *Antisense Nucleic Acid Res*, **7**:431–437.

Grigoriev, M., Praseuth, D., Guieysse, A., Robin, P., Thuong, N.T., Hélène, C., and Harel-Bellan, A., 1993, Inhibition of gene expression by triple helix directed DNA cross-linking at specific sites, *Proc Natl Acad Sci U S A*, **90**:3501–3505.

Grigoriev, M., Praseuth, D., Robin, P., Hemar, A., Saison-Behmoaras, T., Dautry-Varsat, A., Thuong, N.T., Hélène, C., and Harel-Bellan, A., 1992, A triple helix-forming oligonucleotide-intercalator conjugate acts as a transcriptional repressor via inhibition of NF kappa B binding to interleukin-2 receptor alpha-regulatory sequence, *J Biol Chem*, **267**:3389–3395.

Guieysse, A., Praseuth, D., Francois, J., and Hélène, C., 1995, Inhibition of replication initiation by triple helix-forming oligonucleotides, *Biochem Biophys Res Commun*, **217**:186–194.

Guo, Y., Wu, M., Chen, H., Wang, X., Liu, G., Li, G., Ma, J., and Sy, M.S., 1994, Effective tumor vaccine generated by fusion of hepatoma cells with activated B cells, *Science*, **263**:518–520.

Hacia, J., Dervan, P., and Wold, B., 1994, Inhibition of Klenow fragment DNA polymerase on double-helical templates by oligonucleotide-directed triple-helix formation, *Biochemistry*, **33**:6192–6200.

Han, V., and Hill, D., 1992, The insulin-like growth factors: structure and biological functions., In: (P.N. Shofield, Eds.), Oxford University Press, pp. 11–30.

Hausheer, F.H., Singh, U., Saxe, J., Flory, J., and Tufto, K., 1992, Thermodynamic and conformational characterization of 5-methylcytosine-versus cytosine-substituted oligomers in DNA triple helices: Ab initio quantum mechanical and free energy perturbation studies, *J Am Chem Soc*, **114**: 5356–5362.

Havre, P., and Glazer, P., 1993, Targeted mutagenesis of simian virus 40 DNA mediated by a triple helix-forming oligonucleotide, *J. Virol*, **67**:7324–7331.

Hélène, C., 1993, Sequence-specific recognition and cleavage of double helical DNA, *Current opinion in Biotechnology*, **4**:29–36.

Hélène, C., 1994, Control of oncogene expression by antisense nucleic acids, *Eur J Cancer*, **30A**:1721–1726.

Hélène, C., Garestier, T., Giovannangeli, C., and Sun, J.S., 1998, Sequence-specific control of gene expression by antigene and clamp oligonucleotides, *Gene Therapy and Molecular Biology*, **1**:467–474.

Hobbs, C.A., and Yoon, K., 1994, Differential regulation of gene expression *in vivo* by triple helix-forming oligonucleotides as detected by a reporter enzyme, *Antisense Res Dev*, **4**:1–8.

Huang, Y., Snyder, R., Kligshteyn, M., and Wickstrom, E., 1995, Prevention of tumor formation in a mouse model of Burkitt's lymphoma by 6 weeks of treatment with anti-c-myc DNA phosphorothioate, *Mol Med*, **1**:647–658.

Ido, A., Nakata, K., Kato, Y., Nakao, K., Murata, K., Fujita, M., Ishii, N., Tamaoki, T., Shiku, H., and Nagataki, S., 1995, Gene therapy for hepatoma cells using a retrovirus vector carrying herpes simplex virus thymidine kinase gene under the control of human alpha-fetoprotein gene promoter, *Cancer Res*, **55**:3105–3109.

Imagawa, S., Izumi, T., and Miura, Y., 1994, Positive and negative regulation of the erythropoietin gene, *J Biol Chem*, **269**:9038–9044.

Ing, N.H., Beekman, J.M., Kessler, D.J., Murphy, M., Jayaraman, K., Zendegui, J.G., Hogan, M.E., O'Malley, B.W., and Tsai, M.J., 1993, *In vivo* transcription of a progesterone-responsive gene is specifically inhibited by a triplex-forming oligonucleotide, *Nucleic Acids Res*, **21**:2789–2796.

Joseph, J., Kandala, J.C., Veerapanane, D., Weber, K.T., and Guntaka, R.V., 1997, Antiparallel polypurine phosphorothioate oligonucleotides form stable triplexes with the rat alpha1(I) collagen gene promoter and inhibit transcription in cultured rat fibroblasts, *Nucleic Acids Res*, **25**:182–188.

Kiess, W., Lee, L., Graham, D.E., Greenstein, L., Tseng, L.Y., Rechler, M.M., and Nissley, S.P., 1989, Rat C6 glial cells synthesize insulin-like growth factor I (IGF-I) and express IGF-I receptors and IGF-II/mannose 6-phosphate receptors, *Endocrinology*, **124**:1727–1736.

Kim, H.G., and Miller, D.M., 1998, A novel triplex-forming oligonucleotide targeted to human cyclin D1 (bcl-1, proto-oncogene) promoter inhibits transcription in Hela cells, *Biochemistry*, **37**:2666–2672.

Kim, H.G., Reddoch, J.F., Mayfield, C., Ebbinghaus, S., Vigneswaran, N., Thomas, S., Jones, D.E., and Miller, D.M., 1998, Inhibition of transcription of the human c-myc protooncogene by intermolecular triplex, *Biochemistry*, **37**:2299–2304.

Kool, E.T., 1996, Topological modification of oligonucleotides for potential inhibition of gene expression, In: *Antisense therapeutics: Progress and prospects*, 4, (G. Trainor, Eds.), pp. 61–75.

Kovacs, A., Kandala, J., Weber, K., and Guntaka, R., 1996, Triple helix-forming oligonucleotide corresponding to the polypyrimidine sequence in the rat alpha1(I) collagen promoter specifically inhibits factor binding and transcription, *J Biol Chem*, **271**:1805–1812.

Krasilnikov, A., Panyutin, I., Samadashwily, G., Cox, R., Lazurkin, Y., and Mirkin, S., 1997, Mechanisms of triplex-caused polymerization arrest, *Nucleic Acids Res.*, **25**:1339–1346.

Lafarge-Frayssinet, C., Duc, H.T., Frayssinet, C., Sarasin, A., Anthony, D., Guo, Y., and Trojan, J., 1997, Antisense IGF-1 transfer into a rat hepatoma cell line inhibits tumorigenesis by modulating MHC-I cell surface expression, *Cancer Gene Therapy*, **5**:276–285.

Leonetti, C., D'Agnano, I., Lozupone, F., Valentini, A., Geiser, T., Zon, G., Calabretta, B., Citro, G.C., and Zupi, G., 1996, Antitumor effect of c-myc antisense phosphorothioate oligodeoxynucleotides on human melanoma cells *in vitro* and and in mice, *J Natl Cancer Inst*, **88**:419–429.

Maher III, L.J., 1992, Inhibition of T7 RNA polymerase initiation by triple-helical DNA complexes: a model for artificial gene repression, *Biochemistry*, **31**:7587–7594.

Maher III, L.J., 1996, Prospects for the therapeutic use of antigene oligonucleotides, *Cancer Investigation*, **14**:66–82.

Maher III, L.J., Dervan, P.B., and Wold, B., 1992, Analysis of promoter-specific repression by triple-helical DNA complexes in a eukaryotic cell-free transcription system, *Biochemistry*, **31**:70–81.

Mayfield, C., Ebbinghaus, S., Gee, J., Jones, D., Rodu, B., Squibb, M., and Miller, D., 1994, Triplex formation by the human Ha-ras promoter inhibits Sp1 binding and *in vitro* transcription, *J Biol Chem*, **269**:18232–18238.

Mergny, J., Sun, J., Rougee, M., Montenay-Garestier, T., Barcelo, F., Chomilier, J., and Hélène, C., 1991, Sequence specificity in triple-helix formation: experimental and theoretical studies of the effect of mismatches on triplex stability, *Biochemistry*, **30**:9791–9798.

Miller, A.D., 1992, Human gene therapy comes of age, *Nature*, **357**:455–460.

Morishita, R., Higaki, J., Tomita, N., and Ogihara, T., 1998, Application of transcription factor "decoy" strategy as means of gene therapy and study of gene expression in cardiovascular disease, *Circ Res*, **82**:1023–1028.

Moser, H., and Dervan, P., 1987, Sequence-specific cleavage of double helical DNA by triple helix formation, *Science*, **238**:645–650.

Musso, M., Nelson, L., and Van Dyke, M., 1998, Characterization of purine-motif triplex DNA-binding proteins in HeLa extracts, *Biochemistry*, **37**:3086–3095.

Okabe, M., Kunieda, Y., Miyagishima, T., Kobayashi, M., Kurosawa, M., Itaya, T., and Miyazaki, T., 1993, BCR/ABL oncoprotein-targeted antitumor activity of antisense oligodeoxynucleotides complementary to bcr/abl mRNA and herbimycin A, an antagonist of protein tyrosine kinase: inhibitory effects on *in vitro* growth of Ph1-positive leukemia cells and BCR/ABL oncoprotein-associated transformed cells, *Leuk Lymphoma*, **10**:307–316.

Porumb, H., Gousset, H., Letellier, R., Salle, V., Briane, D., Vassy, J., Amor-Gueret, M., Israel, L., and Taillandier, E., 1996, Temporary ex vivo inhibition of the expression of the human oncogene HER2 (NEU) by a triple helix-forming oligonucleotide, *Cancer Res*, **56**:515–522.

Postel, E.H., Flint, S.J., Kessler, D.J., and Hogan, M.E., 1991, Evidence that a triplex-forming oligodeoxyribonucleotide binds to the c-myc promoter in HeLa cells, thereby reducing c-myc mRNA levels, *Proc Natl Acad Sci U S A*, **88**:8227–8231.

Rininsland, F., Johnson, T., Chernicky, C., Schulze, E., Burfeind, P., and Ilan, J., 1997, Suppression of insulin-like growth factor type I receptor by a triple-helix strategy inhibits IGF-I transcription and tumorigenic potential of rat C6 glioblastoma cells, *Proc Natl Acad Sci U S A*, **94**:5854–5859.

Ririe, S., and Guntaka, R., 1998, An RNA oligonucleotide corresponding to the polypyrimidine region of the rat alpha1(I) procollagen promoter forms a stable triplex and inhibits transcription, *Biochem Biophys Res Commun*, **249**:218–221.

Roy, C., 1994, Triple-helix formation interferes with the transcription and hinged DNA structure of the interferon-inducible 6–16 gene promoter, *Eur J Biochem*, **220**:493–503.

Samadashwily, G., Dayn, A., and Mirkin, S., 1993, Suicidal nucleotide sequences for DNA polymerization, *EMBO J*, **12**:4975–4983.

Samadashwily, G., and Mirkin, S., 1994, Trapping DNA polymerases using triplex-forming oligodeoxyribonucleotides, *Gene*, **149**:127–136.

Sandor, Z., and Bredberg, A., 1994, Repair of triple helix directed psoralen adducts in human cells, *Nucleic Acids Res*, **22**:2051–2056.

Scaggiante, B., Morassutti, C., Tolazzi, G., Michelutti, A., Baccarani, M., and Quadrifoglio, F., 1994, Effect of unmodified triple helix-forming oligodeoxyribonucleotide targeted to human multidrug-resistance gene mdr1 in MDR cancer cells, *FEBS Lett*, **352**:380–384.

Semerad, C., and Maher, L.E., 1994, Exclusion of RNA strands from a purine motif triple helix, *Nucleic Acids Res*, **22**:5321–5325.

Shevelev, A., Burfeind, P., Schulze, E., Rininsland, F., Johnson, T., Trojan, J., Chernicky, C., Hélène, C., Ilan, J., and Ilan, J., 1997, Potential triple helix-mediated inhibition of IGF-I gene expression significantly reduces tumorigenicity of glioblastoma in an animal model, *Cancer Gene Ther*, **4**: 105–112.

Skoog, J., and Maher III, L.J., 1993a, Relief of triple-helix-mediated promoter inhibition by elongating RNA polymerases, *Nucleic Acids Res*, **21**:4055–4058.

Skoog, J., and Maher III, L.J., 1993b, Repression of bacteriophage promoters by DNA and RNA oligonucleotides, *Nucleic Acids Res*, **21**:2131–2138.

Skorski, T., Nieborowska-Skorska, M., Campbell, K., Iozzo, R., Zon, G., Darzynkiewicz, Z., and Calabretta, B., 1995, Leukemia treatment in severe combined immunodeficiency mice by antisense oligodeoxynucleotides targeting cooperating oncogenes, *J Exp Med*, **182**:1645–1653.

Sokol, D.L., and Gewirtz, A.M., 1996, Gene therapy: basic concepts and recent advances, *Crit Rev in Eukaryotic gene expression*, **6**:29–57.

Song, C., Jung, M., Supakar, P., Chen, S., Vellanoweth, R., Chatterjee, B., and Roy, A., 1995, Regulation of androgen action by receptor gene inhibition, *Ann N Y Acad Sci*, **761**:97–108.

Stein, C.A., 1996, Exploiting the potential of antisense:beyond phosphorothioate oligodeoxynucleotides, *Chemistry & Biology*, **3**:319–323.

Strobel, S.A., and Dervan, P.B., 1990, Site-specific cleavage of a yeast chromosome by oligonucleotide-directed triple-helix formation, *Science*, **249**:73–75.

Strobel, S.A., Doucette-Stamm, L., Riba, L., Housman, D.E., and Dervan, P.B., 1991, Site-specific cleavage of human chromosome 4 mediated by triple-helix formation, *Science*, **254**:1639–1642.

Svinarchuk, F., Debin, A., Bertrand, J., and Malvy, C., 1996, Investigation of the intracellular stability and formation of a triple helix formed with a short purine oligonucleotide targeted to the murine c-pim-1 proto-oncogene promotor, *Nucleic Acids Res*, **24**:295–302.

Thomas, T., Faaland, C., Gallo, M., and Thomas, T., 1995, Suppression of c-myc oncogene expression by a polyamine-complexed triplex forming oligonucleotide in MCF-7 breast cancer cells, *Nucleic Acids Res*, **23**:3594–3599.

Trojan, J., Blossey, B., Johnson, T., Rudin, S., Tykocinski, M., Ilan, J., and Ilan, J., 1992, Loss of tumorigenicity of rat glioblastoma directed by episome-based antisense cDNA transcription of insulin-like growth factor I, *Proc Natl Acad Sci U S A*, **89**:4874–4878.

Trojan, J., Duc, H., Upegui-Gonzalez, L., Hor, F., Guo, Y., Anthony, D., and Ilan, J., 1996, Presence of MHC-I and B-7 molecules in rat and human glioma cells expressing antisense IGF-I mRNA, *Neurosci Lett*, **212**:9–12.

Trojan, J., Johnson, T.R., Rudin, S.D., Blossey, B.K., Kelley, K.M., Shevelev, A., Abdul-Karim, F.W., Anthony, D.D., Tykocinski, M.L., Ilan, J., and Ilan, J., 1994, Gene therapy of murine teratocarcinoma: separate functions for insulin-like growth factors I and II in immunogenicity and differentiation, *Proc Natl Acad Sci U S A*, **91**:6088–6092.

Trojan, J., Johnson, T.R., Rudin, S.D., Ilan, J., Tykocinski, M.L., and Ilan, J., 1993, Treatment and prevention of rat glioblastoma by immunogenic C6 cells expressing antisense insulin-like growth factor I RNA, *Science*, **259**:94–97.

Tu, G., Cao, Q., and Israel, Y., 1995, Inhibition of gene expression by triple helix formation in hepatoma cells, *J Biol Chem*, **270**:28402–28407.

Upegui-Gonzalez, L.C., Duc, H.T., Buisson, Y., Arborio, M., Lafarge-Frayssinet, C., Jasmin, C., Guoa, Y., and Trojan, J., 1998a, Use of the IGF-I antisense strategy in the treatment of the hepatocarcinoma, In: *Proceeding of Gene therapy of cancer 1997*, 6, (P. Walden, Eds.), pp. 35–44, Plenum PCL, London.

Upegui-Gonzalez, L.C., Duc, H.T., Swiercz, B., Ly, A., Hor, F., Arborio, M., Buisson, Y., Jasmin, C., Pan, Y., Anthony, D., Hénin, D., and Trojan, J., 1998b, IGF-I antisense and triple helix treatment in hepatoma treatment, *Front Biosci.*, in press.

Vasquez, K.M., and Wilson, J.H., 1998, Triplex-directed modification of genes and gene activity, *Trends In Biochem. Sci.*, **23**:4–9.

Wang, G., Levy, D., Seidman, M., and Glazer, P., 1995, Targeted mutagenesis in mammalian cells mediated by intracellular triple helix formation, *Mol Cell Biol*, **15**:1759–1768.

Young, S., Krawczyk, S., Matteucci, M., and Toole, J., 1991, Triple helix formation inhibits transcription elongation *in vitro*, *Proc Natl Acad Sci U S A*, **88**:10023–10026.

VIII

IMMUNO-MODULATION

28

DENDRITIC CELL-BASED IMMUNIZATION FOR CANCER THERAPY

Michael A. Morse[1] and H. Kim Lyerly[2]

[1]Department of Medicine and
[2]Department of Surgery
Duke University Medical Center
Box 2606, Durham, North Carolina, 27710

1. INTRODUCTION

Immunotherapy is showing renewed promise as a modality for treating malignancies. Although the past clinical and experimental experiences of immunotherapy with non-specific immunogens (such as BCG or KLH) and systemic or local administration of various cytokines (such as IL-2 and interferon) have had mixed results (Rosenberg, 1987; Rosenberg, 1989; Robinson, 1977; Bulbul, 1989), an emerging understanding of the biologic basis of antigen-specific cellular recognition and the demonstration in experimental animal studies of an anti-tumor effect mediated through the cellular immune system provide compelling evidence to explore the role of specific cellular immunotherapy in human malignancies. Among the most potent agents for actively stimulating specific cellular immune responses are dendritic cells (DC) loaded with tumor antigen. This chapter will discuss the preclinical support for DC-based immunotherapy, methods for obtaining DC for clinical studies, strategies for delivering antigens to DC, and results of early human studies of DC vaccines.

2. BASIS FOR DENDRITIC CELL-BASED IMMUNOTHERAPY

Most tumors are thought to express rejection antigens (van Pel, 1982) that are recognizable by CD8+ cytotoxic T cells (CTL) as peptide fragments bound in the groove of the tumor cell MHC class I molecule (Darrow, 1989). Such antigens found on human tumors include carcinoembryonic antigen (CEA) (gastrointestinal, breast, lung) (Thompson, 1991), MAGE-1 and -3 (melanoma, breast, ovarian) (Van der Bruggen, 1991), HER2/neu (breast, ovarian) (Disis, 1997), MART-1 (melanoma) (Kawakami, 1994), MUC1 (breast, gastrointestinal) (Barnd, 1989), mutated Ki-ras (pancreatic) (Bos, 1989), and E6 and E7 (cervical) (Halbert, 1991). Unfortunately, tumors may fail to stimulate primary immune responses because they lack the required costimulatory molecules

Cancer Gene Therapy: Past Achievements and Future Challenges, edited by Habib
Kluwer Academic/Plenum Publishers, New York, 2000.

(such as CD80 and CD86) and thus, tumor antigen-specific T cells are usually absent, or present at low frequencies, in patients with malignancies (Van den Eynde, 1989; Fisk, 1995; Disis, 1994; Tsang, 1995, Coulie, 1992; Mazzocchi, 1994).

Based on the hypothesis that once stimulated, naive T cells gain the capacity to recognize and destroy tumors for which they are specific, numerous approaches to inducing primary immune responses have been developed including gene-modified tumor vaccines, tumor lysates, naked DNA, peptide fragments of tumor proteins, and viral vectors carrying the genes for antigens of interest. Recent observations suggest that these modalities may indirectly stimulate immune responses by a process called cross-priming (Matzinger, 1994) in which the delivered antigen is released by inflammatory reactions at the injection site and is taken up, processed, and presented to naive T cells by professional antigen-presenting cells (APC) (Bannerji, 1994; Huang, 1994) such as dendritic cells (DC). This is due to the central role DC play in initiating all T cell responses *in vivo*, including $CD4^+$ T-helper and $CD8^+$ CTL responses (Inaba, 1987; Steinman, 1991; Young, 1990). Furthermore, DC have been thought to participate in *in vivo* anti-tumor immunity based on reports of increased numbers near tumors (Tazi, 1993) and the correlation between the number of tumor-infiltrating antigen-presenting cells and clinical outcome (Azizi, 1987; Becker, 1992; Becker, 1993). Thus, there has been increasing interest in harnessing DC for vaccine strategies.

3. DEVELOPING DC-BASED IMMUNOTHERAPIES

As is typical for a burgeoning field, the increasing understanding of DC development and function requires numerous decisions regarding the type of DC to be used, the method for obtaining the DC of interest, the form of the antigen and method of DC loading, and the protocol for delivery of the vaccine.

3.1. Selecting the Optimal DC for Anti-Tumor Vaccine Strategies

A variety of cells that possess similar morphologies and ability to present antigens have been called "dendritic cells;" but, it is increasingly apparent that there are at least 2 different subgroups with different lineages (myeloid versus non-myeloid (lymphoid)-derived) and functions, and that within the subgroups there are cells that can be further separated by phenotype and functional activity.

Myeloid-lineage DC include blood DC, Langerhans cells, veiled DC in the afferent lymphatics, and germinal center DC (Steinman, 1991; Grouard, 1996). The bone marrow-derived precursors of these DC circulate in the peripheral blood to most tissues where they acquire and process antigen. Subsequently, they migrate to the regional lymphatics where they present the antigen to T cells (Steinman, 1991). When isolated from these locations, the DC are large cells with elongated, stellate processes, that lack lineage-specific markers such as CD3 (T cell), CD19 (B-cell), CD16 and CD56 (NK cell), and CD14 (monocytes) and express high levels of HLA DR, HLA class I, CD11c, CD18, CD40, CD54 (ICAM-1), and CD58 (Steinman, 1991). Coincident with maturation, other markers such as CD83 (Zhou, 1995), CD80 (B7.1), CD86 (B7.2), and CMRF-44 (Hock, 1994) are upregulated. Some markers are expressed by only certain sub-populations such as CD1a, expressed by Langerhans cells. In contrast, the lymphoid-lineage thymic DC (Galy, 1995; Saunders, 1996) express $CD8\alpha$ (Vremee, 1992) and high levels of Fas ligand, and are capable of destroying Fas+ CD4+ T cells, suggesting an immunomodulatory role

(Suss, 1996). Other non-myeloid (lymphoid)-derived DC include the T-zone associated plasmacytoid cells of lymph nodes (Grouard, 1997), for which a peripheral blood precursor has been identified (Strobl, 1998). Follicular dendritic cells, residing in primary and secondary follicles and expressing a unique isoform of CD21, have a different function, being responsible for presenting antigen in the form of immune complexes to B cells (Liu, 1996).

A rapidly expanding body of work has demonstrated the *in vitro* (McKinney, 1989; Macatonia, 1988; Nair, 1992; Mehta-Damani, 1994) and *in vivo* (Takahashi, 1993; Nair, 1993; Rouse, 1994) potency of antigen-loaded, myeloid-derived DC at inducing antigen-specific cytotoxic T cells (CTL) from naive precursors and generating effective tumor-specific immunity (Flamand, 1994; Paglia, 1996; Porgador, 1996; Porgador, 1995; Ossevoort, 1995; Mayordomo, 1995; Celluzzi, 1996). This data has served as the background for testing myeloid-derived DC as a treatment for human malignancies. The possible utilization of lymphoid-derived DC awaits a more comprehensive understanding of these cells.

3.2. Methods for Obtaining DC for Human Studies

Although at least two reported studies of DC-based vaccinations of humans have employed DC isolated from peripheral blood (Hsu, 1996; Valone, 1998), the low frequency of circulating or bone marrow resident DC (0.15% of circulating mononuclear cells (Zhou, 1995)) and the complex procedures required for their isolation (Young, 1988; Knight, 1986; Williams, 1994; Thomas, 1993; Van Voorhis, 1982) have resulted in the development of approaches for generating large numbers of DC from their myeloid precursors *in vitro*. There are also indications from animal studies that DC isolated from tumor-bearing hosts have a diminished capacity to stimulate anti-tumor immunity compared with DC generated *in vitro* (Gabrilovich, 1996; Gabrilovich, 1996).

Intensive experimentation is now ongoing to determine the ideal conditions for generating DC *in vitro*. Although four precursor populations have been identified (CD34+ cells, CD34- depleted peripheral blood mononuclear cells, CD14+ cells, and bulk peripheral blood mononuclear cells [PBMC]), it is likely that the latter three contain the same cells that develop into DC *in vitro*. Culturing CD34+ cells over a 14 day period in the presence of GM-CSF and TNF-α may give rise directly to typical Langerhans DC or may proceed through a CD14+ myelomonocytic stage with the potential to differentiate into either macrophages or DC depending on the cytokine milieu (Caux, 1996; Reid, 1992; Santiago-Schwartz, 1995; Szabolcs, 1994). Modifications to these basic conditions including the addition of c-kit ligand and Flt3 ligand (Szabolcs 1995, Sienna, 1995) at the initiation of the culture, or CD40-ligand and IL-4 near the end of culture, have increased the yield and modulated the phenotype of the resulting DC. Because of the proliferation possible in these cultures, it has been suggested (Sienna, 1995) that massive numbers of DC may be generated by this approach from a single cytokine-mobilized leukapheresis, although this has not actually been demonstrated. As an alternative to CD34+ cells, the cells remaining after selection for CD34+ from a mobilized leukapheresis or bone marrow product, PBMC, or CD14+ monocytes can be driven to DC *in vitro* by the addition of exogenous GM-CSF and IL-4 with or without TNF-α (Tarte, 1997; Zhou, 1996; Romani, 1994; Bender, 1996; Romani, 1996; Morse, 1997). Neither the DC nor their progenitors proliferate in these culture conditions and the yield of DC derived in this manner cannot be increased beyond the number of starting monocytes (Cavanagh,

1998). Nonetheless, because of the large number of monocytoid cells in the peripheral blood, large yields are possible.

In order to be applicable to clinical studies, further modifications to the protocol for DC generation have been made, particularly the use of autologous plasma-containing or serum-free media (Bender, 1996; Romani, 1996; Romani, 1996; Morse, 1997) in order to avoid the undesirable possibility of sensitization to heterologous antigens. Starting with a leukapheresis collection of 5–15 × 10^9 cells, we have been able to generate a heterogeneous population of cells containing 0.25–1.0 × 10^9 DC which represents 5–7% of the original number of PBMC (Morse, unpublished results) and the DC generated under these conditions have the capacity to stimulate naive T cells *in vitro* (Wong, 1998). It is not known whether DC cultured from CD34+ cells or CD14+/PBMC are the optimal delivery system for tumor antigens. Recent data suggests different capacities to stimulate low avidity antigen-specific T cells (Mortarini, 1997). In addition, concern has been raised as to whether DC generated in serum-free medium in the presence of GM-CSF and IL-4 are "unstable" (Romani, 1996; Bender; 1996) and will revert to the phenotype and functional activity of monocytes when removed from the high levels of *in vitro* cytokines and infused into the patient. Possible solutions to the maturation of DC and maintenance of the DC phenotype and function include the addition of monocyte-conditioned medium (Romani, 1996; Bender, 1996), TNF-α (Nair, 1998), or CD40 ligand (Cella, 1996; Morse, 1998).

In order to avoid the manipulations required for *in vitro* DC generation, there is increasing interest in mobilizing DC. Mice that received Flt3 ligand experienced a significant increase of DC in lymphoid tissue and peripheral blood (Maraskovsky, 1996) with at least 5 different populations of DC observed. If this is duplicated in human studies, a large number of DC may be obtained from a single leukapheresis or it may be possible to administer tumor antigens during DC mobilization to avoid the need for *ex vivo* manipulation entirely.

3.3. Delivery of Antigens to DC

Because DC have the remarkable ability to take up various types of molecules from their environment (Steinman, 1995; Austyn, 1996) and because their cellular machinery is adapted towards processing proteins and presenting constituent peptides on the cell surface, numerous approaches for loading DC with antigens are possible (Fig. 1). This has resulted in a debate regarding whether defined or undefined antigens should be used, and whether antigen should be delivered as purified peptides or proteins, genetic material, or tumor extracts. The use of defined antigens (e.g. CEA, HER-2/neu, and MART) reduces the possibility of inducing undesirable immune responses, allows for more accurate measurement of the T cell specificity, and permits manufacturing of large quantities of the antigen so that it is readily available for clinical studies. One disadvantage to this approach is the need to determine whether the antigen is presented in a recognizable form by the targeted tumor. For example, studies of tumor-specific T cells isolated from melanoma patients have indicated that a majority did not recognize well-known melanoma antigens, suggesting that the true tumor rejection antigens may vary from patient to patient or are currently undefined (Anichini, 1996).

Undefined antigens are advantageous because they might supply the most immunogenic epitopes especially if obtained form the patient's own tumor tissue. A drawback to this approach is the need for an adequate amount of tumor tissue for the extraction of the antigenic material. Also, the concentration of any one antigen is likely to be low in a pool of multiple undefined antigens. Further, some undefined antigens may be found

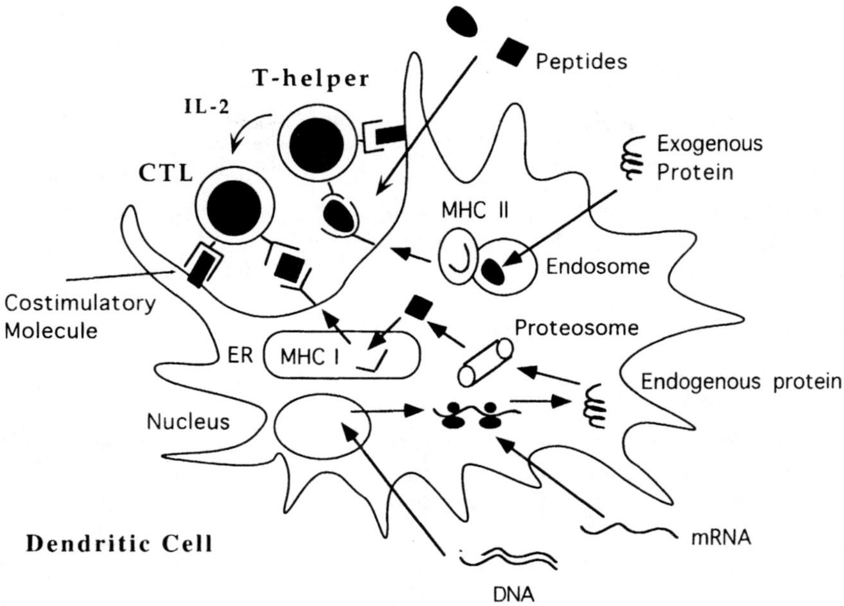

Figure 1. Methods for delivering antigens to DC. Exogenously applied peptides likely bind directly to the MHC molecules during coincubation. Transfected or transduced tumor antigen-encoding DNA is transcribed to mRNA which, in common with exogenously applied mRNA, is translated into proteins that are then degraded by proteosomes. The resulting peptides are transported to the endoplsmic reticulum (ER) and complexed with MHC Class I molecules and β2-microglobulin for transportation to the cell surface. Exogenously applied proteins are degraded in endosomes and complexed with MHC Class II molecules for transport to the cell surface. Some exogenous perotein may instead be degraded by proteosomes and their peptides complexed with MHC Class I molecules. Tumor extracts and fusions likely contain a combination of proteins and genetic material that are channeled into both pathways.

on normal tissues raising the possibility of induction of autoimmunity. Immunization of tumor-bearing mice with tumor extract loaded DC, though, has not been shown to result in major organ autoimmunity (Nair, 1997).

The optimal form of antigen is also under intensive study. Loading of DC by coincubation with peptides, whether defined or undefined, is technically easy to use and has been well-validated *in vitro*. Both defined, manufactured peptide fragments of known tumor antigens (Mayordomo, 1995; Ossevoort, 1995; Porgador, 1995; Celluzzi, 1996; Wong, 1998; Alters, 1998) and undefined tumor eluates (Zitvogel, 1996) have been successfully used to induce T cell responses *in vitro* and *in vivo*. The major drawback to this approach is MHC restriction, which limits the number of patients who could be treated with a particular peptide. Also, if a single, defined class I-restricted peptide is chosen, as is the case in many current studies, tone would fail to take advantage of subdominant and cryptic epitopes or class II epitopes within the native protein that could be recognized by T cells. Antigen in the form of protein is inefficient as it requires large scale purification from significant amounts of tumor tissue for pulsing DC (Nair, 1992). Undefined antigens may also be delivered in the form of tumor lysates (Nestle, 1998) or by fusing DC with tumors (Gong, 1997), although both approaches require a significant amount of tumor tissue. It is also possible to transfect or transduce DC with tumor-antigen encoding DNA (Alijagic, 1995; Reeves, 1996; Specht, 1997; Song, 1997) or RNA (Boczkowski, 1996; Ashley, 1997) by either coincubation or use of viral vectors. This permits the delivery of multiple epitopes, both MHC class I and class II, and requires minimal amounts of material since it is easier to amplify genetic molecules than proteins.

Our preference has been for total tumor mRNA as this fraction is likely to contain the largest number of authentic antigens being presented by the tumor. In addition, there is no risk of integration of genetic material into the genome of the dendritic cell. One theoretical drawback is the possibility that autoimmunity may result from the presentation of epitopes homologous with normal tissue protein. It may be possible to reduce this risk by subtraction hybridization using normal tissue RNA to eliminate common sequences. Whether this method for delivering tumor antigen is superior to the array of existing techniques will ultimately require a direct *in vivo* comparison.

4. THE CURRENT STATUS OF CLINICAL TRIALS OF DENDRITIC CELLS IN HUMANS

The number of clinical trials of DC-based vaccines is increasing rapidly. The earliest study tested antigen-presenting cells generated with GM-CSF *in vitro* and pulsed with a peptide fragment of MAGE-1 for patients with metastatic melanoma. Following vaccination, antigen-specific T cells were demonstrated at the vaccination site and distant metastases, although there was no clinical efficacy (Mukherji, 1995). In a pilot study of 4 patients with B-cell lymphoma (Hsu, 1996), DC were partially isolated from peripheral blood, allowed to mature *in vitro* without added cytokines, and were incubated with the patients' purified idiotype protein. All patients developed cellular proliferative responses against the tumor idiotype and two had complete responses. A similar idiotype-specific CTL response was demonstrated following immunization of a myeloma patient with idiotype protein-loaded, autologous DC (Reichardt, 1996). DC isolated from peripheral blood and loaded with the tumor antigen fusion protein PA2024 (Dendreon) were administered to prostate cancer patients, all of whom developed TH1 responses to the PA2024 (Valone, 1998). In one of the largest phase I clinical trials, DC generated in serum free media supplemented with GM-CSF and IL-4 and pulsed with HLA A2-restricted, prostate tumor antigen peptides were administered to patients with prostate cancer (Murphy, 1996). Cellular immune responses and an average PSA decrement were detected in the group of HLA A2+ patients. A recent phase I series describes the induction of tumor-specific DTH responses and regression of disease in some patients treated with melanoma lysate or peptide loaded dendritic cells (Nestle, 1998). Finally, immunologic responses were described following administration of tumor lysate loaded DC to patients with renal cell carcinoma (Hinkle, 1998). We have been conducting phase I studies of DC generated *in vitro* with GM-CSF and IL-4 and loaded with CEA peptide or mRNA. Preliminary results suggest that the immunizations were well tolerated. Further clinical and immunologic data are pending.

5. CONCLUSIONS

Dendritic cell-based vaccines are currently the most potent method for stimulating antigen-specific cellular immune responses and preliminary results in phase I studies suggest that they are well tolerated and are capable of inducing modest immune and clinical responses *in vivo*. Phase II studies have also been initiated to determine the efficacy of DC vaccines, particularly in the setting of minimal residual disease after surgery or chemotherapy-induced remissions. Many questions regarding the most effective utilization of DC-based vaccines remain to be studied (Fig. 2). For example, strategies for

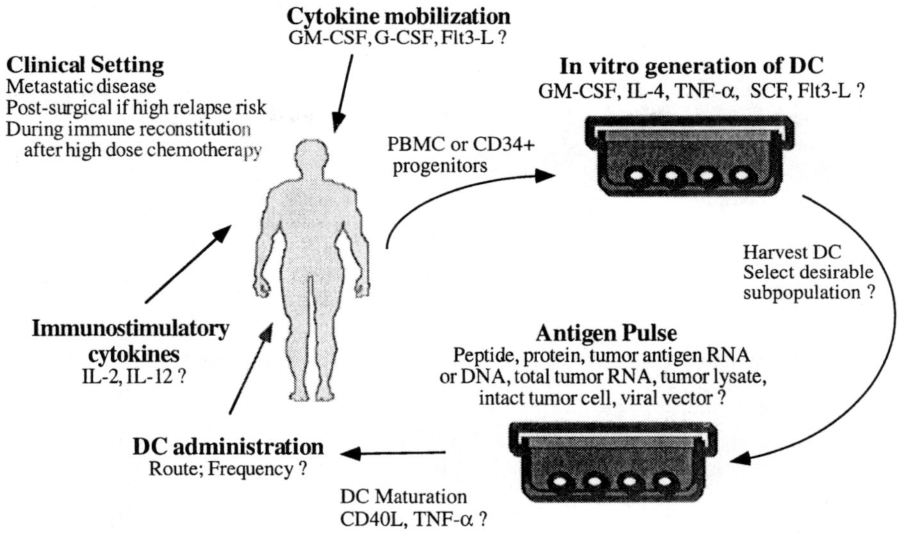

Figure 2. The schema for generation and administration of tumor antigen-pulsed dendritic cells is depicted along with suggested modifications for increasing efficacy. Those requiring further study are followed by question marks. Reprinted from Morse, M.A., and Lyerly, H.K., 1998, Immunotherapy of cancer using dendritic cells. *Cytokines Cell. Mole. Ther.* **4**:35–44.

mobilizing larger numbers of DC or their precursors and automation of the culture process may increase the efficiency of generating DC vaccines. Alternative uses for DC, other than as vaccines, are being developed. Zitvogel (1998) has described the ability of exosomes, antigen-presenting vesicles secreted by DC, to induce immunologic responses *in vivo*. DC may also be used to stimulate antigen-specific T cells *in vitro* for infusion as a form of passive immunotherapy. The number of defined tumor-associated antigens and methods for obtaining undefined antigens from minimal amounts of tumor material continue to expand permitting the application of this technology to more patients. Details of the timing and site of DC injections remain to be clarified. For example, in a murine model, immunizations as frequently as every 4–5 days, was necessary to prevent progression of established tumors (Gabrilovich, 1996). The utility of co-administration of DC vaccines and systemic cytokines such as IL-2 and IL-12 has not been established. Because immune responses are likely to be a prerequisite for clinical responses, sensitive methods for monitoring them must be developed in parallel with DC technology. We are now evaluating flow cytometric and ELISA-based assays for analyzing T cell responses to DC immunizations.

ACKNOWLEDGMENTS

We thank Merrilie Brown for her editorial review of this manuscript.

REFERENCES

Alijagic, S., Moller, P., Artuc, M., Jurgovsky, K., Czarnetzki P.M., and Schadendorf, D., 1995, Dendritic cells generated from peripheral blood transfected with human tyrosinase induce specific T cell activation, *Eur. J. Immunol.* **25**:3100–3107.

Alters, S.E., Gadea, J.R., Sorich, M., O'Donoghue, G., Talib, S., and Philip, R., 1998, Dendritic cells pulsed with CEA peptide induce CEA-specific CTL with restricted TCR repertoire, *J. Immunother.* **21**:17–26.

Anichini, A., Mortarini, R., Maccalli, C., Squarcina, P., Fleischhauer, K., Mascheroni, L., and Parmiani, G., 1996, Cytotoxic T cells directed to tumor antigens not expressed on normal melanocytes dominate HLA-A2.1-restricted immune repertoire to melanoma, *J. Immunol.* **156**:208–217.

Asley, D.M., Faiola, B., Nair, S. Hale, L.P., Bigner, D., and Gilboa, E., 1997, Bone marrow-generated dendritic cells pulsed with tumor extracts or tumor RNA induce antitumor immunity against central nervous system tumors, *J. Exp. Med.* **7**:1177–1182.

Austyn, J., 1996, New insights into the mobilization and phagocytic activity of dendritic cells, *J. Exp. Med.* **183**:1287–1292.

Azizi, E., Bunana, C., Goldberg, L., and Kripke, M.L., 1987, Perturbations of epidermal Langerhans cells in basal cell carcinoma, *Am. J. Dermatopath.* **9**:465–473.

Bannerji, R., Arroyo, C.D., Cordon-Cardo, C., and Gilboa, E., 1994, The role of IL-2 secreted from genetically modified tumor cells in the establishment of antitumor immunity, *J. Immunol.* **152**:2324–2332.

Barnd, D.L., Lan, M.S., Metzgar, R.S., and Finn, O.J., 1989, Specific MHC-unrestricted recognition of tumor associated mucins by human cytotoxic cells, *Proc. Natl. Acad. Sci. USA* **86**:7159–63.

Becker, Y., 1993, Dendritic cell activity against primary tumors: an overview, *In Vivo* **7**:187–191.

Becker, Y., 1992, Anticancer role of dendritic cells (DC) in human and experimental cancers—a review, *Anticancer Res.* **12**:511–520.

Bender, A., Sapp, M., Schuler, G., Steinman, R.M., and Bhardwaj, N., 1996, Improved methods for the generation of dendritic cells from nonproliferating progenitors in human blood, *J. Immunol. Methods* **196**:121–135.

Boczkowski, D., Nair, S., Snyder, D., and Gilboa, E., 1996, Dendritic cells pulsed with RNA are potent antigen presenting cells in vitro and in vivo, *J. Exp. Med.* **184**:465–472.

Bos, J.L., 1989, *ras* Oncogenes in human cancer: a review, *Cancer Res.* **49**:4682.

Bulbul, M.A., Huben, R.P., and Murpy, G.P., 1989, Interferon-β treatment of metastatic prostate cancer, *J. Surg. Oncol.* **33**:231–233.

Caux, C., Vandervliet, B., Massacrier, C., Dezutter-Dambuyant, C., de Saint-Vis, B., Jacquet, C., Yoneda, K., Imamura, S., Schmitt, D., and Banchereau, J., 1996, CD34+ hematopoietic progenitors from human cord blood differentiate along two independent dendritic cell pathways in response to GM-CSF+TNFα, *J. Exp. Med.* **184**:695–706.

Cavanagh, L.L., Saal, R.J., Grimmett, K.L., and Thomas, R., 1998, Proliferation in monocyte-derived dendritic cell cultures is caused by progenitor cells capable of myeloid differentiation, *Blood* **92**:1598–1607.

Cella, M., Scheidegger, D., Palmer-Lehmann, K., Lane, P., Lanzavecchia, A., and Alber, G., 1996, Ligation of CD40 on dendritic cells triggers production of high level of interleukin-12 and enhances T cell stimulator capacity: T-T help via APC activation, *J. Exp. Med.* **184**:747–752.

Celluzzi, C.M., Mayordomo, J.I., Storkus, W.J., Lotze, M.T., and Falo, J.L.D., 1996, Peptide-pulsed dendritic cells induce antigen-specific, CTL-mediated protective tumor immunity, *J. Exp. Med.* **183**:283–287.

Coulie, P.G., Somville, M., Lehmann, F., Hainaut, P., Brasseur, F., Devos, R., and Boon,T., 1992, Precursor frequency analysis of human cytolytic T lymphocytes directed against autologous melanoma cells, *Int. J. Cancer* **50**:289–297.

Disis, M.L, and Cheever, M.A., 1997, HER-2/neu protein: A target for antigen-specific immunotherapy of human cancer, *Adv. Cancer Res.* **71**:343–371.

Darrow, T., Slingluff, C., and Seigler, H, 1989, The role of HLA class I antigens in recognition of melanoma cells by tumor-specific cytotoxic T lymphocytes. Evidence for shared tumor antigens, *J. Immunol.* **42**:3329–3335.

Disis, M.L., Calenoff, E., McLaughlin, G., Murphy, A.E., Chen, W., Groner, B., Jeschke, M., Lydon, N., McGlynn, E., Livingston, R.B., and Cheever, M., 1994, Existent T cell and antibody immunity to HER-2/neu protein in patients with breast cancer, *Cancer Res.* **54**:1620.

Fisk, B., Blevins,T.L., Wharton, J.T., and Ioannides, C.G., 1995, Identification of an immunodominant peptide of HER-2/neu proto-oncogene recognized by ovarian tumorspecific cytotoxic T lymphocyte lines, *J. Exp. Med.* **181**:2109–2117.

Flamand,V., Sornasse, T., Thielemans, K., Demanet, C., Bakkus, M., Bazin, H., Tielemans, F., Leo, O., Urbain, J., and Moser, M., 1994, Murine dendritic cells pulsed in vitro with tumor antigen induce tumor resistance in vivo, *Eur. J. Immunol.* **24**:605–610.

Gabrilovich, D.I., Ciernik, I.F., and Carbone, D.P., 1996, Dendritic cells in antitumor immune responses. I. Defective antigen presentation in tumor-bearing hosts, *Cell. Immunol.* **170**:101–110.

Gabrilovich, D.I., Nadaf, S., Corak, J., Berzofsky, J.A., and Carbone, D.P., 1996, Dendritic cells in antitumor immune responses. II Dendritic cells grown from bone marrow precursors, but not mature DC from tumor bearing mice, are effective antigen carriers in the therapy of established tumors, *Cell. Immunol.* **170**:111–119.

Galy, A., Travis, M., Cen, D., and Chen, B., 1995, Human T, B, natural killer, and dendritic cells arise from a common bone marrow progenitor cell subset, *Immunity* **3**:459–73.

Gong, J, Chen, D., Kashiwaba, M., and Kufe, D., 1997, Induction of antitumor activity by immunization with fusions of dendritic and carcinoma cells, *Nature Med.* **3**:558–61.

Grouard, G., Durand, I., Filgueira, L., Banchereau, J., and Liu, Y.J., 1996, Dendritic cells capable of stimulating T cells in germinal centers, *Nature* **384**:364–367.

Grouard, G., Rissoan, M.C., Filgueira, L., Durand, I. Banchereau, J., and Liu, Y.J., 1997, The enigmatic plasmacytoid T cells develop into dendritic cells with interleukin (IL)-3 and CD40-ligand, *J. Exp. Med.* **185**:1101–1111.

Halbert, C., Demers, G.W., and Galloway, D.A., 1991, The E7 gene of human papillomavirus type 16 is sufficient for immortalization of human epithelial cells, *J. Virol.* **65**:473–478.

Hinkel, A., Gitlitz, B., Mulders, P., Moldawer, N., Tso, C., Kaboo, R., Belldegrun, A., and Figlin, R., 1998, Dendritic cell therapy for metastatic renal cell carcinoma-A translational phase I clinical trial, *Proc. Am. Soc. Clin. Oncol.* **17**:432a.

Hock, B.D., Starling, G.C., Daniel, P.B., and Hart, D.N.J., 1994, CMRF-44, a novel monoclonal antibody to an activation antigen expressed by the allostimulatory cells within peripheral blood, including dendritic cells, *Immunol.* **83**:573–581.

Hsu, F.J., Benike, C., Fagnoni, F., Liles, T.M., Czerwinski, D., Taidi, B., Engleman, E.G., and Levy, R., 1996, Vaccination of patients with B-cell lymphoma using autologous antigen-pulsed dendritic cells, *Nature Med.* **2**:52–58.

Huang, A.Y.C., Golumbek, P., Ahmadzedah, M., Jaffee, E., Pardoll, D., and Levitsky, H., 1994, Role of bone marrow-derived cells in presenting MHC class I-restricted tumor antigens, *Science* **264**:961–965.

Inaba, K., Young, J.W., and Steinman, R.M., 1987, Direct activation of CD8+ cytotoxic T lymphocytes by dendritic cells, *J Exp Med.* **166**:182–194.

Kawakami, Y., Eliyahu, S., Delgado, C.H., Robbins, P.F., Rivoltini, L., Topalian, S.L., Miki, T., and Rosenberg, S.A., 1994, Cloning of the gene coding for a shared human melanoma antigen recognized by autologous T cells infiltrating into tumor, *Proc. Natl. Acad. Sci. USA* **91**:3515–3519.

Knight, S.C., Farrant, J., Bryant, A., Edwards, A.J., Burman, S., Lever, A., Clarke, J., and Webster, A.D., 1986, Non-adherent, low-density cells from human peripheral blood contain dendritic cells and monocytes, both with veiled morphology, *Immunol.* **57**:595–603.

Liu, Y.J., Grouard, G., De Boutteillier, O., and Banchereau, J., 1996, Follicular dendritic cells and germinal centers, *Int. Rev. Cytol.* **166**:139–179.

Macatonia, S.E., Taylor, P.M., Knight, S.C., and Askonas, B.A., 1988, Primary stimulation by dendritic cells induces antiviral proliferative and cytotoxic T cell responses *in vitro*, *J. Exp. Med.* **169**:1255–1264.

Maraskovsky, E., Brasel, K., Teepe, M., Roux, E.R., Lyman, S.D., Shortman, K., and McKenna, H.J., 1996, Dramatic increase in the number of functionally mature dendritic cells in Flt3 ligand-treated mice: multiple dendritic cell subpopulations identified, *J. Exp. Med.* **184**:1953–1962.

Matzinger, P., 1994, Tolerance, danger, and the extended family, *Ann. Rev. Immunol.* **12**:991–1045.

Mayordomo, J.I., Zorina, T., Storkus, W.J., Zitvogel, L., Celluzzi, C., Falo, L.D., Melief, C.J., Ildstad, S.T., Kast, W.M., Deleo, A.B., and Lotze, M.T., 1995, Bone marrow-derived dendritic cells pulsed with synthetic tumour peptides elicit protective and therapeutic antitumour immunity, *Nature Med.* **12**:1297–1302.

Mazzocchi, A., Belli, F., Mascheroni, L., Vegetti, C., Parmiani, G., and Anichini, A., 1994, Frequency of cytotoxic T lymphocyte precursors (CTLp) interacting with autologous tumor via the T-cell receptor: limiting dilution analysis of specific CTLp in peripheral blood and tumor invaded lymph nodes of melanoma patients, *Int. J. Cancer.* **58**:330–339.

McKinney, E.C., and Streilein, E.W., 1989, On the extraordinary capacity of allogeneic epidermal langerhans cells to prime cytotoxic T cells in vivo, *J. Immunol.* **143**:1560–1564.

Mehta-Damani, A., Markowicz, S., and Engleman, E.G., 1994, Generation of antigen-specific CD8+ CTLs from naive precursors, *J. Immunol.* **153**:996–1003.

Morse, M.A., Zhou, L.-J., Tedder, T.F., Lyerly, H.K., and Smith, C., 1997, Generation of dendritic cells *in vitro* from peripheral blood mononuclear cells with GM-CSF, IL-4, and TNF-α for use in cancer immunotherapy, *Ann. Surg.* **226**:6–16.

Morse, M.A., Lyerly, H.K., Gilboa, E., Thomas, E., and Nair, S.K, 1998, The role of CD40-ligand in induction of dendritic cell maturation, *Cancer Res.* (In Press).

Mortarini, R., Anichini, A., Di Nicola, M., Siena, S., Bregni, M., Belli, F., Molla, A., Gianni, A.M., and Parmiani, G., 1997, Autologous dendritic cells derived from CD34+ progenitors and from monocytes are not functionally equivalent antigen-presenting cells in the induction of Melan-A/Mart-1_{27-35}-specific CTLs from peripheral blood lymphocytes of melanoma patients with low frequency of CTL precursors, *Cancer Res.* **57**:5534–5541.

Mukherji, B., Chakraborty, N.G., Yamasaki, S., Okino, T., Yamase, H., Sporn, J.R., Kurtzman, S.K., Ergin, M.T., Ozols, J., and Meehan, J., 1995, Induction of antigen-specific cytolytic T cells *in situ* in human melanoma by immunization with synthetic peptide-pulsed autologous antigen presenting cells, *Proc. Natl. Acad. Sci. USA* **92**:8078–8082.

Murphy, G., Tjoa, B., Ragde, H., Kenny, G., and Boynton, A., 1996, Phase I clinical trial: T-cell therapy for prostate cancer using autologous dendritic cells pulsed with HLA-A0201-specific peptides from prostate-specific membrane antigen, *The Prostate* **29**:371–380.

Nair, S., Zhou, F., Reddy, R., Huang, L., and Rouse, B.T., 1992, Soluble proteins delivered to dendritic cells via pH-sensitive liposomes induce primary cytotoxic T lymphocyte responses *in vitro*, *J. Exp. Med.* **175**:609–612.

Nair, S., Babu, J.S., Dunham, R.G., Kanda, P., Burke, R.L., and Rouse, B.T., 1993, Induction of primary, antiviral cytotoxic, and proliferative responses with antigens administered via dendritic cells, *J. Virol.* **67**:4062–4069.

Nair, S.K., Bocczowski, D., Morse, M.A., Cumming, R.I., Lyerly, H.K., and Gilboa, E., 1998, Induction of primary carcinoembryonic antigen (CEA)-specific cytotoxic T lymphocytes in vitro using human dendritic cells transfected with RNA, *Nature Biotech.* **16**:1–8.

Nair, S.K., Snyder, D., Rouse, B.T., and Gilboa, E., 1997, Regression of tumors in mice vaccinated with professional antigen-presenting cells pulsed with tumor extracts, *Int. J. Cancer* **70**:706–715.

Nestle, F.O., Alijagic, S., Gilliet, M., Sun,Y., Grabbe, S., Dummer, R., Burg, G., and Schadendorf, D., 1998, Vaccination of melanoma patients with peptide- or tumor lysate-pulsed dendritic cells, *Nature Med.* **4**:328–332.

Ossevoort, M.A., Feltkamp, M.C.W., Van Veen, K.J.H., Melief, C.J., and Kast, W.M., 1995, Dendritic cells as carriers for a cytotoxic T-lymphocyte epitope-based peptide vaccine in protection against a human papillomavirus type 16-induced tumor, *J. Immunother.* **18**:86–94.

Paglia, P., Chiodoni, C., Rodolfo, M., and Colombo, M.P., 1996, Murine dendritic cells loaded in vitro with soluble protein prime cytotoxic T lymphocytes against tumor antigen in vivo, *J. Exp. Med.* **183**:317–322.

Porgador, A., Snyder, D., and Gilboa, E., 1996, Induction of antitumor immunity using bone marrow-generated dendritic cells, *J. Immunol.* **156**:2918–2926.

Porgador, A., and Gilboa, E., 1995, Bone marrow-generated dendritic cells pulsed with a class I-restricted peptide are potent inducers of cytotoxic T lymphocytes, *J. Exp. Med.* **182**:255–260.

Reichardt, V., Okada, C., Benike, C., Long, G., Engleman, E., Blume, K., and Levy, R., 1996, Idiotypic vaccination using dendritic cells for multiple myeloma patients after autologous peripheral blood stem cell transplantation, *Blood* **88**:481 (Abstract).

Reid, C.D.L., Stackpoole, A., Meager, A., and Tikerpae, J., 1992, Interactions of tumor necrosis factor with granulocyte-macrophage colony-stimulating factor and other cytokines in the regulation of dendritic cell growth in vitro from early bipotent CD34+ progenitors in human bone marrow, *J. Immunol.* **149**:2681–2688.

Reeves, M.E., Royal, R.E., Lam, J.S., Rosenberg, S.A., and Hwu, P., 1996, Retroviral transduction of human dendritic cells with a tumor-associated antigen gene, *Cancer Res.* **56**:5672–5677.

Robinson, M.R., Rigby, C.C., Pugh, C.B., and Dumonde, D.C., 1977, Adjuvant immunotherapy with BCG in carcinoma of the prostate, *Br. J. Urol.* **49**:221–226.

Romani, N., Gruner, S., Brang, D., Kampgen, E., Lenz, A., Trockenbacher, B., Konwalinka, G., Fritsch, P.O., Steinman, R.M., and Schuler, G., 1994, Proliferating dendritic cell progenitors in human blood, *J. Exp. Med.* **180**:83–93.

Romani, N., Reider, D., Heuer, M., Ebner, S., Kampgen, E., Eibl, B., Niederwieser, D., and Schuler, G., 1996, Generation of mature dendritic cells from human blood. An improved method with special regard to clinical applicability, *J. Immunol. Methods* (1996) **196**:137–151.

Rosenberg, S.A., Lotze, M.T., Muul, L.M., Chang, A.E., Avis, F.P., Leitman, S., Linehan, W.M., Robertson, C.N., Lee, R.E., and Rubin, J.T., 1987, A progress report on the treatment of 157 patients with advanced cancer using lymphokine activated killer cells and interleukin-2 or high-dose interleukin-2 alone, *N. Engl. J. Med.* **316**(15):889–897.

Rosenberg, S.A., Lotze, M.T., Yang, J.C., Aebersold, P.M., Linehan, W.M., Seipp, C.A., and White, D.E., 1989, Experience with the use of high-dose interleukin-2 in the treatment of 652 patients with cancer, *Ann. Surg.* **210**:474–485.

Rouse, R.J.D., Nair, S.K., Lydy, S.L., Bowen, J.C., and Rouse, B.T., 1994, Induction in vitro of primary cytotoxic T-lymphocytes responses with DNA encoding herpes simplex virus proteins, *J. Virol.* **68**:5685–5689.

Santiago-Schwarz, F., Belilos, E., Diamond, B., and Carsons, S.E., 1992, TNF in combination with GM-CSF enhances the differentiation of neonatal cord blood stem cells into dendritic cells and macrophages, *J. Leuk. Biol.* **52**:274–281.

Saunders, D., Lucas, K., Ismaili, J., Wu, L., Maraskovsky, E., Dunn, A., and Shortman, K., 1996, Dendritic cell development in culture from thymic precursor cells in the absence of granulocyte/macrophage colony-stimulating factor, *J. Exp. Med.* **184**:2185–2196.

Siena, S., Di Nicola, M., Bregni, M., Mortarini, R., Anichini, A., Lombardi, L., Ravagnani, F., Parmiani, G., and Gianni, A.M., 1995, Massive ex vivo generation of functional dendritic cells from mobilized CD34+ blood progenitors for anti-cancer therapy, *Exp. Hematol.* **23**:1463–1471.

Song, W., Kong, H.L., Carpenter, H., Torii, H., Granstein, R., Rafii, S., Moore, M.A.S., and Crystal, R.G., 1997, Dendritic cells genetically modified with an adenovirus vector encoding the cDNA for a model antigen induce protective and therapeutic antitumor immunity, *J. Exp. Med.* **186**:1247–1256.

Specht, J.M., Wang, G., Do, M.T., Lam, J.S., Royal, R.E., Reeves, M.E., Rosenberg, S.A., and Hwu, P., 1997, Dendritic cells retrovirally transduced with a model antigen gene are therapeutically effective against established pulmonary metastases, *J. Exp. Med.*, **186**:1213–1221.

Steinman, R.M., and Swanson, J., 1995, The endocytic activity of dendritic cells, *J. Exp. Med.* **182**:283–88.

Steinman, R.M., 1991, The dendritic cell system and its role in immunogenicity, *Ann. Rev. Immunol.* **9**:271–296.

Strobl, H., Scheinecker, C., Riedl, E., Csmarits, B., Bello-Fernandez, C., Pickl, W.F., Majdic, O., and Knapp, W., 1998, Identification of CD68+lin- peripheral blood cells with dendritic precursor characteristics, *J. Immunol.* **161**:740–748.

Suss, G., and Shortman, K., 1996, A subclass of dendritic cells kills CD4 T cells via Fas/Fas-ligand-induced apoptosis, *J. Exp. Med.* **183**:1789–1796.

Szabolcs, P., Moore, M.A., and Young, J.W, 1995, Expansion of immunostimulatory dendritic cells among the myeloid progeny of human CD34+ bone marrow precursors cultured with c-kit ligand, granulocyte-macrophage colony-stimulating factor, and TNF-α, *J. Immunol.* **154**:5851–5861.

Szabolcs, P., Avigan, D., Gezelter, S., Ciocon, D.H., Moore, M.A., Steinman, R.M., and Young, J.W., 1996, Dendritic cells and macrophages can mature independently from a human bone marrow-derived, post-colony forming unit intermediate, *Blood* **87**:4520–4530.

Tazi, A., Bouchonnet, F., Grandsaigne, M., Boumsell, L., Hance, A.J., and Soler, P., 1993, Evidence that granulocyte-macrophage-colony stimulating factor regulates the distribution and differentiated state of dendritic cells/Langerhans cells in human lung and lung cancers, *J. Clin. Invest.* **91**:566–576.

Takahashi, H., Nakagaway, Y., Yokomuro, K., and Berzofsky, J.A., 1993, Induction of CD8+ cytotoxic T lymphocytes by immunization with syngeneic irradiated HIV-1 envelope derived peptide-pulsed dendritic cells, *Int. Immunol.* **5**:849–857.

Tarte, K., Lu, Z.Y., Fiol, G., Legouffe, E., Rossi, J.F., and Klein, B., 1997, Generation of virtually pure and potentially proliferating dendritic cells from non-CD34 apheresis cells from patients with multiple myeloma, *Blood* **90**:3482–3495.

Thomas, R., Davis, L., and Lipsky, P., 1993, Isolation and characterization of human peripheral blood dendritic cells, *J. Immunol.* **150**:821–834.

Thompson, J.A., Gruner, T.F., and Zimmermann, W., 1991, Carcinoembryonic antigen gene family: molecular biology and clinical prospectives, *J. Clin. Lab. Anal.* **5**:344–366.

Tsang, K.Y., Zaremba, S., Nieroda, C.A., Zhu, M.Z., Hamilton, J.M., and Schlom, J., 1995, Generation of human cytotoxic T cells specific for human carcinoembryonic antigen epitopes from patients immunized with recombinant vaccinia-CEA vaccine, *J. Natl. Cancer Inst.* **87**:982–990.

Valone, F.H., Small, E.J., Whisenant, S., Peshwa, M., Strang, G., Laus, R., Ruegg, C., and van Schooten, W., 1998, Immunotherapy of hormone refractory prostate cancer with antigen-pulsed dedritic cells: A Phase I/II study, *Proc. Am. Soc. Clin. Oncol.* **17**:346a.

Van den Eynde, B., Hainut, P., Herin, M., Knuth, A., Lemoine, C., Weynants, P., van der Bruggen, P.C., Fauchet, R., and Boon, T., 1989, Presence on a human melanoma of multiple antigens recognized by autologous CTL, *Int. J. Cancer* **44**:634–640.

Van der Bruggen, P.C., Traversari, C., Chomez, P., Lurquin, C., De Plaen, E., Van den Eynde, B., Knuth, A., and Boon, T., 1991, A gene encoding an antigen recognized by cytotoxic T lymphocytes on a human melanoma, *Science* **254**:1643–1647.

Van Pel, A., and Boon, T, 1982, Protection against a nonimmunogenic mouse leukemia by an immunogenic variant obtained by mutagenesis, *Proc. Natl. Acad. Sci. USA* **79**:4718–4722.

Van Voorhis, W.C., Hair, L.S., Steinman, R.M., and Kaplan, G., 1982, Human dendritic cells: Enrichment and characterization from peripheral blood, *J. Exp. Med.* **155**:1172–1187.

Vremee, D., Zorbas, M., Scollay, R., Saunders, D., Ardavin, C., Wu, L., and Shortman, K., 1992, The surface phenotype of dendritic cells purified from mouse thymus and spleen: investigation of the CD8 expression by a subpopulation of dendritic cells, *J. Exp. Med.* **176**:47–58.

Williams, L.A., Egner, W., and Hart, D.N.J., 1994, Isolation and function of human dendritic cells, *Int. Rev. Cytol.* **153**:41–103.

Wong, C., Morse, M.A., and Nair, S.K., 1998, Induction of primary, human antigen-specific cytotoxic T lymphocytes *in vitro* using dendritic cells pulsed with peptides, *J. Immunother.* **21**:32–40.

Young, J.W., and Steinman, R.M., 1990, Dendritic cells stimulate primary human cytolytic lymphocyte responses in the absence of CD4+ helper T cells, *J. Exp. Med.* **171**:1315–1332.

Young, J.W., and Steinman, R.M., 1988, Accessory cell requirements for the mixed-leukocyte reaction and polyclonal mitogens, as studied with a new technique for enriching blood dendritic cells. *Cell. Immunol.* **111**:167–182.

Zhou, L.-J., and Tedder, T.F., 1995, Human blood dendritic cells selectively express CD83, a member of the immunoglobulin superfamily, *J. Immunol.* **154**:3821–3835.

Zhou, L.-J., and Tedder, T.F., 1996, CD14+ Blood monocytes can differentiate into functionally mature CD83+ dendritic cells, *Proc. Natl. Acad. Sci. USA* **93**:2588–2592.

Zitvogel, L., Mayordomo, J.I., Tjandrawan, T., DeLeo, A.B., Clarke, M.R., Lotze, M.T., and Storkus, W.J., 1996, Therapy of murine tumors with tumor peptide pulsed dendritic cells: dependence on T cells, B7 costimulation, and Th1-asssociated cytokines, *J. Exp. Med.* **183**:87–97.

Zitvogel, L., Regnault, A., Lozier, A., Wolfers, J., Flament, C., Tenza, D. Ricciardi-Castagnoli, P., Raposo, G., and Amigorena, S., 1998, Eradication of established murine tumors using a novel cell-free vaccine; dendritic cell-derived exosomes, *Nature Med.* **4**:594–600.

29

HYBRID CELL VACCINATION FOR CANCER IMMUNOTHERAPY

Peter Walden

Department of Dermatology
Medical Faculty Charité
Humboldt University
Berlin, Germany

1. INTRODUCTION

1.1. Tumour Immunology

The immune surveillance hypothesis proposed by Macfarlane Burnet in 1970 (Burnet, 1970) summarised nearly a hundred years of research on the role of the immune system in cancer. This research had cumulated a substantial body of evidence for the major contribution of the immune system to the control and eradication of cancer. Pathological studies suggested a much more frequent incidence of initial tumour development than what becomes clinically manifest. Burnet's studies had indicated that at a thymus dependent, hence, T cell mediated mechanisms control tumour, may stop its growth and prevent cancer. Since then, tumour associated antigens for T cells were identified, tumour specific antibodies and T cell were isolated from peripheral blood and tumour of cancer patients, and T cell growth and differentiation factors were successfully employed for the treatment of cancer. Tumour infiltrating T lymphocytes were shown to exhibit at least in parts tumour specificity. All these observations confirmed the central role of the immune system and especially of T lymphocytes in tumour biology, and prove that the immune system has the capacity to identify and destroy tumour cells.

Notwithstanding these evidences for the high efficiency of the immune surveillance system, cancer exists among the major health problems of mankind. The failure of the immune system in cases of cancer is the hub of tumour immunology, and a focus of all endeavours towards new strategies for the treatment of cancer. The cause of this failure or, in immunological terms, the reason for the insufficient immunogenicity of the tumour may be in the tumour cells themselves. A lack of antigenicity or a loss of antigens under immune selection pressure renders tumours invisible to the immune system. There may be defects in the immune system such as antigen presentation defects or a lack of suitable tumour specific T cells in the patient's T cell receptor repertoire because of a

Cancer Gene Therapy: Past Achievements and Future Challenges, edited by Habib
Kluwer Academic/Plenum Publishers, New York, 2000.

non-permissive immune genetics or because of immune tolerance. The interaction of tumour and immune systems may result in immune suppression or clonal exhaustion, i.e. a deletion of the tumour specific T cell clones induced by their anti tumour reactions. Finally, there may be deficits in the immune regulatory support for the development of tumour specific cytotoxic T lymphocytes that prevents their induction.

Spontaneous tumour regression has been reported for different cancers, and central regression zones in melanoma lesions with simultaneous tumour progression in the border zones can be very frequent. These observations strongly suggest that the relationship of tumour and immune system in cancer is a dynamic relationship with an ongoing struggle between the two sides. Tumour progression indicates a gradual shift of this balance in favour of the tumour and usually is accompanied by an increasing degree of immune dysfunction. The aim of immune therapeutic intervention is to shift this balance in favour of the immune system by strengthening and supporting its active components.

1.2. Tumour Associated Antigens

A number of tumour associated antigens have been identified in recent years, most for melanoma. The majority of these antigens were identified by T cells assays in combination with molecular cloning (van-den-Eynde and Boon, 1997). These antigens therefore not only define structural differences between tumour and normal tissue, they are also recognised by the immune system and, thus, can be a focus for anti tumour immunity. Detailed analysis has revealed that only a minority of these antigens are tumour specific, most of them are expressed in normal tissues as well. The tumour specific epitopes are mutations occurring in an individual patient only. This group of tumour antigens includes epitopes derived from clonotypic idiotype sequences of lymphomas. The public epitopes fall into two categories: Some are found in the tumour and in testis, an immune privileged organ. MAGE and GAGE are examples for this group of tumour associated antigens. Others, however, are expressed in all cells of the tumour's tissue type. Examples for this category are tyrosinase, gp100 and gp75 which are present in all pigmented cell types. The existence of T cells with specificity for normal tissue antigens in patients with malignant melanoma raises very fundamental questions about the mechanism of self tolerance, questions which have not been resolved yet. The localised vitiligo, which occasionally occurs in melanoma patients, is indicative of cytolytic T cell responses to public T cell epitopes and provides further evidence for the anti tumour response capacity of the cellular immune system. However, these occurrences also show that there is a thin line between immune tolerance of self, anti tumour immunity and autoimmune reactivity.

1.3. Vaccination for Cancer Therapy

Vaccination in the case of infectious diseases is priming of naïve individuals or boosting of individuals with low residual immunity. Therapeutic vaccination as in cancer, on the other hand, aims at the modulation of an ongoing immune response. To what extent this difference requires different vaccination strategies still needs to be investigated. However, the goal of therapeutic vaccination to cancer is a specific shift in the quality and quantity, not priming of the immune response. Therefore, the development of oncologic vaccination strategies must be concerned with both the identification and utilisation of suitable target antigens and the systematic and controllable modulation of the immune regulatory context and requirements for the induction of effective anti tumour cytolytic T cell responses.

2. CELLULAR, MOLECULAR AND GENETIC BASIS OF CELLULAR IMMUNITY

The induction of cytotoxic effector T cells, just like the induction of specific humoral immune responses, requires T cell help (Borges *et al.*, 1994). The basic regulatory units in this case are three cell clusters composed of antigen presenting cells (APC), helper T cells and precursor cytotoxic T cells (Mitchison and O'Malley, 1987; Mitchison, 1990; Stuhler and Walden, 1993). This situation differs from the two cell cluster of T and B cells that control the induction and development of specific antibody mediated immunity. The APC is the organiser of the regulatory three cell cluster systems (Fig. 1). The key to the formation of these clusters is antigen presentation by the APC. Two different antigens must be presented by the same APC on MHC class I and class II molecules for the cytotoxic precursor cell and the helper T cell, respectively (Stuhler and Walden, 1993). Thus, there is a twofold MHC restriction that controls the induction of cytotoxic T cell responses (Juretic *et al.*, 1985) in contrast to T-B cell collaboration and antibody formation which is solely controlled by MHC class II molecules. However, in both cases the two epitopes for the two interacting antigen specific lymphocytes must be physically linked. The physical basis for epitope linkage is in case of the T-T cell collaboration, however, is a cell, the APC, not a molecule as in the case of T-B cell collaboration. The twofold MHC restriction control for the induction of cytolytic effector T cell responses provides additional protection against potential auto-immune reactions. However, the antigens for the two interacting T cells need not be related on the molecular level. Cancer

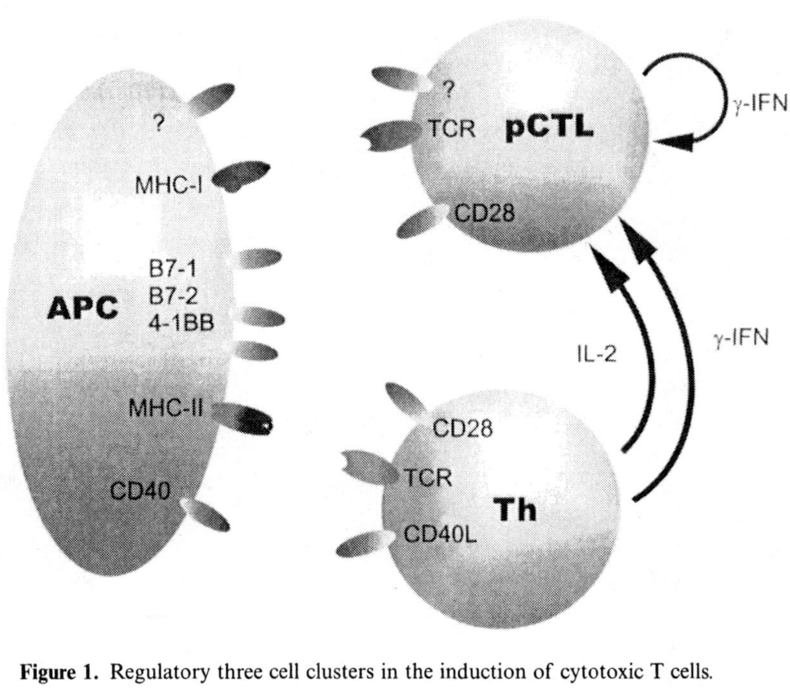

Figure 1. Regulatory three cell clusters in the induction of cytotoxic T cells.

immunology is highly individualised because of the MHC restriction of T cell mediated immunity, i.e. protective anti tumour immunity depends on the perfect combination of the expression of suitable tumour associated antigens presented by permissive MHC allomorphs and the presence of the corresponding T cells in the immune receptor repertoire of the patients.

Recognition of cognate peptides together with their specific MHC class I and class II molecules on the surface of the APC by the two T cell types induces a complex array of molecular interaction between the cells. Both T cells communicate with the APC and with each other via cell surface molecules and soluble factors (Stuhler and Walden, 1993). These molecules are either constitutively expressed or may be modulated. They may be a result of the maturation of the participating cells as CD28, B7 and 4-1BB (Chen *et al.*, 1992), or they may be activation induced as CD40 on the APC and CD40L on the T cells (Bennet *et al.*, 1998; Ridge *et al.*, 1998; Schoenberger *et al.*, 1998). The expression of lymphokines interleukin-2 as growth and differentiation factor for CTL provided by the helper T cell or γ-IFN, a major differentiation factor of CTL which can be produced by both T cells types is strictly activation induced. The latter two soluble factors are major means of the communication between the two T cells and act on short distance. Recent reports demonstrate that additional activation induced factors regulate the interaction of the APC and the precursor cytotoxic T cell (Lanzavecchia, 1998). These factors have not yet been identified and it also remains to be seen whether cell surface molecules are involved in direct T-T cell contacts and communications. These molecular interactions provide signals for co-stimulation, modulation, survival, growth or differentiation and are essential for the development of cellular immunity. New players continue to be discovered and are added to the list or implicated by indirect evidence.

Although several MHC class II expressing cells can serve as APC in these three cell interactions, the dendritic cell (DC) is clearly the most potent cell for antigen presentation, for organisation of the regulatory clusters and for co-stimulation (Banchereau and Steinman, 1998; Reid, 1998). The influence of the specific architecture of the lymphoid organs on the efficiency of the interactions and of the induction of effector T cell is largely unexplored. However, non-lymphoid tissue are far less potent promoters of immune responses. In contrast to these multiple regulatory interactions during the induction of cytotoxic T cells, is the execution of the effector functions, i.e. the cytolysis of the target cells, only dependent on MHC peptide complexes and adhesion molecules.

3. HYBRID CELL VACCINATION

3.1. Conceptual Basis

The hybrid cell vaccination approach to cancer immune therapy aims at the induction of tumour specific cytotoxic T cells and was developed to meet the requirements for effective antigen presentation, co-stimulation and T cell help (Stuhler and Walden, 1994; Guo *et al.*, 1994). While MHC class I restricted CTL epitopes are needed for the antitumour immunity and, thus, are determined by the tumour's antigenicity, MHC class II restricted epitopes can be chosen freely to avoid limitations in helper T cell frequencies and response capacities. While most tumours express MHC class I molecules, MHC class II expression is restricted to a few cell types of the immune system. The probability for simultaneous expression of MHC class I and II on the same tumour cell together with suitable tumour antigens that can address the helper and precursor cytotoxic T cells of the patients at the same time is low. Although any self MHC restricted antigen can be used to induce T cell help for the induction of the tumour specific CTL, the frequency and competence of the helper T cell should not limit the possibility for CTL induction. Therefore, the notoriously vigorous allo MHC specific T cell responses, which are among the most dependable immune reactions, were chosen as a bountiful source for T cell help. Combining all these considerations, fusion of tumour cells and allo MHC class II bearing cells should render antigenic tumour cells immunogenic. The hybrid cells can attract and activate T cell help as well as tumour specific cytotoxic T cells. Moreover, fusion of the tumour cell with an APC may complement potential co-stimulatory deficiencies of the tumour cell. Using the tumour itself as source and carrier of the tumour antigens ensures the broadest possible representation of the tumour's antigenicity in the vaccine. These antigens need not to be identified. Taken together, the above special features of hybrid cell vaccination make this therapeutic strategy well suited for individualised therapy.

In summary, hybrid cell vaccination was designed

- to recruit and activate T cell help for the induction of tumour specific cytotoxic T cells,
- to correct defects in co-stimulatory signaling,
- to utilise a large variety of usually unidentified tumour associated antigens,
- for individualised therapy that can be applied instantly without long preparations.

3.2. Generation of the Hybrid Cell Vaccine

Classical hybridoma technology aims at long term lines that are selected primarily on the basis of their resistance to cytostatic drugs and, subsequently, according to the specific functionality desired. Usually both fusion partners cells are of the same tissue origin and are controlled by very similar genetic programmes. One of the fusion partners is a tumour cell selected for good growth properties. The hybrid cells needed for cancer immune therapy are very different. First, the fusion partners are usually heterogeneous, i.e. they are of very different tissue origin controlled by different genetic programmes and, second, the tumour cell has in most cases not been adapted for stable in culture growth. On the other hand, long term survival and growth of the hybrid cell is not required for the vaccination effect. Just the opposite, it must be ensured that the hybrid cells can not survive after injection into the patient. The vaccines are irradiated prior to inoculation to prevent the tumour from spreading and growing in the vaccinee.

Since the hybrid cells can not be expanded the fusion technology must be designed to yield high fusion efficiencies independent from the specific types of cells used for hybridisation (Fig. 2). Electrofusion was chosen as the most suitable technique for this purpose (Zimmermann, 1986; Chang *et al.*, 1994). Fusion is possible with about any cell type that is available as single cell suspension and the fusion efficiencies range between 10 to 30% for heterogeneous fusion corresponding to 30 to 90% of overall fusion including the homogeneous combinations. Electrofusion requires an initial alignment step to bring the cells in to close contact. This is readily achieved by applying a low voltage (about 25 V/cm) spatially inhomogeneous electric field. The alignment is followed by a short millisecond fusion pulse of about 1,000 V/cm. The optimal parameters for the alignment voltage, the alignment time and the fusion pulse need to established for every cell type combination. However, once established reproducible fusion results can be obtained.

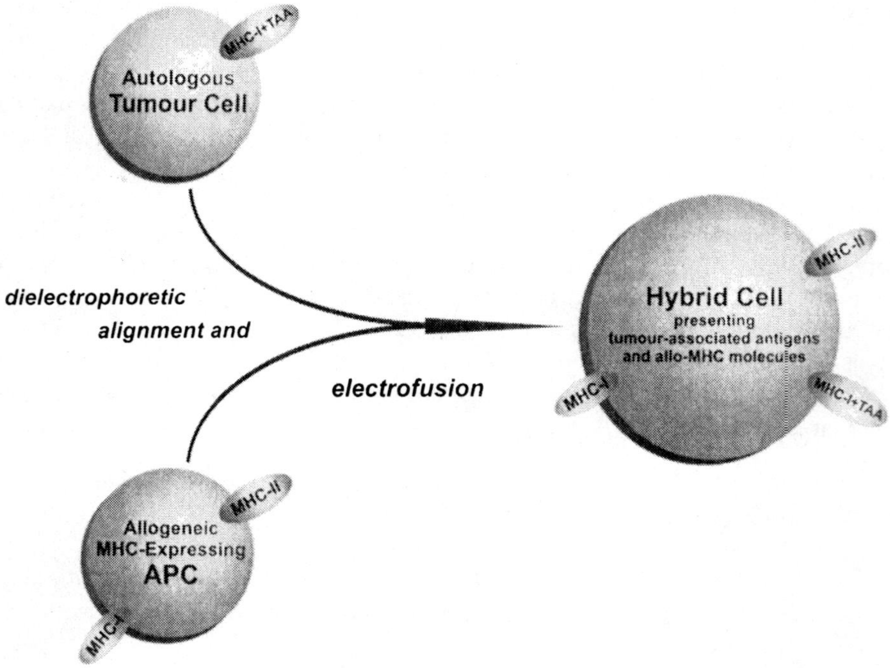

Figure 2. Generation of the hybrid cell vaccines.

4. HYBRID CELL VACCINATION IN PRECLINICAL STUDIES

The hybrid cell vaccination concept has been tested in three different animal transplantation tumour models. These models were liver cell carcinoma in two cases (Guo *et al.*, 1994; Gong *et al.*, 1996) and one example of an intradermal thymoma (Stuhler and Walden, 1994). The fusion partners in the two earlier works were B lymphoma, in contrast to the dendritic cells in the recent report. In all three sets of experiments therapeutic vaccinations as well as primary immunisations with the hybrid cells were tested. In all three cases eradication of established tumours and induction of specific long lasting anti tumour immunity was induced by single *intra peritoneal* injections of the vaccine. The thymoma study provided evidence for the need of cell hybridisation, for the involvement of allo MHC specific helper T cells in the induction of anti tumour immunity and of tumour specific, self-MHC restricted cytotoxic T cell in the effector phase (Stuhler and Walden, 1994). Moreover, the initial liver cell carcinoma study could show that the co-stimulatory molecule B-7.1 is required for effective induction of tumour rejection *in vivo* (Guo *et al.*, 1994). The reported efficiencies in all the sets of experiments were very similar and no difference was seen between B cells and dendritic cells as fusion partners. The inferior APC capacity of B cells (Young and Inaba, 1996) might be overcome by the vigour of anti allo MHC T cell responses.

5. CLINICAL RESULTS WITH HYBRID CELL VACCINATION

The first two clinical cancer immune therapy trials, performed with patients suffering from malignant melanoma or renal cell carcinoma, have yielded evidence for the induction of anti tumour T cell responses in these patients. The main objective of these initial clinical trials was to test efficacy, toxicity, feasibility and anti tumour activity of this treatment. The patients were therefore mostly stage IV patients and had no treatment options left. Both trials yielded cases of complete remission. Stable disease was established in a number of patients who had suffered from rapid tumour progression before. Some of these patients remained in stable disease with periodic hybrid cell vaccinations for years. Immune anti tumour response capacity of the patients is strictly dependent on a functionally intact cellular immune system which can be detected as capacity to mount DTH responses to at least one of a set of common recall antigens. DTH tests thus permit identification of patients who are suited for vaccination therapy.

The side effects seen in the trials were all signs of vaccination induced immune responses. They included erythema at the sites of inoculation, some cases of fever and others of strong but temporal perspiration. In no case did these side effects exceed grade one toxicity. In some of the malignant melanoma patients locally restricted vitiligo occurred after vaccination which again is indicative of the activity of cytotoxic T cells with specificity for public melanoma associated antigens. No other signs of vaccination induced auto-immune reactivity were observed. This low toxicity associated with the hybrid cell vaccination is very encouraging and common to all vaccination therapy strategies tested so far.

6. CONCLUSIONS AND PROSPECTS

Hybrid cell vaccination is a feasible strategy for the treatment of cancer and well suited for individualised therapy. The efficacy of this therapeutic approach will be investigated in the above mentioned as well as in other cancers. These future trials will have to establish the criteria for the selection of malignancies and patients suitable for this type of therapy.

The responses achieved in advanced stage cancer patients with hybrid cell vaccination as well as with other vaccination, immunogene or immune therapy attempts are very encouraging in that they show that there are still tumour specific cytotoxic T cells in these patients that can be activated, that maybe expanded and that can penetrate the tumour lesions and destroy the malignant tissue (Pardoll, 1998). Vaccination therapy for long term maintenance of a stable disease state in patients with metastatic cancers is a new aspect of cancer immune therapy that opens new prospects in the field. Obviously, a repeated reactivation of the cellular immune system is possible without causing clonal exhaustion. However, the silencing of the tumour specific immune reactions after each booster vaccination and vaccination induced immune response indicates that strong immune suppressive mechanisms are effective in these patients. Clearly, more attention needs to be paid to strategies to counteract these immune suppressive mechanisms.

Unlike peptide vaccination (Porgador and Gilboa, 1995; Nestle *et al.*, 1998), for hybrid cell vaccination the tumour antigens need not be known. Using the tumour cell itself as immunogen ensures that a multitude of tumour antigens can be utilised for the induction of anti tumour immunity. While the previous clinical trials have used activated peripheral blood B cells, the more recent protocols use dendritic cells which are now readily accessible. It is conceivable to combine hybrid cell vaccination with other treatment modalities that can help to reduce the tumour burden and to provide relieve from some of the immune suppressive activities of the tumours. While allo MHC molecules ensure a very effective induction of T cell responses and thus of T cell help, there is the disadvantage that allogeneic material has to be transferred into the patient. Gene transfer approach might replace the hybridoma technology and would avoid problems associated with the allogeneic material from human source. However, multiple genes need to be identified and transferred into the tumour cells in order to convert them into efficient antigen presenting cells. Maybe transferring the tumour's antigenicity into dendritic cells is the more promissing approach to the generation of effective tumour vaccines. For now, hybrid cell vaccination appears to be one of the most potent new therapeutic strategies to cancer, well adapted to clinical application.

REFERENCES

Banchereau J., and Steinman R.M.: Dendritic cells and the control of immunity. Nature 1998, 392:245–252.

Bennett S.R., Carbone F.R., Karamalis F., Flavell R.A., Miller J.F., and Heath W.R.: Help for cytotoxic-T-cell responses is mediated by CD40 signalling. Nature 1998, 393:478–480.

Borges E., Wiesmuller K.H., Jung G., and Walden P.: Efficacy of synthetic vaccines in the induction of cytotoxic T lymphocytes. Comparison of the costimulating support provided by helper T cells and lipoamino acid. J Immunol Methods. 1994, 173:253–263.

Burnet F.M.: The concept of immunological surveillance. Prog Exp Tumor Res. 1970, 13:1–27.

Chang D.C., Chassy B.M., Saunders J.A., and Sowers A.E. 1994. Guide to electroporation and electrofusion. Academic Press, San Diego.

Chen L., Ashe S., Brady W.A., Hellstrom I., Hellstrom K.E., Ledbetter J.A., McGowan P., and Linsley P.S.: Costimulation of antitumor immunity by the B7 counterreceptor for the T lymphocyte molecules CD28 and CTLA-4. Cell 1992, 71:1093–1102.

Gong J., Chen D., Kashiwaba M., and Kufe D.: Induction of antitumor activity by immunization with fusions of dendritic and carcinoma cells. Nat Med. 1997, 3:558–561.

Guo Y., Wu M., Chen H., Wang X., Liu G., Li G., Ma J., and Sy M.S.: Effective tumor vaccine generated by fusion of hepatoma cells with activated B cells. Science 1994, 263:518–520.

Juretic A., Malenica B., Juretic E., Klein J., and Nagy Z.A.: Helper effects required during in vivo priming for a cytolytic response to the H-Y antigen in nonresponder mice. J Immunol. 1985, 134:1408–1414.

Lanzavecchia A.: Licence to kill. Nature 1998, 393:413–414.

Mitchison N.A., and O'Malley C.: Three-cell-type clusters of T cells with antigen-presenting cells best explain the epitope linkage and noncognate requirements of the in vivo cytolytic response. Eur J Immunol. 1987, 17:1579–1583.

Mitchison N.A.: An exact comparison between the efficiency of two- and three-cell-type clusters in mediating helper activity. Eur J Immunol. 1990, 20:699–702.

Nestle F.O., Alijagic S., Gilliet M., Sun Y,. Grabbe S., Dummer R., Burg G., and Schadendorf D.: Vaccination of melanoma patients with peptide- or tumor lysate-pulsed dendritic cells. Nat Med. 1998, 4:328–332.

Pardoll D.M.: Cancer vaccines. Nat Med. 1998, 4:525–531.

Porgador A., and Gilboa E.: Bone marrow-generated dendritic cells pulsed with a class I restricted peptide are potent inducers of cytotoxic T lymphocytes. J Exp Med. 1995, 182:255–260.

Reid C.D.: The biology and clinical applications of dendritic cells. Transfus Med. 1998, 8:77–86.

Ridge J.P., Di-Rosa F., and Matzinger P.: A conditioned dendritic cell can be a temporal bridge between a CD4+ T-helper and a T-killer cell. Nature 1998, 393:474–478.

Schoenberger S.P., Toes R.E., van-der-Voort E.I., Offringa R., and Melief C.J.: T-cell help for cytotoxic T lymphocytes is mediated by CD40-CD40L interactions. Nature 1998, 393:480–483.

Stuhler G., and Walden P.: Collaboration of helper and cytotoxic T lymphocytes. Eur J Immunol. 1993, 23:2279–2286.

Stuhler G., and Walden P.: Recruitment of helper T cells for induction of tumour rejection by cytolytic T lymphocytes. Cancer Immunol Immunother. 1994, 39:342–345.

Van-den-Eynde B.J., and Boon T.: Tumor antigens recognized by T lymphocytes. Int J Clin Lab Res. 1997, 27:81–86.

Young J.W., and Inaba K.: Dendritic cells as adjuvants for class I major histocompatibility complex-restricted antitumor immunity. J Exp Med. 1996, 183:7–11.

Zimmermann U.: Electrical breakdown, electropermeabilization and electrofusion. Rev Physiol Biochem Pharmacol. 1986, 105:176–256.

30

MODULATION OF THE IMMUNE RESPONSE THROUGH 4-1BB

Gabriel Sica and Lieping Chen

Department of Immunology, Mayo Clinic
200 First Street, Southwest
Rochester, Minnesota 55905, USA

1. INTRODUCTION

4-1BB (CD137) is a 30 kDa type I (N terminus extracellular) membrane glycoprotein receptor that belongs to the tumor necrosis factor receptor (TNFR) superfamily which also include molecules such as CD27, CD40, Fas (CD95), OX40, and TNFR I/II. Members of the TNFR superfamily are important molecules in the activation, enhancement, and/or downregulation of immune responses, especially cellular immune responses, and are characterized by the presence 2–4 cysteine rich extracellular modules. 4-1BB was originally cloned from stimulated cytolytic T lymphocyte and helper T lymphocyte clone cDNA and its expression is highly limited to lymphoid organs (Kwon *et al.*, 1989). Surface expression of 4-1BB can be detected by flow cytometry on activated CD8+ and CD4+ T cells, activated thymocytes, intraepithelial lymphocytes, activated natural killer (NK) cells, and eosinophils (Goodwin *et al.*, 1993; Melero *et al.*, 1998b; Pinto *et al.*, 1997; Zhou *et al.*, 1994). 4-1BB is not detected on the surface of resting T cells or resting/activated B cells (DeBenedette *et al.*, 1995; Garni-Wagner *et al.*, 1996; Pollok *et al.*, 1995). 4-1BB mRNA is up-regulated approximately 3 hours after T cell activation and can be detected on the surface as early as 24 hours after T cell activation (Garni-Wagner *et al.*, 1996; Pollok *et al.*, 1993). The natural ligand for 4-1BB(4-1BBL) is a type II (C terminus extracellular) membrane glycoprotein of 34 kDa that is a member of the tumor necrosis factor (TNF) superfamily which includes molecules such as OX40 ligand, CD27 ligand, CD40 ligand (CD154), and lymphotoxin α/β (Alderson *et al.*, 1994; Goodwin *et al.*, 1993). 4-1BBL can be detected on the surface of activated T cells, macrophages, monocytes, dendritic cells, B cells, and some murine B lymphomas (Goodwin *et al.*, 1993).

2. ROLE OF 4-1BB AND 4-1BBL IN THE IMMUNE RESPONSE

T cells require two signals for activation, signal one is provided by the interaction of the T cell receptor (TCR) with the appropriate MHC/peptide and signal two is

Cancer Gene Therapy: Past Achievements and Future Challenges, edited by Habib
Kluwer Academic/Plenum Publishers, New York, 2000.

provided by costimulatory pathways such as the well known B7-CD28 pathway. Interaction of naive T cell TCR with MHC/peptide complex without a costimulatory signal can lead to T cell unresponsiveness (Chen, 1998). Naive T cells are most efficiently activated and induced to proliferate by interacting with professional antigen presenting cells (APC) such as B cells, macrophages, and dendritic cells. Professional APCs are well suited to activate naive T cells because of their ability to supply both signals for T cell activation and the presence of multiple molecules involved in costimulation (e.g. B7-1, B7-2, 4-1BBL, OX40L). Many of the costimulatory molecules are upregulated after activation/maturation of the APC (i.e. crosslinking of CD40 or use of TNFα on dendritic cells).

Many *in vitro* studies have shown that 4-1BB can potently co-stimulate T cells to proliferate and to increase cytolytic activity. When T cells are stimulated with suboptimal doses of anti-CD3 mAb, which provide signal one, in the presence of anti-4-1BB mAb or 4-1BBL results in increased T cell proliferation and production of interleukin 2 (Alderson *et al.*, 1994; Pollok *et al.*, 1995; Schwarz *et al.*, 1996; Shuford *et al.*, 1997). T cells are not induced to proliferate if they are exposed only to 4-1BB stimulation without anti-CD3. Similarly, a mixed lymphocyte reaction (MLR) of mouse splenocytes could be inhibited by a soluble form of 4-1BB, although the inhibition of the MLR by soluble 4-1BB is not as potent as CTLA4Ig, a high affinity counterreceptor for B7-1 and B7-2 (DeBenedette *et al.*, 1995; DeBenedette *et al.*, 1997; Hurtado *et al.*, 1995). The costimulatory activity of 4-1BB is preferential for CD8+ T cells as supported by the observations that CD8+ T cells from mice that have been inoculated with tumor cells or allogeneic spleen cells and then treated with anti-4-1BB mAb displayed increased proliferative capacity and increased cytolytic activity (Melero *et al.*, 1997; Shuford *et al.*, 1997).

The costimulatory signal provided through 4-1BB is mediated possibly by two members of the TNF receptor associated factors (TRAF), TRAF 1 and 2, although the src family kinase $p56^{lck}$ may also play a role in 4-1BB intracellular signaling (Arch and Thompson, 1998; Jang *et al.*, 1998; Kim *et al.*, 1993; Saoulli *et al.*, 1998). The TRAF family of proteins is an adapter protein that links many members of the TNFR intracellular signaling to the jun kinase and NFκB pathways. The importance of TRAF2 association with the intracellular domain of 4-1BB was demonstrated in TRAF 2 dominant negative transgenic mice or TRAF2 deficient mice, in which there is a lack of increased IL-2 production in response to stimulation through 4-1BB (Saoulli *et al.*, 1998).

The costimulatory signal mediated by 4-1BB to T cells can act independently of CD28 signaling. *In vitro* studies using B lymphomas which do not express B7-1 or B7-2, but high levels of 4-1BBL, can induce T cells to produce increased levels of IL-2, indicating that T cells are being co-stimulated (DeBenedette *et al.*, 1997). Studies utilizing T cells from $CD28^{-/-}$ mice confirmed the results that 4-1BB, in the absence of CD28 signaling, can provide sufficient costimulatory signals to T cells (Chu *et al.*, 1997; DeBenedette *et al.*, 1997; Saoulli *et al.*, 1998). However, more detailed studies seem to indicate that although 4-1BB signal may provide a sufficient costimulatory signal to T cells, the presence of both CD28 and 4-1BB signal act synergistically to promote T cell activation and clonal expansion (DeBenedette *et al.*, 1997; Kim *et al.*, 1998). A possible model for the role of 4-1BB in T cell activation may be that CD28 signal acts earlier in T cell activation leading to the up-regulation of 4-1BB on the surface of the T cells. 4-1BB would then have a more important role in the clonal expansion of activated T cells rather than the initial costimulation of T cells.

4-1BBL interaction with 4-1BB initiates a bi-directional signal, resulting in costimulation of T cells, and the antigen-presenting cell. B cells that are incubated with anti-μ chain antibodies and soluble 4-1BB are induced to proliferate to a greater extent than control B cells (Pollok *et al.*, 1994). 4-1BBL signal can also activate monocytes and result in the production of IL-6, IL-8, TNF-α and inhibit IL-10 production (Langstein *et al.*, 1998).

4-1BBL signaling may result in the contribution of the antigen presenting cells to augment the T cell response and play a role in inflammation through the secretion of various cytokines.

These *in vitro* studies indicate that the 4-1BB/4-1BBL pathway is a potent T cell costimulatory pathway and therefore a good candidate pathway that can be manipulated to modulate the immune response in disease states such as cancer, autoimmunity, and infection.

3. MODULATION OF THE ANTITUMOR RESPONSE THROUGH 4-1BB

Tumors have been shown to express an increasing array of tumor associated antigens that are usually derived from developmentally regulated genes which are aberrantly expressed or mutated proteins. These tumor associated antigens can be recognized by CD8+ and CD4+ T cells. However, a sufficient immune response to these antigens is rarely elicited, although there are some cases of immunogenic tumors inducing immunity in standard transplantation and challenge experiments. Even tumors that may elicit an effective immune response to a defined tumor antigen do not succumb to the immune system if the inoculating dose of tumor is a few orders of magnitude larger than needed for the outgrowth of tumor in nonimmune recipients (Hellstrom *et al.*, 1995). Various factors may contribute to the ability of tumor cells to evade detection and/or destruction by the immune system. Mechanisms of immune evasion by tumors include the production of immunosuppressive cytokines such as TGFβ, IL-10, and vascular endothelial growth factor (VEGF), down-regulation of MHC class I and/or decreased antigen processing, and induction of T cell ignorance due to engagement of TCR without a costimulatory signal (Chouaib *et al.*, 1997).

4-1BB is a potent T cell costimulatory molecule that can synergize with the B7-CD28 pathway and possibly substitute for the B7-CD28 pathway in the latter stages of T cell activation. Therefore, like B7, the 4-1BB pathway is an attractive pathway to target for activating T cell responses against tumors. Previous studies have shown that the introduction of B7-1 molecule into tumor cells caused many immunogenic tumors to become rejected with immunity to wild-type tumors being established to subsequent tumor rechallenge (Chen *et al.*, 1992). Transfection of B7-1 or B7-2 gene into tumor cells results in immunogenic tumors becoming more immunogenic and non-immunogenic tumors remaining non-immunogenic (Hellstrom *et al.*, 1995). Therefore, the efficacy of B7 mediated antitumor response is limited to small, immunogenic tumors.

The ability to modulate the immune response to tumors through 4-1BB stimulation was tested *in vivo* in two murine tumor models (Melero *et al.*, 1997). One model utilized an immunogenic P815 mastocytoma that is syngeneic to DBA/2 mice. The other model was the poorly immunogenic, spontaneously arising fibrosarcoma, AG104A, which is syngeneic to C3H/HeN mice. Antibodies to 4-1BB had been generated in rats using a soluble fusion protein that contained the extracellular domain of mouse 4-1BB receptor. The antibodies were able to enhance T cell proliferation *in vitro* in the presence of suboptimal doses of anti-CD3 monoclonal antibodies (mAb). Mice were inoculated with either P815 or AG104A subcutaneously (s.c.) three days prior to two treatments with anti-4-1BB mAb intraperitoneally and assayed for tumor growth. In P815 inoculated mice, approximately 92% of the mice showed tumor regression and remained tumor free. Approximately 80% of the mice that had been inoculated with AG104A sarcoma also showed tumor regression and remained tumor free. In a widespread metastasis model, in which AG104A sarcoma was injected intravenously, 50% of the mice that had been treated with anti-4-1BB mAb were cured. Since neither AG104A nor P815 express 4-1BB on the surface, the regression of the tumors was most likely not due to direct cytotoxic effects of the anti-4-1BB mAb administration. Therefore, stimulation through 4-1BB clearly elicited an effective antitumor response to both P815 and AG104A. Previous studies had indicated that an effective antitumor response to AG104A could not be elicited even in the presence of CD28 costimulation. However, an effective immune response could be elicited to AG104A in the presence of 2 costimulatory signals, CD28 signal and CD2 signal, although this immune response was capable of protecting against only a small dose of tumor on subsequent rechallenge (Li *et al.*, 1996). However, these experiments indicated that even poorly immunogenic tumors could be effectively eradicated if there was a strong enough T cell costimulatory signal present.

We investigated whether anti-4-1BB mAb treatments could lead to effective treatment of well-established tumors (Melero *et al.*, 1997). As previously described, tumors were inoculated s.c. into mice, however, they were allowed to grow for twelve days prior to treatments with anti-4-1BB mAb. At twelve days, inoculation of both P815 and AG104A produce well-established, large tumors. When mice with the tumors were treated with anti-4-1BB mAb, 73% of the P815 tumors and 60% of AG104A tumors regressed, signifying that anti-4-1BB mAb treatments were effective against large well established tumors. The time for rejection of AG104A tumors were longer than compared to P815 due to the lower immunogenicity of AG104A compared to P815. When anti-4-1BB mAb modulated antitumor response was compared to either anti-CD28 or anti-CTLA4 mAb, the anti-4-1BB mAb modulated antitumor response proved to be more effective than either anti-CD28 or anti-CTLA4. The antitumor response induced by anti-4-1BB mAb proved to be long lasting and prevented the outgrowth of tumors in cured mice that were re-challenged with a lethal dose of the same tumors. The antitumor response in the re-challenge experiments was not due to the nonspecific activation of immune response because the antigenically different, syngeneic tumor was able to grow in cured mice.

Anti-4-1BB mAb induced tumor immunity is mediated by both CD8+ and CD4+ cells. Splenocytes from mice that had been injected with P815 and anti-4-1BB mAb proliferated four times better *in vitro* than control, with both CD4+ and CD8+ T cell proliferating. Splenocytes from mice that had been inoculated with either P815 or AG104A with subsequent treatment with anti-4-1BB mAb exhibited a 65-fold and 8-fold increase in cytolytic activity against P815 and AG104A respectively. This increased cytolytic activity is dependent on CD8+ cells because if CD8+ cells are depleted *in vivo* prior to harvest of the spleens, there is no increase in cytolytic activity against tumor cells. If either CD4+ or CD8+ cells are depleted *in vivo* prior to inoculation of tumor, no antitumor response can be detected, indicating that both CD8+ and CD4+ cells are required for the anti-4-1BB mAb modulated antitumor response. Later studies in which NK cells are depleted *in vivo* with either anti-asialo GM1 or anti-NK1.1 mAb, resulted in an ineffective antitumor response even when mice were treated with anti-4-1BB mAb (Melero *et al.*, 1998b). The dependence of the antitumor response was not due to the requirement for direct NK-mediated cytotoxicity as supported by the observation that in BALB/c nude mice that were treated with anti-4-1BB mAb there was outgrowth of P815 tumors. The role of the NK cells in the antitumor response may therefore contribute to the T cell mediated antitumor response by providing help, possibly in the form of cell-cell interactions or secreted cytokines. Therefore it seems likely that in order for anti-4-1BB mAb to modulate the antitumor response, several key components of the immune response must contribute to the antitumor response.

To further study the mechanism of 4-1BB augmented antitumor response, P815 and AG104A were transduced with the gene for 4-1BBL in a retrovirus vector (Melero *et al.*, 1998a). Transduction of 4-1BBL into the tumor cells resulted in constitutive expression of 4-1BBL. This system would mimic the roles that professional antigen presenting cells in providing both signal one and signal two on the surface of the tumor cells without the requirement for cross priming. This system, in which costimulatory molecules are transduced into tumor cells have been shown to effectively activate T cells against silent epitopes of tumor-associated antigens, as well as dominant epitopes. 92% of the 4-1BBL+ P815 that were inoculated subcutaneously into mice were rejected. However, AG104A tumors required the presence of both B7-1 and 4-1BBL in order for 60% of the tumors to be rejected. AG104A that had been transduced with only 4-1BBL or B7-1 did not elicit an effective antitumor response and were able to grow in the hosts. The antitumor effect is not due to increased sensitivity of the tumors to cytolysis mediated by transduction of 4-1BBL into genes because tumor cells that are transduced with 4-1BBL do not have increased sensitivity to cytolysis. Long term immunity was induced by the modified tumors as is supported by the observation that cured mice were protected against wild-type tumor re-challenge with 100% survival and 78% survival in the P815 and AG104A rechallenge experiments respectively. In contrast to the anti-4-1BB mAb mediated antitumor response, the tumor response to modified tumors was dependent on CD8+ cells, but not CD4+ cells. CD4+ cells, in the anti-4-1BB mAb experiments, are probably required to help in the rapid clonal expansion of CD8+ cells with the CD4+ cells being activated by the anti-4-1BB mAb treatments. No therapeutic effect was observed against pre-established wild type tumors with injection of 4-1BBL+ tumors at a different site, although there was an effective antitumor response at the site of the injection of the 4-1BBL tumors that contrast with the anti-4-1BB mAb treatment studies. The difference in efficacy of the two types of treatments is possibly due to weaker ligand—receptor interactions compared to mAb—receptor interactions. 4-1BBL transduced tumors may only provide efficient local costimulation, possibly because the transduced tumors may be unable to provide other necessary help signals to promote further expansion of tumor reactive cytolytic lymphocytes fast enough to reject tumors at distant sites, thus allowing the distant wild-type tumors to grow out. The anti-4-1BB mAb, however, provides a systemic costimulatory signal that would allow faster expansion of reactive T cells and thus rejection of distant tumors.

4. ROLE OF 4-1BB IN TRANSPLANTATION AND AUTOIMMUNITY

The role that 4-1BB plays in autoimmunity has not been extensively studied, however, one study indicates that sera of donors with rheumatoid arthritis had higher levels of soluble 4-1BB compared to sera from healthy donors (Michel *et al.*, 1998). Soluble forms of 4-1BB arise from two different alternatively spliced transcripts of 4-1BB mRNA (Setareh *et al.*, 1995) and has been detected in supernatants of phytohemagglutinin or phorbol myristate acetate/calcium ionophore stimulated peripheral blood mononuclear cells. The role of the soluble 4-1BB in the activation state of blood lymphocytes has not been completely studied, but soluble 4-1BB may act as a decoy receptor to interfere with 4-1BB/4-1BBL interactions, thereby modulating the immune response. Alternatively, soluble 4-1BB may play a role in the activation of monocytes and B cells, possibly enhancing the autoreactivity of the immune response. Further investigations may provide insight into whether or not a nonstimulatory, blocking antibody against 4-1BB can ameliorate the symptoms of some autoimmune disorders through the interruption of the 4-1BB/4-1BBL pathway.

Stimulation through 4-1BB results in faster allograft rejection (Shuford *et al.*, 1997). In a model of acute graft versus host disease (GVHD) in which C57BL/6 splenocytes were injected intravenously to allogeneic BDF_1 and then treated with anti-4-1BB mAb, spleens from anti-4-1BB mAb mice were increased in size compared to spleens of mice that had been injected with an isotype-matched control antibody. When splenocytes from GVHD mice that were treated with anti-4-1BB mAb were assayed for CTL activity, there was a four-fold increase in CTL activity compared to control and a reduction in the numbers of splenocytes. Experiments utilizing heart allograft and skin allograft models showed that transplants in mice that were treated with anti-4-1BB mAb had decreased survival, with rejection of heart allografts occurring 2–3 days earlier than control allografts. However, these studies do not give us insight into what role 4-1BB stimulation has in chronic rejection of allografts. One can envision the use of nonstimulatory blocking antibodies to downregulate chronic rejection reactions and lessen the doses of powerful anti-rejection agents to promote graft survival in transplant patients.

5. CONCLUSION

The mechanisms by which the 4-1BB/4-1BBL pathway modulates the immune response is not completely understood, but current studies clearly indicate that 4-1BB has a prominent role in the activation of the T cell mediated immune response. The modulation of the immune response by manipulating the 4-1BB/4-1BBL pathway, either to activate it against tumors, invading pathogens, or to downregulate it in autoimmune disorders holds great promise. It is clearly another molecule that can be added to the growing arsenal with which to specifically "tweak" the immune system.

REFERENCES

Alderson, M.R., Smith, C.A., Tough, T.W., Davis-Smith, T., Armitage, R.J., Falk, B., Roux, E., Baker, E., Sutherland, G.R., and Din, W.S., 1994, Molecular and biological characterization of human 4-1BB and its ligand, *Eur J Immunol* **24**(9):2219–2227.

Arch, R.H., and Thompson, C.B., 1998, 4-1BB and Ox40 are members of a tumor necrosis factor (TNF)-nerve growth factor receptor subfamily that bind TNF receptor-associated factors and activate nuclear factor kappaB, *Mol Cell Biol* **18**(1):558–565.

Chen, L., 1998, Immunological ignorance of silent antigens as an explanation of tumor evasion, *Immunol Today* **19**(1):27–30.

Chen, L., Ashe, S., Brady, W.A., Hellstrom, I., Hellstrom, K.E., Ledbetter, J.A., McGowan, P., and Linsley, P.S., 1992, Costimulation of antitumor immunity by the B7 counterreceptor for the T lymphocyte molecules CD28 and CTLA-4, *Cell* **71**(7):1093–1102.

Chouaib, S., Asselin-Paturel, C., Mami-Chouaib, F., Caignard, A., and Blay, J.Y., 1997, The host-tumor immune conflict: from immunosuppression to resistance and destruction, *Immunol Today* **18**(10):493–497.

Chu, N.R., DeBenedette, M.A., Stiernholm, B.J., Barber, B.H., and Watts, T.H., 1997, Role of IL-12 and 4-1BB ligand in cytokine production by CD28+ and CD28– T cells, *J Immunol* **158**(7):3081–3089.

DeBenedette, M.A., Chu, N.R., Pollok, K.E., Hurtado, J., Wade, W.F., Kwon, B.S., and Watts, T.H., 1995, Role of 4-1BB ligand in costimulation of T lymphocyte growth and its upregulation on M12 B lymphomas by cAMP, *J Exp Med* **181**(3):985–992.

DeBenedette, M.A., Shahinian, A., Mak, T.W., and Watts, T.H., 1997, Costimulation of CD28– T lymphocytes by 4-1BB ligand, *J Immunol* **158**(2):551–559.

Garni-Wagner, B.A., Lee, Z.H., Kim, Y.J., Wilde, C., Kang, C.Y., and Kwon, B.S., 1996, 4-1BB is expressed on CD45RAhiROhi transitional T cell in humans, *Cell Immunol* **169**(1):91–98.

Goodwin, R.G., Din, W.S., Davis-Smith, T., Anderson, D.M., Gimpel, S.D., Sato, T.A., Maliszewski, C.R., Brannan, C.I., Copeland, N.G., Jenkins, N.A., and *et al.*, 1993, Molecular cloning of a ligand for the inducible T cell gene 4-1BB: a member of an emerging family of cytokines with homology to tumor necrosis factor, *Eur J Immunol* **23**(10):2631–2641.

Hellstrom, K.E., Hellstrom, I., and Chen, L., 1995, Can co-stimulated tumor immunity be therapeutically efficacious?, *Immunol Rev* **145**:123–145.

Hurtado, J.C., Kim, S.H., Pollok, K.E., Lee, Z.H., and Kwon, B.S., 1995, Potential role of 4-1BB in T cell activation. Comparison with the costimulatory molecule CD28, *J Immunol* **155**(7):3360–3367.

Jang, I.K., Lee, Z.H., Kim, Y.J., Kim, S.H., and Kwon, B.S., 1998, Human 4-1BB (CD137) signals are mediated by TRAF2 and activate nuclear factor-kappa B, *Biochem Biophys Res Commun* **242**(3):613–620.

Kim, Y.J., Kim, S.H., Mantel, P., and Kwon, B.S., 1998, Human 4-1BB regulates CD28 co-stimulation to promote Th1 cell responses, *Eur J Immunol* **28**(3):881–890.

Kim, Y.J., Pollok, K.E., Zhou, Z., Shaw, A., Bohlen, J.B., Fraser, M., and Kwon, B.S., 1993, Novel T cell antigen 4-1BB associates with the protein tyrosine kinase p56lck1, *J Immunol* **151**(3):1255–1262.

Kwon, B.S., Kestler, D.P., Eshhar, Z., Oh, K.O., and Wakulchik, M., 1989, Expression characteristics of two potential T cell mediator genes, *Cell Immunol* **121**(2):414–422.

Langstein, J., Michel, J., Fritsche, J., Kreutz, M., Andreesen, R., and Schwarz, H., 1998, CD137 (ILA/4-1BB), a member of the TNF receptor family, induces monocyte activation via bidirectional signaling, *J Immunol* **160**(5):2488–2494.

Li, Y, Hellstrom, K.E., Newby, S.A., and Chen, L., 1996, Costimulation by CD48 and B7-1 induces immunity against poorly immunogenic tumors, *J Exp Med* **183**(2):639–644.

Melero, I., Bach, N., Hellstrom, K.E., Aruffo, A., Mittler, R.S., and Chen, L., 1998a, Amplification of tumor immunity by gene transfer of the co-stimulatory 4-1BB ligand: synergy with the CD28 co-stimulatory pathway, *Eur J Immunol* **28**(3):1116–1121.

Melero, I., Johnston, J.V., Shufford, W.W., Mittler, R.S., and Chen, L., 1998b, *Cell Immunol* ***in press***.

Melero, I., Shuford, W.W., Newby, S.A., Aruffo, A., Ledbetter, J.A., Hellstrom, K.E., Mittler, R.S., and Chen, L., 1997, Monoclonal antibodies against the 4-1BB T-cell activation molecule eradicate established tumors, *Nat Med* **3**(6):682–685.

Michel, J., Langstein, J., Hofstadter, F., and Schwarz, H., 1998, A soluble form of CD137 (ILA/4-1BB), a member of the TNF receptor family, is released by activated lymphocytes and is detectable in sera of patients with rheumatoid arthritis, *Eur J Immunol* **28**(1):290–295.

Pinto, A., Aldinucci, D., Gloghini, A., Zagonel, V., Degan, M., Perin, V., Todesco, M., De Iuliis, A., Improta, S., Sacco, C., Gattei, V., Gruss, H.J., and Carbone, A., 1997, The role of eosinophils in the pathobiology of Hodgkin's disease, *Ann Oncol* **8 Suppl 2**:89–96.

Pollok, K.E., Kim, S.H., and Kwon, B.S., 1995, Regulation of 4-1BB expression by cell-cell interactions and the cytokines, interleukin-2 and interleukin-4, *Eur J Immunol* **25**(2):488–494.

Pollok, K.E., Kim, Y.J., Hurtado, J., Zhou, Z., Kim, K.K., and Kwon, B.S., 1994, 4-1BB T-cell antigen binds to mature B cells and macrophages, and costimulates anti-mu-primed splenic B cells, *Eur J Immunol* **24**(2):367–374.

Pollok, K.E., Kim, Y.J., Zhou, Z., Hurtado, J., Kim, K.K., Pickard, R.T., and Kwon, B.S., 1993, Inducible T cell antigen 4-1BB. Analysis of expression and function, *J Immunol* **150**(3):771–781.

Saoulli, K., Lee, S.Y., Cannons, J.L., Yeh, W.C., Santana, A., Goldstein, M.D., Bangia, N., DeBenedette, M.A., Mak, T.W., Choi, Y., and Watts, T.H., 1998, CD28-independent, TRAF2-dependent costimulation of resting T cells by 4-1BB ligand, *J Exp Med* **187**(11):1849–1862.

Schwarz, H., Blanco, F.J., von Kempis, J., Valbracht, J., and Lotz, M., 1996, ILA, a member of the human nerve growth factor/tumor necrosis factor receptor family, regulates T-lymphocyte proliferation and survival, *Blood* **87**(7):2839–2845.

Setareh, M., Schwarz, H., and Lotz, M., 1995, A mRNA variant encoding a soluble form of 4-1BB, a member of the murine NGF/TNF receptor family, *Gene* **164**(2):311–315.

Shuford, W.W., Klussman, K., Tritchler, D.D., Loo, D.T., Chalupny, J., Siadak, A.W., Brown, T.J., Emswiler, J., Raecho, H., Larsen, C.P., Pearson, T.C., Ledbetter, J.A., Aruffo, A., and Mittler, R.S., 1997, 4-1BB costimulatory signals preferentially induce CD8+ T cell proliferation and lead to the amplification in vivo of cytotoxic T cell responses, *J Exp Med* **186**(1):47–55.

Zhou, Z., Pollok, K.E., Kim, K.K., Kim, Y.J., and Kwon, B.S., 1994, Functional analysis of T-cell antigen 4-1BB in activated intestinal intra-epithelial T lymphocytes, *Immunol Lett* **41**(2–3):177–184.

31

HEAT SHOCK PROTEINS IN CANCER THERAPY

Katalin V. Lukacs[1,2], Olivier E. Pardo[1], M. Jo Colston[3], Duncan M. Geddes[1], and Eric WFW Alton[1]

National Heart and Lung Institute
Imperial College School of Medicine
Manresa Rd., London SW3 6LR[1]
Chester Beatty Laboratories
Institute of Cancer Research
Fulham Rd, London SW3 6JB[2] and
National Institute for Medical Research
The Ridgeway, Mill Hill, NW7 1AA
London[3]

1. HEAT SHOCK PROTEINS

Heat shock proteins (hsps) are ubiquitous in living organisms. Their very high level of interspecies sequence homology underline the essential role hsps play as molecular chaperones in normal and pathological processes. As first observed in Drosophila, heat shock proteins are induced by a sudden increase in temperature (Ritossa, 1962). Since several of these proteins respond to a wider array of stressful conditions (oxidative damage, irradiation, metabolic toxins, starvation), hsps are also called stress proteins. Hsps are grouped into families according to their molecular weight. Hsp90s are the most abundant hsps and are expressed in a constitutive manner. Hsp90 binds to transcription factors which regulate the expression of hsps (Nadeau, Das, and Walsh, 1993) and has important roles in signal transduction, cell cycle regulation and development (Pratt and Toft, 1997; Rutherford and Lindquist, 1998). Hsp70s have transient peptide binding properties reversed by ATP (Bukau and Horwich, 1998). Under physiological conditions, Hsp70s assist folding and intracellular transport of newly synthesised polypeptides (Cyr and Neupert, 1996). Under stress conditions, induction of Hsp70 and subsequent binding, limits protein denaturation and aggregation (Hainaut and Milner, 1992; Yiping Yang *et al.*, 1993). Similarly to the Hsp70s, the Hsp60 chaperonin family is crucial for protein folding, transport and oligomere subunit assembly (Hemmingsen *et al.*, 1988; Agsteribbe *et al.*, 1993; Bukau and Horwich, 1998). Bacterial members of the Hsp60 family, including the E coli GroEL and the mycobacterial hsp65, are highly conserved with over 90% amino acid

Cancer Gene Therapy: Past Achievements and Future Challenges, edited by Habib
Kluwer Academic / Plenum Publishers, New York, 2000.

Table 1. Tumorigenicity of hsp65-transfected (J774-hsp65) and control J774 reticulum sarcoma cells in Balb/c mice

Injected tumour cells	Dose	Tumour incidence
J774	10^6	9/10
J774	10^6	5/5
J774-vector	10^6	14/15
J774-hsp65	10^6	0/10
J774-hsp65	10^6	1/10

Balb/c mice were injected with various number of tumour cells intraperitoneally. Tumour cells were either the parent cell line (J774), or the parent cell line transfected with the mycobacterial hsp65 gene (J774-hsp65) or the vector alone (J774-vector). Tumour incidence was determined by autopsy and histological examination 21 days after injection of the cells.

homology, and are amongst the most immunogenic molecules known (Kaufmann, 1990).

2. GENE THERAPY OF CANCER WITH HEAT SHOCK PROTEIN GENES

2.1. Transfer of Hsp65 Gene Protects Against J774 Histiocytic Sarcoma

Studying the immune response to mycobacterial hsps, we serendipitously found that hsp65 gene transfer has a profound anti-tumour effect (Lukacs *et al.*, 1993). Transfection of J774 histiocytic sarcoma cells with the hsp65 gene resulted in a loss of tumorigenicity (Table 1) and immunisation of mice with hsp65-transfected sarcoma cells protected them against a challenge with non-modified parent tumour cells.

More importantly, treating mice with advanced tumours with hsp65 plasmid DNA complexed to liposomes resulted in tumour regression (Fig. 1). J774 reticulum sarcoma is a tumour vith aggressive in vivo growth characteristics (2×10^6 tumour cells injected intraperitoneally cause terminal tumours by day 22–26). Rapid regression of advanced tumours suggests that hsp65 can be en effective treatment in late cancer. After hsp65 gene transfer, tumour-specific antibodies have been found in the sera of hsp65-treated animals (Fig. 2) and tumour specific T cells capable of lysing both hsp65-transfected and untransfected parent tumour cells can be detected (Lukacs *et al.*, 1996; Lukacs *et al.*, 1997). These results confirm increased recognition of tumour-specific antigenic structures after hsp65 gene trasfer.

2.2. Gene Therapy in Malignant Mesothelioma with Hsp65 Gene

Malignant mesothelioma is a fatal tumour of the pleural cavity with occasional localisation to the peritoneum. Most cases are linked to asbestos exposure (Stanton and Wrench, 1972). The lack of effective treatment, as well as a relatively easy route for gene delivery, make malignant mesothelioma a good candidate for gene therapy. Since radio- and chemotherapy are not effective, patients with malignant mesothelioma generally receive no immunosuppressive treatment and could respond well to immunotherapy delivered by hsp65 gene transfer. The feasibility of this approach has been tested in various pre-clinical models of malignant mesothelioma.

Malignant mesothelioma was induced in BALB/c and CBA mice by intraperitoneal injection of 10^6 AB22 or AC29 asbestos-induced syngenic tumour cells, respectively. Mice

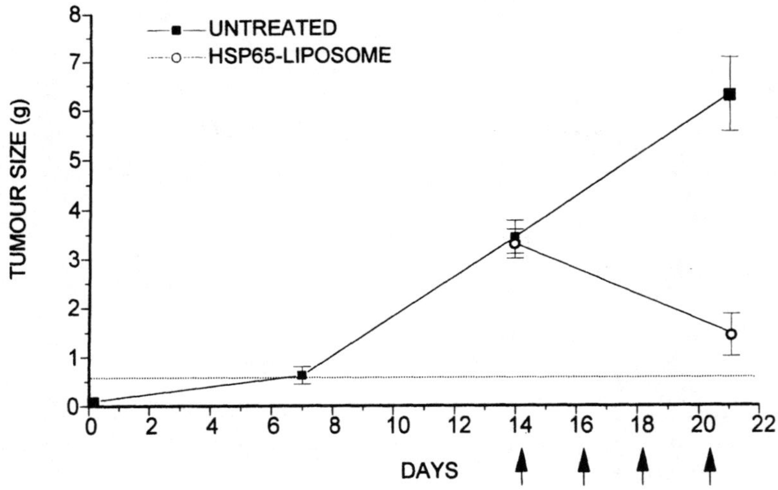

Figure 1. Effect of *in vivo* hsp65 gene therapy on advanced tumors. Balb/c mice were injected with 2×10^6 J774 cells ip. Fourteen days after tumour initiation 10 mice received intraperitoneal injections of hsp65 DNA-liposome complexes repeated at 2 day intervals on four occasions (arrows). On day 21 mice were sacrificed, and tumour sizes were compared to 10 untreated controls (mean ± S.E.).

were treated with intraperitoneal injections of hsp65 plasmid complexed with DC-Chol liposome. DNA was mixed with DC-Chol at a 1:5 weight/weight ratio. Such complexes showed a transfection efficiency of about 5–10% of malignant mesothelioma cells in vitro. Injections of tumour bearing mice with the hsp65-liposome complexes into the peritoneal cavity resulted in increased survival. A single Hsp65 treatment increased average survival from 16.8 ± 0.2 days to 102.5 ± 36.5 days ($p < 0.02$). Repeated injections at one week intervals further increased long term survival (Table 2). In mice surviving one year after

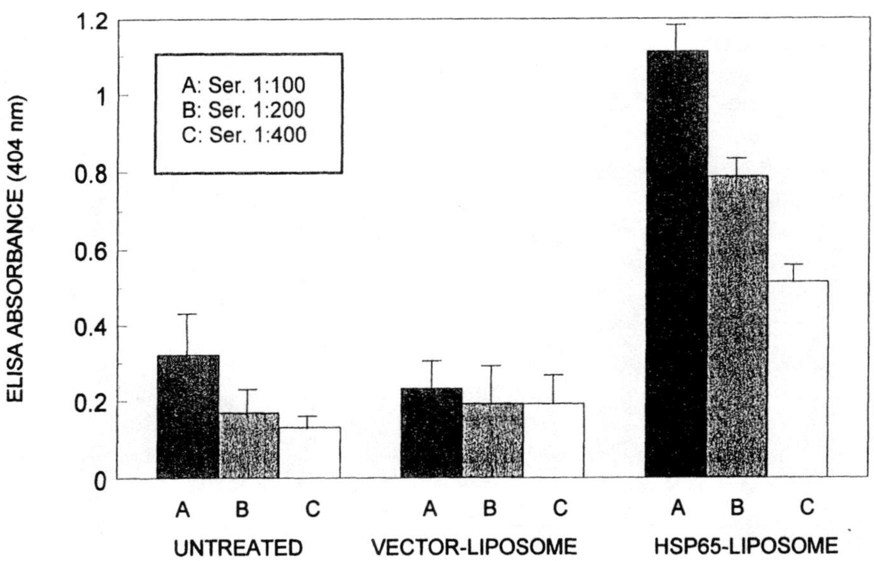

Figure 2. Detection of tumour specific antibodies in the sera of mice after in vivo hsp65 gene therapy. Serum was collected from tumour-bearing Balb/c mice treated with weekly injections of hsp65 gene-liposomes, vector-liposomes or untreated controls started on day 7 and repeated four times. Serum samples in each group were pooled and various dilutions were tested on lysates of nonmodified J774 cells in an ELISA assay.

Table 2. Effect of hsp65-liposome treatment on survival of mice with malignant mesothelioma

	Number of surviving animals 1 year after tumour initiation	
Number of injections	CBA	Balb/c
1 (day 7)	0/7	3/7
2 (day 7, 14)	6/7	ND
3 (day 7, 14, 21)	6/7	47
4 (day 7, 14, 21, 28)	7/7	6/7

Mice were injected with 1 million syngenic malignant mesothelioma cells ip and treated with weekly ip injections of 50 μg hsp65 DNA complexed with 250 μg DC-Chol liposome. Treatment started on day 7.

hsp65 treatment, no tumour was found at autopsy and histological examinations indicating complete eradication of tumours.

Animals treated with a control bacterial gene coding for β-galactosidase, also showed a significant increase in survival (29.7 + 2.6 days) ($p < 0.01$), but none of these animals became tumour free. As β-galactosidase is a bacterial "foreign" antigen, elimination of β-gal transfected tumour cells is not surprising and has been demonstrated (Abina *et al.*, 1996; Lukacs *et al.*, submitted). The immunostimulatory effect of unmethylated CpG motives in bacterial DNA is also likely to contribute to increased survival of both the β-gal-treated and hsp65-treated mice (Roman *et al.*, 1997). Although the development of terminal tumours in β-gal treated mice was delayed, it was not prevented, indicating that the above mechanisms by themselves are not sufficient to eliminate malignant mesothelioma. The ability of the chaperonin Hsp65 to enhance the presentation of tumour specific antigens leading to long term, systemic anti-tumour immune response seems to be crucial for eradication of the tumour in hsp65-treated animals.

2.3. Transfer of the Hsp70 Gene Increases Immunogenicity of Tumours

Recent studies by Melcher and colleagues (1998) suggest an important role of hsp70 in immunogenicity of tumours. They have found that tumour cells transfected with the hsp70 gene are less tumorigenic and are capable of inducing protection against non-modified parent tumour cells. They have also demonstrated that induction of non-apoptotic tumour cell death by HVS-TK/gancyclovir treatment leads to increased expression of hsp70 associated with increased immunogenicity of B16 melanoma. They concluded that tumour immunogenicity is determined by the mechanism of tumour cell death via induction of hsp expression. Increased heat shock protein expression acting as a "danger signal" (Fuchs and Matzinger, 1996) could contribute to break the tolerance to tumour antigens.

3. PROTECTION AGAINST TUMOURS WITH TUMOUR-DERIVED HSPS AND HEAT SHOCK FUSION PROTEINS

The ability of hsps to induce an anti-tumour immune response has also been detected by others using different methods. Srivastava and his colleagues (Suto and Srivastava, 1995; Blachere *et al.*, 1997) found that hsps purified from tumour cells, are associated with a broad range of peptides derived from that particular cell. Hsp-peptide complexes elicit potent T cell responses against the chaperoned peptides. They have shown that immunotherapy of tumour-bearing mice with hsp fractions obtained from autologous tumours resulted in retarded progression of the primary cancer and reduced metastatic load in Lewis lung carcinoma, B16 melanoma and Meth A fibrosarcoma (Yasuaki Tamura *et al.*, 1997). Tumour cells were killed by tumour specific cytotoxic T cells, which were induced because peptides non-covalently associated with the isolated hsps could enter the MHC class I pathway of professional antigen-presenting cells (Suto and Srivastava, 1995). Young and his colleagues obtained similar results with heat shock fusion proteins (Suzue *et al.*, 1997). They showed that a model antigen covalently bound to the mycobacterial hsp70 (ova-hsp70) was successfully delivered to the class I antigen presenting pathway and induced antigen-specific CTL response leading to tumour cell lysis.

4. FUTURE OF HSPS IN CANCER THERAPY

Results originating from various laboratories using entirely different experimental approaches indicate that hsp are powerful inducers of anti-tumour immune responses. We have found that gene therapy with the hsp65 gene is highly effective in various malignant tumours. The use of heterologous hsps (eg. mycobacterial hsp65) add the antigenicity of a foreign protein to the chaperone property of the hsp leading to induction of T cells capable of eliminating both transfected and non-modified tumour cells. Repeated injections of plasmid-liposome preparations coding for the hsp65 gene are well tolerated and achieve one year survival in over 90% of tumour-bearing mice in various pre-clinical models. Hsp65 gene therapy does not require high level gene expression; relatively few hsp65-transfected tumour cells are capable of inducing an effective anti-tumour immune response. Taking into account the limitations of current gene delivery technologies, this is a clear advantage. Safety issues concerning the possible induction of autoimmune diseases cannot be discarded at this stage. However, it is worth noting that none of the mice treated with Hsp65 in our laboratory showed signs of autoimmune disease in long term survival studies and studies from others confirm that hsp65 can be beneficial in preventing autoimmune arthritis (Van den Broek *et al.*, 1989; Yang and Feige, 1992; Wendling and Farine, 1998).

The alternative approach of antitumour vaccines using tumour-derived hsps (Yasuaki Tamura *et al.*, 1997) is also promising but requires individual preparations from the patients own tumours. Heat shock fusion proteins with tumour antigens may also provide an effective cancer vaccine (Suzue *et al.*, 1997), although the availability of defined tumour specific peptides could prove to be a technical limitation. The simplicity and high effectiveness of the hsp65 gene therapy approach makes it a promising candidate. Only the results of clinical trials will decide which of the three approaches will be better tolerated and more effective in patients with cancer. These studies may show that with the help of heat shock proteins tolerance to malignant tumours can be broken.

REFERENCES

Abina, M.A., Lee, M.G., Descamps, V., Cordier, L., Lopez, M., Perricaudet, M., and Haddada, H., 1996, LacZ gene transfer into tumour cells abrogates tumorigenicity and protects mice against the development of further tumors, *Gene Therapy*, **3**:212–216.

Agsteribbe, E., Huckriebe, A., Veenhuis, M., Ruiters, M.H., Niezen-Koning, K.E., Skjeldal, O.H., Skullerud, K., Gupta, R.S., Hallberg, R., van Diggelen, O.P., *et al.*, 1993, A fatal, systemic mitochondrial disease with decreased mitochondrial enzyme activities, abnormal ultrastructure of the mitochondria and deficiency of the heat shock protein 60, *Biochem. Biophy. Res. Comm.* **193**:146–154.

Blachere, N.E., Li, Z., Chandavarkar, R.Y., Suto, R., Jaikaria, N.S., Basu, S., Udono, H., and Srivastava, P.K., 1997, Heat shock protein-peptide complexes, reconstituted in vitro, elicit peptide-specific cytotoxic T lymphocyte response and tumor immunity, *J. Exp. Med.* **186**:1315–1322.

Bohen, S.P., Kralli, A., and Yamamoto, K.R., 1995, Hold'em and fold'em: chaperones and signal transduction, *Science*, **268**:1303–1304.

Van den Broek, M., Hogervorst, E.J.M., van Bruggen, M.C.J., van Eden, W., van der Zee, R., and van der Berg, W.B., 1989, Protection against streptococcal wall-induced arthritis by pretreatment with the 65-KD mycobacterial heat shock protein, *J. Exp. Med.* **170**:449–445.

Bukau, B., and Horwich, A.L., 1998, The Hsp70 and Hsp60 chaperone machines, *Cell.* **92**(3):351–366.

Cyr, D.M., and Neupert, W., 1996, Roles of Hsp70 in protein translocation across membranes of organelles, *EXS*, **77**:25–40.

Fuchs, E.J., and Matzinger, P., 1996, Is cancer dangerous to the immune system? *Sem. Immunol.* **8**:271–280.

Hainaut, P., and Milner, J., 1992, Interaction of heat-shock protein 70 with p53 translated *in vitro*: evidence for interaction with dimeric p53 and for a role in the regulation of p53 conformation, *The EMBO Journal*, **11**:3513–3520.

Hemmingsen, S.M., Woolford, C., van der Vies, S.M., Tilly, K., Dennis, D.T., Georgopoulos, C.P., Hendrix, R.W., and Ellis, R.J., 1988, Homologous plant and bacterial proteins chaperone oligomeric protein assembly, *Nature*, **333**:330–334.

Kauffmann, S.H.E., 1990, Heat shock proteins and the immune response, *Immunol. Today*, **11**:22–27.

Lukacs, K.V., Lowrie, D.B., Stokes, R.W., and Colston, M.J., 1993, Tumor cells transfected with a bacterial heat-shock gene lose tumorigenicity and induce protection against tumors, *J.Exp. Med.* **178**:343–348.

Lukacs, K.V., Lowrie, D.B., and Colston, M.J., 1996, Protection against tumors by stress protein gene transfer, in: *Stress Proteins in Medicine*, (W. van Eden and D.B. Young, eds), pp. 249–265, Marcel Dekker Inc., New York.

Lukacs, L.V., Nakakes, C.J., Lowrie, D.B., and Colston, M.J., 1997, In vivo gene therapy of malignant tumours with heat shock protein-65 gene, *Gene Therapy*, **4**:346–350.

Melcher, A., Todryk, S., Hardwick, N., Ford, M., Jacobson, M., and Vile, R.G., 1998, Tumor immunogenicity is determined by the mechanism of cell death via induction of heat shock protein expression, *Nature Medicine*, **4**:581–587.

Pratt, W.B., and Toft, D.O., 1997, Steroid receptor interactions with heat shock protein and immunophilin chaperones, *Endocrine Reviews*, **18**:306–360.

Ritossa, F.M., 1962, A new puffing pattern induced by a temperature shock and DNP in Drosophila. *Experientia* **18**:571–573.

Roman, M., Martin-Orozco, E., Goodman, J.S., Minh-Duc Nguyen, Yukio Sato, Ronaghi, A., Kornbluth, R.S., Richman, D.D., Carso, D.A., and Raz, E., 1997, Immunostimulatory DNA sequences function as T helper-1-promoting adjuvants, *Nature Medicine*, **3**(8):849–854.

Rutherford, S.L., and Lindquist, S., 1998, Hsp90 as a capacitor for morphological evolution, *Nature*, **396**:336–342.

Stanton, M.F., and Wrench, C., 1972, Mechanism of mesothelioma induction with asbestos and fibrous glass, *J. Natl. Cancer Inst.* **48**:797–821.

Suto, R., and Srivastava, P.K., 1995, A mechanism for the specific immunogenicity of heat shock protein-chaperoned peptides, *Science*, **269**:1585–1588.

Suzue *et al.*, 1997, Heat shock fusion proteins as vehicles for antigen delivery into the Major Histocompatibility Complex class I presentation pathway, *Proc. Natl. Acad. Sci.* **94**:13146–13151.

Yang, X.D., and Feige, U., 1992, Heat shock proteins in autoimmune disease. From causative antigen to specific therapy?, *Experimentia*, **48**(7):650–656.

Yasuaki Tamura, Ping Peng, Kang Liu, Daou, M., and Srivastava, P.K., 1997, Immunotherapy of tumors with autologous tumor-derived heat shock protein preparations, *Science*, **278**:117–120.

Yiping Yang, Janich, S., Cohn, J.A., and Wilson, J.M., 1993, The recombinant variant of cystic fibrosis transmembrane conductance regulator is recognized by Hsp70 and degraded in a pre-Golgi nonlysosomal compartment, *Proc. Natl. Acad. Sci. USA*, **90**:9480–9484.

Wendling, U., and Farine, J.C., 1998, Oral administration of Hsp-containing E. coli extract OM-89 has suppressive effects in autoimmunity. Regulation of autoimmune processes by modulating peripheral immunity towards Hsp's? *Biotherapy*, **10**(3):223–227.

BI-SPECIFIC ANTIBODIES IN CANCER THERAPY

Hao Wang[1,2], Yanjun Liu[1,2], Lixin Wei[1], and Yajun Guo[1,2]

[1]Shanghai International Joint Cancer Institute
The Second Military Medical University
Shanghai, P.R. China
[2]Sidney Kimmel Cancer Center
San Diego, USA

1. PREFACE: NEW MAGIC BULLETS

In the 1980s monoclonal antibodies promised an era of "magic" bullet therapies that could specifically kill tumor cells while leaving healthy tissues intact. But the need to covalently couple the killing agent (toxin, radioisotopes, etc.) limited their clinical potential. Now, new bispecific antibodies (BsAbs, or bi-functional antibodies) have been developed. The prototypic BsAb consists of two different specific binding sites, with one directed to the target antigen on the surface of tumor cells, and the other to a trigger molecule.

2. SCHEME OF BISPECIFIC ANTIBODIES

In the early days of this field, BsAbs were prepared by chemical cross-linking of (mainly Fab fragments of) IgG molecules with different specificities (see Fig. 1) (Brennan *et al.*, 1985; Nolan and O'Kennedy, 1990), or by fusion of two established hybridoma clones secreting antibodies of different specificties, yielding a quadroma (or hybrid hybridoma, tetradroma) (Milstein and Cuello, 1983), or by double transfectomas. However, Chemical modification of antibodies is inefficient and can lead to side reactions that damage the combining sites. In a quadroma it is difficult to separate the bispecific antibodies form the mixed population of heavy and light chains produced by the two parent hybridoma genomes, and even a large variety of mismatch (H/L chains) molecules may emerge. So, large-scale production of BsAb molecules with these methods has always result in batch-to batch variations and low yields. In addition, the rodent-derived Ig molecules usually have to be "humanized" to minimize host antibody response (human anti-mouse antibody response, HAMA) before applied in clinics.

Cancer Gene Therapy: Past Achievements and Future Challenges, edited by Habib
Kluwer Academic/Plenum Publishers, New York, 2000.

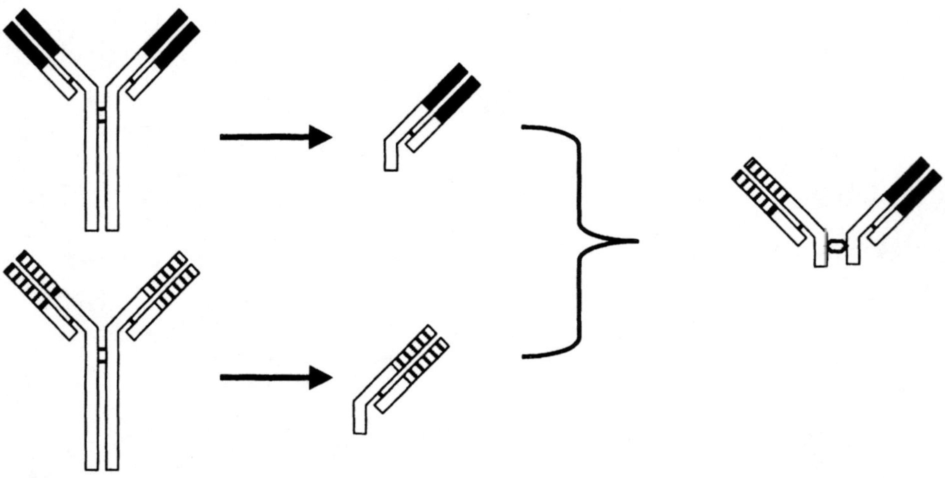

Figure 1. Chemical linked bi-specific antibody F(ab′)$_2$. The monospecific antiobdies are first digested and reduced, yied Fab′ fragments, and are then crosslinked by chemical method.

To address these problems, efforts have been made in improving purification methods for hybrid hybridomas (Allard *et al.*, 1993), for example, selective crosslinking chemistries for bispecific Fab$_2$ (Brennan *et al.*, 1985; Glennie *et al.*, 1987; Carter *et al.*, 1992) and the use of amphipatic helices (leusine-zipper) to drive association (Kostelny *et al.*, 1992), etc. Recently, protein engineering has been adapted to design BsAbs with new formats, in which only the variable domains (Fv) of the antibodies were retained and linked by artificial linkers. These recombinant bispecific Fv fragments include bispecific single chain variable fragment of Ab (BsscFvs) (see Fig. 2) (George *et al.*, 1994; Gruber *et al.*, 1994; Mallender and Voss, 1994; De Jonge *et al.*, 1995; Kurucz *et al.*, 1995; Mack *et al.*, 1995) and diabody (Holliger *et al.*, 1993; Holliger and Winter, 1997a). They could retain the parental specificity, offer several advantages over intact Ab molecules for therapeutic use. While the BsscFv is built up by a single polypeptide chain, the diabody is formed by two polypeptide chains. The smaller Ab molecules have a higher penetrating capacity into solid tumor tissues (Yokota *et al.*, 1992) and a lower immunogenicity due to the lack of the Fc domain.

Diabodies with the antigen binding sites for tumor cell and T cell surface antigens, appear to be as effective at recruiting cytotoxic T cells as are bispecific BsscFv and Fab′$_2$ *in vitro* (see Fig. 3) (Holliger *et al.*, 1996). However, the diabodies have no Fc domain. These kinds of engineered molecules lack the important property of natural antibodies—recruiting the biological effectors of immune response by Fc *in vivo*. Recently, *G Winter's*

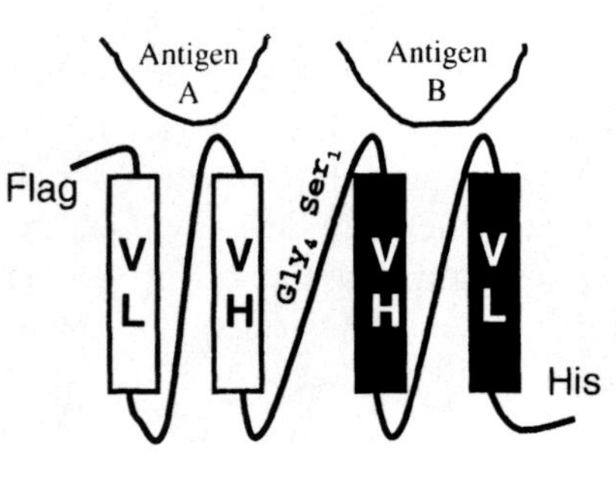

Figure 2. Scheme of the bispecific single-chain Ab. Two scFv fragments were joined by a five amino acids linker. An N-terminal flag-tag was add for easy detection, and a C-terminal histidine tail (His) for protein purification (Mack *et al.*, 1997).

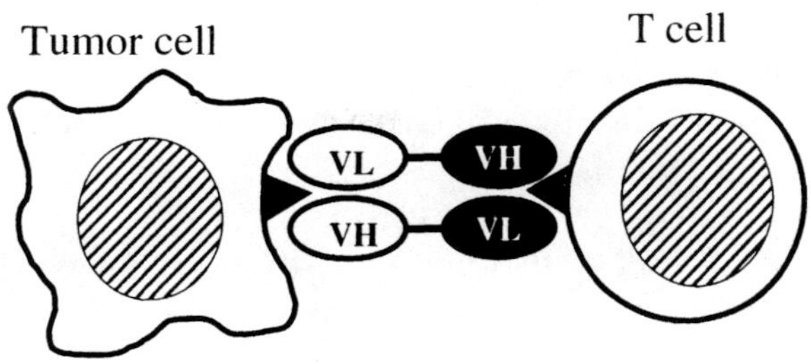

Figure 3. Scheme of bispecific diabody. The linkers are too short to allow pairing between the two domains on the same chain.

group addressed this deficiency by designing diabodies binding specificities for both antigen and an immunological effector protein (Fearon, 1997; Kontermann *et al.*, 1997; Holliger *et al.*, 1997b). For example, the simplest way is to recruit a Fc domain by binding one arm of the diabody to immunoglobulin (on conserve epitopes) itself, making the serum Ig as an intermediate molecular. For the same reason, the binding specificities of the diabodies also can be designed directly to the effector molecules with which Ig Fc interacts.

Since 1990s, efforts were also made for minimizing the immunogenicity of antibodies with the format more resembling the natural human Ab. While the human derived hybridoma manifested unreliable, genetic engineering addressed some of the limitations. It is possible to replace most of the rodent-derived sequence of an antibody with sequences derived from human immunoglobulins without loss of function. The chimeric antibodies, which were combined by the human Ig constant region and rodent Ig variable domain, is the easiest to construct, preserving avidity of the original mAbs. A great portion of the clinical trials performed in the past few years have used this format, although the HAMA response couldn't be eliminated. To further reduce the mouse sequence content of the chimeric antibody, the CDR-grafted antibody had been developed. In this kind of recombinant proteins only the complement-determinant-region (or hypervariable region) of the mouse Ig was retained and grafted to the human Ig framework. However, the process of grafting is tedious and the avidity of the recombinant Ig is sometimes unpredictable (see review by Hayden *et al.*, 1997).

Recently Jakobovits's group (Mendez, 1997) in Abgenis, Inc. in California, USA, reported genome grafting of human Ig loci to the mice, endowing the mice with human antibody response. These mice a likely to be a valuable tool for the generation of completely humanized therapeutic antibodies, even anti-human antigen antibodies.

These smaller Ab fragments can be expressed in bacteria, while the eukaryotic expression systems are necessary for recombinant $F(ab')_2$ or humanized antibodies. The progresses in the protein-engineering field also make it possible to produce sufficient amount of Bs Abs for clinic use.

3. EFFECTOR MOLECULES FOR BsAb THERAPY

The ideal BsAbs combines reactivity with a molecule found exclusively on the surface of a cancer cell and a killing agent. The killing agent can be a toxin or a chelated radioisotope (Nolan *et al.*, 1990; Barbet *et al.*, 1998), but recently major attention was given to the ways to harness optimally the effector capacity of lymphocytes or phagocytes, through engagement of T- or natural killer (NK) cell activation molecules, Ig Fc receptors (FcRs) or complement receptors. These approaches not only provide a physical bridge between the effector cells and tumor cells, but also provide the effector cells primary or costimulating signals for their activation. However, it is not yet clear which effector cell will be the most effective, and this may depend on the type of tumor, and the ultimate aim.

3.1. T Cell Activating Molecular and Costimulating Molecular

A lot of BsAb therapeutic approaches focus on T cell receptor (TCR)/CD3 complex as trigger molecules. Anti-CD3×anti-tumor BsAb have been used to redirect cytotoxic T cells to tumor cells in both MHC-restricted and unrestricted fashion (Weiner *et al.*, 1994; Katayose *et al.*, 1996; Mack *et al.*, 1997; Link *et al.*, 1998).

However, crosslinking of CD3 on T cells with the anti-tumor×anti-CD3 BsAb alone may lead to T-cell unresponsiveness or even apoptosis. The importance of costimulation for optimal T cell activation was also emphasized (Mazzoni *et al.*, 1996; De Jonge *et al.*, 1998). CD28 is the major costimulating factor of naïve T cell, so a lot of immunotherapeutic approaches for treatment of cancer are tested clinically by using the combination of anti-tumor×anti-CD28 BsAb and anti-tumor×anti-CD3 BsAb (Bohlen *et al.*, 1993, 1997; Demanet *et al.*, 1996; Kroesen *et al.*, 1995).

In a SCID mouse xenografted human Hodgkin's lymphoma model, application of anti-tumor×anti-CD3 and anti-tumor×anti-CD28 BsAbs with the patients' T cells could cure the disseminated tumor that were established by injection of tumor cells 2 week before. Treatment results with isolated $CD4^+$ and $CD8^+$ human T cells suggest that both subsets are necessary for the BsAb mediated cure of *in vivo* (Renner *et al.*, 1994; Renner *et al.*, 1996). Whereas both the $CD4^+$ and $CD8^+$ T cell subsets were involved in the growth inhibition, only the $CD8^+$ sub-population accounted for the cytotoxic activity (Mazzoni *et al.*, 1996).

However, several reports found that the CD28 costimulation, although prompt the growth rate of T cells, did not lead to further increase in cytotoxicity of the CTLs (Mack *et al.*, 1997; Katayose *et al.*, 1996). This discrepancy might be due to the difference of the tumor models used in these studies.

Although the anti-CD3 BsAbs proved effective therapeutic effect in mouse xenografted tumor models, clinical trials with anti-CD3 based BsAbs administered systemically resulted in considerable toxic effects. It has been reported that Abs preincubated with lymphocytes *in vitro* and then injected I.V., or given intraperitoneally into the malignant ascites together with the lymphocytes had shown little side effects (Kroesen *et al.*, 1993; Kroesen *et al.*, 1997; Bolhuis *et al.*, 1992; Nitta *et al.*, 1990; Lamers *et al.*, 1997). The systemic administration of these mAbs often resulted in activation of T cells with subsequent release of cytokines such as IFN-γ and TNF-α (Tibben *et al.*, 1993; Kroesen *et al.*, 1994; Tibben *et al.*, 1996), though several well tolerated experience were also reported (De Gast, 1995; De Gast, 1995b). Also, this may be attributed to contamination of bispecific $F(ab')_2$ fragments with intact Abs present after incomplete proteolytic digestion.

Gerstmayer *et al.* (1997) constructed the B7-Ab chimeric protein, which could also direct to the CD28 on the T cell. However, B7 can also conduct a negative signal to the activated T cell by direct to the CTLA-4 molecular on it (Chambers *et al.*, 1996), anti-CD28 BsAbs might have some advantage over this kind of protein.

Other T cell costimulating molecules or adhesion molecules, such as 4-1BB, ICAM-1, VACM-1, LFA-1, CD2, CD5 etc, also can be the candidates of BsAb therapy (Dietsch *et al.*, 1994; Ferrini *et al.*, 1994; Melero *et al.*, 1997).

T cells involved here usually need pre-activation before engaged with the anti-CD3×anti-tumor BsAbs. Some non-specific T cell stimulators, such as staphylococcal enterotoxin B superantigen (SEB), IL-2, IL-12, etc, synergy with these kinds BsAbs (Porter *et al.*, 1997).

3.2. Fc Receptors

The IgG Fc receptors (FcγRs) are expressed primarily on immune effector cells. They are now thought to act as trigger molecules for inflammatory, cytolytic, allergic (hypersensitivity), endocytic and phagocytic activities of immune effector cells (Deo *et al.*, 1997). There are three classes of FcγRs: FcγRI (CD64), FcγRII (CD32) and FcγRIII (CD16), which were expressed by most hematopoietic cells. Their expression can be enhanced by certain inflammatory cytokines such as interferon γ (IFN-γ) and granulocyte colony-stimulating factor (G-CSF). CD64, i.e., is not expressed by neutrophils (PMN) under physical conditions but can be induced by IFN-γ and G-CSF. Tumor specific mAbs can mediate destruction of tumor cells by phagocytosis or ADCC induced via binding to FcγRs. Unlike the TCR/CD3 does in T cells, FcγRs mediate MHC-unrestricted cytotoxicity in the effector cells.

FcγRI and FcγRIII are of particular interest for BsAbs targeting (Deo *et al.*, 1997; van de Winkel *et al.*, 1997). FcRI is expressed on cytotoxic effector cells such as monocytes, macrophages, dendritic cells, and is inducible by IFN-γ and G-CSF on neutrophils (polymorphonucler leukocytes, PMN). It can be most effectively triggered to induce ADCC, phagocytosis and other effector functions by BsAbs that can bind outside the Fc ligand-binding domain. Several such BsAbs have been developed, that have one arm specific for FcγRI and another arm specific for a tumor marker (Elsässer *et al.*, 1996; Goldstein *et al.*, 1997).

Heijnen *et al.* (1997) recently described the capacity of anti-human-CD64 BsAbs to trigger cytotoxic activity of neutrophils (PMN) during G-CSF treatment in the tumor bearing human CD64 transgenic mice, which could effectively provide these phagocytic effector cells tumor specificity. Their results also indicated that under physiologic conditions, serum IgG impairs activation of FcγRI-mediated cytotoxicity by conventional anti-tumor Abs. The IgG blockade can be circumvented with anti-FcγRI BsAbs.

FcγRIII (CD16) is an important Fc receptor triggering ADCC by NK cells and it is also functional on mononuclear phagocytes (van de Winkel *et al.*, 1997). Anti-CD16×anti-tumor BsAbs have shown effective killing of tumor cells *in vivo* (Hombach *et al.*, 1993).

A novel trigger FcR for BsAb-mediated immunotherapy, FcαRI (CD89) was introduced recently. This IgA receptor was expressed on monocytes, macrophages, neutrophils and eosinophils while neutrophils seems to be most important effector cell for FcαRI dependent tumor cell killing (Valerius *et al.*, 1997a).

4. TARGET ANTIGENS

Hematopoietic malignancies usually express the differentiation antigens (CDs) of their normal counterparts. These antigens, sometimes inducible, are commonly selected to be the BsAbs based therapeutic target. Although the parallel cytotoxicity to the normal blood cells, e.g., B cells in B cell lymphoma, could also take place, the new cells could be made by the stem cells in a few weeks. CD19 (Haagen *et al.*, 1994; de Gast *et al.*, 1995a,

1995b; Haagen *et al.*, 1995; Bohlen *et al.*, 1997; Manzke *et al.*, 1997; Kipriyanov *et al.*, 1998), CD20 (Kaminski *et al.*, 1996), CD37 (Ely *et al.*, 1996), MHC-class II (Valerius *et al.*, 1997b), or surface idiotype antigen (Demanet *et al.*, 1994; Honeychurch *et al.*, 1997; De Jonge *et al.*, 1998) have been used for B cell lymphomas and CD30 for Hodgkin's disease (Hombach *et al.*, 1993; Renner *et al.*, 1996a,b; Hartmann *et al.*, 1997), etc.

Honeychurch *et al.* (1997) investigated a panel of BsAbs [IgG and $F(ab)_2$] with dual specificity for T cells and neoplastic B cells antigens, such as CD19, CD22, MHC-II, or surface IgM idiotype (Id), for their anti-tumor effect in murine lymphoma model. Although most of the BsAbs could induce the T cell cytotoxicity *in vitro*, only the anti-Id×anti-T cell BsAbs were effective in prolonging the survival of tumor bearing animals. They concluded that the target cell specificity of a BsAb is critical in determining the therapeutic outcome and that *in vitro* cytotoxicity assays did not predict *in vivo* activity. They hypothesis that the unusual potency of anti-Id BsAb relates to their signaling activity on target lymphoma cells.

Like ordinary mAbs therapy, any tumor specific or tumor associated antigen can be chosen as the specific binding site of the therapeutic BsAbs. For the solid tumors, different tumor associate antigens, such as oncogene products (HER-2/*neu*, etc) (van Ojik *et al.*, 1997; Stockmeyer *et al.*, 1997), folate receptor (FR) (Mazzoni *et al.*, 1996; Roy *et al.*, 1998), EGF receptor (Schmidt *et al.*, 1996), transferrin receptor (Amoroso *et al.*, 1996) etc., are common candidates. But as a matter of fact, no absolute tumor specific antigen recognized by antibody has been found so far. So, it has been suggested that possibly several BsAbs direct to different kinds of tumor antigen should be used together to achieve promising therapeutic result *in vivo*.

To circumvent this problem, we combine the BsAbs with tumor vaccine to treat hepatoma bearing mice, whereas the hepatoma cell line hep1–6 was first treated *in vitro* with the combination cytokines (IFN-γ + TNF-α) to enhance expression of MHC-I and other adhesion molecule on cell surfaces. However, the B7 is detected negative before and after the cytokine treatment, so we used an anti-tumor×anti-CD28 BsAb to arm the treated tumor cell, giving it the co-stimulating signal for activating the T cells. Immunization of the mice with this two-step tumor cell vaccine could prevent and cure the established tumor. One major advantage of this approach is that because the arming of the BsAbs to tumor cell was performed *in vitro*, the antigens recognized by the BsAbs was not necessarily tumor specific. Even the normal tissue antigens could be also chosen as the target antigens as long as they are also expressed on the tumor cells (see Fig. 4) (Guo *et al.*, 1997). Hass *et al.* (1998) reported method of preparing the tumor vaccine by first infecting the tumor cell with New Castle virus, to enhance its expression of primary signal antigens for T cell activation, and the HN antigen of the virus; and next arming the tumor cells with anti-HN×anti-CD3 and anti-HN×anti-CD28 BsAbs. The generation of tumor vaccine also cured the established tumor in mice.

Thorpe's group (Huang *et al.*, 1997) introduced the endothelial cells as a new target for the BsAbs therapy. In their work, a recombinant form of tissue factor (TF) was constructed that contains only the cell surface domain (namely truncated TF, tTF) and conjugated with anti-class II antigen of tumor local vasculature. On injection to the neuroblastoma bearing mice, this antibody-tTF complex homed onto the tumor's blood vessel wall and caused rapid, selective, and complete thrombosis of the tumor's blood supply. In 38% of the mice, the tumors disappeared completely. However, the tumors given to the mice had been engineered to induce the class II marker on the tumor-associated blood vessels, something that could not be done in human cancer patients.

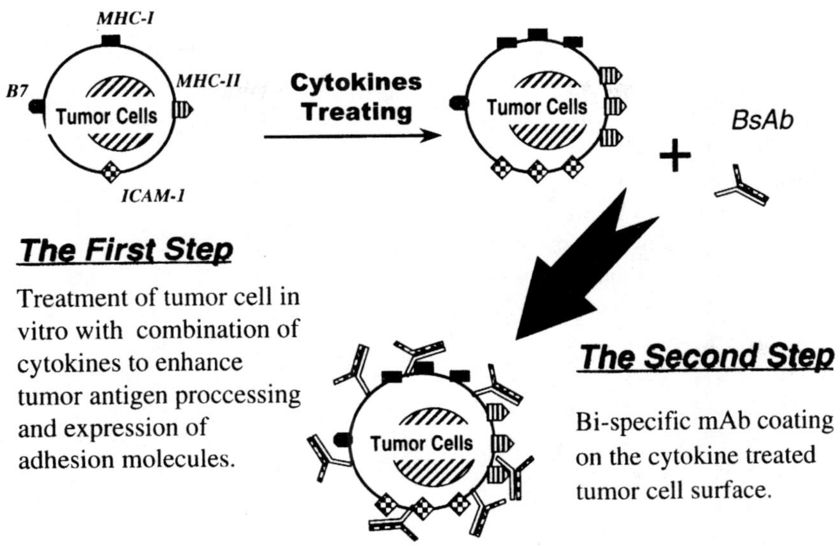

Figure 4. Modification of tumor cells with two-step procedure. The modified tumor cells become immunogenic and are able to elicit tumor-specific cytotoxic T cell response *in vivo.*

But, as the author expected, there are several candidate proteins on newly formed blood vessels that could be used as targets in human.

BsAbs can also be used as vehicles to facilitate of the targeted gene transfer to cells such as endothelium, smooth muscle cells, or T cells. Endothelium expressed receptor that could be induced during either angiogenesis (α_v-integrins) or inflammation (E-selectin) (Wickham *et al.*, 1996, 1997a,b) were proposed to be appropriate targets for gene therapy such as adenovirus-mediated gene transfer in case the target issue lacks sufficient levels of receptors to mediate vector attachment via its fiber coat protein.

5. CLINICAL ASPECTS OF BsAbs

Several phase I and phase II trials have been performed since 1990s on patients with advanced tumor. The results from their trials have showed that these approaches did have limited toxicity. Some efficacy against the tumor has been obtained. However, it is still too early to give a final conclusion for which kinds of BsAb or therapeutic scheme is preferable. Most of the patients in these trials with end-stage tumors are severely immunosuppressed. It is generally assumed that immunotherapy will be more effective in minimal residual disease. So, other therapies to reduce the tumor burden might be needed before the BsAb therapy.

Several anti-CD3 based BsAbs exhibited severe side effect while injected systemically (Mack *et al.*, 1997; Tibben *et al.*, 1996). Most of the anti-CD3 based BsAbs were used together with the *in vitro* activated T cell and showed minor side effects. Riethmüller *et al.* (1994, 1996) reviewed their clinical data on BsAb anti-CD3×anti-17-1A (directed against EpCAM or EGP-2 tumor associated antigen). This antibody had been evaluated for adjuvant therapy of 189 patients with resected Dukes C colorectal carcinoma. Seven-year follow-up data of the prospective randomized trial documented an impressive 32%

reduction in overall death rate of patients. This group, in Munich, Germany, have updated the unmodified murine 17-1A/CD3 antibody with a novel format of single chain Fv, which has same specificity and superior cytotoxic capacity (Mack *et al.*, 1997). Another group experienced limited toxic reaction after intravenously administration of anti-CD3×anti-CD19 BsAb in 4 patients (de Gast 1995a,b).

Canevari *et al.* (1997) reported the intraperitoneal infusion for treatment of 19 case of advanced ovarian carcinoma with autologous T lymphocytes retargeted by a anti-CD3×anti-folate receptor BsAb, Mov-18, and IL-2, resulted 3 complete response, 4 partial response, and 7 stable disease with moderate toxicity.

Two groups reported on the evaluation of BsAbs retarget to NK cell. Hartmann *et al.* (1997) (Universitätskliniken des Saarlandes, Homburg, Germany) reported on the treatment of 15 refractory Hodgkin's disease with HRS-3/A9, a BsAb directed against CD16 and the Hodgkin's associated CD30 antigen. The treatment was well tolerated and induced 1 complete and 1 partial remission (lasting 16 and 3 months, respectively), 3 minor responses (1 to 11+ months), and 1 mixed response. Another group (Weiner *et al.*, Fox Chase Cancer Center, Philadelphia, USA.) reported using 2B1, a anti-CD16 BsAb direct to the c-erbB-2 (product of the HER-2/*neu* proto-oncogene), for treating solid tumor overexpressing this antigen in 27 patients. They sought evidence of an idiotypic antibody induction, indicating a prolonged specific treatment-induced effect on at least one selected target of 2B1 in 50% of the patients. These antibodies were reactive with both intra- and extracellular domains of HER-2/*neu*. The presence of such antibodies reflects induction of active anti-tumor immunity upon FcγR-directed targeting of tumor cells (Clark *et al.*, 1997).

Medarex Inc, a biotech company in New Jersey, USA, developed several BsAbs targeting monocytes and macrophages through CD64. These BsAbs, include MDX-447 (humanized Fab anti-CD64×anti-EGF-receptor) and MDX-H210 (humanized Fab anti-CD64×anti-HER-2/*neu*), were administered intravenously to EGF-receptor or oncogene product HER-2/*neu* overexpressing tumor patients such as renal cell cancer, head and neck, bladder, breast, ovarian, and prostate cancers, and were well tolerated. These results from completed and ongoing clinical trials with the two BsAbs demonstrated excellent tolerability in association with promising anti-tumor effects in tumors that have become refractory to all available therapies (see Review by Curnow, 1997; van Ojik, *et al.*, 1997).

In summary, BsAbs therapy has a broad application clinically. In comparison with other conventional therapies, it may provide a useful strategy for treatment of cancer, especially for prevention of tumor recurrence after operation.

REFERENCES

Allard W.J., Moran C.A., Nagel E., Collins G., and Largen M.T., 1993, Antigen binding properties of highly purified bispecific antibodies. *Mol Immunol* **29**:1219–1227.

Amoroso A.R., Clark J.I., Litwin S., Hsieh-Ma S., Shi T., Alpaugh R.K., Adams G.P., Wolf E.J., Ring D.B., and Weiner L.M., 1996, Binding characteristics and antitumor properties of 1A10 bispecific antibody recognizing gp40 and human transferrin receptor. *Cancer Res* **56**:113–120.

Barbet J., Peltier P., Bardet S., Vuillez J.P., Bachelot I., Denet S., Olivier P., Leccia F., Corcuff B., Huglo D., Proye C., Rouvier E., Meyer P., and Chatal J.F., 1998, Radioimmunodetection of medullary thyroid carcinoma using indium-111 bivalent hapten and anti-CEA×anti-DTPA-indium bispecific antibody. *J Nucl Med* **39**:1172–1178.

Bohlen H., Hopff T., Manzke O., Engert A., Kube D., Wickramanayake P.D., Diehl V., and Tesch H., 1993, Lysis of malignant B cells from patients with B-chronic lymphocytic leukemia by autologous T cells activated with CD3×CD19 bispecific antibodies in combination with bivalent CD28 antibodies. *Blood* **82**:1803–1812.

Bohlen H., Manzke O., Titzer S., Lorenzen J., Kube D., Engert A., Abken H., Wolf J., Diehl V., and Tesch H., 1997, Prevention of Epstein-Barr virus-induced human B-cell lymphoma in severe combined immunodeficient mice treated with CD3×CD19 bispecific antibodies, CD28 monospecific antibodies, and autologous T cells. *Cancer Res* **57**:1704–1709.

Bolhuis R.L., Lamers C.H., Goey S.H., Eggermont A.M., Trimbos J.B., Stoter G., Lanzavecchia A., di Re E., Miotti S., and Raspagliesi F., 1992, Adoptive immunotherapy of ovarian carcinoma with bs-MAb-targeted lymphocytes: A multicenter study. *Int J Cancer* **Suppl 7**:78–81.

Brennan M., Davison P.F., and Paulus H., 1985, Preparation of bispecific antibodies by chemical recombination of monoclonal immunoglobulin G1 fragments. *Science* **229**:81–83.

Canevari S., Mezzanzanica D., Mazzoni A., Negri D.R., Figini M., Ramakrishna V., Bolis G., and Colnaghi M.I., 1997, Approaches to implement bispecific antibody treatment of ovarian carcinoma. *Cancer Immunol Immunother* **45**:187–189.

Carter P., Kelley R.F., Rodrigues M.L., Snedecor B., Covarrubias M., Velligan M.D., Wong W.L., Rowland A.M., Kotts C.E., and Carver M.E., 1992, High level Escherichia coli expression and production of a bivalent humanized antibody fragment. *Biotechnology (N Y)* **10**:163–167.

Chambers C.A., Krummel M.F., Boitel B., Hurwitz A., Sullivan T.J., Fournier S., Cassell D., Brunner M., and Allison J.P., 1996, The role of CTLA-4 in the regulation and initiation of T-cell responses. *Immunol Rev* **153**:27–46.

Clark J.I., Alpaugh R.K., von Mehren M., Schultz J., Gralow J.R., Cheever M.A., Ring D.B., and Weiner L.M., 1997, Induction of multiple anti-c-erbB-2 specificities accompanies a classical idiotypic cascade following 2B1 bispecific monoclonal antibody treatment. *Cancer Immunol Immunother* **44**:265–272.

Curnow R.T., 1997, Clinical experience with CD64-directed immunotherapy. An overview. *Cancer Immunol Immunother* **45**:210–215.

De Gast G.C., Haagen I.A., van Houten A.A., Klein S.C., Duits A.J., de Weger R.A., Vroom T.M., Clark M.R., Phillips J., and van Dijk A.J., 1995a, CD8 T cell activation after intravenous administration of CD3×CD19 bispecific antibody in patients with non-Hodgkin lymphoma. *Cancer Immunol Immunother* **40**:390–396.

De Gast G.C., Van Houten A.A., Haagen I.A., Klein S., De Weger R.A., Van Dijk A., Phillips J., Clark M., and Bast B.J., 1995b, Clinical experience with CD3×CD19 bispecific antibodies in patients with B cell malignancies. *J Hematother* **4**:433–437.

De Jonge J., Brissinck J., Heirman C., Demanet C., Leo O., Moser M., and Thielemans K., 1995, Production and characterization of bispecific single-chain antibody fragments. *Mol Immunol* **32**:1405–1412.

De Jonge J., Heirman C., de Veerman M., Van Meirvenne S., Moser M., Leo O., and Thielemans K., 1998, *In vivo* retargeting of T cell effector function by recombinant bispecific single chain Fv (anti-CD3×anti-idiotype) induces long-term survival in the murine BCL1 lymphoma model. *J Immunol* **161**:1454–1461.

Demanet C., Brissinck J., Leo O., Moser M., and Thielemans K., 1994, Role of T-cell subsets in the bispecific antibody (anti-idiotype×anti-CD3) treatment of the BCL1 lymphoma. *Cancer Res* **54**:2973–2978.

Demanet C., Brissinck J., De Jonge J., and Thielemans K., 1996, Bispecific antibody-mediated immunotherapy of the BCL1 lymphoma: increased efficacy with multiple injections and CD28-induced costimulation. *Blood* **87**:4390–4398.

Deo Y.M., Graziano R.F., Repp R., and van de Winkel J.G.J., 1997, Clinical significance of IgG Fc receptors and FcγR-directed immunotherapies. *Immunol Today* **18**:127–135.

Dietsch M.T., Chan P.Y., Kanner S.B., Gilliland L.K., Ledbetter J.A., Linsley P.S., and Aruffo A., 1994, Coengagement of CD2 with LFA-1 or VLA-4 by bispecific ligand fusion proteins primes T cells to respond more effectively to T cell receptor-dependent signals. *J Leukoc Biol* **56**:444–452.

Elsässer D., Valerius T., Repp R., Weiner G.J., Deo Y., Kalden J.R., van de Winkel J.G., Stevenson G.T., Glennie M.J., and Gramatzki M., 1996, HLA class II as potential target antigen on malignant B cells for therapy with bispecific antibodies in combination with granulocyte colony-stimulating factor. *Blood* **87**:3803–3812.

Ely P., Wallace P.K., Givan A.L., Graziano R.F., Guyre P.M., and Fanger M.W., 1996, Bispecific-armed, interferon γ-primed macrophage-mediated phagocytosis of malignant non-Hodgkin's lymphoma. *Blood* **87**:3813–3821.

Fearon D., 1997, Happy coupling: Recruiting both antigen and effector function. *Nature Biotechnol* **15**:618.

Ferrini S., Sforzini S., Cambiaggi A., Poggi A., Meazza R., Canevari S., Colnaghi M.I., and Moretta L., 1994, The LFA-1/ICAM cell adhesion pathway is involved in tumor-cell lysis mediated by bispecific monoclonal-antibody-targeted T lymphocytes. *Int J Cancer* **56**:846–852.

George A.J., Titus J.A., Jost C.R., Kurucz I., Perez P., Andrew S.M., Nicholls P.J., Huston J.S., and Segal D.M., 1994, Redirection of T cell-mediated cytotoxicity by a recombinant single-chain Fv molecule. *J Immunol* **152**:1802–1811.

Gerstmayer B., Hoffmann M., Altenschmidt U., and Wels W., 1997, Costimulation of T-cell proliferation by a chimeric B7-antibody fusion protein. *Cancer Immunol Immunother* **45**:156–158.

Glennie M.J., McBride H.M., Worth A.T., and Stevenson G.T., 1987, Preparation and performance of bispecific F(ab′γ)$_2$ antibody containing thioether-linked Fab′γ fragments. *J Immunol* **139**:2367–2375.

Goldstein J., Graziano R.F., Sundarapandiyan K., Somasundaram C., and Deo Y.M., 1997, Cytolytic and cytostatic properties of an anti-human FcγRI (CD64)×epidermal growth factor bispecific fusion protein. *J Immunol* **158**:872–879.

Gruber M., Schodin B.A., Wilson E.R., and Kranz D.M., 1994, Efficient tumor cell lysis mediated by a bispecific single chain antibody expressed in *Escherichia coli*. *J Immunol* **152**:5368–5374.

Guo Y.J., Che X.Y., Shen F., Xie T.P., Ma J., Wang X.N., Wu S.G., Anthony D.D., and Wu M.C., 1997, Effective tumor vaccines generated by *in vitro* modification of tumor cells with cytokines and bispecific monoclonal antibodies. *Nat Med* **3**:451–455.

Haagen I.A., Geerars A.J., de Lau W.B., Clark M.R., van de Griend R.J., Bast B.J., and de Gast B.C., 1994, Killing of autologous B-lineage malignancy using CD3×CD19 bispecific monoclonal antibody in end stage leukemia and lymphoma. *Blood* **84**:556–563.

Haagen I.A., Geerars A.J., de Lau W.B., Bast B.J., and de Gast B.C., 1995, The efficacy of CD3×CD19 bispecific monoclonal antibody (BsAb) in a clonogenic assay: the effect of repeated addition of BsAb and interleukin-2. *Blood* **85**:3208–3212.

Haas C., Strauss G., Moldenhauer G., Iorio R.M., and Schirrmacher V., 1998, Bispecific antibodies increase T-cell stimulatory capacity *in vitro* of human autologous virus-modified tumor vaccine. *Clin Cancer Res* **4**:721–730.

Hartmann F., Renner C., Jung W., Deisting C., Juwana M., Eichentopf B., Kloft M., and Pfreundschuh M., 1997, Treatment of refractory Hodgkin's disease with an anti-CD16/CD30 bispecific antibody. *Blood* **89**:2042–2047.

Hayden M.S., Gilliland L.K., and Ledbetter J.A., 1997, Antibody engineering. *Current Opinion in Immunol* **9**:201–212.

Heijnen I.A., Rijks L.J., Schiel A., Stockmeyer B., van Ojik H.H., Dechant M., Valerius T., Keler T., Tutt A.L., Glennie M.J., van Royen E.A., Capel P.J., and van de Winkel J.G.J., 1997, Generation of HER-2/*neu*-specific cytotoxic neutrophils *in vivo*: efficient arming of neutrophils by combined administration of granulocyte colony-stimulating factor and Fcγ receptor I bispecific antibodies. *Immunol* **159**:5629–5639.

Holliger P., and Winter G., 1997a, Diabodies: small bispecific antibody fragments. *Cancer Immunol Immunother* **45**:128–130.

Holliger P., Prospero T., and Winter G., 1993, "Diabodies": small bivalent and bispecific antibody fragments. *Proc Natl Acad Sci USA* **90**:6444–6448.

Holliger P., Brissinck J., Williams R.L., Thielemans K., and Winter G., 1996, Specific killing of lymphoma cells by cytotoxic T-cells mediated by a bispecific diabody. *Protein Eng* **9**:299–305.

Holliger P., Wing M., Pound JD., Bohlen H., and Winter G., 1997b, Retargeting serum immunoglobulin with bispecific diabodies. *Nat Biotechnol* **15**:632–636.

Hombach A., Jung W., Pohl C., Renner C., Sahin U., Schmits R., Wolf J., Kapp U., Diehl V., and Pfreundschuh M., 1993, A CD16/CD30 bispecific monoclonal antibody induces lysis of Hodgkin's cells by unstimulated natural killer cells *in vitro* and *in vivo*. *Int J Cancer* **55**:830–836.

Honeychurch J., Cruise A., Tutt A.L., and Glennie M.J., 1997, Bispecific Ab therapy of B-cell lymphoma: target cell specificity of antibody derivatives appears critical in determining therapeutic outcome. *Cancer Immunol Immunother* **45**:171–173.

Huang X., Molema G., King S., Watkins L., Edgington T.S., and Thorpe P.E., 1997, Tumor infarction in mice by antibody-directed targeting of tissue factor to tumor vasculature. *Science* **275**:547–550.

Kaminski M.S., Zasadny K.R., Francis I.R., Fenner M.C., Ross C.W., Milik A.W., Estes J., Tuck M., Regan D., Fisher S., Glenn S.D., and Wahl R.L., 1996, Iodine-131-anti-B1 radioimmunotherapy for B-cell lymphoma. *J Clin Oncol* **14**:1974–1981.

Katayose Y., Kudo T., Suzuki M., Shinoda M., Saijyo S., Sakurai N., Saeki H., Fukuhara K., Imai K., and Matsuno S., 1996, MUC1-specific targeting immunotherapy with bispecific antibodies: Inhibition of xenografted human bile duct carcinoma growth. *Cancer Res* **56**:4205–4212.

Kipriyanov S.M., Moldenhauer G., Strauss G., and Little M., 1998, Bispecific CD3×CD19 diabody for T cell-mediated lysis of malignant human B cells. *Int J Cancer* **77**:763–772.

Kontermann R.E., Wing M.G., and Winter G., 1997, Complement recruitment using bispecific diabodies. *Nat Biotechnol* **15**:629–631.

Kostelny S.A., Cole M.S., and Tso J.Y., 1992, Formation of a bispecific antibody by the use of leucine zippers. *J Immunol* **148**:1547–1553.

Kroesen B.J., ter Haar A., Spakman H., Willemse P., Sleijfer D.T., de Vries E.G., Mulder N.H., Berendsen H.H., Limburg P.C., and The T.H., 1993, Local antitumour treatment in carcinoma patients with bispecific-monoclonal-antibody-redirected T cells. *Cancer Immunol Immunother* **37**:400–407.

Kroesen B.J., Buter J., Sleijfer D.T., Janssen R.A., van der Graaf WT., The T.H., de Leij L., and Mulder N.H., 1994, Phase I study of intravenously applied bispecific antibody in renal cell cancer patients receiving subcutaneous interleukin-2. *Br J Cancer* **70**:652–661.

Kroesen B.J., Bakker A., van Lier R.A., The H.T., and de Leij L., 1995, Bispecific antibody-mediated target cell-specific costimulation of resting T cells via CD5 and CD28. *Cancer Res* **55**:4409–4415.

Kroesen B.J., Nieken J., Sleijfer D.T., Molema G., de Vries E.G., Groen H.J., Helfrich W., The T.H., Mulder N.H., and de Leij L., 1997, Approaches to lung cancer treatment using the CD3×EGP-2-directed bispecific monoclonal antibody BIS-1. *Cancer Immunol Immunother* **45**:203–206.

Kurucz I., Titus J.A., Jost C.R., Jacobus C.M., and Segal D.M., 1995, Retargeting of CTL by an efficiently refolded bispecific single-chain Fv dimer produced in bacteria. *J Immunol* **154**:4576–4582.

Lamers C.H., Bolhuis R.L., Warnaar S.O., Stoter G., and Gratama J.W., 1997, Local but no systemic immunomodulation by intraperitoneal treatment of advanced ovarian cancer with autologous T lymphocytes re-targeted by a bi-specific monoclonal antibody. *Int J Cancer* **73**:211–219.

Link B.K., Kostelny S.A., Cole M.S., Fusselman W.P., Tso J.Y., and Weiner G.J., 1998, Anti-CD3-based bispecific antibody designed for therapy of human B-cell malignancy can induce T-cell activation by antigen-dependent and antigen-independent mechanisms. *Int J Cancer* **77**:251–256.

Mack M., Riethmüller G., and Kufer P., 1995, A small bispecific antibody construct expressed as a functional single-chain molecule with high tumor cell cytotoxicity. *Proc Natl Acad Sci USA* **92**:7021–7025.

Mack M., Gruber R., Schmidt S., Riethmüller G., and Kufer P., 1997, Biologic properties of a bispecific single-chain antibody directed against 17-1A (EpCAM) and CD3: tumor cell-dependent T cell stimulation and cytotoxic activity. *J Immunol* **158**:3965–3970.

Mallender W.D., and Voss E.W. Jr, 1994, Construction, expression, and activity of a bivalent bispecific single-chain antibody. *J Biol Chem* **269**:199–206.

Mazzoni A., Mezzanzanica D., Jung G., Wolf H., Colnaghi M.I., and Canevari S., 1996, CD3-CD28 costimulation as a means to avoiding T cell preactivation in bispecific monoclonal antibody-based treatment of ovarian carcinoma. *Cancer Res* **56**:5443–5449.

Melero I., Shuford W.W., Newby S.A., Aruffo A, Ledbetter J.A., Hellström K.E., Mittler R.S., and Chen L., 1997, Monoclonal antibodies against the 4-1BB T-cell activation molecule eradicate established tumors. *Nat Med* **3**:682–685.

Mendez M.J., Green L.L., Corvalan J.R., Jia X.C., Maynard-Currie C.E., Yang X.D., Gallo M.L., Louie D.M., Lee D.V., Erickson K.L., Luna J., Roy C.M., Abderrahim H., Kirschenbaum F., Noguchi M., Smith D.H., Fukushima A., Hales J.F., Finer M.H., Davis C.G., Zsebo K.M., and Jakobovits A., 1997, Functional transplant of megabase human immunoglobulin loci recapitulates human antibody response in mice. *Nat Genet* **15**:146–156.

Milstein C., and Cuello A.C., 1983, Hybrid hybridomas and their use in immunohistochemistry. *Nature* **305**:537–540.

Nitta T., Sato K., Yagita H., Okumura K., and Ishii S., 1990, Preliminary trial of specific targeting therapy against malignant glioma. *Lancet* **335**:368–371.

Nolan O., and O'Kennedy R., 1990, Bifunctional antibodies: concept, production and applications. *Biochimica et Biophysica Acta*, **1040**:1–11.

Porter L.E., Nelson H., Ethem Gecim I., Rice D.C., Thibault C., and Chapoval A.I., 1997, T cell activation and retargeting using staphylococcal enterotoxin B and bispecific antibody: an effective *in vivo* antitumor strategy. *Cancer Immunol Immunother* **45**:180–183.

Renner C., Jung W., Sahin U., Denfeld R., Pohl C., Trümper L., Hartmann F., Diehl V., van Lier R., and Pfreundschuh M., 1994, Cure of xenografted human tumors by bispecific monoclonal antibodies and human T cells. *Science* **264**:833–835.

Renner C., Bauer S., Sahin U., Jung W., van Lier R., Jacobs G., Held G., and Pfreundschuh M., 1996a, Cure of disseminated xenografted human Hodgkin's tumors by bispecific monoclonal antibodies and human T cells: The role of human T-cell subsets in a preclinical model. *Blood* **87**:2930–2937.

Renner C., Ohnesorge S., Held G., Bauer S., Jung W., Pfitzenmeier J.P., and Pfreundschuh M., 1996b, T cells from patients with Hodgkin's disease have a defective T-cell receptor ζ-chain expression that is reversible by T-cell stimulation with CD3 and CD28. *Blood* **88**:236–241.

Riethmüller G., Holz E., Schlimok G., Schmiegel W., Raab R., Hoffken K., Gruber R., Pichlmajer H., Hirche H., Pichlmayr R., and G.-A.S. Group., Monoclonal antibody (mAb) adjuvant therapy of Dukes C

colorectal carcinoma: 7 year update of a prospective randomized trial. In *Proceedings of the American Society of Clinical Oncology*, **Vol. 15**. P.444.
Riethmüller G., Schneider-Gädicke E., Schlimok G., Schmiegel W., Raab R., Höffken K., Gruber R., Pichlmaier H., Hirche H., and Pichlmayr R., 1994, Randomised trial of monoclonal antibody for adjuvant therapy of resected Dukes' C colorectal carcinoma. German Cancer Aid 17-1A Study Group. *Lancet* **343**:1177–1183.
Roy E.J., Cho B.K., Rund L.A., Patrick T.A., and Kranz D.M., 1998, Targeting T cells against brain tumors with a bispecific ligand-antibody conjugate. *Int J Cancer* **76**:761–766.
Schmidt M., Hynes N.E., Groner B., and Wels W., 1996, A bivalent single-chain antibody-toxin specific for ErbB-2 and the EGF receptor. *Int J Cancer* **65**:538–546.
Sconocchia G., Titus J.A., and Segal D.M., 1994, CD44 is a cytotoxic triggering molecule in human peripheral blood NK cells. *J Immunol* **153**:5473–5481.
Stockmeyer B., Valerius T., Repp R., Heijnen I.A., Bühring H.J., Deo Y.M., Kalden J.R., Gramatzki M., and van de Winkel J.G., 1997, Preclinical studies with FcγR bispecific antibodies and granulocyte colony-stimulating factor-primed neutrophils as effector cells against HER-2/*neu* overexpressing breast cancer. *Cancer Res* **57**:696–701.
Tibben J.G., Boerman O.C., Claessens R.A., Corstens F.H., van Deuren M., de Mulder P.H., van der Meer J.W., Keijser K.G., and Massuger L.F., 1993, Cytokine release in an ovarian carcinoma patient following intravenous administration of bispecific antibody OC/TR F(ab')$_2$. *J Natl Cancer Inst* **85**:1003–1004.
Tibben J.G., Boerman O.C., Massuger L.F., Schijf C.P., Claessens R.A., and Corstens F.H., 1996, Pharmacokinetics, biodistribution and biological effects of intravenously administered bispecific monoclonal antibody OC/TR F(ab')$_2$ in ovarian carcinoma patients. *Int J Cancer* **66**:477–483.
Valerius T., Stockmeyer B., van Spriel A.B., Graziano R.F., van den Herik-Oudijk I.E., Repp R., Deo Y.M., Lund J., Kalden J.R., Gramatzki M., and van de Winkel J.G., 1997a, FcαRI (CD89) as a novel trigger molecule for bispecific antibody therapy. *Blood* **90**:4485–4492.
Valerius T., Elsässer D., Repp R., Van de Winkel J.G., Gramatzki M., and Glennie M., 1997b, HLA class II antibodies recruit G-CSF activated neutrophils for treatment of B cell malignancies. *Leuk Lymphoma* **26**:261–269.
Van de Winkel J.G.J., Bast B., and de Gast G.C., 1997, Immunotherapeutic potential of bispecific antibodies. *Immunol Today* **18**:562–564.
Van Ojik H.H., Repp R., Groenewegen G., Valerius T., and van de Winkel J.G., 1997, Clinical evaluation of the bispecific antibody MDX-H210 (anti-FcγRI×anti-HER-2/*neu*) in combination with granulocyte-colony-stimulating factor (filgrastim) for treatment of advanced breast cancer. *Cancer Immunol Immunother* **45**:207–209.
Weiner G.J., Kostelny S.A., Hillstrom J.R., Cole M.S., Link B.K., Wang S.L., and Tso J.Y., 1994, The role of T cell activation in anti-CD3×antitumor bispecific antibody therapy. *J Immunol* **152**:2385–2392.
Weiner L.M., Clark J.I., Davey M., Li W.S., Garcia de Palazzo I., Ring D.B., and Alpaugh R.K., 1995, Phase I trial of 2B1, a bispecific monoclonal antibody targeting c-erbB-2 and Fc γ RIII. *Cancer Res* **55**:4586–4593.
Wickham T.J., Segal D.M., Roelvink P.W., Carrion M.E., Lizonova A., Lee G.M., and Kovesdi I., 1996, Targeted adenovirus gene transfer to endothelial and smooth muscle cells by using bispecific antibodies. J Virol **70**:6831–6838.
Wickham T.J., Lee G.M., Titus J.A., Sconocchia G., Bakács T., Kovesdi I., Segal D.M., 1997a, Targeted adenovirus-mediated gene delivery to T cells via CD3. *J Virol* **71**:7663–7669.
Wickham T.J., Haskard D., Segal D., and Kovesdi I., 1997b, Targeting endothelium for gene therapy via receptors up-regulated during angiogenesis and inflammation. *Cancer Immunol Immunother* **45**:149–151.
Yokota T., Milenic D.E., Whitlow M., and Schlom J., 1992, Rapid tumor penetration of a single-chain Fv and comparison with other immunoglobulin forms. *Cancer Res* **52**:3402–3408.

33

B7.1 AND CYTOKINES

Synergy in Cancer Gene Therapy

Marcel Kuiper[1], Raquel Sanches[1], Yves-Jean Bignon[1], and Farzin Farzaneh[2]

[1]Laboratoire de Thérapie Génique Humaine et Expérimentale
INSERM CRI 9502 & EA 2145
Centre Jean Perrin, BP 392
63011 Clermont-Ferrand Cedex 1
France and
[2]Immune Gene Therapy of Cancer Programme
Department of Molecular Medicine
King's College School of Medicine and Dentistry
Rayne Institute, 123 Coldharbour Lane
London SE5 9NU, UK

1. INTRODUCTION

Despite the existence of tumour-specific 'cytotoxic' T lymphocytes (CTL) in cancer patients, an effective immune response against neoplastic cells is rarely detectable. To avoid immune attack, tumour cells often secrete immunosuppressive cytokines like transforming growth factor β (TGF-β), vascular endothelial growth factor (VEGF) and interleukin-(IL-)10, or induce apoptosis in Fas-expressing effector cells by expressing the Fas ligand (FasL) (Chouaib, Asselin-Paturel, Mami-Chouaib, Caignard, and Blay, 1997). However, environmental factors, including, paradoxically, TGF-β, often block the elimination, *via* apoptosis, of these tumour-reactive CTL (Akbar and Salmon, 1997). The challenge of cancer immunotherapy and immune gene therapy is therefore to overcome this immunosuppression and reactivate suppressed tumour-specific CTL.

Tumour cells genetically modified to express immunomodulatory factors are able to stimulate tumour-reactive CTL in a number of animal tumour models (Musiani, Modesti, Giovarelli, Cavallo, Colombo, Lollini, and Forni, 1997; Mackensen, Lindemann, and Mertelsmann, 1997). However, the efficacy of the introduction of such a gene, for example encoding the membrane-bound costimulator B7.1 (see below), may depend on the inherent immunogenicity of the tumour (Chen, McGowan, Ashe, Johnston, Li, Hellström, and Hellström, 1994). One approach to circumvent this problem, is

Cancer Gene Therapy: Past Achievements and Future Challenges, edited by Habib
Kluwer Academic/Plenum Publishers, New York, 2000.

the introduction of combinations of genes encoding immunomodulators, thereby providing a stronger immunostimulatory signal. This Chapter describes pre-clinical cancer gene therapy studies combining the genes encoding two types of immunomodulatory molecules, namely for membrane-bound B7.1 (CD80) and for secreted cytokines such as IL-2 and IL-12. The encouraging results in such animal tumour models have prompted the initiation of a number of clinical trials.

2. SINGLE-GENE CANCER IMMUNE GENE THERAPY

2.1. B7.1 (CD80)

B7.1 is a highly glycosylated, membrane-bound ligand which is expressed by antigen-presenting cells (APC) including B lymphocytes, macrophages and dendritic cells. The receptors for this molecule, CD28 and CTLA4 (CD152) are mainly found on T lymphocytes (Galea-Lauri, Farzaneh, and Gäken, 1996), although it is noteworthy that CD28 is also expressed by murine and human natural killer (NK) cells, and involved in the activation of these cells (Azuma, Cayabyab, Buck, Phillips, and Lanier, 1992; Galea-Lauri, Darling, Kuiper, Gäken, Souberbielle, and Farzaneh, 1998).

Activation of T lymphocytes is dependent on two major signals. Signal 1 is provided by the interaction of peptides bound in the groove of major histocompatibility complex (MHC) molecules with specific T cell receptors (TCR) on T lymphocytes. These interactions differ for $CD4^+$ 'helper' T lymphocytes (Th) and $CD8^+$ CTL, since the MHC class II, the expression of which is normally limited to APC, interacts with CD4 and the TCR complex, whereas MHC class I (expressed by most cells) interacts with CD8 and the TCR complex. Signal 2 is the product of interaction between costimulatory molecules (*e.g.* members of the B7 family) expressed on the APC and their counter-receptors on T lymphocytes. This second exchange of signals is called costimulation. Once activated, effector T cells do not need further costimulation for the lysis of their target cells (Schwartz, 1992).

A number of studies have shown that B7.1 expressed by tumour cells can induce anti-tumour immunity when introduced in immune competent syngeneic hosts (Galea-Lauri *et al.*, 1996). However, analysis of a large panel of tumour models has shown that the efficacy of B7.1 depends on the inherent immunogenicity of the tumour studied; immunogenic tumour cell lines expressing B7.1 do not form tumours and protect against a subsequent challenge with unmodified tumour cells, whereas non-immunogenic tumour cell lines expressing B7.1 form tumours and provide no protection against subsequent challenge with the unmodified tumour cells (Chen *et al.*, 1994).

The mechanism of induction of anti-tumour immunity by B7.1 remains to be fully resolved. The initial reasoning for introducing B7.1 into tumour cells was that it would allow these cells to present their own tumour-associated antigens (TAA) to specific $CD8^+$ precytotoxic T lymphocytes, and simultaneously provide costimulation, thus resulting in activation of these T cells. This would of course depend on the competence in processing and presentation of these antigens by these modified tumour cells. Once activated, the CTL would be able to recognise and destroy not only B7.1-expressing tumour cells, but also their unmodified counterparts (Schwartz, 1992). Although T lymphocytes were shown to be involved in the induced anti-tumour immunity (Galea-Lauri *et al.*, 1996), professional APC are known to be essential for the development of these responses (Huang, Golumbek, Ahmadzadeh, Jaffee, Pardoll, and Levitsky, 1994). Moreover, $B7.1^+$

tumour cells have been shown to be much less efficient than professional APC in stimulating T-cells directed against tumour cells, implying indirect activation of these tumour-reactive T lymphocytes (Huang, Bruce, Pardoll, and Levitsky, 1996). However, if professional APC are necessary for the induction of anti-tumour immunity, the role of B7.1 expressed by tumour cells may seem paradoxical. One explanation is the involvement of other CD28$^+$ and/or CTLA4$^+$ cells, apart from T lymphocytes, specifically killing B7.1$^+$ (not B7.1$^-$) tumour cells. This initial, non-MHC-restricted killing of the tumour cells, may lead to the subsequent uptake of TAA by professional APC, which would then be capable of initiating T lymphocyte-dependent anti-tumour immune responses. As already mentioned, CD28$^+$ NK cells exist and can be activated by B7.1. NK cells may therefore be responsible for the initial killing of B7.1$^+$ tumour cells (Nandi *et al.*, 1994; Azuma *et al.*, 1992; Galea-Lauri *et al.*, 1998). The ability of B7.1/CD28-mediated prevention of apoptosis in CTL (Daniel, Kroidl, Cayeux, Bargou, Blankenstein, and Dörken, 1997) may be important in the maintenance of tumour-specific CTL in a subsequent MHC-restricted phase.

The constitutively expressed B7-2 (CD86), a homologue of B7.1, has a similar function to the activation-induced B7.1, but has been shown to be less potent for cancer gene therapy approaches (Matulonis, Dosiou, Freeman, Lamont, Mauch, Nadler, and Griffin, 1996).

2.2. Interleukin-2

The pleiotropic effects of IL-2 can be explained by the wide variety of cells expressing receptors for this cytokine: APC (mononuclear phagocytes, B lymphocytes), non-MHC-restricted effector cells (NK cells, eosinophils), CTL, and 'helper' T lymphocytes (Foa, Guarini, and Gansbacher, 1992). All of these cell-types may be involved in the development of anti-tumour immunity. Both in cancer patients and tumour-bearing animals, decreased levels of the CD3 ζ chain and other signalling components of the TCR/CD3 complex are frequently observed in unresponsive tumour-specific T lymphocytes, and IL-2 has been shown to restore normal TCR signalling in these T lymphocytes (Zier, Gansbacher, and Salvadori, 1996). These characteristics make IL-2 an attractive cytokine for cancer immunotherapy. The gene for IL-2 has been introduced into tumour cells, with the aim of inducing anti-tumour immunity without the side effects of systemic administration of high doses of IL-2 (Foa *et al.*, 1992). Although effective MHC-restricted anti-tumour responses have been described using IL-2 cancer gene therapy (Foa *et al.*, 1992), introduction of the IL-2 gene alone is not always sufficient to prevent growth of the poorly immunogenic tumour cells, such as the murine NC mammary adenocarcinoma (Gäken, Hollingsworth, Hirst, Buggins, Galea-Lauri, Peakman, Kuiper, Patel, Towner, Patel, Collins, Mufti, Farzaneh, and Darling, 1997).

2.3. Interleukin-12

IL-12 is a heterodimeric cytokine consisting of subunits of 35 and 40 kD (called p35 and p40 respectively). It is produced primarily by activated macrophages, and stimulates activated T lymphocytes and NK cells which possess specific receptors for this cytokine. As a result, the lytic activity of these cells is augmented and their interferon-(IFN-)γ secretion increases. IL-12 has an important role in determining the type of immune response, promoting so-called Th1 responses (cellular immunity) (Tahara and Lotze, 1995). Apart from immunological properties, IL-12 may also have anti-angiogenic activity (Nanni, Rossi, De Giovanni, Landuzzi, Nicoletti, Stoppacciaro, Parenza, Colombo, and Lollini, 1998).

The above-mentioned qualities explain the interest of this cytokine in experimental cancer therapies. Both injection of recombinant IL-12, and IL-12 gene therapy, have given rise to the development of anti-tumour immune responses in murine models of malignant melanoma, sarcomas and mammary carcinoma (Tahara and Lotze, 1995; Mackensen *et al.*, 1997). For the cancer gene therapy protocols, the technical difficulty of delivering the genes for both subunits using a single vector has been resolved by using internal ribosome entry site (IRES) sequences between subunits (Tahara, Zitvogel, Storkus, Zeh, McKinney, Schreiber, Gubler, Robbins, and Lotze, 1995) or constructing a fusion gene encoding a single-chain IL-12 (Lieschke, Rao, Gately, and Mulligan, 1997). However, as for B7.1 and IL-2, gene therapy with IL-12 alone is not 100% effective in all models studied. For example, possibly as a result of stimulation by IL-12 of helper T lymphocytes with inhibitory or suppressive activity, tumour formation of IL-12-secreting CT26 murine colon carcinoma cells was only delayed, not completely prevented (Martinotti, Stoppacciaro, Vagliani, Melani, Spreafico, Wysocka, Parmiani, Trinchieri, and Colombo, 1995).

3. CANCER GENE THERAPY WITH A COMBINATION OF COSTIMULATORY AND CYTOKINE GENES

As desribed above, introduction of single immunomodulatory genes does not always result in the sufficiently potent anti-tumour immune responses, possibly depending on the intrinsic immunogenicity of the tumour model studied (Chen *et al.*, 1994). Therefore, combination of these genes have been employed.

3.1. B7.1 and Interleukin-2

A number of publications have described the combined use of murine B7.1 and IL-2 expressed by murine tumour cells, namely in the fibrosarcoma cell line CMS5 (Salvadori, Gansbacher, Wernick, Tirelli, and Zier, 1995), the mammary adenocarcinoma cell line NC (Gäken *et al.*, 1997) and the plasmocytoma cell line J558L (Cayeux, Richter, Becker, Beck, Aicher, Pezzutto, Dörken, and Blankenstein, 1997). The effect of combined expression of murine B7.1 and human IL-2 was studied in a murine breast adenocarcinoma model based on the PyMT cell line (expressing the polyoma virus middle T antigen) (Emtage, Wan, Bramson, Graham, and Gauldie, 1998). In three of these studies, synergy between B7.1 and IL-2 expression in the stimulation of anti-tumour immune responses could be demonstrated comparing animals inoculated with tumour cells expressing one gene with those possessing both genes (Salvadori *et al.*, 1995; Gäken *et al.*, 1997), or tumour treatment with adenoviruses encoding the single immunomodulators, or both (Emtage *et al.*, 1998). However, in the J588L model, combined B7.1 and IL-2 expression, although resulting in a lower percentage of tumour-bearing animals than cells expressing B7.1 alone, led to a higher percentage of animals forming a tumour relative to IL-2 alone (Cayeux *et al.*, 1997).

Clinically relevant experiments (*i.e.* experiments where treatment follows inoculation with non-modified tumour cells) showed that after resection of a subcutaneous (s.c.) wild-type NC tumour, s.c. vaccination on the other flank with irradiated NC IL-2/B7.1 cells showed the best protection against recurrence of the primary tumour, although even vaccination with non-modified irradiated tumour cells resulted in a high level of protection against recurrence (Gäken *et al.*, 1997). In the PyMT model, s.c. tumours were allowed to develop, after which they were injected intratumorally with adenoviral particles containing the gene(s) for mB7.1, hIL-2, or both mB7.1 and hIL-2. Treatment with the mB7.1/hIL-2 virus resulted in complete regression in all treated tumours, in contrast to mB7.1 (37.5%) or hIL-2 (42%) viruses alone. All animals which rejected their tumours (independent of the virus used) were protected against a subsequent challenge with tumour cells, and exhibited polyoma virus middle T antigen-specific CTL activity (Emtage *et al.*, 1998). These results are in agreement with *in vitro* studies published by Salvadori *et al.* (1995). They showed that mB7.1/mIL-2-expressing CMS5 cells could not induce a response from naive syngeneic spleen cells, but that these modified cells were capable of re-stimulating cytotoxicity of T lymphocytes isolated from CMS5 tumour-bearing mice, in contrast to singly modified or unmodified CMS5 cells (Salvadori *et al.*, 1995).

Cayeux *et al.* (1997) have explained the lack of synergy in their model (J588L) by "hyperstimulation" of the immune system. They hypothesised that tumour cells expressing both transgenes lead to rapid rejection, which prevents the development of anti-tumour immunity (Cayeux *et al.*, 1997). This hypothesis is supported by the results of Emtage *et al.* (1998), showing more rapid rejection of tumours treated with mB7.1/hIL-2 adenovirus, compared to mB7.1 or hIL-2 adenovirus alone, in those animals that rejected their tumours after adenovirus administration.

Taken together, these publications, although not consistent in all aspects, indicate that vaccination with B7.1/IL-2 expressing tumour cells, or B7.1/IL-2 containing adenoviruses may be capable of re-stimulating tumour-specific CTL to attack the patient's tumour cells. This is further supported by the observation that signal transduction through the IL-2 receptor in combination with CD28 signalling has been shown to be superior to either signal alone in reversing TGF-β-induced anergy of lymphokine activated killer (LAK) cells *in vitro* (Koberda, Grimm, and Moser, 1993). This suggests that this combination could be capable of overcoming immunosuppression *in vivo* as well. As summarised in Table 1, on the basis of these promising pre-clinical results, a phase I clinical trial utilising an irradiated vaccine of allogeneic tumour cells expressing hB7.1 mixed with allogeneic fibroblasts releasing hIL-2 has been initiated for colorectal cancer, whereas a phase I clinical trial for malignant melanoma using HLA-A2$^+$ allogeneic melanoma cells expressing both B7.1 and IL-2 is planned (Parmiani, 1998).

3.2. B7.1 and Interleukin-12

Cancer gene therapy using the genes for IL-12 and B7.1 has been studied in a number of experimental murine cancer models. These ranged from CMT93 colorectal cancer and K1735 malignant melanoma (Chong, Todryk, Hutchinson, Hart, and Vile, 1998), B16-BL6 melanoma (Nishimura, Watanabe, Yahata, Uede, Saiki, Herrmann, Kobayashi, and Habu, 1996), MT1A2 mammary adenocarcinoma (Pützer, Hitt, Muller, Emtage, Gauldie, and Graham, 1997), 3LL lung carcinoma (Kato, Okumura, and Yagita, 1997), P815 mastocytoma (Chen, Geer, Podack, and Ksander, 1996) and A20 B cell lymphoma (Pizzoferrato, Chu, Hawley, Lieu, Barber, Hawley, Watts, and Berinstein, 1997).

In the CMT93 and K1735 models, synergy between B7.1 and IL-12 in the prevention of tumour formation was observed. However, this resulted only in protective immunity against re-challenge with unmodified tumour cells in case of the CMT93 model, not in the K1735 model. The authors explain this result by hypothesising that in the former model T lymphocytes are involved, leading to immunological memory, whereas the rejection in the latter is dependent on NK cells without T cell involvement, therefore not resulting in memory. Similar results were obtained with the B16-BL6 model, in which intravenous injection of these cells transduced with both B7.1 and IL-12 genes resulted in lower numbers of tumour metastasis to the lungs than when transduced with either gene alone (Nishimura *et al.*, 1996). In the P815 model, s.c. injection of unmodified or IL-12-secreting cells resulted in 100% tumour take, whereas B7.1 or B7.1/IL-12-expressing cells were rejected by all animals. However, mice who had rejected P815 B7.1/IL-12 cells induced protective immunity, whereas mice inoculated with P815 B7.1 did not (Chen *et al.*, 1996). Pizzoferrato *et al.* (1997) found that 100% of animals injected either with A20 cells expressing both B7.1 and IL-12 or IL-12 alone survived up to 60 days, in contrast to animals receiving A20 cells containing the empty vector or expressing B7.1, of which 0% and 10% survived for 60 days respectively. If unmodified A20 cells were injected s.c. in one flank, and gene-modified A20 cells in the other at the same time, at least some animals in all groups succumbed to their cancers, although only animals receiving A20 B7.1/IL-12 cells showed a statistically significant improvement relative to animals receiving A20 cells containing empty vectors. Finally, if gene-modified and unmodified A20 cells were mixed before s.c. injection, again some animals in all groups formed lethal tumours, however, animals receiving A20 B7.1/IL-12 cells were clearly superior to other groups. Depletion of lymphocyte subsets and experiments in nude mice suggested that the induced anti-tumour immune response depended on both $CD4^+$ and $CD8^+$ cells (Pizzoferrato *et al.*, 1997). In the 3LL lung carcinoma model, CTL activity of lymphocytes isolated from mice inoculated with 3LL B7.1/IL-12 cells was increased 10-fold as compared to lymphocytes isolated from mice injected with unmodified 3LL cells. The therapeutic efficacy was examined by counting lung metastases after intravenous injection of unmodified 3LL cells at day 0, followed by intraperitoneal treatment at days 3, 10 and 17 with irradiated gene-modified 3LL cells. Treatment with 3LL B7.1/IL-12, and to a lesser extent 3LL B7.1, resulted in a statistically significant decrease compared to mock treatment with phosphate-buffered saline (12.2 ± 2.3, 23.8 ± 2.1 and 55.2 ± 6.3 metastatic colonies respectively; Kato *et al.*, 1997). Pützer *et al.* (1997) studied treatment of established s.c. MT1A2 mammary tumours. Intratumoral injection of a B7.1 or IL-12 adenovirus resulted only in growth delay (in 14/15 and 13/15 animals respectively), whereas injection of both these viruses resulted in growth delay in 6/10, a partial response (complete regression for at least 2 weeks) in 1/10 and a complete response (complete regression without recurrence) in 3/10 animals. Interestingly, injection with a high dose of an adenovirus containing both the IL-12 genes and the B7.1 gene resulted in a complete response in 9/10 and a partial response in 1/10 animals, without obvious toxic effects. These clinically relevant experiments show that the combined therapeutic effect of B7.1 and IL-12 is stronger when expressed by the same tumour cell (transduction by the B7.1/IL-12 combination virus), as compared to a situation where the majority of transduced neoplastic cells express only one of these immunomodulators, as a result of injection of a mixture of an adenovirus encoding B7.1 and another encoding IL-12.

In summary, the results of the above-mentioned experiments where gene therapy with B7.1 and IL-12 followed inoculation of an unmodified tumour, support the hypothesis that tumour cells expressing B7.1 and IL-12 (comparable to B7.1 and IL-2; Salvadori *et al.*, 1995) are capable of reactivating suppressed tumour-reactive T lymphocytes. A first clinical trial has been initiated using a combination of recombinant

Table 1. Clinical trials of B7.1/cytokine cancer gene therapy

Malignancy	Description (status)	Principle investigator and affiliation
Colorectal cancer	Subcutaneous injection of irradiated allogeneic hB7.1⁺ tumour cells mixed with allogeneic fibroblasts secreting hIL-2 (open)	Robert E. Sobol, Sidney Kimmel Cancer Center, San Diego, California, USA
Malignant melanoma	*In situ* administration of a recombinant canarypox virus encoding hB7.1, or a combination of the former virus with a canarypox virus encoding hIL-12 (open)	Robert M. Conry, University of Alabama at Birmingham, Birmingham, Alabama, USA
Malignant melanoma	Subcutaneous injection of HLA-A2⁺ allogeneic melanoma cell lines expressing hB7.1 and releasing hIL-2 in HLA-A2⁺ melanoma patients (in preparation)	Giorgio Parmiani, Istituto Nazionale per lo Studio e la Cura dei Tumori, Milan, Italy

canarypox viruses encoding hIL-12 and hB7.1 to treat malignant melanoma *in situ* (Table 1).

3.3. B7.1 and Other Cytokines

Apart from IL-2 and IL-12, a number of other cytokines have been combined with B7.1 in experimental cancer gene therapy. In the J558L plasmocytoma model, no tumours were induced when syngeneic mice were injected with J558L IL-4/B7.1 cells, whereas J558L IL-4, J558L B7.1 and J558L mixed with the adjuvant *Corynebacterium parvum* formed tumours in about 20% of animals. J558L cells formed tumours in 100% of animals (Cayeux, Beck, Dörken, and Blankenstein, 1996). The same scientists also studied B7.1 and IL-7 in the TS/A mammary adenocarcinoma model, with similar results; B7.1 in combination with IL-7 resulted in rejection in all syngeneic animals, whereas TS/A B7.1, TS/A IL-7 and TS/A + *C. parvum* were rejected by a subset of animals. Unmodified TS/A alone formed tumours in all animals. Analysis of the tumour-infiltrating lymphocytes (TIL) showed an increased percentage of $CD4^+$ and $CD8^+$ cells in animals inoculated with TS/A IL-7/B7.1 as compared with the other groups. Both the $CD4^+$ and the $CD8^+$ TIL in the TS/A IL-7/B7.1 tumours showed higher levels of the activation markers CD25 and CD28 as compared to the other groups (Cayeux, Beck, Aicher, Dörken, and Blankenstein, 1995). Finally, B7.1 combined with granulocyte/macrophage-stimulating factor (GM-CSF) induced better rejection in the murine CMT93 colorectal and K1735 melanoma models, as compared to single transfectants. In a separate set of experiments, syngeneic mice were inoculated with unmodified and gene-modified tumour cells. If present, the primary tumour was excised, followed by a re-challenge with unmodified tumour cells. In the CMT93 model, initial inoculation with B7.1/GM-CSF-expressing cells protected about 70% against a wild-type CMT93 challenge, whereas initial inoculation with the single gene-modified cells protected only 30–40%. However, in the K1735 model, initial injection of GM-CSF-secreting or unmodified cells was more effective against a subsequent challenge (about 90% of animals protected) than K1735 B7.1 or K1735 B7.1/GM-CSF (about 55% of animals protected) (Chong *et al.*, 1998). So far, these combinations have been described for one or two tumour models, and animal experiments modelling the therapeutic application of such gene therapy approaches have not yet been performed. Further studies in these and other tumour models will better define their potential for clinical application.

4. FUTURE CHALLENGES

The results of the pre-clinical studies described in this Chapter have formed the basis of numerous clinical trials for cancer gene therapy with the single immunomodulatory molecules B7.1, IL-2 and IL-12 (National Institutes of Health, 1998), and three clinical trials using combinations of B7.1 and IL-2 or IL-12 (Table 1). The positive experimental results with intratumoral injection of viral vectors in murine tumours using combinations of B7.1 and cytokine genes (Emtage *et al.*, 1998; Pützer *et al.*, 1997) have lead to the initiation of an *in situ* B7.1/IL-12 gene therapy trial for malignant melanoma, using a canarypox vector system. The other two trials use cellular vaccines expressing B7.1 and IL-2, the first for colorectal cancer comprises of s.c. injection of irradiated allogeneic hB7.1$^+$ tumour cells mixed with allogeneic fibroblasts secreting hIL-2, whereas the other proposes the use of a mixture of allogeneic cell lines expressing both hB7.1 and hIL-2 (Table 1).

Apart from combinations of B7.1 with cytokines, other combinations have equally demonstrated synergy in cancer gene therapy: two cytokine genes, e.g. *IL-2* and *IL-4* (Hollingsworth, Darling, Gäken, Hirst, Patel, Kuiper, Towner, Humphreys, Farzaneh, and Mufti, 1996), a cytokine gene and a pro-drug convertase ('suicide') gene, for example *IL-2* and the herpes simplex virus thymidine kinase gene (*HSV-TK*) (O'Malley, Cope, Chen, Li, Schwartz, and Woo, 1996), two 'suicide' genes, like *HSV-TK* and the cytosine deaminase gene (*CD*) (Aghi, Kramm, Chou, Breakefield, and Chiocca, 1998), and other associations of genes exist. Most of these studies have been performed on a limited number of tumour models, and it is therefore difficult to establish which combinations are the most promising for clinical application. It would therefore be desirable to initiate comparative studies which would allow the selection of potent single genes or gene combinations applicable to single neoplastic diseases, or for many cancers. The results of these studies would allow rational development of clinical cancer gene therapy trials.

In addition to the choice of the therapeutic gene(s), the mode of gene delivery and control of expression remain important issues (Anderson, 1998). Hence, several Chapters in this book have been dedicated to different delivery systems. Progress in this field is being made, and this, taken together with the promising results of several pre-clinical and clinical studies, some of which have been discussed in this Chapter, might allow cancer gene therapy to develop as an addition to existing cancer treatments.

ACKNOWLEDGMENTS

M.K. is supported by the Centre Médico-Chirurgical de Tronquières (Aurillac, France). R.S. is supported by the Praxis XXI programme, Fundação para a Ciência e a Tecnologia (Lisbon, Portugal).

REFERENCES

Aghi, M., Kramm, C.M., Chou, T.-C., Breakefield, X.O., and Chiocca, E.A., 1998, Synergistic anticancer effects of ganciclovir/thymidine kinase and 5-fluorocytosine/cytosine deaminase gene therapies, *J. Natl. Cancer Inst.* **90**:370–380.

Akbar, A.N., and Salmon, M., 1997, Cellular environments and apoptosis: tissue microenvironments control activated T-cell death, *Immunol. Today* **18**:72–76.

Anderson, W.F., 1998, Human gene therapy, *Nature* **392**:25–30.

Azuma, M., Cayabyab, M., Buck, D., Phillips, J.H., and Lanier, L.L., 1992, Involvement of CD28 in MHC-unrestricted cytotoxicity mediated by a human natural killer leukemia cell line, *J. Immunol.* **149**: 1115–1123.

Cayeux, S., Beck, C., Aicher, A., Dörken, B., and Blankenstein, T., 1995, Tumor cells cotransfected with interleukin-7 and B7.1 genes induce CD25 and CD28 on tumor-infiltrating T lymphocytes and are strong vaccines, *Eur. J. Immunol.* **25**:2325–2331.

Cayeux, S., Beck, C., Dörken, B., and Blankenstein, T., 1996, Coexpression of interleukin-4 and B7.1 in murine tumor cells leads to improved tumor rejection and vaccine effect compared to single gene transfectants and a classical adjuvant, *Hum. Gene Ther.* **7**:525–529.

Cayeux, S., Richter, G., Becker, C., Beck, C., Aicher, A., Pezzutto, A., Dörken, B., and Blankenstein, T., 1997, Lack of correlation between rejection of tumor cells co-expressing interleukin-2 and B7.1 and vaccine efficiency, *Eur. J. Immunol.* **27**:1657–1662.

Chen, L., McGowan, P., Ashe, S., Johnston, J., Li, Y., Hellström, I., and Hellström, K.E., 1994, Tumor immunogenicity determines the effect of B7 costimulation on T cell-mediated tumor immunity, *J. Exp. Med.* **179**:523–532.

Chen, P.W., Geer, D.C., Podack, E.R., and Ksander, B.R., 1996, Tumor cells transfected with B7-1 and interleukin-12 cDNA induce protective immunity, *Ann. N. Y. Acad. Sci.* **795**:325–327.

Cheng, H., Todryk, S., Hutchinson, G., Hart, I.R., and Vile, R.G., 1998, Tumour cell expression of B7 costimulatory molecules and interleukin-12 or granulocyte-macrophage colony-stimulating factor induces a local antitmour response and may generate systemic protective immunity, *Gene Ther.* **5**:223–232.

Chouaib, S., Asselin-Paturel, C., Mami-Chouaib, F., Caignard, A., and Blay, J.Y., 1997, The host-tumor immune conflict: from immunosuppression to resistance and destruction, *Immunol. Today* **18**:493–497.

Daniel, P.T., Kroidl, A., Cayeux, S., Bargou, R., Blankenstein, T., and Dörken, B., 1997, Costimulatory signals through B7.1/CD28 prevent T cell apoptosis during target cell lysis, *J. Immunol.* **159**:3808–3815.

Emtage, P.C.R., Wan, Y., Bramson, J.L., Graham, F.L., and Gauldie, J., 1998, A double recombinant adenovirus expressing the costimulatory molecule B7-1 (murine) and human IL-2 induces complete tumor regression in a murine breast adenocarcinoma model, *J. Immunol.* **160**:2531–2538.

Foa, R., Guarini, A., and Gansbacher, B., 1992, IL2 treatment for cancer: from biology to gene therapy, *Br. J. Cancer* **66**:992–998.

Gäken, J.A., Hollingsworth, S.J., Hirst, W.J.R., Buggins, A.G.S., Galea-Lauri, J., Peakman, M., Kuiper, M., Patel, P., Towner, P., Patel, P.M., Collins, M.K.L., Mufti, G.J., Farzaneh, F., and Darling, D.C., 1997, Irradiated NC adenocarcinoma cells transduced with both B7.1 and interleukin-2 induce CD4$^+$-mediated rejection of established tumors, *Hum. Gene Ther.* **8**:477–488.

Galea-Lauri, J., Farzaneh, F., and Gäken, J., 1996, Novel costimulators in the immune gene therapy of cancer, *Cancer Gene Ther.* **3**:202–213.

Galea-Lauri, J., Darling, D., Kuiper, M., Gäken, J., Souberbielle, B., and Farzaneh, F., 1998, Expression of a variant CD28 on a sub-population of human NK cells: implications for B7-mediated stimulation of NK cells, *J. Immunol.*

Hollingsworth, S.J., Darling, D., Gäken, J., Hirst, W., Patel, P., Kuiper, M., Towner, P., Humphreys, S., Farzaneh, F., and Mufti, G.J., 1996, The effect of combined expression of interleukin 2 and interleukin 4 on the tumorigenicity and treatment of B16F10 melanoma, *Br. J. Cancer* **74**:6–15.

Huang, A.Y.C., Golumbek, P., Ahmadzadeh, M., Jaffee, E., Pardoll, D., and Levitsky, H., 1994, Role of bone marrow-derived cells in presenting MHC class I-restricted tumor antigens, *Science* **264**:961–965.

Huang, A.Y.C., Bruce, A.T., Pardoll, D.M., and Levitsky, H.I., 1996, Does B7-1 expression confer antigen-presenting cell capacity to tumors in vivo? *J. Exp. Med.* **183**:769–776.

Kato, K., Okumura, K., and Yagita, H., 1997, Immunoregulation by B7 and IL-12 gene transfer, *Leukemia* **11**:572–576.

Koberda, J., Grimm, E.A., and Moser, R.P., 1993, Effect of anti-CD3/anti-CD28/interleukin-2 stimulation of mononuclear cells on transforming growth factor β inhibition of lymphokine-activated killer cell generation, *J. Cancer Res. Clin. Oncol.* **119**:131–136.

Lieschke, G.J., Rao, P.K., Gately, M.K., and Mulligan, R.C., 1997, Bioactive murine and human interleukin-12 fusion proteins which retain antitumor activity in vivo, *Nat. Biotechnol.* **15**:35–40.

Mackensen, A., Lindemann, A., and Mertelsmann, R., 1997, Immunostimulatory cytokines in somatic cells and gene therapy of cancer, *Cytokine Growth Factor Rev.* **8**:119–128.

Martinotti, A., Stoppacciaro, A., Vagliani, M., Melani, C., Spreafico, F., Wysocka, M., Parmiani, G., Trinchieri, G., and Colombo, M.P., 1995, CD4 T cells inhibit *in vivo* the CD8-mediated immune response against murine colon carcinoma cells transduced with interleukin-12 genes, *Eur. J. Immunol.* **25**:137–146.

Matulonis, U., Dosiou, C., Freeman, G., Lamont, C., Mauch, P., Nadler, L.M., and Griffin, J.D., 1996, B7-1 is superior to B7-2 costimulation in the induction and maintenance of T cell-mediated antileukemia immunity. Further evidence that B7-1 and B7-2 are functionally distinct. *J. Immunol.* **156**:1126–1131.

Musiani, P., Modesti, A., Giovarelli, M., Cavallo, F., Colombo, M.P., Lollini, P.L., and Forni, G., 1997, Cytokines, tumour-cell death and immunogenicity: a question of choice, *Immunol. Today* **18**:32–36.

Nandi, D., Gross, J.A., and Allison, J.P., 1994, CD28-mediated costimulation is necessary for optimal proliferation of murine NK cells, *J. Immunol.* **152**:3361–3369.

Nanni, P., Rossi, I., De Giovanni, C., Landuzzi, L., Nicoletti, G., Stoppacciaro, A., Parenza, M., Colombo, M.P., and Lollini, P.-L., 1998, Interleukin 12 gene therapy of MHC-negative murine melanoma metastases, *Cancer Res.* **58**:1225–1230.

National Institutes of Health, 1998, Bethesda (July 20, 1998); http://www.nih.gov/od/orda/protocol.htm.

Nishimura, T., Watanabe, K., Yahata, T., Uede, T., Saiki, I., Herrmann, S.H., Kobayashi, M., and Habu, S., 1996, The application of IL-12 to cytokine therapy and gene therapy for tumors, *Ann. N. Y. Acad. Sci.* **795**:375–378.

O'Malley, B.W., Cope, K.A., Chen, S.-H., Li, D., Schwartz, M.R., and Woo, S.L.C., 1996, Combination gene therapy for oral cancer in a murine model, *Cancer Res.* **56**:1737–1741.

Parmiani, G., 1998, Gene modified human melanoma cell vaccines: new strategies, in: *Eurocancer 98. Compte rendu du XIe congrès*, (M. Boiron, and M. Marty, eds.), p. 299, John Libbey Eurotext, Paris.

Pizzoferrato, E., Chu, N.R., Hawley, T.S., Lieu, F.H.L., Barber, B.H., Hawley, R.G., Watts, T.H., and Berinstein, N.L., 1997, Enhanced immunogenicity of B cell lymphoma genetically engineered to express both B7-1 and interleukin-12, *Hum. Gene Ther.* **8**:2217–2228.

Pützer, B.M., Hitt, M., Muller, W.J., Emtage, P., Gauldie, J., and Graham, F.L., 1997, Interleukin 12 and B7-1 costimulatory molecule expressed by an adenovirus vector act synergistically to facilitate tumor regression, *Proc. Natl. Acad. Sci. USA* **94**:10889–10894.

Salvadori, S., Gansbacher, B., Wernick, I., Tirelli, S., and Zier, K., 1995, B7-1 amplifies the response to interleukin-2-secreting tumor vaccines *in vivo*, but fails to induce a response by naive cells *in vitro*, *Hum. Gene Ther.* **6**:1299–1306.

Schwartz, R.H., 1992, Costimulation of T lymphocytes: the role of CD28, CTLA-4, and B7/BB1 in interleukin-2 production and immunotherapy, *Cell* **71**:1065–1068.

Tahara, H., Zitvogel, L., Storkus, W.J., Zeh, H.J., McKinney, T.G., Schreiber, R.D., Gubler, U., Robbins, P.D., and Lotze, M.T., 1995, Effective eradication of established murine tumors with IL-12 gene therapy using a polycistronic retroviral vector, *J. Immunol.* **154**:6466–6474.

Tahara, H., and Lotze, M.T., 1995, Antitumor effects of interleukin-12 (IL-12): applications for the immunotherapy and gene therapy of cancer, *Gene Ther.* **2**:96–106.

Zier, K., Gansbacher, B., and Salvadori, S., 1996, Preventing abnormalities in signal transduction of T cells in cancer: the promise of cytokine gene therapy, *Immunol. Today* **17**:39–45.

INTRALESIONAL VACCINIA/GM-CSF RECOMBINANT VIRUS IN THE TREATMENT OF METASTATIC MELANOMA

Michael J. Mastrangelo, Henry C. Maguire, and Edmund C. Lattime*

Jefferson Medical College of
Thomas Jefferson University
1025 Walnut Street
Philadelphia, Pennsylvania 19107, USA

1. INTRODUCTION

Requisite to the development of an immune response is an array of cytokines which provide co-stimulatory signals to facilitate T lymphocyte activation. Approaches to enrichment of the cytokine milieu have included systemic administration (Maguire, 1995) as well as direct injection at the immunization site (Harris *et al.*, 1998; Russo *et al.*, 1989). We and others hypothesized that higher, and perhaps more effective, cytokine levels could be achieved at immunization sites by active local production rather than passive transfer. The demonstration by Golumbeck *et al.* (1991) that immunization with tumor cells engineered to secrete interleukin-4 could induce regression of an established syngeneic renal cell cancer (RENCA) in mice has led to clinical trials of the methodology. However, the application of this technique in the clinic is severely limited by the requirements that autologous tumor be removed, grown in tissue culture, successfully transfected and prepared for safe re-introduction into the host. Our over-all goal has been to circumvent this cumbersome procedure by using a recombinant virus to insert genes of interest into tumor cells in vivo so that cytokines will be produced in situ, thus facilitating the generation of a therapeutically effective systemic anti-tumor immune response.

In the present study, vaccinia virus was used as a vector for the transfer and expression of the granulocyte-macrophage colony stimulating factor (GM-CSF) cDNA, as well as the cDNA of the reporter gene (beta glalactosidase), to melanoma cells in vivo by direct intratumoral injection. Parameters assessed included toxicity, cytokine and passenger gene expression, anti-vaccinia immunity and tumor regression.

*Current Address: The Cancer Institute of New Jersey 195 Little Albany Street New Brunswick, NJ 08901, USA

Cancer Gene Therapy: Past Achievements and Future Challenges, edited by Habib
Kluwer Academic/Plenum Publishers, New York, 2000.

1.1. Gene Transfer Vector

Mastrangelo *et al.* (1996) reviewed the attributes of various viral and non-viral vectors for gene transfer. Vaccinia virus seemed well suited for intralesional therapy (Moss, 1991). It is highly infectious and targets many cell types. Viral lysis of infected tumor cells directly reduces tumor burden. The wild-type virus alone will trigger a cytokine cascade and recruit helper T lymphocytes to the injection site (Schimizu *et al.*, 1988).

Vaccinia has been successfully employed for gene insertion in patients at risk for AIDs (Cooney *et al.*, 1991) and in colon cancer patients (Cole *et al.*, 1996). The large size of the virus allows it to serve as a carrier for multiple genes as needed. Encoded genes are delivered into and function in the cytoplasm of the host cell. The lytic property of the virus as well as the host's anti-vaccinia immune response prevents unrestricted (and potentially dangerous) cytokine production. Most importantly, vaccinia virus can be given safely (Mastrangelo *et al.*, 1995; Vaccinia, 1985).

1.2. Cytokine

GM-CSF is a hematopoietic growth factor that has been shown to have significant effects on the immune response by 1) increasing class II MHC antigen expression, 2) enhancing dendritic cell maturation and migration, 3) inducing localized inflammation at the site of injection, and 4) exerting immune modulatory effects by its systemic actions on the hematopoietic cytokine network (Jones *et al.*, 1994). Recognition of these properties led to exploration of the possible role of GM-CSF as a vaccine adjuvant. Important to our selection of GM-CSF were the findings of Dranoff *et al.* (1993). These investigators included GM-CSF in a panel of ten molecules, that were engineered to be expressed by tumor cells, to compare their ability to enhance the immunogenicity of the tumor cells. In the B16 melanoma model, nontransduced cells possessed little ability to stimulate systemic anti-tumor immunity. Immunization of syngeneic (C57BL/6) female mice with irradiated B16 melanoma cells expressing murine GM-CSF protected almost all animals against a subsequent challenge with 10^6 nontransduced B16 cells. GM-CSF transduced B16 cells were dramatically more effective (10/10 protected) than wild-type B16 cells (0/10) or cells transduced with genes for interleukins 1,2,3,4,5 or 6, gamma interferon, TNF alpha, ICAM-1 or CD2. Only IL4- and IL6-transduced B16 cells afforded any protection (2/10 and 2/10 respectively) to a challenge with 10^6 nontransduced B16 cells.

2. MATERIALS

2.1. Vaccinia/GM-CSF Recombinant Virus

The procedures used to generate the human GM-CSF producing recombinant virus have been described in detail (Mastrangelo *et al.*, 1999). In brief, the wild-type virus was derived from the New York City Board of Health strain of vaccinia (Center for Disease Control, Atlanta, GA). The cDNA for GM-CSF was obtained from the American Type Culture Collection (Rockville, Maryland). Insertion into the vaccinia thymidine kinase gene was by homologous recombination. The beta galactosidase gene (Lac Z) was included as a reporter gene. Assay for biological activity utilized the GM-CSF dependent

Table 1. Patient population

Patient	Age/Sex	Metastatic Sites	Prior Treatment
#1	81/F	Dermal, Lymph Nodes	Radiation
#2	68/M	Subcutaneous, Lymph Nodes, Lungs	BCDT, TAXOL
#3	32/F	Dermal, Breast	Radiation, BCDT, DCV + IL-2 + IFN, TAXOL
#4	61/F	Dermal, Subcutaneous, Lymph Nodes, Lungs	BCDT, IFN
#5	71/F	Dermal, Lymph Nodes	Limb Perfusion, GP100, MART-1, IL-12, BCDT
#6	67/M	Subcutaneous, Lymph Nodes, Lungs	BCDT
#7	75/M	Dermal	None

BCDT, BCNU (Carmustine) + Cisplatin + Dacarbazine + Tamoxifen; DCV, Dacarbazine + Cisplatin + Vinblastine; IL, Interleukin; IFN, Interferon Alpha; GP100 & MART-1, Melanoma Associated Antigens
Reprinted with permission from Mastrangelo *et al.*, 1999.

Mo7e cell line with recombinant GM-CSF as a standard. Infection of 10^6 143B (human osteosarcoma) cells with a multiplicity of ten plaque forming units (PFU) per cell generated approximately 10^5 units of GM-CSF or 0.1 u/cell.

2.2. Patient Population

Seven patients, immunocompetent as determined by a positive cutaneous delayed type hypersensitivity contact skin response to dinitroflurobenzene, with histologically documented and surgically incurable melanoma were treated. The clinical characteristics of these patients are summarized in Table 1. Only one patient (#3) was less than sixty years of age. Six of seven patients had prior non-surgical treatment and five of these patients had been aggressively treated with combination chemotherapy. Only patient (#7) had intact and clinically uninvolved regional lymph nodes.

3. METHODS

3.1. Overall Treatment Plan

As a prophylaxis for progressive/generalized vaccinia, these patients, with incurable melnaoma, who had had a primary vaccination decades earlier, were re-vaccinated (day 0) with the wild-type vaccinia using a standard scratch technique. The vaccination site was evaluated visually on day 4 to determine whether a "major reaction" was in progress. If such were the case, intralesional therapy commenced on day 4 and was repeated twice weekly with dose escalations based on the local (erythema/inflammation) and systemic (clinical symptoms, physical signs and clinical laboratory values) toxicity from the preceding injections. Toxicity was graded using the NCI Common Toxicity Criteria.

The induction phase of treatment lasted approximately six weeks. Stable and responding patients were continued on treatment until disease resolution or progression.

Doses of intratumorally injected vaccinia ranged from 10^4 to 2×10^7 PFU/lesion and 10^4 to 8×10^7 PFU/treatment session.

3.2. Gene Product Expression

3.2.1. Reverse Transcriptase-Polymerase Chain Reaction (RT-PCR) for Vaccinia (VAC) GM-CSF Message. Total RNA was isolated from melanoma biopsies that had been snap-frozen and stored in liquid nitrogen. The RNA was quantified spectrophotometrically prior to reverse transcription. The synthesized cDNA was amplified by polymerase chain reaction. The primers were designed to amplify only the vaccinia encoded GM-CSF by including flanking regions of the viral plasmid in the primers. Beta actin served as a standard for comparison.

3.2.2. Reporter Beta Galactosidase Gene Expression. Host production of anti-beta galactosidase antibodies served as an index of beta galactosidase production. Serum titers were measured by Elisa using 96 well plates coated with beta galactosidase. Anti-beta galactosidase antibody titers were read as the reciprocal serum dilution yielding 50% maximum absorbance in the assay.

3.3. Cellular Reaction at Injection Sites

Punch biopsies were performed pre-treatment and at various times during the course of treatment. The specimens were processed for routine histology (H&E staining) and immunohistochemistry to elucidate infiltrating cell subsets.

3.4. Humoral Immunity to Vaccinia

The procedure was similar to that used for anti-beta galactosidase antibody except that the wells were coated with vaccinia obtained from cultures of human osteosarcoma cell line 143B infected with the CR-19 strain of vaccinia virus.

4. RESULTS

The results of this trial have been published in detail (Mastrangelo *et al.*, 1999) and are summarized here. The clinical outcome data are updated.

4.1. Toxicity

Inflammation developed at sites injected with the vaccinia/GM-CSF recombinant virus. The intensity and frequency of the inflammation were directly related to dose, being a consistent finding in lesions injected with 10^7 PFU or greater per treatment. Important to the assessment of the potential utility of this approach as a cancer treatment was our ability to infect cells (as evidenced by the development of inflammation) in newly injected lesions late (28 or more days) in the course of treatment despite the presence of anti-vaccinia antibodies. In approximately 10% of instances where patients were treated with a total dose of 4×10^7 PFU or more recombinant virus at one session, mild flu-like symptoms developed within 4–6 hours and resolved spontaneously within 24 hours. No other clinical toxicity and no laboratory toxicity was noted.

4.2. Cellular Reaction at Injection Sites

Typically dermal and subcutaneous melanoma metastases are microscopically devoid of acute or chronic inflammation and lack infiltration by T lymphocytes. Our program of twice weekly injections of 10^7 PFU or more of recombinant virus induced complete necrosis of most lesions that were less than or equal to about 1 cm in diameter. In instances where there was sufficient cellular detail to permit microscopic assessment, lesions were acutely inflamed and densely infiltrated with T lymphocytes (CD3+) of both the CD4+ and CD8+ subsets as well as by tissue macrophages (CD68+). T lymphocytes displaying the activation marker CD45RO as well as activated B lymphocytes (CD30+) were present.

4.3. Humoral Immune Response

4.3.1. Anti-Vaccinia Antibody Response. Our observation that with re-vaccination all seven patients developed major local reactions equivalent to a primary take with a two week interval to resolution indicates that there was essentially no residual immunity from prior vaccination. Indeed, no patient had detectable antibody titers pre-treatment. Representative post-treatment titers are presented in Table 2. Despite continuing treatment with recombinant vaccinia virus, titers tended to plateau at 4 to 6 weeks. Repeated administration of recombinant vaccinia virus failed to boost antibody titers.

4.3.2. Reporter Gene Expression. All seven patients developed a humoral immune response to the reporter gene protein product, beta galactosidase (Table 2). By comparison with the immune response to vaccinia, the anti-beta galactosidase antibody response was slower to develop (3–6 weeks) and seemingly of lesser magnitude. This likely relates to the relative doses and the timing of availability of the respective antigens. Nonetheless, this humoral immune response is clear evidence that the reporter gene was expressed.

Table 2. Humoral immune response to vaccinia virus and the reporter gene product

	Vaccinia Virus		Beta Galactosidase	
Patient	Day*	Antibody Titer⁺	Day**	Antibody Titer
#1	21	1:180	41	1:200
#2	86	1:80	82	1:30
#3	28	1:200	55	1:100
#4	25	1:400	21	1:200
#5	39	1:200	38	1:75
#6	11	>1:300	36	1:80
#7	12	1:100	29	1:50

*Days from re-vaccination with wild-type vaccinia.
**Days from initiation of treatment with recombinant vaccinia.
⁺Titers are the serum dilution yielding 50% of maximum absorbance in the assay.

4.4. Passenger Cytokine (VAC GM-CSF) Gene Function

RT-PCR for VAC GM-CSF mRNA was technically satisfactory in only three patients (#3, #6, #7) who had amelanotic or poorly pigmented tumors. In patients #6 and #7 biopsies were obtained eighteen hours after a single injection of recombinant virus (5 or 7×10^6 PFU/lesion) on day 4 (most likely prior to the development of significant levels of anti-vaccinia antibodies). In these same patients biopsies were obtained in week 5 (anti-vaccinia antibodies present), eighteen hours after a single injection of 20 or 30×10^6 PFU of recombinant virus into a previously untreated lesion. VAC GM-CSF mRNA was found in all samples, convincingly demonstrating passenger cytokine gene function. Patient #3 had been under treatment for thirty-one weeks at the time of biopsy. One lesion was sampled eighteen hours after the last 5×10^6 (of a total of 40×10^6) PFU of recombinant virus were injected. Two additional lesions were sampled eighteen hours after a single injection of 5×10^6 PFU of recombinant virus. In all three biopsies virus derived GM-CSF mRNA was found.

An increased number of eosinophils have been demonstrated in the cutaneous reactions sometimes associated with systemic administration of GM-CSF (Mehreyan *et al.*, 1992). In an effort to determine whether or not functionally significant amounts of VAC GM-CSF were being produced at treatment sites, tumor biopsies were assessed for eosinophilia. Marked infiltration by eosinophils was seen in three of three patients so studied. In comparable specimens injected with wild-type vaccinia there was no eosinophilia (Mastrangelo *et al.*, 1995).

4.5. Tumor Responses

The responses of treated and untreated metastases by location and patient are summarized in Table 3. Patient's #2 and #6 had by far the greatest tumor burdens and neither injected nor uninjected lesions regressed. Patients #1, #4 and #5 had mixed responses characterized by regression of injected dermal lesions in all patients and uninjected

Table 3. Tumor response overall and by type and location of metastasis

Patient	Injected	Uninjected		Overall
		Regional	Distant	Response
#2	Subcutaneous-No Lymph Node-No	Lymph Node-No	Lung-No	None
#6	Subcutaneous-No Lymph Node-No	Subcutaneous-No	Lung-No	None
#1	Dermal-Yes Lymph Node-No	Dermal-Yes	—	Mixed
#4	Dermal-Yes Subcutaneous-Yes	Lymph Node-No	Lung-No	Mixed
#5	Dermal-Yes	Dermal-Yes	Lymph Node-No	Mixed
#3	Dermal-Yes Breast-No	Dermal-Yes	—	PR (14 + MOS)
#7	Dermal-Yes	Dermal-Yes	—	CR (4 + MOS)

Modified with permission from Mastrangelo *et al.* 1999.

dermal lesions in both patients (#1, #5) in whom such were present. In patient #4 injected subcutaneous metastases regressed. Despite this favorable response in dermal and subcutaneous lesions, regional or distant nodal or lung metastases inexorably progressed in these three patients.

Patient #3 is a remarkable young woman who prior to entry onto this study underwent successful treatment for oral, lymph node, lung, bone and brain metastases from a melanoma of unknown primary site (Table 1). She went on to develop parenchymal metastases of the right breast and dermal metastases on the left leg (mostly thigh). Intratumoral treatment commenced in August of 1996. She was treated approximately twice weekly during the follow intervals: 8/18/96-9/16/96, 10/7/96-11/4/96, 12/12/96-12/18/96, 1/15/97-2/13/97, 2/27/97-3/27/97 and 9/18/97-10/16/97. One to eight lesions were injected at each treatment with lesional doses ranging from 10^6 to 2×10^7 PFU. In November of 1997 she was convincingly free of dermal metastases. The appearance of the left thigh is shown pre-treatment, Day +81 and Day +600 in Fig. 1. An uninjected lesion biopsied on Day +31 was infiltrated with CD8+ lymphocytes but devoid of mRNA for vaccinia thymidine kinase or VAC GM-CSF gene product nor was eosinophilia noted. Signal dermal lesions were left uninjected and these regressed as well. The parenchymal breast metastases also were injected. Clinical examination and imaging proved ineffective in distinguishing treatment related swelling from tumor progression. Thus, these lesions were surgically excised converting the treatment induced partial response to a surgical complete remission. Microscopically there were foci of tumor necrosis with minimal inflammation possibly related to the injections of recombinant vaccinia three months before. This patient remained clinically tumor free as of 01/01/99.

Patient #7 was a seventy-five year old male when he presented with approximately twenty-four, 2–3 mm dermal metastases at the perimeter of an area from which a primary melanoma of the scalp had been excised. The regional lymph nodes were intact and clinically free of melanoma. He had had no prior treatment other than for surgery. All injected and uninjected lesions regressed completely after an interval of eight months. Subsequently a pigmented lesion appeared within the skin graft at the primary site. Biopsy revealed melanoma. Five injections of recombinant virus over thirty-four days induced complete remission and the patient was disease free four months later.

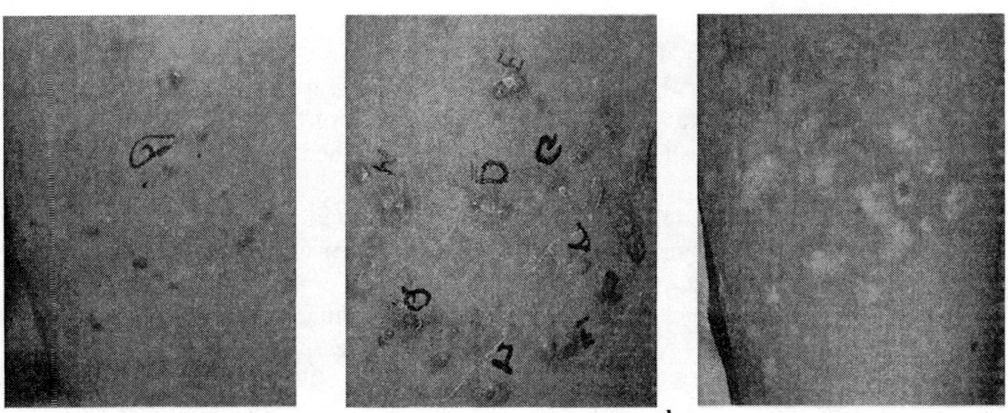

b c

Figure 1. Patient #3, a 32 year old female with extensive dermal metastases of the left thigh: Anterior aspect (a) pretreatment, (b) treatment day +81 and (c) Day +600, 150 days post treatment. (Reproduced with permission from Mastrangelo *et al.*, 1999.)

5. DISCUSSION

Intralesional wild-type vaccinia virus has induced regression of dermal and subcutaneous melanoma metastases. In these same patients, similar but uninjected metastases failed to regress (Mastrangelo *et al.*, 1995). Wild-type vaccinia virus also has been employed as an adjunct to the surgical treatment of melanoma. Everall *et al.* (1975) injected primary melanomas with vaccinia virus two weeks prior to definitive surgical therapy. At four years, 85% of these twenty-three patients were disease free. Only 60% of twenty-five patients alternately allocated to receive only standard excision were similarly disease free. These studies suggest a modest degree of anti-tumor activity. Most importantly there was no serious morbidity from the treatment.

Systemically administered GM-CSF is widely used to reconstitute the bone marrow after intensive chemotherapy. Side effects are well defined and modest (Leukine, 1995).

Despite the favorable toxicity profiles of systemically administered GM-CSF and intracutaneously/intratumorally delivered wild-type vaccinia virus, safety concerns regarding the virulence of the recombinant virus and its use in potentially immunocompromised patients prompted revaccination of study subjects as a precautionary measure. Unfortunately, the resulting anti-vaccinia immunity limits virus proliferation. We hypothesized that in order to maximize the opportunity to generate an immune response against weakly immunogenic tumor cells, cytokine production needed to be maintained at treatment sites for a protracted period, at least, four weeks. However, our experience with wild-type vaccinia indicated that in the face of anti-vaccinia immunity, the persistence of injected virus was almost certain to be less than one week. Further, in designing the recombinant vaccinia, the reporter and cytokine genes were inserted so as to disrupt the continuity of the thymidine kinase gene and attenuate the virulence of the virus (Moss, 1991). To circumvent this problem, a twice weekly treatment schedule was adopted and was successful. Clinical evidence of infection at intratumoral injection sites late in the course of therapy, the generation of antibody to the reporter gene product (beta galactosidase), the presence of RNA message for VAC GM-CSF 5 and 31 weeks after initiation of therapy and eosinophilia at injection sites all support the conclusion that cytokine function could be maintained over a protracted period and seemingly indefinitively. Toxicity was minimal and almost exclusively limited to injection sites.

Although we have demonstrated protracted cytokine and reporter gene function with modest morbidity, anti-tumor activity was narrowly focused to regression of injected dermal and subcutaneous metastases and uninjected regional dermal metastases. Indeed, intralesional treatment with the vaccinia/GM-CSF recombinant virus may be curative in patients with dermal metastases limited to a single region. However, we have yet to observe regression of bulky regional lymph nodes or subcutaneous metastases or of distant disease.

One approach to enhancing therapeutic efficacy is to facilitate virus replication at injection sites. Our clinical impression that the recombinant virus was less virulent that the wild-type virus when used similarly (Mastrangelo *et al.*, 1995) should permit the eventual elimination of the revaccination step, thus, allowing more robust reactions at treatment sites, at least, early on. A second approach is to increase the dosage administered. This is not practical with the current formulation (1.75×10^7 PFU per 0.1 ml). Volume restricts dose escalation in smaller lesions and limited supply precludes saturation of larger lesions. A higher titer formulation is in production. A third possibility under consideration is the use of an immunopotentiating agent such as low dose cyclophosphamide. We have had success with this approach when used in conjunction with an autologous melanoma cell vaccine (Berd *et al.*, 1986).

This approach lends itself to the treatment of a variety of other tumors, most notably intravesical treatment of patients who have failed BCG for superficial bladder cancer. In preclinical studies, Lee *et al.* (1994) instilled influenza hemagglutinin and nuclear protein encoding vaccinia recombinants intravesically in vaccinia-naïve CB7BL/6 mice bearing the MB49 tumor and demonstrated infection/transfection of tumor cells. Similar results were achieved with vaccinia-immune mice. Thus systemic immunity to vaccinia, which would be expected to develop rapidly in bladder cancer patients as it did in the melanoma patients, did not prevent intravesical tumor infection/transfection. Early clinical trials with wild-type vaccinia have similarly demonstrated the ability to infect bladder mucosal/tumor cells in vaccinia-immune patients (Gomella *et al.*, 1997). Collectively our data suggest that this approach will have application for a wide variety of cancers.

6. CONCLUSION

Using vaccinia as a vector we have successfully transferred the GM-CSF and the beta galactosidase cDNAs to melanoma cells in vivo. With twice weekly injection, gene function was sustained over a protracted period despite the presence of anti-vaccinia immunity. Most often, injected metastases regressed and in some patients uninjected metastases regressed as well. Patients with only regional dermal metastases can be placed into complete remission. Toxicity was minimal and virtually limited to injection sites.

REFERENCES

Berd, D., Maguire, H.C., Jr., and Mastrangelo, M.J., 1986, Induction of cell-mediated immunity to autologous melanoma cells and regression of metastases after treatment with a melanoma cell vaccinia preceded by cylophosphamide, *Cancer Res.* **46**:2572–2577.

Cole D.J., Wilson, M.C., Baron, P.L., O'Brien, P., Reed, C., Tsang, K.Y., and Schlom, J., 1996, Phase I study of recombinant CEA vaccinia virus vaccine with post vaccination CEA peptide challenge, *Hum. Gen. Ther.* **7**:1381–1394.

Cooney, E.L., Collier, A.C., Greenberg, P.D., Coombs, R.W., Zarling, J., Arditti, D.E., Hoffman, M.C., Hu, S.L., and Corey, L., 1991, Safety of and immunologic response to a recombinant vaccinia virus vaccine expressing HIV envelope glycoprotein, *Lancet* **337**:567–572.

Dranoff, G., Jaffee, E., Lazenby, A., Golumbek, P., Levitsky, H., Brose, K., Jackson, V., Hamada, H., Pardoll, D.M., and Mulligan, H.C., 1993, Vaccination with irradiated tumor cells engineered to secrete murine granulocyte-macrophage colony stimulating factor stimulated potent, specific and long-lasting anti-tumor immunity, *Proc. Natl. Acad. Sci. U.S.A.*, **90**:3539–3543.

Everall, J.D., O'Doherty, C.J., Ward, J., and Dowd, P.M., 1975, Treatment of primary melanoma by intralesional vaccinia before excision, *Lancet* **2**:583–586.

Golumbek, P.T., Lanzenby, A.J., Levitsky, H.I., Jaffee, L.M., Karasuyama, H., Baker, M., and Pardoll, D.M., 1991, Treatment of established renal cancer by tumor cells engineered to secrete interleukin-4, *Science* **254**:713–716.

Gomella, L.G., Mastrangelo, M.J., Eisenlohr, L.C., McCue, P.A., Monken, C.E., Kovatich, A.J., Mulholland, S.G., Maguire, H.C. Jr., and Lattime, E.C., 1997, Phase I study of intravesical vaccinia virus as a vector for gene therapy of bladder cancer, *Proc. Am. Assoc. Cancer Res.* **38**:9.

Harris, D.T., Matyas G., Mastrangelo, M.J., Muderhwa, J.M., Alving, C., Maida, A., and Spitler, L.E., 1998, OncoVax-P[tm] Vaccine: Pilot studies in prostate cancer patients, *Proc. Am. Soc. Clin. Oncol.* **17**:428a.

Jones, T., Stern, A., and Lin. R., 1994, Potential role of granulocyte-macrophage colony-stimulating factor as vaccine adjuvant, *Eur. J. Clin. Microbiol. Infect. Dis.* **13 (Suppl. 2)**:S47–S53.

Lee, S.S., Eisenlohr, L.C., McCue, P.A., Mastrangelo, M.J., and Lattime, E.C., 1994, Intravesical gene therapy: In-vivo gene transfer using vaccinia vectors, Cancer Res. **54**:3325–3328.

Leukine, in: *Physician's Desk Reference*, 49th Edition, 1995, pp 1161–1164, Medical Economics Data Production Co., Montvale, N.J., U.S.A.

Maguire, H.C., Jr., 1995, Murine recombinant interleukin-12 increases the acquisition of allergic contact dermatitis in the mouse, *Int. Arch. Allergy Immunol.* **106**:166–168.

Mastrangelo, M.J., Maguire, H.C., Jr., McCue, P., Lee, S.S., Alexander, A.A., Nazarian, L.N., Eisenlohr, L.C., Nathan, F., Berd, D., and Lattime, E.C., 1995, A pilot study demonstrating the feasibility of using intratumoral vaccinia injections as a vector for gene transfer, *Vaccine Res.* **4**:55–69.

Mastrangelo, M.J., Berd, D., Nathan, F.E., and Lattime, E.C., 1996, Gene therapy for human cancer: An essay for clinicians, *Semin. Oncol.* **23**:4–21.

Mastrangelo, M.J., Maguire, H.C., Jr., Eisenlohr, L.C., Laughlin, C.E., Monken, C.E., McCue, P.A., Kovatich, A.J., and Lattime E.C., 1999, Intratumoral recombinant GM-CSF encoding virus as gene therapy in patients with cutaneous melanoma, *Cancer Gen. Ther.* (in press).

Mehregan, D.R., Fransway, A.F., Edmonson, J.H., and Leiferman, K.M., 1992, Cutaneous reactions to granulocyte-macrophage colony-stimulating factor, *Arch. Dermatol.* **128**:1055–1059.

Moss, B., 1991, Vaccinia virus: A tool for research and development, *Science* **252**:1660–1667.

Russo, D.M., Sundy, J.S., Young, J.F., Maguire, H.C., Jr., and Weidanz, W.P., 1989, Cell mediated immune responses to vaccine peptides derived from the circumsporozoite protein of Plasmodium falciparum, *J. Immunol.* **143**:655–659.

Shimizu, Y., Hasumi, K., Masubuchi, K., and Okudaira, Y., 1988, Immunotherapy of tumor-bearing mice utilizing virus help, *Cancer Immunol. Immunother.* **27**:223–227.

Vaccinia (Smallpox) vaccine. Recommendations of the Immunization Practices Committee, 1985, *MMWR* **34**:341–342.

IX

SUICIDAL GENES

35

APPROACHES TO GENE-DIRECTED ENZYME PRODRUG THERAPY (GDEPT)

Caroline J. Springer* and Ion Niculescu-Duvaz

CRC Centre for Cancer Therapeutics at the Institute of Cancer Research
15 Cotswold Road
Sutton, Surrey, SM2 5NG

1. INTRODUCTION

Gene therapy for cancer may be broadly defined as a genetic technology aimed at modifying either malignant or non malignant cells for therapeutic gain. There are two "suicide" gene therapy approaches. In toxin gene therapy, genes are introduced into malignant cells and the translated protein directs killing of the tumour cells. The second is termed GDEPT (gene-directed enzyme prodrug therapy) (Niculescu-Duvaz *et al.*, 1998) or VDEPT (virally-directed enzyme prodrug therapy) (Huber *et al.*, 1995). The only difference between these acronyms is that the former involves both viral and non-viral vectors. This review covers the GDEPT approach to make tumour cells more sensitive to chemotherapeutic agents.

2. PRINCIPLES OF GDEPT

GDEPT is a two-step treatment for solid tumours. Foreign enzymes are delivered to, and expressed in target cells where they can activate subsequently administered non-toxic prodrugs to active drugs. In the first step, a gene expressing the foreign enzyme is delivered. In the second step, a prodrug is administered that can be activated to a toxic drug by the enzyme that has been expressed in the tumour. There are a wide range of enzyme/prodrug systems which have been investigated (see Table 1).

The foreign enzyme gene should be expressed exclusively, or with a relatively high ratio in tumour cells compared to normal tissues and blood, and should achieve a sufficient concentration for a clinical benefit. After gene delivery, prodrug administration must be delayed to permit protein expression in the targeted cells. The catalytic activity of the expressed protein should be sufficient for activation of the prodrug.

---

* Author to whom correspondence should be addressed, Tel: 44-181-643 8901, Fax: 44-181-770 7899

Cancer Gene Therapy: Past Achievements and Future Challenges, edited by Habib
Kluwer Academic / Plenum Publishers, New York, 2000.

Table 1. Enzyme/prodrug GDEPT systems[a]

No	Enzyme	Prodrug	Metabolite
1	HSV-TK	GCV, ACV, 1(2′-deoxy-2′-fluoro-β-D-arabinofuranosyl)-5-iodouracil, (E)-5-(2-bromovinyl)- and (E)-5-(2-iodovinyl)-2′-deoxyuridine, 9-(1,3-dihydroxy-2-propoxymethyl)-guanine	nondiffusible
	VZV-TK	6-methoxypurine arabinonucleoside	nondiffusible
2	CD	5-fluorocytosine	diffusible
3	CYP	CP, IF, ipomeanol, 2-aminoanthracene[b]	diffusible
4	PNP	6-methyl-purine, 2-fluoro-adenosine	diffusible
5	TP	5′-DFUR, 2-chloro-2′-deoxyadenosine, 2-fluoro-9-(b-D-arabinofuranosyl)-cytosine, 2′,2′-difluoro-deoxycytidine	nondiffusible
6	dCK	AraC	nondiffusible
7	CPG2	CMDA	diffusible
8	NR	CB 1954	diffusible
9	XGPRT	6-thioguanine, 6-thioxanthine	nondiffusible
10	CE	CPT-11[c]	diffusible

[a]For reference see (Niculescu-Duvaz *et al.*, 1998) and references therein; [b](Rainov *et al.*, 1998); [c](Kojima *et al.*, 1998).

Since expression of the foreign enzymes will not occur in all cells of a targeted tumour *in vivo*, a bystander cytotoxic effect is required whereby the prodrug is cleaved to an active drug that kills not only tumour cells but also neighbouring nonexpressing tumour cells. This means that expression in less than 100% of tumour cells can still give total tumour cell kill.

The foreign enzyme is usually expressed intracellularly (Niculescu-Duvaz *et al.*, 1998) but by expressing the activating enzyme tethered to the outer cell surface of mammalian cells, potential advantages for GDEPT prodrug design are realised (Marais *et al.*, 1997). Since activation in this case occurs in the interstitial spaces of the tumour, the released drug should be taken up by all cells equally, whether or not they express the activating enzyme. Thus, the bystander effect may be increased. The bacterial carboxypeptidase G2 (CPG2) has been expressed intracellularly in colorectal and ovarian tumour cell lines (Marais *et al.*, 1996). Treatment of the enzyme expressing cells with the prodrug (2-chloroethyl)(2-mesyloxyethyl)aminobenzoyl-L-glutamic acid (CMDA) led to enhanced cytotoxicities and bystander effects. A breast tumour line, however, was resistant to the activating effects of intracellularly expressed CPG2 because CMDA prodrug was not cell membrane permeable. Expression of a mutated CPG2, tethered to the outer cell surface led to enhanced cytotoxicity and bystander effect with CMDA (Marais *et al.*, 1996).

3. DELIVERY SYSTEMS AND EFFICIENCIES

A large number of vector systems has been proposed for gene therapy. These include retrovirus (Vile *et al.*, 1995), adenovirus (Roth and Cristiano, 1997), parvovirus (Kremer and Perricaudet, 1995), herpes simplex virus (Glorioso *et al.*, 1995), cationic lipids (Miller, 1998), peptides (Wadha *et al.*, 1997) and naked DNA (Spooner *et al.*, 1995). Targeted delivery of DNA by antibodies, ligands or peptides has been demonstrated (Ebbinghaus *et al.*, 1996).

Transfer efficiency is of paramount importance in gene therapy. *Ex vivo* transfer efficiencies of almost 100% have been achieved with viral carriers. However, *in vivo* transfer efficiencies are lower. The most commonly used vectors have been retrovirus and adenovirus. Retroviruses mediate stable gene transfer to dividing cells only with low potential immunogenicity. Direct injection of retrovirus into brain tumours resulted in low efficiency expression (1–3%). However, 10–55% transfer efficiencies were reported, with direct implantation of retrovirus producer cells in such systems (Culver, 1996). Adenoviruses generate infection regardless of the proliferative status of the cells. Efficiencies greater than 50% (at virus titre, 10^{10} pfu.mL^{-1}). have been observed in mouse hepatocytes after an i.v. injection of adenovirus (Kanai *et al.*, 1996).

4. ENZYME-PRODRUG/DRUG SYSTEMS

The enzymes used in GDEPT must be able to catalyse scission or other types of reactions (e.g.: redox reactions) and should be different from any circulating endogenous enzyme accessible to prodrug. They should be expressed in sufficient amounts in the tumour target cells and able to effect high catalytic activity. It is possible to improve wild type enzymes. For example, mutants have been created in herpes simplex virus thymidine-kinase (HSV-TK) that improved the kinetics of the enzyme for the prodrugs ganciclovir (GCV) and acyclovir (Black *et al.*, 1996). Also bacterial CPG2 has been engineered to express an enzyme tethered to the outer membrane of mammalian cells (see above) (Marais, *et al.*, 1997). Alternatively, the prodrug may be modified to make it a better substrate for the enzyme. (Friedlos *et al.*, 1997, Niculescu-Duvaz *et al.*, 1998)

The enzyme/prodrug systems proposed for GDEPT are summarised in Table 1.

As described earlier the prodrug must be tumour cell membrane permeable for intracellular activation by the enzyme. The cytotoxicity differential between the prodrug and its corresponding drug should be as large as possible and the active drug should be highly diffusible or be actively taken up by neighbouring cells to achieve the bystander effect.

5. *IN VIVO* MODELS

The HSV-TK/GCV system is the most intensively studied. It was demonstrated that the cytotoxicity toward tumour cells was proportional to the level of TK expression and dose dependent (Borelli *et al.*, 1988). In these *in vivo* studies numerous long-term or tumour regressions have been documented. HSV-TK has been used in animal models of hepatocellular carcinoma, using the α-fetoprotein promoter and in brain tumours (Culver *et al.*, 1992). It has been used in treatment of metastatic liver disease with retroviral delivery and for melanoma with the tyrosinase promoter (Vile and Hart, 1993b).

Colorectal carcinoma models have been developed to test 5-fluorocytidine (FC)/cytosine deaminase (CD) as a therapy for metastatic disease in the liver. Different rates of regression were observed (6–47%), depending on the parental tumour line used, indicating that tumour cells differ greatly in their sensitivity to the 5-FC/CD system. The animals that survived the experiment were found to be resistant to subsequent re-challenge with the unmodified parental cell line used, indicating that long-term immune responses had been induced (Huber *et al.*, 1995).

In xenograft models in rats, tumours expressing cytochrome 2B1 (CYP2B1) were found to be sensitive to subsequent cyclophosphamide (CP) administration. After

8 weeks no control animals were alive, whereas 19 out of 20 carrying the CYP2B1-expressing tumours survived (Wei *et al.*, 1994). In model therapy experiments in nude mice with pre-transfected breast carcinoma xenografts expressing the surface tethered CPG2, treatment with CMDA led to tumours regressions and a cure. The bacterial enzyme nitroreductase has been expressed in NIH3T3 cells (Bridgewater *et al.*, 1995) and gives a large IC_{50} differential with the prodrug CB1954 and other prodrugs (Friedlos *et al.*, 1997) Ovarian and colorectal tumours xenografts were made to exppress nitroeductase from an adenoviral vector, followed by CB1954 (McNeish *et al.*, 1998) and significant regressions were obtained.

6. CLINICAL DATA

HSV-TK/GCV is considered to be safe for clinical trials with retroviral producer cells since the cells should be destroyed by GCV treatment. More than 22 clinical studies are had begun by 1996 in glioblastoma, melanoma, head and neck tumours, mesothelioma, multiple myeloma and ovarian carcinoma (Roth and Cristiano, 1997, Sterman *et al.*, 1998).

7. THE BYSTANDER EFFECT AND ACQUIRED IMMUNITY

It was demonstrated, *in vivo,* that when as few as 2% of tumour cells expressed the cytidine deaminase enzyme, subsequent treatment with the 5-FC prodrug gave long term animal survivors. It is considered that an expression efficiency of 10–20% should be enough to achieve 100% cell kill in tumours and efficiencies of 1–5% are sufficient for a therapeutic gain (Huber *et al.*, 1994).

The systems reported have exhibited a significant bystander effect. However, the mechanisms are probably different for individual drugs. For instance, for the toxic nucleoside phosphates resulting from the activation of purines such as GCV, ACV or pyrimidine nucleosides such as arabinosylcytosine (Ara-C), 5′-deoxy-5-fluorouridine (5-FUDR), which do not diffuse across cell membranes, cell to cell contact is required to obtain the bystander effect (Freeman *et al.*, 1993). By contrast, for diffusible drugs such as 5-fluorouracil (5-FU) nitrogen or phosphoramide mustards, 6-mercaptopurine or 2-fluoradenosine, no such requirement is necessary (Chen and Waxman, 1995; Huber *et al.*, 1994).

It has also been demonstrated that different tumour cell types give different results even within the same prodrug/enzyme system. For example, using intracellular expressed CPG2 *in vitro* in the colorectal carcinoma cell line WiDr, 12% transfected cells were required for total cell kill, whereas with the colorectal carcinoma cell line LS174T only 4% were needed to give 100% cytotoxicity (Marais *et al.* 1996).

The bystander effect *in vivo* can be mediated through the release of cytokines in immunocompetent animals. Coinjection of mouse colon carcinoma cells with HSV-TK in retroviral packaging cells followed by GCV, resulted in almost total tumour ablation in immunocompetent BALB/c mice but not in immunocompromised athymic BALB/c animals (Gagandeep *et al.*, 1996). In another study, HSV-TK expressing 3T3 cells transformed by Ki-Ras V12G were sensitive to GCV. Tumour growth with these cells was inhibited for up 50 days and then regrew following GCV treatment in immunocompromised nude mice. By contrast, in immunocompetent BALB/c mice, similar GCV treatment led to the development of a long lasting immunity (Pavlovic *et al.*, 1996). These studies suggest strongly that an intact immune system enhances the therapeutic gain of GDEPT.

8. FUTURE PERSPECTIVES

The major hurdle is for improved gene delivery systems. Advances in the development of better and more efficient delivery systems for exogenous genes together with restricted expression in the tumour alone are being sought and in combination with the appropriate enzyme/prodrug systems will hopefully lead to improved responses for cancer.

ACKNOWLEDGMENT

We thank Dr. R. Marais and Dr. R. Spooner for their helpful comments. We are also grateful to Prof. P. Workman for support and Cancer Research Campaign for funding.

ABBREVIATIONS LIST

HSV-TK: herpes simplex virus thymidine kinase;
VZV-TK: varicella zoster virus thymidine kinase;
GCV: ganciclovir;
ACV: acyclovir;
CD: cytosine deaminase;
CYP: cytochrome P-450;
PNP: purine nucleoside phosphorylase;
TP: thymidine phosphorylase;
dCK: deoxycytidine kinase;
CPG2: carboxypeptidase G2;
NR: nitroreductase;
XGPRT: xanthine-guanine phosphoribosyl transferase;
CE: carboxyl esterase;
CF: cyclophosphamide;
IF: ifosphamide;
CMDA: (2-chloroethyl)(2-mesyloxyethyl)aminobenzoyl-L-glutamic acid
CB1954: 5-aziridinyl-2,4-dinitrobenzamide;
CPT-11: irinotecan.

REFERENCES

Black, M.E., Newcomb, T.G., *et al.*, 1996, Creation of drug-specific herpes simplex virus type 1 thymidine kinase mutants for gene therapy, *Proc. Natl. Acad. Sci., USA*, **93**:3525–3529.

Borelli, E., Heyman, R., *et al.*, 1988, Targeting of an inducible toxic phenotype in animal cell., *Proc. Natl. Acad. Sci., USA*, **85**:7572–7576.

Bridgewater, G., Springer, C.J., *et al.*, 1995, Expression of the bacterial nitroreductase enzyme in mammalian cells renders them selectively sensitive to killing by the prodrug CB1954, *Eur. J. Cancer*, **31A**:2362–2370.

Chen, L., and Waxman, D.J., 1995, Intratumoral activation and enhanced chemotherapeutic effect of oxazaphosphorines following cytochrome P-450 gene transfer: Development of a combined chemotherapy/cancer gene therapy strategy, *Cancer Res.*, **55**:581–589.

Culver, K.W., 1996, The role of herpes simplex thymidine kinase gene transfer in the drug treatment of brain tumors, *CNS Drugs*, **6**:1–11.

Culver, K.W., Ram, Z., *et al.*, 1992, In vivo gene transfer with retroviral vector-producer cells for treatment of experimental brain tumors, *Science*, **256**:1550–1552.

Ebbinghaus, S.W., Vigneswaran, N., *et al.*, 1996, Efficient delivery of triplex forming oligonucleotides to tumour cells by adenovirus-polylysine complexes, *Gene Ther.*, **3**:287–297.

Freeman, S.M., Abboud, C.N., *et al.*, 1993, The "bystander effect": Tumor regression when a fraction of the tumor mass is genetically modified, *Cancer Res.*, **53**:5274–5283.

Friedlos, F., Denny, W.A., *et al.*, 1997, Mustard prodrugs for activation by Escherichia coli nitroreductase in gene-directed enzyme prodrug therapy, *J. Med. Chem.*, **40**:1270–1275.

Gagandeep, S., Brew, R., *et al.*, 1996, Prodrug-activated gene therapy: Involvement of an immunological component in the "bystander effect", *Gene Ther.*, **3**:83–88.

Glorioso, J.C., DeLuca, N.A., *et al.*, 1995, Development and application of herpes simplex virus vectors for human gene therapy, *Annu. Rev. Microbiol.*, **49**:675–710.

Huber, B.E., Austin, E.A., *et al.*, 1994, Metabolism of 5-fluorocytidine to 5-fluorouracil in human colorectal tumor cells transduced with the cytosine deaminase gene: Significant antitumor effects when only a small percentage of tumor cells express cytosine deaminase, *Proc. Natl. Acad. Sci., USA*, **91**: 8302–8306.

Huber, B.E., Richards, C.A., *et al.*, 1995, VDEPT; An enzyme/prodrug gene therapy approach for the treatment of metastatic colorectal cancer, *Advanced Drug Delivery Reviews*, **17**:279–292.

Kanai, F., Shiratori, Y., *et al.*, 1996, Gene therapy for α-fetoprotein producing human hepatoma cells by adenovirus mediated transfer of the herpes simplex virus thymidine kinase gene, *Hepatology*, **23**: 1359–1368.

Kojima, A., Hackett, N.R., *et al.*, 1998, In vivo human carboxylesterase cDNA gene transfer to activate the prodrug CPT-11 for local treatment of solid tumours, *Journal of Clinical Investigations*, **101**:1789–1796.

Kremer, E.J., and Perricaudet, M., 1995, Adenovirus and AAV mediated gene transfer, *Br. Med. Bull*, **51**: 31–44.

Marais, R., Spooner, R.A., *et al.*, 1996, Gene-directed enzyme prodrug therapy with a mustard prodrug/carboxypeptidase G2 combination, *Cancer Res.*, **56**:4735–4742.

Marais, R., Spooner, R.A., *et al.*, 1997, A cell surface tethered enzyme improves efficiency in gene-directed enzyme prodrug therapy, *Nature Biotechnology*, **15**:1373–1377.

McNeish, I.A., Green, N.K., *et al.*, 1998, Virus directed enzyme prodrug therapy for ovarian and pancreatic cancer using retrovirally delivered *E. coli* nitroreductase and CB1954, *Gene Ther.*, **5**:1061–1069.

Miller, A.D., 1998, Cationic liposomes for gene therapy, *Angewandte Chemie, International Edition*, **37**:1768–1785.

Niculescu-Duvaz, D., Niculescu-Duvaz, I., *et al.*, 1998, Self-immolative mustard prodrugs for suicide gene therapy, *J. Med. Chem.*

Niculescu-Duvaz, I., Spooner, R., *et al.*, 1998, Gene-directed enzyme prodrug therapy, *Bioconj. Chem.*, **9**: 4–22.

Pavlovic, J., Nawrath, M., *et al.*, 1996, Anti-tumor immunity is involved in the thymidine kinase-mediated killing of tumors induced by activated Ki-ras (G12V), *Gene Ther.*, **3**:635–643.

Rainov, N.G., Dobberstein, K.-U., *et al.*, 1998, New prodrug activation gene therapy for cancer using cytochrome P450 4B1 and 2-aminoanthracene/4-ipomeanol, *Human Gene Ther.*, **9**:1261–1273.

Roth, J.A., and Cristiano, R.J., 1997, Gene therapy for cancer: What have we done and where are we going?, *Journal of the National Cancer Institute*, **89**:21–39.

Spooner, R.A., Deonarain, M.P., *et al.*, 1995, DNA vaccination for cancer treatment, *Gene Ther.*, **2**:173–180.

Sterman, D., H., Treat, J., *et al.*, 1998, Adenovirus-mediated herpes simplex virus thymidine kinase/ganciclovir gene therapy in patients with localized malignancy: Results of phase I clinical trial in malignant mesothelioma, *Human Gene Ther.*, **9**:1083–1092.

Vile, R.G., Diaz, R.M., *et al.*, 1995, Tissue-specific gene expression from Mo-MLV retroviral vectors with hybrid LTRs containing the murine tyrosinase enhancer/promoter, *Virol.*, **214**:307–313.

Vile, R.G., and Hart, I.R., 1993b, Use of tissue-specific expression of the herpes simplex virus thymidine kinase gene to inhibit growth of established murine melanomas following direct intratumoral injection of DNA, *Cancer Res.*, **53**:3860–3864.

Wadha, M.S., Collard, W.T., *et al.*, 1997, Peptide-mediated gene delivery: influence of peptide structure on gene expression, *Bioconj. Chem.*, **8**:81–88.

Wei, M.X., Tamiya, T., *et al.*, 1994, Experimental tumor therapy in mice using the cyclophosphamide-activating cytochrome P450–2B1 gene, *Human Gene Ther.*, **5**:969–978.

36

SUICIDE GENE THERAPY

Scott M. Freeman

Schering-Plough Research Institute
Clinical Research Oncology
Kenilworth, New Jersey

INTRODUCTION

The field of cancer therapy has made significant advances in the last three decades, but the overall cancer mortality remains virtually unchanged since the 1960's. Further, although approximately 70% of cancer patients achieve a remission, 50% of cancer patients relapse and become non-responsive to conventional therapeutic regimens. Patients with advanced cancers, particularly those in replaces, are the least responsive to therapy and have a median life expectancy of only one and a half years (Feldman, 1986; Boring, 1992). The progression of disease is often a result of growth of tumors cells that are resistant to any combination of chemotherapy, surgery, and radiation (Goldstein, 1991; Goldstein, 1992). Thus, effective treatment will depend on the development of new strategies that attack cancers that are resistant to conventional approaches.

Advances in molecular biology have allowed the elucidation of many cellular functions involved in tumorigenesis. These tumor cell-specific properties are being exploited in a number of novel therapies in which the cancerous cells themselves are targeted in ways that circumvent their resistance to standard agents. Gene therapy is one approach that allows direct treatment of tumor cells by genetically modifying the malignant cells to alter their drug sensitivity phenotype. A suicide gene is a gene encoding a protein which under appropriate conditions (i.e., exposure to a prodrug) is toxic to the genetically modified cell (i.e., tumor cell).

The treatment of cancer is different from treatment of genetic diseases, since an effective cancer therapy requires complete eradication of all tumor cells. Although an effective cancer therapy requires that all tumor cells be killed, a number of studies using suicide genes have shown that when less than 100% of the tumor was genetically modified, the entire tumor mass could be eradicated. Thus, this approach could potentially eliminate a cancer even though the technology to genetically modify an entire tumor is not available.

Genetic modification of tumor cells could impart a new drug sensitivity phenotype to chemotherapy resistant tumor cells which would then allow killing of the cells by exposure to a prodrug (Moolten, 1986; Moolten, 1990; Moolten, 1994). Initially, chemosensitization was developed for a theoretical prophylactic clinical application because gene

Cancer Gene Therapy: Past Achievements and Future Challenges, edited by Habib
Kluwer Academic / Plenum Publishers, New York, 2000.

therapy technology could not be used to genetically modify all tumor cells in a tumor mass, a pre-requisite for a therapeutic application. However, further investigation led to the fundamental discovery, termed the "bystander effect", which allowed a therapeutic application of suicide gene therapy (Freeman, 1992; Freeman, 1992; Freeman, 1993; Freeman, 1995). The "bystander effect" demonstrated that a tumor population could be killed if only a fraction of the tumor expressed the HSV-TK gene, and thus was a seminal finding that overcame the limitations of gene therapy technology, i.e., the inability to genetically modify an entire tumor mass (Freeman, 1992; Freeman, 1992; Freeman, 1993; Freeman, 1995).

The "bystander effect" allowed gene therapists to overcome the limitation in technology through a biological mechanism. Understanding the mechanism of the "bystander effect" will allow enhancement of this effect, which should improve the therapeutic potential of suicide gene therapy. Over two dozen clinical gene therapy trials based on the suicide gene model have been initiated in the last five years with two phase I trials complete and a phase III trial nearing completion (Marcel, 1996; Roth, 1997).

NEGATIVE SELECTABLE MARKERS

One of the first approaches to cancer gene therapy was developed by genetically modifying tumor cells with the Herpes Simplex Virus thymidine kinase gene (HSV-TK) and showing that the gene-modified tumor cells were killed by the anti-viral drug ganciclovir (GCV) (Moolten, 1990; Freeman, 1995). GCV is a derivative of acyclovir, which was developed in the 1970's to treat Herpes Simplex Virus infections (Nishiyama, 1975; Field, 1983; Oliver, 1985). These drugs are nucleoside analogues that can be phosphorylated by the viral thymidine kinase gene into a monophosphate form. Normal cellular enzymes can then di- and triphosphorylate the monophosphate form into a toxic drug, which functions as a DNA chain terminator and an inhibitor of the viral DNA polymerase (Elion, 1977; Elion, 1980; Davidson, 1981). The HSV-TK enzyme is almost one thousand fold more efficient at mono-phosphorylating GCV than the cellular thymidine kinase (Elion, 1977). Thus, GCV is virtually non-toxic to uninfected cells at therapeutic concentrations of the drug (1–10 uM), although neutropenia can be a clinical manifestation of long term drug usage (Elion, 1980; Field, 1983; Shepp, 1985). Since the toxic metabolite of GCV is phosphorylated, its ability to cross cell membranes is impaired and thus its half life (t1/2) within the cell is six times longer (18–24 hours) than unmodified GCV (Elion, 1977; Elion, 1980). The increased cellular t1/2 of the phosphorylated GCV is an important feature of the anti-cancer effects of HSV-TK gene-modified tumors, as discussed below. One advantage to this mechanism of action is that chemotherapy resistant tumor cells are sensitive to GCV when they are genetically modified with the HSV-TK gene (Freeman, 1992; Freeman, 1995).

CHEMOSENSITIZATION

Initial studies demonstrated that the HSV-TK gene could be transferred into tumor cells and would confer sensitivity of the gene-modified cells to GCV (Moolten, 1986). Further, the drug kinetics in gene-modified cells were similar to that of virally infected cells (Sheep, 1985; Moolten, 1986; Freeman, 1993; Freeman, 1995). The HSV-TK genetically modified tumor cells were killed in culture at a concentration of 0.5 uM GCV, which

Table 1. Suicide Genes

Gene	Toxic Metab.	Tumoricidal	Bystander
HSV-TK	GCV → GCV-TP	Yes	Yes
CD	5-FC → 5-FU	Yes	Yes
XGPRT (gpt)	6-TX → 6-TX-TP	Yes	?
VZV-TK	AraM → AraM-MP	Yes	?
DeoD	MeP-dr → 6-MeP	Yes	?
β-glucosidase	amygdalin-cyanide	?	?
β-lactamase	vinca-ceph. → vinca alk.	?	?

can be achieved clinically. Furthermore, since the chemotherapy resistant tumor cells were sensitive to HSV-TK/GCV, the mechanism of drug resistance to cancer chemotherapeutic agents developed by tumor cells does not interfere with drug sensitivity to the antiviral agent ganciclovir in the HSV-TK gene-modified tumor cells.

Further development of this concept was initially limited for the following reasons. *In vivo* gene transfer into cells of a tumor is very inefficient, especially when retroviral vectors are used and thus it is not possible to genetically modify an entire tumor. Recently, adenoviral vectors have been developed that transduce >90% of cells *in vivo*, but this vector remains episomal and is lost when the genetically modified cell divides. Thus, these gene-modified cells would not maintain the transgene upon malignant transformation and cell division. Second, *in vivo* gene expression has been relatively short term, with only a small percentage of the transduced cells expressing the recombinant protein after a few months. The inability to maintain long term expression severely limits gene therapy approaches. Third, even in HSV-TK gene-modified cancer cells in which 100% of the cells contain the HSV-TK gene, a small percentage of these cells are resistant to GCV. Thus, as with virtually every drug therapy, anti-viral drug resistant cells develop posing a potential limitation for this technology.

For suicide gene therapy to become clinically feasible, the above limitations in gene therapy technology must be overcome. Solutions to some of these limitations will be required before suicide genes can be used in clinical gene therapy trials.

THERAPEUTIC CLINICAL APPLICATION OF HSV-TK SUICIDE GENE THERAPY: "THE BYSTANDER EFFECT"

Treatment of an existing tumor with suicide gene therapy had a different set of requirements, which were overcome in 1991 because of a seminal discovery by Freeman and colleagues (Table 2) (Freeman, 1992; Freeman, 1992; Freeman, 1993; Freeman, 1995; Freeman, 1996). The first clinical application of the HSV-TK gene for cancer gene therapy was based on the observation that HSV-TK modified tumor cells were toxic to nearby unmodified tumor cells when the HSV-TK modified cells were exposed to

Table 2. Benefits of Tumor Therapy Using Suicide Genes

1	The Toxic Metabolite Can Kill Chemotherapy Resistant Tumors
2	Requires short-term gene expression
3	Requires transduction of only a fraction of the tumor cells
4	Generation of a "bystander effect"

Table 3. The Three Phases of the "Bystander Effect"

Type	Description	In Vitro	In Vivo
1. Chemosensitization and transfer of drug sensitivity	Gene modification of the tumor with a suicide gene confers a new drug phenotype to the tumor cells and nearby unmodified tumor cells.	+	+
2. Hemorrhagic Tumor Necrosis (HTN)	Release of pro-inflammatory cytokines by the dying HSV-TK gene-modified cells leads to HTN and influx of immune cells into the tumor mass.	–	+
3. Immuno-stimulatory tumor microenvironment	The release of cytokines by tumor infiltrating lymphocytes within the tumor leads to alteration of the tumor microenvironment with upregulation of immune regulatory molecules (ICAM, B7).	–	+

GCV (Freeman, 1992; Freeman, 1993; Freeman, 1995; Freeman, 1996). An important feature of this effect, subsequently termed the "bystander effect", was that when as few as 10% of the cells in a tumor contained the HSV-TK gene-modified cells, the entire population of cells could be killed (Freeman, 1992; Freeman, 1992; Freeman, 1993; Freeman, 1995). This discovery allowed the development of a therapeutic approach to the treatment of cancer because it overcame the need to genetically modify the entire tumor (transduction efficiency). Since only a fraction of the tumor mass needed to be genetically modified to cause tumor regression, the state-of-the-art gene transfer technology could be used to transduce a sufficient number of tumor cells in the tumor mass to allow for the "bystander effect". In addition, long term gene expression was not required since tumor cells expressing the HSV-TK suicide gene are killed after only a short exposure to GCV (Table 2).

The discovery of the "bystander effect" has led a intensive study of this phenomenon to understand its mechanisms (Kolberg, 1994; Seachrist, 1994). There have been a number of mechanisms proposed that included gap junction intercellular communication (GJIC), apoptotic cell death, activation of an antitumor immune response, release of cytokines, and disruption of tumor vasculature. Studies have shown that the "bystander effect" is a complex biological process involving several mechanisms, which may involve different mechanisms *in vitro* and *in vivo* (see Table 3). *In vitro* studies suggested that the "bystander effect" resulted from transfer of toxic GCV metabolites (phosphorylated GCV) from HSV-TK gene-modified cells to nearby unmodified tumor cells because tumor cell proximity was required to observe the effect (Freeman, 1993; Wu, 1994; Kuriyama, 1995; Vrionis, 1995). Further studies using tritiated ganciclovir demonstrated that toxic GCV metabolites were transferred from HSV-TK modified tumor cells to unmodified tumor cells (Bi, 1993). Thus, the role of transfer of toxic ganciclovir metabolites from HSV-TK gene-modified to nearby unmodified tumor cells is the predominant mechanism of the *in vitro* "bystander effect" and one of the *in vivo* mechanisms, as will be discussed below (Table 3).

IN VIVO MECHANISM OF ACTION OF THE BYSTANDER EFFECT

The "bystander effect" is a complex biological process that allows killing of a tumor mass when only a fraction of the tumor is genetically modified. The *in vivo* mechanism of the "bystander effect" appears to encompass a broader spectrum of tumoricidal activity to allow killing of nearby tumor cells by gene-modified tumor cells. This is due to the limited transfer of toxic GCV metabolites to nearby tumor cells and a functional immune system is required to demonstrate the *in vivo* "bystander effect". As described above, *in vitro* studies have demonstrated that the "bystander effect" results from transfer of toxic GCV metabolites from HSV-TK tumor cells to nearby unmodified tumor cells (Bi, 1993). Although a few studies suggest that transfer of toxic GCV metabolites occurs *in vivo* and contributes to the *in vivo* "bystander effect", a number of studies demonstrate that an additional mechanism(s) is operational *in vivo* to account for a number of diverse findings. Studies analyzing the *in vivo* "bystander effect" have shown hemorrhagic tumor necrosis, anti-tumor immunity, and release of cytokines are important (Freeman, 1992; Freeman, 1992; Freeman, 1993; Ram, 1994; Freeman, 1994; Vile, 1994; Consalvo, 1995; Colombo, 1995; Freeman, 1995; Freeman, 1996; Ramesh, 1996; Ramesh, 1996; Freeman, 1997; Vile, 1997; Kruse, 1997; Bi, 1997).

Generation of an *in vivo* "bystander effect" is necessary for an effective treatment, since even in a population of 100% HSV-TK gene-modified tumor cells, resistant cells are present. The unique feature of the "bystander effect" is that a tumor mass consisting of drug resistant and sensitive cells will undergo regression when exposed to the drug. Different processes may be active *in vivo* and *in vitro*, because *in vitro* cell death by bystander killing occurred by apoptosis over a three day time period, but *in vivo*, a hemorrhagic tumor necrosis (HTN) occurred within 24 hours of treatment (Seachrist, 1994; Ram, 1994). This demonstrated that a more rapid mechanism of bystander killing was occurring *in vivo* (Freeman, 1992; Freeman, 1992; Freeman, 1993; Freeman, 1994; Freeman, 1995; Freeman, 1996). The occurrence of HTN was also reported by other investigators using the HSV-TK system (Ram, 1994). The observation of a relatively rapid and centralized hemorrhagic tumor necrosis after HSV-TK/GCV treatment was unexpected and suggested a mechanism(s) in addition to the transfer of toxic metabolites was responsible for the *in vivo* "bystander effect". Another striking observation was the demonstration that the "bystander effect" was diminished or abrogated in athymic nude mice (Table 4) (Freeman, 1992; Freeman, 1992; Freeman, 1993; Vile, 1994; Colombo, 1995; Consalvo, 1995; Freeman, 1995; Freeman, 1996). This indicated that the immune system played an integral role in the "bystander effect" *in vivo* (Ramesh, 1996; Freeman, 1997; Vile, 1997; Kruse, 1997).

These findings suggested that a complex biological process leads to the *in vivo* "bystander effect" (Table 3). First, the centralized hemorrhagic necrosis suggests that a soluble factor may be acting on the central tumor vasculature. Second, as previously mentioned, the HSV-TK gene-modified cells die by apoptosis over a three day period and release soluble factors while HTN occurred within twenty-four hours of treatment with GCV. Third, the host immune/inflammatory system is required. Thus, the *in vivo* "bystander effect" apparently relies on three processes to cause tumor regression: chemosensitization of tumor cells to GCV, hemorrhagic tumor necrosis, and anti-tumor

Table 4. Role of Immunity

Tumor Mass % HSV-TK	Percent Tumor Regression	
	Immune competent	Immune Deficient
100	100	25
50	100	0
0	0	0

immunity (Table 3). The process of chemosensitization leads to *in vivo* killing of the HSV-TK gene-modified tumor cells and some nearby unmodified tumor cells through transfer of toxic GCV metabolites from the GCV exposed HSV-TK gene-modified tumor cells to unmodified tumor cells. Pro-inflammatory cytokines (TNF and IL-1) are involved in generating the centralized HTN observed after the inoculation of HSV-TK tumor cells and GCV because of the known effects of these cytokines on the tumor vasculature (Tracey, 1986; Freeman, 1995; Ramesh, 1996). Their role was suggested by the presence of TNF, IL-1, and IL-6 mRNA in tumors within 24 hours after treatment of HSV-TK gene-modified cells with GCV (Tracey, 1986; Freeman, 1995; Ramesh, 1996; Kruse, 1997; Vile, 1997). Each of the first two phases, chemosensitization and hemorrhagic tumor necrosis leads to killing of *in situ* tumor. The former by a chemical process, GCV toxicity on nearby tumor cells, and the latter through disruption of the tumor vasculature and thus depletion of nutrients to the tumor mass. The majority of the tumor mass can be killed by these two process, but residual tumor remains (Freeman, 1992; Freeman, 1993; Freeman, 1995; Freeman, 1996).

The hemorrhagic tumor necrosis can then lead to a cellular infiltrate within the tumor, especially since the expression of cytokines within the tumor leads to up-regulation of adhesion molecules and the binding of lymphocytes to cells within the tumor (Ramesh, 1996; Freeman, 1997). Studies have shown an increase in T cells and macrophages in the tumor after treatment with HSV-TK gene-modified cells (Freeman, 1992; Freeman, 1993; Freeman, 1995; Ramesh, 1996; Freeman, 1997). The release of cytokines within the tumor also affects the microenvironment of the tumor by causing the up-regulation of adhesion and immune regulatory molecules. Increased expression of immune regulatory molecules (ICAM-1, B7-1, B7-2, MHC) within the tumor have been shown after inoculation of the HSV-TK gene-modified cells (Ramesh, 1996). It has been hypothesized that the tumor infiltrating cells can become activated to develop an anti-tumor cytotoxic response within this new environment (Freeman, 1992; Freeman, 1993; Freeman, 1995; Ramesh, 1996; Freeman, 1997).

Taken together, these findings suggest that treatment of tumors with HSV-TK gene modified cells and GCV leads to an alteration of the tumor microenvironment from one that is immuno-inhibitory to one that is stimulatory, and that the stimulatory environment serves to generate an anti-tumor immune response (Freeman, 1992; Freeman, 1993; Freeman, 1996; Ramesh, 1996; Freeman, 1997). This anti-tumor immunity can be generated because GCV is not immunosuppressive, which may account for the reason why the "bystander effect" has not been observed with cancer chemotherapeutic agents. Thus, the "bystander effect" results from a complex biological process consisting of: chemosensitization, hemorrhagic tumor necrosis, and an anti-tumor immune response. Furthermore, the immune response can be divided into two parts, an early and late. In the early component, as described above, cytokine release leads to alteration of the tumor microenvironment to one that is immuno-stimulatory. This immuno-stimulatory microenvironment by itself cannot alter tumor growth but has profound implications. First, it leads to the "late" component of the immune response: the development of an anti-tumor immune response over several weeks (Freeman, 1992; Freeman, 1995; Consalvo, 1995; Ramesh, 1996; Bi, 1997; Vile, 1997; Freeman, 1997), although this anti-tumor immunity had accounted for an immediate anti-tumor effect observed in nearby untreated tumors, in some cases (Bi, 1997). However, the development of anti-tumor immunity occurs over several weeks (the late component, delayed hypersensitivity immune response) which allows regrowth of residual tumor cells and outpacing of the anti-tumor immune response. Thus, mechanisms to enhance or take advantage of the early immune component (i.e., alteration of the tumor microenvironment) may enhance the "bystander effect".

One major method to enhance the "bystander effect" relies on using the "early phase" of immune activation as an adjuvant to Immunotherapy. During this early phase there is an alteration of the tumor microenvironment from one that is immunosuppressive to one that is immuno-stimulatory, thus generating a microenvironment for immune effector cells to more effectively kill the tumor. Immunotherapy for tumors largely consists of adoptive and active immunotherapy. In the former, immune effector cells are grown *ex vivo* prior to inoculation into the tumor bearing host. In the latter, tumor bearing hosts are immunized to their tumor through vaccination. In both cases, the immune effector cells are in the peripheral circulation. One reason that immunotherapy has met with limited success may be because the tumor microenvironment is generally immunosuppressive due to a variety of tumor immunosuppressive factors. Thus, when peripheral blood "activated" immune cells generated by adoptive or active immunotherapy traffic into the immunosuppressive tumor microenvironment, they become inactivated. Immunotherapy can be potentiated by vaccinating tumor bearing mice with tumor cells prior to injection of the HSV-TK cells (Freeman, 1992; Freeman, 1995; Ramesh, 1996; Freeman, 1996). In this case, the immunized mice have anti-tumor immune effector cells in their peripheral circulation so that when the mice are treated with HSV-TK therapy, the hemorrhagic tumor necrosis allows the immune effector cells to enter the tumor. The altered immuno-stimulatory tumor microenvironment provides the immune effector cells an environment to efficiently function and kill residual tumor cells. The combination of suicide gene therapy and tumor vaccine allows synergy of these two therapies. Thus, HSV-TK suicide gene therapy may provide the second event necessary for successful immunotherapy, an immuno-stimulatory tumor microenvironment that expresses the immune regulatory molecules (e.g., ICAM, B7) needed for an anti-tumor immune response to occur when immune effector cells in the peripheral circulation enter the tumor.

Another method to enhance the "bystander effect", immunomodulation with cytokines, is also based on the discovery that the "bystander effect" is dependent on an intact immune system, although these early studies have met with varying degrees of success. Some studies have focused on using cytokines to enhance the "bystander effect" (Vile, 1997). The combination of HSV-TK therapy and cytokines may lead to an enhanced "bystander effect" because cytokines are involved in this effect. Initial studies used IL-2 secreting cells which were administered in conjunction with the injected HSV-TK therapy but failed to demonstrate an enhanced "bystander effect" (Chen, 1995). It did, however, show an enhanced development of tumor immunity. This is consistent with some studies which have shown that these dying HSV-TK tumor cells are antigenic and in some cases more antigenic than the unmodified irradiated tumor cells (Freeman, 1992; Vile, 1994; Freeman, 1995). Although the combination of HSV-TK modified tumor cells and IL-2 modified tumor cells did not enhance the "bystander effect" when compared to HSV-TK tumor cells alone, the addition of the IL-2 secreting tumor cells increased the "late phase" of tumor immunity in the surviving animals (Chen, 1995).

The first study to demonstrate enhancement of the "bystander effect" using cytokines showed that IFN-a could enhance the effect (Santodonato, 1996). This study's findings fit the model previously proposed (Freeman, 1997), which showed that inflammatory cytokines (i.e., TNF, IL-1, IL-6) were generated by HSV-TK therapy. Therefore, interferon appeared to be a more appropriate choice than IL-2 to enhance the "bystander effect". A second study using IL-2 and GM-CSF in combination with HSV-TK therapy also had some effect on enhancing the "bystander effect" (Chen, 1996).

OTHER NEGATIVE SELECTABLE MARKERS

Other genes have also been used as negative selectable markers, including cytosine deaminase (CD), which converts the nontoxic prodrug 5-fluorocytosine (5-FC) to the toxic metabolite 5-fluorouracil (5-FU) (Table 1). Initial studies demonstrated that CD could kill tumor cells when a capsule containing the enzyme was implanted into a tumor and the animals were treated with 5-FC. Diffusion of the enzyme into the tumor cells allowed conversion of 5-FC into 5-FU (Nishiyama, 1985; Sakai, 1985; Katsuragi, 1987). Genetic modification of tumor cells became possible after the CD gene was cloned (Austin, 1992) and tumor cells genetically modified with the CD gene were shown to be sensitive to 5-FC (Huber, 1994; Consalvo, 1995). Although not initially observed with CD, the "bystander effect" was clearly shown in subsequent studies (Huber, 1994). It has been proposed that the soluble toxic metabolite (5-FU) causes a more powerful "bystander effect" because of its ability to readily diffuse to nearby tumor cells and directly kill them. This mechanism has not been confirmed, and further studies will be required to determine the mechanism and efficiency of the "bystander killing" observed in this system.

The issue to remember, as described above, is that the bystander effect" has three unique characteristics: chemosensitization, hemorrhagic tumor necrosis, and the requirement of a functional immune system (Table 3). The "bystander effect" differentiates itself by these three features from other negative selectable marker genes which rely on the killing of nearby cells only through the diffusion of toxic metabolites. Some suicide genes may produce a more potent toxic metabolite from a prodrug, but without the other characteristic features would not be classified as generating a "bystander effect" and thus may have a weaker anti-tumor effect. Therefore, each suicide gene must be evaluated for its ability to generate the "bystander effect" and to date only HSV-TK and CD have shown these characteristic features.

Other genes that can serve as negative selectable markers are shown in Table 1 (Moolten, 1994). However, they have not been well characterized as the HSV-TK or CD systems and the existence of the "bystander effect" has not been extensively studied.

CLINICAL TRIALS

Approximately 200 clinical gene therapy trials have been approved with over 50% of them related to cancer (Marcel, 1996; Roth, 1997). Suicide genes are employed in a significant number of the cancer trials, because of the several advantages described above (Table 5). First, suicide genes and their prodrug are toxic to chemotherapy resistant tumors. Second, only short term gene expression is required. Third, only a fraction of the tumor cells within the tumor mass (>10%) needs to express the suicide gene to kill the entire tumor, i.e., "bystander effect" (Table 2 and Table 3). The first clinical gene therapy trial using the HSV-TK gene was approved in 1991 to treat ovarian cancer (Table 5) (Freeman, 1993; Freeman, 1995). Ovarian cancer is well-suited for this therapeutic approach for a numbers of reasons. The disease is confined to the peritoneal cavity in over 80% of patients. The CA125 tumor marker allows for the detection of relapse at an early stage, when the residual tumor burden is small and the peritoneal cavity is easily

Table 5. HSV-TK Clinical Protocols

Rac Review Date	P. Investigator	Delivery System	Disease
7/91	S. FREEMAN **TULANE UNIV.**	HSV-TK TRANSDUCED OVARIANCANCER CELLS	OVARIAN CANCER
6/92	E. OLDFIELD **NIH**	HSV-TK VECTOR PRODUCER CELLS (VPC)	BRAIN TUMORS
3/93	J. VAN GILDER M. BERGER M. PRADOW **UNIV. OF IOWA**	HSV-TK VPC	BARIN TUMORS
6/93	C. RAFFEL **CHILD. HOSP. LA**	HSV-TK VPC	BRAIN TUMORS
9/93	L. KUN *et al.* **ST. JUDE HOSP**	HSV-TK VPC	BRAIN TUMORS
12/93*	E. OLDFIELD Z. RAM **NIH**	HSV-TK VPC	LEPTOMENINGEAL CANCER
9/94	S. ECK J. ALAVI **UNIV. OF PENN**	HSV-TK/ ADENOVIRUS	CNS MALIGNANCY
9/94	S. ALBELDA **UNIV. OF PENN**	HSV-TK/ ADENOVIRUS	MESOTHELIOMA
12/94	R. GROSSMAN S. WOO **BAYLOR UNIV.**	HSV-TK ADENOVIRUS	CNS MALIGNANCY
2/95	M. FETELL **COLUMBIA UNIV.**	HSV-TK VPC	GLIOMA
3/95	C. LINK D. MOORMAN **IOWA METHODIST**	HSV-TK VPC	OVARIAN CA
12/95	R. ALVAREZ D. CURIEL **UNIV. OF ALABAMA**	HSV-TK ADENOVIRUS	OVARIAN CANCER

Table 5 shows the first dozen HSV-TK gene therapy clinical trials approved by the NIH Recombinant DNA Committee. An additional several dozen clinical trials have been subsequently approved in the past several years.

accessible. As described above, these studies are based on the premise that the gene-modified cells will "home" to *in situ* ovarian tumor deposits after i.p. infusion so that at least 10% of the *in situ* tumor will be comprised of gene-modified cells. This approach may have a further advantage as serving as a tumor vaccine since GCV is non-immunosuppressive, the dying tumor mass can stimulate an anti-tumor immune response, and the allogeneic HSV-TK gene-modified ovarian tumor cells may express common ovarian tumor antigens (Ioannides, 1993; Kuiper, 1995). Of the eighteen patients treated in this seminal trial, there were four responses as evidenced by either decreased CA125 or tumor with 1 partial and 3 complete responses. Further, average patient survival was similar to other salvage therapies for this patient group.

There have been subsequent trials to treat ovarian cancer that are currently in progress which use other delivery methods including HSV-TK producer cells and adenoviral vectors to deliver the HSV-TK gene. Similarly, brain tumors, particularly glioblastoma, have been an early focus of clinical trials using suicide gene therapy.

CONCLUSIONS

Since conventional cancer treatments are not able to cure approximately 50% of cancers in patients, novel therapies must be developed. A number of different cancer gene therapy protocols are currently underway, including those using suicide genes. Gene therapy provides one approach by which tumors that are resistant to conventional agents may be treated. Inserting a suicide gene into tumor cells will lead to tumor cell death upon exposure to a prodrug, even in chemotherapy resistant tumor cells. One major limitation to using gene therapy for cancer is that modification of an entire tumor is not possible and untreated cells can grow back into tumors. However, the HSV-TK/GCV system has a critical advantage, since the gene-modified cells are toxic to nearby unmodified cells when exposed to the prodrug. This has been termed the "bystander effect". Although the mechanism of this "bystander effect" is complex, the *in vitro* mechanism relies on transfer of toxic metabolites from gene-modified to unmodified cells, while the *in vivo* mechanism appears to be more complex and related to chemosensitization, hemorrhagic tumor necrosis, and activation of immune response.

Clinical trials are underway using the HSV-TK gene in localized tumors including ovarian cancer and brain tumors. To date, it appears that the HSV-TK suicide gene therapy clinical approach is relatively non-toxic and preliminary results show that some patients have achieved partial and complete clinical remission. Future studies will involve better systems to more efficiently and specifically modify the tumor with the suicide gene and focus on methods to enhance the immune aspect of the "bystander effect".

REFERENCES

Austin, E.A., and Huber, B.E., *A first step in the development of gene therapy for colorectal carcinoma: cloning, sequencing, and expression of escherichia coli cytosine deaminase.* Mol Pharmacol, 1992. **43**: p. 380–387.

Barba, D., Hardin, J., Sadelain, M., and Gage, F.H., *Development of anti-tumor immunity following thymidine kinase-mediated killing of experimental brain tumors.* Proc Natl Acad Sci, 1994. **91**: p. 4348–4352.

Bi, W.L., Parysek, L.M., Warnick, R., and Stambrook, P.J., *In vitro evidence that metabolic cooperation is reponsible for the Bystander Effect observed with HSV tk retroviral gene therapy.* Human Gene Therapy, 1993. **4**: p. 725–731.

Bi, W., Kim, Y.G., Feliciano, E.S., Pavelic, L., Wilson, K.M., Pavelic, Z.P., and Stambrook, P.J., *An HSVtk-mediated local and distant antitumor bystander effect in tumors of head and neck origin in athymic mice.* Cancer Gene Therapy, 1997. **4**(4): p. 246–252.

Boring, C.C., Squires, T.S., and Tong, T., *Cancer statistics 1992.* Ca. A Cancer Journal for Clinicians, 1992. **42**: p. 19.

Chen, S.H., Li Chen, X.H., Wang, Y., Kosai, K.I., Finegold, M.J., Rich, S.S., and Woo, S.L.C., *Combination gene therapy for liver metastasis of colon carcioma in vivo.* Proc Natl Acad Sci USA, 1995. **92**(March): p. 2577–2581.

Chen, S.H., Kosai, K., Xu, B., Pham-Nguyen, K., Contanț, C., Finegold, M.J., and Woo, S.L., *Combination suicide and cytokine gene therapy for hepatic metastases of colon carcinoma: sustained antitumor immunity prolongs animal survival.* Cancer Research, 1996. **56**(16): p. 3758–3762.

Colombo, B.M., Benedetti, S., Ottolenghi, S., Mora, M., Pollo, B., Poli, G., and Finocchiaro, G., *The "bystander effect": association of U87 cell death with ganciclovir-mediated apoptosis of nearby cells and lack of effect in athymic mice.* Human Gene Therapy, 1995. **6**: p. 763–772.

Consalvo, M., Mullen, C.A., Modesti, A., Piero, M., Allione, A., Cavallo, F., Giovarelli, M., and Forni, G., *5-fluorocytosine-induced eradication of murine adenocarcinomas engineered to express the cytosine deaminase suicide gene requires host immune competence and leaves an efficient memory*. J Immunol, 1995. **154**: p. 5302–5312.

Davidson, R.L, Kaufman, E.R., Crumpacker, C.S., and Schnipper, L.E., *Inhibition of herpes simplex virus transformed and nontransformed cells by acycloguanosine: mechanisms of uptake and toxicity*. Virology, 1981. **113**: p. 9–19.

Elion, G.B., Furman, P.A., Fyfe, J.A., deMiranda, P., Beauchamp, L., and Schaeffer, H.J., *Selectivity of action of an antiherpetic agent, 9-(2-hydroxyethoxymethyl) guanine*. Proc Natl Acad Sci USA, 1977. **74**: p. 5716–5720.

Elion, G.B., *The chemotherapeutic exploitation of virus-specified enzymes*. Adv Enzyme Regul, 1980. **18**: p. 53–60.

Feldman, A.R., Kessler, L., Myers, M.H., and Naughton, M.D., *The prevalence of cancer. Estimates based on the Connecticut Tumor Registry*. N Engl J Med, 1986. **315**: p. 1394–1397.

Field, A.K., Davies, M.E., DeWitt, C., Perry, H.D., Liou, R., Germershausen, J., Karkas, J.D., Ashton, W.T., Johnston, D.B.R., and Tolman, R.L., *9-{[2-hydroxy-1-(hydroxy-methyl) ethoxy] methyl} guanine: a selective inhibitor of herpes group virus replication*. Proc Natl Acad Sci USA, 1983. **80**: p. 4139–4143.

Freeman, S.M., Whartenby, K.A., Keoplin, D.S., Moolten, F.L., Abboud, C.N., and Abraham, G.N., *Tumor regression when a fraction of the tumor mass contains the HSV-TV gene*. J Cell Biochem, 1992. **16F**: p. 47.

Freeman, S.M., McCune, C., Angel, C., Abraham, G.N., and Abboud, C.N., *Treatment of ovarian cancer using HSV-TK gene-modified vaccine-regulatory issues*. Human Gene Therapy, 1992. **3**: p. 342—349.

Freeman, S.M., Abboud, C.N., Whartenby, K.A., and Abraham, G.N., *The bystander effect: Tumor regression when a fraction of the tumor mass is genetically modified*. Cancer Res, 1993. **53**: p. 5274–5284.

Freeman, S.M., Ramesh, R., Marrogi, A.J., Jensen, A., and Aboud, C.N., *In vivo studies on the mechanism of the "Bystander Effect"*. Cancer Gene Therapy, 1994. **4**: p. 326.

Freeman, S.M., Ramesh, R., Shastri, M., Munshi, A., Jensen, A.K., and Marrogi, A.J., *The role of cytokines in mediating the bystander effect using HSV-TK xenogeneic cells*. Cancer Letters, 1995. **92**(2): p. 167–174.

Freeman, S.M., McCune, C., Robinson, W., Abboud, C.N., Angel, C., Abraham, G.N., and Marrogi, A.J., *Treatment of ovarian cancer using a gene-modified vaccine*. Human Gene Therapy, 1995. **6**: p. 927–939.

Freeman, S.M., Whartenby, K.A., Freeman, J.L, Abboud, C.N., and Marrogi, A.J., *In situ use of suicide genes for cancer therapy*. Seminars in Oncology, 1996. **23**(1): p. 31–45.

Freeman, S.M., Ramesh, R., and Marrogi, A.J., *Immune system in suicidegene therapy*. Lancet, 1997. **349**(9044): p. 2–3.

Freeman, S.M., Ramesh, R., Munshi, A., and Marrogi, A.J., *Tumor killing using the HSV-TK suicide gene*. Cancer Gene Therapy, 1997. **4**: p. 542.

Goldstein, L.J., Pastan, I., and Gottesman, M.M., *Multidrug resistance in human cancer*. Drit Rev Oncol Hematol, 1992. **12**: p. 243–253.

Goldstein, L.J., Gottesman, M.M., and Pastan, I., *Expression of the MDR1 gene in human cancers*. Cancer Treat Res, 1991. **57**: p. 101–119.

Huber B.E., Austin, E.A., Richards, C.A., Davis, S.T., and Good, S.S., *Metabolism of 5-fluorocytosine to 5-fluorouracil in human colorectal tumor cells transduced with the cytosine deaminase gene: significant anti-tumor effects when only a small percentage of tumor cells express cytosine deaminase*. Proc Natl Acad Sci USA, 1994. **91**: p. 8302–8306.

Ioannides, C.G., Fisk, B., Jerome, K.R., Irimura, T. Wharton, J.T., and Finn, O.J., *Cytotoxic T cells from ovarian malignant tumors can recognize polymorphic epithelia mucin core peptides*. Journal of Immunology, 1993. **151**: p. 3693–3703.

Katsuragi, T., Sakai, T., and Tonomura, K., *Implantable enzyme capsules for cancer chemotherapy from bakers yeast cytosine deaminase immobilized on epoxy-acrylic resin and urethane prepolymer*. Appl Biochem Biotechnol, 1987. **16**: p. 61—69.

Kolberg, R., *The bystander effect in gene therapy: Great, but how does it work*. J NIH Res, 1994. **6**: p. 62–64.

Kruse, C.A., Roper, M.D., Kleinschmidt-DeMasters, B.K., Banuelos, S.J., Smiley, W.R., Robbins, J.M., and Burrows, F.J., *Purified herpes simplex thymidine kinase Retrovector particles. I. In vitro characterization, in situ transduction efficiency, and histopathological analyses of gene therapy-treated brain tumors*. Cancer Gene Therapy, 1997. **4**(2): p. 118–128.

Kuiper, M., Peakman, M., and Farzaneh, F., *Ovarian tumour antigens as potential targets for immune gene therapy*. Gene Therapy, 1995. **2**: p. 7–15.

Kuriyama, S., Nakatani, T., Masui, K., Sakamoto, T., Tominaga, K., Yoshikawa, M., Fukui, H., Ikenaka, K., and Tsujii, T., *Bystander effect caused by suicide gene expression indicates the feasibility of gene therapy for hepatocellular carcinoma*. Hepatology, 1995. **22**(6): p. 1838–1946.

Marcel, T., and Grausz, J.D., *The TMC Worldwide Gene Therapy Enrollment Report (June 1996)*. Human Gene Therapy, 1996. **7**(16): p. 2025–2046.

Moolten, F.L., *Drug sensitivity ("suicide") genes for selective cancer chemotherapy*. Cancer Gene Therapy, 1994. **4**: p. 279–287.

Moolten, F.L., and Wells, J.M., *Curability of tumors bearing herpes thymidine kinase genes transferred by retroviral vectors*. Journal of the National Cancer Institute, 1990. **82**: p. 297–300.

Moolten, F.L., Wells, J.M., Heyman, R.A., and Evans, R.M., *Lymphoma regression induced by ganciclovir in mice bearing a herpesthymidine kinase transgene*. Human Gene Therapy, 1990. **1**: p. 125–134.

Moolten, F.L., *Tumor chemosensitivity conferred by inserted herpes thymidine kinase genes: Paradigm for a prospective cancer control strategy*. Cancer Res, 1986. **46**: p. 5276–5281.

Nishiyama, Y., and Rapp, F., *Anticellular effects of 9-(2-hydroxy-ethoxymethyl) guanine against herpes simplex virus-transformed cells*. J Gen Virol, 1979. **45**: p. 227–230.

Nishiyama, T., Kawamura, Y., Kawamoto, K., Matsumura, H., Yamamoto, N., Ito, T., Ohyama, A., Katsuragi, T., and Sakai, T., *Antineoplastic effects in rats of 5-fluorocytosine in combination with cytosine deaminase capsules*. Cancer Research, 1985. **45**: p. 1753–1761.

Oliver, S., Bubley, G., and C., C., *Inhibition of HSV-transformed murine cells by nucleoside analogs, 2'-NDG and 2'-nor-cGMP: mechanisms of inhibition and reversal by exogenous nucleosides*. Virology, 1985. **145**: p. 84–93.

Ram, Z., Walbridge, S., Shawker, T., Culver, K.W., Blaese, R.M., and Oldfield, E.H., *The effect of thymidine kinase transduction and ganciclovir therapy on tumor vasculature and growth of 9L gliomas in rats*. J Neurosurg, 1994. **81**: p. 256–260.

Ramesh, R, Marrogi, A.J., Munshi, A., Abboud, C.N., and Freeman, S.M., *In vivo analysis of the "bystander effect": a cytokine cascade*. Experimental Hematology, 1996. **24**(7): p. 829–838.

Ramesh, R., Munshi, A., Abboud, C.N., Marrogi, A.J., and Freeman, S.M., *Expression of costimulatory molecules: B7 and ICAM up-regulation after treatment with a suicide gene*. Cancer Gene Therapy, 1996. **3**(6): p. 373–384.

Roth, J.A., and Cristiano, R.J., *Gene therapy for cancer: What have we done and where are we going*? J Natl Cancer Inst, 1997. **89**: p. 21–29.

Sakai, T., Katsuragi, T., Tonomura, K., Nishiyama, T., and Kawamura, Y., *Implantable encapsulated cytosine deaminase havine 5-fluorocytosine deaminating activity*. Journal of Biotechnology, 1985. **2**: p. 13–21.

Santodonato, L., Ferrantini, M., Gabriele, L., Proietti, E., Venditti, M., Musiani, A., Modesti, A., Modica, A., Lupton, S.K., and Belardelli, F., *Cure of mice with established metastatic friend leukemia cell tumors by a combined therapy with tumor cells expressing both intereron-alpha 1 and herpes simplex thymidine kinase followed by ganciclovir*. Hum Gen Ther, 1996. **7**: p. 1–10.

Seachrist, L., *Successful gene therapy has researchers looking for the bystander effect*. J Natl Cancer Inst, 1994. **86**: p. 82–83.

Shepp, D.H., Dandiliker, P., DeMiranda, P., Burnette, T.C., Cederberg, D.M., Kirk, L.E., and Meyers, J.D., *Activity of 9-[2-hydroxy-1-(hydroxymethyl) ethoxymethyl] guanine in the treatment of cytomegalovirus pneumonia*. Ann Intern Med, 1985. **103**: p. 368–373.

Tracey, K.J., Beutler, B., Lowry, S.F., and *et al.*, *Shock and tissue injury induced by recombinant human cachectin*. Science, 1986. **234**: p. 470–474.

Vrionis, F.D., Wu, J.K., Qi, P., Cano, W., and Cherington, V., *A more potent bystander cytocidal effect elicited by tumor cells expressing the herpes simples virus-thymidine kinase gene than by fibroblast virus-producer cells in vitro*. Journal of Neurosurgery, 1995. **83**(4): p. 698–704.

Vile, R.G., Nelson, J.A., Castleden, S., Chong, H., and Hart, I.R., *Systemic gene therapy of murine melanoma using tissue specific expression of the HSVtk gene involves an immune component*. Cancer Research, 1994. **54**(23): p. 6228–6234.

Vile, R.G., Castleden, S., Marshall, J., Camplejohn, R., Upton, C., and Chong, H., *Generation of an anti-tumour immune response in a non-immunogenic tumour: HSVtk killing in vivo stimulates a monocuclear cell infiltrate and a th1-like profile of intratumoural cytokine expression*. Intl J Cancer, 1997. **71**: p. 267–274.

Vile, R.G., Diaz R.M., Castleden, S., and Chong, H., *Targeted gene therapy for cancer: herpes simplex virus thymidine kinase gene-mediated cell killing leads to anti-tumour immunity that can be augmented by co-expression of cytokines in the tumour cells*. Biochemical Society Transactions, 1997. **25**(2): p. 717–722.

Wu, J.K., Cano, W.G., Meylaerts, S.A., Qi, P., Vrionis, F., and Cherington, V., *Bystander tumoricidal effect in the treatment of experimental brain tumors*. Neurosurgery, 1994. **35**(6): p. 1094–1102; discussion 1103–1103.

37

ADENOVIRUSES AS GENE DELIVERY VECTORS

Anu-Maaria Sandmair[1,2]*, Matti Vapalahti[2,4], and Seppo Ylä-Herttuala[1,3,4]

[1]A.I.Virtanen Institute
University of Kuopio
P.O.B. 1627, FIN-70211 Kuopio
Finland
[2]Department of Neurosurgery
[3]Department of Medicine and
[4]Gene Therapy Unit
University Hospital of Kuopio
P.O.B. 1777, FIN-70110 Kuopio
Finland

1. DEFINITION

Adenoviral gene transfer vector is a genetically modified adenovirus in which a portion of the normal viral genome has been deleted to prevent viral replication and replaced with a new gene or genes to be delivered into target cells (Culver, Rigolet and Raffel, 1997).

2. HISTORY

Adenoviruses were first cultured and reported as distinct viral agents in 1953 (Rowe *et al.*, 1953). Similar viral agents were isolated from febrile military personnel with a variety of respiratory illnesses and it was soon realized that there were multiple serotypes of adenoviruses. Acute respiratory disease syndromes caused by adenoviruses had been defined for 100 years before their association with this group of viruses (Horwitz, 1990). Adenovirus serotypes used for gene transfer are not associated with human malignancies. Graham *et al.* constructed an adenovirus complementing 293 cell line, which was the basis for adenovirus—mediated gene therapy (Graham *et al.*, 1977). Furthermore, Spector *et al.* showed that in 293 cell line the replication of E1 deleted viruses was possible via complementation of integrated sequencies (Spector *et al.*, 1980).

* Fax: +358-17-163030

Cancer Gene Therapy: Past Achievements and Future Challenges, edited by Habib
Kluwer Academic/Plenum Publishers, New York, 2000.

3. WILD TYPE ADENOVIRUSES

Adenovirus infections occure worldwide in humans as well as in a variety of animals. With a few exceptions, human adenovirus serotypes are generally not pathogenic to animals and animal adenoviruses are only pathogenic within their own species. Approximately 5% of the acute respiratory disease in children under the age of 5 is due to adenoviruses (Horwitz, 1990). Antibodies to serotypes 1, 2 and 5 are most common and are present in 40% to 60% of children (Horwitz, 1990). More than 10 million people have been immunized safely with serotypes 4 and 7 (Horwitz, 1990). The usual symptoms include nasal congestion, cough, fever, myalgia and headache. Some serotypes can cause conjunctivitis, pneumonia, gastroenteritis, acute hemorrhagic cystitis and rarely meningoencephalitis. Specific antiviral therapy for adenovirus infections have been attempted but were unsuccessful. Effective oral vaccines exist to prevent disease by some adenoviruses and have been used for over 20 years in military recruits (Brody *et al.*, 1994).

4. ADENOVIRUS AS A GENE THERAPY TOOL

Adenoviruses have been used in modern molecular biology for in vivo delivery of cloned DNA for therapy of cancer and various other diseases. Adenoviruses can be designed to enter cells and express genes of interest. However, they are made replication-deficient in order to prevent production of infectious viruses. Adenovirus vectors commonly used for gene therapy are based on serotypes 2 or 5. The majority of viruses have E1A and E1B deletions in order to prevent replication (Graham *et al.*, 1977). The E1 and E3 regions can be deleted to provide additional space for the insertion of 7.5 kb of exogenous DNA allowing the packaging of approximately 105% of the normal adenoviral genome. These viruses can be grown only in cells such as the 293 cell line, a transformed human kidney cell line, which has been stably transfected with E1A and E1B sequences that complement the defect in the recombinant virus. However, these recombinant adenoviruses are fully competent to enter human cells, which is mediated via an adenovirus receptor (CAR) (Bergelson *et al.*, 1997).

A vector is made by inserting an expression cassette containing a gene or genes of interest into the E1 portion of the adenovirus genome. The cassette contains inverse terminal repeats (ITR), promoters, a gene and a polyadenylation signal. Nuclear localizion signal within the expression cassette of the lacZ marker gene leads to the better identification of transduced cells.

Adenoviral vectors have several biological characteristics that make them attractive candidates for somatic gene therapy of cancer. They are large, double-stranded DNA viruses and can carry large fragments of foreign DNA. Adenoviruses have a known tropism for lung and intestine epithelia, they are not neurotoxic and are linked to only minor diseases in humans. Replication-defective adenoviruses have a broad host and cell range and can be produced in high titers up to 1×10^{12} particles/ml. Therefore, they are capable of high-efficiency gene delivery into a variety of organs, including lung, sceletal muscle, heart, liver, blood vessels and central nervous system. Moreover, in contrast to retroviruses, adenoviruses are able to transfer genes to nonproliferating cells, which may be preferable in cancer gene therapy, given the heterogenity of tumor cell cycling. For tumor cell specificity, the selectivity must come form the transferred genes, either by the use of tissue specific promoters or by use of a gene that specifically targets dividing cells (e.g. HSVtk, Herpes simplex thymidine kinase gene, Moolten, 1994).

Adenoviruses enter cells via receptor-mediated endocytosis. Unlike retroviruses, which integrate into the host genome, adenoviruses exist extrachromosomally in the cell, although they migrate to the cell nuclei. Therefore the risks associated with insertional mutagenesis, inactivation of tumor suppressor genes, and tumorigenesis are minimal. Adenoviruses lead to a short term gene expression. This is not a disadvantage in cancer gene therapy, since the target is to destroy the transfected cells, not to have a long term gene expression. The transient nature of gene expression and the lack of a significant potential for integration and insertional mutagenesis after adenovirus gene delivery allow selection of a molecular intervention that may cause tumor cell lysis.

Adenoviruses usually raise immunologic reactions which can limit repeated gene transfers, but also can be advantageous by increasing immunologic reactions against the tumor (Kajiwara *et al.*, 1997). Specifically, early cytokine responses (IL-1, IL-6 and TNF-α) and appearance of monocytes and NK cells followed by specific T- and B-lymphocytes have been described (Byrnes *et al.*, 1995). Stimulation of cel-mediated immunoresponses has been considred for the induction of antitumor immunity. Activated T-cells and macrophages have the capacity to sensitize and kill target cells, for example tumor cells. The activation of the immune system has been seen before as a limitation of repeated administration of adenoviruses, but, on the other hand, it can also be seen as a possible stimulation of antitumor activity. Brain is considered as a relatively priviledged site for immuno responses. However, the blood-brain barrier is partly interrupted due to the growth of malignant glioma. As we have shown before, cellular immunoreactivity is raised after TK and GCV-induced cell death in rat brain (Puumalainen *et al.*, 1999) and humoral immune system is activated after adenovirus—mediated gene transfer (Puumalainen *et al.*, 1998). Moreover, it has been shown previously, that survival of rats can be significantly improved by β-galactosidase marker gene transfer alone, which may be due immunoresponse against β-galactosidase transfected cells (Abina *et al.*, 1996). All these features and the high gene transfer efficiency make adenoviruses one of the most promising gene delivery vectors for the treatment of cancer.

In order to combine efficiency of adenovirus vector and stable transduction of retroviruses Curiel *et al.* have developed a chimeric vector (Feng *et al.*, 1997). Gene transfer is achived by adenoviral delivery of retroviral vector and packaging functions which leads to transient production of retrovirus producers in vivo (Feng *et al.*, 1997). They have also shown that simultaneous addition of replication-deficient adenovirus and plasmid containing replication-enabling sequences results in the production of additional recombinant adenoviruses (Goldsmith *et al.*, 1994). Furthermore, they are developing an adenovirus vector which would replicate only in tumor cells providing the spread of the virions into the tumor mass (Rancourt *et al.*, 1997). Several constructs of adenovirus vectors with targeting elements have been developed: for T-cells (Wickham *et al.*, 1997), endothelial and smooth muscles cells (Wickham *et al.*, 1996), malignant cells via folate receptor (Douglas *et al.*, 1996) and for a variety of other cells to increase the efficacy via heparan-containing cellular receptors (Wickham *et al.*, 1996).

Further more, second and third generation adenoviruses have been developed by deleting large portions of adenovirus genome in order to decrease cytopathic effects, to reduce immuno responses againts adenoviral proteins and to increase capacity to accomodate large DNA fragments. Deletions of E4 and E2 areas have been described (Wang and Finer, 1996). Adenovirus vector consisting only one or two adenovirus proteind instead of the 11 proteins and containing no viral genome have also been developed for potentially safer adenovirus gene therapy (Fender *et al.*, 1997). Using adenovirus vectors with all viral coding sequences deleted, a long term expression of transgene and reduced toxicity have been observed in vivo in mice (Schiedner, 1998).

5. ANIMAL AND HUMAN STUDIES

Previous gene therapy studies have shown positive responses for the treatment of various cancers in animal models. Tumor regression have shown after adenovirus—mediated IL-2 (Addison *et al.*, 1995) and combinations of IL-2 and p53 (Putzer *et al.*, 1998) gene therapy in transgenic mouse mammary adenocarcinoma model. Pancreatic cells have shown to be sensitive for Adv-HSVtk gene therapy when injected into the liver (Block *et al.*, 1997) or intraperitoneal cavity of a nude mice (Rosenfeld *et al.*, 1997). Tumor regression and prolongation of survival was shown in a syngeneic mouse model of intrahepatic implantation of lung cancer cells (Kwong *et al.*, 1997). Treatment effect of Adv-tk in glioma models has been demonstrated in various studies (Vincent *et al.*, 1997; Chen *et al.*, 1994; Viola *et al.*, 1995; Colak *et al.*, 1995; Maron *et al.*, 1996; Perez-Cruet *et al.*, 1994; Tyynelä *et al.*, 1998). On the other hand, no suppression of tumor growth were seen after Adv-p53 treatment of hepatocellular carcinoma in mice after gene transfer via portal vein. Instead, severe toxicity was seen after this treatment (Bao *et al.*, 1996).

According to animal studies 5% to 50% transfection efficiencies have been observed with adenoviruses (Boviatsis *et al.*, 1994; Viola *et al.*, 1995) and <0,01 to 11% in human malignant glioma (Puumalainen *et al.*, 1998). The maximal gene expression is reached at day 7 and at that time point approximately 40–50% gene transfer efficiency can be reached in 9L—gliosarcoma tumors in Fischer rats (Viola *et al.*, 1995). This is in line with our experience of 65% gene transfer efficiency in BT4C rat glioma model (Puumalainen *et al.*, 1999). On the other hand Boviatsis *et al.* reported 5% expression at day 6 and only 1% expression at day 12 after adenovirus—mediated β-galactosidase gene transfer in 9L/Fischer rat model (Boviatsis *et al.*, 1994). Gene expression after adenovirus—mediated gene transfer is transient with 0.1% efficiency 30 days after the gene transfer (Puumalainen *et al.*, 1999).

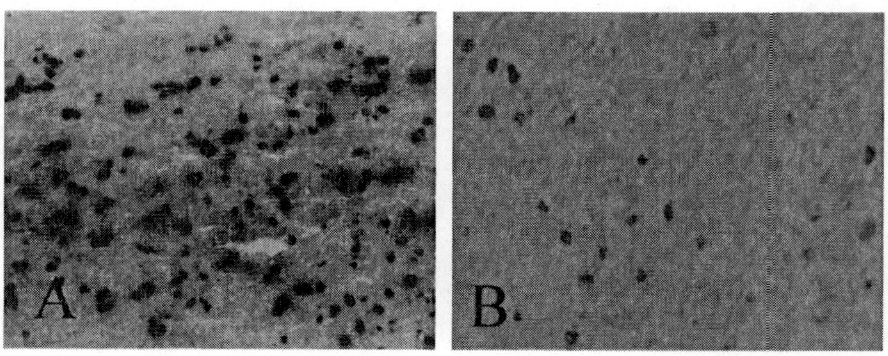

Figure 1A. Adenovirus-ediated gene transfer leads to a high efficiency gene expression in a target tissue. Nuclear targeted β-galactosidase gene expression is seen in BT4C rat brain tissue after a single stereotactic injection of Adv/Lac Z (2.4×10^8) into the right corpus callosum of rat 7 days after the gene transfer. **B.** Adenovirus gene therapy leads to an immunological response in the transduced tissue. Macrophages are seen in human malignant glioma after β-galactosidase gene transfer and CD68 immunostaining.

Table 1. Advantages of adenoviruses as gene delivery vectors for the treatment of cancer

High efficiency gene transfer
Wide host cell range
Transient gene expression
Lack of integration
Increased immunologic responses towards target tissue
Low pathogenity in humans

Published studies from clinical trials are still limited from adenovirally treated patients. We have shown previously that LacZ adenoviruses can transfect human malignant glioma in vivo (Puumalainen *et al.*, 1998). Gene transfers were safe and well tolerated. Lung cancer patients have shown promising responses when treated with Adv-p53 gene therapy (Roth *et al.*, 1998). According to their series no severe adverse events were detected. Adenoviruses expressing CFTR have been used for the treatment of cystic fibrosis (Bellon *et al.*, 1997). Also in this study the treatment was safe and expression of transgene was detected. Adv-p53 was also used safely for the treatment of head and neck squamous cell carsinoma with objective antitumor activity (Clayman *et al.*, 1998). Phase I and II adenovirus—mediated HSVtk studies for the treatment of malignant glioma are currently underway (Puumalainen *et al.*, 1999) and the preliminary results show the safety and efficacy of the method.

6. CONCLUSIONS

Adenovirus—mediated gene therapy has in a short period of time become one of the most promising novel treatment modalities for cancer. Results from several preclinical studies have been promising and the first experiences from clinical studies are available. While clinical trials are under way in order to examine a true treament effect using randomized protocols with the first generation adenoviruses further development is needed to improve gene transfer efficiency of adenovirus vectors.

ACKNOWLEDGMENTS

This study was supported by grants from Finnish Academy and Kuopio University Hospital (EVO grants 5021 and 5118).

REFERENCES

Abina, M.A., Lee, M.G., Descamps, V., Cordiel, L., Lopez, M., Perricaudet, M., and Haddada, H., 1996, LacZ gene transfer into tumor cells abrogates tumorigenicity and protects mice against the development of further tumors. *Gene Ther.* **3**:212–216.

Addison, C.L., Braciak, T., Ralston, R., Muller, W.J., Gauldie, J., and Graham, F.L., 1995, Intratumoral injection of an adenovirus expressing interleukin-2 induces regression and immunity in a murine brest cancer model. *Proc Natl Acad Sci USA* **18**:8522–8526.

Bao, J.J., Zhang, W.W., and Kuo, M.T., 1996, Adenoviral delivery of recombinant DNA into transgenic mice bearing hepatocellular carcinomas. *Hum Gene Ther.* **3**:355–365.

Bellon, G., Michel-Calemard, L., Thouvenot, D., Jagneaux, V., Poitevin, F., Malcus, C., Accart, N., Layani, M.P., Aymard, M., Bernon, H., Bienvenu, J., Courtney, M., Döring, G., Gilly, B., Gilly, R., Lamy, D., Levrey, H., Morel, Y., Paulin, C., Perraud, F., Rodillon, L., Sene, C., So, S., Touraine-Moulin, F., Schatz, C., and Pavirani, A., 1997, Aerosol Administration of a Recombinant Adenovirus Expressing CFTR to Cystic Fibrosis Patients: A Phase I Clinical Trial. *Human Gene Therapy* **8**:15–25.

Bergelson, J.M., Cunningham, J.A., Droguett, G., Kurt-Jones, E.A., Krithivas, A., Hong, J.S., Horwitz, M.S., Crowell, R.L., and Finberg, R.W., 1997, Isolation of a common receptor for Coxsackie B viruses and adenoviruses 2 and 5. *Science* **275**:1320–1323.

Block, A., Chen, S.H., Kosai, K., Finegold, M., and Woo, S.L., 1997, Adenoviral-mediated herpes simplex virus thymidine kinase gene transfer: regression of hepatic metastasis of pancreatic tumors. *Pancreas* **1**:25–34.

Boviatsis, E., Chase, M., Wei, M.X., Tamiya, T., Hurford, R.K., Kowall, N.W., Tepper, R.I., Breakfield, X.O., and Chiocca, E.A. 1994, Gene transfer into experimental brain tumors mediated by adenovirus, herpes simplex virus, and adenovirus vectors. *Human Gene Therapy* **5**:183–191.

Brody, S.L., and Crystal, R.G., 1994, Adenovirus-Mediated in Vivo Gene Transfer. *Annals New York Academy of Sciences*. **716**:90–103.

Byrnes, A.P., Rusby, J.E., Wood, M.J.A., Charlton, H.M., 1995, *Neuroscience*, **66**:1015–1024.

Chen, S.-H., Shine, H.D., Goodman, J.C., Grossman, R.G., and Woo, S.L.C., 1994, Gene therapy for brain tumors: Regression of experimental gliomas by adenovirus-mediated gene transfer in vivo. *Proc Natl Acad Sci USA* **91**:3054–3057.

Clayman, G.L., El-Naggar, A.K., Lippman, S.M., Henderson, Y.C., Frederick, M., Merritt, J.A., Zumstein, L.A., Timmons, T.M., Liu, T.J., Ginsberg, L., Roth, J.A., Hong, W.K., Bruso, P., and Goepfert, H., 1998, Adenovirus—mediated p53 gene transfer in patients with advanced recurrent head and neck squamous cell carcinoma. *J. Clin. Oncol.* **16**:2221–2232.

Colak, A., Goodman, J.C., Chen, S.H., Woo, S.L.C., Grossman, R.G., and Shine, H.D., 1995, Adenovirus-mediated gene therapy for experimental spinal cord tumors: tumoricidal efficacy and functional outcome. *Brain Research*. **691**:76–82.

Culver, K.W., Rigolet, M., and Raffel, C., 1997, Gene Therapy Strategies for Brain Cancer. *Encyclopedia of Cancer*. Vol II. Pp. 704–717, Academic Press.

Douglas, J.T.,Rogers, B.E., Rosenfeld, M.E., Michael, S.I., Feng, M, and Curiel, D.T., 1996, Targeted gene delivery by tropism-modified adenoviral vectors. *Nat Biotechnol.* **14**:1574–1578.

Fender, P., Ruigrok, R.W., Gout, E., Buffet, S., and Chroboczek, J., 1997, Adenovirus dodecahedron, a new vector for human gene transfer. *Nat Biotechnol.* **15**:52–56.

Feng, M., Jackson, W.H.Jr, Goldman, C.K., Rancourt, C., Wang, M., Dusing, S.K., Siegal, G., and Curiel, D.T., 1997, Stable in vivo gene transduction via a novel adenoviral/retroviral chimeric vector. *Nature Biotech*, **15**:866–870.

Goldsmith, K.T., Curiel, D.T., Engler, J.A., and Garver, R.I. Jr, 1994, Trans Complementation of an E1A-Deleted Adenovirus with Codeliverd E1A Sequences to Make Recombinant Adenoviral Producer Cells. *Hum Gene Ther*. **5**:1341–1348.

Graham, F.L., Smiley, J., Russell, W.C., and Nair, R., 1977, Characteristics of a human cell line transformed by DNA from human adenovirus type 5. *J. Gen. Virol.* **36**:59–74.

Horwitz, M.S., 1990, Adenoviruses. In: *Fields Virology*, volume 2. (Fields and Knipe eds.), pp. 1723–1742, Raven Press.

Kajiwara, K., Byrnes, A.P., Charlton, H.M., Wood, M.J.A., and Wood, K.J., 1997, Immune Responses to Adenoviral Vectors During Gene Transfer in the Brain. *Hum. Gene Ther*. **8**:253–265.

Kwong, Y.L., Chen, S.H., Kosai, K., Finegold, M., and Woo, S.L., 1997, Combination therapy with suicide and cytokine genes for hepatic metastases of lung cancer. *Chest* **5**:1332–1337.

Moolten, F.L., 1994, Drug sensitivity ("suicide") genes for selective cancer chemotherapy. *Cancer Gene Ther.* **1**:279–287.

Perez-Cruet, M.J., Trask, T.W., Chen, S.H., Goodman, J.C., Woo, S.L.C., Grossman, R.G., and Shine, H.D. 1994, Adenovirus-Mediated Gene Therapy of Experimental Gliomas. *Journal of Neuroscience Research*. **39**:506–511.

Putzer, B.M., Bramson, J.L., Addison, C.L., Hitt, M., Siegel, P.M., Muller, W.J., Graham, F.L., 1998, Combination therapy with interleukin-2 and wild-type p53 expressed by adenoviral vectors potentiates tumor regression in a murine model of breast cancer. *Hum Gene Ther.* **5**:707–718.

Puumalainen, A.M., Puranen, P., Loimas, S., Puranen, M., Hurskainen, H., Kossila, M., Tyynelä, K., Turunen, M., Vanninen, R., Lehtolainen, P., Paljärvi, L., Johansson, R., Vapalahti, M., and Ylä-Herttuala, S. 1999, Thymidine kinase gene therapy for human malignant glioma using replication-deficient retroviruses or adenoviruses. Phase I safety and efficacy trial. Submitted.

Puumalainen, A.M., Turunen, M., Loimas, S., Vainio, P., Tyynelä, K., Vanninen, R., Vapalahti, M., Bjerkvig, R., Jänne, J., and Ylä-Herttuala, S. 1999, Herpes Simplex Virus Thymidine Kinase Gene Therapy in Experimental Rat BT4C Glioma Model. Submitted.

Puumalainen, A.M., Vapalahti, M., Agrawal, R., Laukkanen, J., Kossila, M., Viita, H., Vanninen, R., Paljärvi, L., and Ylä-Herttuala, S., 1998, B-galactosidase gene transfer to human malignant glioma using retro- and adenovirus vectors. *Hum. Gene Ther.* **9**:1769–1774.

Rancourt, C., and Curiel, D.T., 1997, Conditionally replicative adenoviruses for cancer gene therapy. *Advanced Drug Delivery Reviews.* **27**:67–81.

Rosenfeld, M.E., Vickers, S.M., Raben, D., Wang, M., Sampson, L., Feng, M., Jaffee, E., and Curiel, D.T., 1997, Pancreatic carcinoma cell killing via adenoviral mediated delivery of the herpes simplex virus thymidine kinase gene. *Ann Surg.* **5**:609–620.

Roth, J.A., Swisher, S.G., Merrit, J.A., Lawrence, D.D., Kemp, B.L., Carrasco, C.H., El-Naggar, A.K., Fossella, F.V., Glisson, B.S., Hong, W.K., Khurl, F.R., Kurie, J.M., Nesbitt, J.C., Pisters, K., Putnam, J.B., Schrump, D.S., Shin, D.M., and Walsh, G.L., 1998, Gene therapy for non-small cell lung cancer: a preliminary report of a phase I trial of adenoviral p53 gene replacement. *Semin-Oncol.* **25**:33–37.

Rowe, W.P., Huebner, R.J., Gillmore, L.K., Parrott, R.H., and Ward, T.G., 1953, Isolation of a cytopathogenic agent from human adenoids undergoing spontaneous degeneration in tissue culture. *Proc Soc Exp Biol Med.* **84**:570–573.

Schiedner, G., Morral, N., Parks, R.J., Wu, Y., Koopmans, S.C., Langston, C., Graham, F.L., Beaudet, A.L., and Kochanek, S., 1998, Genomic DNA transfer with a high-capacity adenovirus vector results in improved in vivo gene expression and decreased toxicity. *Nat Genetics* **18**:180–183.

Spector, D.J., Halbert, D.N., and Raskas, H.J., 1980, Regulation of integrated adenovirus sequencies during adenovirus infection of transformed cells. *J. Virol.* **36**:860–871.

Tyynelä, K., Puumalainen, A.M., Turunen, M., Hakumäki, J., Kauppinen, R., Johansson, R., and Ylä-Herttuala, S., 1998, Adenovirus-mediated Herpes simplex virus -thymidine kinase gene therapy in BT4C rat glioma in vivo. Submitted.

Vincent, A.J.P.E., Esandi, M.C., Avezaat, C.J.J., Vecht, C., Smitt, P.S., Bekkum, D.W., Valerio, D., Hoogerbrugge, P.M., and Bout, A. 1997, Preclinical Testing of Recombinant Adenoviral Herpes Simplex Virus-Thymidine Kinase Gene Therapy for Central Nervous System Malignancies. *Neurosurgery.* **41**:442–452.

Viola, J.J., Ram, Z., Walbridge, S., Oshiro, E.M., Trapnell, B., Tao-Cheng, J.H., and Oldfield, E.H. 1995, Adenovirally mediated gene transfer into experimental solid brain tumors and leptomeningeal cancer cells. *J Neurosurg* **82**:70–76.

Wang, Q., and Finer, M.H., 1996, Second-generation adenovirus vectors. *Nat Medicine,* **2**:714–716.

Wickham, T.J., Lee, G.M., Tutus, J.A., Sconoccia, G., Bakacs, T, Kovesdi, I., and Segal, D.M., 1997, Targeted adenovirus-mediated gene delivery to T cells via CD3. *J Virol.* **71**:7663–7669.

Wickham, T.J., Roelvink, P.W., Brough, D.E., and Kovesdi, I., 1996, Adenovirus targeted to heparan-containing receptors increases its gene delivery efficiency to multiple cell types. *Nat Biotechnol,* **14**:1570–1573.

Wickham, T.J., Segal, D.M., Roelvink, P.W., Carrion, M.E., Lizonova, A., Lee, G.M., and Koveski, I. 1996, Targeted adenovirus gene transfer to endothelial and smooth muscle cells by using bispecific antibodies. *J Virol.* **70**:6831–6838.

Wilson, J., 1995, Adenoviruses as Gene-Delivery Vehicles. *N Engl J Med* **334**:1185–1187.

Ylä-Herttuala, S., 1998, Gene therapy for cancer. In: *Tumor Invasion: Biological, Clinical and Therapeutical Considerations.* (Mikkelsen, Bjerkvig, Laerum, Rosenblum eds.), pp. 435–445, John Wiley & Sons.

X

ANGIOGENESIS CONTROL

EVALUATION OF AN INHIBITOR OF DNA METHYLATION, 5-AZA-2′-DEOXYCYTIDINE, FOR THE TREATMENT OF LUNG CANCER AND THE FUTURE ROLE OF GENE THERAPY

Richard L. Momparler, Nicoletta Eliopoulos, and Joseph Ayoub

Département de pharmacologie
Université de Montréal, Centre de recherche pédiatrique
Hôpital Ste-Justine
3175 Côte Ste-Catherine
Montréal, Québec H3T 1C5 and Centre d'Oncologie
Hôpital Notre-Dame
1560 rue Sherbrooke est
Montréal, Québec H2L 1R7 Canada

1. INTRODUCTION

The antineoplastic activity and pharmacology of 5-aza-2′-deoxycytidine (5-AZA-CdR) have been under investigation in our laboratory for some time (Momparler, 1985). The chemical structures of 5-AZA-CdR and its related natural nucleoside, deoxycytidine, are shown in Fig. 1. The only structural difference between these two compounds is the presence of a nitrogen at position 5 of the cytosine ring in 5-AZA-CdR as compared to a carbon at this position for deoxycytidine. In this article we will review the pharmacology of 5-AZA-CdR, its preliminary antitumor activity in patients with advanced non-small cell lung cancer (NSCLC), and the possible use of gene therapy to improve its anticancer effect.

1.1. Pharmacodynamics of 5-AZA-CdR

5-AZA-CdR is a prodrug that must first be phosphorylated to be converted to its active form. The rate limiting enzyme for this metabolic activation is deoxycytidine kinase (Fig. 2). The affinity of 5-AZA-CdR for the catalytic site of this enzyme is similar to that of the natural substrate, deoxycytidine (Momparler, 1985). Deoxycytidine kinase is

Cancer Gene Therapy: Past Achievements and Future Challenges, edited by Habib
Kluwer Academic/Plenum Publishers, New York, 2000.

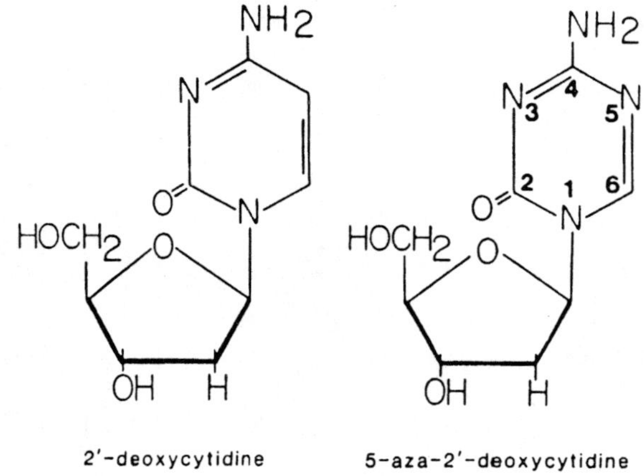

Figure 1. Chemical structures of 2′-deoxycytidine and 5-aza-2′-deoxycytidine.

primarily synthesized during the S phase of the cell cycle. Cells deficient in this enzyme are completely resistant to 5-AZA-CdR.

Another key enzyme involved in the action of 5-AZA-CdR is DNA polymerase. 5-AZA-CdR, after conversion to its triphosphate form, 5-AZA-dCTP, is incorporated into replicating DNA at a rate similar to that of the natural substrate, dCTP (Bouchard and Momparler, 1983). The major pharmacological characteristics of 5-AZA-CdR are summarized in Table 1. The antineoplastic action of 5-AZA-CdR is S phase specific (Momparler, Samson, Momparler, and Rivard, 1984). This analogue produces a marked loss of clonogenicity for many types of leukemic and tumor cell lines, as determined by colony assay. 5-AZA-CdR has been demonstrated to induce terminal differentiation of both embryonic and neoplastic cells (Momparler, Bouchard, and Samson, 1985a; Jones and Taylor, 1980). The plasma half-life of 5-AZA-CdR in man is short, in the range of 15 to 20 min (Momparler, Rivard, and Gyger, 1985b), and this is due primarily to deamination of this analogue by the high levels of cytidine deaminase in the liver (Ho, 1973). In animal

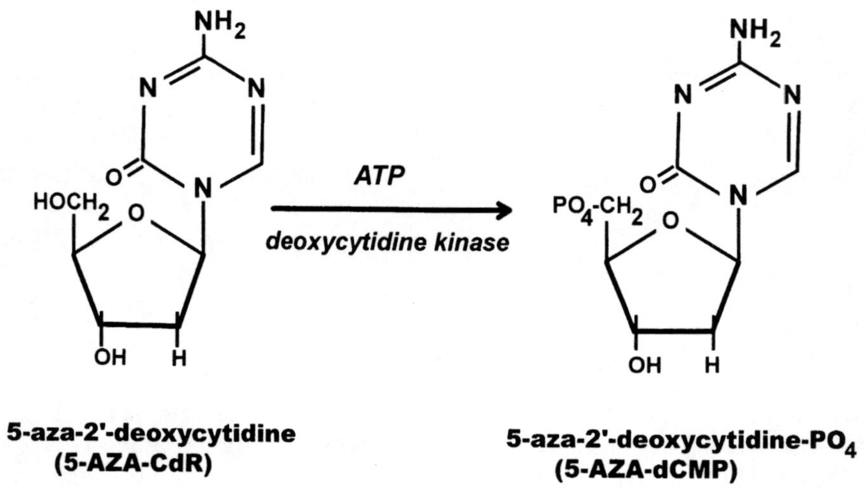

Figure 2. Activation of 5-AZA-CdR by deoxycytidine kinase.

Table 1. Pharmacological characteristics of 5-AZA-CdR

1. S phase specific agent
2. specific inhibitor of DNA methylation
3. inducer of cellular differentiation
4. short *in vivo* half-life
5. excellent tissue distribution
6. hematological toxicity
7. antineoplastic activity very dependent on dose-schedule

models, 5-AZA-CdR shows potent antileukemic and antitumor activity (Momparler, 1985). Its *in vivo* antineoplastic activity is very dose-schedule dependent.

1.2. Inhibition of DNA Methylation by 5-AZA-CdR

Methylation of cytosine to 5-methylcytosine occurs at the level of DNA. In this reaction catalyzed by DNA methylase, a methyl group is covalently linked to the 5 position of cytosine using S-adenosyl-methionine as the methyl donor. Each class of differentiated cells has its own distinct methylation pattern. After chromosomal duplication, in order to conseve this pattern of methylation, the 5-methylcytosine on the parental strand serves to direct methylation on the complementary daughter DNA strand. Substituting the carbon at the 5 position of the cytosine for a nitrogen interferes with this normal process of DNA methylation. The presence of 5-AZA-CdR in place of 5-methylcytosine at a specific site of methylation produces an irreversible inactivation of DNA methylase (Fig. 3). As a result, all the methylation sites downstream of its incorporation will not be methylated. The inhibition of DNA methylation correlates with the antineoplastic action of 5-AZA-CdR (Wilson, Jones, and Momparler, 1983). As illustrated in Fig. 4, one consequence of DNA hypomethylation is the transcriptional activation of genes that were previously silenced by methylation (Jones, 1996; Razin and Cedar, 1991).

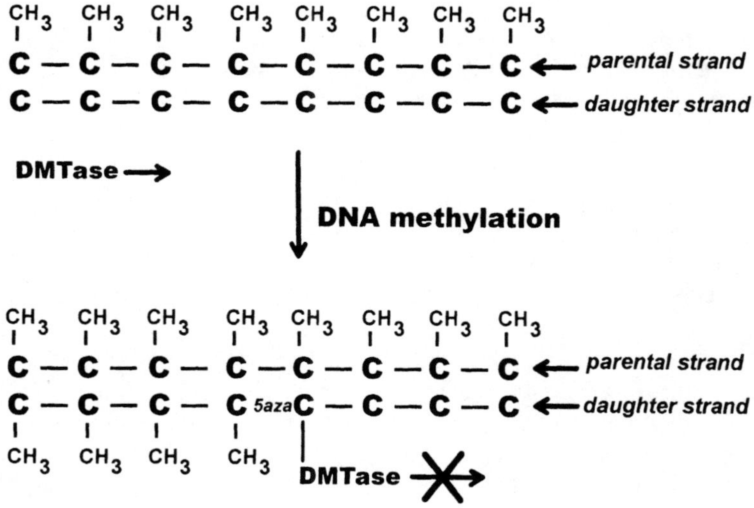

Figure 3. Inactivation of DNA methylase by the presence of 5-AZA-CdR in DNA.

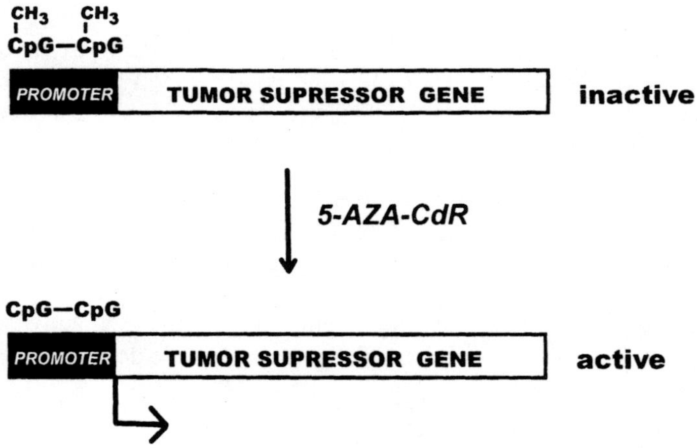

Figure 4. Activation of a tumor suppressor gene by demethylation of the promoter region by 5-AZA-CdR.

Recent laboratory investigations have shown that DNA methylation of some tumor suppressor genes can silence their expression (Baylin, Herman, Graff, Vertino, and Issa, 1998). This methylation occurs primarily in the CpG islands located in the promoter region of tumor suppressor genes, such as p16 (Merlo, Herman, Mao, Lee, Gabrielson, Burger, Baylin, and Sidranski, 1995; Otterson, Khleif, Chen, Coxon, and Kaye, 1995; Herman, Merlo, Mao, Lapidus, Issa, Davidson, Sidranski, and Baylin, 1995), p15 (Herman, Jen, Merlo, and Baylin, 1996b) and VHL (Herman, Latif, Weng, Lerman, Zbar, Liu, Samid, Duan, Gnarra, Linehan, and Baylin, 1994). Other cancer-related genes that have also been demonstrated to be silenced by methylation include E-cadherin which is an invasion suppressor gene (Yoshiura, Kanai, Ochiai, Shimoyama, Sugimura, and Hirohashi, 1995) and thrombospondin-1, an angiogenesis inhibitor (Ahuja, Mohan, Li, Stolker, Herman, Hamilton, Baylin, and Issa, 1997). The expression of the repair genes, hMLH1 and the methylguanine DNA methyltransferase, have been also shown to be blocked by DNA methylation (Herman, Umar, Polyak, Graff, Ahuja, Issa, Markowitz, Willson, Hamilton, Kinzler, Kane, Kolodner, Vogelstein, Kunkel, and Baylin, 1998; Kane, Loda, Gaida, Lipman, Mishra, Goldman, Jessup, and Kolodner, 1997; Qian and Brent, 1997). The silencing of these genes can lead to tumor progression and growth by increased metastasis and angiogenesis and to a more malignant phenotype due to additional mutations. Genes related to breast cancer, such as the estrogen receptor (Ferguson, Lapidus, Baylin, and Davidson, 1995), the mammary-derived growth inhibitor (Huynh, Alpert, and Pollak, 1996) and BRCA1 (Dobrovic and Simpfendorfer, 1997), have also been reported to be methylated.

Considerable interest in 5-AZA-CdR was provided by *in vitro* studies showing that many of the genes that have been silenced by DNA methylation can be reactivated by treatment of tumor cells with this analogue (Table 2). 5-AZA-CdR exerts a growth inhibitory effect on human tumor cell lines by its demethylation action which reactivates the expression of growth regulatory genes (Bender, Pao, and Jones, 1998). Genes reported to have their expression activated by the 5-AZA-CdR-induced block in methylation include p16 in lung cancer (Merlo *et al.*, 1995; Otterson *et al.*, 1995), bladder carcinoma (Gonzalgo, Hayashida, Bender, Pao, Tsai, Gonzales, Nguyen, Nguyen, and Jones, 1998) and gliomas (Costello, Berger, Huang, and Cavenee, 1996), VHL in renal cancer (Herman *et al.*, 1994), mammary-derived growth inhibitor in breast cancer (Huynh *et al.*, 1996),

Table 2. Genes silenced by methylation and activated by 5-AZA-CdR

Gene	Function	Activated by 5-AZA-CdR
p16 (cyclin kinase inhibitor)	tumor suppressor	+
von Hippel Lindau (VHL)	tumor suppressor	+
Retinoic acid receptor β (RARβ)	tumor suppressor	+
Mammary-derived Growth Inhibitor	tumor suppressor	+
BRCA1	tumor suppressor	nd
E-Cadherin	tumor invasion inhibitor	+
Thrombospondin 1 (TSP-1)	angiogenesis inhibitor	+
hMLH1	mismatch DNA repair	+
O^6-methylguanine methyltransferase	DNA repair	+

nd, not done.

and E-cadherin in primary tumors of breast and prostate (Graff, Herman, Lapidus, Chopra, Xu, Jarrard, Isaacs, Pitha, Davidson, and Baylin, 1995). A rapid PCR assay has been developed to evaluate the methylation status of genes (Herman, Graff, Myohanen, Nelkin, and Baylin, 1996a).

Our laboratory has been investigating the tumor suppressor gene retinoic acid receptor beta (RARβ). We observed that treatment of human DLD-1 colon carcinoma cells with 5-AZA-CdR results in demethylation of RARβ and activation of its expression (Côté and Momparler, 1997; Côté, Sinnett, and Momparler, 1998). One interesting outcome of the 5-AZA-CdR exposure is that the tumor cells become more sensitive to the inhibitory effects of retinoic acid, resulting in a synergistic interaction between these two agents (Côté and Momparler, 1995). Retinoic acid and its analogues are under investigation as cancer chemoprevention agents. It will be interesting to test retinoic acid combined with 5-AZA-CdR in patients with tumors that show methylation of RARβ. It is interesting to note that RARβ expression is often decreased in NSCLC (Xu, Sozzi, Lee, Lee, Pastorino, Pilotti, Kurie, Hong, and Lotan, 1997). In human lung cancer cells, gene transfer with RARβ produced a tumor inhibitory action (Houle, Rochette-Egly, and Bradley, 1993).

2. CLINICAL ACTIVITY OF 5-AZA-CdR IN CANCER

2.1. Hematological Malignancies

Our group in Montreal was the first to perform a clinical trial on 5-AZA-CdR. We observed that this cytosine nucleoside analogue could induce complete remissions in some children with relapsed leukemia (Rivard, Momparler, Demers, Benoit, Raymond, Lin, and Momparler, 1981). In addition, 5-AZA-CdR showed activity against adult leukemia (Momparler *et al.*, 1985b; Richel, Colly, Kluin-Nelemans, and Willemze, 1991). One group demonstrated that leukemic blasts from adult patients were induced due to differentiate by *in vitro* treatment with 5-AZA-CdR (Pinto, Attadia, Fusco, Ferrara, Spada, DiFiore, 1984). In recent investigations, 5-AZA-CdR demonstrated interesting activity against chronic myeloid leukemia in blastic phase (Kantarjian, O'Brien, Estey, Giralt, Beran, Rios, Keating, de Vos, and Talpaz, 1997) and against myelodysplastic syndrome, a preleukemic disease (Zagonel, Lo Re, Marotta, Babare, Sardeo, Gattei, De Angelis, Monfardini, and Pinto, 1993).

2.2. Non-Small Cell Lung Cancer (NSCLC)

DNA methylation analysis and clonogenic assays on human tumor cell lines (Côté and Momparler, 1995), and animal studies in nude mice with human tumors (Braakhuis, van Dongen, van Walsum, Leyva, and Snow, 1988; Bender *et al.*, 1998) indicate that 5-AZA-CdR may be an active antineoplastic agent in patients with malignant tumors. In order to investigate this possibility, we performed a pilot clinical trial on 5-AZA-CdR in patients with stage IV non-small cell lung cancer (NSCLC) (Momparler, Bouffard, Momparler, Dionne, Bélanger, and Ayoub, 1997). We chose this tumor type because most patients with advanced metastatic NSCLC respond poorly to conventional chemotherapy. For example, one clinical study compared the efficacy of chemotherapy using a combination of cisplatin, cyclophosphamide and mitomycin against that of supportive therapy alone in patients with stage IV NSCLC (Cartei, Cartei, Cantone, Causarano, Genco, Tobaldin, Interlandi, and Giraldi, 1993). The patients that received chemotherapy survived only a few months longer that the supportive therapy group. With chemotherapy, less than 50% of the patients survived more than 10 months and only 10% of the these patients survived 24 months.

In our clinical trial, patients with stage IV NSCLC were administered an 8 hour i.v. infusion of 5-AZA-CdR at the indicated total dose for one or more cycles at intervals of 5 to 7 weeks, depending on the recovery of the granulocyte count. These patients did not receive any prior chemotherapy. Except for patient #4, the survival time increased with the number of cycles of 5-AZA-CdR administered (Table 3). Most interesting was the response of patient #7 who survived more than 6 years, an outcome in our clinic never seen before with this tumor type at an advanced stage. This patient, a heavy smoker with a diagnosis of adenocarcinoma which is a form of lung cancer with a very poor prognosis, received 5 cycles of 5-AZA-CdR, more than any other patient.

One phenomena we encountered in this clinical study was the delayed action of 5-AZA-CdR on tumor growth. Most conventional cytotoxic drugs used in tumor therapy, when administered at the maximum tolerated dose, produce an initial rapid reduction in tumor size. On the other hand, 5-AZA-CdR has a delayed action on tumor growth that begins to occur after one or more cell divisions (Fig. 5). This can result in a slight increase in tumor size after 5-AZA-CdR treatment, followed by a delayed arrest in tumor growth. Patient #7, our best responder, was actually removed from the study after the 5th cycle of 5-AZA-CdR, about 12 months from the initial treatment, due to signs of tumor

Table 3. Response to 5-AZA-CdR in evaluable patients with stage IV non-small cell lung cancer*

Pt #	Diagnosis	Age/sex	Dose per cycle (mg/m^2)	Cycles treatment	Survival time (months)
1	adeno	59/F	660	1	3.3
2	adeno	47/M	660	2	6.7
3	squam	62/F	660	2	9.3
4	adeno	62/M	200–400	3	4.3
5	adeno	51/F	660	3	16.0
6	mixed	62/M	200–400	4	15.3
7	adeno	64/F	495–660	5	81

5-AZA-CdR administered as 8 hr i.v. infusion; Interval between cycles 5–7 weeks.
*modified from Momparler *et al.*, Anti-Cancer Drugs 8:358–368 (1997).

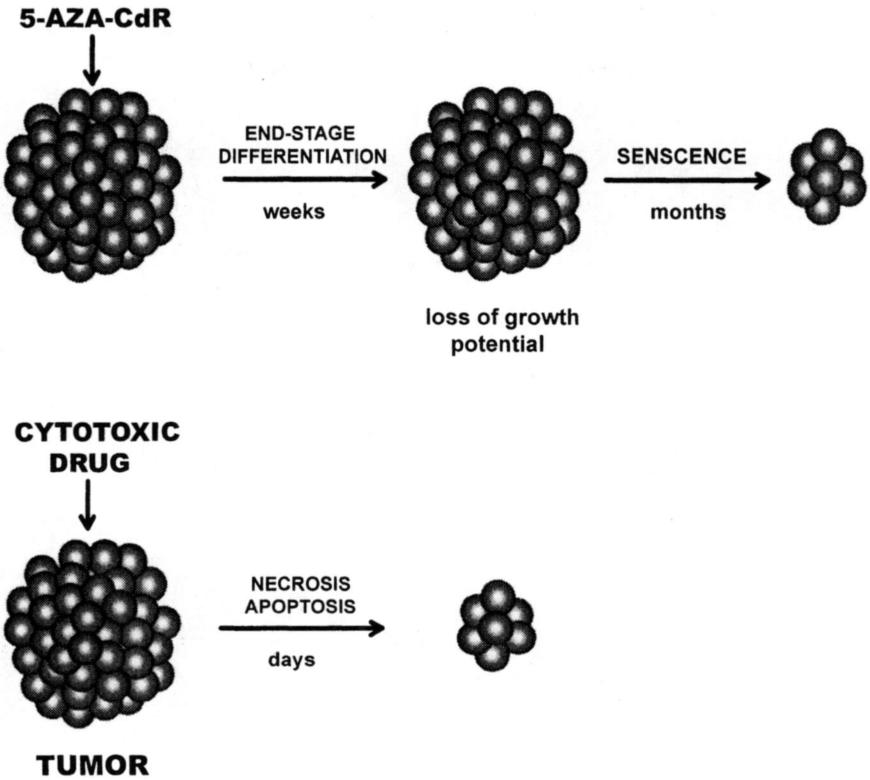

Figure 5. Model showing delayed action of 5-AZA-CdR on tumor size.

progression. This signifies that different end points have to be used for 5-AZA-CdR in clinical trials to evaluate its antitumor activity. The key end point for this analogue should be patient survival. Several patients with NSCLC may have been removed prematurely from this study due to signs of tumor progression since the delayed action of 5-AZA-CdR was not taken into account and thus our data are possibly an underestimate of its "true" clinical effect. The antitumor activity of 5-AZA-CdR may require 3 or more cycles of treatment before it becomes evident. It should be pointed out that we had also observed in patients with leukemia a delayed action of 5-AZA-CdR on the clearance of leukemic blasts from the bone marrow as compared to the clearance produced by conventional cytotoxic drugs (Momparler *et al.*, 1985b).

2.3. Hematological Toxicity

Cancer chemotherapy with cytosine nucleoside analogues produces myelosuppression as its main side effect. The major toxicity produced by 5-AZA-CdR in patients with lung cancer was leukopenia. The hemogram of a patient with lung cancer treated with an 8 hour i.v. infusion of 5-AZA-CdR at a dose of 660 mg/m^2 is shown in Fig. 6.

Treatment with 5-AZA-CdR caused a considerable drop in the leukocyte count which lasted from about day 15 to day 30. The hematopoietic toxicity produced by 5-AZA-CdR was acceptable in most patients and not life-threatening. The patients were monitored after each cycle of 5-AZA-CdR on an outpatient basis and did not require

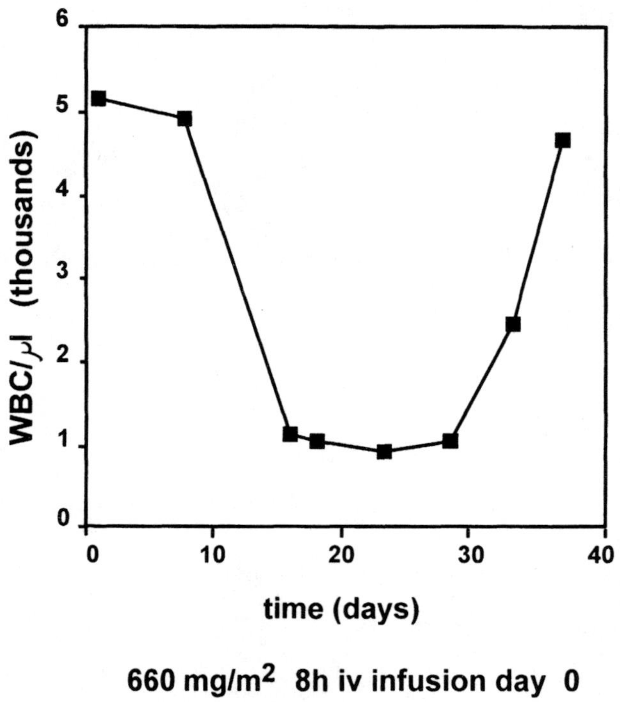

Figure 6. Representative hematopoietic toxicity of 5-AZA-CdR in patients with NSCLC.

hospitalization during the nadir of the white blood cell count. In future clinical trials in patients with advanced cancer, it may be necessary to use a longer 5-AZA-CdR infusion time to prevent, with the more rapidly proliferating tumors, marked tumor progression which may occur in patients between cycles of treatment.

However, longer infusions may produce more pronounced granulocytopenia which can result in severe infections. A new approach may be needed to allow patients to tolerate dose intensification without experiencing marked myelosuppression.

3. GENE THERAPY OF CANCER WITH 5-AZA-CdR

3.1. Rationale for Chemoprotection

Gene therapy can be used to overcome the problem of hematopoietic toxicity produced by chemotherapeutic drugs used for the treatment of cancer. Gene transfer and expression of drug resistance genes in hematopoietic cells has been utilized to protect these cells from the toxic effects of different classes of anticancer agents (Koç, Allay, Keunmyoung, Davis, Reese, and Gerson, 1996). For example, retroviral vectors have been used to insert the P-glycoprotein gene (MDR) into blood cells to protect them from the myelosuppression produced by MDR drugs (Sorrentino, Brandt, Bodine, Gottesman, Pastan, Cline, and Nienhuis, 1992; Ward, Pioli, Ayello, Reiss, Urzi, Richardson, Hesdorffer, and Bank, 1996). Clinical trials using the MDR gene are under investigation in patients with breast cancer (Deisseroth, Holmes, Hortobagyi, and Champlin, 1996). Moreover, other drug resistance genes, such as dihydrofolate reductase (Zhao, Banerjee, Mineishi, and Bertino, 1997), O_6-methylguanine-DNA-methyltransferase (Allay,

Dumenco, Koç, Liu, and Gerson, 1995) and glutathione-S-transferase (Létourneau, Greenbaum, and Cournoyer, 1996) have been used to protect hematopoietic cells from the toxicity of antifolates and alkylating agents, respectively.

3.2. Cytidine Deaminase

The drug resistance gene cytidine deaminase (CD) catalyzes the deamination of cytosine nucleosides and related analogues. The deamination by CD of cytosine nucleoside analogues, such as cytosine arabinoside and 5-AZA-CdR results in a loss of their pharmacological acitivity (Fig. 7). We have cloned the cDNA for human CD (Laliberté and Momparler, 1994) and shown that the coding region consists of 441bp, a desirable small size facilitating its manipulation into mammalian expression vectors. We inserted the CD cDNA into a retroviral vector and transferred this gene into fibroblast and marrow cells. CD gene-modified cells demonstrated *in vitro* increased expression of CD and drug resistance to cytosine nucleoside analogues (Momparler, Eliopoulos, Bovenzi, Létourneau, Greenbaum, and Cournoyer, 1996a; Momparler, Laliberté, Eliopoulos, Beauséjour, and Cournoyer, 1996b).

In order to show that the acquired drug resistance phenotype was due to increased expression of CD, we treated CD-transduced fibroblast cells with tetrahydrouridine, a potent competitive inhibitor of this enzyme. As shown in Fig. 8, the transduced cells showed drug resistance to 5-AZA-CdR which was reversed by the addition of THU, confirming that this resistance was the result of elevated CD enzyme activity (Eliopoulos, Cournoyer, and Momparler, 1998b).

In animal studies, we observed that about one year after transplantation of CD-transduced marrow cells into mice it was possible by PCR to detect the presence of the CD transgene in hematopoietic tissues (Eliopoulos, Bovenzi, Le, Momparler, Greenbaum, Létourneau, Cournoyer, and Momparler, 1998a). At this time interval we also demonstrated in some mice a significantly increased CD expression in blood and marrow cells and, using a clonogenic assay, signs of *in vitro* drug resistance to cytosine nucleoside analogues. These preclinical studies suggest that it may be possible to use gene therapy with CD to confer chemoprotection in patients with cancer.

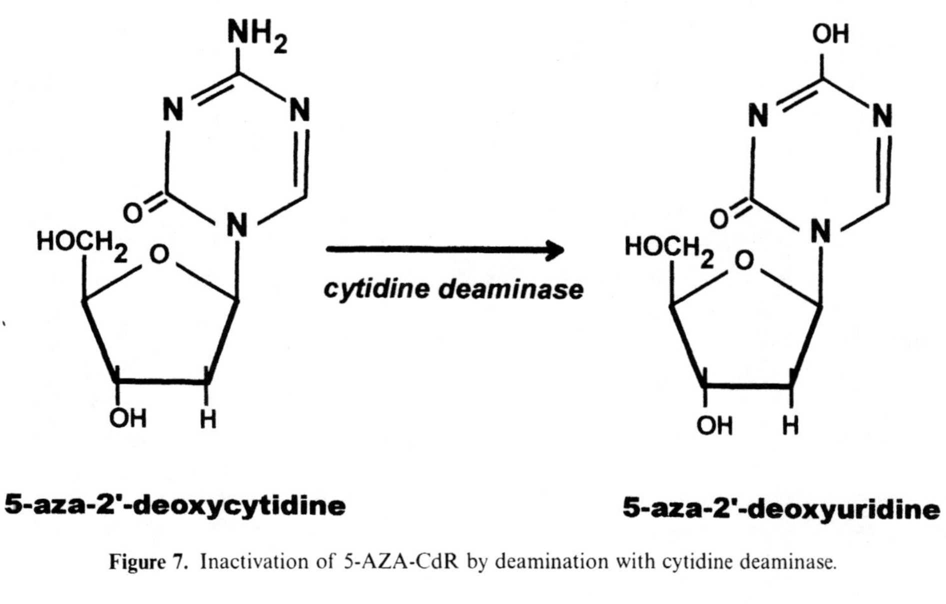

Figure 7. Inactivation of 5-AZA-CdR by deamination with cytidine deaminase.

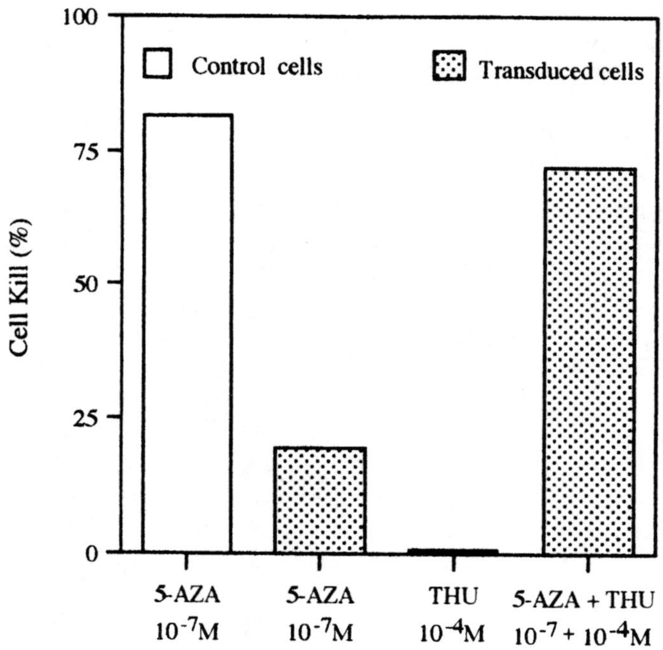

Figure 8. Drug resistance to 5-Aza-2′-deoxycytidine (5-AZA) in cells transduced with cytidine deaminase (CD) and its reversal with tetrahydrouridine (THU).

3.3. Model for Clinical Therapy Using Chemoprotection

As stated earlier, the major dose-limiting toxicity produced by 5-AZA-CdR is bone marrow suppression which limits its curative potential in cancer therapy. One approach to overcome this problem is to confer to the hematopoietic stem cells drug resistance to cytosine nucleoside analogues using the CD gene. The schema for the clinical use of this form of gene therapy is shown in Fig. 9. Marrow or blood cells are harvested from the patient. The CD34+ antigen-expressing hematopoietic cells are isolated by centrifugation and affinity chromatography and placed in tissue culture with growth factors and retroviral particles containing the CD gene. The transduced cells are reinfused back into the patient who is then ready for intensive chemotherapy with 5-AZA-CdR. It should be noted that expression of the drug resistance gene is only necessary during chemotherapy and not for the entire lifespan of the patient, as is the case with gene therapy of genetic diseases. A single transplantation of CD-transduced cells from an autologous marrow should be enough to offer protection against the hematotoxicity for several cycles of drug treatment.

There are several potential problems that must be addressed in order to utilize this type of gene therapy in patients. One important feature of this approach is that the hematopoietic cells to be transfused should not be contaminated with tumor stem cells. Only patients that do not have metastasis of tumor cells to the bone marrow can be selected for a clinical trial on chemoprotection. The significant advances in the techniques used for separating tumor cells from normal hematopoietic cells will allow malignant cells to be detected, even in cases where the patient does not show any histological signs of tumor metastasis. These cell isolation procedures will eliminate the possibility of having contaminating tumor cells in the transplanted hematopoietic cells. Another limitation to

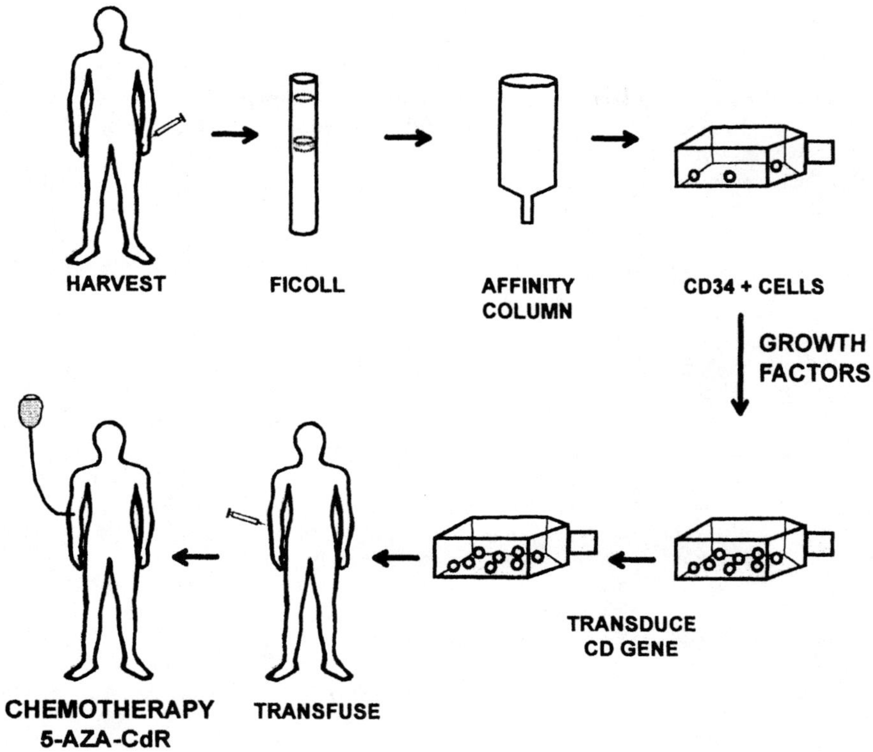

Figure 9. Schema for clinical gene therapy for chemoprotection.

the gene therapy approach is the necessity to express sufficient levels of the drug resistance gene in order to obtain the required chemoprotection. Improved vectors for gene expression and better packaging cells can be used to overcome the problem of inadequate transgene expression.

In conclusion, the development of drug resistance by tumor cells has always been a major obstacle that limits the effectiveness of chemotherapy. Gene therapy has the potential to use this same process to reverse the tables, as stated by Bertino (1990), to protect the normal cells against the toxic effects of anticancer drugs. This approach will permit the use of higher doses of antineoplastic agents so as to increase their ability to eradicate all the tumor cells. The end result will be the improvement of the survival of patients with cancer. The successful use of gene therapy to confer chemoprotection may permit the realization of the full potential in cancer therapy of the drug, 5-AZA-CdR, which has such an interesting mechanism of action.

REFERENCES

Allay, J.A., Dumenco, L.L., Koç, O.N., Liu, L., and Gerson, S.L., 1995, Retroviral transduction and expression of the human alkyltransferase cDNA provides nitrosourea resistance to hematopoietic cells, *Blood* **85**:3342–3351.

Ahuja, N., Mohan, A.L., Li, Q., Stolker, J.M., Herman, J.G., Hamilton, S.R., Baylin, S.B., and Issa, J.-P.J., 1997, Association between CpG island methylation and microsatellite instability in colorectal cancer, *Cancer Res.* **57**:3370–3374.

Baylin, S.B., Herman, J.G., Graff, J.R., Vertino, P.M., and Issa, J.-P., 1998, Alterations in DNA methylation: A fundamental aspect of neoplasia, *Adv. Cancer Res.* **72**:141–196.

Bender, C.M., Pao, M.M., and Jones, P.A., 1998, Inhibition of DNA methylation by 5-aza-2′-deoxycytidine suppresses the growth of human tumor cell lines, *Cancer Res.* **58**:95–101.

Bertino, J.R., 1990, "Turning the Tables"—Making normal marrow resistant to chemotherapy, *J. Natl. Cancer I.* **82**:1234–1235, 1990.

Bouchard, J., and Momparler, R.L., 1983, Incorporation of 5-aza-2′-deoxycytidine 5′-triphosphate into DNA Interaction with mammalian DNA polymerase α and DNA methylase, *Mol. Pharmacol.* **24**:109–114.

Braakhuis, B.J.M., van Dongen, G.A.M.S., van Walsum, M., Leyva, A., and Snow, G.B., 1988, Preclinical antitumor activity of 5-aza-2′-deoxycytidine against human head and neck cancer xenografts, *Invest. New Drugs* **6**:299–304.

Cartei, G., Cartei, F., Cantone, A., Causarano, D., Genco, G., Tobaldin, A., Interlandi, G., and Giraldi, T., 1993, Cisplatin-cyclophosphamide-mitomycin combination chemotherapy with supportive care versus suppotive care alone for treatment of metastatic non-small-cell lung cancer, *J. Natl. Cancer I.* **85**:794–800.

Costello, J.F., Berger, M.S., Huang, H.S., and Cavenee, W.K., 1996, Silencing of p16/CDKN2 expression in human gliomas by methylation and chromatin condensation. *Cancer Res.* **56**:2405–2410.

Côté, S., and Momparler, R.L., 1995, Antineoplastic action of all-trans retinoic acid and 5-aza-2′-deoxycytidine on human DLD-1 colon carcinoma cells, *Cell. Pharmacol.* **2**:221–228.

Côté, S., and Momparler, R.L., 1997, Activation of the retinoic acid receptor β gene by 5-aza-2′-deoxycytidine in human DLD-1 colon carcinoma cells, *Anti-Cancer Drugs* **8**:56–61.

Côté, S., Sinnett, D., and Momparler, R.L., 1998, Demethylation by 5-aza-2′-deoxycytidine of specific 5-methylcytosine sites in the promoter region of the retinoic acid receptor beta gene in human colon carcinoma cells, *Anti-Cancer Drugs* **9** (in press).

Deisseroth, A.B., Holmes, F., Hortobagyi, G., and Champlin, R., 1996, Use of safety-modified retroviruses to introduce chemotherapy resistance sequences into normal hematopoietic cells for chemoprotection during the therapy of breast cancer cancer: A pilot trial, *Hum. Gene Ther.* **7**:401–416.

Dobrovic, A., and Simpfendorfer, D., 1997, Methylation of the BRCA1 gene in sporadic breast cancer, *Cancer Res.* **57**:3347–3350.

Eliopoulos, N., Bovenzi, V., Le, N.L.O., Momparler, L.F., Greenbaum, M., Létourneau, S., Cournoyer, D., and Momparler, R.L., 1998a, Retroviral transfer and long-term expression of human cytidine deaminase cDNA in hematopoietic cells following transplantation in mice, *Gene Ther.* **5** (in press).

Eliopoulos, N., Cournoyer, D., and Momparler, R.L., 1998b, Drug resistance to 5-aza-2′-deoxycytidine, 2′,2′-difluorodeoxycytidine and cytosine arabinoside conferred by retroviral-mediated transfer of human cytidine deaminase cDNA into murine cells, *Cancer Chemother. Pharmacol.* **42**:373–378.

Ferguson, A.T., Lapidus, R.G., Baylin, S.B., and Davidson, N.E., 1995, Demethylation of the estrogen receptor gene in estrogen-negative breast cancer cells can reactivate estrogen gene expression, *Cancer Res.* **55**:2279–2283.

Gonzalgo, M.L., Hayashida, T., Bender, C.M., Pao, M.M., Tsai, Y.C., Gonzales, F.A., Nguyen, H.D., Nguyen, T.T., and Jones, P.A., 1998, The role of DNA methylation in expression of the p19/p16 locus in human bladder cancer cell lines. *Cancer Res.* **58**:1245–1252, 1998.

Graff, J.R., Herman, J.G., Lapidus, R.G., Chopra, H., Xu, R., Jarrard, D.F., Isaacs, W.B., Pitha, P.M., Davidson, N.E., and Baylin, S.B., 1995, E-cadherin expression is silenced by DNA hypermethylation in human breast and prostate carcinomas. *Cancer Res.* **55**:5195–5199.

Herman, J.G., Latif, F., Weng, Y., Lerman, M.I., Zbar, B., Liu, S., Samid, D., Duan, D-S.R., Gnarra, J.R., Linehan, W.M., and Baylin, S.B., 1994, Silencing of the VHL tumor-suppressor gene by DNA methylation in renal carcinoma, *Proc. Natl. Acad. Sci. USA* **91**:9700–9704.

Herman, J.G., Merlo, A., Mao, L., Lapidus, R.G., Issa, J-P.J., Davidson, N.E., Sidranski, K., and Baylin, S.B., 1995, Inactivation of the CDKN2/p16/MTS1 gene is frequently associated with aberrant DNA methylation in all common human cancers, *Cancer Res.* **55**:4525–4530.

Herman, J.G., Graff, J.R., Myohanen, S., Nelkin, B.D., and Baylin, S.B., 1996a, Methylation-specific PCR: a novel PCR assay for methylation status of CpG islands, *Proc. Natl. Acad. Sci. USA* **93**:9821–9826.

Herman, J.G., Jen, J., Merlo, A., and Baylin, S.B., 1996b, Hypermethylation-associated inactivation indicates a tumor suppressor role for p15INK4B1, *Cancer Res.* **56**:722–727.

Herman, J.G., Umar, A., Polyak, K., Graff, J.R., Ahuja, N., Issa, J-P., Markowitz, S., Willson, J.K.V., Hamilton, S.R., Kinzler, K.W., Kane, M.F., Kolodner, R.D., Vogelstein, B., Kunkel, T.A., and Baylin, S.B., 1998, Incidence and functional consequences of *hMLH1* promoter hypermethylation in colorectal cancer, *Proc. Natl. Acad. Sci. USA* **95**:6870–6875.

Ho. D.H.W., 1973, Distribution of kinase and deaminase of 1-β-D-arabinofuranosylcytosine in tissues of man and mouse, *Cancer Res.* **33**:2816–2820.

Houle, B., Rochette-Egly, C., and Bradley, W.E.C., 1993, Tumor-suppressive effect of the retinoic acid receptor β in human epidermoid lung cancer cells, *Proc. Natl. Acad. Sci. USA* **90**:985–989.

Huynh, H., Alpert, L., and Pollak, M., 1996, Silencing of the mammary-derived growth inhibitor (MDGI) gene in breast neoplasms is associated with epigenetic changes, *Cancer Res.* **56**:4865–4870.

Issa, J-P.J., Ottaviano, Y.L., Celano, P., Hamilton, S.R., Davidson, N.E., and Baylin, S.B., 1994, Methylation of the oestrogen receptor CpG island links ageing and neoplasia in human colon, *Nat. Genet.* **7**:536–540.

Jones, P.A., 1996, DNA methylation errors and cancer, *Cancer Res.* **56**:2463–2467.

Jones, P.A., and Taylor, S.M., 1980, Cellular differentiation, cytidine analogs and DNA methylation, *Cell* **20**:85–93.

Kane, M.F., Loda, M., Gaida, G.M., Lipman, J., Mishra, R., Goldman, H., Jessup, J.M., and Kolodner, R., 1997, Methylation of the hMLH1 promoter correlates with lack of expression of hMLH1 in sporadic colon tumors and mismatch repair-defective human tumor cell lines, *Cancer Res.* **57**:808–811.

Kantarjian, H.M., O'Brien, S.M., Estey, E., Giralt, S., Beran, M., Rios, M.B., Keating, M., de Vos, D., and Talpaz, M., 1997, Decitabine studies in chronic and acute myelogenous leukemia, *Leukemia* **11**:S35–36.

Koç, O.N., Allay, J.A., Keunmyoung, L., Davis, B.M., Reese, J.S., and Gerson, S.L., 1996, Transfer of drug resistance genes into hematopoietic progenitors to improve chemotherapy tolerance, *Semin. Oncol.* **23**:46–65.

Laliberté, J., and Momparler, R.L., 1994, Human cytidine deaminase: Purification of enzyme, cloning and expression of its cDNA, *Cancer Res.* **54**:5401–5407.

Létourneau, S., Greenbaum, M., and Cournoyer, D., 1996, Retrovirus-mediated gene transfer of rat glutathione S-transferase Yc confers in vitro resistance to alkylating agents in human leukemia cells and in clonogenic mouse hematopoietic progenitor cells, *Hum. Gene Ther.* **7**:831–840.

Merlo, A., Herman, J.G., Mao, L., Lee, D.J., Gabrielson, E., Burger, P.C., Baylin, S.B., and Sidranski, D., 1995, 5′ CpG island methylation is associated with transcriptional silencing of the tumour suppressor gene p16/CDKN2/MTS1 in human cancers, *Nat. Med.* **1**:686–692.

Momparler, R.L., Samson, J., Momparler, L.F., and Rivard, G.E., 1984, Cell cycle effects and cellular pharmacology of 5-aza-2′-deoxycytidine, Cancer Chemother. Pharmacol. **13**:191–194, 1984.

Momparler, R.L., 1985, Molecular, cellular and animal pharmacology of 5-Aza-2′-deoxycytidine, *Pharmacol. Ther.* **30**:287–299.

Momparler, R.L., Bouchard, J., and Samson, J., 1985a, Induction of differentiation and inhibition of DNA methylation in HL-60 myeloid leukemic cells by 5-AZA-2′-deoxycytidine, *Leukemia Res.* **9**:1361–1366.

Momparler, R.L., Rivard, G.E., and Gyger, M., 1985b, Clinical trial on 5-Aza-2′-deoxycytidine in patients with acute leukemia, *Pharmacol. Ther.* **30**:277–286.

Momparler, R.L., Laliberté, J., Eliopoulos, N., Beauséjour, C., and Cournoyer, D., 1996b, Transfection of murine fibroblast cells with human cytidine deaminase cDNA confers resistance to cytosine arabinoside, *Anti-Cancer Drugs* **7**:266–274.

Momparler, R.L., Eliopoulos, N., Bovenzi, V., Létourneau, S., Greenbaum, M., and Cournoyer D., 1996a, Resistance to cytosine arabinoside by retroviral-mediated gene transfer of human cytidine deaminase into murine fibroblast and hematopoietic cells, *Cancer Gene Ther.* **3**:331–338.

Momparler, R.L., Bouffard, D.Y., Momparler, L.F., Dionne, J., Bélanger, K., and Ayoub, J., 1997, Pilot phase I-II study on 5-Aza-2′-deoxycytidine (Decitabine) in patients with metastatic lung cancer, *Anti-Cancer Drugs* **8**:358–368.

Otterson, G.A., Khleif, S.N., Chen, W., Coxon, A.B., and Kaye, F.J., 1995, CDKN2 gene silencing in lung cancer by DNA hypermethylation and kinetics of p16INK4 protein induction by 5-Aza 2′ deoxycytidine, *Oncogene* **11**:1211–1216.

Pinto, A., Attadia, V., Fusco, I.A., Ferrara, F., Spada, O.A., DiFiore, P., 1984, 5-Aza-2′-deoxycytidine induces terminal differentiation of leukemic blasts from patients with acute myeloid leukemia, *Blood* **64**:922–929.

Qian, X.C., and Brent, T.P., 1997, Methylation hot spots in the 5′ flanking region denote silencing of the O^6-methylguanine-DNA methyltransferase gene, *Cancer Res.* **57**:3672–3677.

Razin, A., and Cedar, H., 1991, DNA methylation and gene expression, *Microbiol. Rev.* **55**:451–458, 1991.

Richel, D.J., Colly, L.P., Kluin-Nelemans, J.C., and Willemze, R., 1991, The antileukaemic activity of 5-Aza-2′-deoxycytidine (Aza-dC) in patients with relapsed and resistant leukaemia, *Brit. J. Cancer* **64**:144–148.

Rivard, G.E., Momparler, R.L., Demers, J., Benoit, P., Raymond, R., Lin, K.T., and Momparler, L.F., 1981, Phase I study on 5-Aza-2′-deoxycytidine in children with acute leukemia, *Leukemia Res.* **5**:453–462.

Sorrentino, B.P., Brandt, S.J., Bodine, D., Gottesman, M., Pastan, I., Cline, A., and Nienhuis, A.W., 1992, Selection of drug-resistant bone marrow cells in vivo after retroviral transfer of human MDR-1, *Science* **257**:99–103.

Ward, M., Pioli, P., Ayello, J., Reiss, R., Urzi, G., Richardson, C., Hesdorffer, C., and Bank, A., 1996, Retroviral transfer and expression of the human multiple drug resistance (MDR) gene in peripheral blood progenitor cells, *Clin. Cancer Res.* **2**:873–876.

Wilson, V.L., Jones, P.A., and Momparler, R.L., 1983, Inhibition of DNA methylation in L1210 leukemic cells by 5-aza-2′-deoxycytidine as a possible mechanism for chemotherapeutic action, *Cancer Res.* **43**:3493–3497.

Xu, X.-C., Sozzi, G., Lee, J.S., Lee, J.J., Pastorino, U., Pilotti, S., Kurie, J.M., Hong, W.K., and Lotan, R., 1997, Suppression of retinoic acid receptor β in non-small-cell lung cancer in vivo: implications for lung cancer development, *J. Natl. Cancer I.* **89**:624–629.

Yoshiura, K., Kanai, Y., Ochiai, A., Shimoyama, Y., Sugimura, T., and Hirohashi, S., 1995, Silencing of the E-cadherin invasion-suppressor gene by CpG methylation in human carcinomas, *Proc. Natl. Acad. Sci. USA* **92**:7416–7419.

Zagonel, V., Lo Re, G., Marotta, G., Babare, R., Sardeo, G., Gattei, V., De Angelis, V., Monfardini, S., and Pinto, A., 1993, 5-Aza-2′-deoxycytidine (Decitabine) induces trilineage response in unfavorable myelodysplastic syndromes, *Leukemia* **7**: suppl. 1, 30–35.

Zhao, S.C., Banerjee, D., Mineishi, S., and Bertino, J.R., 1997, Post-transplant methotrexate administration leads to improved curability of mice bearing a mammary tumor transplanted with marrow transduced with a mutant human dihydrofolate reductase cDNA, *Hum. Gene Ther.* **8**:903–909.

39

VASCULAR ENDOTHELIAL GROWTH FACTOR AS A TARGET FOR CANCER GENE THERAPY

Josephine Tuong Nguyen

Department of Diagnostic Radiology
Beth Israel Deaconess Medical Center Boston
Massachusetts, 02115, U.S.A.

1. INTRODUCTION TO VASCULAR ENDOTHELIAL GROWTH FACTOR

Vascular endothelial growth factor (VEGF), also known as vascular permeability factor (VPF), has drawn increasing attention as a candidate biological target for cancer therapy because it is one of the most important factors promoting angiogenesis, the process by which tumors recruit and expand the blood supply necessary to support their growth (Dvorak, H. F., Brown, L. F., Detmar, Dvorak, 1995; Folkman, J., 1995). Besides being a potent angiogenic factor, VEGF is the only one which acts specifically on endothelial cells and which increases microvascular hyperpermeability (Connolly, Heuvelman, Nelson, Olander, Eppley, Delfino, Siegel, Leimgruber, and Feder, 1989), which can lead to significant morbidity in certain cancer settings and may facilitate tumor invasion. Abundant data exists that inhibiting VEGF leads to dose-dependent suppression of tumor growth and metastasis. That VEGF is expressed and secreted by almost all solid tumors underscores its fundamental role in neoplasia and implies the broad applicability of its inhibition as an anti-cancer strategy.

1.1. The Upregulation of VEGF in Neoplasia

VEGF is a homodimer of two identical 23 kDa subunits linked by disulfide bridges (reviewed by Neufeld, Cohen, Gitay-Goren, Poltorak,Tessler, Sharon, Gengrinovitch, and Levi, 1996). Alternative splicing of one mRNA transcript yields four products, leading to proteins of 121, 165, 189, and 206 amino acids (Tischer, Mitchell, Hartmann, Silva, Gosodarowicz, Tiddles, and Abraham, 1991). Isoforms 121 and 165 have mitogenic and permeabilizing activities and are secretable, the 165 isoform being more abundant, while the two larger isoforms, also mitogenic, remain mostly bound to cell surface heparan sulfate proteoglycans. The crystal structure of VEGF has recently been elucidated (Muller, Li, Christinger, Wells, Cunningham, and de Vos, 1997). Three

Cancer Gene Therapy: Past Achievements and Future Challenges, edited by Habib
Kluwer Academic / Plenum Publishers, New York, 2000.

surface loops are situated symmetrically at either end of the dimeric molecule to form two receptor binding interfaces. Binding of VEGF to each of its tyrosine kinase receptors causes receptor dimerization and activation. Cell-surface heparan-sulfates may help VEGF-165 bind to its receptors, especially under oxidizing conditions, such as inflammation.

VEGF is induced by the microenvironments caused by poor perfusion and acts primarily on postcapillary venules and small veins, but in other respects, the VEGF signalling system is distinctly different in benign as opposed to neoplastic conditions. In normal tissues, such as in kidney, heart, lung, and adrenal cortex, VEGF receptors are barely detectable, possibly all that is needed to carry out maintenance of the microvasculature (Plate, Breier, Weich, and Risau, 1992; Dvorak, Brown, Detmar, and Dvorak, 1995; Hatva, Kaipainen, Mentule, Haltia, and Alitalo, 1995). By contrast, in the endothelium of growing tumors and other pathologic conditions such as delayed hypersentivity, rheumatoid arthritis, and proliferative retinopathy, VEGF receptors are abundant (reviewed Dvorak *et al.,* 1995). In tumors, a potent synergism in inducing angiogenesis exists between VPF and basic fibroblast growth factor (bFGF) (Goto, Goto, Weindel, and Folkman, 1993; Pepper, Ferrara, Orci, and Montesano, 1992). Unlike in normal settings, VEGF expression in neoplasia is constitutive and can be drastically further upregulated by several characteristics peculiar to tumors.

In neoplasia, the VEGF signalling system is enhanced in order to mediate the angiogenesis necessary for tumor growth. A strong inducer is hypoxia, which can increase the level of VEGF mRNA by 2–16 fold above normoxia in many tumor types (reviewed by Claffey and Robinson, 1996). A binding site for hypoxia-inducible factor 1 (HIF-1) has been identified in the 5′ untranslated region of the human VEGF gene (Forsythe, Jiang, Iyer, Agani, Leung, Koos, and Semenza, 1996). Hypoxia also stabilizes the induced VEGF mRNA, probably by binding to a site within the 3′ untranslated region of the gene (Levy, Levy, and Goldberg, 1996; Damert, Machein, Breier, Fujita, Hanahan, Risau, and Plate, 1997; Ikeda, Achen, Breier, and Risau, 1995). Hypoglycemia may also stabilize VEGF mRNA (Stein, Neeman, Shweiki, Itin, and Keshet, 1995). Platelet derived growth factor (PDGF) also upregulates VEGF, via activation of Protein Kinase C (Rinkenzeller, Marme, Weich, and Hug, 1992). The activation of several oncogenes, such as V-Ha-*ras,* v-*raf,* and *src*, which is activated by hypoxia, have also been found to increase VEGF (Rak, Mithuhashi, Bayko, Filmus, Shirasawa, Sasazuki, and Kerbel, 1995; Grugel, Finkenzeller, Weindel, Barleon, and Marme, 1995; Mukhopadhyay, Tsiokas, Zhou, Foster, Brugge, and Sukhatme, 1995). The inactivation of tumor-suppressor genes such as *p53* and *VHL* has been found to upregulate VEGF expression (Kieser, Brandner, Weich, Marme, and Kolch, 1994; Volpert, Dameron, and Bouch, 1997; Siemeister, Weindel, Mohrs, Barleon, Mariny-Baron, and Marme, 1996). It is interesting that although different tumors may express different amounts of VEGF at baseline, they tend to increase expression to similar levels under hypoxia (Leith and Michelson, 1995); this may imply that a threshold level is needed to cause the switch to the angiogenic phenotype, the onset of aggressive growth.

1.2. The Role of VEGF in Angiogenesis

Angiogenesis requires endothelial cell proliferation as well as adhesion, migration, and tube formation. To mediate these distinct functions, VEGF acts via two major endothelial cell receptors found exclusively on endothelial cells, VEGF-R1 (flt-1) and VEGF-R2 (KDR in humans; flk-1 in mice) (Millauer, Wizigmann-Voos, Schnurca,

Martinez, Mollet, Risau, and Ullrich, 1993). The binding of VEGF to each receptor causes receptor dimerization, autophosphorylation, and activation.

KDR transduces a proliferative signal on endothelial cells, encouraging them to elongate and divide (Keyt, Nguyen, Berleau, Duarte, Park, Chen, and Ferrara, 1996; Waltenberger, Claesson-Welsh, Siegbahn, Shibuya, and Heldin, 1994). Downregulated in normal adult tissues, KDR exists abundantly in tumor endothelium and is further upregulated by its own ligand (Brogi, Schatteman, Wu, Kim, Varticoski, Keyt, and Isner, 1996; Barleon, Totzke, Herzog, Blanke, Kremmer, Siemeister, Marme, and Martiny-Baron, 1997).

Flt-1 mainly facilitates the migration of endothelial and monocyte cells by inducing conditions which favor the deposition of an extracellular fibrin gel (Waltenberger, Claesson, Liegbahn, shibuya, and Heldin, 1994). It induces the expression of tissue factor (TF), a transmembrane glycoprotein which functions as the initiator in a major pathway of coagulation ending in the formation of cross-linked fibrin clots (Barleon, Sozzani, Zhou, Weich, Mantovani, and Marme, 1996; Clauss, Weich, Breier, Knies, Rockl, Waltenberger, and Risau, 1996). TF is found on endothelial cells, more often in malignant than in benign tumors (Dvorak, Dvorak, Manseau, Wiberg, and Churchill, 1979; Contrino, Hair, Kreutzer, and Rickles, 1996). TF and VEGF upregulate each other, conspiring to lay down the angiogenesis-engendering fibrin matrix found abundantly in many tumors (Zhang, Deng, Luther, Muller, Ziegler, Walkherr, Stern, and Nawroth, 1994; Shoji, Dillehay, Casper, Micko, Danave, and Rickles, 1996; Lanir, Ciano, Van de Water, McDonagh, Dvorak, and Dvorak, 1988; Brown, Lanir, McDonagh, Czamecki, Estrella, Dvorak, and Dvorak, 1993). The laying down of matrix is further helped along by the vascular permeabilizing effect of VEGF, which acts within minutes with a potency 50,000 times that of histamine and which results in extravasation of circulating fibrinogen and thrombin, components needed in the formation of an angiogenesis supporting matrix (Senger, Connolly, Van de Water, Feder, and Dvorak, 1990; Fong, Rossant, Gertsenstein, and Breitman, 1990). In intracerebral tumors, the edema which is caused by this hyperpermeabilizing action is responsible for neuronal damage, leading to significant morbidity, and threatens fatal brain herniation. VEGF further induces endothelial cells to express osteopontin and avb3integrin, which stimulate endothelial cell migration and neovascularization (reviewed by Claffey, Brown, del Aguila, Tognazzi, Yeo, Manseau, and Dvorak, 1996). The end cumulative influence of VEGF on angiogenesis is comprehensive.

1.3. The Role of VEGF in Invasion and Metastasis

Many of the same actions of VEGF which mediate angiogenesis may also encourage tumor metastasis. The process of metastasis requires stepwise invasion of the tumor cells into the circulation, followed by intravasation into the vasculature, survival within the circulatory blood, extravasation from the microvasculature, and tumor cell growth within the distant site. While abundant evidence shows that VEGF supports the last step, its role in tumor invasion is implied by many of the effects it mediates. For instance, VEGF induces urokinase and its receptor, tissue plasminogen activator, both of which convert plasminogen to plasmin, which dissolves cross-linked fibrin polymers. In the presence of the opposing polymerizing action of thrombin, the net result is an extracellular fibrin gel permissive to endothelial cell invasion during angiogenesis. Just as a fibrin matrix helps endothelial cells to attach and invade, it may also help tumor cells to do the same; indeed, it is well known that coating culture dishes with extracellular matrix-like gels helps certain tumor cell lines to attach and grow and that co-injecting such gels along with tumor cells during innoculation into animals helps the cells to establish tumors at the implanted site. VEGF also increases the release of collagen digesting matrix metalloproteinases (MMPs) such as gelatinase A, interstitial collagenase (MMP-1), and MMP-2 (Unemori, Ferrara, Bauer, and Amento, 1992; Kurizaki, Toi, and Tominaga, 1998), while it inhibits endothelial cells from releasing tissue inhibitors of metalloproteases (Lamoreaux, Fitzgerald, Reiner, Hasty, and Charles, 1998). The cumulative effect favors digestion of the extracellular collagen, which dissolves a barrier to cell invasion through basement membranes, thus implicating VEGF in the early steps in tumor metastasis. Indeed, VEGF is overexpressed by ductal carcinomas in situ of the breast before evidence of invasion and is a prognostic indicator of invasiveness and metastatic potential in many types of solid tumor (Brown, Berse, Jackman, Tognazzi, Guidi, Dvorak, Senger, Connolly, Schnitt, 1995).

2. ANTI-VEGF STRATEGIES

Because of the diverse influences of VEGF on cancer progression, it is important to consider VEGF signalling in cancer therapy. Inhibiting such a central angiogenic factor can significantly tip the scale away from angiogenesis and towards tumor dormancy, a state in which angiogenesis and tumor growth are inhibited (O'Reilly, Holmgren, Chen, Folkman, 1996). Although different tumors may express different amounts of VEGF at baseline, they tend to increase expression to similar levels under hypoxia, and so for a wide array of tumors, only a partial reduction in VEGF concentration may be effective, as long as the ambient level is suppressed below that required for angiogenesis, the ED_{50} for endothelial cell proliferation being about 54 nM (Goto, *et al.*, 1993). Strategies such as this whose end action is on endothelial cells may be less vulnerable than conventional chemotherapy to the development of resistance to treatment. This is because endothelial cells are much more genetically stable than tumor cells, the targets of conventional chemotherapy (Folkman, 1996). To inhibit the influence of VEGF, various points along the gene expression and signal transduction cascade, from the VEGF genes to the VEGF receptors, have been inhibited with some degree of success, as will be described. Gene therapy may be a viable way to implement many of these strategies.

2.1. Antibodies

Antibodies comprise a major anti-VEGF modality comprising strategies which target both VEGF and its receptors. The growth and metastasis of colon carcinoma, epidermoid carcinoma, rhabdomyosarcoma, glioblastoma multiforme, leiosarcoma, and fibrosarcoma were inhibited by intraperitoneal or intravenous injections of various antibodies against VEGF, often in a dose-dependent manner (Warren, Yuan, Matli, Gillett, and Ferrara, 1995; Melnyk, Shuman, and Kim, 1996; Kim, Li, Winer, Armanini, Gilett, Phillips, and Ferrara, 1993; Asano, Yukita, Matsumoto, Kondo, and Suzuki, 1995; Borgstrom, Hillan, Sriramarao, and Ferrara, 1996). Antibodies against the receptor VEGF-R2 prevented the invasion of malignant cells without reducing tumoral proliferation in a model using human keratinocytes in surface transplants in nude mice (Skobe, Rockwell, Goldstein, Vosseler, and Fusenig, 1997). Antibodies have shown high efficacy. However, they may not be deliverable by gene therapy to mediate long-term administration because the genes coding for immunoglobulins may be too large to package into some vectors.

2.2. Strategies Based on VEGF-receptor Binding

Many other methods to inhibit VEGF which may find a useful administration modality in gene therapy, among which are those designed to disturb the VEGF-receptor interaction. For instance, a mutant VEGF protein has been designed which binds but cannot activate a VEGF receptor. This competitive inhibition required 100x excess of the mutant to completely inhibit VEGF-stimulated proliferation of HUVE cells (Siemeister, Schimer, Reusch, Barleon, Marme, and Martiny-Baron, 1998). In another strategy, infection of endothelial cells with retrovirus encoding a dominant negative mutant of the flk-1 receptor, designed to dimerize to the wild type receptors and to render them inactivatable by their ligand, prevented the growth of C6 glioma in nude mice (Millaur, Shawver, Plate, Risau, and Ullrich, 1994). Yet a different receptor-focused approach features a recombinant soluble Flt-1 receptor which binds VEGF to neutralize its action (Barleon, Totzke, Herzog, Blanke, Kremmer, Siemeister, Marme, Martiny-Baron, 1997). Gene therapy may be applicable to delivering these anti-VEGF strategies, as has been shown with retroviral vectors (Millaur *et al.*, 1994).

2.3. Antisense RNA

So far most anti-VEGF gene therapy strategies have focused on antisense RNA technology, which inhibits the secretion of VEGF. This technology involves transduction of cells by a vector which carries a piece of VEGF cDNA cloned in antisense orientation relative to the promoter. Antisense RNA technology is based on the high affinity of RNA-RNA binding. Within the cell, an mRNA which is complementary to the target VEGF mRNA is transcribed which binds to the target mRNA, prevents its translation, and causes its degradation by double-stranded-specific RNAses (Wu, MacLeod, Lima, and Crooke, 1998; Izant and Weintraub, 1985; Delihas, Rokita, and Zheng, 1997).

The first Investigational New Drug Applications for antisense oligonucleotides were submitted to the Food and Drug Administration in late 1991. To date, several of these have been approved for small scale, phase I/II clinical trials (Black, DeGeorge, Cavagnaro, Jordan, and Ahn, 1993). Several potential problems include cellular toxicity arising from homology to transcripts other than the targets and decreased efficacy from the unwinding of RNA-RNA hybrids by RNA helicases. Usual antisense targets include translational control sites, the poly adenylation signal, and the intron splice sites. For VEGF, other potential targets include exon 3, essential for function, and exons 7 and 8, which both exist in the secretable isoform-165 (Muller, Li, Christinger, Wells, Cunningham, and de Vos, 1997; Carmeliet, Ferreira, Breier, Pollefeyt, Kierkens, Gertsenstein, Fahrig, Vandenhoek, Harpal, Eberhardt, Declerq, Pawlign, Moons, Collen, Risau, and Nagy, 1996; Ferrara, Carver-Moore, Chen, Dowd, Lu, O'Shea, Powell, Braxton, Hillan, and Moore, 1996). An overall 20–25% amino acid homology to PDGF, a potent mitogen for vascular smooth muscle which plays an important role in wound healing, may result in side effects of unclear consequence (Wagner,1995; Nellen and Lichtenstein, 1993). For antisense technology to have maximal effect, as many target cells as possible must be transduced; multiple cycles of transduction may increase the yield but only if not prevented by immune reactions, such as those elicited by adenoviral vectors (Roth, 1993).

Despite these drawbacks, strategies such as these, which affect the ambient concentration of a secreted cytokine, enjoy the advantage that efficacy may result despite transduction of only a fraction of the total cell target population.

Almost all published reports using antisense VEGF have relied on *in-vitro* transfection of cells with plasmids. In one study, SK-MEL-2 melanoma cells, which inherently secreted relatively little VEGF, were transfected *in-vitro* with a plasmid carrying antisense mouse VEGF-164 cDNA. The selected transfectant cells, once xenografted into mice, grew small, minimally vascularized subcutaneously tumors and formed no visible lung metastases after intravenous injection, in contrast to cells transfected with sense mouse VEGF cDNA (Claffey, Brown, del Aguila, Tognazzi, Yeo, Manseau, and Dvorak, 1996). Other experiments similarly used plasmids to transfect *in-vitro* glioblastoma and glioma cells, which were then inoculated into nude mice. The implanted stable transfectants failed to sustain growth and formed tumors with reduced vascular density (Cheng, Huang, Nagane, Wang, Shih, Arap, Huang, and Cavenee, 1996; Saleh , Stacker, and Wilks, 1996). While these experiments successfully demonstrated that inhibiting VEGF expression inhibited tumor growth and vascularization, these models are not applicable to clinical use, where it is the VEGF in endogenous tumors which must be repressed and where the *in vivo* environment render plasmids poor gene therapy vectors, with very low efficiency, marked instability, and transient expression of transgene (Wolff, Malone, Williams, Chong, Acsadi, Jani, and Felgner, 1990).

The only antisense VEGF RNA gene transfer study employing a vector other than a plasmid used adeno-associated viral (AAV) vectors. In this study, recombinant AAV transduced an mRNA directed against VEGF-165 cDNA into glioblastoma, colorectal carcinoma, and hepatocelllular carcinoma cells and dramatically inhibited these cells from secreting endogenous VEGF (Nguyen, Wu, Clouse, Hlatky, and Terwilliger, 1998). Strategies to inhibit VEGF delivered via vectors which have qualities such as those found in AAV may have a significant chance for clinical success.

3. GENE THERAPY

Gene therapy may be useful, if not necessary, for implementing many of the above anti-VEGF strategies. Because tumor dormancy depends on a sustained, favorable balance between pro- and anti-angiogenic factors, the chief pro-angiogenic factor being VEGF, the suppression of VEGF must persist undiminished over the long periods. Direct, chronic administration of the proteins and RNA agents described above may be technically difficult because of the chemical instability of such agents and the expense involved in their large scale purification. Gene therapy may offer a simplified alternative by mediating continuous, *in situ* manufacture of anti-VEGF substances.

Many vector systems are currently available, including adenoviral, retroviral, lentiviral, AAV, and plasmid vectors, and are more fully described in section 10.3. Although only partial reduction in VEGF may effectively suppress angiogenesis below a threshold concentration of VEGF needed for angiogenesis, a vector is required with a tranduction efficiency high enough and expression enduring enough to suppress and to keep VEGF secretion at a level sufficiently low to maintain tumor dormancy. One vector which has the potential to effect both efficient and long-lasting *in vivo* gene transfer is AAV, a nonpathogenic, replication deficient vector which is capable of long-term expression and can transduce a wide variety of both nondividing and dividing cells (Muzyczka, 1992; Berns, 1990). Because the targets of anti-VEGF approaches are endothelial cells, which have very low mutation rates, resistance to treatment is not likely to develop, but this quality can be enjoyed only if the vector used for delivery can be effective on repeat administrations. AAV has low immunogenicity, which improves the efficacy of repeated deliveries. To maintain the advantage of the broad applicability of anti-VEGF strategies to many tumors and because different cell types within tumors can secrete VEGF, the chosen gene transfer vector must also have tropism for many cell types. Vectors such as AAV (Xiao, Li, and Samulski, 1996), which are capable of efficiently transducing a wide variety of both dividing and nondividing cell types, and which are also nonimmunogenic and can lead to lasting expression, may have real potential for success in carrying out anti-VEGF approaches to cancer therapy.

4. CONCLUSIONS

Solid tumors require angiogenesis to grow beyond microscopic size (Folkman, 1971). VEGF influences all aspects of angiogenesis and plays a central role in cancer progression. Anti-VEGF strategies may be especially useful under specific circumstances, such as in intracerebral vascular tumors, where blocking VEGF may ameliorate the considerable morbidity caused by the edema resulting from the hyperpermeability effects of VEGF. Anti-VEGF strategies employ biological agents such as proteins and RNA, which are unstable and difficult to manufacture, especially in the large quantities for administration over the long-periods of time necessary to maintain tumor dormancy. Gene therapy may offer a simplified way to chronically administer these substances. Unlike many other vectors currently in use, AAV shows high efficacy on repeat administrations, has wide tropism for many types of dividing and nondividing cells, and has been shown in many tissue types to mediate long-term expression (Muzyczka, 1992; Xiao *et al.*, 1996). A vector with these characteristics used to target VEGF mRNA has been shown to effectively suppress endogenous tumor cell VEGF (Nguyen *et al.*, 1998) and may stimulate further explorations of the potentials of anti-VEGF cancer therapy.

REFERENCES

Asano, M., Yukita, A., Matsumoto, T., Kondo, S., and Suzuki, H., 1995, Inhibition of tumor growth and metastasis by an immunoneutralizing monoclonal antibody to human VEGF/VPF-121, *Cancer Res.* **55**:5296–531.

Barleon, B., Sozzani, S., Zhou, D., Weich, H.A., Mantovani, A., and Marme, D., 1996, Migration of human monocuytes in response to VEGF is mediated via the VEGF receptor Flt-1, *Blood* **87**:3336–3343.

Barleon, B., Totzke, F., Herzog, C., Blanke, S., Kremmer, E., Siemeister, G., Marme, D., and Martiny-Baron, G., 1997, Mapping of the sites for ligand binding and receptor dimerization at the extracellular domain of the VEGF receptor Flt-1, *J. Biol. Chem.* **272**:10382–10388.

Berns, K.I., 1990, Parvovirus replication. *Microbiol. Rev.* **54**:316–329.

Black, L.E., DeGeorge, J.J., Cavagnaro, J.A., Jordan, A., and Ahn, C.H., 1993, Regulatory considerations for evaluating the pharmacology and toxicology of antisense drugs, *Antisense Res. Dev.* **3**:399–404.

Borgstrom, P., Hillan, K.J., Sriramarao, P., and Ferrara, N., 1996, Complete inhibition of angiogenesis and growth of microtumors by anti-VEGF neutralizing antibody: Novel concepts of angiostatic therapy from intravital videomicroscopy, *Cancer Res.* **56**:4032–4039.

Brogi, E., Schatteman, G., Wu, T., Kim, E.A., Varticorski, L., Kyet, B., and Isner, J.M., 1996, Hypoxia-induced paracrine regulation of VEGF receptor expression, *J. Clin. Invest.* **97**:469–476.

Brown, L., Berse, B., Jackman, R., Tognazzi, K., Guidi, A., Dvorak, H., Senger, D., Connolly, J., and Schnitt, S., 1995, Expression of VPF/VEGF and its receptors in breast cancer, *Hum. Pathol.* **26**:86–91.

Brown, L.F., Lanir, N., McDonagh, J., Czamecki, K., Estrella, P., Dvorak, A.M., and Dvorak, H.F., 1993, Fibroblast migration in fibrin gel matrices, *Am. J. Pathol.* **142**:273–283.

Carmeliet, P., Ferreira, V., Breier, G., Polefeyt, S., Kierkens, L., Gertsenstein, M., Fahrig, M., Vandenhoeck, A., Harpal, K., Eberhardt, C., Declercq, C., Pawlig, J., Moons, L., Collens, D., Risau, W., and Nagy, A., 1996, Abnormal blood vessel development and lethality in embryos lacking a single VEGF allele, *Nature* **380**:435.

Cheng, S.Y., Huang, H.J.S., Nagane, M., Ji, X.D., Wang, D., Shih, C.C.Y., Arap, W., Huang, S.M., and Cavenee, W.K., 1996, Suppression of glioblastoma angiogenicity and tumorigenicity by inhbition of endogenous expression of VEGF, *Proc. Nat. Acad. Sci. USA* **93:**8502–8507.

Claffey, K.P., Brown, L.F., del Aguila, L.F., Tognazzi, K., Yeo, K.-T., Manseau, E.J., and Dvorak, H.F., 1996, Expression of VPF/VEGF by melanoma cells increases tumor growth, angiogenesis, and experimental metastasis, *Cancer Res.* **56**:172–181.

Clauss, M., Weich, H., Breier, G., Knies, U., Rockl, W., Waltenberger, J., and Risau, W., 1996, The VEGF receptor Flt-1 mediates biological activities, *J. Biol. Chem.* **271:**17629–17634.

Connolly, D.T., Heuvelman, D.M., Nelson, R., Olander, J.V., Eppley, B.L., Delfino, J.J., Siegel, N.R., Leimgruber, R.M., and Feder, J., 1989, Tumor VPF stimulates endothelial cell growth and angiogenesis. *J. Clin. Invest.* **84**:1470–1478.

Contrino, J., Hair, G., Kreutzer, D., and Rickles, F.R., 1996, In situ detection of expression of tissue factor in vascular endothelial cells: Correlation with the malignant phenotype of human breast tissue, *Nature Med.* **2**:209–215.

Damert, A., Machein, M., Breier, G., Fujita, M.Q., Hanahan, D., Risau, W., and Plate, K.H., 1997, Upregulatio of VEGF expression in a rat glioma is conferred by two distinct hypoxia-driven mechanisms, *Cancer Res.* **57**:3860–3864.

Delihas, N., Rokita, S.E., and Zheng, P., 1997, Natural antisense RNA/target RNA interactions: Possible models for antisense oligonucleotide drug design, *Nature Biotech.* **15**:751–753.

Dvorak, H.F., Brown, L.F., Detmar, M., and Dvorak, A.M., 1995, VPF/VEGF, microvascular hyperpermeability, and angiogenesis, *Am. J. Path.* **146**:1029–1039.

Dvorak, H.F., Dvorak, A.M., Manseau, E.J., Wiberg, L., and Churchill, W.H., 1979, Fibrin gel investment associated with line 1 and line 10 solid tumor growht, angiogenesis, and fibroplasia in guinea pigs. Role of cellular immunity, myofibroblasts, microvascular damage, and infarction in line 1 tumor regression. *J. Natl. Cancer Inst.* **62**:1459–1472.

Ferrara, N., Carver-Moore, K., Chen, H., Dowd, M., Lu, L., O'Shea, K.S., Powell-Braxton, L., Hillan, K.J., and Moore, M.W., 1996, Heterozygous embryonic lethality induced by targeted inactivation of the VEGF gene, *Nature* **380:**439–443.

Folkman, J., 1971, Tumor angiogenesis: Therapeutic implications, *New Engl. J. Med.*, **285**:1182–1186.

Folkman, J., 1995, Clinical applications of research on angiogenesis, *New Engl. J. Med.*, **333**:1757–1763.

Folkman, J., 1996, Tumor angiogenesis and tissue factor. *Nature Med.* **2**:167–168.

Fong, G.H., Rossant, J., Gertsenstein, M., and Breitman, M.L., 1995, Role of the Flt-1 receptor tyrosine kinase in regulationg the assembly of vascular endothelium, *Nature* **376**:66–70.

Forsythe, J.A., Jiang, B.H., Iyer, N.V., Agani, F., Leung, S.W., Koos, R.D., and Semenza, G.L., 1996, Activation of VEGF gene transcription by hypoxia-inducible factor 1, *Mol. Cell Biol.* **16**:4604–4613.

Goto, F., Goto, K., Weindel, K., and Folkman, J., 1993, Synergistic effects of VEGF and bFGF on the proliferation and cord formation of BCE cells within collagen gels, *Lab. Invest.* **69**:508–517.

Grugel, S., Finkenzeller, G., Weindel, K., Barleon, B., and Marme, D., 1995, Both V-Ha-*ras* and v-*raf* induce expression of the vascular endothelial growth factor in NIH 3T3 cells, *J. Biol. Chem.* **270**: 25915–15919.

Hatva, E., Kaipainen, A., Mentule, P., Haltia, M., and Alitalo, K., 1995, Expression of endothelial cell-specific receptor tyrosine kinases and growth factors in human brain tumors, *Am. J. Pathol.* **146**:368–378.

Ikeda, E., Achen, M.G., Breier, G., and Risau, W., 1995, Hypoxia-induced transcriptional activation and increased mRNA stability of VEGF in C6 glioma cells., *J. Biol. Chem.* **270**:19761–19766.

Izant, J.G., and Weintraub, H., 1985, Constitutive and conditional suppression of exogenous and endogenous genes by antisense RNA, *Science* **229**:345–352.

Keyt, B.A., Nguyen, H.V., Berleau, L.T., Duarte, C.M., Park, J., Chen, H., and Ferrara, N., 1996, Identification of VEGF determinants for binding KDR and Flt-1 receptors. Generation of recptor selective VEGF variants by site-directed mutagenesis, *J. Biol. Chem.* **271**:5638–5646.

Kieser, A., Brandner, G., Weich, H.A., Marme, D., and Kolch, W., 1994, Mutant p53 potentiates protein kinase C induction of VEGF expression, *Oncogene* **9**:963–969.

Kim, K.J., Li, B., Winer, J., Armanini, M., Gillett, N., Phillips, H.S., and Ferrara, N., 1993, Inhibition of VEGF-induced angiogenesis suppresses tumour growth in vivo. *Nature* **362**:841–844.

Kurizaki, T., Toi, M., and Tominaga, T., 1998, Relationship between matrix metalloproteinase expression and tumor angiogenesis in human breast carcinoma, *Oncol. Rep.* **5**:673–677.

Lamoreaux, W.J., Fitzgerald, N.E., Reiner, A., Hasty, K.A., and Charles, S.T., 1998, VEGF increases release of gelatinase A and decreases release of tissue inhibitor of metalloproteinases by microvascular endothelial cells in vitro, *Microvasc. Res.* **55**:29–42.

Lanir, N., Ciano, P.S., Van De Water, L., McDonagh, J., Dvorak, A.M., and Dvorak, H.F., 1988, Macrophage migration in fibrin gel matrices. II. Effects of clotting factor XIII, fibronectin, and glycosaminoglycan content on cell migration. *J. Immunol.* **140**:2340–2349.

Leith, J.T., and Michelson, S., 1995, Levels of selected growth factors in viable and necrotic regions of xenografted HCT-8 human colon tumours, *Cell Prolif.* **28**:279–286.

Levy, A.P., Levy, N.S., and Goldberg, M.A., 1996, Post-transcriptional regulation of VEGF by hypoxia, *J. Biol. Chem.* **271**:2746–2753.

Melnyk, O., Shuman, M.A., and Kim, K.J., 1996, Vascular endothelial growth factor promotes tumor dissemination by a mechanism distinct from its effect on primary tumor growth, *Cancer Res.* **56**:921–924.

Millauer, B., Shawver, L.K., Plate, K.H., Risau, W., and Ullich, A., 1994, Glioblastoma growth inhibited in vivo by a dominant-negative Flk-1 mutant, *Nature* **367**:576–579.

Millauer, B., Wizigmann-Voos, S., Schnurca, H., Martinez, R., Mollet, M.P., Risau, W., and Ullrich, A., 1993, High affinity VEGF binding and developmental expresssion suggest Flk-1 as a major regulator of vasculogenesis and angiogenesis, *Cell* **72**:835–846.

Mukhopadhyay, D., Tsiokas, L., Zhou, X., Foster, D., Brugge, J.S., and Sukhatme, V.P., 1995, Hypoxic induction of human VEGF expression through c-Src activation, 1995, *Nature* **375**:577–581.

Muller, Y.A., Li, B., Christinger, H.W., Wells, J.A., Cunningham, B.C., de Vos, A.A., and VEGF: Crystal structure and functional mapping of the kinase receptor binding site, 1997, *Proc. Nat. Acad. Sci. USA* **94**:7192–7197.

Muzyczka, N., 1992, Use of AAV as a general transduction vector for mammalian cells. *Curr. Top. Microbiol. Immunol.* **158**:97–131.

Nellen, W., and Lichtenstein, C., 1993, What makes an mRNA anti-sense-itive? *TIBS* **18**:419–423.

Neufeld, G., Cohen, T., Gitay-Goren, H., Poltorak, Z., Tessler, S., Sharon, R., Gengrinovitch, S., and Levi, B.-Z., 1996, Similarities and differenced between the VEGF splice variants, *Cancer Met. Rev.* **15**: 153–158.

Nguyen, J.T., Wu, P., Clouse, M.E., Hlatky, L., and Terwilliger, E.F., 1998, Adeno-associated virus-mediated delivery of anti-angiogenic factors as an anticancer strategy, *Cancer Res.* **58**:5673–5677.

O'Reilly, M.S., Holmgren, L., Chen, C., and Folkman, J., 1996, Angiostatin induces and sustains dormancy of human primary tumors in mice, *Nature Med.* **2**:689–692.

Pepper, M.S., Ferrara, N., Orci, L., and Montesano, R., 1992, Potent synergism between VEGF and bFGF in the induction of angiogenesis *in-vitro*, *Biochem. Biophys. Res. Commun.* **182**:824–831.

Plate, K.H., Breier, G., Weich, H.A., and Risau, W., 1992, VEGF is a potential tumour angiogenesis factor in human gliomas in vivo, *Nature* **359**:845–848.

Rak, J., Mithuhashi, Y., Bayko, L., Filmus, J., Shirasawa, S., Sasazuki, T., and Kerbel, R.S., 1995, Mutant ras oncogenes upregulate VEGF/VPF expression: Implications for induction and inhibition of tumor angiogenesis, *Cancer Res.* **55**:4575–4580.

Rinkenzeller, G., Marme, D., Weich, H., and Hug, H., 1992, Platelet-derived growth factor-induced transcription of the VEGF gene is mediated by protein kinase C, *Cancer Res.* **52**:4821–4823.

Roth, J.A., 1993, Targeting cancer genes: A novel approach to cancer prevention and therapy. *Cancer Bulletin* **45**:171–177.

Saleh, M., Stacker, S.A., and Wilks, A.F., 1996, Inhibition of growth of C6 glioma cells in vivo by expression of antisense VEGF sequence. *Cancer Res.* **56**:393–401.

Senger, D.R., Connolly, D.T., Van De Water, L., Feder, J., and Dvorak, H.F., 1990, Purification and NH2-terminal amino-acid sequence of guinea pig tumor-secreted vascular permeability factor, *Cancer Res.* **50**:1774–1778.

Shoji, M., Dillehay, D., Casper, K., Micko, C., Danave, I., Nawroth, P.P., and Rickles, F.R., 1996, TF regulates the expression of VEGF in vitro and angiogenesis in vivo in human breast cancer and melanoma, *Blood* **88**:514–518.

Siemeister, G., Schirner, M., Reusch, P., Barleon, B., Marme, D., and Martiny-Baron, G., 1998, An antagonistic VEGF variant inhibits VEGF-stimulated receptor autophosphorylation and proliferation of human endothelial cells, *Proc. Nat. Acad. Sci. USA* **95**:4625–4629.

Siemeister, G., Weindel, K., Mohrs, K., Barleon, B., Mariny-Baron, G., and Marme, D., 1996, Reversion of deregulated expression of VEGF in human renal carcinoma by von Hippel-Lindau tumor suppressor protein, *Cancer Res.* **56**:2299–2301.

Skobe, M., Rockwell, P., Goldstein, N., Vosseler, S., and Fusenig, N.E., 1997, Halting angiogenesis suppresses carcinoma cell invasion, *Nature Med.* **3**:1222–1227.

Stein, I., Neeman, M., Shweiki, D., Itin, A., and Keshet, E., 1995, Stabilization of VEGF mRNA by hypoxia and hypoglycemia and coregulation with other ischemia-induced gene, *Mol. Cell. Biol.* **15**:5363–5368.

Tischer, E., Mitchell, R., Hartmann, T., Silva, M., Gosodarowicz, D., Tiddles, J.D., and Abraham, J.A., 1991, The human gene for VEGF. *J. Biol. Chem.* **266**:11947–11954.

Toi, M., Inada, K., Suzuki, H., and Tominaga, T., 1995, Tumor angiogenesis in breast cancer: Its importance as a prognostic indicator and the association with VEGF expression, *Breast Cancer Res. & Treat.* **36**:193–204.

Unemori, E.N., Ferrara, N., Bauer, E.A., and Amento, E.P., 1992, VEGF induces interstitial collagenase expression in human endothelial cells, *J. Cell. Phys.* **153**:557–562.

Volpert, O.V., Dameronm K.M., and Bouch, N., 1997, Sequential development of an angiogenic phenotype by human fibroblasts progressing to tumorigenicity, *Oncogene* **14**:1495–1502.

Wagner, R.W., 1995, The state of the art in antisense research, *Nature Med.* **1**:1116–1118.

Waltenberger, J., Claesson-Welsh, L., Siegbahn, A., Shibuya, M., and Heldin, C.H., 1994, Different signal transduction properties of KDR and Flt-1, two receptors for VEGF, *J. Biol. Chem.* **269**:26988–26995.

Warren, R.S., Yuan, H., Matli, M.R., Gillett, N.A., and Ferrara, N., 1995, Regulation by VEGF of human colon cancer tumorigenesis in a mouse model of experimental liver metastasis. *J. Clin. Invest.* **95**:1789–1797.

Wolff, J.A., Malone, R.W., Williams, P., Chong, W., Acsadi, G., Jani, A., and Felgner, P.L., 1990, Direct gene transfer into mouse muscle in vivo, *Science* **247**:1465–1468.

Wu, H., MacLeod, A.R., Lima, W.F., and Crooke, S.T., 1998, Identification and partial purification of human dsRNAse activity. A novel terminating mechanism for oligoribonucleotide antisense drugs, *J. Biol. Chem.* **273**:2532–2542.

Xiao, X., Li, J., and Samulski, R.J., 1996, Efficient long-term gene transfer into muscle tissue of immunocompetent mice by adeno-associated virus vector, *J. Virol.* **70**:8098–8108.

Zhang, Y., Deng, Y., Luther, T., Muller, M., Ziegler, R., Waldherr, R., Stern, D.M., and Nawroth, P.P., 1994, Tissue factor controls the balance of angiogenic and antiangiogenic properties of tumor cells in mice. *J. Clin. Invest.* **94**:1320–1327.

ADENO-ASSOCIATED VIRUS AND OTHER POTENTIAL VECTORS FOR ANGIOSTATIN AND ENDOSTATIN GENE THERAPY

Josephine Tuong Nguyen

Department of Diagnostic Radiology
Beth Israel Deaconess Medical Center Boston
Massachusetts, 02115, U.S.A.

1. INTRODUCTION

Recently, evidence has gathered that inhibiting angiogenesis, the process by which new vessels are generated from existing ones, may have significant advantages over conventional cancer chemotherapy, including nontoxicity, lack of drug resistance, and broad-spectrum efficacy against tumors of varied origins. To date, the demonstration of these benefits has relied solely on the administration of purified anti-angiogenic proteins into mouse models. A major hindrance to the clinical applicability of anti-angiogenic proteins lies in the current lack of the means to generate and administer these proteins in large enough quantities to achieve an effective dose over the lengths of time that would be needed to keep tumors suppressed in human subjects. Gene transfer using the appropriate vector may offer a single modality which can both generate and simply administer these proteins over the entire course needed to achieve anti-angiogenic cancer therapy.

2. ANTI-ANGIOGENESIS CANCER THERAPY

The inhibition of angiogenesis can control tumor growth, as was first proposed many years ago (Folkman, 1971) and recently demonstrated in many animal studies. This feat requires the subversion of autocrine and paracrine forces which act on tumors to generate angiogenic factors as growth threatens to deplete energy supply. Whether angiogenesis proceeds depends on the relative intratumoral levels of negative and positive regulators (Folkman, 1995). To meet the energy demand of growth, many tumors expand their blood supply by secreting angiogenic factors, among which include vascular endothelial growth factor (VEGF), basic fibroblast growth factor (bFGF), platelet-derived growth factor (PDGF), insulin-like growth factor-1 (IGF-1). When the relative concentrations of anti- and pro-angiogenesis substances within a tumor change so that

Cancer Gene Therapy: Past Achievements and Future Challenges, edited by Habib
Kluwer Academic/Plenum Publishers, New York, 2000.

the pro-angiogenesis forces prevail, the switch of a tumor's phenotype from dormant to angiogenic is said to occur, whereby blood vessel formation, tumor growth, and metastasis accelerate (Folkman, 1995). Indeed, angiogenesis correlates with growth and metastatic potential in many types of tumor. By contrast, in dormancy, sufficient ambient concentrations of angiogenesis inhibitors keep tumor growth in check. Among the most potent of these natural angiogenesis inhibitors are angiostatin and endostatin. Anti-angiogenesis cancer therapy is achieved by raising the concentration of inhibitors such as these to a high enough level to inhibit angiogenesis and tumor growth.

Anti-angiogenesis therapy complements conventional chemotherapy because each attacks a different major cellular compartment in a tumor. The tumor cell and the endothelial cell compartments are known to exert paracrine forces to stimulate each other's growth (Nicosia, Tchao, Leighton, 1986; Folkman, 1996). Combined anti-angiogenic and chemotherapy resulted in 42% longer survival of tumor-bearing mice (Teicher, Holden, Ara, Sotomayor, Dong, 1994).

When considered alone, each approach has specific strengths and weaknesses. Conventional chemotherapy targets the tumor cells, directly killing them. The toxicity which renders these drugs so effective also damages or kills normal cells, and the patient suffers significant systemic toxic side effects in hopes of remission, which occurs only if the right combination of chemotherapy is chosen to which the particular tumor is sensitive. Often, however, drug resistance inevitably develops within the rapidly mutating tumor cells, and another drug combination must be chosen.

In contrast, angiogenic substances target endothelial cells, which are much more genetically stable than tumor cells and much less prone to developing drug resistance (Folkman, 1996). Because they act specifically on proliferating endothelial cells, the risk of unwanted toxicity is low. Many animal experiments with angiostatin and endostatin have documented no detectable systemic toxicity in treated animals (O'Reilly, Holmgren, Chen, Folkman, 1996; Parangi, O'Reilly, Christofori, Holmgren, Grosfeld, Folkman, Hanahan, 1996). Lastly, since proliferating endothelial cells are ubiquitous in growing solid tumors, anti-angiogenic therapy promises to be effective for many types of solid tumor, unlike other biological targets, such as p53 mutations and oncogenes, which are not present in all tumors. This broad spectrum action has been evident in many studies to date.

A significant difficulty in translating these promising results to clinical use has been that dormancy therapy can work only as long as enough of the agent is present to inhibit angiogenesis. In addition, frequent re-administrations would be required because of the short half-lives of proteins administered systemically. It has been prohibitively expensive and difficult to generate enough active, purified angiostatin and endostatin protein to demonstrate efficacy in human subjects. Therefore the greatest problem to solve before anti-angiogenic cancer therapy is clinically applicable is how to effect long-term drug administration. In this section, angiostatin and endostatin, the two most potent, specific angiogenic inhibitors discovered to date, will be introduced with regard to their origins and mechanisms of action, and the studies performed so far regarding their administration via gene therapy will be discussed in view of the relative strengths and weaknesses of the major vector systems available.

3. ANGIOSTATIN

Angiostatin is a 38 kD protein with more than 98% identity to the first four Krinkle domains of plasminogen (Cao, Ji, Davidson, Schaller, Marti, Sohndel, McCance,

O'Reilly, Llinas, and Folkman, 1996) and was originally purified from serum and urine of mice bearing Lewis lung carcinoma (O'Reilly, Holmgren, Shing, Chen, Rosenthal, Moses, Lane, Cao, Sage, and Folkman, 1994). It is not secreted by the tumor cells but results from elastase digestion of plasminogen. It inhibits the *in-vitro* bFGF-stimulated proliferation of microvascular endothelial cells but not of other cell types. The human form of angiostatin similarly inhibited bovine capillary endothelial (BCE) cells, inhibited angiogenesis in *in vivo* models, and induced tumor dormancy, defined as a balance between tumor cell apoptosis, such as mediated by starvation or hypoxia, and proliferation. No systemic toxicity or drug resistance was evident for the duration of therapy, up to 60 days (O'Reilly, Holmgren, Chen, Folkman, 1996).

More information about the origin and mechanism of action of angiostatin has surfaced since its discovery. Angiostatin is probably generated not directly from plasminogen but from plasmin, itself a serine protease. Chinese hamster ovary and the human fibrosarcoma cell line HT1080 secrete plasmin reductases which reduce and thereby activate plasmin, causing it to autolyse to form an angiostatin which, like the endogenous form, specifically inhibits micro- and not macrovascular endothelial cells. S-carbamidomethylation of the free sulfhydryl group(s) in angiostatin ablated its inhibitory activity, which implies that the free sulfhydryl groups are essential for activity (Stathakis *et al.*, 1997). Annexin II, a receptor on endothelial cells, binds to both plasminogen and tPA and may generate plasmin on the surface of endothelial cells, leading to the generation of angiostatin near its target (Cesarman, Guevara, Hajjar, 1994; Hajjar, Nachman, 1988). Both recombinant and native human angiostatin are produced in a glycosylated and nonglycosylated forms; glycosylation may be important in receptor recognition (Sim *et al.*, 1997).

The survival and proliferation of endothelial cells depends on its attachment to the extracellular matrix (ECM). Endothelial cell mitogens such as VEGF and bFGF induce an ordered turnover of focal adhesion contacts. Components of the ECM, such as fibronectin, induce the clustering of β1-integrins when these integrins bind to the fibronectin tripeptide RGF. Clustering of β1-integrins activates endothelial cell expression of focal adhesion kinase (FAK), which then targets the endothelial cells towards focal adhesions. This normal activity is disturbed by angiostatin, which leads to abnormal FAK induction which is not integrin-dependent. This may reroute FAK to sites within endothelial cells not accessible for binding to focal adhesion contacts, thus hampering endothelial cell survival.

A recent experiment found that while angiostatin had no effect the rate of DNA synthesis in BCE cells, it increased their rate of apoptosis (Claesson, Welsh, Ito, Anand-Apte, Soker, Zetter, O'Reilly, and Folkman, 1998) and inhibited the migration of human umbilical vein endothelial cells (HUVECs) toward VEGF and bFGF. Others, however, did not find increased apoptosis in HUVECs treated with angiostatin, perhaps because HUVEC medium contained endothelial growth factors. Rather than inducing apoptosis, angiostatin was found to induce clonogenic killing in human aortic endothelial cells and HUVECs (Mauceri, Hanna, Beckett, Gorski, Staba, Stellato, Gigelow, Heimann, Gately, Dhanabal, Soff, Sukhatme, Kufe, and Weichselbaum, 1988). Decreasing the mitotic rate of endothelial cells has also been suggested as a mechanism of inhibition, evidenced when endothelial cells infected with adenovirus which expressed angiostatin lacked M-phase promoting factor (Griscelli, Li, Bennaceur-Griscelli, Soria, Opolon, Soria, Perricaudet, Yeh, and Lu, 1998). However, how much of this was due to a cytotoxic effect of adenovirus has not been addressed.

Although angiostatin has strong affinity for heparin (O'Reilly, 1994), and heparan sulfate proteoglycans may assist the association of bFGF and VEGF with their receptors or promote bFGF oligomerization (Yalyon, Klagsbrun, Esko, Leder, and Omitz, 1991; Moy, Safran, Seddon, Ap, Kitchen, Bohlen, Aviezer, Yayon, Powers, 1997; Friesel, and Maciaq, 1995), angiostatin has no effect on VEGF or bFGF binding to their receptors, on the subsequent Shb protein tyrosine phosphorylation, or on the *Ras* pathway (Claesson *et al.*, 1998). Other evidence that angiostatin interacts directl with endothelia is that intraperitoneal injection inhibited intracranial glioma growth, suggesting an endovascular route of action (Kirsh, Strasser, Allende, Bello, Zhang, and Black, 1998).

Preliminary demonstrations that gene transfer of angiostatin leads to generation of active protein include the following. Plasmid DNA encoding mouse angiostatin cDNA was used to transfect a murine T241 fibrosarcoma cell line, and stably transfected cells were implanted subcutaneously into C57BL6/J mice. Angiostatin transfected cells grew larger tumors than untransfected cells and generated fewer lung metastases (Cao, O'Reilly, Marshall, Flynn, and Folkman, 1998). Adenoviral vectors expressing angiostatin, when used to infect C6 glioma cells *in-vitro*, slowed the growth of these tumors for up to 20 days. When injected intravenously 24 hours before implantation of C6 glioma cells, the vectors slowed the establishment and growth of these cells for up to 20 days. When injected directly into preestablished breast carcinoma and C6-glioma tumors, the vectors significantly inhibited their growth to 45 days after injection (Griscelli *et al.*, 1998). Retroviral-mediated delivery of mouse angiostatin cDNA *in-vitro* into glioma cells led to the formation of small and poorly vascularized tumors after these cells were implanted. Because retroviruses proved to be poor vectors *in vivo*, adenoviral vectors for angiostatin cDNA were then co-injected with glioma cells, which led to tumors which had lower vascular densities; the survival was also significantly increased to 35 days, compared to 30 days for negative controls (Tanaka, Cao, Folkman, and Fine, 1998). Adeno-associated virus (AAV) carrying angiostatin cDNA and transduced into glioblastoma, colorectal carcinoma, and hepatocellular carcinoma mediated secretion of active angiostatin protein by these tumor cells (Nguyen, Wu, Clouse, Hlatky, and Terwilliger, 1998).

4. ENDOSTATIN

Endostatin is a 20 kDa C-terminal globular domain of the collagen XVIII which corresponds to the last 184 of some 300 residues of the non-collagenous C-terminal domain of this collagen (Hohenester, 1998). It was first isolated from the culture media from mouse hemangioendothelioma (O'Reilly, Boehm, Shing, Fukai, Vasios, Lane, Flynn, Birkhead, Olsen, and Folkman, 1997). Sera from healthy human donors contains 120–300 ng/mL endostatin, similar to concentration which efficiently inhibited endothelial cell proliferation in vitro. Repeated cycles of administration of purified endostatin into mice carrying tumors followed by a period allowed for tumor regrowth did not generate resistance, as repeated administrations showed equally effective inhibition (Boehm, Folkman, Browder, and O'Reilly, 1997).

Collagen XVIII, localized in the perivascular basement membranes and in vessel walls, may function as the sequestration site for generation of endostatin (Hanahan and Folkman, 1996; Muragaki, Timmons, Griffith, Oh, Fadel, Quertemous, and Olsen, 1996). Tissue extract (0.1–0.2 g/mL) from mouse brain, skeletal muscle, heart, kidney, testis, and liver had increasing concentrations from 0.3–2 μg/mL, suggesting that circulating endostatin is derived from tissue sources (Sasaki, Fukai, Mann, Gohring, Olsen, and Timple,

1998). What regulates the release of endostatin around tumors and normal cells is unknown.

The three-dimensional structure of endostatin has been elucidated (Hohenester, Sasaki, Olsen, and Timpl, 1998) and consists of a single globular domain of 35 × 30 × 30 angstroms. The X-ray structure reveals similarity to C-type lectin and E-selectin, an endothelial-cell specific membrane adhesion glycoprotein but without the calcium binding site of selectins. It is also structurally similar to tetranectin, which binds plasminogen, a dissolver of crosslinked fibrin, and which co-localizes with plasminogen in the invasive fronts of melanomas (De Vries, De Wit, Clemensen, Verspaget, Weidle, Brocker, Ruiter, VanMuijen, 1996). A potential heparin-binding site is thought to competitively inhibit bFGF or VEGF from binding to cell surface heparan sulfate proteoglycans (Honehester *et al.*, 1998). A zinc binding site, most closely resembling that of the matrix metalloproteinases, is thought important for activation of proteolytic activity (Ding, Javaherian, Lo, Chopra, Boehm, Lanciotti, Harris, Li, Shapiro, Hohenester, Timple, Folkman, and Wiley, 1998). The only study so far on gene therapy of endostatin employed an AAV vector to transduce glioblastoma, colorectal carcinoma, and hepatocellular carcinoma and effected the secretion of active endostatin by these cells (Nguyen *et al.*, 1998).

5. GENE THERAPY VECTORS

The appeal of gene therapy as a method to chronically deliver drug agents lies in the potential for simplification of delivery, whereby only one or infrequent administrations lead to continuously *in-situ* manufacture of fresh, active protein. To be applicable, the gene therapy vector needs to express the drug enduringly while retaining all the advantages which make anti-angiogenesis therapy so desirable, that is, nontoxicity, absence of drug resistance, and broad-spectrum action.

A variety of gene therapy vectors are available for consideration, including naked plasmid, plasmid packaged in liposomes, adenovirus, retrovirus, lentivirus, herpesvirus, and adeno-associated virus (AAV). Each of the major available vectors will be considered in light of the characteristics desirable in the transfer of anti-angiogenic proteins.

Adenoviral vectors are the most popular vectors in the United States and have many strengths, such as high-titer preparations, the capacity for inclusion of transgenes as large as 35 kilobases, and the ability to infect both dividing and nondividing cells. The disadvantages include inflammatory and immune responses which lead to loss of the transduced cell and result in transient gene expression and limited efficacy of repeat deliveries (Tripathy, Black, Goldwasser, and Leiden, 1996). Immune responses have been observed to be generated against the transgene product itself when adenovirus is used, thus clearing the transgene product from the circulation. The adenoviral virion itself may be immunogenic because it is composed of many different proteins (Tripathy *et al.*, 1996). Even if the immune reactions could be minimized, the full potential for long-term expression is not realizable because adenovirus cannot integrate. Moreover, the risk of recombination with wild-type and generation of replicative mutant virus has not been eliminated.

Retroviruses were historically the first viruses used as gene therapy vectors. Having an RNA genome, they require host DNA synthesis enzymes for transduction and therefore efficiently integrate into and transduce only dividing cells, a characteristic which may allow selective transduction of tumor cells (Gordon and Anderson, 1994; Salmons and Gunzber, 1993). This selectiveness may not apply in the setting of anti-angiogenesis therapy, where the tumor cells are not the exclusive targets. Although retroviral vectors can integrate and thus mediate long-term transgene expression, integration is inefficient. In addition, retroviral long-terminal repeats (LTRs) contain dominant transcriptional regulatory sequences which can interfere with internally placed promoter or enhancer elements, thereby inadvertently shutting down transgene expression. The retroviral LTRs also have transcriptional regulatory elements which facilitate insertional activation of protooncogenes (Samulski, Chang, and Shenk, 1989; Zlokovic and Apuzzo, 1997). Reverse transcriptase errors lead to mutation rates of 5% during reverse transcription, leading to genetic variability of vector. Other drawbacks include low titers of preparations and inactivation in plasma (Zlokovic *et al.*, 1997). The use of mouse producer cells to overcome inefficient titers can provoke immune toxic reactions to xenogeneic cells (Ram, Culver, Walbridge, Frank, Blaese, and Oldfield, 1993).

Lentivirus vectors are retroviral transduction systems, the human form of which is based on HIV-1. Except for being able to transduce nondividing as well as dividing cells, this vector system suffers from the same disadvantages as for retroviruses. Moreover, full testing to exclude the danger of generating a replication-competent recombinant virus is not currently feasible (Naldini *et al.*, 1996). Feline (ungulate) form may be more safely tested, as they cannot infect human cells, but are narrowly limited in their use because of highly restricted tropisms (Poeschla, Wong-Staal, and Looney, 1998).

Herpesvirus vectors can infect post-mitotic cells and pose no danger of insertional mutagenesis. However, shortcomings include the presence of many viral genes of unknown function, short-term expresssion of transgene (Geller, During, Jaycock, Freese, and Neve, 1993), low titer preparations of replication-defective preparations, and possible recombination with wild type virus, which could lead to lytic infection (Zlokovic *et al.*, 1997).

Naked plasmid vectors are simple to prepare. However, many impracticalities exist, namely transient expression, very low *in vivo* transfection efficiency, instability in extracellular environments, immunogenicity and toxicity due to acidity (Wolff, Malone, Williams, Chong, Acsadi, Jani, and Felgner, 1990; Zlokovic *et al.*, 1997). Encapsulation of plasmid DNA within liposomes protects the plasmid from the extracellular environment, eliminates toxicity, and somewhat improves transduction efficiency. However, most liposomes are inactivated by plasma and therefore still yield low transduction efficiency *in vivo*. Moreover, transgene expression is very transient, among the longest expression documented being about 9 weeks (Zhu, Liggitt, Liu, and Debs, 1993).

Adeno-associated virus (AAV) is an alternative vector system which has some characteristics which may be very important for anti-angiogenic therapy. The first is low toxicity; it has never been shown to cause any pathology in humans or animals. A parvovirus, it is naturally replication defective, requiring coinfection with adenovirus or herpesvirus for replication, and thus has no risk for generating replication-competent recombinants. For generating virus stock, the AAV vector plasmid is transfected along with a helper E1A-adenovirus and a helper plasmid which provides the *rep* and *cap* proteins into a cell line expressing the E1A gene. The helper plasmid itself cannot be packaged because it does not contain any recognition signals for replication and packaging (Samulski *et al.*, 1989). All viral genes can be deleted in vectors to minimize any immune reaction because the only signal needed for replication and packaging lie in the two inverted terminal repeats (ITRs), between which a transgene cassette is inserted. Unlike retroviral long terminal repeats, these ITRs appear to contain no dominant transcriptional regulatory sequences interfering with internally placed promoter or enhancer elements. The ITRs appear to lack any transcriptional regulatory elements, reducing the risk for insertional activation of protooncogenes, a recognized mechanism of oncogenesis by murine retroviruses. While the adenovirus virion is composed of fourteen different proteins, the AAV virion consists of only one major polypeptide, Vp3 (Berns and Bohenzky, 1987). Thus AAV offers much less for stimulation of an immune reaction and therefore has more potential for long-term expression than the latest generation of adenoviruses. In one experiment, in contrast to adenoviral vectors, AAV elicited no inflammatory reaction and resulted in much longer expression, as adenovirus-delivered transgene ceased expression by 3–4 weeks after administration (Li, Ueno, Pan, Tomita, Yamamoto, Kanegae, Saito, and Takeshita, 1995). Readministration of AAV may be more efficacious than with other vectors, as seen when anti-AAV antibodies mounted against AAV capsid proteins diminished but did not block the readministration of AAV vectors. Immune responses mounted against β-galactosidase were strong after adenovirus vector was injected but were not mounted after AAVlacZ was injected, implying that the vector modulates the immune response to the transgene product (Fisher, Jooss, Alston, Yang, Haecker, High, Pathak, Raper, and Wilson, 1997).

Several weaknesses also exist for AAV vectors. They transduce some primary cell types much less efficiently than immortalized cells (Halbert, Alexander, Wolgamot, and Miller, 1995). For instance, transduction of CMV-LacZ into human bone marrow stem cells resulted in strong expression for up to 21 days, but this disappeared by 5 weeks, with no evidence for integration (Gardern, Zhu, Colosi, Kurtzman, and Scadden, 1997). Integration is not directed to chromosome 19 when *rep* is absent (Kumar and Leffak, 1991). The nonselective targeting may be a problem in certain settings. Perhaps most important, packaging systems, although highly efficient among vectors, can nevertheless use much improvement in standardizability and efficiency.

Despite these weaknesses, the long-term transgene expression of AAV has been well documented in many organ systems. A single peroral administration of AAV expressing β-galactosidase into rats led to transgene expression lasting undiminished for 6 months (During *et al.*, 1998). AAV-mediated galactosidase transgene expression lasted in the brain for three months and tyrosine hydroxylase expression in the striatum and glia for four months (McCown, Xiao, Li, Breese, Samulski, 1996; Kaplitt, Leone, Samulski, Xiao, Pfaff, O'Malley, and During, 1994). Infusion of rAAV carrying β-galactosidase into myocardium resulted in transgene expression lasting for 6 months (Kaplitt, Xiao, Samulski, Li, Ojamaa, Klein, Makimura, Kaplitt, Strumpf, and Diethrich, 1996). AAV carrying the erythropoeitin gene administered as a single injection into mouse hindlimb resulted in at least 40 weeks of increased RBC production (Kessler, Podsakoff, Chen, McQuiston, Colosi, Metelis, Kurtzman, and Byrne, 1996). AAV also mediated β-galactosidase expression lasting for at least 18 months after one injection into mouse skeletal muscle (Xiao, Li, and Samulski, 1996). AAV-LacZ or AAV β-glucuronidase injection into mouse muscle yielded high level transduction which lasted at least 240 days. Significant levels of transgene expression were also achieved with similar injections into the rhesus monkey (Fisher *et al.*, 1997). AAV carrying clotting factor IX injected in mice resulted in 200–350 ng/ml of the protein plasma for at least 6 months (Herzog, Hagstrom, Kung, Tai, Wilson, Fisher, and High, 1997). AAV carrying the alkaline phosphatase or clotting factor IX gene injected into mouse via the tail vein or directly into the liver led to prolonged expression in hepatocytes (Koeberl, Alexander, Halbert, Russell, and Miller, 1997). AAV-Factor IX via the portal vein yielded high serum levels for at least 36 weeks (Snyder, Miao, Patijn, Spratt, Danos, Nagy, Gown, Winther, Meuse, Cohen, Thompson, and Kay, 1997). Ample evidence therefore exists that AAV may possess the qualities needed in a successful anti-angiogenic gene therapy vector.

6. CONCLUSIONS

Nontoxic proteins have recently surfaced as potential cancer therapeutic agents which are very broad in spectrum of tumor targets and seem impervious to drug resistance. Among the most effective of these substances include the recently discovered angiostatin and endostatin, specific inhibitors of angiogenesis. The idea introduced in the 1970's by Judah Folkman that tumors could not grow beyond a size of 2–3 mm^3 without the ability to develop new blood vessels, has been supported by multiple animal experiments with purified angiostatin and endostatin which demonstrated the principle of dormancy therapy. Preliminary reports have shown that gene transfer can generate active anti-angiogenic proteins. The strength of gene therapy in anti-angiogenic treatment for cancer may depend on the ability of the chosen vector to preserve the considerable advantages of the specific angiogenesis inhibitors while achieving high-level, long-lasting expression.

REFERENCES

Berns, K.L., and Bohenzky, R.A., 1987, Adeno-associated viruses: An update (review), *Adv. Virus Res.* **32**: 243–306.

Boehm, T., Folkman, J., Browder, T., and O'Reilly, M.S., 1997, Antiangiogenic therapy of experimental cancer does not induce acquired drug resistance, *Nature* **390**:404–407.

Cao, Y., Ji, R.W., Davidson, D., Schaller, J., Marti, D., Sohndel, S., McCance, S.G., O'Reilly, M.S., Llinas, M., and Folkman, J., 1996, Kringle domains of human angiostatin, *J. Biol. Chem.* **271**:29461–29467.

Cesarman, G.M., Guevara, C.A., and Hajjar, K.A., 1994, An endothelial cell receptor for plasminogen/t-PA, *J. Biol. Chem.* **269**:21198–21203.

Claesson, L., Welsh, M., Ito, N., Anand-Apte, B., Soker, S., Zetter, B., and O'Reilly, M., Folkman., 1998, Angiostatin induces endothelial cell apoptosis and activation of focal adhesion kinase independely of the integrin-binding motif RGF, *Proc. Nat. Acad. Sci. USA* **95**:5579–5583.

De Vries, T.J., De Wit, P.E.J., Clemensen, I., Verspaget, H.W., Weidle, U.H., Brocker, E.B., Ruiter, D.J., and Van Muijen, G.N.P., 1996, *J. Path.* **179**:260–265.

Ding, Y.H., Javaherian, K., Lo, K.M., Chopra, R., Boehm, T., Lanciotti, J., Harris, B.A., Li, Y., Shapiro, R., Hohenester, E., Timpl, R., Folkman, J., and Wiley, D.C., 1998, Zinc-dependent dimers observed in crystals of human endostatin, *Proc. Nat. Acad. Sci. USA* **95**:10443–10448.

During, M.J., Xu, R., Young, D., Kaplitt, M.G., Sherwin, R.S., and Leone, P., 1998, Peroral gne therapy of lactose intoleranceusing an AAV vector, *Nature Med.* **4**:1131–1135.

Fisher, K.J., Jooss, K., Alston, J., Yang, Y., Haecker, S.E., High, K., Pathak, R., Raper, S.E., and Wilson, J.M., 1997, Recombinant adene-associated viurs for muscle directed gene therapy, *Nature Med.* **3**:306–312.

Fisher, K.J., Jooss, K., Alston, J., Yang, Y., Haecker, S.E., High, K., Pathak, R., Raper, S.E., and Wilson, J.M., 1997, Recombinant adene-associated viurs for muscle directed gene therapy, *Nature Med.* **3**:306–312.

Folkman, J., 1971, Tumor angiogenesis: Therapeutic implications, *New Engl. J. Med.* **285**:1182–1186.

Folkman, J., 1995, Clinical applications of research on angiogenesis, *New Eng. J. Med.* **333**:1757–1801.

Folkman, J., 1996, Tumor angiogenesis and tissue factor. *Nature Med.* **2**:167–168.

Friesel, R.E., and Maciaq, T., 1995, Molecular mechanisms of angiogenesis: FGF signal, *FASEB J.* **9**:919–925.

Gardner, J.P., Zhu, H., Colosi, P.C., Kurtzman, G.J., and Scadden, D.T., 1997, Robust, but transient expression of AAV-transduced genees during human T lymphopoiesis, *Blood* **90**:4354–4864.

Geller, A.I., During, M.J., Jaycock, J.W., Freese, A., Neve, R., 1993, Long-term increases in neurotransmitter release from neuronal cells expressing a constitutively active adenylate cyclase from a HSV-type 1 vector. *Proc. Nat. Acad. Sci. USA* **90**:7603–7607.

Gordon, E.M., Anderson, W.F., 1994, Gene therapy using retroviral vecttors, *Curr. Opin. Biotech.* **5**:611–616.

Griscelli, F., Li, H., Bennaceur-Griscelli, A., Soria, J., Opolon, P., Soria, C., Perricaudet, M., Yeh, P., and Lu, H., 1998, Angiostatin gene transfer: Inhibition of tumor growth *in vivo* by blockage of endothelial cell proliferation associated with a mitosis arrest, *Proc. Nat. Acad. Sci. USA* **95**:6367–6372.

Hajjar, K.A., and Nachman, R.L., 1988, Endothelial cell-mediated conversion of Glu-plasminogen to Lys-plasminogen. Further evidence for assembly of the fibrinolytic system on the endothelial cell surface, *J. Clin. Invest.* **82**:1769–1778.

Halbert, C.L., Alexander, I.E., Wolgamot, G.M., and Miller, A.D., 1995, Adeno-associated virus vectors tranduce primary cells uch less efficiently than immortalized cells, *J. Virol.* **69**:1473–1479.

Hanahan, D., and Folkman, J., 1996, Patterns and emerging mechanisms of the angiogenic switch during tumor growth, *Cell* **86**:353–364.

Herzog, R.L., Hagstrom, J.N., Kung, S.H., Tai, S.J., Wilson, J.M., Fisher, K.J., and High, K.A., 1997, Stable gne transfer and expression of human blood coagulation factor IX after intramuscular injection of recombinant adeno-associated virus, *Proc. Nat. Acad. Sci. USA* **94**:5804–5809.

Hohenester, E., Sasaki, T., Olsen, B.R., and Timpl, R., 1998, Crystal structure of the angiogenesis inhibitor endostatin at 1.5A resolution, *EMBO* **17**:1656–1664.

Kaplitt, M.G., Xiao, X., Samulski, R.J., Li, J., Ojamaa, K., Klein, I.L., Makimura, H., Kaplitt, M.J., Strumpf, R.K., and Diethrich, E.B., 1996, Long-term gene transfer in procine myocardium after coronary infusion of an adeno-associated virus vector, *Ann. Thorac. Surg.* **62**:1669–1676.

Kessler, P.A., Podsakoff, G.M., Chen, X., McQuiston, S.A., Colosi, P.C., Matelis, L.A., Kurtzman, G.J., and Byrne, B.J., 1996, Gene delivery into skeletal muscle results in sustained expression and systemic delivery of a therapeutic protein, *Proc. Nat. Acad. Sci. USA* **93**:14082–14087.

Kirsh, M., Strasser, J., Allende, R., Bello, L., Zhang, J., and Black, P McL., 1998, Angiostatin suppresses malignant glioma growth *in vivo*, *Cancer Res.* **58**:4654–4659.

Koeberl, D.D., Alexander, I.E., Halbert, C.L., Russell, D.W., and Miller, D., 1997, Persistent expression of human clotting factor IX from mouse liver after intravenous injection of adeno-associated virus vectors, *Proc. Nat. Acad. Sci. USA* **94**:1426–1431.

Kumar, S., and Leffak, M., 1991, Conserved chromatin structure in c-myc 5′-flanking DNA after viral transduction, *J. Mol. Biol.* **222**:45–57.

Li, J.J., Ueno, H., Pan, Y., Tomita, H., Yamamoto, H., and Kanegae, Saito, Takeshita, 1995, Percutaneous transluminal gene transfer into canine myocardium in vivo by replication-defective adenovirus. *Cardiovasc. Res.* **30**:97–105.

Mauceri, H.J., Hanna, N.N., Beckett, M.A., Gorski, D.H., Staba, M.J., Stellato, K.A., Bigelow, K., Heimann, R., Gately, S., Dhanabal, M., Soff, G.A., Sukhatme, V.P., Kufe, D.W., and Weichselbaum, R.R., 1998, Combined effects of angiostatin and ionizing radiation in antitumour therapy, *Nature* **394**:287–291.

McCown, T.J., Xiao, X., Li, J., Breese, G.R., and Samulski, R.J., 1996, Differential and persistent expression patterns of CNS gene transfer by an adeno-associated virus (AAV) vector, *Brain Res.* **713**:99–107.

Meuse, L., Cohen, L.K., Thompson, A.R., and Kay, M.A., 1997, Persistent and therapeutic concentrations of human factor IX in mice after hepatic gene transfer of recombinant AAV vectors, *Nature Genet.* **16**:270–276.

Moy, F.J., Safran, M., Seddon, Ap.P., Kitchen, D., Bohlen, P., Aviezer, D., Yayon, A., and Powers, R., 1997, Properly oriented heparin desaccharide-induced dimers are the biologically active form of bFGF, *Biochemistry* **36**:4782–4791.

Muragaki, Y., Timmons, S., Griffith, C.M., Oh, S.P., Fadel, B., Quertermous, T., and Olsen, B.R., 1995, Mouse col18a1 is expressed in a tissue-specific manner as three alternative variants and is localized in basement membrane zones, *Proc. Nat. Acad. Sci. USA* **92**:8763–8767.

Naldini, L., Blomer, U., Gallay, P., Ory, D., Mulligan, R., Gage, F.H., Verma, I.M., and Trono, D., 1996, In vivo gene delivery and stable transductionof nondividing cells by a lentiviral vector, *Science* **272**:263–267.

Nguyen, J.T., Wu, P., Clouse, M.E., Hlatky, L., and Terwilliger, E.F., 1998, Adeno-associated virus-mediated delivery of anti-angiogenic factors as an anticancer strategy, *Cancer Res.* **58**:5673–5677.

Nicosia, R.F., Tchao, R., and Leighton, J., 1986, Interactions between newly formed endothelial channels and carcinoma cells in plasma clot culture, *Clin. Exp. Metas.* **4**:91–104.

O'Reilly, M.S., Boehm, T., Shing, Y., Fukai, N., Vasios, G., Lane, W.S., Flynn, E., Birkhead, J.R., Olsen, B.R., and Folkman, J., 1997, Endostatin: an endodenous inhibitor of angiogenesis and tumor growth, *Cell* **88**:277–285.

O'Reilly, M.S., Holmgren, L., Chen, C., and Folkman, J., 1996, Angiostatin induces and sustains dormancy of human primary tumors in mice, *Nature Med.* **2**:689–692.

O'Reilly, M.S., Holmgren, L., Shing, Y., Chen, C., Rosenthal, R.A., Moses, M., Lane, W.S., Cao, Y., Sage, E.H., and Folkman, J., 1994, Angiostatin: A novel angiogenesis inhibitor that mediates the suppression of metastases by a Lewis lung carcinoma, *Cell* **79**:315–328.

Olsen, B.R., and Folkman, J., 1997, Endostatin: an endodenous inhibitor of angiogenesis and tumor growth, *Cell* **88**:277–285.

Parangi, S., O'Reilly, M., Christofori, G., Holmgren, L., Grosfeld, J., Folkman, J., and Hanahan, D., 1996, Antiangiogenic therapy of transgenic mice impairs de novo tumor growth, *Proc. Nat. Acad. Sci. USA* **93**:2002–2005.

Poeschla, E.M., Wong-Staal, F., and Looney, D.J., 1998, Efficient transduction of nondividing cells by feline immunodeficiency virus lentiviral vectors, *Nature Medicine* **4**:354–357.

Ram, Z., Culver, K.W., Walbridge, S., Frank, J.A., Blaese, R.M., and Oldfield, E.H., 1993, Toxicity studies of retroviral-mediated gene transfer for the treatment of brain tumors, *J. Neurosurg.* **79**:400–407.

Salmons, B., and Gunzber, W.H., 1993, Targeting of retroviral vectors for gene therapy, *Hum. Gene Ther.* **4**:129–141.

Samulski, R., Chang, L., and Shenk, T., 1989, Helper-free stocks of recombinant adeno-associated viruses: Normal integration does not require viral gene expression, *J. Virol.* **63**:3822–3828.

Sasaki, T., Fukai, N., Mann, K., Gohring, W., Olsen, B.R., and Timple, R., 1998, Structure, function and tissue forms of the C-terminal globular domain of collagen XVIII containg the angiogenesis inhibitor endostatin, *EMBO* **17**:4249–4256.

Sim, B.K.L., O'Reilly, M.S., Liang, H., Fortier, A.H., He, W., Madsen, J.W., Lapcevich, R., and Nacy, C.A., 1997, A recombinant human angiostatin protein inhibits experimental primary and metastatic cancer, *Cancer Res.* **57**:1329–1334.

Snyder, R.O., Miao, C.H., Patijn, G.A., Spratt, S.K., Danos, O., Nagy, D., Gown, A.M., Winther, B., Meuse, L., Cohen, L.K., Thompson, A.R., and Kay, M.A., 1997, Persistent and therapeutic concentrations of human factor IX in mice after hepatic gene transfer of recombinant AAV vectors, *Nature Genet.* **16**:270–276.

Stathakis, P., Fitzgerald, M., Matthias, L.J., Chesterman, C.N., and Hogg, P.J., 1997, Generation of angiostatin by reduction and proteolysis of plasmin, *J. Biol. Chem.* **272**:20641–20645.

Tanaka, T., Cao, Y., Folkman, J., and Fine, H.A., 1998, Viral vector-targeted antiangiogenic gene therapy utilizing an angiostatin complementary DNA, *Proc. Nat. Acad. Sci. USA* **58**:3362–3369.

Teicher, B.A., Holden, S.A., Ara, G., Sotomayor, E.A., and Dong, H.Z., 1994, Potentiation of cytotoxic cancer therapies by TNP-470 alone and with other anti-angiogenic agents, *Int. J. Cancer* **57**:1–6.

Tripathy, S.K., Black, H.B., Goldwasser, E., and Leiden, J.M., 1996, Immune responses to transgene-encoded proteins limit the stability of gene expression after injection of replication-defective adenovirus vectors, *Nature Med.* **2**:545–550.

Wolff, J.A., Malone, R.W., Williams, P., Chong, W., Acsadi, G., Jani, A., and Felgner, P.L., 1990, Direct gene transfer into mouse muscle in vivo, *Science* **247**:1465–1468.

Xiao, X., Li, J., and Samulski, R.J., 1996, Efficient long-term gene transfer into muscle tissue of immunocompetent mice by adeno-associated virus vector, *J. Virol.* **70**:8098–8108.

Yalyon, A., Klagsbrun, M., Esko, J.D., Leder, P., and Ornitz, D.M., 1991, Cell surface, heparin-like molecules are required for binding of bFGF to its high affinity receptor, *Cell* **64**:841–848.

Zhu, N., Liggitt, D., Liu, Y., and Debs, R., 1993, Systemic gene expression after IV DNA delivery into adult mice, *Science* **261**:209–211.

Zlokovic, B.V., and Apuzzo, M.L.J., 1997, Cellular and molecular neurosurgery: Pathways from concept to reality-Part II: Vector systems and delivery methodologies for gene therapy of the central nervous system, *Neurosurg.* **40**:805–813.

XI

MATRIX METALLO PROTEINASE

41

POTENTIAL APPLICATIONS OF TISSUE INHIBITOR OF METALLOPROTEINASE (TIMP) OVEREXPRESSION FOR CANCER GENE THERAPY

Andrew H. Baker*, Matti Ahonen[1], and Veli-Matti Kähäri[1]

Department of Surgery
University of Bristol, Bristol BS2 8HW
United Kingdom
[1]Department of Dermatology
Turku University Central Hospital and
Department of Medical Biochemistry and
MediCity Research Laboratory
University of Turku
FIN-20520 Turku
Finland

1. INTRODUCTION

Invasion and metastasis of malignant cells is a multistep process which involves detachment of cells from primary tumor, degradation of structural barriers such as basement membrane and collagenous extracellular matrix (ECM), as well as migration of cells through the degraded matrix (see MacDougal and Matrisian, 1995; Johnsen *et al.*, 1998). Tumor invasion *in vivo* involves interplay between tumor cells, stromal cells and inflammatory cells, all of which can express a variable set of proteolytic enzymes and contribute to degradation of peritumoral stromal matrix. Matrix metalloproteinases (MMPs) are a family of zinc-dependent neutral endopeptidases collectively capable of degrading essentially all components of the ECM (see Birkedal-Hansen *et al.*, 1993). Up-regulation of MMP expression has been documented in a number of malignant tumors and has been associated with invasive capacity of tumors. In addition, MMPs apparently play a role in tumor-induced angiogenesis. The activity of MMPs is inhibited specifically by tissue inhibitors of metalloproteinases (TIMPs). Here we discuss the role of MMPs

* Tel: +44 1179 283586/8, Fax: +44 1179 283581, Email: A.H.Baker@bris.ac.uk

Cancer Gene Therapy: Past Achievements and Future Challenges, edited by Habib
Kluwer Academic / Plenum Publishers, New York, 2000.

and TIMPs in tumor growth and invasion, with emphasis on potential applicability of TIMPs in targeted gene therapy of cancer.

2. MATRIX METALLOPROTEINASES

At present, the MMP gene family consists of 17 structurally related human members, which can be divided into subgroups of collagenases, gelatinases, stromelysins and stromelysin-like MMPs, membrane type MMPs (MT-MMPs) and novel MMPs, according to their substrate specificity and primary structure (Fig. 1) (see Kähäri

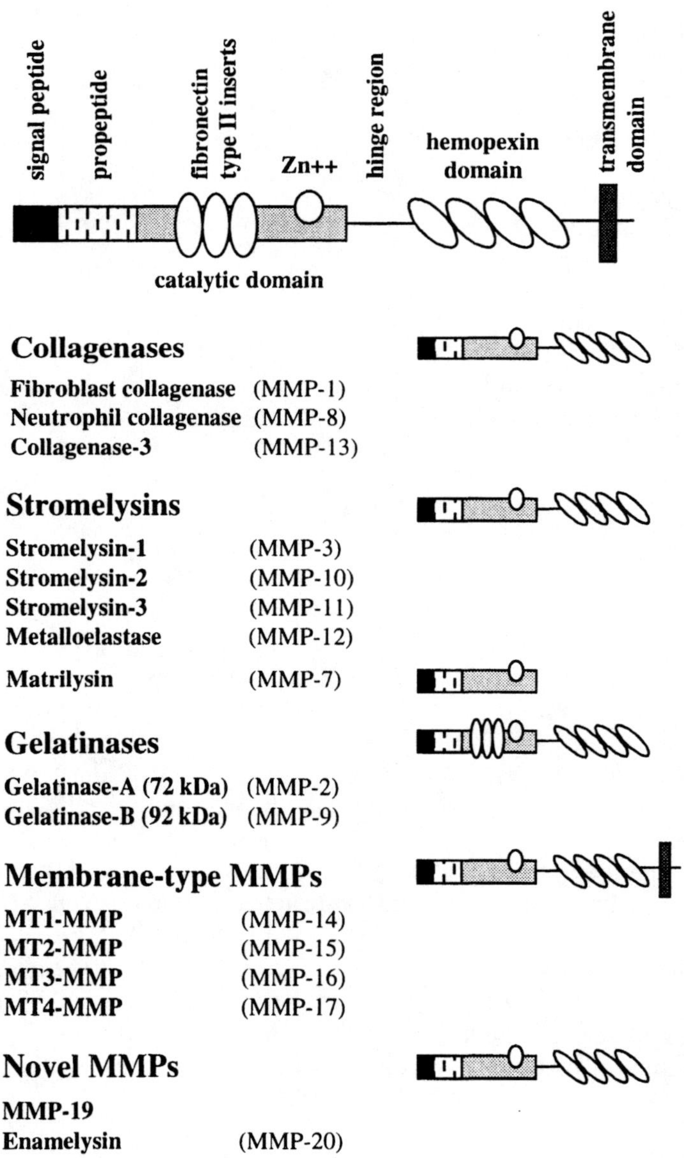

Figure 1. Human MMPs. The basic structure of individual sub-group MMPs are shown. See text for details.

and Saarialho-Kere, 1997; Shapiro, 1998). MMPs contain distinct functional domains: a signal peptide, a propeptide, a catalytic domain with the highly conserved zinc-binding site, and a hemopexin-like domain linked to the catalytic domain by a short hinge region (Fig. 1). Gelatinase-A (MMP-2) and gelatinase-B (MMP-9) also contain fibronectin type II-like inserts within the catalytic domain. Membrane-type MMPs (MT-MMPs) contain a transmembrane domain at the carboxy-terminal end of the hemopexin-like domain (Fig. 1). In contrast, the smallest MMP, matrilysin (MMP-7) lacks the hemopexin domain. For details of substrate specificity and expression, please see (Birkedal-Hansen, 1993; Kähäri and Saarialho-Kere, 1997; Shapiro, 1998) and references therein.

2.1. Collagenases

The collagenase subgroup consists of collagenase-1 (fibroblast collagenase, MMP-1), collagenase-2 (neutrophil collagenase, MMP-8), and collagenase-3 (MMP-13), which are the principal secreted neutral proteinases capable of initiating degradation of native fibrillar collagens of type I, II, III, and V. All three collagenases cleave fibrillar collagens at a specific site generating amino-terminal 3/4 and carboxy-terminal 1/4 fragments, which rapidly denature at body temperature and are further degraded by other MMPs, e.g. gelatinases. In addition, human MMP-13 cleaves type I collagen at the amino-terminal non-helical telopeptide. MMP-13 cleaves fibrillar collagens with preference for type II collagen over type I and III collagens, and displays over 40-fold stronger gelatinase activity than MMP-1 and MMP-8 (Freije *et al.*, 1994). MMP-13 degrades type IV, X and XIV collagens, tenascin, fibronectin and aggrecan core protein. MMP-1 is expressed by various types of cells in culture and *in vivo*, whereas MMP-8 is produced by neutrophils and chondrocytes (see Birkedal-Hansen, 1993; Kähäri and Saarialho-Kere, 1997). Apparently due to its ability to degrade an exceptionally wide range of ECM components, the physiologic expression of MMP-13 is limited to situations in which rapid and effective remodeling of collagenous ECM is required, i.e. fetal bone development and postnatal bone remodelling. On the other hand, MMP-13 is implicated in excessive degradation of collagenous ECM in osteoarthritic cartilage, rheumatoid synovium, chronic cutaneous ulcers, intestinal ulcerations, and periodontitis (see Kähäri and Saarialho-Kere, 1997; Shapiro, 1998), as well as in malignant tumors, i.e. breast carcinomas (Freije *et al.*, 1994; Heppner *et al.*, 1996; Uría *et al.*, 1997), squamous cell carcinomas (SCCs) of the head, neck and vulva (Johansson *et al.*, 1997; 1999), cutaneous basal cell carcinomas (Airola *et al.*, 1998) and chondrosarcomas (Uría *et al.*, 1998).

2.2. Gelatinases

This subgroup of MMPs contains only two members, gelatinase-A (72-kDa gelatinase, MMP-2) and gelatinase-B (92-kDa gelatinase, MMP-9), also called 72-kDa and 92 kDa type IV collagenases, respectively. MMP-9 is produced by epithelial cells, e.g. keratinocytes and stored in the secretory granules of neutrophils and eosinophils. MMP-2 and MMP-9 apparently play an important role in the final degradation of denatured fibrillar collagens after initial cleavage by collagenases. MMP-2 also cleaves native type I collagen to aminoterminal 3/4 and carboxyterminal 1/4 fragments identical to those generated by collagenases. In addition, MMP-9, which is expressed by osteoclasts during murine fetal development and in adult human bone, has been shown to cleave type I, II and V collagen in the N-terminal non-helical telopeptide. It is therefore possible that MMP-2 and MMP-9 also play a role in the remodelling of collagenous ECM.

2.3. Stromelysins

The principal members of the stromelysin subgroup, stromelysin-1 (MMP-3) and stromelysin-2 (MMP-10) are closely related to each other with respect to structure and substrate specificity. MMP-3 and MMP-10 are both expressed by epithelial cells and fibroblasts, as well as by various types of malignant cells and they are able to degrade a wide range of substrates, e.g. fibronectin, type IV, V, IX and X collagens, elastin, laminins, gelatin, and proteoglycan core proteins (see Kähäri and Saarialho-Kere, 1997; Shapiro, 1998). Stromelysin-3 (MMP-11), matrilysin (MMP-7) and metalloelastase (MMP-12) are often included in the stromelysin subgroup, although they are less closely related to MMP-3 and MMP-10. Stromelysin-3 (MMP-11) has not been shown to degrade any ECM components, but degrades serine-proteinase inhibitors, α2-antitrypsin and α1-proteinase inhibitor (Pei *et al.*, 1994). Matrilysin (MMP-7) is the smallest MMP, as it lacks the hemopexin-like domain. This is probably responsible for the wide substrate specificity of MMP-7, it is able to degrade elastin, fibronectin, laminins, nidogen, type IV collagen, and proteoglycan core proteins. Macrophage metalloelastase (MMP-12) is expressed by macrophages and stromal cells at sites of rapid matrix turnover during murine fetal development (see Shapiro, 1998). MMP-12 can degrade elastin, type IV collagen, fibronectin, vitronectin and laminin.

2.4. MT-MMPs

This subgroup consists of four members, all of which contain a transmembrane sequence within the carboxy-terminal hemopexin domain. MT1-MMP (MMP-14) is expressed by fibroblastic cells and malignant epithelial cells. MT1-MMP, MT2-MMP (MMP-15), and MT3-MMP (MMP-16) have been shown to proteolytically activate latent MMP-2, and MT1- and MT2-MMP also activate latent MMP-13. In addition, MT1-MMP degrades native type I collagen, gelatin, fibronectin, laminin-1 and cartilage proteoglycan core protein. MT2-MMP has been shown to degrade laminin, fibronectin and tenascin (d'Ortho *et al.*, 1997) and MT3-MMP has been shown to degrade gelatin, type III collagen and fibronectin. The substrate specificity of MT4-MMP is not known.

2.5. New MMPs

Two recently cloned members of the MMP family, MMP-19 (Pendás *et al.*, 1997) and enamelysin (MMP-20) (Llano *et al.*, 1997) can not be classified in any of the MMP subgroups based on their structure and substrate specificity. MMP-19 is expressed by various normal human tissues, but its ability to degrade native ECM components is not known. MMP-20 is expressed during tooth development and degrades amelogenin.

3. REGULATION OF MMP EXPRESSION AND ACTIVITY

In general, MMPs are not stored by cells in culture or *in vivo*, but their expression is rapidly induced in response to exogenous stimuli, e.g. cytokines, growth factors, cell-matrix interactions, and altered cell-cell contacts. Exceptions to this rule are MMP-8 and MMP-9, which are both stored in secretory granules of neutrophils and eosinophils, and MMP-7, which is stored in secretory epithelial cells of exocrine glands in skin, gastrointestinal tract, and airways (Saarialho-Kere *et al.*, 1995; Dunsmore *et al.*, 1998).

The expression of distinct MMPs is induced at transcriptional level. After translation and secretion in latent form, their proteolytic capacity is regulated at the level of pro-enzyme activation and inhibition of enzyme activity. The expression of inducible MMPs (MMP-1, MMP-3, MMP-7, MMP-9, MMP-10, MMP-12, and MMP-13) in unstimulated cultured cells is low, and is induced at transcriptional level by a variety of stimuli, e.g. cytokines and growth factors (see Birkedal-Hansen *et al.*, 1993; Kähäri and Saarialho-Kere, 1997). The 5′-flanking regulatory regions of inducible MMPs contain an AP-1 element at position -70 upstream of the transcription initiation site. Extracellular signals result in activation of nuclear AP-1 transcription factor dimers, composed of the members of the Jun and Fos gene families, which bind to the AP-1 element and activate transcription of the corresponding MMP gene.

Another *cis*-element, PEA3, which binds members of the ETS family of transcription factors, is present in the promoter regions of several MMPs (MMP-1, MMP-3, MMP-7, MMP-9, MMP-10, MMP-11, MMP-12, and MMP-13), and has been shown to co-operate with the AP-1 motif for maximal activation of the MMP-1, MMP-3 and MMP-9 promoters (Shindoh *et al.*, 1996; Westermarck *et al.*, 1997). Expression of ETS-1 has been demonstrated in stromal fibroblasts adjacent to invading tumor cells and in endothelial cells during tumor vascularization (Wernert *et al.*, 1992; 1994). It has also been shown that transcription factor NF-κB plays a pivotal role in expression of MMP-9 in fibroblasts and vascular smooth muscle cells (Bond *et al.*, 1998).

As MMPs are secreted as latent precursors or zymogens, which are proteolytically activated in the extracellular space. The latent MMPs are retained in the latent form by a "cysteine switch" formed by interaction between a conserved cysteine in the propeptide and the catalytic zinc blocking the access of the catalytic site to substrates (van Wart and Birkedal-Hansen, 1990). Activation of the latent MMP by chaotropic agents (e.g. organomercurials) or partial proteolytic cleavage of the propeptide dissociates the covalent bond between the cysteine and the catalytic zinc and results in exposure of the catalytic site (van Wart and Birkedal-Hansen, 1990). The ability of MMPs to activate each other may create a network of proteases in the extracellular space enabling effective activation of latent MMPs produced (see Birkedal-Hansen *et al.*, 1993; Kähäri and Saarialho-Kere, 1997). Furthermore, activation of latent MMP-2 by MT1-MMP and interaction of MMP-2 with $\alpha_v\beta_3$ integrin at the cell membrane also provides a mechanism for directing proteolytic activity to focal areas in the pericellular space. In contrast to other MMPs, stromelysin-3 (MMP-11) and MT-MMPs have a cleavage site for Golgi-associated furin proteinases between the propeptide and the catalytic domain providing basis for furin-dependent activation of these latent MMPs prior to secretion.

4. THE TIMP GENE FAMILY

The activity of MMPs *in vivo* is tightly regulated in the extracellular space by TIMPs, a gene family currently consisting of four members, TIMPs 1-4. TIMPs are small secreted proteins with a molecular weight of between 20 and 30 kDa and thus are highly suited for gene therapy as significant bystander effects are expected. Each TIMP has 12 conserved cysteine residues, which contribute to the secondary structure and MMP-inhibitory activity of TIMPs (see Birkedal-Hansen *et al.*, 1993 and Fig. 2). For the purpose of application of TIMPs in gene therapy it is important to define how inhibition of MMPs takes place. TIMPs can bind to both the latent and active forms of MMPs. Binding of TIMP to latent MMPs occurs in a 1:1 stoichiometry at the

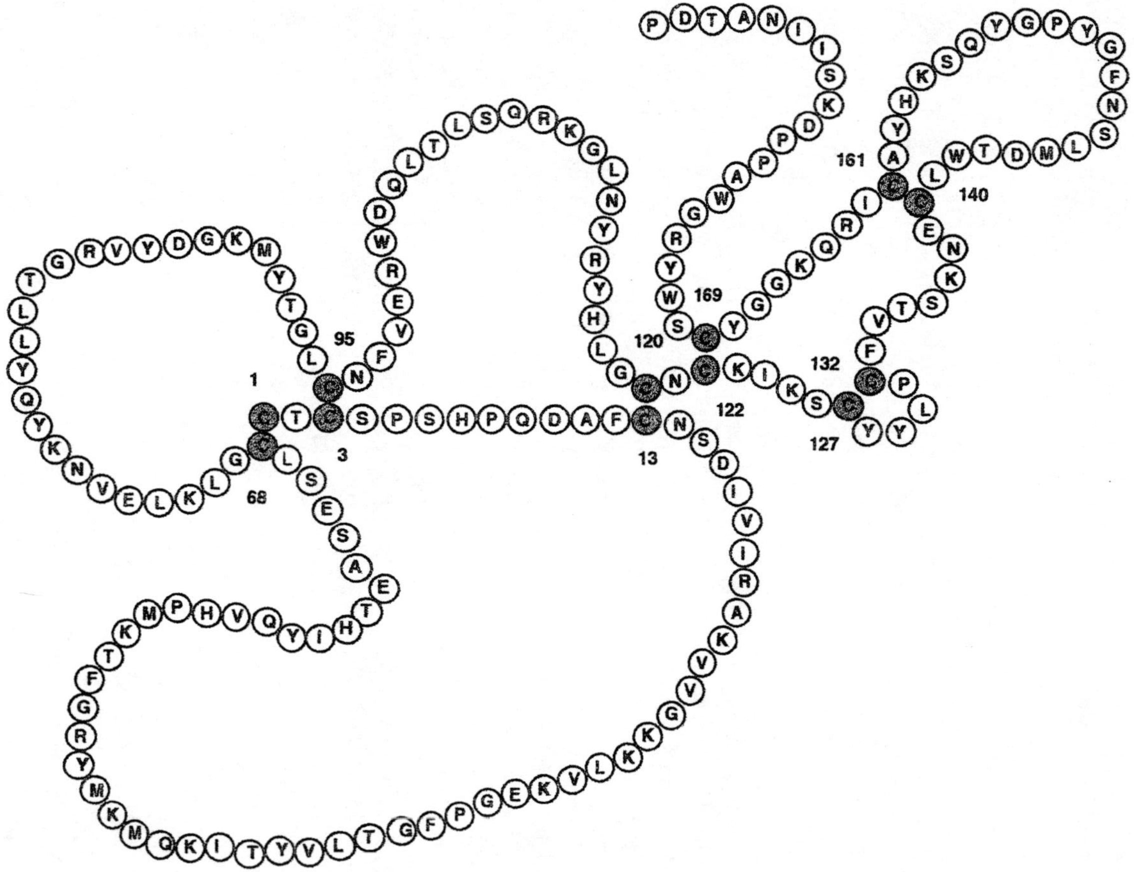

Figure 2. Secondary structure of TIMPs. 12 cysteine residues located throughout the protein (dark circles) form 6 bonds generating a uniquely folded structure required for MMP-inhibition.

C-terminal region of individual MMPs (Bode *et al.*, 1994). This action stabilizes the latent MMP:TIMP complex in the extracellular matrix and retards activation of the MMP, thus slowing proteolysis of the matrix substrate(s). However, the latent MMP:TIMP complex can still be activated, albeit delayed. A further molecule of TIMP can bind to the activated complex, this time at the active site of the MMP thereby inhibiting matrix proteolysis (Howard *et al.*, 1991; Willenbrock and Murphy, 1994). This second binding of TIMP therefore directly prevents local matrix destruction.

Individual TIMPs have variable affinities for different MMPs. For example, TIMP-1 binds to the latent form of MMP-9 with a higher affinity than TIMP-2. Conversely, TIMP-2 binds with higher affinity to latent MMP-2 than TIMP-1. The role of TIMP-2 in the activation of latent MMP-2 is a complex process that has only recently been defined. TIMP-2 binds latent MMP-2 and engages this complex with a molecule of MT1-MMP on the cell surface of MT1-MMP expressing cells (Zucker *et al.*, 1998; Butler *et al.*, 1998). A second molecule of MT1-MMP is required for activation to proceed, leading to the release of active MMP-2. Thus, TIMP-2 is required for activation of latent MMP-2. However, high concentrations of TIMP-2 or -3, but not TIMP-1 inhibit activity of MT1-MMP, thus preventing latent MMP-2 activation (Murphy *et al.*, 1992). It is therefore important to ensure that high levels of TIMP-2 would be produced locally in a potential gene therapy strategy. Thus, it is of importance to define and understand, which MMPs are implicated in the progression of different malignant tumors. This basic knowledge will therefore be useful in determining which TIMP(s) may be most beneficial for potential gene therapies.

5. ROLE OF MMPS AND TIMPS IN TUMOR GROWTH AND INVASION

As mentioned above, proteolytic degradation of stromal ECM is essential in cancer cell migration and invasion, as well as in tumor-induced angiogenesis. At present, most of the evidence for the role of MMPs in tumor invasion is based on the observations of expression of distinct MMPs in invasive primary tumors or in their metastases, whereas there is little evidence on the actual activity of distinct MMPs in tumor tissues. As the expression of most MMPs is low or undetectable in most normal tissues, activation of their expression in malignant tumors *in vivo* has been taken to indicate that they are produced and activated *in situ*. This notion is supported by observations that expression of certain MMPs is associated with invasion capacity in certain tumors. For example, expression of collagenase-1 (MMP-1) is associated with poor prognosis in colorectal cancer and oesophageal cancer (Murray *et al.*, 1996; 1998) and expression of MMP-2 and MMP-3 is closely related to lymph node metastasis and vascular invasion in SCC of esophagus (Shima *et al.*, 1992). Similarly, abundant expression of collagenase-3 (MMP-13) in SCCs of the head and neck and vulva is associated with their metastatic capacity (Johansson *et al.*, 1997; 1999). Stromelysin-3 has also been suggested to be associated with increased local invasiveness in head and neck SCCs (Muller *et al.*, 1993). It should be noted, that all MMPs, the expression of which has been documented in malignant tumors, can also be expressed by non-neoplastic cells. However, MMP-13, MMP-7, and MT1-MMP can be regarded as transformation specific MMPs at least in keratinocytes. Their expression is not detectable in normal keratinocytes, but are readily expressed by malignantly transformed keratinocytes in SCCs, indicating that their expression serves as a marker for transformation and invasion capacity (Johansson *et al.*, 1997; 1999).

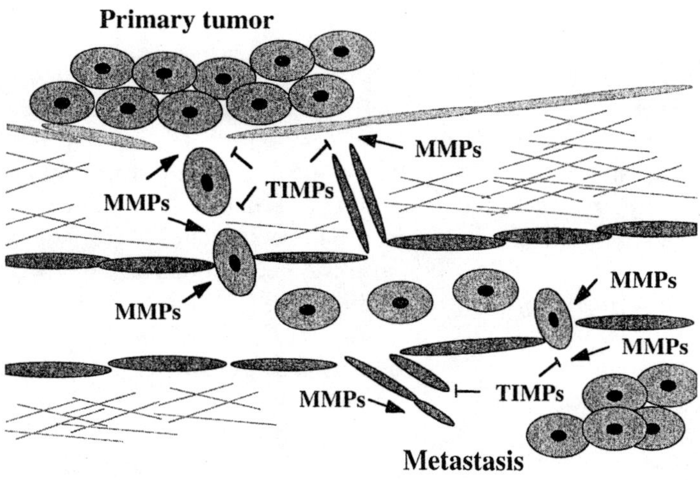

Figure 3. The role of MMPs and TIMPs in tumor growth, invasion, and metastasis. MMPs potentially act at different stages of tumour growth, invasion, and extravasation. The potential inhibition by TIMPs is indicated.

Invasion of malignant tumors involves interplay between tumor cells, stromal cells, and inflammatory cells and it is likely that all these cells express distinct, although somewhat overlapping patterns of MMPs (see MacDougal and Matrisian, 1995; Basset *et al.*, 1997 and Fig. 3). For example, in SCCs MMP-13 is mainly expressed by tumor cells, whereas MMP-1 is expressed mainly by stromal fibroblasts. matrilysin (MMP-7) is only expressed by tumor cells, MMP-2 by stromal fibroblasts, MT1-MMP by tumor and stromal cells and MMP-9 by tumor cells and inflammatory cells (Johansson *et al.*, 1997; Airola *et al.*, 1997; Johansson *et al.*, 1999). In SCCs, it is likely that co-localization of cells expressing MT1-MMP and MMP-2 with tumor cells producing MMP-13 creates optimal conditions for pericellular activation of tumor cell-derived latent MMP-13 and stromal-cell derived MMP-2. In addition, expression of MMP-3 and MMP-7 by some SCC tumor cells provides further activity for a tumor driven proteolytic cascade, in which latent MMP-13 can be activated by MT1-MMP or MMP-3, both expressed by tumor cells. Furthermore, expression of MMP-9 by invading tumor cells and adjacent tumor infiltrating inflammatory cells adds a further link to this MMP cascade, as MMP-13 can activate latent MMP-9. As the substrate specificity of the MMPs present in the peritumoral environment of SCCs is different, it can be proposed that each of these MMPs plays a distinct role in SCC invasion. Therefore, it is possible that specific inhibition of the expression or activity of even one of these MMPs may disturb the balance in this complex network of proteases and markedly inhibit invasion of SCC cells. An interesting observation is that in SCCs of skin, expression of TIMP-3 is detected in stromal fibroblasts, but not in tumor cells (Airola *et al.*, 1998). In this context it should be noted that TIMP-3 is the most potent inhibitor of MMP-13 activity among the TIMP family members.

Up-regulation of the expression of collagenase-1 (MMP-1) and gelatinase-A (MMP-2) has been detected in melanoma cells *in vivo*, and together with gelatinase-B (MMP-9), they have been implicated in the invasion and migration of melanoma cells *in vitro* (Woolley *et al.*, 1980; Väisänen *et al.*, 1996; Ray *et al.*, 1995; MacDougall *et al.*,

1995). $\alpha_v\beta_3$ integrin also appears to play an important role in melanoma cell invasion by mediating activation of MMP-2 expression by vitronectin and by directing the proteolytic activity of MMP-2 to the pericellular space of invading cells (Brooks *et al.*, 1996; Bafetti *et al.*, 1998).

Further evidence for the role of MMPs in tumor growth and invasion has been provided by studies on MMP-deficient transgenic mice (see Shapiro, 1998). Matrilysin (MMP-7) knock-out mice show reduced intestinal tumorigenesis (Wilson *et al.*, 1997) while gelatinase-A (MMP-2) knock-out mice show reduced angiogenesis and tumor progression (Itoh *et al.*, 1998). Stromelysin-3 knock-out mice display reduced tumorigenesis in response to chemical mutagenesis (Masson *et al.*, 1998).

6. TIMPS AS MODULATORS OF TUMOR INVASION, GROWTH AND APOPTOSIS

A number of studies have demonstrated expression of TIMPs in tumor stroma and tumor tissue, but the prognostic value of TIMP expression is not established. For example, in breast cancer TIMP-2 expression correlated with tumor recurrence (Visscher *et al.*, 1994). Similarly, in malignant breast cancer TIMP-1 has been shown to be overexpressed compared to non-malignant breast cancer (Yoshiji *et al.*, 1996). However, the MMP:TIMP ratio is elevated in cervical carcinomas with poor prognosis, indicating that evaluation of either MMP or TIMP expression alone is not always sufficient for prognostication of malignancies (Nuovo *et al.*, 1995). Although expression patterns of TIMPs by tumor and stromal cells do not always correlate with the degree of malignancy, there is a significant body of evidence to suggest that overexpression of TIMPs by cancer cells, or by the host reduces invasive and metastatic capacity of tumor cells. This indicates that disturbance of proteolytic activity cripples invasive cancer cells. This hypothesis has been confirmed using synthetic MMP inhibitors (see Brown, 1998). However, gene therapy by local overexpression of TIMPs would provide an opportunity to halt the progression of cancer cell invasion without the undesirable systemic side effects elicited by synthetic MMP inhibitors.

Inhibition of cancer cell invasion by overexpression of either TIMP-1, -2, -3 or -4 has been achieved using *in vitro* systems following delivery of TIMPs using recombinant protein or a wide variety of gene delivery vehicles. Albini *et al.* (1991) showed that recombinant TIMP-2 inhibited the invasion of HT1080 fibrosarcoma cells *in vitro*. This effect was further defined and shown to be dependent on the inhibition of MMP-2 activity, as a specific antisera against MMP-2 also inhibited HT1080 cell invasion. These findings were in agreement with another study by DeClerk *et al.* (1991) who found that rTIMP-2 inhibited degradation of matrices by MMPs leading to inhibition of invasion. Interestingly, rTIMP-2 had no effect on cell growth indicating that, at least in the *in vitro* system used, MMPs were important for invasion but not for growth of tumor cells. However, a subsequent *in vivo* study demonstrated an effect of TIMP-2 gene transfer on tumor cell growth and invasion (DeClerk *et al.*, 1992). In this study, overexpression of TIMP-2 in stably transfected cell lines inhibited their invasion and suppressed tumor growth in nude mouse lung colonies following injection of cell clones. The capacity of TIMP-2 to inhibit invasion was further defined in the study by Imren *et al.* (1996), in which tumors derived from injected cells overexpressing TIMP-2 from a retrovirus showed a thick connective tissue capsule surrounding the tumor, presumably through inhibition of MMP activity *in vivo*. Recently, combination therapy using TIMP-2 to inhibit invasion and a bisphosphonate to inhibit bone resorption was found to show a beneficial effect in a murine model (Yoneda *et al.*, 1997).

Similar findings have been reported for other TIMPs. Overexpression of TIMP-1 in gastric cancer cells reduced metastasis but only in clones expressing significantly high levels of TIMP-1 (Tsuchiya *et al.*, 1993). TIMP-1 can also reduce the growth rate and inhibit invasion of astrocytoma cells (Matsuzawa *et al.*, 1996), prevent metastasis of gastric cancer cells (Watanabe *et al.*, 1996) and inhibit invasion and tumor growth in mammary carcinoma cells (Alonso *et al.*, 1998). The ability of TIMP-1 to alter tumor development at different stages has been highlighted by the use of transgenic mice. Mice which a constitutively overexpress TIMP-1 in the liver had the ability to hinder tumor initiation, growth and angiogenesis when crossed with transgenics expressing the SV40 T antigen, which develop hepatocellular carcinomas (Martin *et al.*, 1996). The effects on tumor initiation may represent a protective effect conveyed to the liver by TIMP-1 overexpression. Further evidence for this hypothesis was recently produced in a study, in which TIMP-1 overexpression in the brain prevented tumor formation (Kruger *et al.*, 1998).

Studies in melanoma cells have also highlighted the potential applications and limitations for gene therapeutic approaches using TIMPs. As mentioned earlier, melanoma cells abundantly express MMPs and have a high metastatic capacity (Montgomery *et al.*, 1993). Overexpression of TIMP-2 in tumor cells reduced MMP activity and suppressed tumor growth in the skin of immunodeficient mice (Montgomery *et al.*, 1994). TIMP-2 did not, however, affect the rate of spontaneous metastasis. Conversely, Oku *et al.* (1997) demonstrated that melanoma cells overexpressing TIMP-2 had a reduced metastatic activity *in vivo*. Using a different melanoma cell line, Koop *et al.* (1994) showed that melanoma cells overexpressing TIMP-1 had a reduced metastatic capacity due to inhibition of tumor growth following extravasation. These studies highlight both similar, and diverse effects that overexpression of individual TIMPs may have on tumor phenotypes *in vivo*. Although TIMP-4 has been cloned more recently than TIMPs-1, -2 and -3, its ability to inhibit cell invasion has been established, as TIMP-4 overexpression inhibits invasion *in vitro* and tumor growth *in vivo*, and results in reduction in axillary lymph node and lung metastasis (Wang *et al.*, 1997).

Further studies have highlighted the different effects of individual TIMPs on cell death. Valente *et al.* (1998) reported that overexpression of TIMP-2 using an expression vector reduced invasion and angiogenesis, but also protected the melanoma cells from apoptosis although it increased necrosis. This phenotype, while scientifically intriguing, would diminish the usefulness of TIMP-2 for melanoma gene therapy due to the potential increase in cancer cell survival. Furthermore, TIMP-1 has recently been shown to promote survival of B-cells through modulation of CD40 levels (Guedez *et al.*, 1998). However, we have recently shown, using an adenoviral-mediated gene delivery approach that TIMP-3 efficiently promotes apoptosis of a number of cell types including human melanoma, HeLa, MCF-7 and HT1080 cells, while TIMP-1 and -2 had no effect on cell viability (Ahonen *et al.*, 1998; Baker *et al.*, 1998; 1999). This apoptotic effect was linked to a reduced capacity of TIMP-3 transduced cells to bind matrix components implicating disruption of a cell-matrix survival pathway. Additionally, stable colon carcinoma cell lines overexpressing TIMP-3 show reduced tumor growth *in vivo* (Bian *et al.*, 1996). Additionally, TIMP-3 overexpression in this system resulted in apoptosis of tumor cells through stabilization of tumor necrosis factor-α receptors on the cell surface (Smith *et al.*, 1997).

The involvement of TIMPs in the regulation of cell survival and cell death is therefore intriguing both in physiological conditions and in cancer gene therapy. Individual TIMPs may modulate the levels of death proteins at the cell surface. For example, release of FAS from the cell surface is mediated by MMP-induced cleavage and is inhibited by synthetic MMP inhibitors (Kayagaki *et al.*, 1995). Furthermore, TIMP-3, but not TIMP-1 or -2 has recently been shown to specifically inhibit tumor necrosis factor-α convertase (TACE or ADAM-17) providing evidence for highly complex regulation of death ligands and receptors by MMPs and TIMPs (Amour *et al.*, 1998).

It is therefore clear from early studies that TIMPs can inhibit the invasion, and sometimes the growth of cancer cells through the ECM. Recent evidence has demonstrated that TIMPs are involved in the complex regulation of proteins at the cell surface that can either promote or inhibit cell death. Further definition of these cellular events in both physiological and pathological states will require clarification before conclusions regarding the mechanisms of action can be drawn.

7. CONCLUDING REMARKS

The involvement of MMPs in invasion and growth of a variety of malignant tumors makes them an attractive target for therapeutic intervention, either pharmacological or gene-based with the aim to inhibit tumor growth, invasion, and angiogenesis. Due to the high affinity of TIMPs for MMPs in the pericellular environment, overexpression of TIMPs may be one possible route to cancer gene therapy. TIMPs have many qualities important for efficient gene therapy including their secretion (therefore inducing significant bystander effects), are highly diffusible (TIMPs-1, -2 and -4) and inhibit the activity of multiple MMPs. Furthermore, induction of apoptosis (TIMP-3) may be an additional attribute in attempts to ablate tumor mass. However, as yet the role of MMPs and TIMPs in the highly complex regulation of both physiological and pathological conditions *in vivo* remains to be fully elucidated. Further experimentation will fully define the clinical potential for TIMP gene therapy of cancer.

ACKNOWLEDGMENTS

The original work of the authors has been supported by grants from the Medical Research Council of Great Britain, the Academy of Finland, Sigrid Jusélius Foundation, Cancer Foundation of Finland, and Turku University Central Hospital. M. A. is a student in Turku Graduate School in Biomedical Sciences (TuBS).

REFERENCES

Ahonen, M., Baker, A.H., and Kähäri, V.-M., 1998, Adenovirus-mediated gene delivery of tissue inhibitor of metalloproteinases-3 inhibits invasion and induces apoptosis in melanoma cells. *Cancer Res.* **58**:2310–2315.

Airola, K., Johansson, N., Kariniemi, A.-L., Kähäri, V.-M., and Saarialho-Kere, U., 1997, Human collagenase-3 is expressed in malignant squamous epithelium of the skin. *J. Invest. Dermatol.* **109**:225–231.

Airola, K., Ahonen, M., Johansson, N., Heikkilä, P., Kere, J., Kähäri, V.-M., and Saarialho-Kere, U.K., 1998, Human TIMP-3 is expressed during fetal development, hair growth cycle and cancer progression. *J. Histochem. Cytochem.* **46**:437–448.

Albini, A., Melchiori, A., Santi, L., Liotta, L.A., Brown, P.D., and Stetler-Stevenson, W.G., 1991, Tumor cell invasion inhibited by TIMP-2. *J. Natl. Cancer Inst.* **83**:775–779.

Amour, A., Slocombe, P.M., Webster, A., Butler, M., Knight, C.G., Smith, B.J., Stephens, P.E., Shelley, C., Hutton, M., Knauper, V., Docherty, A.J., and Murphy, G., 1998, TNF-α converting enzyme (TACE) is inhibited by TIMP-3. *FEBS Lett.* **435**:39–44.

Alonso, D.F., Skilton, G., DeLorenzo, M.S., Scursoni, A.M., Yoshiji, H., and Gomez, D.E., 1998, Histopathological findings in a highly invasive mouse mammary carcinoma transfected with human tissue inhibitor of metalloproteinases-1. *Oncology Reports.* **5**:1083–1087.

Bafetti, L.M., Young, T.N., Itoh, Y., and Stack, M.S., 1998, Intact vitronectin induces matrix metalloproteinase-2 and tissue inhibitor of metalloproteinases-2 expression and enhanced cellular invasion by melanoma cells. *J. Biol. Chem.* **273**:143–149.

Baker, A.H., Zaltsman, A.B., George, S.J., and Newby, A.C., 1998, Divergent effects of tissue inhibitor of metalloproteinase-1, -2 or -3 overexpression on rat vascular smooth muscle cell invasion, proliferation and death *in vitro*: TIMP-3 promotes apoptosis. *J. Clin. Invest.* **101**:1478–1487.

Baker, A.H., George, S.J., Zaltsman, A.B., Murphy, G., and Newby, A.C., 1999, Inhibition of invasion and induction of apoptotic cell death of cancer cell lines by overexpression of TIMP-3. *Br. J. Cancer*, in press.

Basset, P., Okada, A., Chenard, M.P., Kannan, R., Stoll, I., Anglard, P., Bellocq, J.P., and Rio, M.C., 1997, Matrix metalloproteinases as stromal effectors of human carcinoma progression: therapeutic implications. *Matrix Biol* **15**:535–541.

Bian, J., Wang, Y., Smith, M.R., Kim, H., Jacobs, C., Jackman, J., Kung, H.F., Colburn, N.H., and Sun, Y., 1996, Suppression of in vivo tumor growth and induction of suspension cell death by tissue inhibitor of metalloproteinases (TIMP)-3. *Carcinogenesis* **17**:1805–1811.

Birkedal-Hansen, H., Moore, W.G.I., Bodden, M.K., Windsor, L.J., Birkedal-Hansen, B., De Carlo, A., and Engler, J.A., 1993, Matrix metalloproteinases: A review. *Crit. Rev. Oral Biol. Med.* **4**:197–250.

Bode, W., Reinemer, P., Huber, R., Kleine, T., Schnierer, S., and Tschesche, H., 1994, The X-ray crystal structure of the catalytic domain of human neutrophil collagenase inhibited by a substrate analogue reveals the essentials for catalysis and specificity. *EMBO J.* **13**:1263–1269.

Bond, M., Fabunmi, R.P., Baker, A.H., and Newby, A.C., 1998, Synergistic upregulation of metalloproteinase-9 by growth factors and inflammatory cytokines: an absolute requirement for transcription factor NF-κB. *FEBS Lett.* **435**:29–34.

Brooks, P.C., Stromblad, S., Sanders, L.C., von Schalscha, T.L., Stetler-Stevenson, W.G., Quigley, J.P., and Cheresch, D.A., 1996, Localization of matrix metalloproteinase MMP-2 to the surface of invasive cells by interaction with integrin $\alpha_v\beta_3$. *Cell* **85**:683–693.

Brown, P.D., 1998, Synthetic inhibitors of matrix metalloproteinases. In: *Matrix metalloproteinases*. W.C. Parks, and R.P. Mecham, eds. Academic Press, San Diego. 243–261.

Butler, G.S., Butler, M.J., Atkinson, S.J., Will, H., Tamura, T., van Westrum, S.S., Crabbe, T., Clements, J., d'Ortho, M.P., and Murphy, G., 1998, The TIMP2 membrane type 1 metalloproteinase "receptor" regulates the concentration and efficient activation of progelatinase A. A kinetic study. *J. Biol. Chem.* **273**:871–880.

Declerck, Y.A., Yean, T.D., Chan, D., Shimada, H., and Langley, K.E., 1991, Inhibition Of Tumor Invasion Of Smooth-Muscle Cell-Layers By Recombinant Human Metalloproteinase Inhibitor. *Cancer Res.* **51**:2151–2157.

DeClerck, Y.A., Perez, N., Shimada, H., Boone, T.C., Langley, K.E., and Taylor, S.M., 1992, Inhibition of invasion and metastasis in cells transfected with an inhibitor of metalloproteinases. *Cancer Res.* **52**:701–708.

d'Ortho, M.P., Will, H., Atkinson, S., Butler, G., Messent, A., Gavrilovic, J., Smith, B., Timpl, R., Zardi, L., and Murphy, G., 1997, Membrane type matrix metalloproteinases 1 and 2 exhibit broad-spectrum proteolytic capacities comparable to many matrix metalloproteinases. *Eur. J. Biochem.* **250**: 751–757.

Dunsmore, S.E., Saarialho-Kere, U.K., Roby, J.D., Wilson, C.L., Matrisian, L.M., Welgus, H.G., and Parks, W.C., 1998 Matrilysin expression and function in airway epithelium. *J. Clin. Invest.* **102**:1321–1331.

Freije, J.M.P., Díez-Itza, I., Balbín, M., Sánchez, L.M., Blasco, R., Tolivia, J., López, and Otín, C., 1994, Molecular cloning and expression of collagenase-3, a novel human matrix metalloproteinase produced by breast carcinomas. *J. Biol. Chem.* **269**:16766–16773.

Guedez, L., Courtemanch, L., and StetlerStevenson, M., 1998, Tissue inhibitor of metalloproteinase (TIMP)-1 induces differentiation and an antiapoptotic phenotype in germinal center B Cells. *Blood.* **92**:1342–1349.

Heppner, K.J., Matrisian, L.M., Jensen, R.A., and Rodgers, W.H., 1996, Expression of most matrix metalloproteinases family members in breast cancer represents a tumor-induced host response. *Am. J. Pathol.* **149**:273–282.

Howard, E.W., Bullen, E.C., and Banda, M.J., 1991, Preferential inhibition of 72- and 92-kDa gelatinases by tissue inhibitor of metalloproteinases-2. *J. Biol. Chem.* **266**:13070–13075.

Imren, S., Kohn, D.B., Shimada, H., Blavier, L., and DeClerck, Y.A., 1996, Overexpression of tissue inhibitor of metalloproteinases-2 retroviral-mediated gene transfer in vivo inhibits tumor growth and invasion. *Cancer Res.* **56**:2891–2895.

Itoh, T., Tanioka, M., Yoshida, H., Yoshioka, T., Nishimoto, H., and Itohara, S., 1998, Reduced angiogenesis and tumor progression in gelatinase A-deficient mice. *Cancer Res.* **58**:1048–1051.

Johansson, N., Airola, K., Grenman, R., Kariniemi, A.-L., Saarialho-Kere, U., and Kähäri, V.-M., 1997, Expression of collagenase-3 (matrix metalloproteinase-13) in squamous cell carcinomas of the head and neck. *Am. J. Pathol.* **151**:499–508.

Johansson, N., Vaalamo, M., Grénman, S., Hietanen, S., Klemi, P., Saarialho-Kere, U., and Kähäri V.-M., 1999 Collagenase-3 (MMP-13) is expressed by tumor cells in invasive vulvar squamous carcinomas. *Am. J. Pathol.*, in press.

Johnsen, M., Lund, L.R., Romer, J., Almholt, K., and Dano, K., 1998, Cancer invasion and tissue remodeling: common themes in proteolytic matrix degradation. *Curr. Opin. Cell. Biol.* **10**:667–671.

Kähäri, V.-M., and Saarialho-Kere, U., 1997, Matrix metalloproteinases in skin. *Exp. Dermatol.* **6**:199–213.

Kayagaki, N., Kawasaki, A., Ebata, T., Ohmoto, H., Ikeda, S., Inoue, S., Yoshino, K., Okumra, K., and Yagita, H., 1995, Metalloproteinase-Mediated Release Of Human Fas Ligand. *J. Exp. Med.* **182**:1777–1783.

Kruger, A., SanchezSweatman, O.H., Martin, D.C., Fata, J.E., Ho, A.T., Orr, F.W., Ruther, U., and Khokha, R., 1998,. Host TIMP-1 overexpression confers resistance to experimental brain metastasis of a fibrosarcoma cell line. *Oncogene.* **16**:2419–2423.

Koop, S., Khokha, R., Schimdt, E.E., MacDonald, I.C., Morris, V.L., Chambers, A.F., and Groom, A.C., 1994, Overexpression of metalloproteinase inhibitor in B16F10 cells does not affect extravasation but reduces tumor growth. *Cancer Res.* **54**:4791–4797.

Llano, E., Pendas, A.M., Knäuper, V., Sorsa, T., Ding, Y., Santavirta, S., Ceponis, A., and López-Otín, C., 1997, Identification and characterization of human enamelysin. *Biochemistry* **36**:15101–15108.

MacDougall, J.R., and Matrisian, L.M., 1995, Contributions of tumor and stromal matrix metalloproteinases to tumor progression, invasion and metastasis. *Cancer Metastasis Rev.* **14**:351–362.

MacDougall, J.R., Bani, M.R., Lin, Y., Rak, J., and Kerbel, R.S., 1995, The 92-kDa gelatinase B is expressed by advanced stage melanoma cells: suppression by somatic cell hybridization with early stage melanoma cells. *Cancer Res.* **55**:4174–4181.

Martin, D.C., Ruther, U., Sanchez-Sweatman, O.H., Orr, F.W., and Khokha, R., 1996, Inhibition of SV40 T antigen-induced hepatocellular carcinoma in TIMP-1 transgenic mice. *Oncogene* **13**:569–576.

Masson, R., Lefebvre, O., Noel, A., Fahime, M.E., Chenard, M.P., Wendling, C., Kebers, F., LeMeur, M., Dierich, A., Foidart, J.M., Basset, P., and Rio, M.C., 1998, In vivo evidence that the stromelysin-3 metalloproteinase contributes in a paracrine manner to epithelial cell malignancy. *J. Cell Biol.* **140**: 1535–41.

Matsuzawa, K., Fukuyama, K., Hubbard, S.L., Dirks, P.B., and Rutka, J.T., 1996, Transfection of an invasive human astrocytoma cell line with a TIMP-1 cDNA: Modulation of astrocytoma invasive potential. *J Neuropath. Exp. Neurol.* **55**:88–96.

Montgomery, A.M., De Clerck, Y.A., Langley, K.E., Reisfeld, R.A., and Mueller, B.M., 1993, Melanoma-mediated dissolution of extracellular matrix: contribution of urokinase-dependent and metalloproteinase-dependent proteolytic pathways. *Cancer Res.* **53**:693–700.

Montgomery, A.M., Mueller, B.M., Reisfeld, R.A., Taylor, S.M., and DeClerck, Y.A., 1994, Effect of tissue inhibitor of the matrix metalloproteinases-2 expression on the growth and spontaneous metastasis of a human melanoma cell line. *Cancer Res.* **54**:5467–5473.

Muller, D., Wolf, C., Abecassis, J., Millon, R., Engelmann, A., Bronner, G., Rouyer, N., Rio, M.C., Eber, M., Methlin, G., Chambon, P., and Basset, P., 1993, Increased stromelysin 3 gene expression is associated with increased local invasiveness in head and neck squamous cell carcinomas. *Cancer Res.* **53**:165–169.

Murray, G.I., Duncan, M.E., O'Neil, P., Melvin, W.T., and Fothergill, J.E., 1996, Matrix metalloproteinase-1 is associated with poor prognosis in colorectal cancer. *Nat. Med.* **2**:461–462.

Murray, G.I., Duncan, M.E., O'Neil, P., McKay, J.A., Melvin, W.T., and Fothergill, J.E., 1998, Matrix metalloproteinase-1 is associated with poor prognosis in oesophageal cancer. *J. Pathol.* **185**:256–261.

Murphy, G., Willenbrock, F., Ward, R.V., Cockett, M.I., Eaton, D., and Docherty, A.J.P., 1992, The C-terminal domain of 72 kDa gelatinase A is not required for catalysis, but is essential for membrane activation and modulates interactions with tissue inhibitors of metalloproteinases. *Biochem. J.* **283**: 637–641.

Nuovo, G.J., MacConnell, P.B., Simsir, A., Valea, F., and French, D.L., 1995, Correlation of the in situ detection of polymerase chain reaction amplified metalloproteinase complementary cDNAs and their inhibitors with prognosis in cervical carcinoma. *Cancer Res.* **55**:267–275.

Oku, T., Ata, N., Yonezawa, K., Tokai, H., Fujii, H., Shinagawa, A., Ohuchi, E., and Saiki, I., 1997, Antimetastatic and antitumor effect of a recombinant human tissue inhibitor of metalloproteinases-2 in murine melanoma models. *Biol. Pharm. Bull.* **20**:843–849.

Pei, D., Majmudar, G., and Weiss, S.J., 1994. Hydrolytic inactivation of a breast carcinoma cell-derived serpin byhuman stromelysin-3. *J. Biol. Chem.* **269**:25849–25855.

Pendas, A.M., Knäuper, V., Puente, X.S., Llano, E., Mattei, M.G., Apte, S., Murphy, G., and López-Otín, C., 1997 Identification and characterization of a novel human matrix metalloproteinase with unique structural characteristics, chromosomal localization and tissue distribution. *J. Biol. Chem.* **272**:4281–4286.

Ray, J.M., and Stetler-Stevenson, W.G., 1995, Gelatinase A activity directly modulates melanoma cell adhesion and spreading. *EMBO J.* **14**:908–917.

Saarialho-Kere, U.K., Crouch, E.C., and Parks, W.C., 1995, Matrix metalloproteinase matrilysin is constitutively expressed in adult human exocrine epithelium. *J. Invest. Dermatol.* **105**:190–196.

Shapiro, S.D., 1998, Matrix metalloproteinase degradation of extracellular matrix: biological consequences. *Curr. Opin. Cell. Biol.* **10**:602–608.

Shima, I., Sasapuri, Y., Kusukawa, J., Fujita, H., Kakegava, T., and Morimatsu, M., 1992, Production of matrix metalloproteinase-2 and metalloproteinase-3 related to malignant behavior of esophageal carcinoma. A clinicopathological study. *Cancer* **70**:2747–2753.

Shindoh, M., Higashino, F., Kaya, M., Yasuda, M., Funaoka, K., Hanzawa, M., Hida, K., Kohgo, T., Amemiya, A., Yoshida, K., and Fujinaga, K., 1996 Correlated expression of matrix metalloproteinases and ets family transcription factor E1A-F in invasive oral squamous-cell-carcinoma-derived cell lines. *Am. J. Pathol.* **148**:693–700.

Smith, M.R., Kung, H.F., Durum, S.K., Colburn, N.H., and Sun, Y., 1997, TIMP-3 induces cell death by stabilizing TNF-alpha receptors on the surface of human colon carcinoma cells. *Cytokine.* **9**:770–780.

Tsuchiya, Y., Sato, H., Endo, Y., Okada, Y., Mai, M., Sasaki, T., and Seiki, M., 1993, Tissue Inhibitor Of Metalloproteinase-1 Is a Negative Regulator Of the Metastatic Ability Of a Human Gastric-Cancer Cell-Line, Kkls, In the Chick-Embryo. *Cancer Res.* **53**:1397–1402.

Uría, A., Ståhle-Bäckdahl, M., Seiki, M., Fueyo, A., and López-Otín, C., 1997, Regulation of collagenase-3 expression in human breast carcinomas is mediated by stromal-epithelial cell interactions. *Cancer Res.* **57**:4882–4888.

Uría, J.A., Balbin, M., López, J.M., Alvarez, J., Vizoso, F., Takigawa, M., and López-Otín, C., 1998, Collagenase-3 (MMP-13) expression in chondrosarcoma cells and its regulation by basic fibroblast growth factor. *Am. J. Pathol.* **153**:91–101.

Valente, P., Fassina, G., Melchiori, A., Masiello, L., Cilli, M., Vacca, A., Onisto, M., Santi, L., Stetler-Stenvenson, W.G., and Albini, A., 1998, TIMP-2 over-expression reduces invasion and angiogenesis and protects B16F10 melanoma cells from apoptosis. *Int. J. Cancer* **75**:246–253.

Väisänen, A., Tuominen, H., Kallioinen, M., and Turpeenniemi-Hujanen, T., 1996, Matrix metalloproteinase-2 (72 kD type IV collagenase) expression occurs in the early stage of human melanocytic tumour progression and may have prognostic value. *J. Pathol.* **180**:283–289.

van Wart, H.E., and Birkedal-Hansen, H., 1990, The cystein switch: A principle of regulation of metalloproteinase activity with potential applicability to the entire matrix metalloproteinase family. *Proc. Natl. Acad. Sci. USA* **87**:5578–5582.

Visscher, D.W., Hoyhtya, M., Ottosen, S.K., Liang, C.M., Sarkar, F.H., Crissman, J.D., and Fridman, R., 1994, Enhanced expression of tissue inhibitor of metalloproteinase-2 (TIMP-2) in the stroma of breast carcinomas correlates with tumor recurrence. *Int. J. Cancer* **59**:339–344.

Wang, M., Liu, Y.E., Greene, J., Sheng, S., Fuchs, A., Rosen, E.M., and Shi, Y.E., 1997, Inhibition of tumor growth and metastasis of human breast cancer cells transfected with tissue inhibitor of metalloproteinase 4. *Oncogene* **14**:2767–2774.

Watanabe, M., Takahashi, Y., Ohta, T., Mai, M., Sasaki, T., and Seiki, M., 1996, Inhibition of metastasis in human gastric cancer cells transfected with Tissue inhibitor of metalloproteinase 1 gene in nude mice. *Cancer.* **77**:1676–1680.

Wernert, N., Raes, M.B., Lassalle, P., Dehouck, M.P., Gosselin, B., Vandenbunder, B., and Stehelin, D., 1992, c-ets1 proto-oncogene is a transcription factor expressed inendothelial cells during tumor vascularization and other forms of angiogenesis in humans. *Am. J. Pathol.* **140**:119–127.

Wernert, N., Gilles, F., Fafeur, V., Bouali, F., Raes, M.B., Pyke, C., Dupressoir, T., Seitz, G., Vandenbunder, B., and Stehelin, D., 1994, Stromal expression of c-Ets1 transcription factor correlates with tumor invasion. *Cancer Res.* **54**:5683–5688.

Westermarck, J., Seth, A., and Kähäri, V.-M., 1997, Differential regulation of interstitial collagenase (MMP-1) gene expression by ETS transcription factors. *Oncogene* **14**:2651–2660.

Willenbrock, F., and Murphy, G., 1994, Structure-function relationships in the tissue inhibitors of metalloproteinases. *Am. J. Respir. Crit. Care Med.* **150**:S165–S170.

Wilson, C.L., Heppner, K.J., Labosky, P.A., Hogan, B.L., and Matrisian, L.M., 1997, Intestinal tumorigenesis is suppressed in mice lacking the metalloproteinase matrilysin. *Proc. Natl. Acad. Sci. USA* **94**:1402–1407.

Woolley, D.E., and Grafton, C.A., 1980, Collagenase immunolocalization studies of cutaneous secondary melanomas. *Br. J. Cancer* **42**:260–265.

Yoneda, T., Sasaki, A., Dunstan, C., Williams, P.J., Bauss, F., De Clerck, Y.A., and Mundy, G.R., 1997, Inhibition of osteolytic bone metastasis of breast cancer by combinedtreatment with the bisphosphonate ibandronate and tissue inhibitor of the matrix metalloproteinase-2. *J. Clin. Invest.* **99**:2509–2517.

Yoshiji, H., Gomez, D.E., and Thorgeirsson, U.P., 1996, Enhanced RNA expression of tissue inhibitor of metalloprotenases-1 (TIMP-1) in human breast cancer. *Int. J. Cancer* **69**:131–134.

Zucker, S., Drews, M., Conner, C., Foda, H.D., DeClerck, Y.A., Langley, K.E., Bahou, W.F., Docherty, A.J., and Cao, J., 1998, Tissue inhibitor of metalloproteinase-2 (TIMP-2) binds to the catalytic domain of the cell surface receptor, membrane type 1-matrix metalloproteinase 1 (MT1-MMP). *J. Biol. Chem.* **273**:1216–1222.

CONTRIBUTORS

Matti Ahonen
Department of Dermatology
Turku University Central Hospital
University of Turku
FIN-20520 Turku
Finland

Eric WFW Alton
National Heart and Lung Institute
Imperial College School of Medicine
Emmanuel Kaye Building
Manresa Road
London SW3 6LR
UK

Joseph Ayoub
Centre de Recherche Pédiatrique
Hôpital Ste-Justine
3175 Chemin Côte Ste-Catherine
Montréal
Québec H3T 1C5
Canada

Ichiro Azuma
Institute of Immunological Science
Hokkaido University
Kita-ku
Sapporo 060
Japan

Andrew H. Baker
Department of Surgery
Bristol Heart Institute
University of Bristol
Bristol Royal Infirmary
Bristol BS2 8HW
UK

Nasim A. Begum
Dept of Immunology
Osaka Medical Center for Cancer and Cardiovascular Disease
Higashinari-ku
Osaka 537
Japan

David Bermudes
Vion Pharmaceuticals, Inc
4 Science Park
New Haven, CT 06511
USA

Yves-Jean Bignon
Laboratoire de Thérapie Génique Humaine at Expérimentale
INSERM CR1 9502 & EA 2145
Centre Jean perrin
BP 392
63011 Clermont-Ferrand Cedex 1
France

Lieping Chen
Mayo Clinic
Department of Immunology
Room 342B, Guggenheim Building
200 First Street, S.W.
Rochester, MN 55905
USA

Gary Clawson
Department of Pathology C7768
Penn State University
500 University Drive
Hershey, PA 17033
USA

Susan J. Cleator
Department of Clinical Oncology
Imperial College School of Medicine
Hammersmith Hospital
Du Cane Road
London W12 0NN,
UK

Stephen Coe
The Royal Free and University College
Medical School
Dept. Anatomy & Developmental
Biology & Dept Surgery
University of London
Royal Free Campus
Rowland Hill Street
London NW3 2PF
UK

M. Jo Colston
National Institute for Medical Research
The Ridgeway
Mill Hill
London NW7 1AA
UK

Michael Dahm
Gene Therapy and Biotechnology
HEXAL AG
Industriestr, 25
D-83607 Holzkirchen
Germany

Wafik El-Deiry
HHMI/U. Penn CRB 437A
415 Curie Boulevard
Philadelphia, PA 19104
USA

Nicoletta Eliopoulos
Centre de Recherche Pédiatrique
Hôpital Ste-Justine
3175 Chemin Côte Ste-Catherine
Montréal
Québec H3T 1C5
Canada

Farzin Farzaneh
Immune Gene Therapy of Cancer
Programme
Department of Molecular Medicine
King's College School of Medicine and
Dentistry
Rayne Institute
123 Coldharbour Lane
London SE5 9NU
UK

Paul Fisher
Herbert Irving Comprehensive Cancer
Centre
Departments of Pathology and Urology
Columbia University
College of Physicians and Surgeons
BB1501
630 West 168th Street
New York, NY 10032
USA

Jean-Christophe François
Laboratoire de Biophysique
INSERM U201
CNRS UA481
Muséum National d'Histoire Naturelle
43 rue Cuvier 75231
Paris Cedex 05
France

Scott Freeman
Schering-Plough Research Institute
2015 Galloping Hill Road (K-15-3 3200)
Kenilworth
New Jersey, 07033
USA

Duncan M. Geddes
National Heart and Lung Institute
Imperial College School of Medicine
Emmanuel Kaye Building
Manresa Road
London SW3 6LR
UK

Ruth A Gjerset
Sidney Kimmel Cancer Center
10835 Altman Row
San Diego, CA 92121
USA

Professor Geoffrey Goldspink
The Royal Free and University College Medical School
Dept. Anatomy and Developmental Biology & Dept. Surgery
University of London
Royal Free Campus
Rowland Hill Street
London NW3 2PF
UK

Neil I. Goldstein
GenQuest Incorporated
New York, NY 10032
USA

Beverley E. Griffin
Department of Infectious Diseases (Virology)
7th Floor, Commonwealth Building
Imperial College School of Medicine
Hammersmith Hospital
Du Cane Road
London W12 0NN
UK

Yajun Guo
Sidney Kimmel Cancer Center
10835 Altman Row
San Diego, CA 92121
USA

Nagy A. Habib
Head of Liver Surgery Section
Imperial College School of Medicine
Hammersmith Hospital
Du Cane Road
London W12 0NN
UK

Paul L. Hallenbeck
Genetic Therapy
21 Firstfield Road
Gaithersburg, MD 20878
USA

Michael Harron
The Royal Free and University College Medical School
Dept. Anatomy and Developmental Biology & Dept. Surgery
University of London
Royal Free Campus
Rowland Hill Street
London NW3 2PF
UK

Roman Havlik
Imperial College School of Medicine
Hammersmith Hospital
Du Cane Road
London W12 0NN
UK

Akira Hayashi
Organisation for Pharmaceutical Safety and Research
Tokyo 113
Japan

Daniel J. Hellrung
Human Gene Therapy Research Institute
1415 Woodlands Avenue
Des Moines
Iowa 50309-3203
USA

Phil W. Hinds
Howard Hughes Medical Institute
Department of Pathology
Harvard Medical School
200 Longwood Avenue
Boston, MA 02115, USA

Brian Hoel
Department of Microbiology and Immunology
Medical University of South Carolina
171 Ashley Avenue
Charleston, SC 29425
USA

Gabriel N. Hortobagyi
Dept. of Breast Medical Oncology
The University of Texas
M. D. Anderson Cancer Centre
1515 Holcombe Boulevard, Box 079
Houston, Texas 77030
USA

Mien-Chie Hung
Section of Molecular Cell Biology
Dept. of Cancer Biology
The University of Texas
M. D. Anderson Cancer Centre
1515 Holcombe Boulevard, Box 079
Houston, Texas 77030
USA

Veli-Matti Kähäri
Department of Dermatology
Turku University Central Hospital
University of Turku
FIN-20520 Turku
Finland

Kunghong Kim
HHMI/U. Penn CRB 437A
415 Curie Boulevard
Philadelphia, PA 19104
USA

Junzo Kigawa
Departments of Obstetrics and Gynecology
Tottori University School of Medicine
36-1 Nishimachi
Yonago 6838504
Japan

Nina Krauzewicz
Department of Infectious Diseases (Virology)
7th Floor, Commonwealth Building
Imperial College School of Medicine
Hammersmith Hospital
Du Cane Road
London W12 0NN
UK

Marcel Kuiper
Laboratoire de Thérapie Génique Humaine et Expérimentale
INSERM CR1 9502 & EA 2145
Centre Jean Perrin
BP 392
63011 Clermont-Ferrand Cedex 1
France

Edmund C. Lattime
The Cancer Institute of New Jersey
195 Little Albany Street
New Brunswick, NJ 08901
USA

Charles J. Link Jr
Gene Therapy Program
Human Gene Therapy Research Institute
1415 Woodlands Avenue
Des Moines
Iowa 50309-3203
USA

Yanjun Liu
Sidney Kimmel Cancer Center
10835 Altman Row
San Diego, CA 92121
USA

Brooks Low
Vion Pharmaceuticals, Inc
4 Science Park
New Haven, CT 06511
USA

Katalin V. Lukacs
National Heart and Lung Institute
Imperial College School of Medicine
Emmanuel Kaye Building
Manresa Road
London SW3 6LR
USA

Adama Ly
INSERM—CRJ 9701
Lab. de Neurologie due Développement
Hôpital Robert Debre
48, Bd Sérurier
75019 Paris
France

H. Kim Lyerly
Department of Surgery
Duke University Medical Center
Box 2606
Durham, NC 27710
USA

Malavi T. Madireddi
Department of Urology
Columbia University
College of Physicians and Surgeons
BB1501
630 West 168th Street
New York, NY 10032
USA

Frideriki Maggouta
Department of Microbiology and Immunology
Medical University of South Carolina
171 Ashley Avenue
Charleston, SC 29425
USA

Henry C. Maquire, Jr
Division of Medical Oncology
Jefferson Medical College
Room 1024 Curtis
1015 Walnut Street
Philadelphia, PA 19107
USA

Raphael J. Mannino
Biodelivery Sciences Inc.,
UMDNJ, MSB A691
185 South Orange Avenue
Newark, New Jersey 07103
USA

Marc R. Mansour
Imperial College School of Medicine
Hammersmith Hospital
De Cane Road
London W12 0NN
UK

Michael J. Mastrangelo
Division of Medical Oncology
Jefferson Medical College
Room 1024 Curtis
1015 Walnut Street
Philadelphia, PA 19107
USA

Misako Matsumoto
Dept of Immunology
Osaka Medical Center for Cancer and Cardiovascular Disease
Higashinari-ku
Osaka 537
Japan

Dan Mercola
Sidney Kimmel Cancer Center
10835 Altman Row
San Diego, Ca 92121
USA

Ragai R. Mitry
Imperial College School of Medicine
Hammersmith Hospital
Du Cane Road
London W12 0NN
UK

Richard L. Momparler
Centre de Recherche Pédiatrique
Hôpital Ste-Justine
3175 Chemin Côte Ste-Catherine
Montréal
Québec H3T 1C5
Canada

Michael A. Morse
Department of Medicine
Duke University Medical Center
Box 2606
Durham, NC 27710
USA

Josephine T. Nguyen
Department of Diagnostic Radiology
Beth Israel Deaconess Medical Center
3rd Floor, One Deaconess Road
Boston, MA 02115
USA

Ion Nicolescu-Duvaz
CRC Centre for Cancer Therapeutics
CRC Labs, ICR
15 Cotswold Road
Sutton
Surrey SM2 5NG
UK

Midori Nomura
Dept of Immunology
Osaka Medical Center for Cancer and Cardiovascular Disease
Higashinari-ku
Osaka 537
Japan

James S. Norris
Department of Microbiology and Immunology
Medical University of South Carolina
171 Ashley Avenue
Charleston, SC 29425
USA

Mathieu Noteborn
P.O. Box 9503
2300 RA
Leiden
The Netherlands

Yaw Ohene-Abuakwa
Department of Histopathology
Division of Investigative Sciences
Imperial College School of Medicine
Hammersmith Hospital
Du Cane Road
London W12 0NN
UK

Weihua Pan
Department of Pathology C7768
Penn State University
500 University Drive
Hershey, PA 17033
USA

Olivier E. Pardo
National Heart and Lung Institute
Imperial College School of Medicine
Emmanuel Kaye Building
Manresa Road
London SW3 6LR
UK

John M. Pawelek
Vion Pharmaceuticals, Inc
4 Science Park
New Haven, CT 06511
USA

Alexandra Pietersen
P.O. Box 9503
2300 RA
Leiden
The Netherlands

Massimo Pignatelli
Department of Histopathology
Division of Investigative Sciences
Imperial College School of Medicine
Hammersmith Hospital
Du Cane Road
London W12 0NN
UK

Massimo Pizzato
Institute of Cancer Research
Chester Beatty Laboratories
237 Fullham Road
London SW3 6JB
UK

Pat Price
Department of Clinical Oncology
Imperial College School of Medicine
Hammersmith Hospital
Du Cane Road
London W12 0NN
UK

Raquel Sanches
Laboratoire de Thérapie Génique Humaine et Expérimentale
INSERM CR1 9502 & EA 2145
Centre Jean Perrin
BP 392
63011 Clermont-Ferrand Cedex 1
France

Anu-Maaria Sandmair
Department of Medicine and Gene Therapy Unit
A.I. Virtanen Institute
University of Kuopio
P. O. B. 1627
Fin-70211 Kuopio
Finland

Tatiana Seregina
Gene Therapy Program
Human Gene Therapy Research Institute
1415 Woodlands Avenue
Des Moines
Iowa 50309-3203
USA

Prem Seth
Human Gene Therapy Research Institute
1415 Woodland Avenue
Suite 218
Des Moines
IA 50309
USA

Tsukasa Seya
Department of Immunology
Osaka Medical Center for Cancer and
Cardiovascular Diseases
Higashinari-ku
Osaka 537
Japan

Gabriel Sica
Mayo Clinic
Department of Immunology
Room 342B, Guggenheim Building
200 First Street, S.W.
Rochester, MN 55905
USA

Caroline J. Springer
CRC Centre for Cancer Therapeutics
CRC Labs, ICR
15 Cotswold Road
Sutton
Surrey SM2 5NG
UK

Susan C. Stevenson
Genetic Therapy
21 Firstfield Road
Gaithersburg, MD 20878
USA

Zao-Zhong Su
Departments of Urology
Columbia University
College of Physicians and Surgeons
BB1501
630 West 168th Street
New York, NY 10032
USA

Matoko Sunamura
Department of Surgery 1
Tohoku University School of Medicine
1-1 Seiryo-machi, Aoba-ku
Sendai 1-2 980-77
Japan

Yasuhiro Takeuchi
Chester Beatty Laboratories
Institute of Cancer Research
237 Fulham Road
London SW3 6JB
UK

Taiki Tamaoki
The University of Calgary
Department of Biochemistry &
Molecular Biology
Health Sciences Centre
3330 Hospital Drive N.W.
Calgary, Alberta T2N 4N1
Canada

Vivien J. Tannoch
Howard Hughes Medical Institute
Department of Pathology
Harvard Medical School
200 Longwood Avenue
Boston, MA 02115
USA

Naoki Terakawa
Departments of Obsstetrics and
Gynecology
Tottori University School of Medicine
36-1 Nishimachi
Yonago 6838504
Japan

Kumao Toyoshima
Dept of Immunology
Osaka Medical Center for Cancer and Cardiovascular Disease
Higashinari-ku
Osaka 537
Japan

Jerzy Trojan
INSERM—CRJ9701
Lab. de Neurologie de Développement
Hôpital Robert Debre
48, Bd Sérurier
75019 Paris
France

Li-Huei Tsai
Howard Hughes Medical Institute
Department of Pathology
Harvard Medical School
200 Longwood Avenue
Boston, MA 02115
USA

Shoutaro Tsuji
Dept of Immunology
Osaka Medical Center for Cancer and Cardiovascular Disease
Higashinari-ku
Osaka 537
Japan

Philip C. Turner
School of Biological Sciences
University of Liverpool
Crown Street
Liverpool L69 7ZB
UK

Naoto T. Ueno
Department of Blood and Marrow Transplantation
The University of Texas
M. D. Anderson Cancer Centre
1515 Holcombe Boulevard, Box 65
Houston, Texas 77030
USA

Lia C. Upegui-Gonzalez
INSERM—CRJ 9701
Lab. de Neurologie du Développement
Hôpital Robert Debre
48, Bd Sérurier, 75019 Paris
France

Matti Vapalahti
Department of Neurosurgery
A.I. Virtanen Institute
University of Kuopio
P. O. B. 1777
Fin-70110 Kuopio
Finland

Dale Voeks
Department of Microbiology and Immunology
Medical university of South Carolina
171 Ashley Avenue
Charleston, SC 29425
USA

Peter Walden Department of Dermatology
Medical Faculty Charité
Humboldt University
Schumannstrasse 20/21
D-10089 Berlin
Germany

Hao Wang
Sidney Kimmel Cancer Center
10835 Altman Row
San Diego, CA 92121
USA

Suming Wang
Gene Therapy Program
Human Gene Therapy Research Institute
1415 Woodlands Avenue
Des Moines
Iowa 50309-3203
USA

Lixin Wei
Sidney Kimmel Cancer Center
10835 Altman Row
San Diego, CA 92121
USA

Marc Winslet
The Royal Free and University College Medical School
Dept. Anatomy and Developmental Biology & Dept Surgery
University of London
Royal Free Campus
Rowland Hill Street
London NW3 2PF
UK

Gen Sheng Wu
HHMI/U. Penn CRB 437A
415 Curie Boulevard
Philadelphia, PA 19104
USA

Seppo Ylä-Herttuala
Department of Medicine and Gene Therapy Unit
A.I. Virtanen Institute
University of Kuopio
P. O. B. 1627
Fin-70211 Kuopio
Finland

Charles S. H. Young
Departments of Microbiology
Columbia University
College of Physicians and Surgeons
BB1501
630 West 168th Street
New York, NY 10032
USA

Leila Zarif
BioDelivery Sciences Inc.,
C/o UMDNJ, MSB A691
185 South Orange Avenue
Newark, New Jersey 07103
USA

INDEX